Introduction to Optics

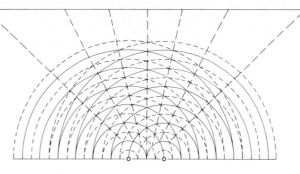

Second Edition

Introduction to Optics

FRANK L. PEDROTTI, S.J.

Marquette University
Milwaukee, Wisconsin

Vatican Radio,
Rome

LENO S. PEDROTTI

Center for Occupational
Research and Development
Waco, Texas

Emeritus Professor of Physics
Air Force Institute of Technology
Dayton, Ohio

Prentice-Hall International, Inc.

This edition may be sold only in those countries to which
it is consigned by Prentice-Hall International. It is not to
be re-exported and it is not for sale in the U.S.A., Mexico,
or Canada.

© 1996 by Prentice Hall PTR
Prentice-Hall, Inc.
A Simon & Schuster Company
Upper Saddle River, New Jersey 07458

Printed in the United States of America

10 9   8   7   6

ISBN 0-13-016973-0

Prentice-Hall International (UK) Limited, *London*
Prentice-Hall of Australia Pty. Limited, *Sydney*
Prentice-Hall Canada Inc., *Toronto*
Prentice-Hall Hispanoamericana, S.A., *Mexico*
Prentice-Hall of India Private Limited, *New Delhi*
Prentice-Hall of Japan, Inc., *Tokyo*
Simon & Schuster Asia Pte., Ltd., *Singapore*
Editora Prentice-Hall do Brasil, Ltda., *Rio de Janeiro*
Prentice-Hall, Inc., *Englewood Cliffs, New Jersey*

*This book is dedicated to the memory of our parents,
immigrants from beautiful Dambel*

Contents

List of Tables   *xv*

Physical Constants   *xvii*

Preface   *xix*

1   **Nature of Light**   **1**

        Introduction   *1*
        A Brief History   *2*
        Problems   *6*
        References   *7*

2   **Production and Measurement of Light**   **8**

        Introduction   *8*
2-1   Electromagnetic Spectrum   *8*
2-2   Radiometry   *10*
2-3   Photometry   *13*
2-4   Blackbody Radiation   *15*
*2-5   Sources of Optical Radiation   *17*
*2-6   Detectors of Radiation   *23*
        Problems   *28*
        References   *29*

3   *Geometrical Optics*   **30**

         Introduction   *30*

3-1   Huygens' Principle   *31*

3-2   Fermat's Principle   *34*

3-3   Principle of Reversibility   *36*

3-4   Reflection in Plane Mirrors   *37*

3-5   Refraction through Plane Surfaces   *38*

3-6   Imaging by an Optical System   *40*

3-7   Reflection at a Spherical Surface   *43*

3-8   Refraction at a Spherical Surface   *47*

3-9   Thin Lenses   *50*

3-10   Vergence and Refractive Power   *55*

3-11   Newtonian Equation for the Thin Lens   *57*

         Problems   *58*

         References   *60*

4   *Matrix Methods in Paraxial Optics*   **62**

         Introduction   *62*

4-1   The Thick Lens   *62*

4-2   The Matrix Method   *65*

4-3   The Translation Matrix   *66*

4-4   The Refraction Matrix   *66*

4-5   The Reflection Matrix   *67*

4-6   Thick-Lens and Thin-Lens Matrices   *68*

4-7   System Ray-Transfer Matrix   *71*

4-8   Significance of System Matrix Elements   *72*

4-9   Location of Cardinal Points for an Optical System   *74*

4-10   Examples Using the System Matrix and Cardinal Points   *77*

*4-11   Ray Tracing   *79*

         Problems   *84*

         References   *86*

5   *Aberration Theory*   **87**

         Introduction   *87*

5-1   Ray and Wave Aberrations   *88*

*5-2   Third-Order Treatment of Refraction at a Spherical Interface   *89*

5-3   Spherical Aberration   *93*

5-4   Coma   *96*

5-5   Astigmatism and Curvature of Field   *98*

5-6   Distortion   *100*

5-7   Chromatic Aberration   *102*

         Problems   *106*

         References   *107*

6 **Optical Instrumentation   109**

Introduction   *109*

6-1   Stops, Pupils, and Windows   *109*

6-2   Prisms   *116*

6-3   The Camera   *125*

6-4   Simple Magnifiers and Eyepieces   *131*

6-5   Microscopes   *135*

6-6   Telescopes   *139*

Problems   *146*

References   *149*

7 **Optics of the Eye   151**

Introduction   *151*

7-1   Biological Structure of the Eye   *152*

7-2   Optical Representation of the Eye   *153*

7-3   Functions of the Eye   *155*

7-4   Errors of Refraction and Their Correction   *158*

*7-5   Laser Therapy for Ocular Defects   *163*

Problems   *169*

References   *171*

8 **Wave Equations   172**

Introduction   *172*

8-1   One-Dimensional Wave Equation   *172*

8-2   Harmonic Waves   *174*

8-3   Complex Numbers   *177*

8-4   Harmonic Waves as Complex Numbers   *178*

8-5   Plane Waves   *178*

8-6   Spherical Waves   *180*

8-7   Electromagnetic Waves   *181*

*8-8   Doppler Effect   *183*

Problems   *184*

References   *186*

9 **Superposition of Waves   187**

Introduction   *187*

9-1   Superposition Principle   *187*

9-2   Superposition of Waves of the Same Frequency   *188*

9-3   Random and Coherent Sources   *191*

9-4   Standing Waves   *192*

9-5   Phase and Group Velocities   *194*

Problems   *197*

References   *199*

Contents

10   *Interference of Light*   **200**

Introduction   *200*

10-1   Two-Beam Interference   *200*

10-2   Young's Double-Slit Experiment   *205*

10-3   Double-Slit Interference with Virtual Sources   *209*

10-4   Interference in Dielectric Films   *211*

10-5   Fringes of Equal Thickness   *215*

10-6   Newton's Rings   *217*

*10-7   Film-Thickness Measurement by Interference   *219*

Problems   *221*

References   *223*

11   *Optical Interferometry*   **224**

Introduction   *224*

11-1   The Michelson Interferometer   *225*

11-2   Applications of the Michelson Interferometer   *228*

11-3   Variations of the Michelson Interferometer   *230*

*11-4   Stokes Relations   *232*

*11-5   Multiple-Beam Interference in a Parallel Plate   *233*

*11-6   Fabry-Perot Interferometer   *236*

*11-7   Fringe Profiles: The Airy Function   *238*

*11-8   Resolving Power   *239*

*11-9   Free Spectral Range   *242*

Problems   *244*

References   *246*

12   *Coherence*   **247**

Introduction   *247*

12-1   Fourier Analysis   *248*

12-2   Fourier Analysis of a Finite Harmonic Wave Train   *251*

12-3   Temporal Coherence and Natural Line Width   *253*

*12-4   Partial Coherence   *254*

12-5   Spatial Coherence   *259*

12-6   Spatial Coherence Width   *260*

Problems   *263*

References   *265*

13   *Holography*   **266**

Introduction   *266*

13-1   Conventional versus Holographic Photography   *266*

13-2   Hologram of a Point Source   *267*

13-3   Hologram of an Extended Object   *269*

13-4   Hologram Properties   *273*

13-5   White-Light Holograms   *273*

13-6   Other Applications of Holography   *274*
Problems   *277*
References   *279*

14   *Matrix Treatment of Polarization*   *280*

Introduction   *280*
14-1   Mathematical Representation of Polarized Light: Jones Vectors   *281*
14-2   Mathematical Representation of Polarizers: Jones Matrices   *289*
Problems   *295*
References   *297*

15   *Production of Polarized Light*   *298*

Introduction   *298*
15-1   Dichroism: Polarization by Selective Absorption   *298*
15-2   Polarization by Reflection from Dielectric Surfaces   *301*
15-3   Polarization by Scattering   *303*
15-4   Birefringence: Polarization with Two Refractive Indices   *306*
15-5   Double Refraction   *310*
*15-6   Optical Activity   *313*
*15-7   Photoelasticity   *317*
Problems   *319*
References   *322*

16   *Fraunhofer Diffraction*   *323*

Introduction   *323*
16-1   Diffraction from a Single Slit   *324*
16-2   Beam Spreading   *329*
16-3   Rectangular and Circular Apertures   *330*
16-4   Resolution   *335*
16-5   Double-Slit Diffraction   *338*
16-6   Diffraction from Many Slits   *341*
Problems   *346*
References   *348*

17   *The Diffraction Grating*   *349*

Introduction   *349*
17-1   The Grating Equation   *349*
17-2   Free Spectral Range of a Grating   *351*
17-3   Dispersion of a Grating   *352*
17-4   Resolution of a Grating   *353*
*17-5   Types of Gratings   *355*

*17-6   Blazed Gratings   *356*
*17-7   Grating Replicas   *359*
*17-8   Interference Gratings   *359*
*17-9   Grating Instruments   *361*
        Problems   *363*
        References   *365*

18   *Fresnel Diffraction*   *366*

        Introduction   *366*
18-1   Fresnel-Kirchhoff Diffraction Integral   *367*
18-2   Criterion for Fresnel Diffraction   *369*
18-3   The Obliquity Factor   *370*
18-4   Fresnel Diffraction from Circular Apertures   *370*
18-5   Phase Shift of the Diffracted Light   *374*
18-6   The Fresnel Zone Plate   *374*
18-7   Fresnel Diffraction from Apertures with Rectangular Symmetry   *376*
18-8   The Cornu Spiral   *378*
18-9   Applications of the Cornu Spiral   *382*
18-10   Babinet's Principle   *388*
        Problems   *388*
        References   *390*

19   *Theory of Multilayer Films*   *391*

        Introduction   *391*
19-1   Transfer Matrix   *392*
19-2   Reflectance at Normal Incidence   *396*
19-3   Two-Layer Antireflecting Films   *398*
19-4   Three-Layer Antireflecting Films   *401*
19-5   High-Reflectance Layers   *402*
        Problems   *405*
        References   *406*

20   *Fresnel Equations*   *407*

        Introduction   *407*
20-1   The Fresnel Equations   *407*
20-2   External and Internal Reflections   *411*
20-3   Phase Changes on Reflection   *414*
20-4   Conservation of Energy   *417*
*20-5   Evanescent Waves   *419*
*20-6   Complex Refractive Index   *420*
*20-7   Reflection from Metals   *422*
        Problems   *423*
        References   *425*

21   *Laser Basics*   **426**

          Introduction   *426*
- **21-1**   Einstein's Quantum Theory of Radiation   *427*
- **21-2**   Essential Elements of a Laser   *431*
- **21-3**   Simplified Description of Laser Operation   *434*
- **21-4**   Characteristics of Laser Light   *440*
- **21-5**   Laser Types and Parameters   *451*
          Problems   *453*
          References   *454*

22   *Characteristics of Laser Beams*   **456**

          Introduction   *456*
- **22-1**   Three-Dimensional Wave Equation and Electromagnetic Waves   *457*
- **22-2**   Phase Variation of Spherical Waves Along a Transverse Plane   *459*
- **22-3**   Basis for Defining Laser-Beam Mode Structures   *459*
- **22-4**   Gaussian Beam Solution for Lasers   *461*
- **22-5**   Spot Size and Radius of Curvature of a Gaussian Beam   *464*
- **22-6**   Laser Propagation through Arbitrary Optical Systems   *468*
- ***22-7**   Higher-Order Gaussian Beams   *476*
          Problems   *480*
          References   *483*

23   *Laser Applications*   **484**

          Introduction   *484*
- **23-1**   Lasers and Interaction   *485*
- **23-2**   Lasers and Information   *491*
- **23-3**   More Recent Developments   *496*
          Problems   *498*
          References   *500*

24   *Fiber Optics*   **501**

          Introduction   *501*
- **24-1**   Applications   *501*
- **24-2**   Communications System Overview   *502*
- **24-3**   Bandwidth and Data Rate   *504*
- **24-4**   Optics of Propagation   *505*
- **24-5**   Allowed Modes   *507*
- **24-6**   Attenuation   *509*
- **24-7**   Distortion   *512*
          Problems   *519*
          References   *521*

Contents   **xiii**

25   *Fourier Optics*   *522*

Introduction   *522*
25-1   Optical Data Imaging and Processing   *523*
25-2   Fourier-Transform Spectroscopy   *536*
Problems   *539*
References   *540*

26   *Nonlinear Optics and the Modulation of Light*   *541*

Introduction   *541*
26-1   The Nonlinear Medium   *542*
26-2   Second Harmonic Generation   *543*
26-3   Frequency Mixing   *546*
26-4   The Pockels Effect   *547*
26-5   The Kerr Effect   *551*
26-6   The Faraday Effect   *553*
26-7   The Acousto-Optic Effect   *556*
26-8   Nonlinear Optical Phase Conjugation   *559*
Problems   *565*
References   *567*

27   *Optical Properties of Materials*   *568*

Introduction   *568*
27-1   Polarization of a Dielectric Medium   *568*
27-2   Propagation of Light Waves in a Dielectric   *571*
27-3   Conduction Current in a Metal   *575*
27-4   Propagation of Light Waves in a Metal   *576*
27-5   Skin Depth   *577*
27-6   Plasma Frequency   *578*
Problems   *579*
References   *580*

Suggestions for Further Reading   *581*

Articles on Optics from *Scientific American*   *581*

Answers to Selected Problems   *585*

Index   *597*

List of Tables

Table 2-1	Radiometric and Photometric Terms	*10*
Table 3-1	Summary of Gaussian Mirror and Lens Formulas	*54*
Table 4-1	Summary of Some Simple Ray-Transfer Matrices	*70*
Table 4-2	Cardinal Point Locations in Terms of System Matrix Elements	*77*
Table 4-3	Meridional Ray-Tracing Equations	*82*
Table 5-1	Sample of Optical Glasses	*104*
Table 6-1	Fraunhofer Lines	*120*
Table 6-2	Standard Relative Apertures and Irradiance Available on Cameras	*128*
Table 7-1	Constants of a Schematic Eye	*154*
Table 14-1	Summary of Jones Vectors	*288*
Table 14-2	Summary of Jones Matrices	*293*
Table 15-1	Refractive Indices for Several Materials	*308*
Table 15-2	Specific Rotation of Quartz	*314*
Table 15-3	Refractive Indices for Quartz	*316*
Table 18-1	Fresnel Integrals	*380*
Table 19-1	Refractive Indices for Several Coating Materials	*400*
Table 19-2	Reflectance of a High-Low Quarter-Wave Stack	*403*
Table 21-1	Comparison of Linewidths	*441*
Table 21-2	Laser Parameters for Several Common Lasers	*452*

Table 23-1   A Classification of Laser Applications   *485*

Table 23-2   Laser Types in Medicine   *489*

Table 23-3   Medical Fields Involved with Lasers   *490*

Table 24-1   Characterization of Several Optical Fibers   *506*

Table 26-1   Linear and Nonlinear Processes   *546*

Table 26-2   Linear Electro-optic Coefficients for Representative Materials   *549*

Table 26-3   Kerr Constant for Selected Materials   *553*

Table 26-4   Verdet Constant for Selected Materials   *554*

PHYSICAL CONSTANTS

Speed of light	$c = 2.998 \times 10^8$ m/s
Electron charge	$e = 1.602 \times 10^{-19}$ C
Electron rest mass	$m = 9.109 \times 10^{-31}$ kg
Planck constant	$h = 6.626 \times 10^{-34}$ Js
Boltzmann constant	$k = 1.3805 \times 10^{-23}$ J/K
Permittivity of vacuum	$\epsilon_0 = 8.854 \times 10^{-12}$ C^2/N-m^2
Permeability of vacuum	$\mu_0 = 4\pi \times 10^{-7}$ T-m/A

Preface

Optics is today perhaps the most flourishing area of both theoretical and applied physics. Since the 1960s the parallel emergence and development of lasers, fiber optics, and a variety of semiconductor sources and detectors have revitalized the field. The need for a variety of updated optics texts with different approaches and emphases is therefore apparent, both for the student of optics and for the laborer in the field who needs an occasional review of the basics.

With *Introduction to Optics* we propose to teach introductory modern optics at an intermediate level. Except for several of the final chapters (19, 20, 22, 25, 27), which are written at a somewhat higher level, the text assumes as background a good course in introductory physics, at the level usually given to physics and engineering majors, and at least two semesters of calculus. The book is written at the level of understanding appropriate to the average sophomore physics major, who has usually completed the necessary physics and mathematics prerequisites as a freshman. Encompassing the traditional areas of college optics, as well as several rather new ones, the text can be designed for either a half- or a full-year course. We believe that the vitality and importance of optics today warrant readjustment of curricula to provide for a full year of optics early in the program.

For those who are familiar with the first edition, it may be helpful to summarize the major changes introduced in this second edition. Two entirely new chapters dealing with laser-beam characteristics and nonlinear optics have been added. The new laser chapter now appears, together with the two earlier laser chapters, toward the end of the book, where the three function as a unit. In addition, the chapter on fiber optics has been greatly expanded and moved to a later chapter. Several new sections have been introduced. They are Ray Tracing and The Thick Lens (Chapter 4), Doppler Effect (Chapter 8), and Evanescent Waves (Chapter 20). Worked examples are now highlighted within the text, and 175 new problems have been added to the chapter exercises.

Specific features of the text, in terms of coverage beyond the traditional areas, include extensive use of 2×2 matrices in dealing with ray tracing, polarization, and multiple thin-film interference; three chapters devoted to lasers; a separate chapter on the eye, including laser treatments of the eye; and individual chapters on holography, coherence, fiber optics, interferometry, Fourier optics, nonlinear optics, and Fresnel equations. A final chapter provides a brief introduction to the optical constants of dielectrics and metals. We have attempted to make many of the more specialized chapters independent of the others so that they can be omitted without detriment to the remainder of the book. This should be helpful in designing shorter versions of the course.

Organization of the material in three major parts follows essentially traditional lines. The first part of the book deals with geometrical optics, presented as a limiting form of wave optics. The middle part develops wave optics in detail, and the final part treats topics generally referred to as *modern optics*. In the first part, Chapter 1 presents a brief historical review of the theories of light, including wave, particle, and photon descriptions. In Chapter 2, we describe a variety of common sources and detectors of light, as well as the radiometric and photometric units of measurement that are used throughout the book. In this chapter and the remainder of the text, the rationalized MKS system of units is employed. Chapter 3 reviews the geometrical optics covered by introductory physics courses, deriving the usual reflection and refraction relations for mirrors and lenses. Chapter 4 shows how one can extend paraxial optics to systems of arbitrary complexity through the use of 2×2 matrices. Also in this chapter we include an introduction to the ray-tracing techniques that are widely applied using computer programming. Chapter 5 presents a semiquantitative treatment of third-order aberration theory. Chapter 6 discusses the principles of geometrical optics and aberration theory as applied to apertures and to several optical devices: the prism, the camera, the eyepiece, the microscope, and the telescope. The importance of the eye as the final optical instrument in many optical systems is recognized in a separate chapter (7). This chapter explains the functions and the defects of the eye and discusses some of the treatments of these defects that make use of the unique properties of laser light.

The next section of the text introduces wave or physical optics with two chapters (8 and 9) that discuss the wave equations and the superposition of waves. Interference phenomena are then treated in Chapters 10 and 11, the second dealing with both Michelson and Fabry-Perot interferometers in some detail. Although the concept of coherence is handled in general terms in preceding discussions, it receives a more precise and quantitative treatment in Chapter 12. After a brief explanation of Fourier series and the Fourier integral, the chapter deals with both temporal and spatial coherence and presents a quantitative discussion of partial coherence. Chapter 13 presents, as a special application of interference, an introduction to holography, including some current applications.

Chapters 14 and 15 treat the polarization of light. We first give a mathematical presentation using 2×2 matrices to represent the electric field vector (Chapter 14), before examining in detail the physical mechanisms responsible for the production of polarized light (Chapter 15). Thus Chapter 14 uses matrices to describe the various modes of polarized light and types of polarizers without reference to the physics of its production. Although the order of these chapters can be reversed, we feel this choice is pedagogically more effective. Diffraction is discussed in the following three chapters (16, 17, 18). Since an adequate treatment of Fraunhofer diffraction is too long for a single chapter, we have included a separate chapter (17) on the diffraction grating and grating instruments immediately following the discussion of multiple-slit diffraction in Chapter 16. Fresnel diffraction is then taken up in Chapter 18.

The final chapters are generally more demanding in mathematical sophistication. Chapter 19 employs 2×2 matrices to treat reflectance of multilayer thin films. Chapter 20 derives the Fresnel equations in an examination of reflection from both dielectric and metallic surfaces. The basic elements of a laser and the basic characteristics of laser light are treated in Chapter 21, followed by a rather demanding chapter (22) that describes the principal features of laser beams. The divergence and mode structure of laser beams are dealt with here in a quantitative manner. Chapters 21 and 22 are best taken in sequence, and together with Chapter 23, essentially an essay on laser applications, form a suitable three-chapter unit for a minicourse on lasers. The other chapters in this final part of the book are self-contained in the sense that no particular sequence is required.

Chapter 24 presents a survey of the basic features of optical fibers with special attention given to communication applications. Thus topics of bandwidth, allowed modes, and mechanisms of attenuation and distortion are treated here. Chapter 25 introduces the subject of Fourier optics in a discussion of optical data processing and Fourier-transform spectroscopy. Chapter 26 presents a variety of physical effects under the umbrella of nonlinear optics. The final chapter (27) considers the propagation of a light wave in both dielectric and metallic media and shows how the optical constants arise.

Each of the 27 chapters contains a limited bibliography related to the chapter contents and referred to at times within the text using square brackets. In addition, at the end of the book, we have included a chronological listing of articles related to optics that have appeared in *Scientific American* over the last 40 years or so. It is hoped that this list of excellent articles will prove helpful, especially to the undergraduate student.

This text is intended to be adaptable for either one or two semester sequences. The precise selection of material will depend on the particular goals of both teacher and student. As a rough guide, however, a typical one-semester course might include the basic sequence:

Chapter	1	Nature of Light
	3	Geometrical Optics
	6	Optical Instrumentation
	8	Wave Equations
	9	Superposition of Waves
	10	Interference of Light
	12	Coherence
	13	Holography
	15	Production of Polarized Light
	16	Fraunhofer Diffraction
	18	Fresnel Diffraction
	21	Laser Basics

As a further aid to selection, those sections that could be omitted in abbreviated versions of the course are marked with an asterisk. See the Contents.

We wish to thank the many teachers who have inspired us with an interest in optics and in teaching and the many students who have motivated us to teach with clarity and efficiency. For their very helpful reading of portions of the manuscript for the first edition, we are indebted to Hugo Weichel, James Tucci, Hajime Sakai, Arthur H. Guenther, and Thomas B. Greenslade. For their suggestions for improving the second edition, which we have considered very seriously, we wish to thank the team of reviewers selected by Prentice Hall: Joel Blatt, Florida Institute of Technology; James Boger, Oregon Institute of Technology; Harry Daw, New Mexico State University; Edward Eyler, University of Delaware; and Daniel Wilkins, University of Nebraska. We are grateful to Leno M. Pedrotti for his critical reading of most of the new material added in this second edition. For his review and suggestions in the chapter on the eye, we are also pleased to acknowledge and thank Dr. Michael Pedrotti, O.D. We also wish to thank Judy Lawson for her sketch of Einstein that graces page 1. Finally, we express our gratitude to the editorial and production staff of Prentice Hall. In particular, we are indebted to our acquisitions editors, Holly Hodder for the first edition and Ray Henderson for the second edition, and to our production editor, Kathleen Lafferty, who helped us with both editions of this text.

Frank L. Pedrotti, S.J.
Leno S. Pedrotti

1

Nature of Light

INTRODUCTION

The evolution in our understanding of the physical nature of light forms one of the most fascinating accounts in the history of science. Since the dawn of modern science in the sixteenth and seventeenth centuries, light has been pictured either as particles or waves—incompatible models—each of which enjoyed a period of prominence among the scientific community. In the twentieth century it became clear that somehow light was both wave and particle, yet it was precisely neither. For some time this perplexing state of affairs, referred to as the *wave-particle duality*, motivated the greatest scientific minds of our age to find a resolution to apparently contradictory models of light. The solution was achieved through the creation of *quantum electrodynamics*, one of the most successful theoretical structures in the annals of physics.

In what follows, we will be content to sketch briefly a few of the high points of this developing understanding.[1] Certain areas of physics once considered to be disciplines apart from optics—electricity and magnetism, and atomic physics—are very much involved in this account. This alone suggests that the resolution achieved also constitutes one of the great unifications in our understanding of the physical world. The final result is that light and subatomic particles, like electrons, are both consid-

---

[1] A more in-depth historical account may be found, for example, in [1] (see References at the end of the chapter).

ered to be manifestations of matter or energy under the same set of formal principles.

In the seventeenth century the most prominent advocate of a particle theory of light was Isaac Newton, the same creative giant who had erected a complete science of mechanics and gravity. In his treatise *Optics*, Newton clearly regarded rays of light as streams of very small particles emitted from a source of light and traveling in straight lines. Although Newton often argued forcefully against positing hypotheses that were not derived directly from observation and experiment, here he adopted a particle hypothesis, believing it to be adequately justified by the phenomena. Important in his considerations was the observation that light can cast sharp shadows of objects, in contrast to water and sound waves, which bend around obstacles in their paths. At the same time, Newton was aware of the phenomenon now referred to as *Newton's rings*. Such light patterns are not easily explained by viewing light as a stream of particles traveling in straight lines. Newton maintained his basic particle hypothesis, however, and explained the phenomenon by endowing the particles themselves with what he called "fits of easy reflection and easy transmission," a kind of periodic motion due to the attractive and repulsive forces imposed by material obstacles. Newton's eminence as a scientist was such that his point of view dominated the century that followed his work.

A BRIEF HISTORY

Christian Huygens, a Dutch scientist contemporary with Newton, championed the view (in his *Treatise on Light*) that light is a wave motion, spreading out from a light source in all directions and propagating through an all-pervasive elastic medium called the ether. He was impressed, for example, by the experimental fact that when two beams of light intersected, they emerged unmodified, just as in the case of two water or sound waves. Adopting a wave theory, Huygens was able to derive the laws of reflection and refraction and to explain double refraction in calcite as well.

Within two years of the centenary of the publication of Newton's *Optics*, the Englishman Thomas Young performed a decisive experiment that seemed to demand a wave interpretation, turning the tide of support to the wave theory of light. It was the double-slit experiment, in which an opaque screen with two small, closely spaced openings was illuminated by monochromatic light from a small source. The "shadows" observed formed a complex interference pattern like those produced with water waves.

Victories for the wave theory continued up to the twentieth century. In the mood of scientific confidence that characterized the latter part of the nineteenth century, there was little doubt that light, like most other classical areas of physics, was well understood. We mention a few of the more significant confirmations.

In 1821 Augustin Fresnel published results of his experiments and analysis, which required that light be a transverse wave. On this basis, double refraction in calcite could be understood as a phenomenon involving polarized light. It had been assumed that light waves in an ether were necessarily longitudinal, like sound waves in a fluid, which cannot support transverse vibrations. For each of the two components of polarized light, Fresnel developed the *Fresnel equations*, which give the amplitude of light reflected and transmitted at a plane interface separating two optical media.

Working in the field of electricity and magnetism, James Clerk Maxwell synthesized known principles in his set of four *Maxwell equations*. The equations yielded a prediction for the speed of an electromagnetic wave in the ether that turned out to be the measured speed of light, suggesting its electromagnetic character. From then on, light was viewed as a particular region of the electromagnetic spectrum of radiation. The experiment (1887) of Albert Michelson and Edward Morley, which attempted to detect optically the earth's motion through the ether, and the special theory of relativity (1905) of Albert Einstein were of monumental importance. Together they led inevitably to the conclusion that the assumption of an ether was superfluous. The problems associated with transverse vibrations of a wave in a fluid thus vanished.

If the nineteenth century served to place the wave theory of light on a firm foundation, this foundation was to crumble as the century came to an end. The wave-particle controversy was resumed with vigor. Again, we mention only briefly some of the key events along the way. Difficulties in the wave theory seemed to show up in situations that involved the interaction of light with matter. In 1900, at the very dawn of the twentieth century, Max Planck announced at a meeting of the German Physical Society that he was able to derive the correct blackbody radiation spectrum only by making the curious assumption that atoms emitted light in discrete energy chunks rather than in a continuous manner. Thus *quanta* and *quantum mechanics* were born. According to Planck, the energy E of a quantum of electromagnetic radiation is proportional to the frequency of the radiation, ν,

$$E = h\nu \tag{1-1}$$

where the constant of proportionality, *Planck's constant*, has the very small value of 6.63×10^{-34} J-s. Five years later, in the same year that he published his theory of special relativity, Albert Einstein offered an explanation of the photoelectric effect, the emission of electrons from a metal surface when irradiated with light. Central to his explanation was the conception of light as a stream of photons whose energy is related to frequency by Planck's equation (1-1). Then in 1913 the Danish physicist Niels Bohr once more incorporated the quantum of radiation in his explanation of the emission and absorption processes of the hydrogen atom, providing a physical basis for understanding the hydrogen spectrum. Again in 1922, the photon model of light came to the rescue for Arthur Compton, who explained the scattering of X-rays from electrons as particlelike collisions between photons and electrons in which both energy and momentum were conserved.

All such victories for the photon or particle model of light indicated that light could be treated as a particular kind of matter, possessing both energy and momentum. It was Luis de Broglie who saw the other side of the picture. In 1924 he published his speculations that subatomic particles are endowed with wave properties. He suggested, in fact, that a particle with momentum p had an associated wavelength of

$$\lambda = \frac{h}{p} \tag{1-2}$$

where h was, again, Planck's constant. Experimental confirmation of de Broglie's hypothesis appeared during the years 1927–1928, when Clinton Davisson and Lester Germer in the United States and Sir George Thomson in England performed experiments that could only be interpreted as the diffraction of a beam of electrons.

Thus the wave-particle duality came full circle. Light behaved like waves in its propagation and in the phenomena of interference and diffraction; it could, however, also behave as particles in its interaction with matter, as in the photoelectric effect. On the other hand, electrons usually behaved like particles, as observed in the pointlike scintillations of a phosphor exposed to a beam of electrons; in other situations they were found to behave like waves, as in the diffraction produced by an electron microscope.

Photons and electrons that behaved both as particles and as waves seemed at first an impossible contradiction, since particles and waves are very different entities indeed. Gradually it became clear, to a large extent through the reflections of Niels Bohr and especially in his *principle of complementarity*, that photons and electrons were neither waves nor particles, but something more complex than either.

In attempting to explain physical phenomena, it is natural that we appeal to well-known physical models like waves and particles. As it turns out, however, the full intelligibility of a photon or an electron is not exhausted by either model. In certain situations, wavelike attributes may predominate; in other situations, particlelike attributes stand out. We can appeal to no simpler physical model that is adequate to handle all cases.

Quantum mechanics, or wave mechanics, as it is often called, deals with all particles more or less localized in space, and so describes both light and matter. Combined with special relativity, the momentum p, wavelength λ, and speed v for both material particles and photons are given by the same general equations:

$$p = \frac{\sqrt{E^2 - m^2c^4}}{c} \tag{1-3}$$

$$\lambda = \frac{h}{p} = \frac{hc}{\sqrt{E^2 - m^2c^4}} \tag{1-4}$$

$$v = \frac{pc^2}{E} = c\sqrt{1 - \frac{m^2c^4}{E^2}} \tag{1-5}$$

In these equations, m is the *rest mass* and E is the total energy, the sum of rest-mass energy mc^2 and kinetic energy, that is, the work done to accelerate the particle from rest to its measured speed. The relativistic mass is given by γm, where γ is the ratio $1/\sqrt{1 - (v/c)^2}$. The proper expression for kinetic energy is no longer simply $\frac{1}{2}mv^2$, but $mc^2(\gamma - 1)$. The relativistic expression for kinetic energy approaches $\frac{1}{2}mv^2$ for $v \ll c$.[2]

A crucial difference between particles like electrons and neutrons and particles like photons is that the latter have zero rest mass. Equations (1-3) to (1-5) then take the simpler forms for photons:

$$p = \frac{E}{c} \tag{1-6}$$

$$\lambda = \frac{h}{p} = \frac{hc}{E} \tag{1-7}$$

$$v = \frac{pc^2}{E} = c \tag{1-8}$$

Thus, while nonzero rest-mass particles like electrons have a limiting speed of c, Eq. (1-8) shows that zero rest-mass particles like photons must travel with the constant speed c. The energy of a photon is not a function of its speed but of its frequency, as expressed in Eq. (1-1) or in Eqs. (1-6) and (1-7), taken together. Notice that for a photon, because of its zero rest mass, there is no distinction between its total energy and its kinetic energy. The following example illustrates the preceding equations.

---

[2] This discussion is not meant to be a condensed tutorial on relativistic mechanics, but, with the help of Eqs. (1-3) to (1-8), a summary of some basic relations that unify particles of matter and light.

Example

An electron is accelerated to a kinetic energy of 2.5 MeV. Determine its relativistic momentum, de Broglie wavelength, and speed. Also determine the same properties for a photon having the same energy as the electron.

Solution The electron's total energy E must be the sum of its rest mass energy and its kinetic energy, E_k:

$$E = mc^2 + E_k = 0.511 \text{ MeV} + 2.5 \text{ MeV} = 3.011 \text{ MeV}$$

or

$$E = 3.011 \times 10^6 \text{ eV} \times (1.602 \times 10^{-19} \text{ J/eV}) = 4.82 \times 10^{-13} \text{ J}$$

The other quantities are then calculated in order. From Eq. (1-3):

$$p = 1.58 \times 10^{-21} \text{ kg-m/s}$$

From Eq. (1-4):

$$\lambda = 41.8 \times 10^{-12} \text{ m} = 41.8 \text{ pm}$$

From Eq. (1-5):

$$v = 2.95 \times 10^8 \text{ m/s}$$

For the photon, with $m = 0$, we get instead From Eq. (1-6):

$$p = 1.61 \times 10^{-21} \text{ kg-m/s}$$

From Eq. (1-7):

$$\lambda = 0.412 \text{ pm}$$

From Eq. (1-8):

$$v = c = 3.00 \times 10^8 \text{ m/s}$$

Another important distinction between electrons and photons is that electrons obey Fermi statistics whereas photons obey Bose statistics. A consequence of Fermi statistics is the restriction that no two electrons in the same interacting system be in the same *state*, that is, have precisely the same physical properties. Bose statistics impose no such prohibition, so that identical photons with the same energy and momentum can occur together in large numbers. Because light beams can possess so many similar photons in proximity, the granular structure of the beam is not ordinarily experienced, and the beam can be adequately represented by a continuous electromagnetic wave. From this point of view, electromagnetic fields appear as a special manifestation of photons.

A profound consequence of the wave nature of particles is embodied in the Heisenberg principle of indeterminacy. As a result of this principle, particles do not obey deterministic laws of motion. Rather, the theory predicts only probabilities. Wave functions are associated with the particles through the fundamental wave equation of quantum mechanics. The wave amplitudes, or, better, the square of the wave amplitudes assigned to these particles, provide a means of expressing the probability that a particle will be found within a region of space during an interval of time. Thus the *irradiance* (power/area) of these waves at some intercepting surface, also proportional to the square of the wave amplitudes, provides a measure of this probability. When large numbers of particles are involved, probabilities approach certainties, so that the irradiance E_e of light at a location is proportional to the number of photons passing through the location per second.

$$n \text{ (photons/m}^2\text{-s)} = \frac{E_e}{h\nu} \tag{1-9}$$

In this way, the interference and diffraction patterns previously explained by waves can be interpreted as manifestations of particles. Particle wave amplitudes predict the probabilities of their locations in the same patterns.

In the theory called quantum electrodynamics, which combines the principles of quantum mechanics with those of special relativity, photons are assumed to interact only with charges. An electron, for example, is capable of both absorbing and emitting a photon, with a probability that is proportional to the square of the charge. There is no conservation law for photons as there is for the charge associated with particles. In this theory the wave-particle duality becomes reconciled. Essential distinctions between photons and electrons are removed. Both are considered subject to the same general principles. Through this unification, light is viewed as basically just another form of matter. Nevertheless, the complementary aspects of particle and wave descriptions of light remain, justifying our use of one or the other description when appropriate. The wave description of light will be found adequate to describe most of the optical phenomena treated in this text.

PROBLEMS

1-1. Calculate the de Broglie wavelength of (a) a golf ball of mass 50 g moving at 20 m/s and (b) an electron with kinetic energy of 10 eV.

1-2. The threshold of sensitivity of the human eye is about 100 photons per second. The eye is most sensitive at a wavelength of around 550 nm. For this wavelength determine the threshold in watts of power.

1-3. What is the energy, in electron volts, of light photons at the ends of the visible spectrum, that is, at wavelengths of 400 and 700 nm?

1-4. Determine the wavelength and momentum of a photon whose energy equals the rest-mass energy of an electron.

1-5. Show that the rest-mass energy of an electron is 0.511 MeV.

1-6. Show that the relativistic momentum of an electron, accelerated through a potential difference of 1 million volts, can be conveniently expressed as 1.422 MeV/c, where c is the speed of light.

1-7. Show that the wavelength of a photon, measured in angstroms, can be found from its energy, measured in electron volts, by the convenient relation,

$$\lambda(\text{Å}) = \frac{12,400}{E(\text{eV})}$$

1-8. Show that the relativistic kinetic energy,

$$E_k = mc^2(\gamma - 1)$$

reduces to the classical expression, $\frac{1}{2}mv^2$, when $v \ll c$.

1-9. A proton is accelerated to a kinetic energy of 2 billion electron volts (2 BeV). Find (a) its momentum; (b) its de Broglie wavelength; (c) the wavelength of a photon with the same total energy.

1-10. Solar radiation is incident at the earth's surface at an average of 1353 W/m² on a surface normal to the rays. For a mean wavelength of 550 nm, calculate the number of photons falling on 1 cm² of the surface each second.

1-11. Two parallel beams of electromagnetic radiation with different wavelengths deliver the same power to equivalent surface areas normal to the beams. Show that the numbers of photons striking the surfaces per second for the two beams are in the same ratio as their wavelengths.

REFERENCES

[1] Ronchi, Vasco. *The Nature of Light*. Cambridge: Harvard University Press, 1970.

[2] Hoffmann, Banesh. *The Strange Story of the Quantum*. New York: Dover Publications, 1959.

[3] Guillemin, Victor. *The Story of Quantum Mechanics*. New York: Charles Scribner's Sons, 1968.

[4] Feinberg, Gerald. "Light." In *Lasers and Light*, pp. 4–13. San Francisco: W. H. Freeman and Company Publishers, 1969.

[5] Cantor, G. N. *Optics after Newton*. Dover, N.H.: Manchester University Press, 1983.

[6] Strong, John. *Concepts of Classical Optics*. San Francisco: W. H. Freeman and Company Publishers, 1958. Appendix G.

2

Production and Measurement of Light

INTRODUCTION

Electromagnetic radiation may vary in wavelength (or frequency) and in "strength." Classification due to variation in wavelength is summarized in the *electromagnetic spectrum*. Variations in strength are described in more precise physical terms, which have developed in the areas called *radiometry* and *photometry*. *Sources* and *detectors* of electromagnetic radiation can be classified on the basis of their spectral range and the strength of signal produced (sources) or detected (detectors). These considerations are essential to the production and measurement of electromagnetic radiation and are discussed in this chapter.

2-1 ELECTROMAGNETIC SPECTRUM

An electromagnetic disturbance that propagates through space as a wave may be *monochromatic*, that is, characterized for practical purposes by a single wavelength, or *polychromatic*, in which case it is represented by many wavelengths, either discrete or in a continuum. The distribution of energy among the various constituent waves is called the *spectrum* of the radiation, and the adjective *spectral* implies a dependence on wavelength. Various regions of the *electromagnetic spectrum* are referred to by particular names, such as radio waves, cosmic rays, light, and ultraviolet radiation, because of differences in the way they are produced or detected. Most of the common descriptions are given in Figure 2-1, in which the electromagnetic spectrum is displayed in terms of both wavelength (λ) and frequency (ν). The two quantities are related, as with all wave motion, through the wave velocity (c):

$$c = \lambda\nu \qquad (2\text{-}1)$$

The radiation described in Figure 2-1 is assumed to propagate in free space, for which, approximately, $c = 3 \times 10^8$ m/s. Common units for wavelength shown are the *angstrom* (1 Å $= 10^{-10}$ m), the *nanometer* (1 nm $= 10^{-9}$ m), and the *micrometer* (1 $\mu m = 10^{-6}$ m). The regions ascribed to various types of waves, as shown, are not precisely bounded. Regions may overlap, as in the case of the continuum from X-rays to gamma rays. The choice of label will depend on the manner in which the

Figure 2-1 Electromagnetic spectrum, arranged by wavelength in meters and frequency in hertz. The narrow portion occupied by the visible spectrum is highlighted.

radiation is either produced or used. The narrow range of electromagnetic waves from approximately 380 to 770 nm is capable of producing a visual sensation in the human eye and is properly referred to as "light." This *visible region* of the spectrum, which includes the spectrum of colors from red (long-wavelength end) to violet (short-wavelength end) is bounded by the invisible *ultraviolet* and *infrared* regions, as shown. The three regions taken together comprise the *optical spectrum*, that region of the electromagnetic spectrum of special interest in a textbook of optics.

2-2 RADIOMETRY

Radiometry is the science of measurement of electromagnetic radiation. In the discussion we present the *radiometric quantities* or physical terms used to characterize the energy content of radiation. Later we briefly discuss some of the more common principles used in the instruments designed to measure radiation. Many radiometric terms have been introduced and used; however we include here only approved International System (SI) units. These terms and their units are summarized in Table 2-1.[1]

TABLE 2-1 RADIOMETRIC AND PHOTOMETRIC TERMS

Term	Symbol (units)	Defining equation	Term	Symbol (units)	Defining equation
Radiant energy	Q_e (J = W-s)	—	Luminous energy	Q_v (lm-s) (talbot)	—
Radiant energy density	w_e (J/m³)	$w_e = dQ_e/dV$	Luminous energy density	w_v (lm-s/m³)	$w_v = dQ_v/dV$
Radiant flux	Φ_e (W)	$\Phi_e = dQ_e/dt$	Luminous flux	Φ_v (lm)	$\Phi_v = dQ_v/dt$
Radiant exitance	M_e (W/m²)	$M_e = d\Phi_e/dA$	Luminous exitance	M_v (lm/m²)	$M_v = d\Phi_v/dA$
Irradiance	E_e (W/m²)	$E_e = d\Phi_e/dA$	Illuminance	E_v (lm/m²) or (lx)	$E_v = d\Phi_v/dA$
Radiant intensity	I_e (W/sr)	$I_e = d\Phi_e/d\omega$	Luminous intensity (candlepower)	I_v (lm/sr) or (cd)	$I_v = d\Phi_v/d\omega$
Radiance	L_e (W/sr-m²)	$Le = dI_e/dA \cos \theta$	Luminance	L_v (cd/m²)	$L_v = dI_v/dA \cos \theta$

Abbreviations: J, joule; W, watt; m, meter; lm, lumen; lx, lux; sr, steradian; cd, candela.

Radiometric quantities appear either without subscripts or with the subscript *e* (*electromagnetic*) to distinguish them from similar photometric terms, to be described afterwards. The terms *radiant energy*, Q_e (J = joules), *radiant energy density*, w_e (J/m³), and *radiant flux*, Φ_e (W = watts = J/s), need no further explanation. *Radiant flux density* at a surface, measured in watts per square meter, may be either emitted (scattered, reflected) from a surface, in which case it is called *radiant exitance*, M_e, or incident onto a surface, in which case it is called *irradiance*, E_e. The radiant flux (Φ_e) emitted per unit of solid angle (ω) by a point source in a given direction (Figure 2-2) is called the *radiant intensity*, I_e. This quantity, often confused with *irradiance*, is given by

$$I_e = \frac{d\Phi}{d\omega} \text{ (W/sr)} \tag{2-2}$$

---

[1] The introduction "in the abstract" of so many new units, some rarely used and others misused, is not very palatable pedagogically. Table 2-1 is meant to serve as a convenient summary that can be referred to when needed.

Figure 2-2 The radiant intensity is the flux through the cross section dA per unit of solid angle. Here the solid angle $d\omega = dA/r^2$.

where sr = steradian. The radian intensity I_e from a sphere radiating Φ_e W of power uniformly in all directions, for example, is $\Phi_e/4\pi$ W/sr, since the total surrounding solid angle is 4π sr.

The familiar inverse-square law of radiation from a point source, illustrated in Figure 2-3, is now apparent by calculating the irradiance of a point source on a spherical surface surrounding the point, of solid angle 4π sr and surface area $4\pi r^2$. Thus

$$E_e = \frac{\Phi_e}{A} = \frac{4\pi I_e}{4\pi r^2} = \frac{I_e}{r^2} \qquad (2\text{-}3)$$

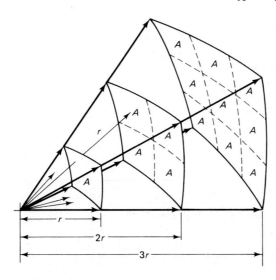

Figure 2-3 Illustration of the inverse square law. The flux leaving a point source within any solid angle is distributed over increasingly larger areas, producing an irradiance that decreases inversely with the square of the distance.

The *radiance*, L_e, describes the radiant intensity per unit of projected area, perpendicular to the specified direction, and is given by

$$L_e = \frac{dI_e}{dA \cos \theta} = \frac{d^2\Phi_e}{d\omega \, (dA \cos \theta)} \; (\text{W/m}^2\text{-sr}) \qquad (2\text{-}4)$$

The importance of the radiance is suggested in the following considerations. Suppose a plane radiator or reflector is perfectly *diffuse*, by which we mean that it radiates uniformly in all directions. The radiant intensity is measured for a fixed solid angle defined by the fixed aperture A_p at some distance r from the radiating surface, shown in Figure 2-4. The aperture might be the input aperture of a detecting instrument measuring all the flux that so enters. When viewed at $\theta = 0°$, along the normal to the surface, a certain maximum intensity $I(0)$ is observed. As the aperture is moved along the circle of radius r, thereby increasing the angle θ, the cross section of radiation presented by the surface decreases in such a way that

$$I(\theta) = I(0) \cos \theta \qquad (2\text{-}5)$$

a relation called *Lambert's cosine law*. If the radiance is determined at each angle θ, it is found to be constant, because the intensity must be divided by the projected

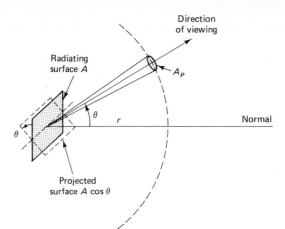

Direction
of viewing

Radiating
surface A

A_P

θ

r

Normal

θ

Projected
surface $A \cos \theta$

Figure 2-4 Radiant flux collected along a direction making an angle θ with the normal to the radiating surface. The projected area of the surface is shown shaded.

area $A \cos \theta$ such that the cosine dependence cancels:

$$L_e = \frac{I(\theta)}{A \cos \theta} = \frac{I(0) \cos \theta}{A \cos \theta} = \frac{I(0)}{A} = \text{constant} \qquad (2\text{-}6)$$

Thus when a radiating (or reflecting) surface has a radiance that is independent of the viewing angle, the surface is said to be perfectly diffuse or a *Lambertian surface*.

We show next that the radiance has the same value at any point along a ray propagating in a uniform, nonabsorbing medium. Figure 2-5 illustrates a narrow beam of radiation in such a medium, including a central ray and a small bundle of surrounding rays (not shown) that pass through the elemental areas dA_1 and dA_2 situated at different points along the beam. The central ray makes angles of θ_1 and θ_2,

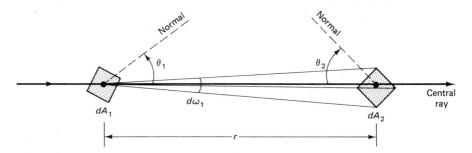

Normal

θ_1

Normal

θ_2

Central
ray

dA_1

$d\omega_1$

dA_2

r

Figure 2-5 Geometry used to show the invariance of the radiance in a uniform, lossless medium.

respectively, relative to the area normals, as shown. The solid angle $d\omega_1 = dA_2 \cos \theta_2/r^2$, where $dA_2 \cos \theta_2$ represents the projection of area dA_2 normal to the central ray. According to Eq. (2-4), the radiance L_1 at dA_1 is given by

$$L_1 = \frac{d^2\Phi_1}{d\omega_1(dA_1 \cos \theta_1)} = \frac{d^2\Phi_1}{(dA_2 \cos \theta_2/r^2)(dA_1 \cos \theta_1)} \qquad (2\text{-}7)$$

By a similar argument, in which we reverse the roles of dA_1 and dA_2 in the figure,

$$L_2 = \frac{d^2\Phi_2}{d\omega_2(dA_2 \cos \theta_2)} = \frac{d^2\Phi_2}{(dA_1 \cos \theta_1/r^2)(dA_2 \cos \theta_2)} \qquad (2\text{-}8)$$

For a nonabsorbing medium, the power associated with the radiation passing through the continuous bundle of rays remains constant, that is, $d\Phi_1 = d\Phi_2$, so that we can conclude from Eqs. (2-7) and (2-8) that $L_1 = L_2$. It follows that the radiance of the beam is also the radiance of the source, at the initial point of the beam, or $L_1 = L_2 = L_0$.

Suppose, referring to Figure 2-6, that we wish to know the quantity of radiant power reaching an element of area dA_2 on surface S_2 due to the source element dA_1 on surface S_1. The line joining the elemental areas, of length r_{12}, makes angles of θ_1

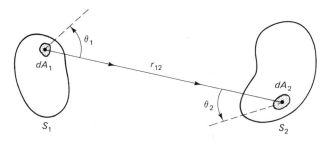

Figure 2-6 General case of the illumination of one surface by another radiating surface. Each elemental radiating area dA_1 contributes to each elemental irradiated area dA_2.

and θ_2 with the respective normals to the surfaces, as shown. The radiant power is $d^2\Phi_{12}$, a second-order differential because both the source and receptor are elemental areas. By Eq. (2-7) or Eq. (2-8),

$$d^2\Phi_{12} = \frac{L\,dA_1\,dA_2 \cos \theta_1 \cos \theta_2}{r_{12}^2}$$

and the total radiant power at the entire second surface due to the entire first surface is, by integration,

$$\Phi_{12} = \int_{A_1}\int_{A_2} \frac{L \cos \theta_1 \cos \theta_2\, dA_1\, dA_2}{r_{12}^2} \tag{2-9}$$

By adding powers rather than amplitudes in this integration, we have tacitly assumed that the radiation source emits incoherent radiation. We shall say more about coherent and incoherent radiation later.

2-3 PHOTOMETRY

Radiometry applies to the measurement of all radiant energy. *Photometry*, on the other hand, applies only to the visible portion of the optical spectrum. Whereas radiometry involves purely physical measurement, photometry takes into account the response of the human eye to radiant energy at various wavelengths and so involves psycho-physical measurements. The distinction rests on the fact that the human eye, as a detector, does not have a "flat" spectral rsponse; that is, it does not respond with equal sensitivity at all wavelengths. If three sources of light of equal radiant power but radiating blue, yellow, and red light, respectively, are observed visually, the yellow source will appear to be far brighter than the others. When we use photometric quantities, then, we are measuring the properties of visible radiation as they appear to the normal eye, rather than as they appear to an "unbiased" detector. Since not all human eyes are identical, a standard response has been determined by the International Commission on Illumination (CIE) and is reproduced in Figure 2-7. The relative response or sensation of brightness for the eye is plotted versus wavelength, showing that peak sensitivity occurs at the "yellow-green" wavelength of 555 nm. Actually the curve shown is the luminous efficiency of the eye for *photopic vision*, that is, when adapted for day vision. For lower levels of illumination, when adapted for night or *scotopic vision*, the curve shifts toward the green, peaking at 510 nm. It is interesting to note that human color sensation is a function of illumi-

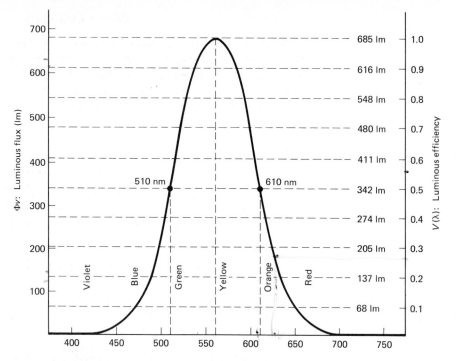

Figure 2-7 CIE luminous efficiency curve. The luminous flux corresponding to 1 W of radiant power at any wavelength is given by the product of 685 lm and the luminous efficiency at the same wavelength: $\Phi_v(\lambda) = 685V(\lambda)$ for each watt of radiant power.

nation and is almost totally absent at lower levels of illumination. One way to confirm this is to compare the color of stars, as they appear visually, to their photographic images made on color film using a suitable time exposure. Another, very dramatic way to demonstrate human color dependence on illumination is to project a 35-mm color slide of a scene onto a screen with a low current in the projector bulb. At sufficiently low currents, the scene appears black and white. As the current is increased, the full colors in the scene gradually emerge. On the other hand, very intense radiation may be visible beyond the limits of the CIE curve. The reflection of an intense laser beam of wavelength 694.3 nm from a ruby laser is easily seen. Even the infrared radiation around 900 nm from a gallium-arsenide semiconductor laser can be seen as a deep red.

Radiometric quantities are now related to photometric quantities through the luminous efficiency curve of Figure 2-7 in the following way: Corresponding to a radiant flux of 1 W at the peak wavelength of 555 nm, where the luminous efficiency is maximum, the *luminous flux* is defined to be 685 lm. Then, for example, at $\lambda = 610$ nm, in the range where the luminous efficiency is 0.5 or 50%, 1 W of radiant flux would produce only 0.5×685 or 342 lm of luminous flux. The curve shows that again at $\lambda = 510$ nm, in the blue-green, the brightness has dropped to 50%.

Photometric units, in terms of their definitions, parallel radiometric units. This is amply demonstrated in the summary and comparison provided in Table 2-1. In general, analogous units are related by the following equation:

$$\text{photometric unit} = K(\lambda) \times \text{radiometric unit} \qquad (2\text{-}10)$$

where $K(\lambda)$ is called the *luminous efficacy*. If $V(\lambda)$ is the *luminous efficiency*, as given on the CIE curve, then

$$K(\lambda) = 685V(\lambda) \qquad (2\text{-}11)$$

Photometric terms are preceded by the word *luminous* and the corresponding units are subscripted with the letter v (*visual*); otherwise the symbols are the same. Notice that the SI unit of luminous energy is the *talbot*, the unit of luminous incidence is the *lux* (lx), and the unit of luminous intensity is the *candela* (cd). Notice also the distinction between the analogous terms *irradiance* (radiometric) and *illuminance* (photometric).

Example

A light bulb emitting 100 W of radiant power is positioned 2 m from a surface. The surface is oriented perpendicular to a line from the bulb to the surface. Calculate the irradiance at the surface. If all 100 W is emitted from a red bulb at $\lambda = 650$ nm, calculate also the illuminance at the surface.

Solution

$$\text{irradiance } E_e = P/A = 100 \text{ W}/4\pi\,(2 \text{ ft})^2 \cong 2 \text{ W/m}^2$$

From the CIE curve, $V(650 \text{ nm}) = 0.1$. Thus

$$\text{illuminance } E_v = K(\lambda) \times \text{irradiance} = 685\, V(\lambda) \times E_e$$
$$E_v = 685 \times 0.1 \times 2 = 137 \text{ lm/m}^2 \text{ or lux}$$

Thus, whereas a *radiometer* with aperture at the surface measures 2 W/m², a *photometer* in the same position would be calibrated to read 137 lx.

When the radiation consists of a spread of wavelengths, the radiometric and the photometric terms may be functions of wavelength. This dependence is noted by preceding the term with the word *spectral* and by using a subscript λ or adding the λ in parentheses. For example, *spectral radiant flux* is denoted by $\Phi_{e\lambda}$ or $\Phi_e(\lambda)$. The total radiant flux is then determined by integration over the wavelength region of interest:

$$\Phi_e = \int_{\lambda_1}^{\lambda_2} \Phi_e(\lambda)\, d\lambda$$

2-4 BLACKBODY RADIATION

A *blackbody* is an ideal absorber: All radiation falling on a blackbody, irrespective of wavelength or angle of incidence, is completely absorbed. It follows that a blackbody is also a perfect emitter: No body at the same temperature can emit more radiation at any wavelength or into any direction than a blackbody. Blackbodies are approached in practice by blackened surfaces and by tiny apertures in radiating cavities. An excellent example of a blackbody is the surface formed by the series of sharp edges of a stack of razor blades. The array of blade edges effectively traps the incident light, resulting in almost perfect absorption.

The spectral radiant exitance M_λ of a blackbody can be derived on theoretical grounds. It was first so derived by Max Planck, who found it necessary to postulate quantization in the process of radiation and absorption by the blackbody. The result of this calculation [1] is given by

$$M_\lambda = \frac{2\pi hc^2}{\lambda^5} \left(\frac{1}{e^{hc/\lambda kT} - 1} \right) \qquad (2\text{-}12)$$

where the physical constants h, c, and k represent the Planck constant, the speed of light in vacuum, and the Boltzmann constant, respectively. When the known values of these constants are used, the result is

$$M_\lambda = \frac{3.745 \times 10^8}{\lambda^5} \left(\frac{1}{e^{14388/\lambda T} - 1} \right) (\text{W/m}^2\text{-}\mu\text{m})$$

where λ is in micrometers and T is in Kelvin. The quantity M_λ is plotted in Figure 2-8 for different temperatures. The spectral radiant exitance is seen to increase with

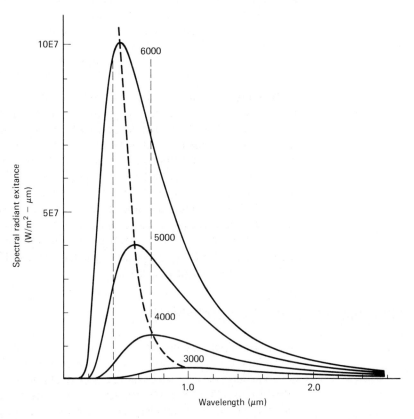

Figure 2-8 Blackbody radiation spectral distribution at four Kelvin temperatures. The vertical dashed lines mark the visible spectrum, and the dashed curve connecting the peaks of the four curves illustrates the Wien displacement law ($5E7 = 5 \times 10^7$).

absolute temperature at each wavelength. The peak exitance also shifts toward shorter wavelengths with increasing temperature, falling into the visible spectrum (dashed vertical lines) at $T = 5000$ and 6000 K. The variation of λ_{max}, the wavelength at which M_λ peaks, with the temperature can be found by differentiating M_λ with respect to λ and setting this equal to zero. The result is the *Wien displacement law*,[2] given by

$$\lambda_{\text{max}} T = \frac{hc}{5k} = 2.88 \times 10^3 \ (\mu\text{m-K}) \tag{2-13}$$

---

[2] Although Wien's law is often found written in this form, the number 5 is an approximation to 4.965 that, when used, gives a more accurate Wien constant of $2.898 \times 10^3 \ \mu\text{m-K}$.

and is indicated in Figure 2-8 by the dashed curve. If, on the other hand, the spectral exitance of Eq. (2-12) is integrated over all wavelengths, the total radiant exitance or area under the blackbody radiation curve at temperature T is

$$M = \sigma T^4 \qquad (2\text{-}14)$$

known as the *Stefan-Boltzmann law*, with σ as the Stefan-Boltzmann constant, equal to 5.67×10^{-8} W/m^2-K^4.

The radiation from real surfaces is always less than that of the blackbody or *Planckian source* and is accounted for quantitatively by the *emissivity* ϵ. Distinguishing now between the radiant exitance M of a measured specimen and that of a blackbody M_{bb} at the same temperature, we define

$$\epsilon(T) = \frac{M}{M_{bb}} \qquad (2\text{-}15)$$

If the radiant exitance of the blackbody and the specimen are compared in various narrow wavelength intervals, a spectral emissivity is calculated, which is not in general a constant. In those special cases where the emissivity is independent of wavelength, the specimen is said to be a *graybody*. In this instance the spectral exitance of the specimen is proportional to that of the blackbody and their curves are the same except for a constant factor. The spectral radiation from a heated tungsten wire, for example, is close to that of a graybody with $\epsilon = 0.4$–0.5.

Blackbody radiation is used to establish a color scale in terms of absolute temperature alone. The *color temperature* of a specimen of light is then the temperature of the blackbody with the closest spectral energy distribution. In this way, a candle flame can be said to have a color temperature of 1900 K, whereas the sun has a typical color temperature of 5500 K.

2-5 SOURCES OF OPTICAL RADIATION

Sources of light may be natural, as in the case of sunlight and skylight, or artificial, as in the case of incandescent or discharge lamps. Light from various sources may also be classified as monochromatic, spectral line, or continuous. The way in which energy is distributed in the radiation determines the color of the light and, consequently, the color of surfaces seen under the light. Anyone who has used a camera is aware that the actual color response of film depends on the type of light used to illuminate the subject.

The following brief survey of sources of light cannot hope to be comprehensive; rather it is intended to direct attention to an extensive area of practical information. For the purposes of this limited survey, we classify a number of sources as follows:

    A. Sunlight, skylight

    B. Incandescent sources
       1. Blackbody sources
       2. Nernst glower and globar
       3. Tungsten filament

    C. Discharge lamps
       1. Monochromatic and spectral sources
       2. High-intensity sources
          a. Carbon arc
          b. Compact short arc

         c. Flash
         d. Concentrated zirconium arc
      3. Fluorescent lamps
   D. Semiconductor light-emitting diodes (LEDs)
   E. Coherent source—laser

Daylight is a combination of sunlight and skylight. Direct light from the sun has a spectral distribution that is clearly different from that of skylight, which has a predominantly blue hue. A plot of spectral solar irradiance is given in Figure 2-9.

Figure 2-9 Solar spectral irradiation above the atmosphere and on a horizontal surface at sea level: clear day, sun at zenith.

Extraterrestrial solar radiation indicates that the sun behaves approximately as a blackbody with a temperature of 6000 K at its center and 5000 K at its edge, but the radiation received at the earth's surface is modified by absorption in the earth's atmosphere. The annual average of total irradiance just outside the earth's atmosphere is the *solar constant*, 1350 W/m^2. Although solar radiation is not routinely used as a light source in the laboratory, high-pressure xenon lamps, with appropriate filters, provide an excellent artificial source for solar simulation and are available commercially.

Artificial optical sources that use light produced by a material heated to incandescence by an electric current are called *incandescent lamps*. Radiation arises from the de-excitation of the atoms or molecules of the material after they have been thermally excited. The energy is emitted over a broad continuum of wavelengths. Commercially available *blackbody sources* consist of cavities equipped with a small hole. Radiation from the small hole has an emissivity that is essentially constant and equal to unity. Such sources are available at operating temperatures from that of liquid nitrogen (−196°C) to 3000°C. Incandescent sources particularly useful in the infrared include the *Nernst glower*. This source is a cylindrical tube or rod of refractory material (zirconia, yttria, thoria), heated by an electric current and useful from the visible to around 30 μm. The Nernst glower behaves like a graybody with an emissivity greater than 0.75. When the material is a rod of bonded silicon carbide, the source is called a *globar*, approximating a graybody with an average emissivity of 0.88 (see Figure 2-10).

The tungsten filament lamp is the most popular source for optical instrumentation designed to use continuous radiation in the visible and into the infrared region. The lamp is available in a wide variety of filament configurations and of bulb and base shapes. The filament is in coil or ribbon form, the ribbon providing a more uni-

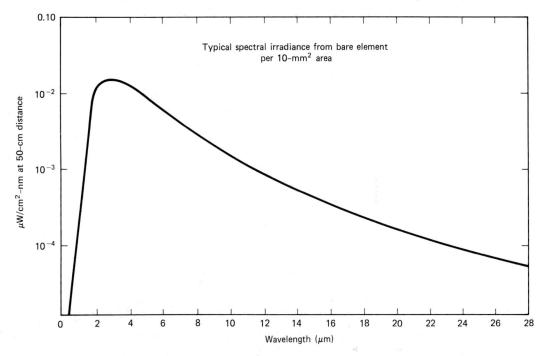

Figure 2-10 Globar infrared source, providing continuous usable emission from 1 to over 25 μm at a temperature variable up to 1000 K. The source is a 6.2-mm diameter silicon carbide resistor. (Oriel Corp., General Catalogue, Stratford, Conn.)

form radiating surface. The bulb is usually a glass envelope, although quartz is used for higher-temperature operation. Radiation over the visible spectrum approximates that of a graybody, with emissivities approaching unity for tightly coiled filaments. Lumen output depends both on the filament temperature and the electrical power input (wattage). During operation, tungsten gradually evaporates from the filament and deposits on the inner bulb surface, leaving a dark film that can decrease the flux output by as much as 18% during the life of the lamp. This process also weakens the filament and increases its electrical resistance. The presence of an inert gas, usually nitrogen or argon, introduced at around 0.8 atmospheric pressure, helps to slow down the evaporation. More recently this problem has been minimized by the addition of a halogen vapor (iodine, bromine) to the gas in the *quartz-halogen* or *tungsten-halogen* lamp. The halogen vapor functions in a regenerative cycle to keep the bulb free of tungsten. Iodine reacts with the deposited tungsten to form the gas tungsten iodide, which then dissociates at the hot filament to redeposit the tungsten and free the iodine for repeated operation. A typical spectral irradiance curve for a 100-W quartz-halogen filament source is given in Figure 2-11. The lamp approximates a 3000-K graybody source, providing a useful continuum from 0.3 to 2.5 μm. In *tungsten arc lamps*, an arc discharge between two tungsten electrodes heats the

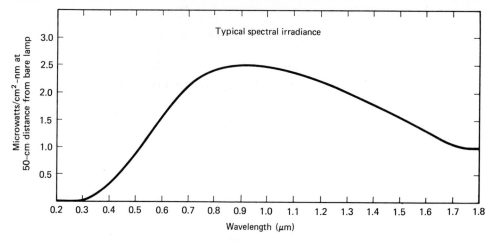

Figure 2-11 Spectral irradiance from a 100-W quartz halogen lamp, providing continuous radiation from 0.3 to 2.5 μm. (Oriel Corp., General Catalogue, Stratford, Conn.)

electrodes to incandescence in an atmosphere of argon, providing a spectral distribution of radiation like that of tungsten lamps at 3100 K.

The *discharge lamp* depends for its radiation output on the dynamics of an electrical discharge in a gas. A current is passed through the ionized gas between two electrodes, sealed in a glass or quartz tube. (Glass envelopes absorb ultraviolet radiation below about 300 nm, whereas quartz transmits down to about 180 nm.) An electric field accelerates electrons sufficiently to ionize the vapor atoms. The source of the electrons may be a heated cathode (thermionic emission), a strong field applied at the cathode (field emission), or the impact of positive ions on the cathode (secondary emission). De-excitation of the excited vapor atoms provides a release of energy in the form of photons of radiation. High-pressure and high-current operation generally results in a continuous spectral output, in addition to spectral lines characteristic of the vapor. At lower pressure and current, sharper spectral lines appear, and the background continuum is minimal. When sharp spectral lines are desired, as in *monochromatic sources*, the lamp is designed to operate at low temperature, pressure, and current. The *sodium arc lamp*, for example, provides radiation almost completely confined to a narrow "yellow" band due to the spectral lines at 589.0 and 589.6 nm. The low-pressure *mercury discharge tube* is often used to provide, with the help of isolating filters, strong monochromatic radiation at wavelengths of 404.7 and 435.8 nm (violet), 546.1 nm (green), and 577.0 and 579.1 nm (yellow). Other gases or vapors may be used to provide spectral lines of other desired wavelengths. For the highest spectral purity, particular isotopes of the gas are used.

When high intensity rather than spectral purity is desired, other designs become available. Perhaps the oldest source of this kind is the *carbon arc*, still widely used in searchlights and motion picture projectors. The high-current arc is formed between two carbon rods in air. A 200-A carbon arc lamp may have a peak luminance of 1600 cd/mm². The source has a spectral distribution close to that of a graybody at 6000 K. A wide range of spectral outputs is possible by using different materials in the core of the carbon rod. When the arc is enclosed in an atmosphere of vapor at high pressure, the lamp is a *compact short-arc source* and the radiation is divided between line and continuous spectra. See Figure 2-12 for a sketch of this type of lamp and its housing. The most useful of these lamps, designed to operate from 50 W to 25 kW, are the high-pressure *mercury arc lamp*, with comparatively weak background radiation but strong spectral lines and a good source of ultraviolet; the *xenon arc lamp*, with practically continuous radiation from the near-ultraviolet through the visible and into the near-infrared; and the *mercury-xenon arc lamp*, providing essentially the mercury spectrum but with xenon's contribution to the continuum and its own strong spectral emission in the 0.8- to 1-μm range. As mentioned earlier, the color quality of the xenon lamp is similar to that of sunlight at 6000-K color temperature. Spectral emission curves for Xe and Hg-Xe lamps are shown in Figures 2-13 and 2-14. The *hydrogen* and *deuterium arc lamps* are ideal

Figure 2-13 Spectral emission for xenon compact arc lamp. (Canrad-Hanovia, Inc.)

Figure 2-14   Spectral emission for Hg-Xe arc lamp. (Canrad-Hanovia, Inc.)

for ultraviolet spectroscopy because they produce a high radiance with a continuous background in the ultraviolet region. Figure 2-15 shows the typical spectral output of a deuterium lamp, which produces a line-free continuum from below 180 nm to 400 nm.

Flash tubes represent a high output source of visible and near infrared radiation, produced by a rapid discharge of stored electrical energy through a gas-filled tube. The gas is most often xenon. The *photoflash* tube, in contrast, provides high-intensity, short-duration illumination by the rapid combustion of metallic (aluminum or zirconium) foil or wire in a pure oxygen atmosphere.

When a high-intensity *point source* of radiation is desired in optical instrumentation, a useful lamp is the *concentrated zirconium arc lamp*, with wattages varying from 2 W to 300 W. Zirconium vapor is formed by evaporation from an oxide-

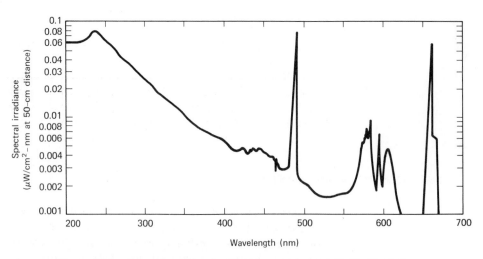

Figure 2-15   Spectral emission for deuterium arc lamp at 50 W. (Oriel Corp., General Catalogue, Stratford, Conn.)

coated cathode in an argon atmosphere. Radiation originates both from the incandescence of a molten cathode surface and from the excited zirconium vapor and argon gas. It is viewed through a small hole, 0.13 to 2.75 mm in diameter, in a metallic anode, and through an optically flat window sealed into the tube. The spectral distribution approximates that of a 3200-K graybody source.

The familiar *fluorescent lamps* use low-pressure, low-current electrical discharges in mercury vapor. The ultraviolet radiation from excited mercury atoms is converted to visible light by stimulating fluorescence in a phosphor coating on the inside of the glass-envelope surface. Spectral outputs depend on the particular phosphor used. "Daylight" lamps, for example, use a mixture of zinc beryllium silicate and magnesium tungstate.

A very different type of light source is the low-intensity light-emitting diode (LED), a solid-state device employing a *p-n* junction in a semiconducting crystal. The device is hermetically sealed in an optically centered package. When a small bias voltage is applied in the forward direction, optical energy is produced by the recombination of electrons and holes in the vicinity of the junction. Popular LEDs include the infrared GaAs device, with a peak output wavelength near 900 nm, and the visible SiC device, with peak output at 580 nm. LEDs provide narrow spectral emission bands, as evident in Figure 2-16. Solid solutions of similar compound semiconductor materials produce output in a variety of spectral regions when the composition of the alloy is varied.

Figure 2-16 Spectral output from a GaAs light-emitting diode.

The *laser* is a very important modern source of coherent and extremely monochromatic radiation, capable of very high intensity. Lasers emit radiation in the ultraviolet, visible, and infrared regions of the spectrum. Because of the central role that lasers play in optical instrumentation, they are treated in a separate chapter.

2-6 DETECTORS OF RADIATION

Any device that produces a measurable physical response to incident radiant energy is a detector. The most common detector is, of course, the eye. Whereas the eye provides a qualitative and subjective response, the detectors discussed here provide a quantitative and objective response. In view of the unique role played by the eye in human vision, it is treated separately in another chapter.

The most common detectors may be classified as follows:

A. Thermal detectors
   1. Thermocouples and thermopiles
   2. Bolometers and thermistors
   3. Pyroelectric
   4. Pneumatic or Golay
B. Quantum detectors
   1. Photoemissive—phototubes and photomultipliers
   2. Photoconductive
   3. Photovoltaic
   4. Photographic

When the primary measurable response of a detector to incident radiation is a rise in temperature, the device is a *thermal detector*. The receptor is usually a blackened metal strip or flake, which efficiently absorbs at all wavelengths. Such a device, in which an increase in temperature at a junction of two dissimilar metals or semiconductors generates a voltage, is called a *thermocouple* (Figure 2-17a). When the

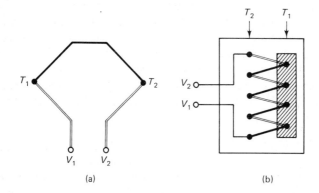

Figure 2-17 (a) Thermocouple made of dissimilar materials (dark and light lines) joined at points T_1 and T_2, where a difference in temperature produces an emf between terminals V_1 and V_2. (b) Thermocouple made of couples in series. Radiation is absorbed at the junctions T_1 in thermal contact with a black absorber and thermally insulated from the junctions T_2.

effect is enhanced by using an array of such junctions in series, the device is called a *thermopile* (Figure 2-17b). Thermal detectors also include bulk devices that respond to a rise in temperature by a significant change in resistance. Such an instrument may employ as its sensitive element either a metal (*bolometer*) or, more commonly, a semiconductor (*thermistor*). Typically, two blackened sensitive elements are used in adjacent arms of a bridge circuit, one of which is exposed to the incident radiation. The imbalance in the circuit, due to the change in resistance, is indicated by a galvanometer deflection or current. In the *pyroelectric detector*, the temperature change causes a change in the surface charge of certain materials, like lithium tantalate or triglycine sulfate (TGS) that exhibit the pyroelectric effect. The detector behaves like a capacitor whose charge is a function of the temperature. The *Golay cell* measures instead the thermal expansion of a gas. Heat absorbed by a blackened membrane is transmitted to the gas in an airtight chamber. The pressure rise in the gas is usually detected optically, by the deflection of a mirror. A schematic of the cell is shown in Figure 2-18. Thermal detectors are generally characterized by a slow response to changes in the incident radiation. If the detector is expected to follow a changing input signal, such as a pulse, thermal detectors are not as desirable as the faster-responding quantum detectors to be discussed next. The speed of response is described by a *time constant*, a measure of the time required to regain equilibrium in output after a change in input. Thus quantum detectors are better suited to high-frequency operation.

Figure 2-18 Golay pneumatic infrared detector. (Oriel Corp., General Catalogue, Stratford, Conn.)

Quantum detectors respond to the rate of incidence of photons of the radiation rather than to thermal energy. Photons interact directly with the electrons in the detector material. When the measurable effect is the release of electrons from an illuminated surface, the device is called a *photoemissive detector*. A photosensitive surface, typically containing alkali metals, absorbs incident photons that transfer enough energy to enable some electrons to overcome the work function and escape from the surface. If the photoemitted electrons are simply collected by a positive-biased anode in an evacuated tube, enabling a current to be drawn into an external circuit, the detector is called a *diode phototube*. When the signal is internally amplified by secondary electron emission, the detector is a *photomultiplier*; see Figure 2-19. In this case the primary photoelectrons are accelerated, so that as a result of a sequence of collisions, each multiplying the current by the addition of secondary electrons, an avalanche of electrons becomes available at the output corresponding to each primary photoelectron.

Figure 2-19 One type of photomultiplier tube structure. Electrons photoemitted from the cathode are accelerated along zigzag paths down the tube so as to strike each of the curved dynode surfaces, each time producing additional secondary electrons. The multiplied current is collected at the anode.

Another means of amplification, used in the *gas-filled photocell*, allows the generation of additional electrons by ionization of the residual gas. In the case of energetic photons ($\lambda < 550$ nm), the sensitivity of photoemissive detectors is sufficient to allow the counting of individual photons. Such detectors possess superior sensitivity in the visible and ultraviolet spectral ranges. For wavelengths in the infrared, over 1 μm, photoemitters are not available and *photoconducting detectors* are used. In these detectors, photons absorbed into thin films or bulk material produce additional free charges in the form of electron-hole pairs. Both the negative (electrons) and positive (holes) charges increase the electrical conductivity of the sample. Without illumination, a bias voltage across such a material with high intrinsic resistivity produces a small or "dark" current. The presence of illumination and the extra free-charge carriers so produced effectively lower the resistance of the material, and a larger photocurrent results. Semiconducting compounds CdS and CdSe are often used in the visible and near-infrared regions (see Figure 2-20), while farther out in the near-infrared region, the compounds PbS (0.8 to 3 μm) and PbSe (1 to 5 μm) are popular.

The most common *photovoltaic detector* is a *p-n* junction, the *semiconductor photodiode*. This device consists of a junction between doped *p*-type (rich in posi-

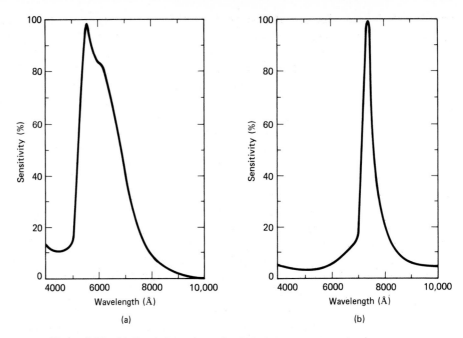

Figure 2-20 (a) Spectral response of a CdS photoconducting cell. The peak response at 5500 Å closely matches the response of the human eye. The cell is useful with incandescent, fluorescent, or neon lamps. (b) Spectral response of a CdSe photoconducting cell with peak at 7350 Å. The cell is sensitive to the near infrared and is useful with incandescent or neon lamps.

tive charge carriers) and doped *n*-type (rich in negative charge carriers) materials, most often silicon. Doping involves adding small amounts of an impurity to the semiconductor to provide either an excess (*n*-type) or deficiency (*p*-type) of conduction electrons. In a narrow region between these materials, a built-in electric field occurs as a consequence of current equilibrium. When photons are absorbed in the vicinity of the junction, the electron-hole pairs are separated by the field, causing a change in voltage, the *photovoltaic effect*. The solar cell and the photographic exposure meter are perhaps the best known applications of this detector. A variation of the photovoltaic cell, the *avalanche diode*, provides an internal mechanism of amplification that results in enhanced sensitivity out to around 1.5 μm. In the region of 1 to 8 μm, the semiconductor compounds PbS, PbSe, and PbTe possess a large photovoltaic effect and greater sensitivity than the thermocouple or the ordinary bolometer. As with other detectors that are designed to operate at longer wavelengths, photovoltaic detectors are often cooled to enable operation at greater sensitivity.

Two-dimensional arrays or panels of photodiodes allow the detection of images. Each photodiode, or *MOS* (metal-oxide-semiconductor) device, responds to the incident radiation to provide one *pixel* (picture element) of output. On exposure to light, each of the discrete devices fabricated on a silicon chip, say, stores photo-induced charge in a potential well created by an applied gate voltage. The stored charge contributed by each pixel, a measure of the local irradiance, is scanned to produce an electronic record of the image. Scanning and readout is performed by charge transfer along rows of such devices (*CCD*, or *charge-coupled devices*) or by injection of the charge into the underlying bulk semiconductor material (*CID*, or *charge-injection devices*) [2]. By sequential reading of the stored charge, the set of *x-y* coordinates of the original irradiance distribution is electronically reconstructed to form the image. Such photodiode arrays are used in television cameras and have been used to replace photographic plates in astronomical telescopes and spectographs.

Finally, a detector made popular by the widespread use of the camera is the *photographic film* or *plate*. Such photographic emulsions are available with spectral sensitivity that extends from the X-ray region into the near infrared at around 1.2 μm. The sensitive material is an emulsion of silver halide crystals or grains. An incident photon imparts energy to the valence electron of a halide ion, which can then combine with the silver ion, producing a neutral silver atom. Even before developing, the emulsion contains a latent image, a distribution of reduced silver atoms determined by the variations in radiant energy received. The latent image is then "amplified," so to speak, by the action of the developer. The resulting chemical action provides further free electrons to continue the reduction process, with the latent image acting as a catalytic agent to further action. The density of the silver atoms, and thus the opacity of the film, is a measure of both the irradiance and the time of exposure, so that photographic film, unlike other detectors, has the advantage of light-signal integration. Even weak radiation can be detected by the cumulative effect of a long exposure.

In addition to a knowledge of the spectral range over which a particular detector is effective, it is important to know the actual sensitivity or, more precisely, the *responsivity S* of the detector, defined as the ratio of output to input:

$$S = \frac{\text{output}}{\text{input}} \qquad (2\text{-}16)$$

Input may be radiant flux or irradiance. Output is almost always a current or voltage. For the responsivity to be a useful specification of a detector, it should be constant over the useful range of the instrument. In other words, the detector, together with its associated amplifier and circuits, should provide a linear response, with output proportional to input. In general, however, responsivity is not independent of wavelength. Curves of responsivity versus wavelength are provided with commercial detectors. When the responsivity is a function of λ, the detector is said to be *selective*. A *nonselective* detector is one that depends only on the radiant flux, not on the wavelength. Thermal detectors using a blackened strip as a receptor may be nonselective; however, entrance windows to such devices may well make them selective. The *detectivity D* of a detector is the reciprocal of the minimum detectable power, called the *noise equivalent power*, Φ_N, of the detector:

$$D = \frac{1}{\Phi_N}$$

The minimum detectable power is limited by the *noise* inherent in the operation of the detector. The noise is that part of the signal or output not related to the desired input. Many sources of noise exist, including the statistical fluctuations of photons, or *radiation noise*, and the thermal agitation of current carriers, or *Johnson noise*, inherent in all detectors; the generation and recombination noise due to statistical fluctuations of current carriers in photoconductors; the *shot noise* due to random emission of electrons in photoemissive detectors; and the noise due to temperature fluctuations in thermal detectors. Mere amplification of a signal is not useful when it does not distinguish between signal and noise and results in the same signal-to-noise ratio, just as the mere magnification of an optical image is not useful in clarifying its details.

2-1. Calculate the frequencies of electromagnetic radiation capable of producing a visual sensation in the normal eye.

2-2. A small, monochromatic light source, radiating at 500 nm, is rated at 500 W.
   (a) If only 2% of its total power is perceived by the eye as luminous power, what is its luminous flux output?
   (b) If the source radiates uniformly in all directions, determine its radiant and luminous intensities.
   (c) If the surface area of the source is 50 cm², determine the radiant and luminous exitances.
   (d) What are the irradiance and illuminance on a screen situated 2 m from the source, with its surface normal to the radiant flux?
   (e) If the screen contains a hole with diameter 5 cm, how much radiant and luminous flux get through?

2-3. (a) A 50-mW He-Cd laser emits at 441.6 nm. A 4-mW He-Ne laser emits at 632.8 nm. Using Figure 2-7, compare the relative brightness of the two laser beams of equal diameter when projected side by side on a white piece of paper. Assume photopic vision.
   (b) What power argon laser emitting at 488 nm is required to match the brightness of a 0.5-mW He-Ne green laser at 543.5 nm, under the conditions of (a)?

2-4. A lamp 3 m directly above a point P on the floor of a room produces at P an illuminance of 100 lm/m².
   (a) What is the luminous intensity of the lamp?
   (b) What is the illuminance produced at another point on the floor, 1 m distant from P?

2-5. A lot is illuminated at night by identical lamps at the top of two poles 30 ft high and 40 ft apart. Assuming the lamps radiate equally in all directions, compare the illuminance at ground level for points directly under one lamp and midway between them.

2-6. A small source of 100 cd is situated at the focal point of a spherical mirror of 50-cm focal length and 10-cm diameter. What is the average illuminance of the parallel beam reflected from the mirror, assuming an overall reflectance of about 80%?

2-7. (a) The sun subtends an angle of 0.5° at the earth's surface, where the illuminance is about 10^5 lx at normal incidence. Determine the luminance of the sun.
   (b) Determine the illuminance of a horizontal surface under a hemispherical sky with uniform luminance L.

2-8. A circular disc of radius 20 cm and uniform luminance of 10^5 cd/m² illuminates a small plane surface area of 1 cm², 1 m distant from the center of the disc. The small surface is oriented such that its normal makes an angle of 45° with the axis joining the centers of the two surfaces. The axis is perpendicular to the circular disc. What is the luminous flux incident on the small surface?

2-9. Derive the Wien displacement law from the Planck blackbody spectral radiance formula.

2-10. Derive the Stefan-Boltzmann law from the Planck blackbody spectral radiance formula. (*Hint:* Use a substitution of $x = hc/\lambda kT$ to facilitate the integration.)

2-11. The peak of the solar spectrum falls at about 500 nm. Determine the sun's surface temperature, assuming that it radiates like a blackbody.

2-12. (a) At what wavelength does a blackbody at 6000 K radiate the most per unit wavelength?
   (b) If the blackbody is a 1-mm diameter hole in a cavity radiator at this temperature, find the power radiated through the hole in the narrow wavelength region 5500–5510 Å.

2-13. At a given temperature, $\lambda_{max} = 550$ nm for a blackbody cavity. The cavity temperature is then increased until its total radiant exitance is doubled. What is the new temperature and the new λ_{max}?

2-14. What must be the temperature of a graybody with emissivity of 0.45 for it to have the same total radiant exitance as a blackbody at 5000 K?

REFERENCES

[1] Resnick, Robert. *Basic Concepts in Relativity and Early Quantum Mechanics*. New York: John Wiley and Sons, 1972. Ch. 4.

[2] Barbe, D. F. "Imaging Devices Using the Charge-Coupled Concept." *Proceedings of the IEEE* 63 (Jan. 1975): 38.

[3] Malacara, Zacarias H., and Morales R. Arquimedes. "Light Sources." In *Geometrical and Instrumental Optics*, edited by Daniel Malacara. Boston: Academic Press, 1988.

[4] Nussbaum, Allen. *Geometrical Optics: An Introduction*. Reading, Mass.: Addison-Wesley Publishing Co., 1968. Ch. 6.

[5] Budde, W. *Physical Detectors of Optical Radiation*. Optical Radiation Measurements Series, vol. 4. New York: Academic Press, 1983.

[6] Grum, Franc, and Richard J. Becherer. *Radiometry*. Optical Radiation Measurements Series, vol. 1. New York: Academic Press, 1979.

[7] *IES Lighting Handbook*. New York: Illuminating Engineering Society, 1981.

[8] *Handbook of Optics*, edited by Walter G. Driscoll and William Vaughan. Sponsored by the Optical Society of America. New York: McGraw-Hill Book Company, 1978: Stephen F. Jacobs, "Nonimaging Detectors." Jay F. Snell, "Radiometry and Photometry." George J. Zissis and Anthony J. Larocca, "Optical Radiators and Sources."

[9] Kingston, R. H. *Detection of Optical and Infrared Radiation*. New York: Springer-Verlag, 1978.

[10] Stimson, A. *Photometry and Radiometry for Engineers*. New York: Wiley-Interscience, 1974.

[11] James, T. H. "Photographic Development." *Scientific American* (Nov. 1952): 30.

[12] Morehead, Fred F., Jr. "Light-Emitting Semiconductors." *Scientific American* (May 1967): 108.

[13] Clark Jones, R. "How Images are Detected." In *Lasers and Light. Readings from Scientific American*, pp. 81–88. San Francisco: W. H. Freeman and Company Publishers, 1968.

[14] Fitch, James Marston. "The Control of the Luminous Environment." In *Lasers and Light. Readings from Scientific American*, pp. 131–39. San Francisco: W. H. Freeman and Company Publishers, 1968.

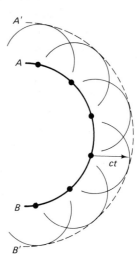

3

Geometrical Optics

INTRODUCTION

The treatment of light as wave motion allows for a region of approximation in which the wavelength is considered to be negligible compared with the dimensions of the relevant components of the optical system. This region of approximation is called *geometrical optics*. When the wave character of the light may not be so ignored, the field is known as *physical optics*. Thus geometrical optics forms a special case of physical optics in a way that may be summarized as follows:

$$\lim_{\lambda \to 0} \{\text{physical optics}\} = \{\text{geometrical optics}\}$$

Since the wavelength of light is very small compared to ordinary objects, early unrefined observations of the behavior of a light beam passing through apertures or around obstacles in its path could be handled by geometrical optics. Recall that the appearance of distinct shadows influenced Newton to assert that the apparent rectilinear propagation of light was due to a stream of light corpuscles rather than a wave motion. Wave motion characterized by longer wavelengths, such as those in water waves and sound waves, was known to give distinct bending around obstacles. Newton's model of light propagation, therefore, seemed not to allow for the existence of a wave motion with very small wavelengths. There was in fact already evidence of some degree of bending, even for light waves, in the time of Isaac Newton. The Jesuit Francesco Grimaldi had noticed the fine structure in the edge of a shadow, a structure not explainable in terms of the rectilinear propagation of light. This bending of light waves around the edges of an obstruction came to be called *diffraction*.

Within the approximation represented by geometrical optics, light is understood to travel out from its source along straight lines or *rays*. The ray is then simply the path along which light energy is transmitted from one point to another in an optical system. The ray is a useful construct, although abstract in the sense that a light beam, in practice, cannot be narrowed down indefinitely to approach a straight line. A pencil-like laser beam is perhaps the best actual approximation to a ray of light. (When an aperture through which the beam is passed is made small enough, however, even a laser beam begins to spread out in a characteristic diffraction pattern.) When a light ray traverses an optical system consisting of several homogeneous media in sequence, the optical path is a sequence of straight-line segments. Discontinuities in the line segments occur each time the light is reflected or refracted. The laws of geometrical optics that describe the subsequent direction of the rays are the following:

Law of Reflection. When a ray of light is reflected at an interface dividing two uniform media, the reflected ray remains within the *plane of incidence*, and the angle of reflection equals the angle of incidence. The plane of incidence includes the incident ray and the normal to the point of incidence.

Law of Refraction (Snell's Law). When a ray of light is refracted at an interface dividing two uniform media, the transmitted ray remains within the plane of incidence and the sine of the angle of refraction is directly proportional to the sine of the angle of incidence.

Both laws are summarized in Figure 3-1, in which an incident ray is partially reflected and partially transmitted at a plane interface separating two transparent media.

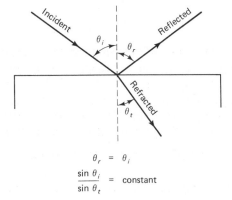

$$\theta_r = \theta_i$$

$$\frac{\sin \theta_i}{\sin \theta_t} = \text{constant}$$

Figure 3-1 Illustration of the law of reflection and refraction.

3-1 HUYGENS' PRINCIPLE

The Dutch physicist Christian Huygens envisioned light as a series of pulses emitted from each point of a luminous body and propagated in relay fashion by the particles of the ether, an elastic medium filling all space. Consistent with his conception, Huygens imagined each point of a propagating disturbance as capable of originating new pulses that contributed to the disturbance an instant later. To show how his model of light propagation implied the laws of geometrical optics, he enunciated a fruitful principle that can be stated as follows: Each point on the leading surface of a wave disturbance—the wave front—may be regarded as a secondary source of spherical waves (or *wavelets*), which themselves progress with the speed of light in the medium and whose envelope at a later time constitutes the new wavefront. Simple applications of the principle are shown in Figure 3-2 for a plane and spherical wave. In each case, AB forms the initial wave disturbance or wavefront, and $A'B'$ is the new wavefront at a time t later. The radius of each wavelet is, accordingly, ct, where c is the speed of light in the medium. Notice that the new wavefront is tangent to each wavelet at a single point. According to Huygens, the remainder of each wavelet is to be disregarded in the application of the principle. Indeed, were the remainder of the wavelet considered to be effective in propagating the light disturbance, Huygens could not have derived the law of rectilinear propagation from his principle. To see this more clearly, refer to Figure 3-3, which shows a spherical wave disturbance originating at O and incident upon an aperture with an opening SS'. According to the notion of rectilinear propagation, the lines OA and OB form the sharp edges of the shadow to the right of the aperture. Some of the wavelets that originate from points of the wavefront (arc SS'), however, overlap into the region of shadow. According to Huygens, however, these are ignored and the new wavefront ends abruptly at points P and P', precisely where the extreme wavelets originating at points S and S' are tangent to the new wavefront. In so disregarding the effective-

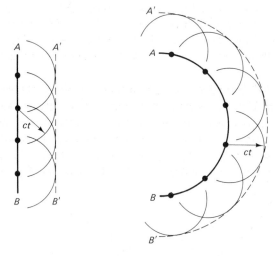

Figure 3-2 Illustration of Huygens' principle for plane and spherical waves.

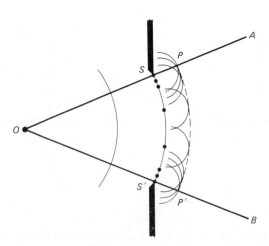

Figure 3-3 Huygens' construction for an obstructed wavefront.

ness of the overlapping wavelets, Huygens avoided the possibility of diffraction of the light into the region of geometric shadow. Huygens also ignored the wavefront formed by the back half of the wavelets, since these wavefronts implied a light disturbance traveling in the opposite direction. Despite weaknesses in this model, remedied later by Fresnel and others, Huygens was able to apply his principle to prove the laws of both reflection and refraction, as we show in what follows.

Figure 3-4a illustrates the Huygens construction for a narrow, parallel beam of light to prove the law of reflection. Huygens' principle must be modified slightly to accommodate the case in which a wavefront, such as AC, encounters a plane interface, such as XY, at an angle. Here the angle of incidence of the rays AD, BE, and CF relative to the perpendicular PD is θ_i. Since points along the plane wavefront do not arrive at the interface simultaneously, allowance is made for these differences in constructing the wavelets that determine the reflected wavefront. If the interface XY were not present, the Huygens construction would produce the wavefront GI at the instant ray CF reached the interface at I. The intrusion of the reflecting surface, however, means that during the same time interval required for ray CF to progress from F to I, ray BE has progressed from E to J and then a distance equivalent to JH *after reflection*. Thus a wavelet of radius JH centered at J is drawn above the reflecting surface. Similarly, a wavelet of radius DG is drawn centered at D to represent the propagation after reflection of the lower part of the beam. The new wavefront, which must now be tangent to these wavelets at points M and N, and include the point I, is shown as KI in the figure. A representative reflected ray is DL, shown perpendicular to the reflected wavefront. The normal PD drawn for this ray is used to define angles of incidence and reflection for the beam. The construction makes clear the equivalence between the angles of incidence and reflection, as outlined in Figure 3-4a.

Similarly, in Figure 3-4b, a Huygens construction is shown that illustrates the law of refraction. Here we must take into account a different speed of light in the upper and lower media. If the speed of light in vacuum is c, we express the speed in the upper medium by the ratio c/n_i, where n_i is a constant that characterizes the medium and is referred to as the *refractive index*. Similarly, the speed of light in the lower medium is c/n_t. The points D, E, and F on the incident wavefront arrive at

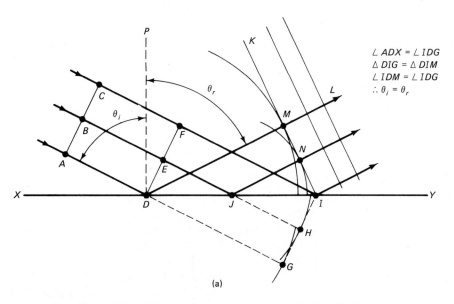

$\angle ADX = \angle IDG$
$\triangle DIG = \triangle DIM$
$\angle IDM = \angle IDG$
$\therefore \theta_i = \theta_r$

(a)

Figure 3-4 (a) Huygens' construction to prove the law of reflection.

$\angle DIM = \theta_t$

$\angle IDF = \theta_i$

$\sin \theta_i = \dfrac{FI}{DI}$ and $\sin \theta_t = \dfrac{DM}{DI}$

$\dfrac{\sin \theta_i}{\sin \theta_t} = \dfrac{FI}{DM} = \dfrac{DG}{DM} = \dfrac{n_t}{n_i} = $ constant

(b)

Figure 3-4 (*continued*) (b) Huygens' construction to prove the law of refraction.

points D, J, and I of the plane interface XY at different times. In the absence of the refracting surface, the wavefront GI is formed at the instant ray CF reaches I. During the progress of ray CF from F to I in time t, however, the ray AD has entered the lower medium, where its speed is, let us say, slower. Thus if the distance DG is $v_i t$, a wavelet of radius $v_t t$ is constructed with center at D. The radius DM can also be expressed as

$$ DM = v_t t = v_t \left(\frac{DG}{v_i} \right) = \left(\frac{n_i}{n_t} \right) DG $$

Similarly, a wavelet of radius (n_i/n_t) JH is drawn centered at J. The new wavefront KI includes point I on the interface and is tangent to the two wavelets at points M and N, as shown. The geometric relationship between the angles θ_i and θ_t, formed by the representative incident ray AD and refracted ray DL, is *Snell's law*, as outlined in Figure 3-4b. Snell's law of refraction may be expressed as

$$ n_i \sin \theta_i = n_t \sin \theta_t \qquad (3\text{-}1) $$

3-2 FERMAT'S PRINCIPLE

The laws of geometrical optics can also be derived, perhaps more elegantly, from a different fundamental hypothesis. The root idea had been introduced by Hero of Alexandria, who lived in the second century B.C. According to Hero, when light is propagated between two points, it takes the shortest path. For propagation between two points in the same uniform medium, the path is clearly the straight line joining the two points. When light from the first point A, Figure 3-5, reaches the second

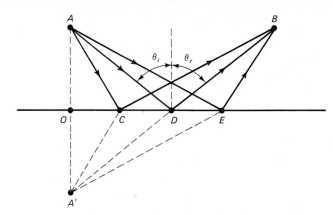

Figure 3-5 Construction to prove the law of reflection from Hero's principle.

point B after reflection from a plane surface, however, the same principle predicts the law of reflection, as follows. Figure 3-5 shows three possible paths from A to B, including the correct one, ADB. Consider, however, the arbitrary path ACB. If point A' is constructed on the perpendicular AO such that $AO = OA'$, the right triangles AOC and $A'OC$ are equal. Thus $AC = A'C$ and the distance traveled by the ray of light from A to B via C is the same as the distance from A' to B via C. The shortest distance from A' to B is obviously the straight line $A'DB$, so the path ADB is the correct choice taken by the actual light ray. Elementary geometry shows that for this path, $\theta_i = \theta_r$. Note also that to maintain $A'DB$ as a single straight line, the reflected ray must remain within the plane of incidence, that is, the plane of the page.

The French mathematician Pierre de Fermat generalized Hero's principle to prove the law of refraction. If the terminal point B lies below the surface of a second medium, as in Figure 3-6, the correct path is definitely not the shortest path or straight line AB, for that would make the angle of refraction equal to the angle of incidence, in violation of the empirically established law of refraction. Appealing to the "economy of nature," Fermat supposed instead that the ray of light traveled the path of least *time* from A to B, a generalization that included Hero's principle as a special case. If light travels more slowly in the second medium, as assumed in Figure 3-6, light bends at the interface so as to take a path that favors a shorter time in the second medium, thereby minimizing the overall transit time from A to B. Mathematically, we are required to minimize the total time,

$$t = \frac{AO}{v_i} + \frac{OB}{v_t}$$

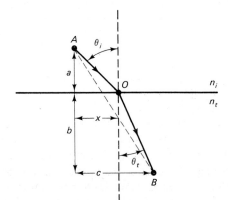

Figure 3-6 Construction to prove the law of refraction from Fermat's principle.

where v_i and v_t are the velocities of light in the incident and transmitting media, respectively. Employing the Pythagorean theorem and the distances defined in Figure 3-6, we have

$$t = \frac{\sqrt{a^2 + x^2}}{v_i} + \frac{\sqrt{b^2 + (c - x)^2}}{v_t}$$

Since other choices of path change the position of point O and therefore the distance x, we can minimize the time by setting $dt/dx = 0$:

$$\frac{dt}{dx} = \frac{x}{v_i \sqrt{a^2 + x^2}} - \frac{c - x}{v_t \sqrt{b^2 + (c - x)^2}} = 0$$

Again from Figure 3-6, the angles of incidence and refraction can be conveniently introduced into this condition, giving

$$\frac{dt}{dx} = \frac{\sin \theta_i}{v_i} - \frac{\sin \theta_t}{v_t} = 0$$

so that $v_t \sin \theta_i = v_i \sin \theta_t$. Introducing the refractive indices of the media through the relations, $v = c/n$, we arrive at Snell's law,

$$n_i \sin \theta_i = n_t \sin \theta_t$$

Fermat's principle, like that of Huygens, required refinement to achieve more general applicability. Situations exist where the actual path taken by a light ray may represent a maximum time or even one of many possible paths, all requiring equal time. As an example of the latter case, consider light propagating from one focus to the other inside an ellipsoidal mirror, along any of an infinite number of possible paths. Since the ellipse is the locus of all points whose combined distances from the two foci is a constant, all paths are indeed of equal time. A more precise statement of Fermat's principle, which requires merely an extremum relative to neighboring paths, may be given as follows: The actual path taken by a light ray in its propagation between two given points in an optical system is such as to make its optical path equal, in the first approximation, to other paths closely adjacent to the actual one.

With this formulation, Fermat's principle falls in the class of problems called *variational calculus*, a technique that determines the form of a function that minimizes a definite integral. In optics, the definite integral is the integral of the time required for the transit of a light ray from starting to finishing points.[1]

3-3 PRINCIPLE OF REVERSIBILITY

Refer again to the cases of reflection and refraction pictured in Figures 3-5 and 3-6. If the roles of points A and B are interchanged, so that B is the source of light rays, Fermat's principle of least time must predict the same path as determined for the original direction of light propagation. In general then, any actual ray of light in an optical system, if reversed in direction, will retrace the same path backward. This principle will be found to be very useful in various applications to be dealt with later.

---

[1] It is of interest to note here that a similar principle, called *Hamilton's principle of least action* in mechanics that calls for a minimum of the definite integral of the Lagrangian function (the kinetic energy minus the potential energy), represents an alternative formulation of the laws of mechanics and indeed implies Newton's laws of mechanics themselves.

Before discussing the formation of images in a general way, we discuss the simplest—and experientially, the most accessible—case of images formed by plane mirrors. In this context it is important to distinguish between *specular reflection* from a perfectly smooth surface and *diffuse reflection* from a granular or rough surface. In the former case, all rays of a parallel beam incident on the surface obey the law of reflection from a plane surface and therefore reflect as a parallel beam; in the latter case, though the law of reflection is obeyed locally, the microscopically granular surface results in reflected rays in various directions and thus a diffuse scattering of the originally parallel rays of light. Every plane surface will produce some such scattering, since a perfectly smooth surface is not attainable in practice. The treatment that follows assumes the case of specular reflection.

Consider the specular reflection of a single light ray OP from the xy-plane in Figure 3-7a. By the law of reflection, the reflected ray PQ remains within the plane of incidence, making equal angles with the normal at P. If the path OPQ is resolved into its x-, y-, and z-components, it is clear that the direction of ray OP is altered by the reflection only along the z-direction, and then in such a way that its z-component is simply reversed. If the direction of the incident ray is described by its unit vector, $\hat{r}_1 = (x, y, z)$, then the reflection causes

$$\hat{r}_1 = (x, y, z) \longrightarrow \hat{r}_2 = (x, y, -z)$$

It follows that if a ray is incident from such a direction as to reflect sequentially from all three rectangular coordinate planes, as in the "corner reflector" of Figure 3-7b,

$$\hat{r}_1 = (x, y, z) \longrightarrow \hat{r}_2 = (-x, -y, -z)$$

and the ray returns precisely parallel to the line of its original approach. A network of such corner reflectors ensures the exact return of a beam of light, a headlight beam from highway reflectors, for example, or a laser beam from the moon.

Figure 3-7 Geometry of a ray reflected from a plane.

Image formation in a plane mirror is illustrated in Figure 3-8a. A point object S sends rays toward a plane mirror, which reflect as shown. The law of reflection ensures that pairs of triangles like SNP and $S'NP$ are equal, so that all reflected rays appear to originate at the *image point* S', which lies along the normal line SN, and at such a depth that the *image distance* $S'N$ equals the *object distance* SN. The eye sees a point image at S' in exactly the same way as it would see a real point object placed there. Since none of the actual rays of light lies below the mirror surface, the image is said to be a *virtual image*. The image S' cannot be projected on a screen as

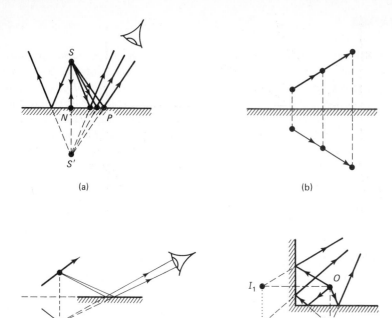

(a)

(b)

(c)

(d)

Figure 3-8 Image formation in a plane mirror.

in the case of a *real image*. All points of an extended object, such as the arrow in Figure 3-8b, are imaged by a plane mirror in similar fashion: Each object point has its image point along its normal to the mirror surface and as far below the reflecting surface as the object point lies above the surface. Note that the image position does not depend on the position of the eye. Further, the construction of Figure 3-8b makes clear that the image size is identical with the object size, giving a *magnification* of unity. In addition, the transverse orientation of object and image are the same. A right-handed object, however, appears left-handed in its image. In Figure 3-8c, where the mirror does not lie directly below the object, the mirror plane may be extended to determine the position of the image as seen by an eye positioned to receive reflected rays originating at the object. Figure 3-8d illustrates multiple images of a point object O formed by two perpendicular mirrors. Images I_1 and I_2 result from single reflections in the two mirrors, but a third image I_3 results from sequential reflections from both mirrors.

3-5 REFRACTION THROUGH PLANE SURFACES

Consider light ray (1) in Figure 3-9a, incident at angle θ_1 at a plane interface separating two transparent media characterized, in order, by refractive indices n_1 and n_2. Let the angle of refraction be the angle θ_2. Snell's law, which now takes the form,

$$n_1 \sin \theta_1 = n_2 \sin \theta_2 \tag{3-2}$$

requires an angle of refraction θ_2 such that the refracted ray bends away from the normal, as in Figure 3-9a, rays 1 and 2, when $n_2 < n_1$. For $n_2 > n_1$, on the other hand, the refracted ray bends toward the normal. The law also requires that ray 3, incident normal to the surface ($\theta_1 = 0$), be transmitted without change of direction ($\theta_2 = 0$), regardless of the ratio of refractive indices. In Figure 3-9a the three rays shown originate at a source point S below an interface and emerge into an upper

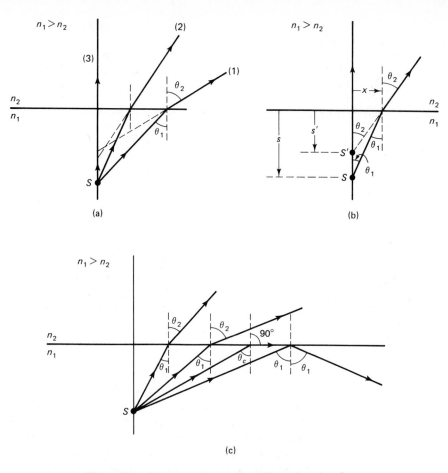

Figure 3-9 Geometry of rays refracted by a plane interface.

medium of lower refractive index, as in the case of light emerging from water ($n_1 = 1.33$) into air ($n_2 = 1.00$). A unique image point is not determined by these rays because they have no common intersection or virtual image point below the surface from which they appear to originate after refraction, as shown by the dashed line extensions of the refracted rays. For rays making a small angle with the normal to the surface, however, a reasonably good image can be located. In this approximation, where we allow only such *paraxial rays*[2] to form the image, the angles of incidence and refraction are both small and the approximation,

$$\sin \theta \cong \tan \theta \cong \theta \text{ (in radians)}$$

is valid. From Eq. (3-2), Snell's law can be approximated by

$$n_1 \tan \theta_1 \cong n_2 \tan \theta_2 \tag{3-3}$$

and taking the appropriate tangents from Figure 3-9b, we have

$$n_1\left(\frac{x}{s}\right) = n_2\left(\frac{x}{s'}\right)$$

---

[2] In general, a paraxial ray is one that remains near the central axis of the image-forming optical system, thus making small angles with the optical axis.

The image point occurs at the vertical distance s' below the surface given by

$$s' = \left(\frac{n_2}{n_1}\right)s \qquad (3\text{-}4)$$

where s is the corresponding depth of the object. Thus objects underwater, viewed from directly overhead, appear to be nearer the surface than they actually are, since in this case $s' = (1/1.33)s = \frac{3}{4}s$. Even when the viewing angle θ_2 is not small, a reasonably good retinal image of an underwater object is formed because the aperture or pupil of the eye admits only a small bundle of rays while forming the image. Since these rays differ very little in direction, they will appear to originate from approximately the same image point. However, the depth of this image will not be $\frac{3}{4}$ the object depth, as for paraxial rays, and in general will vary with the angle of viewing.

Rays from the object that make increasingly larger angles of incidence with the interface must, by Snell's law, refract at increasingly larger angles, as shown in Figure 3-9c. A critical angle of incidence θ_c is reached when the angle of refraction reaches 90°. Thus from Snell's law,

$$\sin \theta_c = \left(\frac{n_2}{n_1}\right)\sin 90 = \frac{n_2}{n_1}$$

or

$$\theta_c = \sin^{-1}\left(\frac{n_2}{n_1}\right) \qquad (3\text{-}5)$$

For angles of incidence $\theta_1 > \theta_c$, the incident ray experiences *total internal reflection*, as shown. This phenomenon is essential in the transmission of light along glass fibers by a series of total internal reflections, as discussed in Chapter 24. Note that the phenomenon does not occur unless $n_1 > n_2$, so that θ_c can be determined from Eq. (3-5).

We return to the nature of images formed by refraction at a plane surface when we deal with such refraction as a special case of refraction from a spherical surface.

3-6 IMAGING BY AN OPTICAL SYSTEM

We discuss now what is meant by an image in general and indicate the practical and theoretical factors that render an image less than perfect. In Figure 3-10, let the region labeled "optical system" include any number of reflecting and/or refracting surfaces, of any curvature, that may alter the direction of rays leaving an *object point O*. This region may include any number of intervening media, but we shall assume that each individual medium is homogeneous and isotropic, and so characterized by its own refractive index. Thus rays spread out radially in all directions from object point O, as shown, in real *object space*, which precedes the first reflecting or refracting

Figure 3-10 Image formation by an optical system.

surface of the optical system. The family of spherical surfaces normal to the rays are the *wavefronts*, the locus of points such that each ray contacting a wavefront represents the same transit time of light from the source. In real object space the rays are diverging and the spherical wavefronts are expanding. Suppose now that the optical system redirects these rays in such a way that on leaving the optical system and entering real *image space*, the wavefronts are contracting and the rays are converging to a common point that we define to be the *image point*, I. In the spirit of Fermat's principle, we can say that since every such ray starts at O and ends at I, every such ray requires the same transit time. These rays are said to be *isochronous*. Further, by the principle of reversibility, if I is the object point, each ray will reverse its direction but maintain its path through the optical system, and O will be the corresponding image point. The points O and I are said to be *conjugate* points for the optical system. In an ideal optical system, every ray from O intercepted by the system—and only these rays—also passes through I. To image an actual object, this requirement must hold for every object point and its conjugate image point.

Nonideal images are formed in practice because of (1) light scattering, (2) aberrations, and (3) diffraction. Some rays leaving O do not reach I due to reflection losses at refracting surfaces, diffuse reflections from reflecting surfaces, and scattering by inhomogeneities in transparent media. Loss of rays by such means merely diminishes the brightness of the image; however, some of these rays are scattered through I from nonconjugate object points, degrading the image. When the optical system itself cannot produce the one-to-one relationship between object and image rays required for perfect imaging of all object points, we speak of system aberrations. Such aberrations are treated later. Finally, since every optical system intercepts only a portion of the wavefront emerging from the object, the image cannot be perfectly sharp. Even if the image were otherwise perfect, the effect of using a limited portion of the wavefront leads to diffraction and a blurred image, which is said to be *diffraction limited*. This source of imperfect images, discussed further in the sections under diffraction, represents a fundamental limit to the sharpness of an image that cannot be entirely overcome. This difficulty rises from the wave nature of light. Only in the unattainable limit of geometrical optics, where $\lambda \to 0$, would diffraction effects disappear entirely.

Reflecting or refracting surfaces that form perfect images are called *Cartesian surfaces*. In the case of reflection, such surfaces are the conic sections, as shown in Figure 3-11. In each of these figures, the roles of object and image points may be reversed by the principle of reversibility. Notice that in Figure 3-11b, the image is virtual. In Figure 3-11c, the parallel reflected rays are said to form an image "at infinity." In each case, one can show that Fermat's principle, requiring isochronous rays between object and image points, leads to a condition that is equivalent to the geometric definition of the corresponding conic section.

Cartesian surfaces that produce perfect imaging by refraction may be more complicated. Let us ask for the equation of the appropriate refracting surface that images object point O at image point I, as illustrated in Figure 3-12. There an arbitrary point P with coordinates (x, y) is on the required surface Σ. The requirement is that every ray from O, like OPI, refracts and passes through the image I. Another such ray is evidently OVI, normal to the surface at its vertex point V. By Fermat's principle, these are isochronous rays. Since the media on either side of the refracting surface are characterized by different refractive indices, however, the isochronous rays are not equal in length. The transit time of a ray through a medium of thickness x with refractive index n is

$$t = \frac{x}{v} = \frac{nx}{c}$$

(a) Ellipsoid

(b) Hyperboloid

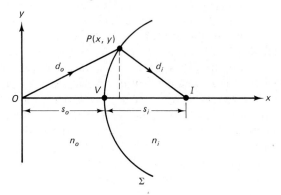

(c) Paraboloid

Figure 3-11 Cartesian reflecting surfaces showing conjugate object and image points.

Figure 3-12 Cartesian refracting surface which images object point O at image point I.

Therefore, equal times imply equal values of the product nx, called the *optical path length*. In the problem at hand then, Fermat's principle requires that

$$n_o d_o + n_i d_i = n_o s_o + n_i s_i = \text{constant} \qquad (3\text{-}6)$$

where the distances are defined in Figure 3-12. In terms of the (x, y)-coordinates of P, the first sum of Eq. (3-6) becomes

$$n_0(x^2 + y^2)^{1/2} + n_i[\, y^2 + (s_o + s_i - x)^2]^{1/2} = \text{constant} \qquad (3\text{-}7)$$

The constant in the equation is determined by the middle member of Eq. (3-6), which can be calculated once the specific problem is defined. Equation (3-7) describes the *Cartesian ovoid* of revolution shown in Figure 3-13a.

In most cases, however, the image is desired in the same optical medium as the object. This goal is achieved by a lens that refracts light rays twice, once at each surface, producing a real image outside the lens. Thus it is of particular interest to determine the Cartesian surfaces that render every object ray parallel after the first refraction. Such rays incident on the second surface can then be refracted again to form an image. The solutions to this problem are illustrated in Figure 3-13b and c. Depending on the relative magnitudes of the refractive indices, the appropriate refracting surface is either a hyperboloid ($n_i > n_o$) or an ellipsoid ($n_o > n_i$), as shown.

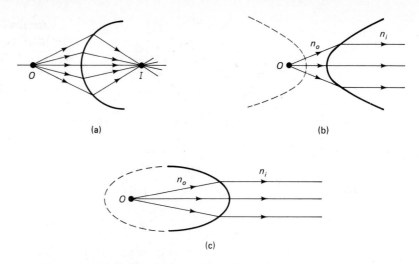

(a)

(b)

(c)

Figure 3-13 Cartesian refracting surfaces. (a) Cartesian ovoid images O at I by refraction. (b) Hyperbolic surface images object point O at infinity when O is at one focus and $n_i > n_o$. (c) Ellipsoid surface images object point O at infinity when O is at one focus and $n_o > n_i$.

The first of these corresponds to the usual case of an object in air. A double hyperbolic lens then functions as shown in Figure 3-14. Note, however, that the aberration-free imaging so achieved applies only to object point O at the correct distance from the lens and on axis. For nearby points, imaging is not perfect. The larger the actual object, the less precise is its image. Because images of actual objects are not free from aberrations and because hyperboloid surfaces are difficult to grind exactly, most optical surfaces are spherical. The *spherical aberrations* so introduced are accepted as a compromise when weighed against the relative ease of fabricating spherical surfaces. In the remainder of our treatment of geometrical optics, we concentrate on the case of spherical reflecting and refracting surfaces with a *radius of curvature* R. Of course, in the limit $R \rightarrow \infty$, we deal with the special case of a plane surface.

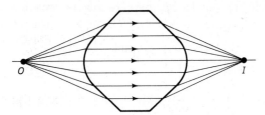

Figure 3-14 Aberration-free imaging of point object O by a double hyperbolic lens.

3-7 REFLECTION AT A SPHERICAL SURFACE

Spherical mirrors may be either concave or convex relative to an object point O, depending on whether the center of curvature C is on the same or opposite side of the surface. In Figure 3-15 the mirror shown is convex, and two rays of light originating at O are drawn, one normal to the spherical surface at its vertex V and the other an arbitrary ray incident at P. The first ray reflects back along itself; the second reflects at P as if from a tangent plane at P, satisfying the law of reflection. The two reflected rays diverge as they leave the mirror. The intersection of the two rays (extended backward) determines the image point I conjugate to O. The image is virtual, located behind the mirror surface. Object and image distances from the vertex are

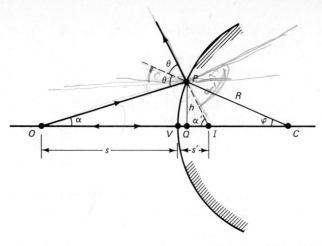

Figure 3-15 Reflection at a spherical surface.

shown as s and s', respectively. A perpendicular of height h is drawn from P to the axis at Q. We seek a relationship between s and s' that depends only on the radius of curvature R of the mirror. As we shall see, such a relation is possible only to first-order approximation of the sines and cosines of the angles made by the object and image rays to the spherical surface. This means that in place of the expansion of

$$\sin \varphi = \varphi - \frac{\varphi^3}{3!} + \frac{\varphi^5}{5!} - \cdots$$

$$\cos \varphi = 1 - \frac{\varphi^2}{2!} + \frac{\varphi^4}{4!} + \cdots \tag{3-8}$$

we consider the first terms only and write

$$\sin \varphi \cong \varphi \quad \text{and} \quad \cos \varphi \cong 1 \tag{3-9}$$

relations that can be accurate enough if the angle φ is small enough. This approximation leads to *first-order*, or *Gaussian*, optics, after Karl Friedrich Gauss, who in 1841 developed the foundations of the subject. Returning now to the problem at hand, notice that two angular relationships may be obtained from Figure 3-15, because the exterior angle of a triangle equals the sum of its interior angles. These are

$$\theta = \alpha + \varphi \quad \text{and} \quad 2\theta = \alpha + \alpha'$$

which combine to give

$$\alpha - \alpha' = -2\varphi \tag{3-10}$$

Using the small-angle approximation, the angles of Eq. (3-10) can be replaced by their tangents, yielding

$$\frac{h}{s} - \frac{h}{s'} = -2\frac{h}{R}$$

where we have also neglected the axial distance VQ, small when angle ϕ is small. Cancellation of h produces the desired relationship,

$$\frac{1}{s} - \frac{1}{s'} = \frac{-2}{R} \tag{3-11}$$

If the spherical surface is chosen to be concave instead, the center of curvature would be to the left. For certain positions of the object point O, it is then possible to find a real image point also to the left of the mirror. In these cases, the resulting geometric relationship analogous to Eq. (3-11) consists of terms that are all positive.

It is possible, employing a sign convention, to represent all cases by the single equation

$$\frac{1}{s} + \frac{1}{s'} = -\frac{2}{R} \tag{3-12}$$

The sign convention to be used in conjunction with Eq. (3-12) is as follows: Assume the light propagates from left to right:

1. The object distance s is positive when O is to the left of V, corresponding to a real object. When O is to the right, corresponding to a virtual object, s is negative.
2. The image distance s' is positive when I is to the left of V, corresponding to a real image, and negative when I is to the right of V, corresponding to a virtual image.
3. The radius of curvature R is positive when C is to the right of V, corresponding to a convex mirror, and negative when C is to the left of V, corresponding to a concave mirror.

These rules[3] can be quickly summarized by noticing that positive object and image distances correspond to real objects and real images and that convex mirrors have positive radii of curvature. Applying Rule 2 to Figure 3-15, we see that the general Eq. (3-12) becomes identical with Eq. (3-11), a special case derived in conjunction with Figure 3-15. Virtual objects occur only with a sequence of two or more reflecting or refracting elements and are considered later.

The spherical mirror described by Eq. (3-12) yields, for a plane mirror with $R \rightarrow \infty$, $s' = -s$, as determined previously. The negative sign implies a virtual image for a real object. Notice also in Eq. (3-12) that object distance and image distance appear symmetrically, implying their interchangeability as conjugate points. For an object at infinity, incident rays are parallel and $s' = -R/2$, as illustrated in Figure 3-16a and b for both concave ($R < 0$) and convex ($R > 0$) mirrors. The image distance in each case is defined as the *focal length* of the mirrors. Thus

$$f = -\frac{R}{2} \begin{cases} > 0, & \text{concave mirror} \\ < 0, & \text{convex mirror} \end{cases} \tag{3-13}$$

and the mirror equation can be written, more compactly, as

$$\frac{1}{s} + \frac{1}{s'} = \frac{1}{f} \tag{3-14}$$

In Figure 3-16c, a construction is shown that allows the determination of the transverse magnification. The object is an extended object of transverse dimension h_o. The image of the top of the arrow is located by two rays whose behavior on reflection is known. The ray incident at the vertex must reflect to make equal angles with the axis. The other ray is directed toward the center of curvature along a normal and so must reflect back along itself. The intersection of the two reflected rays occurs behind the mirror and locates a virtual image there. Because of the equality of the three angles shown, it follows that

$$\frac{h_o}{s_o} = \frac{h_i}{s_i}$$

---

[3] While this set of sign conventions is widely used, the student is cautioned that other schemes exist. No one with a continuing involvement in optics can hope to escape confronting other conventions, nor should the matter be beyond the mental flexibility of the serious student to accommodate.

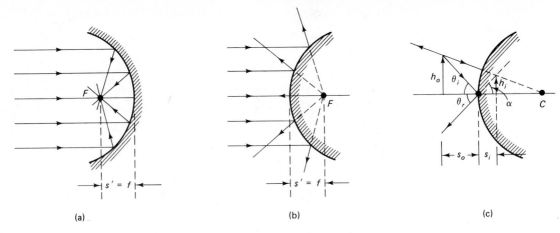

(a)   (b)   (c)

Figure 3-16 Location of focal points (a) and (b) and construction to determine magnification (c) of a spherical mirror.

The lateral magnification is defined by the ratio of lateral image size to corresponding lateral object size, giving

$$m = \frac{h_i}{h_o} = \frac{s_i}{s_o} \tag{3-15}$$

Extending the sign convention to include magnification, we assign a (+) magnification to the case where the image has the same orientation as the object and a (−) magnification when the image is inverted relative to the object. To produce a (+) magnification in the construction of Figure 3-16c, where s' must itself be negative, we modify Eq. (3-15) to give the general form

$$m = -\frac{s_i}{s_o} \tag{3-16}$$

Once the points C and F are located, image formation by a spherical mirror may be determined approximately by graphical methods. Figure 3-17 illustrates several examples that should be examined carefully. The validity of each ray reflection has been established by the discussion above. In each case the image of the top of the arrow is located by the intersection of three reflected rays.

Example

An object 3 cm high is placed 20 cm from (a) a convex and (b) a concave spherical mirror, each of 10-cm focal length. Determine the position and nature of the image in each case.

Solution   (a) Convex mirror: $f = -10$ cm and $s = +20$ cm.

$$\frac{1}{s} + \frac{1}{s'} = \frac{1}{f} \quad \text{or} \quad s' = \frac{fs}{s - f} = \frac{(-10)(20)}{(20) - (-10)} = -6.67 \text{ cm}$$

$$m = -\frac{s'}{s} = -\frac{-6.67}{20} = +0.333 = \frac{+1}{3}$$

The image is virtual (because s' is negative), 6.67 cm to the right of the mirror vertex, and is erect (because m is positive) and $\frac{1}{3}$ the size of the object, or 1 cm high.

(b) Concave mirror: $f = +10$ cm and $s = +20$ cm.

$$s' = \frac{fs}{s - f} = \frac{(10)(20)}{20 - 10} = +20 \text{ cm}$$

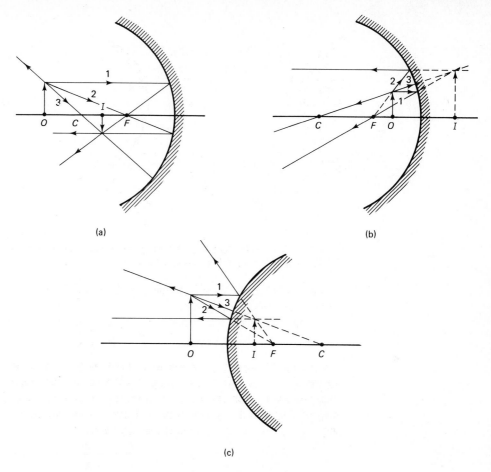

Figure 3-17 Ray diagrams for spherical mirrors. (a) Real image, concave mirror. (b) Virtual image, concave mirror. (c) Virtual image, convex mirror.

$$m = -\frac{s'}{s} = -\frac{20}{20} = -1$$

The image is real (because s' is positive), 20 cm to the left of the mirror vertex, and is inverted (because m is negative) and the same size as the object, or 3 cm high. Image and object happen to be at $2f = 20$ cm, the center of curvature of the mirror.

3-8 REFRACTION AT A SPHERICAL SURFACE

We turn now to a similar treatment of refraction at a spherical surface, choosing in this case the concave surface of Figure 3-18. Two rays are shown emanating from object point O. One is an axial ray, normal to the surface at its vertex and so refracted without change in direction. The other ray is an arbitrary ray incident at P and refracting there according to Snell's law,

$$n_1 \sin \theta_1 = n_2 \sin \theta_2 \tag{3-17}$$

The two refracted rays appear to emerge from their common intersection, the image point I. In triangle CPO, the exterior angle $\alpha = \theta_1 + \varphi$. In triangle CPI, the exte-

Figure 3-18 Refraction at a spherical surface for which $n_2 > n_1$.

rior angle $\alpha' = \theta_2 + \varphi$. Approximating for paraxial rays and substituting for θ_1 and θ_2 in Eq. (3-17), we have

$$n_1(\alpha - \varphi) = n_2(\alpha' - \varphi) \tag{3-18}$$

Next, writing the tangents for the angles by inspection of Figure 3-18, where again we may neglect the distance QV in the small angle approximation,

$$n_1\left(\frac{h}{s} - \frac{h}{R}\right) = n_2\left(\frac{h}{s'} - \frac{h}{R}\right)$$

or

$$\frac{n_1}{s} - \frac{n_2}{s'} = \frac{n_1 - n_2}{R} \tag{3-19}$$

Employing the same sign convention as introduced for mirrors (i.e., positive distances for real and negative distances for virtual objects and images), the virtual image distance $s' < 0$ and the radius of curvature $R < 0$. If these negative signs are understood to apply to these quantities for the case of Figure 3-18, a general form of the refraction equation may be written as

$$\frac{n_1}{s} + \frac{n_2}{s'} = \frac{n_2 - n_1}{R} \tag{3-20}$$

which holds equally well for convex surfaces. When $R \to \infty$, the spherical surface becomes a plane refracting surface, and

$$s' = -\left(\frac{n_2}{n_1}\right)s \tag{3-21}$$

where s' is the apparent depth determined previously. For a real object ($s > 0$), the negative sign in Eq. (3-21) indicates that the image is virtual. The lateral magnification of an extended object is simply determined by inspection of Figure 3-19. Snells law requires, for the ray incident at the vertex and in the small-angle approximation, $n_1\theta_1 = n_2\theta_2$ or, using tangents for angles,

$$n_1\left(\frac{h_o}{s}\right) = n_2\left(\frac{h_i}{s'}\right)$$

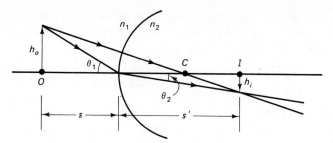

Figure 3-19 Construction to determine lateral magnification at a spherical refracting surface.

The lateral magnification is then

$$m = \frac{h_i}{h_o} = -\frac{n_1 s'}{n_2 s} \qquad (3\text{-}22)$$

where the negative sign is attached to give a negative value corresponding to an inverted image. For the case of a plane refracting surface, Eq. (3-21) may be incorporated into Eq. (3-22), giving $m = +1$. Thus the images formed by plane refracting surfaces have the same lateral dimensions and orientation as the object.

Example

As an extended example of refraction by spherical surfaces, refer to Figure 3-20. In (a), a real object is positioned in air, 30 cm from a convex spherical surface of radius 5 cm. To the right of the interface, the refractive index is that of water. Before constructing representative rays, we first find the image distance and lateral magnification of the image, using Eqs. (3-20) and (3-22). Equation (3-20) becomes

$$\frac{1}{30} + \frac{1.33}{s_1'} = \frac{1.33 - 1}{5}$$

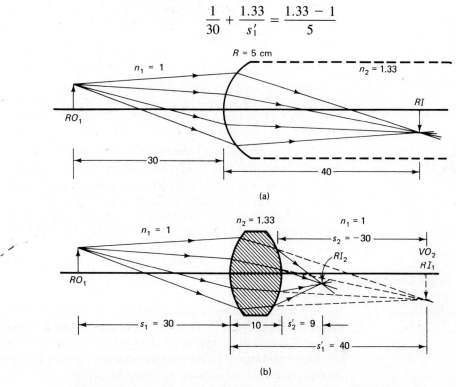

(a)

(b)

Figure 3-20 Example of refraction by spherical surfaces. (a) Refraction by a single spherical surface. (b) Refraction by a thick lens. Subscripts 1 and 2 refer to refractions at the first and second surfaces, respectively.

giving $s_1' = +40$ cm. The positive sign indicates that the image is real and so is located to the right of the surface, where real rays of light are refracted. Equation (3-22) becomes

$$m = -\frac{(1)(+40)}{(1.33)(+30)} = -1$$

indicating an inverted image, equal in size to that of the object. Figure 3-20a shows the image as well as several rays, which are now determined. In this example we have assumed that the medium to the right of the spherical surface extends far enough so that the image is formed inside it, without further refraction. Let us suppose now (Figure 3-20b) that the second medium is only 10 cm thick, forming a *thick lens*, with a second, concave spherical surface, also of radius 5 cm. The refraction by the first surface is, of course, unaffected by this change. Inside the lens, therefore, rays are directed as before to form an image 40 cm from the first surface. However, these rays are intercepted and again refracted by the second surface to produce a different image, as shown. Since the convergence of the rays striking the second surface is determined by the position of the first image, its location now specifies the appropriate object distance to be used for the second refraction. We call the real image for surface (1) a *virtual object* for surface (2). Then, by the sign convention established previously, we make the virtual object distance, relative to the second surface, a negative quantity when using Eqs. (3-20) and (3-22). For the second refraction then, Eq. (3-20) becomes

$$\frac{1.33}{-30} + \frac{1}{s_2'} = \frac{1 - 1.33}{-5}$$

or $s' = +9$ cm. The magnification, according to Eq. (3-22), is

$$m = \frac{(-1.33)(+9)}{(1)(-30)} = +\frac{2}{5}$$

The final image is then $\frac{2}{5}$ the lateral size of its (virtual) object and appears with the same orientation. Relative to the original object, the final image is $\frac{2}{5}$ as large and inverted.

In general, whenever a train of reflecting or refracting surfaces is involved in the processing of a final image, the individual reflections and/or refractions are considered in the order in which light is actually incident upon them. The object distance of the nth step is determined by the image distance for the $(n - 1)$st step. If the image of the $(n - 1)$ step is not actually formed, it serves as a *virtual object* for the nth step.

3-9 THIN LENSES

We now apply the preceding method to discover the thin-lens equation. As in the example of Figure 3-20, two refractions at spherical surfaces are involved. The simplification we make is to neglect the thickness of the lens in comparison with the object and image distances, an approximation that is justified in most practical situations. At the first refracting surface, of radius R_1,

$$\frac{n_1}{s_1} + \frac{n_2}{s_1'} = \frac{n_2 - n_1}{R_1} \tag{3-23}$$

and at the second surface, of radius R_2,

$$\frac{n_2}{s_2} + \frac{n_1}{s_2'} = \frac{n_1 - n_2}{R_2} \qquad (3\text{-}24)$$

We have assumed that the lens faces the same medium of refractive index n_1 on both sides. Now the second object distance, in general, is given by

$$s_2 = t - s_1' \qquad (3\text{-}25)$$

where t is the thickness of the lens. Notice that this relationship produces the correct sign of s_2, as in Figure 3-20, and also when the intermediate image falls inside or to the left of the lens. In the thin-lens approximation, neglecting t,

$$s_2 = -s_1' \qquad (3\text{-}26)$$

When this value of s_2 is substituted into Eq. (3-24) and Eqs. (3-23) and (3-24) are added, the terms n_2/s_1' cancel and there results

$$\frac{n_1}{s_1} + \frac{n_1}{s_2'} = (n_2 - n_1)\left(\frac{1}{R_1} - \frac{1}{R_2}\right)$$

Now s_1 is the original object distance and s_2' is the final image distance, so we may drop their subscripts and write simply

$$\frac{1}{s} + \frac{1}{s'} = \frac{n_2 - n_1}{n_1}\left(\frac{1}{R_1} - \frac{1}{R_2}\right) \qquad (3\text{-}27)$$

The *focal length* of the thin lens is defined as the image distance for an object at infinity, or the object distance for an image at infinity, giving

$$\frac{1}{f} = \frac{n_2 - n_1}{n_1}\left(\frac{1}{R_1} - \frac{1}{R_2}\right) \qquad (3\text{-}28)$$

Equation (3-28) is called the *lensmaker's equation* because it predicts the focal length of a lens fabricated with a given refractive index and radii of curvature and used in a medium of refractive index n_1. In most cases, the ambient medium is air, and $n_1 = 1$. The thin-lens equation, in terms of the focal length, is then

$$\frac{1}{s} + \frac{1}{s'} = \frac{1}{f} \qquad (3\text{-}29)$$

Wavefront analysis for plane wavefronts, as shown in Figure 3-21, indicates that a lens thicker in the middle causes convergence, while one thinner in the middle

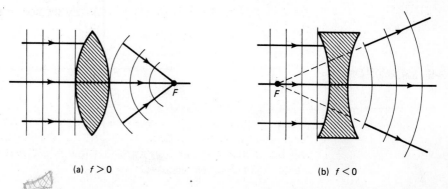

(a) $f > 0$ (b) $f < 0$

Figure 3-21 Action of converging lens (a) and diverging lens (b) on plane wavefronts of light.

causes divergence of the incident parallel rays. The portion of the wavefront that must pass through the thicker region is delayed relative to the other portions. Converging lenses are characterized by positive focal lengths and diverging lenses by negative focal lengths, as is evident from the figure, where the images are real and virtual, respectively.

Sample ray diagrams for converging (or *convex*) and diverging (or *concave*) lenses are shown in Figure 3-22. The thin lenses are best represented, for purposes of ray construction, by a vertical line with edges suggesting the general shape of the lens. Ray (1) from the top of the object is incident parallel to the axis and converges, in Figure 3-22a, through the focal point or diverges, in Figure 3-22b, as if proceeding from the focal point. A second ray is simply the inverse of the first. Although two rays are sufficient to locate the image, a third may also be drawn through the center of the lenses without bending. The midsection of the lens behaves as a parallel plate, which does not alter the direction of the incident ray, and because it is thin, displaces the ray by a negligible amount. In constructing ray diagrams, observe that, except for the central ray, every ray refracted by a convex lens bends toward the axis and every ray refracted by a concave lens bends away from the axis. From either diagram, the angles subtended by object and image at the center of the lens are seen to be equal. For either the real image in (a) or the virtual image in (b), it follows that

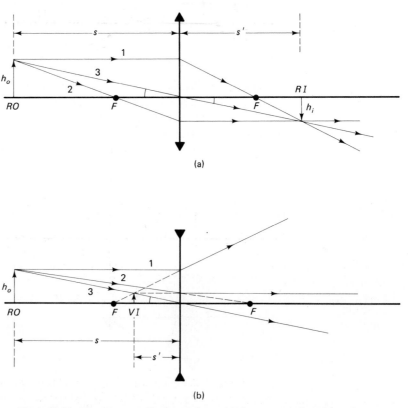

(a)

(b)

Figure 3-22 Ray diagrams for image formation by a convex lens (a) and a concave lens (b).

$$\frac{h_o}{s} = \frac{h_i}{s'}$$

and lateral magnification

$$m = \frac{h_i}{h_o} = \frac{s'}{s}$$

In accordance with the sign convention adopted here, a negative sign should be added to this expression. In case (a), $s > 0$, $s' > 0$, and $m < 0$ because the image is inverted; in case (b), $s > 0$, $s' < 0$, and $m > 0$. In either case then,

$$m = -\frac{s'}{s} \tag{3-30}$$

Further ray-diagram examples for a train of two lenses are illustrated in Figure 3-23. Table 3-1 and Figure 3-24 provide a convenient summary of image formation in lenses and mirrors.

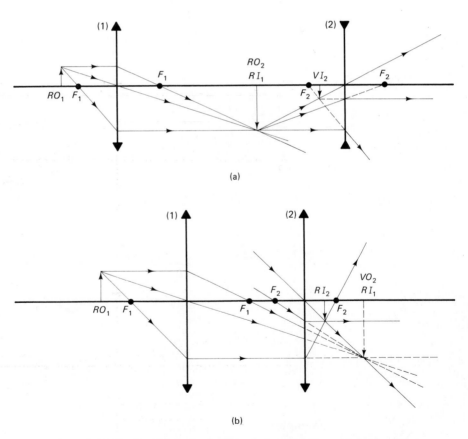

(a)

(b)

Figure 3-23 (a) Formation of a virtual image by a two-element train of a convex lens (1) and concave lens (2). (b) Formation of a real image RI_2 by a train of two convex lenses. The intermediate image RI_1 serves as a virtual object VO_2 for the second lens.

TABLE 3-1 SUMMARY OF GAUSSIAN MIRROR AND LENS FORMULAS

	Spherical surface	Plane surface
Reflection	$$\frac{1}{s} + \frac{1}{s'} = \frac{1}{f}, \quad f = -\frac{R}{2}$$ $$m = -\frac{s'}{s}$$ Concave: $f > 0, R < 0$ Convex: $f < 0, R > 0$	$s' = -s$ $m = +1$
Refraction Single surface	$$\frac{n_1}{s} + \frac{n_2}{s'} = \frac{n_2 - n_1}{R}$$ $$m = -\frac{n_1 s'}{n_2 s}$$ Concave: $R < 0$ Convex: $R > 0$	$s' = \frac{n_2}{n_1} s$ $m = +1$
Refraction Thin lens	$$\frac{1}{s} + \frac{1}{s'} = \frac{1}{f}$$ $$\frac{1}{f} = \frac{n_2 - n_1}{n_1}\left(\frac{1}{R_1} - \frac{1}{R_2}\right)$$ $$m = -\frac{s'}{s}$$ Concave: $f < 0$ Convex: $f > 0$	

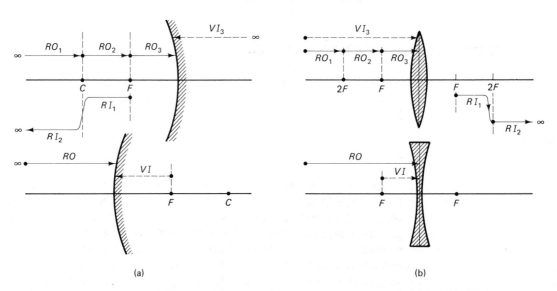

Figure 3-24 Summary of image formation by spherical mirrors and thin lenses. The location, nature, magnification, and orientation of the image are indicated or suggested. (a) Spherical mirrors. (b) Thin lenses.

Example

Find and describe the intermediate and final images produced by a two-lens system such as the one sketched in Figure 3-23a. Let $f_1 = 15$ cm, $f_2 = 15$ cm, and their separation be 60 cm. Let the object be 25 cm from the first lens, as shown.

Solution The first lens is convex: $f_1 = +15$ cm, $s_1 = 25$ cm.

$$\frac{1}{s_1} + \frac{1}{s_1'} = \frac{1}{f} \quad \text{or} \quad s_1' = \frac{s_1 f}{s_1 - f} = \frac{(25)(15)}{25 - 15} = +37.5 \text{ cm}$$

$$m_1 = -\frac{s_1'}{s_1} = -\frac{37.5}{25} = -1.5$$

Thus the first image is real (because s_1' is positive), 37.5 cm to the right of the first lens, inverted (because m is negative), and 1.5 times the size of the object.

The second lens is concave: $f_2 = -15$ cm. Since real rays of light diverge from the first real image, it serves as a real object for the second lens, with $s_2 = 60 - 37.5 = +22.5$ cm to the left of the lens. Then

$$s_2' = \frac{s_2 f}{s_2 - f} = \frac{(22.5)(-15)}{(22.5) - (-15)} = -9 \text{ cm}$$

$$m_2 = -\frac{s_2'}{s_2} = -\frac{-9}{22.5} = +0.4$$

Thus the final image is virtual (because s_2' is negative), 9 cm to the *left* of the second lens, erect *with respect to its own object* (because m is positive), and 0.4 times its size. The *overall* magnification is given by $m = m_1 m_2 = (-1.5)(0.4) = -0.6$. Thus the final image is inverted relative to the *original* object and 6/10 its lateral size. All these features are exhibited qualitatively in the ray diagram of Figure 3-23a.

3-10 VERGENCE AND REFRACTIVE POWER

Another way of interpreting the thin-lens equation is useful in certain applications, including optometry. The interpretation is based on two considerations. In the thin-lens equation,

$$\frac{1}{s} + \frac{1}{s'} = \frac{1}{f} \tag{3-31}$$

notice that (1) the reciprocals of distances in the left member add to give the reciprocal of the focal length, and (2) the reciprocals of the object and image distances describe the curvature of the wavefronts incident at the lens and centered at the object and image positions O and I, respectively. A plane wavefront, for example, has a curvature of zero. In Figure 3-25 spherical waves expand from the object point O and attain a curvature, or *vergence*, V, given by $1/s$, when they intercept the thin lens. On the other hand, once refracted by the lens, the wavefronts contract, in Figure 3-25a, and expand further, in Figure 3-25b, to locate the real and virtual image points shown. The curvature, or vergence, V', of the wavefronts as they emerge from the lens is $1/s'$. The change in curvature from object space to image space is due to the *refracting power P of the lens*, given by $1/f$. With these definitions, Eq. (3-31) may be written

$$V + V' = P \tag{3-32}$$

The units of the terms in Eq. (3-32) are reciprocal lengths. When the lengths are measured in meters, their reciprocals are said to have units of *diopters* (D). Thus the

(a)

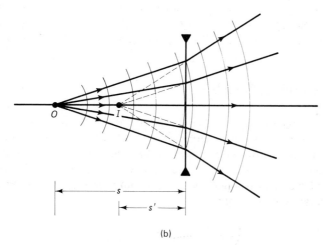

(b)

Figure 3-25 Change in curvature of wavefronts on refraction by a thin lens. (a) Convex lens. (b) Concave lens.

refracting power of a lens of focal length 20 cm is said to be 5 diopters. This alternative point of view emphasizes the degree of wave curvature or ray convergence rather than object and image distances. Accordingly, the degree of convergence V' of the image rays is determined by the original degree of convergence V of the object rays and the refracting power P of the lens. Eq. (3-32) can also be applied to the case of refraction at a single surface, Eq. (3-20), in which case the refractive indices in object and image space need not be 1. In this event, the *power of the refracting surface* is $(n_2 - n_1)/R$.

This approach is useful for another reason. When thin lenses are placed together, the focal length f of the combination, treated as a single thin lens, can be found in terms of the focal lengths $f_1, f_2, \ldots$ of the individual lenses. For example, with two such lenses back to back, we write the lens equations

$$\frac{1}{s_1} + \frac{1}{s_1'} = \frac{1}{f_1} \quad \text{and} \quad \frac{1}{s_2} + \frac{1}{s_2'} = \frac{1}{f_2}$$

Since the image distance for the first lens plays the role of the object distance for the second lens, we may write

$$s_2 = -s_1'$$

and, adding the two equations,

$$\frac{1}{s_1} + \frac{1}{s_2'} = \frac{1}{f_1} + \frac{1}{f_2} = \frac{1}{f}$$

The reciprocals of the individual focal lengths, therefore, add to give the reciprocal of the overall focal length f of the pair. In general, for several thin lenses,

$$\frac{1}{f} = \frac{1}{f_1} + \frac{1}{f_2} + \frac{1}{f_3} + \cdots \tag{3-33}$$

Expressed in diopters, the refractive powers simply add.

$$P = P_1 + P_2 + P_3 + \cdots \tag{3-34}$$

In a nearsighted eye, the refracted (converging) power of the lens is too great, so that a real image is formed in front of the retina. By reducing the convergence with a number of diverging lenses placed in front of the eye until an object is clearly focused, an optometrist can determine the net diopter specification of the single corrective lens needed by simply adding the diopters of these test lenses. In a farsighted eye, the natural converging power of the eye is not strong enough, and additional converging power must be added in the form of spectacles with a converging lens.

3-11 NEWTONIAN EQUATION FOR THE THIN LENS

When object and image distances are measured relative to the focal points of a lens, as by the distances x and x' in Figure 3-26, an alternative form of the thin-lens equation results, called the *Newtonian form*. In the figure, the two rays shown determine two right triangles, joined by the focal point, on each side of the lens. Since each pair constitutes similar triangles, we may set up proportions between sides that represent the lateral magnification:

$$\frac{h_i}{h_o} = \frac{f}{x} \quad \text{and} \quad \frac{h_i}{h_o} = \frac{x'}{f}$$

Introducing a negative sign for the magnification, due to the inverted image,

$$m = -\frac{f}{x} = -\frac{x'}{f} \tag{3-35}$$

The two parts of Eq. (3-35) also constitute the Newtonian form of the thin-lens equation,

$$xx' = f^2 \tag{3-36}$$

This equation is somewhat simpler than Eq. (3-29) and is found to be more convenient in certain applications.

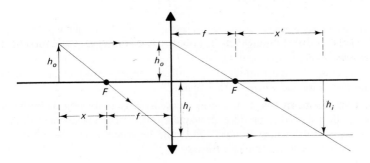

Figure 3-26 Construction used to derive Newton's equations for the thin lens.

3-1. Derive an expression for the transit time of a ray of light that travels a distance x_1 through a medium of index n_1, a distance x_2 through a medium of index n_2, . . . , and a distance x_m through a medium of index n_m. Use a summation to express your result.

3-2. Deduce the Cartesian oval for perfect imaging by a refracting surface when the object point is on the optical x-axis 20 cm from the surface vertex and its conjugate image point lies 10 cm inside the second medium. Assume the refracting medium to have an index of 1.50 and the outer medium to be air. Find the equation of the intersection of the oval with the xy-plane, where the origin of the coordinates is at the object point. Generate a table of (x, y)-coordinates for the surface and plot, together with sample rays.

3-3. A double convex lens has a diameter of 5 cm and zero thickness at its edges. A point object on an axis through the center of the lens produces a real image on the opposite side. Both object and image distances are 30 cm, measured from a plane bisecting the lens. The lens has a refractive index of 1.52. Using the equivalence of optical paths through the center and edge of the lens, determine the thickness of the lens at its center.

3-4. Determine the minimum height of a wall mirror that will permit a 6-ft person to view his or her entire height. Sketch rays from the top and bottom of the person, and determine the proper placement of the mirror such that the full image is seen, regardless of the person's distance from the mirror.

3-5. A ray of light makes an angle of incidence of 45° at the center of the top surface of a transparent cube of index 1.414. Trace the ray through the cube.

3-6. To determine the refractive index of a transparent plate of glass, a microscope is first focused on a tiny scratch in the upper surface, and the barrel position is recorded. Upon further lowering the microscope barrel by 1.87 mm, a focused image of the scratch is seen again. The plate thickness is 1.50 mm. What is the reason for the second image, and what is the refractive index of the glass?

3-7. A small source of light at the bottom face of a rectangular glass slab 2.25 cm thick is viewed from above. Rays of light totally internally reflected at the top surface outline a circle of 7.60 cm in diameter on the bottom surface. Determine the refractive index of the glass.

3-8. Show that the lateral displacement s of a ray of light penetrating a rectangular plate of thickness t is given by

$$s = \frac{t \sin (\theta_1 - \theta_2)}{\cos \theta_2}$$

where θ_1 and θ_2 are the angles of incidence and refraction, respectively. Find the displacement when $t = 3$ cm, $n = 1.50$, and $\theta_1 = 50°$.

3-9. A meter stick lies along the optical axis of a convex mirror of focal length 40 cm, with its nearer end 60 cm from the mirror surface. How long is the image of the meter stick?

3-10. A glass hemisphere is silvered over its curved surface. A small air bubble in the glass is located on the central axis through the hemisphere 5 cm from the plane surface. The radius of curvature of the spherical surface is 7.5 cm, and the glass has an index of 1.50. Looking along the axis into the plane surface, one sees two images of the bubble. How do they arise and where do they appear?

3-11. A concave mirror forms an image on a screen twice as large as the object. Both object and screen are then moved to produce an image on the screen that is three times the size of the object. If the screen is moved 75 cm in the process, how far is the object moved? What is the focal length of the mirror?

3-12. A sphere 5 cm in diameter has a small scratch on its surface. When the scratch is viewed through the glass from a position directly opposite, where does the scratch appear and what is its magnification? Assume $n = 1.50$ for the glass.

3-13. (a) At what position in front of a spherical refracting surface must an object be placed so that the refraction produces parallel rays of light? In other words, what is the focal length of a single refracting surface?

(b) Since real object distances are positive, what does your result imply for the cases $n_2 > n_1$ and $n_2 < n_1$?

3-14. A small goldfish is viewed through a spherical glass fishbowl 30 cm in diameter. Determine the apparent position and magnification of the fish's eye when its actual position is (a) at the center of the bowl and (b) nearer to the eye, halfway from center to glass, along the line of sight. Assume that the glass is thin enough so that its effect on the refraction may be neglected.

3-15. A small object faces the convex spherical glass window of a small water tank. The radius of curvature of the window is 5 cm. The inner back side of the tank is a plane mirror, 25 cm from the window. If the object is 30 cm outside the window, determine the nature of its final image, neglecting any refraction due to the thin glass window itself.

3-16. A plano-convex lens having a focal length of 25.0 cm is to be made with glass of refractive index 1.520. Calculate the radius of curvature of the grinding and polishing tools to be used in making this lens.

3-17. Calculate the focal length of a thin meniscus lens whose spherical surfaces have radii of curvature of magnitude 5 and 10 cm. The glass is of index 1.50. Sketch both positive and negative versions of the lens.

3-18. One side of a fish tank is built using a large-aperture thin lens made of glass $(n = 1.50)$. The lens is equiconvex, with radii of curvature 30 cm. A small fish in the tank is 20 cm from the lens. Where does the fish appear when viewed through the lens? What is its magnification?

3-19. Two thin lenses have focal lengths of -5 and $+20$ cm. Determine their equivalent focal lengths when (a) cemented together and (b) separated by 10 cm.

3-20. Two identical, thin, plano-convex lenses with radii of curvature of 15 cm are situated with their curved surfaces in contact at their centers. The intervening space is filled with oil of refractive index 1.65. The index of the glass is 1.50. Determine the focal length of the combination. (*Hint:* Think of the oil layer as an intermediate thin lens.)

3-21. An eyepiece is made of two thin lenses each of $+20$-mm focal length, separated by a distance of 16 mm.

(a) Where must a small object be positioned so that light from the object is rendered parallel by the combination?

(b) Does the eye see an image erect relative to the object? Is it magnified? Use a ray diagram to answer these questions by inspection.

3-22. A diverging thin lens and a concave mirror have focal lengths of equal magnitude. An object is placed $3f/2$ from the diverging lens, and the mirror is placed a distance $3f$ on the other side of the lens. Using Gaussian optics, determine the *final* image of the system, after two refractions (a) by a three-ray diagram and (b) by calculation.

3-23. A small object is placed 20 cm from the first of a train of three lenses with focal lengths, in order, of 10, 15, and 20 cm. The first two lenses are separated by 30 cm and the last two by 20 cm. Calculate the final image position relative to the last lens and its linear magnification relative to the original object when (a) all three lenses are positive; (b) the middle lens is negative; (c) the first and last lenses are negative. Provide ray diagrams for each case.

3-24. A convex thin lens with refractive index of 1.50 has a focal length of 30 cm in air. When immersed in a certain transparent liquid, it becomes a negative lens with a focal length of 188 cm. Determine the refractive index of the liquid.

3-25. It is desired to project onto a screen an image that is four times the size of a brightly illuminated object. A plano-convex lens with $n = 1.50$ and $R = 60$ cm is to be used.

Employing the Newtonian form of the lens equations, determine the appropriate distance of the object and screen from the lens. Is the image erect or inverted? Check your results using the ordinary lens equations.

3-26. Three thin lenses of focal lengths 10 cm, 20 cm, and −40 cm are placed in contact to form a single compound lens.

(a) Determine the powers of the individual lenses and that of the unit, in diopters.

(b) Determine the vergence of an object point 12 cm from the unit and that of the resulting image. Convert the result to an image distance in centimeters.

3-27. A lens is moved along the optical axis between a fixed object and a fixed image screen. The object and image positions are separated by a distance L that is more than four times the focal length of the lens. Two positions of the lens are found for which an image is in focus on the screen, magnified in one case and reduced in the other. If the two lens positions differ by distance D, show that the focal length of the lens is given by $f = (L^2 − D^2)/4L$. This is *Bessel's method* for finding the focal length of a lens.

3-28. An image of an object is formed on a screen by a lens. Leaving the lens fixed, the object is moved to a new position and the image screen moved until it again receives a focused image. If the two object positions are S_1 and S_2 and if the transverse magnifications of the image are M_1 and M_2, respectively, show that the focal length of the lens is given by

$$f = \frac{(S_2 − S_1)}{(1/M_1 − 1/M_2)}$$

This is *Abbe's method* for finding the focal length of a lens.

3-29. Derive the law of reflection from Fermat's principle by minimizing the distance of an arbitrary (hypothetical) ray from a given source point to a given receiving point.

3-30. Determine the ratio of focal lengths for two identical, thin plano-convex lenses when one is silvered on its flat side and the other on its curved side. Light is incident on the unsilvered side.

3-31. Show that the minimum distance between an object and its image, formed by a thin lens, is $4f$. When does this occur?

3-32. A ray of light traverses successively a series of plane interfaces, all parallel to one another and separating regions of differing thickness and refractive index.

(a) Show that Snell's law holds between the first and last regions, as if the intervening regions did not exist.

(b) Calculate the net lateral displacement of the ray from point of incidence to point of emergence.

3-33. A parallel beam of light is incident on a plano-convex lens that is 4 cm thick. The radius of curvature of the spherical side is also 4 cm. The lens has a refractive index of 1.50 and is used in air. Determine where the light is focused for light incident on each side.

3-34. A spherical interface, with radius of curvature 10 cm, separates media of refractive index 1 and $\frac{4}{3}$. The center of curvature is located on the side of the higher index. Find the focal lengths for light incident from each side. How do the results differ when the two refractive indices are interchanged?

3-35. An airplane is used in aerial surveying to make a map of ground detail. If the scale of the map is to be 1:50,000 and the camera used has a focal length of 6 in., determine the proper altitude for the photograph.

REFERENCES

[1] Feynman, Richard P., Robert B. Leighton, and Matthew Sands. *The Feynman Lectures on Physics.* Vol. 1. Reading, Mass.: Addison-Wesley Publishing Co., 1975. Ch. 26, 27.

[2] Smith, F. Dow. "How Images are Formed." In *Lasers and Light. Readings from Scientific American*, pp. 59–70. San Francisco: W. H. Freeman and Company Publishers, 1968.

[3] Winston, Roland. "Nonimaging Optics." *Scientific American* (March 1991): 76.

[4] Longhurst, R. S. *Geometrical and Physical Optics*, 2d ed. New York: John Wiley and Sons, 1967. Ch. 1, 2.

[5] Rossi, Bruno. *Optics*. Reading, Mass.: Addison-Wesley Publishing Company, 1957. Ch. 1, 2.

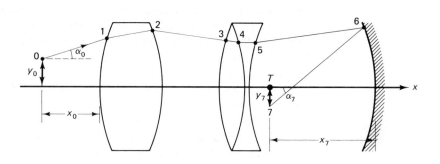

4

Matrix Methods
in Paraxial Optics

INTRODUCTION

This chapter deals with methods of analyzing optical systems when they become complex, involving a number of refracting and/or reflecting elements in trainlike fashion. Beginning with a description of a single *thick lens* in terms of its *cardinal points*, the discussion proceeds to an analysis of a train of optical elements by means of multiplication of 2×2 matrices representing the elementary refractions or reflections involved in the train. In this way, a *system matrix* for the entire optical system can be found that is related to the same cardinal points characterizing the thick lens. Finally, computer ray-tracing methods for tracing a given ray of light through an optical system are briefly described.

4-1 THE THICK LENS

Consider a spherical *thick lens*, that is, a lens whose thickness along its optical axis cannot be ignored without leading to serious errors in analysis. Just when a lens moves from the category of *thin* to *thick* clearly depends on the accuracy required. The thick lens can be treated by the methods of Chapter 3. The glass medium is bounded by two spherical refracting surfaces. The image of a given object, formed by refraction at the first surface, becomes the object for refraction at the second surface. The object distance for the second surface takes into account the thickness of the lens. The image formed by the second surface is then the final image due to the action of the composite thick lens.

The thick lens can also be described in a way that allows graphical determination of images corresponding to arbitrary objects, much like the ray rules for a thin lens. This description, in terms of the so-called *cardinal points* of the lens, is useful also because it can be applied to more complex optical systems, as will become evident in this chapter. Thus, even though we are at present interested in a single thick lens, the following description is applicable to an arbitrary optical system that we can imagine is contained within the outlines of the thick lens.

There are six cardinal points on the axis of a thick lens, from which its imaging properties can be deduced. Planes[1] normal to the axis at these points are called the *cardinal planes*. The six cardinal points (see Figures 4-1 and 4-2) consist of the first and second *system focal points* (F_1 and F_2), which are already familiar; the first and second *principal points* (H_1 and H_2); and the first and second *nodal points* (N_1 and N_2).

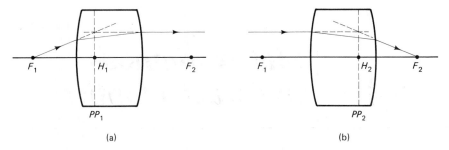

(a)   (b)

Figure 4-1 Illustration of the (a) first and (b) second principal planes of an optical system.

Figure 4-2 Illustration of the nodal points of an optical system.

A ray from the first focal point F_1 is rendered parallel to the axis (Figure 4-1a), and a ray parallel to the axis is refracted by the lens through the second focal point F_2 (Figure 4-1b). The extensions of the incident and resultant rays in each case intersect, by definition, in the *principal planes*, and these cross the axis at the principal points, H_1 and H_2. If the thick lens were a single thin lens, the two principal planes would coincide at the vertical line that is usually drawn to represent the lens. Principal planes in general do not coincide and may even be located outside the optical system itself. Once the locations of the principal planes are known, accurate ray diagrams can be drawn. The usual rays, determined by the focal points, bend at their intersections with the principal planes, as in Figure 4-1. The third ray usually drawn for thin-lens diagrams is one through the lens center, undeviated and negligibly displaced. The nodal points of a thick lens, or of any optical system, permit the correc-

---

[1] These "planes" are actually slightly curved surfaces that can be considered plane in the paraxial approximation.

tion to this ray, as shown in Figure 4-2. Any ray directed toward the first nodal point N_1 emerges from the optical system parallel to the incident ray, but displaced so that it appears to come from the second nodal point on the axis, N_2.

The positions of all six cardinal points are indicated in Figure 4-3. Distances are *directed*, positive or negative by a sign convention that makes distances directed to the left negative and distances to the right, positive. Notice that for the thick lens, the distances r and s determine the positions of the principal points relative to the vertices V_1 and V_2, while f_1 and f_2 determine focal point positions relative to the principal points. Note carefully that these focal points are *not* measured from the vertices of the lens.

Figure 4-3 Symbols used to signify the cardinal points and locations for a thick lens. Axial points include focal points (F), vertices (V), principal points (P), and nodal points (N). Directed distances separating their corresponding planes are defined in the drawing.

We summarize the basic equations for the thick lens without proof. Although the derivations involve simple algebra and geometry, they are rather arduous. We shall be content to await the matrix approach later in this chapter as a simpler way to justify these equations, and even then some of the work is relegated to the problems.

Utilizing the symbols defined in Figure 4-3, the focal length f_1 is given by

$$\frac{1}{f_1} = \frac{n_L - n'}{nR_2} - \frac{n_L - n}{nR_1} - \frac{(n_L - n)(n_L - n')}{nn_L}\frac{t}{R_1R_2} \tag{4-1}$$

and the focal length f_2 is conveniently expressed in terms of f_1 by

$$f_2 = -\frac{n'}{n}f_1 \tag{4-2}$$

Notice that the two focal lengths have the same magnitude if the lens is surrounded by a single refractive medium, so that $n = n'$. The principal planes can be located next using

$$r = \frac{n_L - n'}{n_L R_2}f_1 t \quad \text{and} \quad s = -\frac{n_L - n}{n_L R_1}f_2 t \tag{4-3}$$

The positions of the nodal points are given by

$$v = \left(1 - \frac{n'}{n} + \frac{n_L - n'}{n_L R_2}t\right)f_1 \quad \text{and} \quad w = \left(1 - \frac{n}{n'} - \frac{n_L - n}{n_L R_1}t\right)f_2 \tag{4-4}$$

Image and object distances and lateral magnification are related by

$$-\frac{f_1}{s_o} + \frac{f_2}{s_i} = 1 \quad \text{and} \quad m = -\frac{ns_i}{n's_o} \tag{4-5}$$

as long as the distances s_o and s_i, as well as focal lengths, are measured relative to corresponding principal planes. In the ordinary case of a lens in air, with $n = n' = 1$, notice that $r = v$ and $s = w$: First and second principal points are superimposed over corresponding nodal points. Also, first and second focal lengths are equal in magnitude, and the equations,

$$-\frac{1}{s_o} + \frac{1}{s_i} = \frac{1}{f} \quad \text{and} \quad m = -\frac{s_i}{s_o} \tag{4-6}$$

are valid, with symbols properly reinterpreted. Except for one negative sign, required by the sign convention, these are identical with the thin-lens equations.

Example

Determine the focal lengths and the principal points for a 4-cm thick, biconvex lens with refractive index of 1.52 and radii of curvature of 25 cm, when the lens caps the end of a long cylinder filled with water ($n = 1.33$).

Solution Use the equations for the thick lens in the order given:

$$\frac{1}{f_1} = \frac{1.52 - 1.33}{1(-25)} - \frac{1.52 - 1}{1(+25)} - \frac{(1.52 - 1)(1.52 - 1.33)}{1(1.52)} \frac{4}{(-25)(+25)}$$

or $f_1 = -35.74$ cm to the *left* of the first principal plane. Then

$$f_2 = -\left(\frac{1.33}{1}\right)(-35.74) = 47.53 \text{ cm}$$

to the *right* of the second principal plane, and

$$r = \frac{1.52 - 1.33}{(1.52)(-25)}(-35.74)(4) = 0.715 \text{ cm}$$

$$s = -\frac{1.52 - 1}{(1.52)(+25)}(47.53)(4) = -2.60 \text{ cm}$$

Thus the principal point H_1 is situated 0.715 cm to the *right* of the left vertex of the lens, and H_2 is situated 2.60 cm to the *left* of the right vertex.

4-2 THE MATRIX METHOD

When the optical system consists of several elements—for example, the four or five lenses that constitute a photographic lens—we need a systematic approach that facilitates analysis. As long as we restrict our analysis to paraxial rays, this systematic approach is well handled by the matrix method. We now present a treatment of image formation that employs matrices to describe changes in the height and angle of a ray as it makes its way by successive reflections and refractions through an optical system. We show that, in the paraxial approximation, changes in height and direction of a ray can be expressed by linear equations that make this matrix approach possible. By combining matrices that represent individual refractions and reflections, a given optical system may be represented by a single matrix, from which the essential properties of the composite optical system may be deduced. The method lends itself to computer techniques for tracing a ray through an optical system of arbitrary complexity.

Figure 4-4 shows the progress of a single ray through an arbitrary optical system. The ray is described at distance x_0 from the first refracting surface in terms of its height y_0 and slope angle α_0 relative to the optical axis. Changes in angle occur at each refraction, such as at points 1 through 5, and at each reflection, such as point 6. The height of the ray changes during translations between these points. We look for

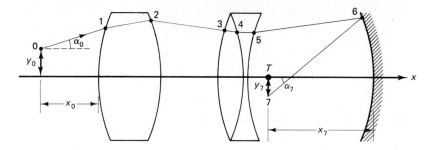

Figure 4-4 Steps in tracing a ray through an optical system. Progress of a ray can be described by changes in its elevation and direction.

a procedure that will allow us to calculate the height and slope angle of the ray at any point in the optical system, for example, at point T, a distance x_7 from the mirror. In other words, given the input data y_0, α_0 at point 0, we wish to predict values of y_7, α_7 at point 7 as output data.

4-3 THE TRANSLATION MATRIX

Consider a simple translation of the ray in a homogeneous medium, as in Figure 4-5. Let the axial progress of the ray be L, as shown, such that at point 1, the elevation and direction of the ray are given by "coordinates" y_1 and α_1, respectively. Evidently,

$$\alpha_1 = \alpha_0 \quad \text{and} \quad y_1 = y_0 + L \tan \alpha_0$$

These equations may be put into an ordered form, where the paraxial approximation $\tan \alpha_0 \cong \alpha_0$ has been used:

$$y_1 = (1)y_0 + (L)\alpha_0$$
$$\alpha_1 = (0)y_0 + (1)\alpha_0 \tag{4-7}$$

In matrix notation, the two equations are written

$$\begin{bmatrix} y_1 \\ \alpha_1 \end{bmatrix} = \begin{bmatrix} 1 & L \\ 0 & 1 \end{bmatrix} \begin{bmatrix} y_0 \\ \alpha_0 \end{bmatrix} \tag{4-8}$$

Evidently, the 2×2 *ray-transfer matrix* represents the effect of the translation on the ray. The input data (y_0, α_0) is modified by the ray-transfer matrix to yield the correct output data (y_1, α_1).

Figure 4-5 Simple translation of a ray.

4-4 THE REFRACTION MATRIX

Consider next the refraction of a ray at a spherical surface separating media of refractive indices n and n', as shown in Figure 4-6. We need to relate the ray coordinates

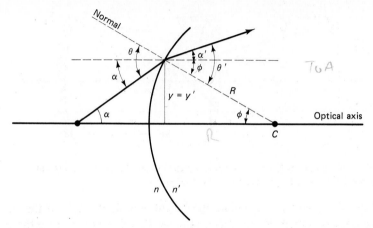

Figure 4-6 Refraction of a ray at a spherical surface.

(y', α') after refraction to those before refraction, (y, α). Since refraction occurs at a point, there is no change in elevation, and $y = y'$. The angle α', on the other hand, is by inspection of Figure 4-6

$$\alpha' = \theta' - \phi = \theta' - \frac{y}{R} \quad \text{and} \quad \alpha = \theta - \phi = \theta - \frac{y}{R}$$

Incorporating the paraxial form of Snell's law,

$$n\theta = n'\theta'$$

we have

$$\alpha' = \left(\frac{n}{n'}\right)\theta - \frac{y}{R} = \left(\frac{n}{n'}\right)\left(\alpha + \frac{y}{R}\right) - \frac{y}{R}$$

or

$$\alpha' = \left(\frac{1}{R}\right)\left(\frac{n}{n'} - 1\right)y + \left(\frac{n}{n'}\right)\alpha$$

The appropriate linear equations are then

$$y' = (1)y + (0)\alpha$$

$$\alpha' = \left[\left(\frac{1}{R}\right)\left(\frac{n}{n'} - 1\right)\right]y + \left(\frac{n}{n'}\right)\alpha \tag{4-9}$$

or, in matrix form,

$$\begin{bmatrix} y' \\ \alpha' \end{bmatrix} = \begin{bmatrix} 1 & 0 \\ \dfrac{1}{R}\left(\dfrac{n}{n'} - 1\right) & \dfrac{n}{n'} \end{bmatrix} \begin{bmatrix} y \\ \alpha \end{bmatrix} \tag{4-10}$$

We use the same sign convention as designed earlier. If the surface is instead concave, R is negative. Furthermore, allowing $R \to \infty$ yields the appropriate refraction matrix for a plane interface.

4-5 THE REFLECTION MATRIX

Finally, consider reflection at a spherical surface, illustrated in Figure 4-7. In the case considered, a concave mirror, R is negative. We need to add a sign convention for the angles that describe the ray directions. Angles are considered positive for all

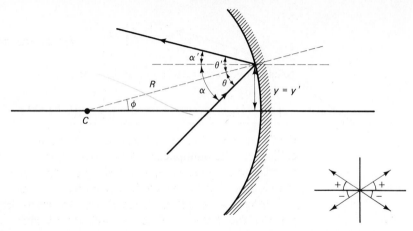

Figure 4-7 Reflection of a ray at a spherical surface. The inset illustrates the sign convention for ray angles.

rays pointing upward, either before or after a reflection; angles for rays pointing downward are considered negative. The sign convention is summarized in the inset of Figure 4-7. From the geometry of Figure 4-7, with both α and α' positive,

$$\alpha = \theta + \phi = \theta + \frac{y}{-R} \quad \text{and} \quad \alpha' = \theta' - \phi = \theta' - \frac{y}{-R}$$

Using these relations together with the law of reflection, $\theta = \theta'$,

$$\alpha' = \theta' + \frac{y}{R} = \theta + \frac{y}{R} = \alpha + \frac{2y}{R}$$

and the two desired linear equations are

$$y' = (1)y + (0)\alpha$$

$$\alpha' = \left(\frac{2}{R}\right)y + (1)\alpha \tag{4-11}$$

In matrix form,

$$\begin{bmatrix} y' \\ \alpha' \end{bmatrix} = \begin{bmatrix} 1 & 0 \\ \dfrac{2}{R} & 1 \end{bmatrix} \begin{bmatrix} y \\ \alpha \end{bmatrix} \tag{4-12}$$

4-6 THICK-LENS AND THIN-LENS MATRICES

We construct now a matrix that represents the action of a thick lens on a ray of light. For generality, we assume different media on opposite sides of the lens, having refractive indices n and n', as shown in Figure 4-8. In traversing the lens, the ray undergoes two refractions and one translation, steps for which we have already derived matrices. Referring to Figure 4-8, where we have chosen for simplicity a lens with positive radii of curvature, we may write, symbolically,

$$\begin{bmatrix} y_1 \\ \alpha_1 \end{bmatrix} = M_1 \begin{bmatrix} y_0 \\ \alpha_0 \end{bmatrix} \qquad \text{for the first refraction}$$

$$\begin{bmatrix} y_2 \\ \alpha_2 \end{bmatrix} = M_2 \begin{bmatrix} y_1 \\ \alpha_1 \end{bmatrix} \qquad \text{for the translation}$$

and

$$\begin{bmatrix} y_3 \\ \alpha_3 \end{bmatrix} = M_3 \begin{bmatrix} y_2 \\ \alpha_2 \end{bmatrix} \qquad \text{for the second refraction}$$

Telescoping these matrix equations results in

$$\begin{bmatrix} y_3 \\ \alpha_3 \end{bmatrix} = M_3 M_2 M_1 \begin{bmatrix} y_0 \\ \alpha_0 \end{bmatrix}$$

Evidently the entire thick lens can be represented by a matrix $M = M_3 M_2 M_1$. Recalling that the multiplication of matrices is associative but not commutative, the descending order must be maintained. The individual matrices operate on the light ray in the same order in which the corresponding optical actions influence the light ray as it traverses the system. Generalizing, the matrix equation representing any number N of translations, reflections, and refractions is given by

$$\begin{bmatrix} y_f \\ \alpha_f \end{bmatrix} = M_N M_{N-1} \cdots M_2 M_1 \begin{bmatrix} y_0 \\ \alpha_0 \end{bmatrix} \tag{4-13}$$

and the ray-transfer matrix representing the entire optical system is

$$M = M_N M_{N-1} \cdots M_2 M_1 \tag{4-14}$$

We apply this result first to the thick lens of Figure 4-8, whose index is n_L and whose thickness for paraxial rays is t. The correct approximation for a thin lens is then made by allowing $t \to 0$. Letting $\mathcal{R}$ represent a refraction matrix and $\mathcal{T}$ represent a translation matrix, the matrix for the thick lens is, by Eq. (4-14), the composite matrix

$$M = \mathcal{R}_2 \mathcal{T} \mathcal{R}_1$$

or

$$M = \begin{bmatrix} 1 & 0 \\ \dfrac{n_L - n'}{n' R_2} & \dfrac{n_L}{n'} \end{bmatrix} \begin{bmatrix} 1 & t \\ 0 & 1 \end{bmatrix} \begin{bmatrix} 1 & 0 \\ \dfrac{n - n_L}{n_L R_1} & \dfrac{n}{n_L} \end{bmatrix} \tag{4-15}$$

We simplify immediately for the case where t is negligible ($t = 0$) and where the lens is surrounded by the same medium on either side ($n = n'$). Then

$$M = \begin{bmatrix} 1 & 0 \\ \dfrac{n_L - n}{n R_2} & \dfrac{n_L}{n} \end{bmatrix} \begin{bmatrix} 1 & 0 \\ 0 & 1 \end{bmatrix} \begin{bmatrix} 1 & 0 \\ \dfrac{n - n_L}{n_L R_1} & \dfrac{n}{n_L} \end{bmatrix} \tag{4-16}$$

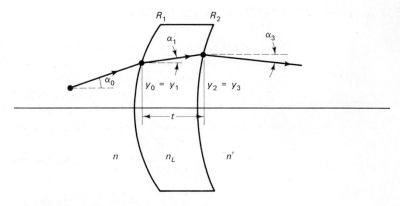

Figure 4-8 Progress of a ray through a thick lens.

TABLE 4-1 SUMMARY OF SOME SIMPLE RAY-TRANSFER MATRICES

Translation matrix:

$$M = \begin{bmatrix} 1 & L \\ 0 & 1 \end{bmatrix}$$

Refraction matrix,
  spherical interface:

$$M = \begin{bmatrix} 1 & 0 \\ \dfrac{n - n'}{Rn'} & \dfrac{n}{n'} \end{bmatrix}$$

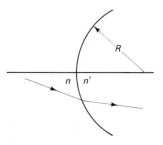

$(+R)$: convex
$(-R)$: concave

Refraction matrix,
  plane interface:

$$M = \begin{bmatrix} 1 & 0 \\ 0 & \dfrac{n}{n'} \end{bmatrix}$$

Thin-lens matrix:

$$M = \begin{bmatrix} 1 & 0 \\ -\dfrac{1}{f} & 1 \end{bmatrix}$$

$$\frac{1}{f} = \frac{n' - n}{n}\left(\frac{1}{R_1} - \frac{1}{R_2}\right)$$

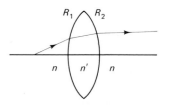

$(+f)$: convex
$(-f)$: concave

Spherical mirror
  matrix:

$$M = \begin{bmatrix} 1 & 0 \\ \dfrac{2}{R} & 1 \end{bmatrix}$$

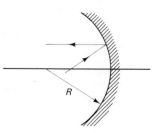

$(+R)$: convex
$(-R)$: concave

When multiplied together,

$$M = \begin{bmatrix} 1 & 0 \\ \dfrac{n_L - n}{n}\left(\dfrac{1}{R_2} - \dfrac{1}{R_1}\right) & 1 \end{bmatrix} \qquad (4\text{-}17)$$

The matrix element in the first column, second row, may be expressed in terms of the focal length of the lens, by the lensmaker's formula,

$$\frac{1}{f} = \frac{n_L - n}{n}\left(\frac{1}{R_1} - \frac{1}{R_2}\right)$$

so that the thin-lens ray-transfer matrix is simply

$$M = \begin{bmatrix} 1 & 0 \\ -\dfrac{1}{f} & 1 \end{bmatrix} \qquad (4\text{-}18)$$

As usual, f is taken as positive for a convex lens and negative for a concave lens. This matrix together with those previously derived are summarized for quick reference in Table 4-1.

4-7 SYSTEM RAY-TRANSFER MATRIX

By combining appropriate individual matrices in the proper order, according to Eq. (4-14), it is possible to express any optical system by a single 2×2 matrix, which we call the *system matrix*.

Example

Find the system matrix for the thick lens of Figure 4-8, whose matrix before multiplication is expressed by Eq. (4-15), and specify the thick lens exactly by choosing $R_1 = 45$ cm, $R_2 = 30$ cm, $t = 5$ cm, $n_L = 1.60$, and $n = n' = 1$.

Solution

$$M = \begin{bmatrix} 1 & 0 \\ \dfrac{1}{50} & 1.6 \end{bmatrix}\begin{bmatrix} 1 & 5 \\ 0 & 1 \end{bmatrix}\begin{bmatrix} 1 & 0 \\ -\dfrac{1}{120} & 1.6 \end{bmatrix} \quad \text{or} \quad M = \begin{bmatrix} \dfrac{23}{24} & \dfrac{25}{8} \\ \dfrac{7}{1200} & \dfrac{17}{16} \end{bmatrix}$$

The elements of this composite ray-transfer matrix, usually referred to in the symbolic form

$$M = \begin{bmatrix} A & B \\ C & D \end{bmatrix}$$

describe the relevant properties of the optical system, as we will see. Be aware that the particular values of the matrix elements of a system depend on the location of the ray at input and output. In the case of the thick lens just calculated, the *input plane* was chosen at the left surface of the lens, and the *output plane* was chosen at its right surface. If each of these planes is moved some distance from the lens, the system matrix will also include an initial and a final translation matrix incorporating these distances. The matrix elements change, and the system matrix now represents this enlarged "system." In any case, the determinant of the system matrix has a very useful property:

$$\text{Det } M = AD - BC = \frac{n_0}{n_f} \qquad (4\text{-}19)$$

where n_0 and n_f are the refractive indices of the initial and final media of the optical system. The proof of this assertion follows upon noticing first that the determinant of all the individual ray-transfer matrices in Table 4-1 have values of either n/n' or unity and then making use of the theorem[2] that the determinant of a product of matrices is equal to the product of the determinants. Symbolically, if $M = M_1 M_2 M_3 \cdots M_N$, then

$$\text{Det } (M) = (\text{Det } M_1)(\text{Det } M_2)(\text{Det } M_3) \cdots (\text{Det } M_N) \qquad (4\text{-}20)$$

In forming this product, using determinants of ray-transfer matrices, all intermediate refractive indices cancel, and we are left with the ratio n_0/n_f, as stated in Eq. (4-19). Most often, as in the case of the thick-lens example, n_0 and n_f both refer to air, and Det (M) is unity. The condition expressed by Eq. (4-19) is useful in checking the correctness of the calculations that produce a system matrix.

4-8 SIGNIFICANCE OF SYSTEM MATRIX ELEMENTS

We examine now the implications that follow when each of the matrix elements is zero. In symbolic form, we have, from Eq. (4-13),

$$\begin{bmatrix} y_f \\ \alpha_f \end{bmatrix} = \begin{bmatrix} A & B \\ C & D \end{bmatrix} \begin{bmatrix} y_0 \\ \alpha_0 \end{bmatrix} \qquad (4\text{-}21)$$

which is equivalent to the algebraic relations

$$y_f = Ay_0 + B\alpha_0$$
$$\alpha_f = Cy_0 + D\alpha_0 \qquad (4\text{-}22)$$

1. $D = 0$. In this case, $\alpha_f = Cy_0$, independent of α_0. Since y_0 is fixed, this means that all rays leaving a point in the input plane will have the same angle α_f at the output plane, independent of their angles at input. As shown in Figure 4-9a, the input plane thus coincides with the first focal plane of the optical system.

2. $A = 0$. This case is much like the previous one. Here $y_f = B\alpha_0$ implies that y_f is independent of y_0, so that all rays departing the input plane at the same angle, regardless of altitude, arrive at the same altitude y_f at the output plane. As shown in Figure 4-9b, the output plane thus functions as the second focal plane.

3. $B = 0$. Then $y_f = Ay_0$, independent of α_0. Thus all rays from a point at height y_0 in the input plane arrive at the same point of height y_f in the output plane. The points are then related as object and image points, as shown in Figure 4-9c, and the input and output planes correspond to conjugate planes for the

---

[2] The theorem can easily be verified for the product of two matrices and generalized by induction to the product of any number of matrices. Formal proofs can be found in any standard textbook on matrices and determinants, for example, E. T. Browne, *Introduction to the Theory of Determinants and Matrices* (Chapel Hill: University of North Carolina, 1958).

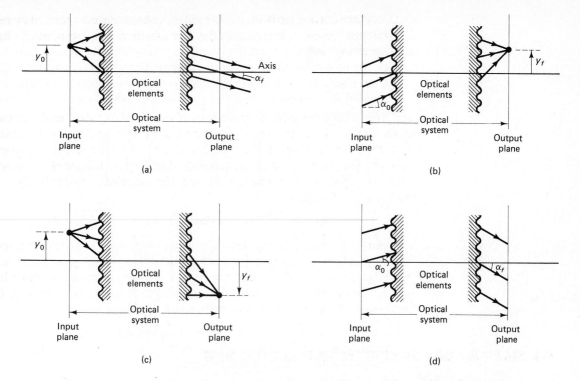

(a)

(b)

(c)

(d)

Figure 4-9 Diagrams illustrating the significance of the vanishing of system matrix elements. (a) When $D = 0$, the input plane corresponds to the first focal plane of the optical system. (b) When $A = 0$, the output plane corresponds to the second focal plane of the optical system. (c) When $B = 0$, the output plane is the image plane conjugate to the input plane, and A is the linear magnification. (d) When $C = 0$, a parallel bundle of rays at the input plane is parallel at the output plane, and D is the angular magnification.

optical system. Furthermore, since $A = y_f/y_0$, the matrix element A represents the linear magnification.

4. $C = 0$. Now $\alpha_f = D\alpha_0$, independent of y_0. This case is analogous to case 3, with directions replacing ray heights. Input rays, all of one direction, now produce parallel output rays in some other direction. Moreover, $D = \alpha_f/\alpha_0$ is the angular magnification. A system for which $C = 0$ is sometimes called a "telescopic system," because a telescope admits parallel rays into its objective and outputs parallel rays for viewing from its eyepiece.

Example

We illustrate case 3 by an example. We place a small object at a distance of 16 cm from the left end of a long, plastic rod with a polished spherical end of radius 4 cm, as indicated in Figure 4-10. The refractive index of the plastic is

Figure 4-10 Schematic defining an example for ray-transfer matrix methods.

1.50 and the object is in air. Let the unknown image be formed at the output reference plane, a distance x from the spherical cap. We desire to determine the image distance x and the lateral magnification. The system matrix consists of the product of three matrices, corresponding to (1) a translation $\mathcal{T}_1$ in air from object to the rod, (2) a refraction $\mathcal{R}$ at the spherical surface, and (3) a translation $\mathcal{T}_2$ in plastic to the image. Remembering to take the matrices in reverse order, we have $M = \mathcal{T}_2\mathcal{R}\mathcal{T}_1$

$$
M = \begin{bmatrix} 1 & x \\ 0 & 1 \end{bmatrix} \begin{bmatrix} 1 & 0 \\ \dfrac{1-1.50}{4(1.50)} & \dfrac{1}{1.50} \end{bmatrix} \begin{bmatrix} 1 & 16 \\ 0 & 1 \end{bmatrix}
$$

or

$$
M = \begin{bmatrix} 1 - \dfrac{x}{12} & 16 - \dfrac{2x}{3} \\ -\dfrac{1}{12} & -\dfrac{2}{3} \end{bmatrix}
$$

with the unknown quantity x incorporated in the matrix elements. According to this discussion, when $B = 0$, the output plane is the image plane, so that the image distance is determined by setting

$$
16 - \frac{2x}{3} = 0 \quad \text{or} \quad x = 24 \text{ cm}
$$

Further, the linear magnification is then given by the value of element A:

$$
m = A = 1 - \frac{x}{12} = -1
$$

We conclude that the image occurs 24 cm inside the rod, is inverted, and has the same lateral size as the object. This illustrates how the system matrix can be used to perform the customary job of finding image locations and sizes, although this may usually be done more quickly by using the Gaussian image formulas derived earlier.

4-9 LOCATION OF CARDINAL POINTS FOR AN OPTICAL SYSTEM

Since the properties of an optical system can be deduced from the elements of the system ray-transfer matrix, it follows that relationships must exist between the matrix elements, A, B, C, and D and the cardinal points of the system. In Figure 4-11, we generalize Figure 4-3 by defining distances locating the six cardinal points relative to the input and output planes that define the limits of an optical system. The focal points are located at distances f_1 and f_2 from the principal planes and at distances p and q from the reference input and output planes, respectively. Further, measured from the input and output planes, the distances r and s locate the principal points and the distances v and w locate the nodal points. Distances measured to the right of their reference planes are considered positive and to the left, negative. The principal and nodal points often occur outside the optical system, that is, outside the region defined by the input and output planes.

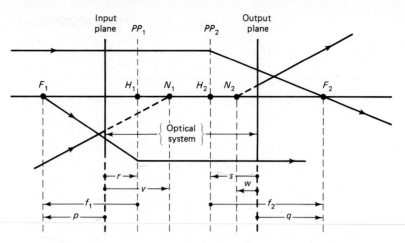

Figure 4-11 Location designations for the six cardinal points of an optical system. Rays associated with the nodal points and principal planes are also shown.

We now derive the relationships between the distances defined in Figure 4-11 and the system matrix elements. Consider Figure 4-12a, which highlights distances p, r, and f_1, as they are determined by the positions of the first focal point and the first principal plane. Input coordinates of the given ray are (y_0, α_0) and output coordinates are $(y_f, 0)$. Thus the ray equations, Eqs. (4-22), become for this ray

$$y_f = Ay_0 + B\alpha_0$$

and

$$0 = Cy_0 + D\alpha_0 \quad \text{or} \quad y_0 = -\left(\frac{D}{C}\right)\alpha_0 \tag{4-23}$$

For small angles, Figure 4-12a shows that

$$\alpha_0 = \frac{y_0}{-p}$$

where the negative sign indicates that F_1 is located a distance p to the left of the input plane. Incorporating Eq. (4-23),

$$p = \frac{-y_0}{\alpha_0} = \frac{D}{C} \tag{4-24}$$

Similarly, $\alpha_0 = y_f/(-f_1)$, and thus

$$f_1 = \frac{-y_f}{\alpha_0} = \frac{-(Ay_0 + B\alpha_0)}{\alpha_0} = \frac{AD}{C} - B$$

$$f_1 = \frac{AD - BC}{C} = \frac{\text{Det }(M)}{C} = \left(\frac{n_0}{n_f}\right)\frac{1}{C} \tag{4-25}$$

Finally, with the help of Eqs. (4-24) and (4-25), the positive distance r can be expressed in terms of p and f_1:

$$r = p - f_1 = \frac{D}{C} - \frac{n_0}{n_f}\frac{1}{C} = \frac{1}{C}\left(D - \frac{n_0}{n_f}\right) \tag{4-26}$$

Using Figure 4-12b, one can similarly discover relations for the output distances q, f_2, and s. The results, together with those just derived for p, f_1, and r, are

Figure 4-12 (a) Construction used to relate distances p, r, and f_1 to matrix elements. (b) Construction used to relate distances q, s, and f_2 to matrix elements. (c) Construction used to relate distances v and w to matrix elements.

listed in Table 4-2. With the help of Figure 4-12c, the nodal plane distances v and w may also be determined. For example, for small angle α,

$$\alpha = \frac{-y_0}{v} \tag{4-27}$$

where the negative sign indicates that the ray intersects the input plane below the axis. Input and output rays make the same angle relative to the axis. From Eq. (4-22), with $\alpha_0 = \alpha_f = \alpha$,

$$\alpha = Cy_0 + D\alpha \quad \text{or} \quad \frac{y_0}{\alpha} = \frac{1 - D}{C} \tag{4-28}$$

Combining Eqs. (4-27) and (4-28),

$$v = \frac{D - 1}{C} \tag{4-29}$$

$$p = \frac{D}{C} \qquad\qquad F_1$$

$$q = -\frac{A}{C} \qquad\qquad F_2$$

$$r = \frac{D - n_0/n_f}{C} \qquad H_1 \qquad \text{Located relative to}$$
$$\qquad\qquad\qquad\qquad\qquad \text{input (1) and output}$$
$$s = \frac{1 - A}{C} \qquad\qquad H_2 \qquad \text{(2) reference planes}$$

$$v = \frac{D - 1}{C} \qquad\qquad N_1$$

$$w = \frac{n_0/n_f - A}{C} \qquad N_2$$

$$f_1 = p - r = \frac{n_0/n_f}{C} \qquad F_1 \qquad \text{Located relative to}$$
$$\qquad\qquad\qquad\qquad\qquad \text{principal planes}$$
$$f_2 = q - s = -\frac{1}{C} \qquad F_2$$

Similarly, one can show that

$$w = \frac{(n_0/n_f) - A}{C} \tag{4-30}$$

again using the fact that Det $(M) = AD - BC = n_0/n_f$. These results are also included in Table 4-2. The relationships listed there can be used to establish the following useful generalizations:

1. Principal points and nodal points coincide, that is, $r = v$ and $s = w$, when the initial and final media have the same refractive indices.
2. First and second focal lengths of an optical system are equal in magnitude when initial and final media have the same refractive indices.
3. The separation of the principal points is the same as the separation of nodal points, that is, $r - s = v - w$.

4-10 EXAMPLES USING THE SYSTEM MATRIX AND CARDINAL POINTS

As an example, consider an optical system that consists of two thin lenses in air, separated by a distance L, as shown in Figure 4-13. The lenses have focal lengths of f_A and f_B, which may be either positive or negative. If input and output reference

Figure 4-13 Optical system consisting of two thin lenses in air, separated by a distance L.

planes are located at the lenses, the system matrix includes two thin-lens matrices, $\mathscr{L}_A$ and $\mathscr{L}_B$, and a translation matrix $\mathscr{T}$ for the distance L between them. The system matrix $M = \mathscr{L}_B \mathscr{T} \mathscr{L}_A$, or

$$
M = \begin{bmatrix} 1 & 0 \\ -\dfrac{1}{f_B} & 1 \end{bmatrix} \begin{bmatrix} 1 & L \\ 0 & 1 \end{bmatrix} \begin{bmatrix} 1 & 0 \\ -\dfrac{1}{f_A} & 1 \end{bmatrix}
$$

$$
M = \begin{bmatrix} 1 - \dfrac{L}{f_A} & L \\[2ex] \dfrac{1}{f_B}\left(\dfrac{L}{f_A} - 1\right) - \dfrac{1}{f_A} & 1 - \dfrac{L}{f_B} \end{bmatrix} \tag{4-31}
$$

We may "derive" the equivalent focal length of any such system, measured relative to principal planes, by simply reading out the C element. Since $f_{eq} = |1/C|$,

$$
\frac{1}{f_{eq}} = \frac{1}{f_A} + \frac{1}{f_B} - \frac{L}{f_A f_B} \tag{4-32}
$$

Furthermore, the first principal and nodal points coincide at a distance given by $(D - 1)/C$ from the first lens, and the second principal and nodal points coincide at a distance given by $(1 - A)/C$ from the second lens. Thus

$$
r = v = \left(\frac{f_{eq}}{f_B}\right)L \quad \text{and} \quad s = w = -\left(\frac{f_{eq}}{f_A}\right)L \tag{4-33}
$$

Example

Let us apply these results to the case of a Huygens eyepiece, which consists of two positive, thin lenses separated by a distance L equal to the average of their focal lengths. Suppose $f_A = 3.125$ cm and $f_B = 2.083$ cm, giving $L = 2.604$ cm and $f_{eq} = 2.5$ cm, by Eq. (4-32). Incidentally, the magnifying power of this eyepiece, given by $25/f$, is therefore $10\times$. From Eq. (4-33), we conclude that $r = +3.125$ cm and $s = -2.083$ cm. The optical system, together with its cardinal points and sample rays, is shown roughly to scale in Figure 4-14. The incident rays determine an image location between the lenses, which acts as a virtual object for the optical system. An enlarged, virtual image (not shown) is formed by the diverging rays leaving the system, as seen by an eye looking into the eyepiece. This eyepiece is discussed further in Chapter 6.

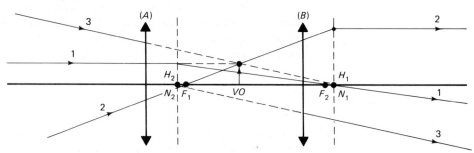

Figure 4-14 Ray construction for a Huygens eyepiece, using cardinal points.

Example

As a final calculation, let us find the cardinal points and sketch a ray diagram for the hemispherical glass lens shown in Figure 4-15. The radii of curvature are $R_1 = 3$ cm and $R_2 \to \infty$, and the lens in air has a refractive index of 1.50.

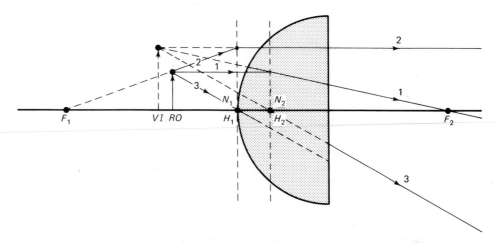

Figure 4-15 Ray construction for a hemispherical lens, using cardinal points.

Solution The system matrix, for input and output reference planes at the two surfaces of the lens, is then $M = \mathcal{R}_2 \mathcal{T} \mathcal{R}_1$,

$$M = \begin{bmatrix} 1 & 0 \\ 0 & 1.5 \end{bmatrix} \begin{bmatrix} 1 & 3 \\ 0 & 1 \end{bmatrix} \begin{bmatrix} 1 & 0 \\ \dfrac{-0.5}{1.5(3)} & 1.5 \end{bmatrix}$$

or

$$M = \begin{bmatrix} \dfrac{2}{3} & 2 \\ -\dfrac{1}{6} & 1 \end{bmatrix}, \qquad \text{with Det } (M) = 1$$

The relations in Table 4-2 then give values of $p = -6$ cm, $q = 4$ cm, $r = 0$, $s = -2$ cm, $f_1 = -6$ cm, and $f_2 = 6$ cm. Principal and nodal points coincide. The cardinal points are located, approximately to scale, in Figure 4-15. Ray diagrams using the principal planes and nodal points are constructed for an arbitrary real object. In this case the emerging rays determine a virtual image near the object, erect, and slightly magnified.

4-11 RAY TRACING

The assumption of paraxial rays greatly simplifies the description of the progress of rays of light through an optical system, because trigonometric terms do not appear in the equations. For many purposes, this treatment is sufficient. In practice, rays of light contributing to an image in an optical system are, in fact, usually rays in the near neighborhood of the optical axis. If the quality of the image is to be improved, however, ways must be found to reduce the ever-present aberrations that arise from the presence of rays deviating, more or less, from this ideal assumption. To determine the actual path of individual rays of light through an optical system, each ray must be *traced*, independently, using only the laws of reflection and refraction together with geometry. This technique is called *ray tracing* because it was formerly done by hand, graphically, with ruler and compass, in a step-by-step process through an accurate sketch of the optical system. Today, with the help of computers, the necessary calculations yielding the progressive changes in a ray's altitude and angle is done more easily and quickly. Graphics techniques are used to actually draw the optical system and to trace the ray's progress through the optical system on the monitor.[3]

Ray-tracing procedures, such as the one to be described here, are often limited to *meridional rays*, that is, rays that pass through the optical axis of the system. Since the law of refraction requires that refracted rays remain in the plane of incidence, a meridional ray remains within the same meridional plane throughout its trajectory. Thus the treatment in terms of meridional rays is a two-dimensional treatment,[4] greatly simplifying the geometrical relationships required. Rays contributing to the image that do not pass through the optical axis are called *skew rays* and require three-dimensional geometry in their calculations. The added complexity is no problem for the computer, once the ray-tracing program is written. Analysis of various aberrations, such as spherical aberration, astigmatism, and coma (described in the following chapter), require knowledge of the progress of selected nonparaxial rays and skew rays. The design of a complex lens system, such as a photographic lens with four or five elements, is a combination of science and skill. By alternating ray tracing with small changes in the positions, focal lengths, and curvatures of the surfaces involved and in refractive indices of the elements, the design of the lens system is gradually optimized.

For our present purposes, it will be sufficient to show how the appropriate equations for meridional ray tracing can be developed and how they can be repeated in stepwise fashion to follow a ray through any number of spherical refracting surfaces that constitute an optical system. The technique is well adapted to reiterative loops handled by computer programs.

Figure 4-16 shows a single, representative step in the ray-tracing analysis. By incorporating a sign convention, the equations developed from this diagram can be made to apply to any ray and to any spherical refracting surface. The ray selected originates (or passes through) point A, making an angle α with the optical axis. The ray passes through the optical axis at O and then intersects the refracting surface at P, where it is refracted into a medium of index n', cutting the axis again at I. The angles of incidence θ and refraction θ' are related by Snell's law. Points O and I are conjugate points with distances s and s' from the surface vertex at V. The radius of the surface is also shown, passing through the center of curvature at C. Other points and lines are added to help in developing the necessary geometrical relationships.

The sign convention is the same as that used previously in this chapter. Distances to the left of the vertex V are negative and to the right, positive. If we use

---

[3] An example of commercial software available, well-adapted for academic purposes, and inexpensive, is BEAM2, a product of Stellar Software, Berkeley, California. It is IBM PC, XT, AT, and PS/2 compatible and supports CGA, EGA, VGA, and Hercules graphics adapters.

[4] The two dimensions are those of the page on which we have been drawing our ray diagrams. Without emphasizing this, we have been using meridional rays in all our diagrams.

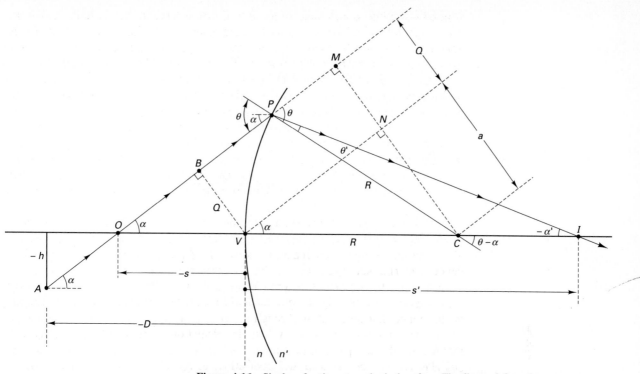

Figure 4-16 Single refraction at a spherical surface. The figure defines the symbols and shows the geometrical relationships that lead to ray-tracing equations for a meridional ray.

light rays progressing from left to right, their angles have the same sign as their slopes. Distances measured above the axis are positive and below, negative. An important quantity in the calculations, also subject to this sign convention, is the parameter Q, the perpendicular distance from the vertex to the ray, as shown.

The input parameters for the ray are assumed to be its elevation h, angle α, and distance D. Figure 4-16 shows that the "object distance," s, is related to D by

$$s = D - \frac{h}{\tan \alpha} \tag{4-34}$$

Also, in $\triangle OBV$:

$$\sin \alpha = \frac{Q}{-s} \tag{4-35}$$

In $\triangle PMC$:

$$\sin \theta = \frac{a + Q}{R}$$

In $\triangle VNC$:

$$\sin \alpha = \frac{a}{R}$$

Eliminating the length a from the last two equations, we get

$$\sin \theta = \frac{Q}{R} + \sin \alpha \tag{4-36}$$

Snell's law at P:

$$n \sin \theta = n' \sin \theta' \tag{4-37}$$

Sec. 4-11   Ray Tracing

81

In $\triangle CPI$:

$$\theta - \alpha = \theta' - \alpha' \qquad (4\text{-}38)$$

The Q parameter for the refracted ray is shown in Figure 4-17a. Analogous to the relations just found, we see that

In $\triangle CMV$:

$$\sin(-\alpha') = \frac{a'}{R}$$

In $\triangle PLC$:

$$\sin \theta' = \frac{Q' - a'}{R}$$

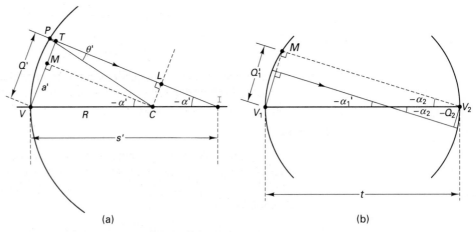

(a)   (b)

Figure 4-17 (a) Geometrical relationship of refracted-ray parameters with the distance Q'. (b) Geometrical relationships illustrating the transfer between Q and α after one refraction and before the next.

TABLE 4-3   MERIDIONAL RAY-TRACING EQUATIONS (INPUT: n, n', R, α, h, D)

General case	Ray parallel to axis: $\alpha = 0$	Plane Surface: $R \Rightarrow \infty$
$s = D - \dfrac{h}{\tan \alpha}$	—	$s = D - \dfrac{h}{\tan \alpha}$
$Q = -s \sin \alpha$	$Q = h$	$Q = -s \sin \alpha$
$\theta = \sin^{-1}\left(\dfrac{Q}{R} + \sin \alpha\right)$	$\theta = \sin^{-1}\left(\dfrac{Q}{R} + \sin \alpha\right)$	—
$\theta' = \sin^{-1}\left(\dfrac{n \sin \theta}{n'}\right)$	$\theta' = \sin^{-1}\left(\dfrac{n \sin \theta}{n'}\right)$	—
$\alpha' = \theta' - \theta + \alpha$	$\alpha' = \theta' - \theta + \alpha$	$\alpha' = \sin^{-1}\dfrac{n}{n' \sin \alpha}$
$Q' = R(\sin \theta' - \sin \alpha')$	$Q' = R(\sin \theta' - \sin \alpha')$	$Q' = Q\dfrac{\cos \alpha'}{\cos \alpha}$
$s' = \dfrac{-Q'}{\sin \alpha'}$	$s' = \dfrac{-Q'}{\sin \alpha'}$	$s' = \dfrac{-Q'}{\sin \alpha'}$

Transfer: Input: t
  $Q = Q' + t \sin \alpha'$
  $\alpha = \alpha'$
  $n = n'$
Input: new n', R
Return: to calculate θ

       Chap. 4   Matrix Methods in Paraxial Optics

As before, when a' is eliminated, there results

$$Q' = R(\sin \theta' - \sin \alpha') \tag{4-39}$$

In $\triangle ITV$:

$$\sin(-\alpha') = \frac{Q'}{s'} \quad \text{or} \quad s' = \frac{-Q'}{\sin \alpha'} \tag{4-40}$$

The relevant equations describing the first refraction are included in Table 4-3 under the first column for the general case. The calculations lead to new values of α, Q, and s (now primed), which prepare for the next refraction in the sequence. The geometrical *transfer* to the next surface, at distance t from the first, is shown in Figure 4-17b, where, in $\triangle V_2 M V_1$,

$$\sin(-\alpha_2) = \frac{V_1 M}{t} = \frac{Q_1' - Q_2}{t}$$

or

$$Q_2 = Q_1' + t \sin \alpha_2 \tag{4-41}$$

Table 4-3 also shows how the equations must be modified for two special cases: (1) when the incident ray is parallel to the axis and (2) when the surface is plane, with an infinite radius of curvature.

Example

Do a ray trace for two rays through a *Rapid landscape* photographic lens of three elements. The rays enter the lens from a distant object at altitudes of 1 and 5 mm above the optical axis. The lens specifications (all dimensions in mm) are as follows:

$$R_1 = -120.8$$
$$R_2 = -34.6 \qquad t_1 = 6 \qquad n_1 = 1.521$$
$$R_3 = -96.2 \qquad t_2 = 2 \qquad n_2 = 1.581$$
$$R_4 = -51.2 \qquad t_3 = 3 \qquad n_3 = 1.514$$

Solution Since the rays are parallel to the axis, the second column of Table 4-3 is used to calculate the progress of the ray. These can be tabulated as follows:

Input:	Results: ray at $h = 1$	Results: ray at $h = 5$
First surface:		
$n = 1$, $n' = 1.521$	$Q = 1$	$Q = 5$
$\alpha = 0$	$\alpha' = 0.1625°$	$\alpha' = 0.8128°$
$h = 1$ or 5	$s' = -352.66$	$s' = -352.53$
$R = -120.8$	$Q' = 1.0000$	$Q' = 5.0010$
Second surface:		
$t = 6$	$Q = 1.0170$	$Q = 5.0861$
$n = 1.581$	$\alpha' = 0.2202°$	$\alpha' = 1.1041°$
$R = -34.6$	$s' = -264.59$	$s' = -264.03$
	$Q' = 1.0170$	$Q' = 5.0876$
Third surface:		
$t = 2$	$Q = 1.0247$	$Q = 5.1261$
$n = 1.514$	$\alpha' = 0.2030°$	$\alpha' = 1.0178°$
$R = -96.2$	$s' = -289.26$	$s' = -288.58$
	$Q' = 1.0247$	$Q' = 5.1260$
Final surface:		
$t = 3$	$Q = 1.0353$	$Q = 5.1793$
$n = 1$	$\alpha' = -0.2883°$	$\alpha' = -1.4520°$
$R = -51.2$	$s' = 205.72$	$s' = 203.91$
	$Q' = 1.0353$	$Q' = 5.1672$

Thus the two rays intersect the optical axis at 205.72 and 203.91 mm beyond the final surface, missing a common focus by 1.8 mm.

PROBLEMS

4-1. A biconvex lens of 5 cm thickness and index 1.60 has surfaces of radius 40 cm. If this lens is used for objects in water, with air on its opposite side, determine its effective focal length and sketch its focal and principal points.

4-2. A double concave lens of glass with $n = 1.53$ has surfaces of 5 D (diopters) and 8 D, respectively. The lens is used in air and has an axial thickness of 3 cm.
  (a) Determine the position of its focal and principal planes.
  (b) Also find the position of the image, relative to the lens center, corresponding to an object at 30 cm in front of the first lens vertex.
  (c) Calculate the paraxial image distance assuming the thin-lens approximation. What is the percent error involved?

4-3. A biconcave lens has radii of curvature of 20 cm and 10 cm. Its refractive index is 1.50 and its central thickness is 5 cm. Describe the image of a 1-in. tall object, situated 8 cm from the first vertex.

4-4. An equiconvex lens having spherical surfaces of radius 10 cm, a central thickness of 2 cm, and a refractive index of 1.61 is situated between air and water ($n = 1.33$). An object 5 cm high is placed 60 cm in front of the lens surface. Find the cardinal points for the lens and the position and size of the image formed.

4-5. A hollow, glass sphere of radius 10 cm is filled with water. Refraction due to the thin glass walls is negligible for paraxial rays.
  (a) Determine its cardinal points and make a sketch to scale.
  (b) Calculate the position and magnification of a small object 20 cm from the sphere.
  (c) Verify your analytical results by drawing appropriate rays on your sketch.

4-6. Light rays enter the plane surface of a glass hemisphere of radius 5 cm and refractive index 1.5.
  (a) Using the system matrix representing the hemisphere, determine the exit elevation and angle of a ray that enters parallel to the optical axis and at an elevation of 1 cm.
  (b) Enlarge the system to a distance x beyond the hemisphere and find the new system matrix as a function of x.
  (c) Using the new system matrix, determine where the ray described above crosses the optical axis.

4-7. Using Figure 4-12b and c, verify the expressions given in Table 4-2 for the distances q, f_2, s, and w.

4-8. A lens has the following specifications: $R_1 = +1.5$ cm $= R_2$, d (thickness) $= 2.0$ cm, $n_1 = 1.00$, $n_2 = 1.60$, $n_3 = 1.30$. Find the principal points using the matrix method. Include a sketch, roughly to scale, and do a ray diagram for a finite object of your choice.

4-9. A positive thin lens of focal length 10 cm is separated by 5 cm from a thin negative lens of focal length -10 cm. Find the equivalent focal length of the combination and the position of the foci and principal planes using the matrix approach. Show them in a sketch of the optical system, roughly to scale, and use them to find the image of an arbitrary object placed in front of the system.

4-10. A glass lens 3 cm thick along the axis has one convex face of radius 5 cm and the other, also convex, of radius 2 cm. The former face is on the left in contact with air and the other in contact with a liquid of index 1.4. The refractive index of the glass is 1.50. Find the positions of the foci, principal planes, and focal lengths of the system. Use the matrix approach.

4-11. (a) Find the matrix for the simple "system" of a thin lens of focal length 10 cm, with input plane at 30 cm in front of the lens and output plane at 15 cm beyond the lens.

(b) Show that the matrix elements predict the locations of the six cardinal points as they would be expected for a thin lens.

(c) Why is $B = 0$ in this case? What is the special meaning of A in this case?

4-12. A gypsy's crystal ball has a refractive index of 1.50 and a diameter of 8 in.

(a) By the matrix approach, determine the location of its principal points.

(b) Where will sunlight be focused by the crystal ball?

4-13. A thick lens presents two concave surfaces, each of radius 5 cm, to incident light. The lens is 1 cm thick and has a refractive index of 1.50. Find (a) the system matrix for the lens when used in air and (b) its cardinal points. Do a ray diagram for some object.

4-14. An achromatic doublet consists of a crown glass positive lens of index 1.52 and of thickness 1 cm, cemented to a flint glass negative lens of index 1.62 and of thickness 0.5 cm. All surfaces have a radius of curvature of magnitude 20 cm. If the doublet is to be used in air, determine (a) the system matrix elements for input and output planes adjacent to the lens surfaces; (b) the cardinal points; (c) the focal length of the combination, using the lensmaker's equation and the equivalent focal length of two lenses in contact. Compare this calculation of f, which assumes thin lenses, with the previous value.

4-15. Enlarge the optical system of Figure 4-15 to include an object space to the left and an image space to the right of the lens. Let the new input plane be located at distance s in object space and the new output plane at distance s' in image space.

(a) Recalculate the system matrix for the enlarged system.

(b) Examine element B to determine the general relationship between object and image distances for the lens. Also determine the general relationship for the lateral magnification.

(c) From the results of (b), calculate the image distance and lateral magnification for an object 20 cm to the left of the lens.

(d) What information can you find for the system by setting matrix elements A and D equal to zero? (See Figure 4-9.)

4-16. Find the system matrix for a Cooke triplet camera lens (see Figure 6-18a). Light entering from the left encounters six spherical surfaces whose radii of curvature are, in turn, r_1 to r_6. The thickness of the three lenses are, in turn, t_1 to t_3, and the refractive indices are n_1 to n_3. The first and second air separations between lens surfaces are d_1 and d_2. Sketch the lens system with its cardinal points. How far behind the last surface must the film plane occur to focus paraxial rays?

Data:	$r_1 = 19.4$ mm	$t_1 = 4.29$ mm	$d_1 = 1.63$ mm	$n_1 = 1.6110$
	$r_2 = -128.3$ mm	$t_2 = 0.93$ mm	$d_2 = 12.90$ mm	$n_2 = 1.5744$
	$r_3 = -57.8$ mm	$t_3 = 3.03$ mm		$n_3 = 1.6110$
	$r_4 = 18.9$ mm			
	$r_5 = 311.3$ mm			
	$r_6 = -66.4$ mm			

4-17. Process the product of matrices for a thick lens, as in Eq. (4-15), without assuming the special conditions, $n = n'$ and $t = 0$. Thus find the general matrix elements A, B, C, and D for a thick lens.

4-18. Using the cardinal point locations (Table 4-2) in terms of the matrix elements for a general thick lens (problem 4-17), verify that f_1 and f_2 are given by Eqs. (4-1) and (4-2).

4-19. Using the cardinal point locations (Table 4-2) in terms of the matrix elements for a general thick lens (problem 4-17), verify that the distances r, s, v, and w are given by Eqs. (4-3) and (4-4).

4-20. Write a computer program that incorporates Eqs. (4-34) to (4-41) for ray tracing through an arbitrary number of refracting, spherical surfaces. The program should allow for the special cases of rays from far-distant objects and for plane surfaces of refraction.

4-21. Trace two rays through the hemispherical lens of Figure 4-15. The rays originate from the same object point, 2 cm above the optical axis and an axial distance of 10 cm from the first surface. One ray is parallel to the axis and the other makes an angle of $-20°$ with the axis.

4-22. Trace a ray originating 7 mm below optical axis and 100 mm distant from a doublet. The ray makes an angle of $+5°$ relative to the horizontal. The doublet is an equiconvex lens of radius 50 mm, index 1.50, and central thickness 20 mm, followed by a matched meniscus lens of radii -50 mm and -87 mm, index 1.8, and central thickness 5 mm. Determine the final values of s, α, and Q.

4-23. Trace two rays, both from far-distant objects, through a *Protor photographic lens*, one at altitude of 1 mm and the other at 5 mm. Determine where and at what angle the rays cross the optical axis. The specifications of this four-element lens, including an intermediate air space of 3 mm, is as follows, with distances in mm:

$$R_1 = 17.5$$
$$R_2 = 5.8 \qquad t_1 = 2.9 \qquad n_1 = 1.6489$$
$$R_3 = 18.6 \qquad t_2 = 1.3 \qquad n_2 = 1.6031$$
$$R_4 = -12.8 \qquad t_3 = 3.0 \qquad n_3 = 1$$
$$R_5 = 18.6 \qquad t_4 = 1.1 \qquad n_4 = 1.5154$$
$$R_6 = -14.3 \qquad t_5 = 1.8 \qquad n_5 = 1.6112$$

REFERENCES

[1] Fincham, W. H. A, and M. H. Freeman. *Optics*, 9th ed. Boston: Butterworth Publishers, 1980. Ch. 8, 9, 19.

[2] Kingslake, Rudolf. *Lens Design Fundamentals*. New York: Academic Press, 1978. Ch. 2, 3, 7.

[3] Smith, Warren J. "Image Formation: Geometrical and Physical Optics." In *Handbook of Optics*, edited by Walter G. Driscoll and William Vaughan. New York: McGraw-Hill Book Company, 1978.

[4] Brouwer, William. *Matrix Methods in Optical Instrument Design*. New York: W. A. Benjamin, 1964.

[5] Gerard, A., and J. M. Burch. *Introduction to Matrix Methods in Optics*. New York: John Wiley and Sons, 1975.

[6] Atneosen, Richard, and Richard Feinberg. "Learning Optics with Optical Design Software." *American Journal of Physics* 59 (March 1991): 242–47.

[7] Nussbaum, Allen. *Geometric Optics: An Introduction*. Reading, Mass.: Addison-Wesley Publishing Company, 1968. Ch. 2–4.

[8] Nussbaum, Allen, and Richard A. Phillips. *Contemporary Optics for Scientists and Engineers*. Englewood Cliffs, N.J.: Prentice-Hall, 1976. Ch. 1.

[9] Blaker, J. Warren. *Geometric Optics: The Matrix Theory*. New York: Marcel Dekker, 1971.

5

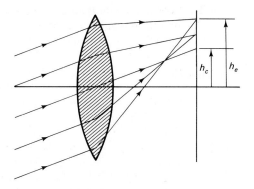

Aberration Theory

INTRODUCTION

The paraxial formulas developed earlier for image formation by spherical reflecting and refracting surfaces are, of course, only approximately correct. In deriving those equations, it was necessary to assume paraxial rays, that is, rays both near to the optical axis and making small angles with it. Mathematically, the power expansions for the sine and cosine functions, given by

$$\sin x = x - \frac{x^3}{3!} + \frac{x^5}{5!} - \cdots$$

$$\cos x = 1 - \frac{x^2}{2!} + \frac{x^4}{4!} - \cdots$$

were accordingly approximated by their first terms. To the extent that these first-order approximations are valid, Gaussian optics implies exact imaging. The inclusion of higher-order terms in the derivations, however, predicts increasingly larger departures from "perfect" imaging with increasing angle. These departures are referred to as "aberrations." When the next term involving x^3 is included in the approximation for $\sin x$, a *third-order aberration theory* results. The aberrations have been studied and classified by the German mathematician Ludwig von Seidel and are referred to as third-order or *Seidel aberrations*. For monochromatic light, there are five Seidel aberrations: *spherical aberration*, *coma*, *astigmatism*, *curvature of field*, and *distortion*. An additional aberration, *chromatic aberration*, results from the wavelength dependence of the imaging properties of an optical system. The details of aberration theory are too formidable to treat in this chapter. We include here a brief, quantitative description of how the various aberrations follow from a third-order treatment and a qualitative description of each aberration, with typical procedures for its elimination.

5-1 RAY AND WAVE ABERRATIONS

The departure from ideal, paraxial imaging may be described quantitatively in several ways. In Figure 5-1 two wavefronts are shown emerging from an optical system. Wavefront $W1$ is a spherical wavefront representing the Gaussian, or paraxial, approximation that produces an image at I. Wavefront $W2$ is an example of the actual wavefront, an aspherical envelope whose shape represents an exact solution of the optical system. This shape could be deduced by precisely tracing a sufficient number of rays through the optical system by the methods described in Section 4-11. Rays from adjacent points A and B, being normal to their respective wavefronts, do not intersect the paraxial image plane at the same point. The "miss" along the optical axis, represented by the distance LI, is called the *longitudinal aberration*, while the miss IS, measured in the image plane, is called the *transverse*, or *lateral*, *aberration*. These are *ray aberrations*. Alternatively, the aberration may be described in terms of the deviation of the deformed wavefront from the ideal at various distances from the optical axis. At the location of point B, shown in Figure 5-1, the wave aberration is given by the distance AB. Notice that rays from both wavefronts, at their point O of tangency on the optical axis, reach the same image point I. Rays from intermediate points of the actual wavefront between O and B intersect the image screen at other points around I, producing a blurred image, the result of aberration. The maximum ray aberration thus indicates the size of the blurred image. The ultimate goal of optical design is to reduce the ray aberrations until they are comparable to the unavoidable blurring due to diffraction itself.

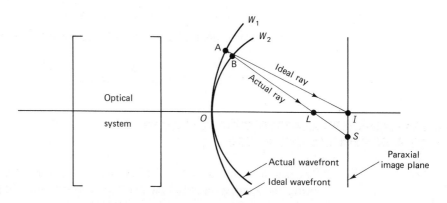

Figure 5-1 Illustration of ray and wave aberrations.

Lateral ray aberrations corresponding to the wave aberration AB may be calculated once the variation in AB with aperture dimension y is known. Referring to Figure 5-2, the angle α between actual and ideal rays from a point P of the wavefront, at elevation y, is the same as the angle between wavefront tangents at P. The wavefronts, having been shaped by the optical system, exist in image space with refrac-

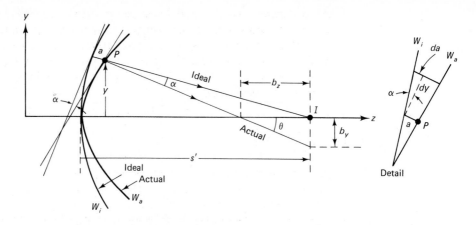

Figure 5-2 Construction used to relate the ray aberrations b_y and b_z to the wave aberration a. The detail shows how to relate a change da in wave aberration to a change dy in the aperture dimension.

tive index n_2. The detail construction of Figure 5-2 then shows that the incremental wave aberration da, expressed as an optical path length in image space, is

$$da = n_2(\alpha\, dy) \qquad (5\text{-}1)$$

The derivative da/dy describes the local curvature of the wavefront at P. The lateral ray aberration b_y due to the rays from the neighborhood of P may then be approximated by

$$b_y = \alpha s' = \frac{s'\, da}{n_2\, dy} \qquad (5\text{-}2)$$

where s' is the paraxial image distance from the wavefront and α has been taken from Eq. (5-1). Similarly, along the other transverse direction, perpendicular to the yz- axes in the plane of the page,

$$b_x = \frac{s'\, da}{n_2\, dx} \qquad (5\text{-}3)$$

The longitudinal ray aberration b_z is related to the lateral ray aberration b_y by

$$b_z = \frac{b_y}{\tan \theta} = \frac{s' b_y}{y} \qquad (5\text{-}4)$$

5-2 THIRD-ORDER TREATMENT OF REFRACTION AT A SPHERICAL INTERFACE

Let us solve now the case of refraction from a single spherical surface, where we improve the approximation to include "third-order" angle effects. In Figure 5-3, an arbitrary ray PQ from an axial object point P is refracted by a spherical surface, centered at C, that separates media of refractive indices n_1 and n_2. The refracted ray locates an axial image at I. To a first approximation, the optical path lengths of rays PQI and POI are identical, according to Fermat's principle. Aberration contributes to the formation of the image because, beyond a first approximation, ray paths PQI differ for different points Q along the spherical surface. Thus we define the aberration at Q as

$$a(Q) = (PQI - POI)_{opd} \qquad (5\text{-}5)$$

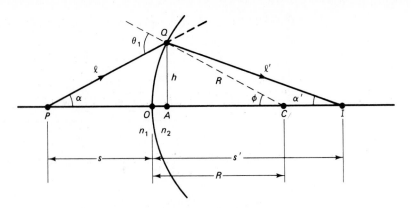

Figure 5-3 Refraction of a ray at a spherical surface.

where *opd* indicates the optical path difference. More precisely,

$$a(Q) = (n_1 \ell + n_2 \ell') - (n_1 s + n_2 s') \tag{5-6}$$

Employing the cosine law, the lengths ℓ and ℓ' may be exactly expressed, in terms of the quantities defined in Figure 5-3, by

$$\ell^2 = R^2 + (s + R)^2 - 2R(s + R) \cos \phi \tag{5-7}$$

$$\ell'^2 = R^2 + (s' - R)^2 + 2R(s' - R) \cos \phi \tag{5-8}$$

Modification of aberration calculation, pp. 90–91, to improve the approximation used there.

Expressing $\cos \phi$ in terms of h, we write,

$$\cos \phi = (1 - \sin^2\phi)^{1/2} = [1 - (h/R)^2]^{1/2} \tag{5-9}$$

where we have used the fact that $\sin \phi = h/R$. The binomial expansion then permits expansion to the fourth power in h:

$$\cos \phi \cong 1 - \frac{h^2}{2R^2} - \frac{h^4}{8R^4} \tag{5-10}$$

Introducing Eq. (5-10) into Eqs. (5-7) and (5-8) and rearranging terms,

$$\ell = s\left(1 + \left[\frac{h^2(R + s)}{Rs^2} + \frac{h^4(R + s)}{4R^3s^2}\right]\right)^{1/2} \tag{5-11}$$

$$\ell' = s'\left(1 + \left[\frac{h^2(R - s')}{Rs'^2} + \frac{h^4(R - s')}{4R^3s'^2}\right]\right)^{1/2} \tag{5-12}$$

Next, representing the quantities enclosed in square brackets by x in Eq. (5-11) and x' in Eq. (5-12), the square roots of the expression in outer parentheses may be approximated, again using the binomial expansion:

$$(1 + x)^{1/2} \cong 1 + \frac{x}{2} - \frac{x^2}{8} \tag{5-13}$$

Thus

$$\ell \cong s\left(1 + \frac{x}{2} - \frac{x^2}{8}\right) \tag{5-14}$$

$$\ell' \cong s'\left(1 + \frac{x'}{2} - \frac{x'^2}{8}\right) \tag{5-15}$$

When all terms of order higher than h^4 are discarded, there remains

$$\ell = s\left[1 + \frac{h^2(R + s)}{2Rs^2} + \frac{h^4(R + s)}{8R^3s^2} - \frac{h^4(R + s)^2}{8R^2s^4}\right] \qquad (5\text{-}16)$$

$$\ell' = s'\left[1 + \frac{h^2(R - s')}{2Rs'^2} + \frac{h^4(R - s')}{8R^3s'^2} - \frac{h^4(R - s')^2}{8Rs'^4}\right] \qquad (5\text{-}17)$$

These expressions for ℓ and ℓ' are introduced into Eq. (5-6) and the result is simplified. In the process, there occurs in both the h^2 and the h^4 terms the expression,

$$\left(\frac{n_1}{s} + \frac{n_2}{s'}\right) - \left(\frac{n_2 - n_1}{R}\right)$$

When this expression is set equal to zero, it reproduces the Gaussian formula for imaging by a spherical surface. The terms therefore represent a first-order approximation to the aberration and vanish, by Fermat's principle. There remains the third-order aberration represented by the term in h^4, given in Eq. (5-18).

$$a(Q) = -\frac{h^4}{8}\left[\frac{n_1}{s}\left(\frac{1}{s} + \frac{1}{R}\right)^2 + \frac{n_2}{s'}\left(\frac{1}{s'} - \frac{1}{R}\right)^2\right] \qquad (5\text{-}18)$$

When h is small enough, the rays are essentially paraxial and the aberration represented by this term may be negligible. In any case, since the square brackets include quantities independent of h, we have shown that third-order theory predicts a wave aberration that is proportional to the fourth power of the aperture h, measured from the optical axis, or

$$a = ch^4 \qquad (5\text{-}19)$$

where c represents the constant of proportionality. This is the principal result of our calculation for *axial* object points. We will use this in generalizing the aberration calculation to include *off-axis* imaging. In this way, the other Seidel aberrations will also appear.

The aberration $a(Q)$ we have calculated as a difference in optical-path lengths between ideal and actual rays must correspond to the wave aberration AB of Figure 5-1. The deviation AB of the actual from the ideal spherical wavefront is clearly a function of the distance from the optical axis at which the ray intersects the wavefront and is referred to as *spherical aberration*.

Before examining spherical aberration in more detail, however, we wish to show how the other third-order aberrations arise. To do this, we need to consider the case of an off-axis object point. Shown in Figure 5-4 are two pencils of rays whose limits are determined by an aperture E_nP serving as the entrance pupil. An axial pencil from the on-axis object point O forms an image at and around the paraxial image point I. This image will be affected by spherical aberration, as discussed earlier, to a degree determined by the displacement y of the extreme rays of the pencil. This pencil is symmetrical about the axis OCI, where C is the center of curvature of the refracting surface. Also shown is an oblique pencil of rays originating at the off-axis point O'. This pencil is certainly not symmetrical about the axis OI; in the absence of the limiting aperture E_nP, its axis of symmetry would be the line $O'CI'$. It is from this axis that the displacement y' of the rays of the oblique pencil would have to be measured to determine the degree of aberration described by Eq. (5-19). Notice that such displacement from the axis of symmetry is much greater in the case

---

[1] The same quantity appears in processing the h^4 term and is also set equal to zero there in arriving at Eq. (5-18).

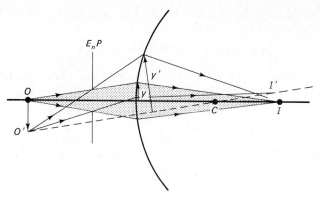

Figure 5-4 Comparison of axial and oblique pencils of rays from an object, defined by passage through entrance aperture E_nP.

of the oblique pencil. Thus an oblique pencil of rays due to off-axis object points is far more susceptible to aberration than corresponding axial points. The position of the aperture is critical in determining the magnitude of y' and is least harmful in this respect when placed at the center of curvature, C. (In this regard, one may recall the use of symmetrical lenses or lens combinations, such as the achromatic double-meniscus objective, where the aperture is placed midway between them.)

Consider then the off-axis pencil of rays from object point P, as shown in Figure 5-5. The aberration function $a'(Q)$ for the point Q on the wavefront may be expressed as

$$a'(Q) = (PQP' - PBP')_{opd} = c(BQ)^4 = c\rho'^4 \qquad (5\text{-}20)$$

In Eq. (5-20) we relate the elevation of the ray PQP' to the axis PBP' and consider points B, O, and Q to lie in a vertical plane approximating the wavefront at O. It can be shown that this approximation does not affect the results of third-order aberration theory. We have also made use of Eq. (5-19) and identified the distance BQ with a quantity ρ'. A section of the plane that includes the relevant points and defines the

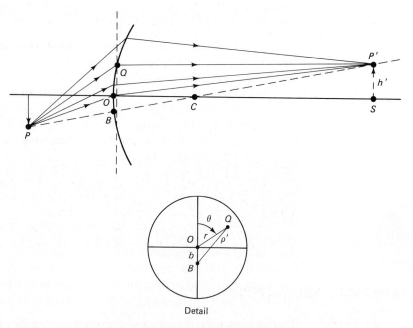

Detail

Figure 5-5 Imaging of off-axis point P. Aberration at an arbitrary point Q on the wavefront may be related to the symmetry axis PBP' or the optical axis OCS. The detail shows a frontal view of a portion of a wavefront.

distances ρ', b, and r is also shown in Figure 5-5 (detail). In a similar manner, we may write, for the wavefront point O,

$$a'(O) = (POP' - PBP')_{opd} = c(BO)^4 = cb^4 \qquad (5\text{-}21)$$

If the point Q is referred to the optical axis OC, an off-axis aberration function $a(Q)$ may be expressed as the difference between the axial aberrations at Q and O found previously.

$$a(Q) = a'(Q) - a'(O) = c\rho'^4 - cb^4 = c(\rho'^4 - b^4) \qquad (5\text{-}22)$$

Applying the cosine law to the geometric detail shown in Figure 5-5, we have

$$\rho'^2 = r^2 + b^2 + 2rb \cos \theta$$

and introducing this expression for ρ' into Eq. (5-22) gives

$$a(Q) = c(r^4 + 4r^2b^2 \cos^2 \theta + 2r^2b^2 + 4r^3b \cos \theta + 4rb^3 \cos \theta) \qquad (5\text{-}23)$$

From similar triangles OBC and SCP' in Figure 5-5, we see that the distance $OB = b$ is proportional to the height h' of the paraxial image P' above the optical axis. This may be expressed by

$$b = kh' \qquad (5\text{-}24)$$

where k is the appropriate proportionality constant. When b in Eq. (5-23) is replaced by kh', we have, lumping all constants into term-by-term coefficients,

$$a(Q) = {}_0C_{40}r^4 + {}_1C_{31}h'r^3 \cos \theta + {}_2C_{22}h'^2r^2 \cos^2 \theta$$
$$+ {}_2C_{20}h'^2r^2 + {}_3C_{11}h'^3r \cos \theta \qquad (5\text{-}25)$$

The C coefficients in Eq. (5-25) are subscripted by numbers that specify the powers of the term dependence on h', r, and $\cos \theta$, respectively. For example, the C coefficient ${}_1C_{31}$ accompanies the term $h'r^3 \cos \theta$, where h' is to the first power, r is cubed, and $\cos \theta$ is to the first power. The individual terms describe wavefront aberrations that contribute to the total aberration at the image. These terms comprise the five monochromatic, or Seidel, aberrations, as follows:

r^4	spherical aberration
$h'r^3 \cos \theta$	coma
$h'^2r^2 \cos^2 \theta$	astigmatism
h'^2r^2	curvature of field
$h'^3r \cos \theta$	distortion

Each aberration is characterized by its dependence on h' (departure from axial imaging), r (aperture of refracting surface), and θ (symmetry around the axis). Notice that the first term for spherical aberration agrees with Eq. (5-19), derived for axial imaging, where h represents the aperture.

We now briefly describe each of these aberrations in terms of their visual effects and indicate some means that are employed to reduce them.

5-3 SPHERICAL ABERRATION

The aberration known as *spherical aberration* results from the first term, ${}_0C_{40}r^4$, in Eq. (5-25). It is the only term in the third-order wave aberration $a(Q)$ that does not depend on h'. Thus spherical aberration exists even for axial object and image points, as illustrated for a single lens in Figure 5-6a. The paraxial image point I is distinct from axial image points, such as E, due to rays refracted by larger lens apertures. The axial miss distance EI, due to rays from the extremities of the lens, provides the usual measure of longitudinal spherical aberration, whereas the distance IG in the paraxial image plane measures the corresponding transverse spherical aberration. These quantities also depend on the object distance. When E is to the left of I, as shown for the case of a positive lens, the spherical aberration is positive; for a negative lens, E falls to the right of I, and the spherical aberration is considered negative. At some intermediate point between E and I, a "best" focus is attained in practice. The broadened image there is called, descriptively, the "circle of least confusion." Using Eqs. (5-2) and (5-4) for lateral and longitudinal aberration, the corresponding spherical ray aberrations may be determined as follows:

$$b_y = \frac{s'}{n_2}\frac{da}{dy} = \frac{s'}{n_2}\frac{da}{dr} = \frac{4_0 C_{40} s'}{n_2}r^3$$

and

$$b_z = \frac{s' b_y}{r} = \frac{4_0 C_{40} s'^2}{n_2}r^2$$

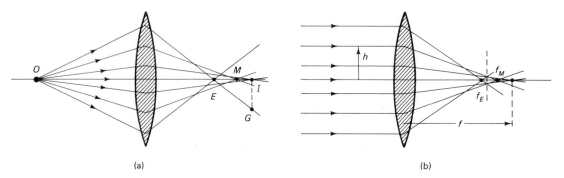

(a)   (b)

Figure 5-6 Spherical aberration of a lens, producing in (a) different image distances and in (b) different focal lengths, depending on the lens aperture.

Example

Axially collimated light enters a glass rod through its end, a convex, spherical surface of radius 4 cm. The glass rod has a refractive index of 1.60. Determine the longitudinal and lateral spherical ray aberrations for light entering at an aperture height of 1 cm.

Solution   According to Eq. (5-18), with object distance s very large, there remains

$$a = -\frac{h^4}{8}\left[\frac{n_2}{s'}\left(\frac{1}{s'} - \frac{1}{R}\right)^2\right]$$

To calculate b_z, one needs the derivative da/dh:

$$\frac{da}{dh} = -\frac{h^3}{2}\left[\frac{n_2}{s'}\left(\frac{1}{s'} - \frac{1}{R}\right)^2\right]$$

The image distance s', also the focal length of the surface, is found from the paraxial equation, giving

$$\frac{1}{\infty} + \frac{1.6}{s'} = \frac{0.6}{4} \quad \text{or} \quad s' = 10.667 \text{ cm}$$

Then da/dh can be calculated:

$$\frac{da}{dh} = -\frac{1}{2}\left[\frac{1.6}{10.67}\left(\frac{1}{10.67} - \frac{1}{4}\right)^2\right] = -0.001831$$

$$b_y = \frac{s'}{n_2}\frac{da}{dy} = \frac{10.667}{1.6}(-0.001831) = -0.0122 \text{ cm}$$

$$b_z = \frac{s'}{r}b_y = \frac{10.667}{1}(-0.0122) = -0.130 \text{ cm}$$

Figure 5-6b shows spherical aberration when the object is at infinity. Various circular zones of the lens about the axis produce different focal lengths, so that f is a function of aperture h. The specified focal length of the lens is due to the intersection of paraxial rays for which $h \to 0$. This focal length is given by the lensmaker's formula,

$$\frac{1}{f} = (n - 1)\left(\frac{1}{r_1} - \frac{1}{r_2}\right) \tag{5-26}$$

for a thin lens of refractive index n and radii of curvature r_1 and r_2, when used in air. From Eq. (5-26) it is obvious that a given f may result from different combinations of r_1 and r_2. Various choices of the radii of curvature, while not changing the focal length, may have a large effect on the degree of spherical aberration of the lens. Figure 5-7 illustrates the "bending," or change in shape, of a lens as its radii of curvature vary but its focal length remains fixed. A measure of this bending is the *Coddington shape factor* σ, defined by

$$\sigma = \frac{r_2 + r_1}{r_2 - r_1} \tag{5-27}$$

where the usual sign convention for r_1 and r_2 is assumed. For example, a thin lens of $n = 1.50$ and $f = 10$ cm may result from an equiconvex lens of $\sigma = 0$ ($r_1 = 10$, $r_2 = -10$ cm); a plano-convex lens of $\sigma = +1$ ($r_1 = 5$ cm); a meniscus lens of $\sigma = +2$ ($r_1 = 3.33$, $r_2 = 10$ cm). These shapes, as well as their mirror images with negative shape factors, are shown in Figure 5-7.

-2   -1   0   +1   +2

Figure 5-7 "Bending" of a single lens into various versions having the same focal length. The Coddington shape factor below each version serves to classify them.

The spherical aberration of a single, spherical refracting surface is given in Eq. (5-18). A thin lens combines two such surfaces, each of which contributes to the total spherical aberration. The total longitudinal spherical aberration of a thin lens with focal length f and index n, where s_h' is the image distance for a ray at elevation h, s_p' is the paraxial image distance, and

$$p = \frac{s' - s}{s' + s}$$

is given by

$$\frac{1}{s_h'} - \frac{1}{s_p'} = \frac{h^2}{8f^3}\frac{1}{n(n - 1)}\left[\frac{n + 2}{n - 1}\sigma^2 + 4(n + 1)p\sigma\right.$$
$$\left. + (3n + 2)(n - 1)p^2 + \frac{n^3}{n - 1}\right] \tag{5-28}$$

One can show further (problem 5-11), that minimum (but not zero!) spherical aberration results when the bending is such that

$$\sigma = -\frac{2(n^2 - 1)}{n + 2}p \tag{5-29}$$

Notice that for an object at infinity, $\sigma \cong 0.7$ for a lens of refractive index $n = 1.50$. This shape factor is close to that of the plano-convex lens with $\sigma = +1$. Accordingly, optical systems often employ plano-convex lenses (with the convex side facing the parallel incident rays) to reduce spherical aberration. In general, a minimum in spherical aberration is associated with the condition of equal refraction by each of the two surfaces, calling to mind the case of minimum deviation in a prism. When lenses are used in combination, the possibility of canceling spherical aberration arises from the fact that positive and negative lenses produce spherical aberration of opposite sign. A common application of this technique is found in the cemented "doublet" lens.

5-4 COMA

Coma is represented by the term $_1C_{31}h'r^3 \cos \theta$, which indicates an off-axial aberration ($h' \neq 0$) that is nonsymmetrical about the optical axis ($\cos \theta \neq$ constant) and increases rapidly with the aperture r. Figure 5-8a illustrates the aberration due to a vertical, or *tangential*, fan of parallel rays refracted by a single lens. Each circular zone of the lens forms a circular image called the *comatic circle*. Zone rays lying in the tangential fan shown form an image at the top of each comatic circle, whereas zone rays lying in a *sagittal* fan, in the horizontal plane, form an image at the bottom of each comatic circle. Every other fan of rays forms images that complete the comatic circle. The combination of all such comatic circles, which increase in radius as the zone radius increases, is the cometlike figure shown in Figure 5-8b, giving this aberration its name. In effect, each zone produces a different magnification, so that h_c due to central rays is not equal to h_e due to extreme rays. Coma, like spherical aberration, may occur as a positive quantity ($h_e > h_c$) or a negative quantity ($h_e < h_c$).

Without the usual paraxial approximation—restricting rays to those making small angles with the axis—one can show that for a small object near the axis, any ray from an object point that is refracted at a spherical interface must satisfy the *Abbe sine condition*,

$$nh \sin \theta + n'h' \sin \theta' = 0 \tag{5-30}$$

Here h and h' are object and image size, respectively, and the angles θ and θ' are the slope angles of the rays in optical media n and n', respectively. These quantities are illustrated in Figure 5-8c. When Eq. (5-30) is rearranged to express the lateral magnification, the condition can be written

$$m = \frac{h'}{h} = -\frac{n \sin \theta}{n' \sin \theta'}$$

To prevent coma, the lateral magnification resulting from refraction by all zones of a lens must be the same. Thus coma is absent when, for all values of θ,

$$\frac{\sin \theta}{\sin \theta'} = \text{constant}$$

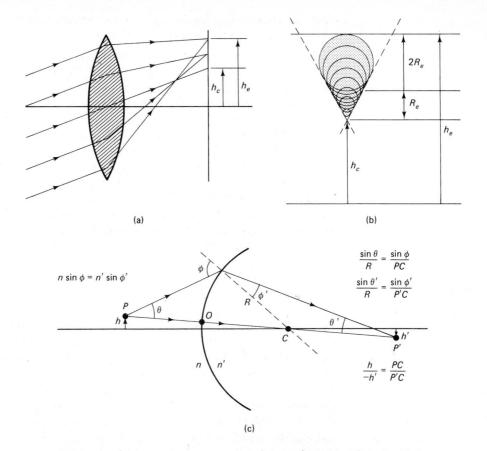

Figure 5-8 (a) Coma due to a tangential fan of parallel rays. When all such azimuthal fans are considered, each image point in the figure becomes the top of a comatic circle of image points. (b) Formation of a comatic image from a series of comatic circles. The shape of the comatic image is such that its maximum extension is three times the radius of the comatic circle formed by rays from the outer zone of the lens. The angle between the dashed lines is 60°. (c) Nonparaxial rays from object point P near the axis form an image at P', subject to the Abbe sine condition. The condition follows from Snell's law and the geometric relationships given in the figure.

The bending of a lens, found useful in reducing spherical aberration, is also useful in reducing coma. The Coddington shape factor, Eq. (5-27), which results in minimum spherical aberration, is close to that producing zero coma, so that both aberrations may be significantly reduced in the same lens by proper bending. One can show that coma is absent in a lens when

$$\sigma = \left(\frac{2n^2 - n - 1}{n + 1}\right)\left(\frac{s - s'}{s + s'}\right) \tag{5-31}$$

For the example of the lens considered previously, with $n = 1.50$ and object at infinity, Eq. (5-31) gives a value of $\sigma = 0.8$, quite close to the value of $\sigma = 0.7$, which yielded minimum spherical aberration. A lens or optical system free of both spherical aberration and coma is said to be *aplanatic*.

5-5 ASTIGMATISM AND CURVATURE OF FIELD

Aplanatic optics is still susceptible to two closely related aberrations whose wave aberration terms can be combined to give $h'^2r^2(_2C_{22}\cos^2\theta + {}_2C_{20})$. The first term produces *astigmatism*, and the second, which is symmetrical about the optical axis, is called *curvature of field*. Both aberrations increase similarly with the off-axis distance of the object and with the aperture of the refracting surface.

Figure 5-9a and b illustrates the astigmatic images of an off-axis point P due to a tangential fan of rays through the section tt' and a sagittal fan of rays through the section ss' of a single lens. Since these perpendicular fans of rays focus at different distances from the lens, the two images are line images, shown as T and S for the tangential and sagittal fans, respectively. The focal line T lies in the sagittal plane, and the focal line S falls in the tangential plane. If a screen held perpendicular to the principal ray is moved from S to T, intermediate images will be elliptical in shape. Approximately midway between S and T, the focus will be circular, the *circle of least confusion*. The locus of the line images S and T for various object points P are paraboloidal surfaces, as illustrated in Figure 5-9c. The deviation between the two surfaces along any principal ray from a given object point measures the magnitude of the astigmatism for this object, approximately proportional to the square of the distance from the optical axis. When the T surface falls to the left of the S surface, as shown, the astigmatic difference is taken as positive; otherwise it is negative.

If points like P fall along a circle in an object plane perpendicular to the optical axis, the corresponding line images in the T surface merge into a well-focused image circle. In the S surface, however, the image of the circle will not be sharp, having everywhere the width of the S focal line. On the other hand, object points along radial lines in the object circle produce sharp radial images only in the S surface, where the elongated radial images merge to produce well-focused radial lines. Thus if the object plane contains both circular and radial elements, the image distance for a good focus will be different for each type of element, with a compromise image somewhere between.

From the point of view of Figure 5-9c, the elimination of astigmatism requires that the tangential and sagittal surfaces be made to coincide. When the curvatures of these surfaces are changed by altering lens shapes or spacings so that they coincide, the resulting surface is called the *Petzval surface*. In this focal surface, for an aplanatic system, point images are formed. If the surface is curved, then, although astigmatism has been eliminated, the associated aberration called *curvature of field* remains. To record sharp images under these conditions, the film must be shaped to fit the Petzval surface. A Petzval surface can be determined for any optical system, even when the T and S surfaces do not coincide. Unlike the T and S surfaces, the Petzval surface is unaffected by lens bending or placement and depends only on the refractive indices and focal lengths of the lenses involved. In third-order theory, the Petzval surface is always situated three times farther from the T surface than from the S surface and always lies on the side of the S surface opposite to that of the T surface. For example, two lenses will have a flat Petzval surface, eliminating curvature of field, if

$$n_1f_1 + n_2f_2 = 0$$

In general the Petzval surface for a number of thin lenses in air satisfies

$$\sum \frac{1}{n_if_i} = \frac{1}{R_p} \tag{5-32}$$

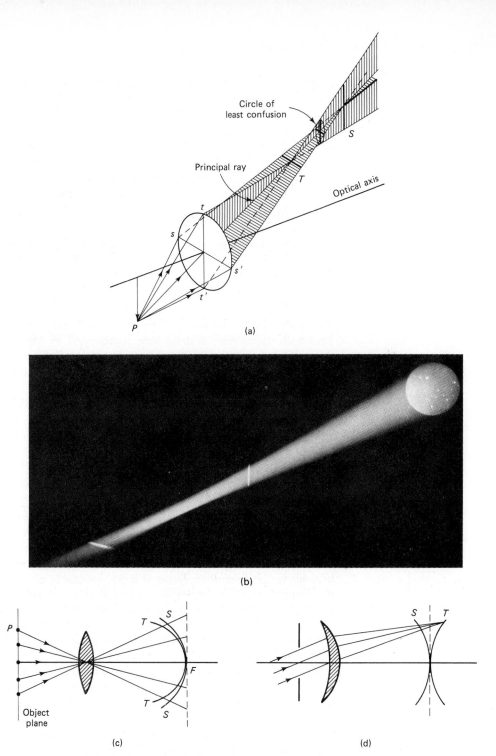

Figure 5-9 (a) Astigmatic line images T and S of an off-axial point P due to tangential (tt') and sagittal (ss') fans of light rays through a lens. (b) Photograph of astigmatic images formed by a lens, as illustrated in Figure 5-9a. The separated line images T and S are revealed as sections of the beam by fluorescent screens. (From M. Cagnet, M. Francon, and J. C. Thrierr, *Atlas of Optical Phenomenon*, Plate 4, Berlin: Springer-Verlag, 1962.) (c) Astigmatic surfaces in the field of a lens. (d) Use of a stop to artificially "flatten" the field of a lens. The compromise surface between the S and T surfaces is indicated by the dashed line.

where R_p is the radius of curvature of the Petzval surface. Field flattening by this condition cannot be accomplished for a single lens, but artificial field flattening may be accomplished by use of an aperture stop positioned as in Figure 5-9d. In this arrangement, oblique chief rays, now determined by the aperture, do not penetrate the lens center. The S and T astigmatic surfaces then appear oppositely curved, and the surface of least confusion is flat, as shown. This inexpensive method for artificially flattening the field has been used in simple box cameras. In more difficult situations, where the Petzval condition cannot be satisfied without sacrificing other requirements, a low-power lens is sometimes used near the image plane. The lens helps to counteract curvature of field without otherwise seriously compromising image quality. Finally, according to fifth-order aberration theory, the T and S surfaces may actually be made to come together again and intersect at some distance from the optical axis. The result is less average astigmatism over the compromise focal plane. The *anastigmat* camera objective is designed to take advantage of this.

5-6 DISTORTION

The last of the five monochromatic Seidel aberrations, present even if all the others have been eliminated, is *distortion*, represented by the term ${}_3C_{11}h'^3 r \cos\theta$. Even though object points are imaged as points, distortion shows up as a variation in the lateral magnification for object points at different distances from the optical axis. If the magnification increases with distance from the axis, the rectangular grid of Figure 5-10a, serving as object, will have an image as shown in Figure 5-10b, descriptively called *pincushion distortion*. On the other hand, if magnification decreases with distance from the axis, the image appears as in Figure 5-10c, with *barrel distortion*. The image in either case is sharp but distorted. Such distortion is often augmented due to the limitation of ray bundles by stops or by elements effectively acting as stops. To see this effect, refer to Figure 5-11a. Shown there is the image of an off-axis point, formed by a single lens. Two pencils of rays are drawn, each limited by the aperture stop when located (1) at some distance from the lens and (2) near the lens. As the aperture approaches the lens, it permits a shorter path to the lens. It will be seen that the effective object to image distance is greater—hence the lateral magnification is smaller—for position 1. This decrease in lateral magnification due to the aperture position is more noticeable as the object point recedes farther from the axis, so that the image suffers from barrel distortion. The effect of placing the aperture stop on the image side of the lens can also be seen from the same figure by reversing all rays and the roles played by object and image. Now the ratio of effective object to image distance is smaller, and pincushion distortion ap-

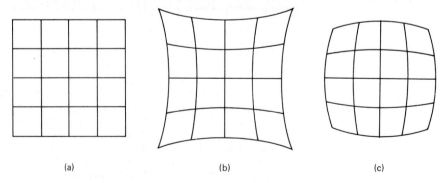

(a)   (b)   (c)

Figure 5-10 Images of a square grid (a) showing pincushion distortion (b) and barrel distortion (c) due to nonuniform magnifications.

(a)

(b)

(c)

(d)

Figure 5-11 (a) Effect of an aperture stop on the distortion of an image by a lens. The aperture in position (1) produces more barrel distortion than it does in position (2). If object and image are interchanged, the same system produces pincushion distortion. (b) Image of a square grid by a positive lens. With the stop located between object (far right) and lens, barrel distortion occurs in the image. (c) Image of a square grid by a positive lens. With the stop located at the lens, the image is free from distortion. (d) Image of a square grid by a positive lens. With the stop located between lens and image, pincushion distortion occurs in the image. (Figures 5-11b, c, and d from M. Cagnet, M. Francon, and J. C. Thrierr, *Atlas of Optical Phenomenon*, Plate 5, Berlin: Springer-Verlag, 1962.)

pears in the image. When the aperture stop is placed at the position of the lens, such distortion does not occur. Also, a symmetric doublet with a central stop, combining both effects, is free from distortion for unit magnification. Photographs of the effects of stop location on distortion are reproduced in Figure 5-11b, c, and d.

5-7 CHROMATIC ABERRATION

The final aberration to be discussed is not one of the Seidel aberrations, which are all monochromatic aberrations. Neither our first-order (Gaussian or paraxial) approximations nor the third-order theory sketched briefly in the preceding sections took into account an important fact of refraction: the variation of refractive index with wavelength, or the phenomenon of dispersion. Because of dispersion, an additional *chromatic aberration* (C.A.) appears, even for paraxial optics, in which images formed by different colors of light are not coincident. In terms of the monochromatic third-order aberrations of Eq. (5-25), we could introduce chromatic effects by considering the wavelength dependence of each of the coefficients of the terms.

The chromatic aberration of a lens is simply demonstrated by Figure 5-12a. Since the focal length f of a lens depends on the refractive index n of the glass, f is also a function of wavelength. The figure shows convergence of parallel incident light rays by the lens to distinct focal points for the red and violet ends of the visible spectrum. Notice that a cone of violet light will form a halo around the red focus at R. If the incident light contains all wavelengths of the visible spectrum, intermediate colors focus between these points on the axis. Just as for a prism, greater refraction of shorter wavelengths brings the violet focus nearer the lens for the positive lens shown. Figure 5-12b illustrates chromatic aberration for an off-axial object point and displays both *longitudinal chromatic aberration* and *lateral chromatic aberration*. Notice that if longitudinal chromatic aberration were absent, the lateral chromatic aberration could be interpreted as a difference in magnification for different colors. The longitudinal chromatic aberration of a convex lens may easily be comparable to its spherical aberration for rays at widest aperture.

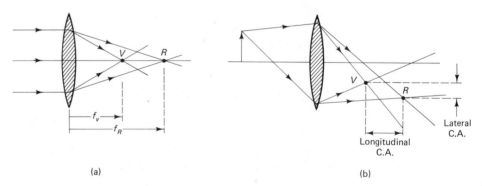

(a)   (b)

Figure 5-12   Chromatic aberration (exaggerated) for a thin lens, illustrating the effect on the focal length (a) and the lateral and longitudinal misses (b) for red (R) and violet (V) wavelengths.

Chromatic aberration is eliminated by making use of multiple refracting elements of opposite power. The most common solution is achieved with the *achromatic doublet*, consisting of a convex and concave lens, of different glasses, cemented together. The focal lengths and powers of the lenses differ, through shaping of their surfaces, to produce a net power of the doublet that may be either positive or

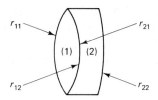

Figure 5-13 Achromatic doublet, consisting of (1) crown glass equiconvex lens cemented to (2) a negative flint glass lens. Notation for the four radii of curvature are shown.

negative. The dispersing powers of the components are, through appropriate selection of glasses, in inverse proportion to their powers. The result is a compound lens that has a net focal length but reduced dispersion over a significant portion of the visible spectrum.

We consider next the quantitative details of this design. The general shape of the achromatic doublet is shown in Figure 5-13. The powers of the two lenses for the yellow center of the visible spectrum, conveniently represented by the *Fraunhofer wavelength*, $\lambda_D = 587.6$ nm, are

$$P_{1D} = \frac{1}{f_{1D}} = (n_{1D} - 1)\left(\frac{1}{r_{11}} - \frac{1}{r_{12}}\right) = (n_{1D} - 1)K_1 \tag{5-33}$$

$$P_{2D} = \frac{1}{f_{2D}} = (n_{2D} - 1)\left(\frac{1}{r_{21}} - \frac{1}{r_{22}}\right) = (n_{2D} - 1)K_2 \tag{5-34}$$

where the radii of curvature are designated in Figure 5-13. Here n_D refers to the refractive index of each glass for the D Fraunhofer line, and we have introduced constants K_1 and K_2 as an abbreviation for the curvatures. We have already shown with Eq. (4-32) that the power of a doublet, with lens separation L, is given by

$$\frac{1}{f} = \frac{1}{f_1} + \frac{1}{f_2} - \frac{L}{f_1 f_2} \tag{5-35}$$

and

$$P = P_1 + P_2 - LP_1 P_2 \tag{5-36}$$

For a cemented doublet of thin lenses, $L = 0$, and the powers of the lenses are simply additive:

$$P = P_1 + P_2 \tag{5-37}$$

Incorporating Eqs. (5-33) and (5-34),

$$P = (n_1 - 1)K_1 + (n_2 - 1)K_2 \tag{5-38}$$

Chromatic aberration is absent at the wavelength λ_D if the power is independent of wavelength, or $(\partial P/\partial \lambda)_D = 0$. Applied to Eq. (5-38), this condition is

$$\frac{\partial P}{\partial \lambda} = K_1 \frac{\partial n_1}{\partial \lambda} + K_2 \frac{\partial n_2}{\partial \lambda} = 0 \tag{5-39}$$

The variation of n with λ in the neighborhood of λ_D may be approximated using the red and blue Fraunhofer wavelengths, $\lambda_C = 656.3$ nm and $\lambda_F = 486.1$ nm, respectively:

$$\frac{\partial n}{\partial \lambda} \cong \frac{n_F - n_C}{\lambda_F - \lambda_C} \tag{5-40}$$

The dispersion constant for the glasses may be introduced by expressing the terms of Eq. (5-39) as

$$K_1 \frac{\partial n_{1D}}{\partial \lambda} = K_1 \left(\frac{n_{1F} - n_{1C}}{\lambda_F - \lambda_C} \right) \left(\frac{n_{1D} - 1}{n_{1D} - 1} \right) = \frac{P_{1D}}{(\lambda_F - \lambda_C)V_1} \qquad (5\text{-}41)$$

$$K_2 \frac{\partial n_{2D}}{\partial \lambda} = K_2 \left(\frac{n_{2F} - n_{2C}}{\lambda_F - \lambda_C} \right) \left(\frac{n_{2D} - 1}{n_{2D} - 1} \right) = \frac{P_{2D}}{(\lambda_F - \lambda_C)V_2} \qquad (5\text{-}42)$$

where we have used Eqs. (5-33) and (5-34) as well as a dispersive constant V, defined as the reciprocal of the *dispersive power* (see pages 119–120) and given by

$$V \equiv \frac{1}{\Delta} = \frac{n_D - 1}{n_F - n_C} \qquad (5\text{-}43)$$

Substituting Eqs. (5-41) and (5-42) into Eq. (5-39), the condition for the absence of chromatic aberration may be written as

$$V_2 P_{1D} + V_1 P_{2D} = 0 \qquad (5\text{-}44)$$

Combining Eqs. (5-37) and (5-44), the powers of the individual elements may be expressed in terms of the desired power P_D of the combination:

$$P_{1D} = P_D \frac{-V_1}{V_2 - V_1} \quad \text{and} \quad P_{2D} = P_D \frac{V_2}{V_2 - V_1} \qquad (5\text{-}45)$$

The K curvature factors expressed in Eqs. (5-33) and (5-34) may then be calculated using

$$K_1 = \frac{P_{1D}}{n_{1D} - 1} \quad \text{and} \quad K_2 = \frac{P_{2D}}{n_{2D} - 1} \qquad (5\text{-}46)$$

Finally, from the values of K_1 and K_2, the four radii of curvature of the lens faces may be determined. For simplicity of construction, the crown glass lens (1) may be chosen to be equiconvex. In addition the curvature of the two lenses must match at their interface. The radii of curvature thus satisfy

$$r_{12} = -r_{11}, \quad r_{21} = r_{12}, \quad \text{and} \quad r_{22} = \frac{r_{12}}{1 - K_2 r_{12}} \qquad (5\text{-}47)$$

In the design of an achromatic doublet, the three indices of refraction for each of the glasses to be used are taken from manufacturer's specifications, like those presented in Table 5-1. One also inputs the desired overall focal length of the achro-

TABLE 5-1 SAMPLE OF OPTICAL GLASSES

Type	Catalog code	V	n_C	n_D	n_F
	$\dfrac{n_D - 1}{10V}$	$\dfrac{n_D - 1}{n_F - n_C}$	656.3 nm	587.6 nm	486.1 nm
Borosilicate crown	517/645	64.55	1.51461	1.51707	1.52262
Borosilicate crown	520/636	63.59	1.51764	1.52015	1.52582
Light barium crown	573/574	57.43	1.56956	1.57259	1.57953
Dense barium crown	638/555	55.49	1.63461	1.63810	1.64611
Dense flint	617/366	36.60	1.61218	1.61715	1.62904
Flint	620/380	37.97	1.61564	1.62045	1.63198
Dense flint	689/312	31.15	1.68250	1.68893	1.70462
Dense flint	805/255	25.46	1.79608	1.80518	1.82771
Fused silica	458/678	67.83	1.45637	1.45846	1.46313

mat. In the series of calculations leading to the four radii of curvature, a calculation that is easily programmed, Eqs. (5-43), (5-45), (5-46), and (5-47) are employed in sequence. For example, if 520/636 crown glass and 617/366 flint glass are used in designing an achromat of focal length 15 cm, these equations lead to lenses with radii of curvature given by

$$r_{11} = 6.6218 \text{ cm}$$

$$r_{12} = -6.6218 \text{ cm}$$

$$r_{21} = -6.6218 \text{ cm}$$

$$r_{22} = -223.29 \text{ cm}$$

With these values, Eqs. (5-33) and (5-34) permit the calculation of focal lengths for each of the Fraunhofer wavelengths. In this case we find

	f_1	f_2	f
λ_D	6.3653 cm	−11.0575 cm	15.0000 cm
λ_C	6.3961 cm	−11.147 cm	15.007 cm
λ_F	6.2966 cm	−10.8485 cm	15.007 cm

For a thin lens, achromatizing renders focal lengths (nearly) equal, eliminating longitudinal and lateral aberration at the same time. In a thick lens or an optical system of lens combinations, the second principal planes for different wavelengths may not coincide as they do in a thin lens. When this is the case, equal focal lengths for two wavelengths, measured as they are from their respective principal planes, do not lead to a single focal point on the axis, and longitudinal chromatic aberration remains (Figure 5-14a). If the focal lengths for red and blue light are made unequal, such that they produce a single focus (Figure 5-14b), the difference in f_B and f_R results in a difference of lateral magnifications, and lateral chromatic aberration remains. Thus the condition for removing lateral chromatic aberration is the coincidence of the principal planes for the two corrected wavelengths.

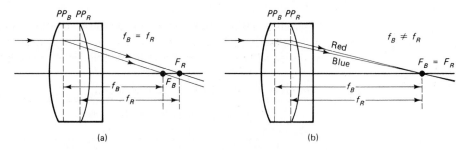

Figure 5-14 Doublet with second principal planes separated for red and blue light. (a) Equal focal lengths result in residual longitudinal chromatic aberration. (b) Equal foci result in residual lateral chromatic aberration.

Another solution for zero longitudinal chromatic aberration results if one uses two separated lenses ($L \neq 0$) of the same glass ($n_1 = n_2 = n$). The condition $\partial P / \partial \lambda = 0$ applied to Eq. (5-36) now gives

$$\frac{\partial P}{\partial \lambda} = \frac{\partial}{\partial \lambda} [(n - 1)(K_1 + K_2) - (n - 1)^2 K_1 K_2 L] = 0$$

Performing the differentiation and canceling $\partial n/\partial \lambda$, there remains

$$L = \frac{f_1 + f_2}{2} \tag{5-48}$$

which is the same result derived for a double-lens eyepiece in the following chapter. Thus two lenses of the same material, separated by a distance equal to the average of their focal lengths, exhibit zero longitudinal chromatic aberration for the wavelength at which the focal lengths are calculated.

PROBLEMS

5-1. Carry out the "rearranging" called for in arriving at Eq. (5-18).

5-2. If image and object distance for a spherical refracting surface—in addition to satisfying Eq. (3-20)—also satisfy the relation $1/s' = (1/s) + (1/R)$, show that
(a) $s' = -(n_1/n_2)s$, and
(b) $a(Q)$ for spherical aberration in Eq. (5-18) vanishes.
(c) Show that $a(Q)$ also vanishes for $s' = R$ and for rays intersecting with the spherical surface vertex. Such image points are called *aplanatic points*.
(d) Find the aplanatic points for a spherical surface of $+8$ cm separating two media of refractive indices 1.36 and 1.70, respectively.

5-3. A collimated light beam is incident on the plane side of a plano-convex lens of index 1.50, diameter 50 mm, and radius 40 mm. Find the spherical wave aberration and the longitudinal and transverse spherical ray aberrations.

5-4. Show that for a spherical concave mirror, a calculation like that done for a refracting surface gives a third-order aberration of

$$a = \frac{h^4}{4R}\left(\frac{1}{s} - \frac{1}{R}\right)^2$$

where R is the magnitude of the radius of curvature.

5-5. Using the result of problem 5-4, determine the wave aberration, transverse aberration, and longitudinal aberration for a spherical mirror of 2-m focal length and 50-cm diameter, when it forms an image of a distant point object.

5-6. A reflecting telescope uses a spherical mirror with a 3-m focal length and an aperture given by $f/3.75$.
(a) Using the results of problem 5-4, determine the magnitude of the spherical wave aberration for the telescope.
(b) If a Schmidt-type correcting plane of refractive index 1.40 were installed to correct the spherical aberration, what would be the required difference in thickness between the center and edge of the plate?

5-7. In forming an image of an axial point object, a $+4.0$-diopter lens with a diameter of 6.0 cm gives a longitudinal spherical aberration of $+1.0$ cm. If the object is 50 cm from the lens, determine (a) the transverse spherical aberration and (b) the diameter of the blur circle in the paraxial focal plane.

5-8. Determine the longitudinal and lateral spherical ray aberration for a thin lens of $n = 1.50$, $r_1 = +10$ cm, and $r_2 = -10$ cm due to rays parallel to the axis and through a zone or radius $h = 1$ cm.

5-9. Using the equation for spherical aberration of a thin lens, given in problem 5-8, find the longitudinal spherical ray aberration of a lens as a function of ray height h. Do this by plotting the longitudinal ray aberration as a function of ray height for $h = 0, 1, 2, 3, 4,$ and 5 cm. The lens has a refractive index of 1.60 and radii $r_1 = 36$ cm and $r_2 = -18$ cm. The incident light rays are parallel to the optical axis.

5-10. An equiconvex thin lens of index 1.50 and radius 15 cm forms an image of an axial object point 25 cm in front of the lens and for rays through a zone of radius $h = 2$ cm. Determine the longitudinal and lateral spherical ray aberration. (See problem 5-8.)

5-11. Show that if $L = (1/s_h') - (1/s_p')$, setting $dL/d\sigma = 0$ produces the condition for minimum spherical aberration:

$$\sigma = -\frac{2(n^2 - 1)p}{n + 2}$$

5-12. A positive lens of index 1.50 and focal length 30 cm is "bent" to produce Coddington shape factors of 0.700 and 3.00. Determine the corresponding radii of curvature.

5-13. A positive thin lens of focal length 20 cm is designed to have minimal spherical aberration in its image plane, 30 cm from the lens. If the lens index is 1.60, determine its radii of curvature.

5-14. A thin, plano-convex lens with 1-m focal length and index 1.60 is to be used in an orientation that produces less spherical aberration while focusing a collimated light beam. Prove that the proper orientation is with light incident on the spherical side by comparing the Coddington shape factor for each orientation with the value giving minimum spherical aberration.

5-15. A positive lens is needed to focus a parallel beam of light with minimum spherical aberration. The required focal length is 30 cm. If the glass has a refractive index of 1.50, determine (a) the required Coddington factor and (b) the radii of curvature of the lens. (c) If the lens is to be used instead to produce a collimated beam, how do these answers change?

5-16. Answer problem 5-15 when the lens is designed to reduce coma.

5-17. A 20-cm focal length positive lens is to be used as an inverting lens, that is, it simply inverts an image without altering its size. What radii of curvature lead to minimum spherical aberration in this application? The lens refractive index is 1.50.

5-18. Answer problem 5-17 when the lens is designed to reduce coma.

5-19. It is desired to reduce the curvature of field of a lens of 20-cm focal length made of crown glass ($n = 1.5230$). For this purpose a second lens of flint glass ($n = 1.7200$) is added. What should be its focal length? Refractive indices are given for sodium light of 589.3 nm.

5-20. A doublet telescope objective is made of a cemented positive lens ($n_1 = 1.5736$, $f_1 = 3.543$ cm) and negative lens ($n_2 = 1.6039$, $f_2 = 5.391$ cm).
(a) Determine the radius of their Petzval surface.
(b) What focal length for the negative lens gives a flat Petzval surface?

5-21. Design an achromatic doublet of 517/645 crown and 620/380 flint glasses that has an overall focal length of 20 cm. Assume the crown glass lens to be equiconvex. Determine the radii of curvature of the outer surfaces of the lens, as well as its resultant focal length for the D, C, and F Fraunhofer lines.

5-22. Design an achromatic doublet of 5-cm focal length using 638/555 crown and 805/255 flint glass. Determine (a) radii of curvature; (b) focal lengths for D, C, and F Fraunhofer lines; (c) powers and dispersive powers of the individual elements. (d) Is Eq. (5-44) satisfied?

5-23. Design an achromatic doublet of -10-cm focal length, using 573/574 and 689/312 glasses. Assume the crown glass lens to be equiconcave. Determine (a) radii of curvature of the lens surfaces; (b) individual focal lengths for the Fraunhofer D line; (c) the overall focal lengths of the lens for the Fraunhofer D, C, and F lines.

REFERENCES

[1] Martin, L. C. *Technical Optics*, Vol. 2, 2d ed. London: Sir Isaac Pitman & Sons, 1960. Ch. V and Appendix V.

[2] Fincham, W. H. A., and M. H. Freeman. *Optics*, 9th ed. Boston: Butterworth Publishers, 1980. Ch. 18.

[3] Welford, W. T. *Geometrical Optics*. Amsterdam: North-Holland Publishing Company, 1962. Ch. 6.

[4] Welford, W. T. *Aberrations of Optical Systems*. Boston: Adam Hilger Ltd., 1986.

[5] Guenther, Robert D. *Modern Optics*. New York: John Wiley and Sons, 1990. Appendix 5-B.

[6] Kingslake, Rudolf. *Lens Design Fundamentals*. New York: Academic Press, 1978.

[7] Conrady, A. E. *Applied Optics and Optical Design*. New York: Dover Publications, 1957.

[8] Smith, Warren J. "Image Formation: Geometrical and Physical Optics." In *Handbook of Optics*, edited by Walter G. Driscoll and William Vaughan. New York: McGraw-Hill Book Company, 1978.

[9] Hopkins, R. E. "Geometrical Optics." In *Geometrical and Instrumental Optics*, edited by Daniel Malacara. Boston: Academic Press, 1988.

[10] Brouwer, William. *Matrix Methods in Optical Instrument Design*. New York: W. A. Benjamin, 1964.

[11] Nussbaum, Allen. *Geometric Optics: An Introduction*. Reading, Mass.: Addison-Wesley Publishing Company, 1968. Ch. 7, 8.

[12] Smith, F. Dow. "How Images Are Formed." *Scientific American* (Sept. 1968): 59–70.

6

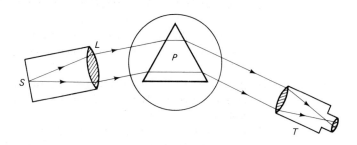

Optical Instrumentation

INTRODUCTION

The principles of geometrical optics, developed earlier, are applied in this chapter in order to discuss several practical optical instruments. The discussion begins with an introduction to the operation of stops, pupils, and windows, of great practical importance to optical instrumentation. The optical instruments treated in the following articles then include the prism, the camera, the eyepiece, the microscope, and the telescope.

6-1 STOPS, PUPILS, AND WINDOWS

We have been studying ways to trace rays through an optical system using the step-by-step application of Gaussian formulas, matrix methods, and ray tracing. However, not every light ray from an object point, directed toward or into an optical system, survives to the final image. Depending on the location of the object point and the ray angle, many of these rays are blocked by the limiting apertures of lenses and mirrors or by actual apertures intentionally inserted into the optical system. In this section, we wish to concentrate on the effects of such spatial limitations of light beams in an optical system.

The apertures dealt with are often purposely inserted into an optical system to achieve various practical purposes. We have seen (Chapter 5) how apertures can be used to modify the effects of spherical aberration, astigmatism, and distortion. In other applications apertures may be introduced to produce a sharp border to the image, like the sharp outline we see looking into the eyepiece of an optical instrument. Apertures may also be used to shield the image from undesirable light scattered from optical components. In any case, apertures are inevitably present because every lens has a finite diameter that effectively puts an aperture into the system.

The presence of apertures in an optical system influences its properties in important ways. We have learned how reducing the useful aperture of a single lens can reduce spherical aberration. An image formed by the lens is obviously also reduced in brightness. In this chapter we examine how the depth of field (or focus) of the lens increases as its effective diameter becomes smaller. In the study of diffraction (Chapter 16) we examine the effect of lens diameter on the resolution of the image. All these properties are also influenced by the appearance of apertures in more complex optical instruments. In addition, apertures play a role in limiting the field of view, the extent of the object field that appears in the final image.

The variety in complexity of practical optical systems, and in the relative placement of apertures with reflecting and refracting surfaces, is obviously very large. In this introduction we must limit ourselves to the definition and explanation of the terms required by this kind of analysis and to their application in a few simple optical systems.

Image Brightness: Aperture Stops and Pupils

Aperture Stop (AS).   The aperture stop of an optical system is the actual optical component that limits the size of the maximum cone of rays from an axial object point that can be processed by the entire system. Thus it controls the brightness of the image. The diaphragm of a camera and the iris of the human eye are examples of aperture stops. Another example is the telescope, in which the first, or objective, lens determines how much light is accepted by the telescope to form a final image on the retina of the eye. In this simple case, the objective lens is the aperture stop of the optical system. However, the aperture stop is not always identical with the first component of an optical system. For example, refer to Figure 6-1a. As shown, the aperture stop (AS) in front of the lens determines the extreme (or *marginal*) rays that can be accepted by the lens. But if the object OO' is moved nearer to AS, at some point the lens rim becomes the limiting aperture. (This point is just the intersection with the axis of a line drawn from the lens rim through the edge of the aperture stop AS.) In this case, the angle subtended by the lens at O becomes smaller than the angle subtended by the aperture, and we designate the lens as the aperture stop.

Entrance Pupil (E_nP).   The entrance pupil is the limiting aperture that the light rays "see," looking into the optical system from the object. In Figure 6-1a, this is simply the aperture stop itself, so in this case, AS and E_nP are identical. To see that this is not always the case, look at Figure 6-1b, where the aperture stop is behind the lens (a *rear stop*), as in most cameras. Which component now limits the cone of light rays? It is that component whose aperture edges limit rays from O to their smallest angle relative to the axis. Looking into the optical system from object space, one sees the lens directly, but sees the AS through the lens. In other words, the effective aperture due to AS is its image formed by the lens, the dashed line marked E_nP. Since rays from O, directed toward this virtual aperture, make a smaller angle than rays directed toward the lens edge, this virtual aperture serves as the effective entrance pupil for the system. Notice that rays from O, directed toward the edges of E_nP, are in fact refracted by the lens so as to pass through the

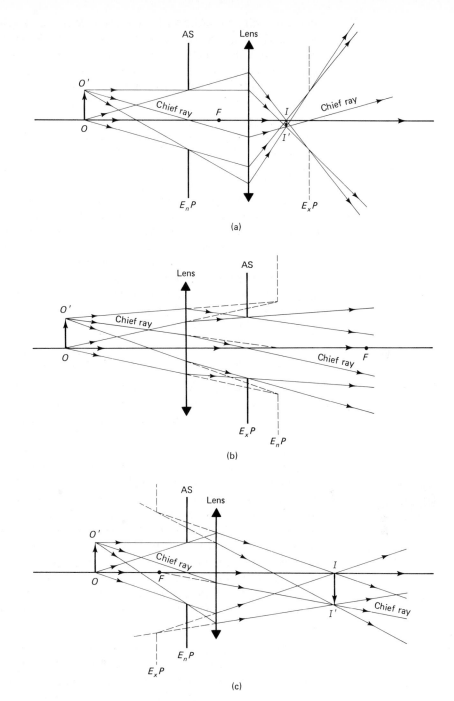

Figure 6-1 Limitation of light rays by various combinations of positive lens and diaphragm.

edges of the real aperture stop. This must be the case, since AS and $E_n P$ are, by definition, conjugate planes: The edges of $E_n P$ are the images of the edges of AS. This example illustrates the general rule: *The entrance pupil is the image of the controlling aperture stop formed by the imaging elements preceding it.*[1] When the con-

---

[1] "Preceding" is used in the sense that light must pass through those imaging elements first. If we always use light rays directed from left to right, we can simply say, "by all imaging elements to its left."

trolling aperture stop is the first such element (a *front stop*), it serves itself as the entrance pupil.

Another example, in which an aperture placed in front of the lens functions as the AS for the system, is shown in Figure 6-1c. It is different from Figure 6-1a in that the aperture is placed *inside* the focal length of the lens. Nevertheless, the aperture is the AS for the system because it, not the lens, limits the system rays to their smallest angle with the axis. Furthermore, it is the $E_n P$ of the system because it is the first element encountered by the light from the object.

Exit Pupil ($E_x P$). We have described the $E_n P$ of an optical system as the image of the AS one sees by looking into the optical system from the object. If one looks into the optical system from image space, another image of the AS can be seen that appears to limit the output beam size. This image is called the *exit pupil* of the optical system. Thus *the exit pupil is the image of the controlling aperture stop formed by the imaging elements following it* (or *to the right* of it, in our figures). The rear stop in Figure 6-1b is automatically the $E_x P$ for the system because it is the last optical component. According to our definition of the $E_x P$, the exit pupil is the optical conjugate of the AS; $E_x P$ and AS are conjugate planes. It follows that $E_x P$ is also conjugate with $E_n P$. In Figure 6-1a, $E_x P$ is the real image of $E_n P$; in Figure 6-1c, it is the virtual image. Notice that in each case, rays intersecting the edges of the entrance pupil also (actually, or when extended) intersect the edges of the exit pupil.

In a system like that of Figure 6-1a, a screen held at the position of the exit pupil receives a sharp image of the circular opening of the aperture stop. If the system represents the eyepiece of some optical instrument, the exit pupil is matched in position and diameter to the pupil of the eye. Notice further that if the screen is moved closer to the lens, it intercepts a sharp image II' of the object OO'. The exit pupil is seen to limit the solid angle of rays forming each point of the image.

Chief Ray. The *chief*, or *principal*, *ray* is a ray from an object point that passes through the *axial point*, in the plane of the entrance pupil. Given the conjugate nature of the entrance pupil with both the aperture stop and the exit pupil, this ray must also pass (actually or when extended) through their axial points. Verify this behavior in all three systems of Figure 6-1. Notice that the chief ray in the pencil of rays from the axial point O coincides with the optical axis.

Before adding to our collection of new concepts that arise from a consideration of apertures in optical systems, we consider a system slightly more complex than those of Figure 6-1. Of course, the single lens in those systems could each represent an entire system whose ray paths are determined by its cardinal points. However, in Figure 6-2, we specify a particular optical system consisting of two lenses $L1$ and $L2$ with an aperture A placed between them, as shown. The first question to be answered is: Which element serves as the effective AS for the whole system? The answer to this question is not always obvious. It can always be answered, however, by determining which of the actual elements in the given system—in this case, A, $L1$, or $L2$—has an entrance pupil that confines rays to their smallest angle with the axis, as seen from the object point. To decide which candidate presents the limiting aperture, it is necessary to find the entrance pupil for each by imaging each through that part of the optical system lying to its left:

$L2$: By ray diagram or by calculation, the image of lens $L2$, formed by $L1$ (as if light went from right to left), is $L2'$. Both its location and size (magnification) are shown.

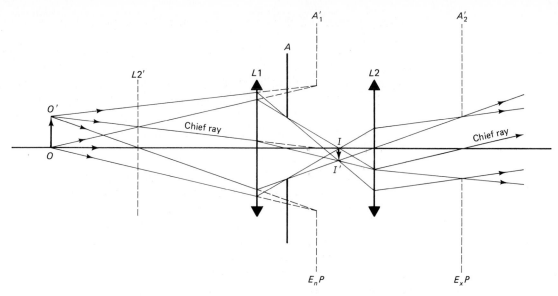

Figure 6-2 Limitation of light rays in an optical system consisting of two positive lenses and a diaphragm.

A: The image of aperture A backward through $L1$ is virtual and is shown as A_1'.

$L1$: Since lens $L1$ is the first element, it acts as its own entrance pupil.

The three entrance pupils, $L2'$, $L1$, and A_1', are next viewed from the axial point O. Since A_1' subtends the smallest angle at O, we conclude that aperture A is the aperture stop of the system.

Once the AS is identified, it is imaged through the optical elements to its right to find the exit pupil. In this case, aperture A is imaged through $L2$ to form $A2'$. The chief ray, together with the two marginal rays, is drawn from each of the extreme points O and O' of the object. Notice that it passes (actually or when extended) through the centers of AS and its conjugate planes, E_nP and E_xP. The chief ray intersects the axis at A, at A_1' (where it is virtual), and at A_2'. The bundle of incident rays from either object point O or O', limited by the size of the entrance pupil A_1', just makes it through the exit pupil A_2'. The image formed by $L1$ is shown as II'; the final image (not shown) is virtual, since the rays from either O or O' diverge on leaving $L2$.

Field of View: Field Stops and Windows. In describing the limitations of a cone of rays from an axial object point, we have seen that entrance and exit pupils are related to the aperture stop and so govern the brightness of the image. Apertures also determine the *field of view* handled by the system. The controlling element in this connection is called the *field stop*, and it is related to an *entrance window* and an *exit window* in the same way that the aperture stop is related to entrance and exit pupils.

The simplest experience of a limitation in the field of view is that of looking through an opening such as a window. The edges of the window determine how much of the outdoors we can see. This field of view can be described in terms of the lateral dimensions of the object viewed, or in terms of the angular extent of the window, relative to the line of sight. Again, one can talk about the field of view in terms of the object or in terms of the image (on the retina, in this example).

To see how an aperture restricts the field of view—using diagrams that could well be applied to the case of window and eye lens, just discussed—look at Figure 6-3. In part (a), the optical system is a single aperture A placed in front of a single lens. Object and image planes are also shown. Rays from an axial point O are limited in angle by the aperture and focused by the lens at point O'. The same is true for the off-axis point T and its image, T'. In both cases, the lens is large enough to intercept the entire cone of rays. If the object plane is uniformly bright and the aperture is a circular hole, then a circle of radius $O'T'$ is uniformly illuminated in the image plane.

Figure 6-3 Referring to the same optical system, diagrams (a) and (b) illustrate both the way in which an aperture limits the field of view and the process by *vignetting*. Diagram (c) is an example of a more complex optical system, showing the angular field of view in object and image space, α and α', respectively.

However, if one considers object points below T, the top rays from such points, passing through the aperture, miss the lens. Such a point, U, is shown in part (b) of the figure, the same optical system as (a), but redrawn for clarity. It is chosen such that the chief or central ray of the bundle just misses the top of the lens. About half the beam is lost so that image point U' receives only about half as much light as points O' and T'. Thus the circular image begins to dim as its radius increases. This partial shielding of the outer portion of the image by the aperture for off-axis object points is called *vignetting*. Excessive vignetting may make the image of a point appear astigmatic. Finally, object point V is chosen such that all its rays through the aperture miss the lens entirely. The lateral field of view processed by this optical system is at most a circle of radius OV. It is often defined as the smaller circle of radius OU if one considers the usable field of view as consisting of all object points that produce image points having at least half the maximum irradiance near the center. One can then also define the angular field of view as twice the angle made with the axis by the chief ray at the center of the aperture representing the entrance pupil. The lens itself, in this case, acts both at the field stop and the entrance window. In other, more complex, systems, the relevant quantities may be described as follows.

Field Stop (FS). The field stop is the aperture that controls the field of view by limiting the solid angle formed by chief rays. As seen from the center of the entrance pupil, the field stop (or its image) subtends the smallest angle. When the edge of the field of view is to be sharply delineated, the field stop should be placed in an image plane so that it is sharply focused along with the final image. A simple example of such a field stop is the opening directly in front of the film that outlines the final image in a camera. Intentional limitation of the field of view using an aperture is desirable when either far-off axis imaging is of unacceptable quality due to aberrations or when vignetting severely reduces the illumination in the outer portions of the image.

Entrance Window ($E_n W$). The entrance window is the image of the field stop formed by all optical elements preceding it (just as the entrance pupil is an image of the aperture stop). The entrance window delineates the lateral dimensions of the object to be viewed, as in the viewfinder of a camera, and its angular diameter determines the angular field of view. When the field stop is located in an image plane, the entrance window lies in the conjugate object plane, where it defines the lateral dimensions of the object field imaged by the optical system.

Exit Window ($E_x W$). The exit window, analogous to the exit pupil, is the image of the field stop formed by all optical elements following it. To an observer in image space, the exit window seems to limit the area of the image in the same way as an outdoor scene appears limited by the window of a room.

In Figure 6-3c, the role of field stops and entrance and exit windows is shown in a more complex optical system consisting of two lenses and two apertures. The first aperture is the AS of the system and, as we have seen, is related to an entrance pupil, its image in $L1$, and an exit pupil, its image in $L2$. The second aperture is the field stop, with corresponding images through the lenses: the entrance window to the left and the exit window to the right. The field of view in object space can then be described by α, the angle subtended by the entrance window at the center of the entrance pupil. Similarly, the field of view in image space can be described by α', the angle subtended by the exit window at the center of the exit pupil. We see that the size of the field imaged by the optical system is effectively determined by the entrance window and, actually, by the size of the field stop. Notice that, since $E_n W$ and $E_x W$ are both images of the FS, they are conjugate planes. Thus the beam that fills the entrance window also fills the field stop and the exit window.

The Summary of Terms is provided as a convenient reference to a subject that requires patience, and experience with many examples, to master.

Brightness

Aperture stop: The real aperture in an optical system that limits the size of the cone of rays accepted by the system from an axial object point.

Entrance pupil: The image of the aperture stop formed by the optical elements (if any) that precede it.

Exit pupil: The image of the aperture stop formed by the optical elements (if any) that follow it.

Field of view

Field stop: The real aperture that limits the angular field of view formed by an optical system.

Entrance window: The image of the field stop formed by the optical elements (if any) that precede it.

Exit window: The image of the field stop formed by the optical elements (if any) that follow it.

6-2 PRISMS

Angular Deviation of a Prism. The top half of a double-convex, spherical lens can form an image of an axial object point within the paraxial approximation, as shown in Figure 6-4. If the lens surfaces are flat, a prism is formed, and paraxial rays can no longer produce a unique image point. It is nevertheless helpful in some cases to think of a prism as functioning approximately like one-half of a convex lens.

Figure 6-4 Focusing due to half of a convex lens approximates the action of a prism.

In the following we derive the relationships that describe exactly the progress of a single ray of light through a prism. The bending that occurs at each face is determined by Snell's law. The degree of bending is a function of the refractive index of the prism and is, therefore, a function of the wavelength of the incident light. The variation of refractive index and light speed with wavelength is called *dispersion* and is discussed later. For the present we assume monochromatic light, which has its own characteristic refractive index in the prism medium. The relevant angles describing the progress of the ray through the prism are defined in Figure 6-5. Angles of incidence and refraction at each prism face are shown relative to the normals constructed at the point of intersection with the light ray. The total angular deviation δ

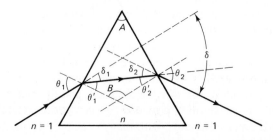

Figure 6-5 Progress of an arbitrary ray through a prism.

of the ray due to the action of the prism as a whole is the sum of the angular deviations δ_1 and δ_2 at the first and second faces, respectively. Snell's law at each prism face requires that

$$\sin \theta = n \sin \theta_1' \tag{6-1}$$

$$n \sin \theta_2' = \sin \theta_2 \tag{6-2}$$

and inspection will show that the following geometrical relations must hold between the angles:

$$\delta_1 = \theta_1 - \theta_1' \tag{6-3}$$

$$\delta_2 = \theta_2 - \theta_2' \tag{6-4}$$

$$B = 180 - \theta_1' - \theta_2' = 180 - A \tag{6-5}$$

$$A = \theta_1' + \theta_2' \tag{6-6}$$

The two members of Eq. (6-5) follow because the sum of the angles of a triangle is 180° and because the sum of the angles of a quadrilateral must be 360°. Notice that the angles A and B and the two right angles formed by the normals with the prism sides constitute such a quadrilateral.

Using Eqs. (6-1) through (6-6), a programmable calculator or computer may easily be programmed to perform the sequential operations that finally determine the angle of deviation, δ. Assuming that the *prism angle A* and refractive index n are given, then the stepwise calculation for a ray incident at an angle θ_1 is as follows:

$$\theta_1' = \sin^{-1}\left(\frac{\sin \theta_1}{n}\right) \tag{6-7}$$

$$\delta_1 = \theta_1 - \theta_1' \tag{6-8}$$

$$\theta_2' = A - \theta_1' \tag{6-9}$$

$$\theta_2 = \sin^{-1}(n \sin \theta_2') \tag{6-10}$$

$$\delta = \theta_1 + \theta_2 - \theta_1' - \theta_2' \tag{6-11}$$

The variation of deviation with angle of incidence for $A = 30°$ and $n = 1.50$ is shown in Figure 6-6. Notice that a minimum deviation occurs for $\theta_1 = 23°$. Refraction by a prism under the condition of minimum deviation is most often utilized in practice. We may argue rather neatly that when minimum deviation occurs, the ray of light passes symmetrically through the prism, making it unnecessary to subscript angles, as shown in Figure 6-7. Suppose this were not the case, and minimum deviation occurred for a nonsymmetrical case, as in Figure 6-5. Then if the ray were reversed, following the same path backward, it would have the same total deviation as the forward ray, which we have supposed to be a minimum. Hence there would be two angles of incidence, θ_1 and θ_2, producing minimum deviation, contrary to experience. The geometric relations simplify in this case:

From Eq. (6-11),

$$\delta = 2\theta - 2\theta' \tag{6-12}$$

and from Eq. (6-6),

$$A = 2\theta' \tag{6-13}$$

Together these allow us to write

$$\theta' = \frac{A}{2} \quad \text{and} \quad \theta = \frac{\delta + A}{2} \tag{6-14}$$

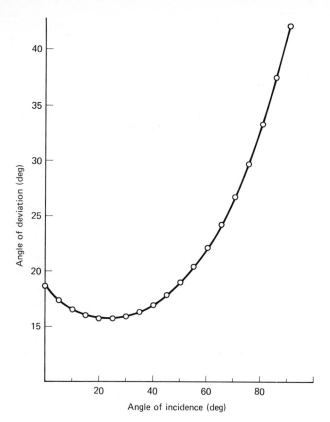

Figure 6-6 Graph of total deviation versus angle of incidence for a light ray through a prism with $A = 30°$ and $n = 1.50$. Minimum deviation occurs for an angle of 23°.

Figure 6-7 Progress of a ray through a prism under the condition of minimum deviation.

so that Eq. (6-1) becomes

$$\sin\left(\frac{A + \delta}{2}\right) = n \sin\left(\frac{A}{2}\right)$$

or

$$n = \frac{\sin[(A + \delta)/2]}{\sin(A/2)} \quad (6\text{-}15)$$

Eq. (6-15) provides a method of determining the refractive index of a material that can be produced in the form of a prism. Measurement of both prism angle and minimum deviation of the sample determines n. An approximate form of Eq. (6-15) follows for the case of small prism angles and, consequently, small deviations. Approximating the sine of the angles by the angles in radians, we may write

$$n \cong \frac{(A + \delta)/2}{A/2}$$

or

$$\delta \cong A(n - 1), \quad \text{minimum deviation, small } A \quad (6\text{-}16)$$

For $A = 15°$, the deviation given by Eq. (6-16) is correct to within about 1%. For $A = 30°$, the error is about 5%.

Figure 6-8 Typical normal dispersion curve and consequent color separation for white light refracted through a prism.

Dispersion. The minimum deviation of a monochromatic beam through a prism is given implicitly by Eq. (6-15) in terms of the refractive index. The refractive index, however, depends on the wavelength, so that it would be better to write n_λ for this quantity. As a result, the total deviation is itself wavelength dependent, which means that various wavelength components of the incident light are separated on refraction from the prism. A typical *normal dispersion* curve and the nature of the resulting color separation are shown in Figure 6-8. Notice that shorter wavelengths have larger refractive indices and, therefore, smaller speeds in the prism. Consequently violet light is deviated most in refraction through the prism. The dispersion indicated in the graph of Figure 6-8 is called "normal" dispersion. When the refracting medium has characteristic excitations that absorb light of wavelengths within the range of the dispersion curve, the curve is monotonically decreasing, as shown, but has a positive slope in the wavelength region of the absorption. When this occurs, the term *anomalous dispersion* is used, although there is nothing anomalous about it. (This subject is treated further in Section 27-2.) The normal dispersion curve shown is typical but varies somewhat for different materials. An empirical relation that approximates the curve, introduced by Augustin Cauchy, is

$$n_\lambda = A + \frac{B}{\lambda^2} + \frac{C}{\lambda^4} + \cdots \qquad (6\text{-}17)$$

where $A, B, C, \ldots$ are empirical constants to be fitted to the dispersion data of a particular material. Often the first two terms are sufficient to provide a reasonable fit, in which case experimental knowledge of n at two distinct wavelengths is sufficient to determine values of A and B that represent the dispersion. The *dispersion*, defined as $dn/d\lambda$, is then approximately, using Cauchy's formula, $dn/d\lambda = -2B/\lambda^3$.

It is important to distinguish dispersion from deviation. Although prism materials of large n produce a large deviation at a given wavelength, the dispersion or separation of neighboring wavelengths need not be correspondingly large. Figure 6-9 depicts extreme cases illustrating the distinction. Historically, dispersion has been characterized by using three wavelengths of light near the middle and ends of the

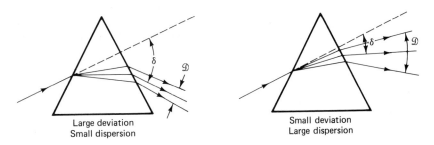

Large deviation
Small dispersion

Small deviation
Large dispersion

Figure 6-9 Extreme cases showing the dispersion $\mathscr{D}$ for three wavelengths and the deviation δ for the intermediate wavelength.

visible spectrum, called *Fraunhofer lines*. These lines were among those that appeared in the solar spectrum studied by J. von Fraunhofer. Their wavelengths, together with refractive indices, are given in Table 6-1. The F and C *dark lines* are due to absorption by hydrogen atoms, and the D dark line is due to absorption by the sodium atoms in the sun's outer atmosphere.[2] Using the thin prism at minimum

TABLE 6-1 FRAUNHOFER LINES

		n	
(nm)	Characterization	Crown glass	Flint glass
486.1	F, blue	1.5286	1.7328
589.2	D, yellow	1.5230	1.7205
656.3	C, red	1.5205	1.7076

deviation for the D line, for example, the ratio of angular spread of the F and C wavelengths to the deviation of the D wavelength, as suggested in Figure 6-9, is

$$\frac{\mathscr{D}}{\delta} = \frac{n_F - n_C}{n_D - 1}$$

This measure of the ratio of dispersion to deviation is defined as the *dispersive power*,

$$\Delta = \frac{n_F - n_C}{n_D - 1} \tag{6-18}$$

Using Table 6-1, we may calculate the dispersive power of crown glass to be $\frac{1}{65}$, while that of flint glass is $\frac{1}{29}$, more than twice as great. The reciprocal of the dispersive power is known as the *Abbe number*.

Prism Spectrometers. An analytical instrument employing a prism as a dispersive element, together with the means of measuring the prism angle and the angles of deviation of various wavelength components in the incident light, is called a *prism spectrometer*. Its essential components are shown in Figure 6-10. Light to be analyzed is focused onto a narrow slit S and then collimated by lens L and refracted by the prism P, which typically rests on a rotatable platform. Rays of light corresponding to each wavelength component emerge mutually parallel after refraction by the prism and are viewed by a telescope focused for infinity. As the telescope is rotated around the prism table, a focused image of the slit is seen for each

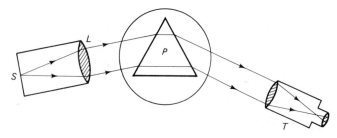

Figure 6-10 Essentials of a spectrometer.

[2] Because the yellow sodium D line is a doublet (589.0 and 589.6 nm), the more monochromatic d line of helium at 587.56 nm is often preferred to characterize the center of the visible spectrum. The green line of mercury at 546.07, lying nearer to the center of the luminosity curve (Figure 2-7), is also used.

wavelength component at its corresponding angular deviation. The deviation δ is measured relative to the telescope position when viewing the slit without the prism in place. When the instrument is used for visual observations without the capability of measuring the angular displacement of the spectral "lines," it is called a *spectroscope*. If means are provided for recording the spectrum, for example, with a photographic film in the focal plane of the telescope objective, the instrument is called a *spectrograph*. When the prism is made of some type of glass, its wavelength range is limited by the absorption of glass outside the visible region. To extend the usefulness of the spectrograph farther into the ultraviolet, for example, prisms made from quartz (SiO_2) and fluorite (CaF_2) have been used. Wavelengths extending further into the infrared can be handled by prisms made of salt (NaCl, KCl) and sapphire (Al_2O_3).

Chromatic Resolving Power.

If the wavelength difference between two components of the light incident on a prism is allowed to diminish, the ability of the prism to resolve them will ultimately fail. The resolving power of a prism spectograph thus represents an important performance parameter, which we shall evaluate in this section. Imagine two spectral lines formed on a photographic film in a prism spectrograph. The lines are images of the slit, so that for precise wavelength measurements the entrance slit should be kept as narrow as possible consistent with the requirement of adequate illumination of the film. Even with the narrowest of slit widths, however, the spectral line image is found to possess a width, directly traceable to the limitation that the edges of the collimating lens or prism face impose on the light beam. The phenomenon is therefore due to the *diffraction* of light, treated later. Since the line images have an irreducible width due to diffraction, as $\Delta\lambda$ decreases and the lines approach one another, a point is reached where the two lines appear as one, and the limit of resolution of the instrument is realized. No amount of magnification of the images can produce a higher resolution or enhancement of the ability to discriminate between such closely spaced spectral lines.

Consider Figure 6-11a, in which a monochromatic parallel beam of light is incident on a prism, such that it fills the prism face. Employing Fermat's principle, the ray *FTW* is isochronous with ray *GX*, since they begin and end on the same plane wavefronts, *GF* and *XW*, respectively. Their optical paths can be equated to give

$$FT + TW = nb$$

(a)

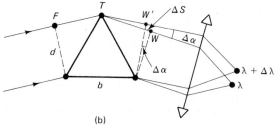

(b)

Figure 6-11 Constructions used to determine chromatic resolving power of a prism. (a) Refraction of monochromatic light. (b) Refraction of two wavelength components separated by $\Delta\lambda$.

where b is the base of the prism and n is the refractive index of the prism, corresponding to the wavelength λ. If a second neighboring wavelength component λ' is now also present in the incident beam, such that $\lambda' - \lambda = \Delta\lambda$, the component λ' will be associated with a different refractive index, $n' = n - \Delta n$. For normal dispersion, Δn will be a small, positive quantity. The emerging wavefronts for the two components, shown in Figure 6-11b, are thus separated by a small angular difference $\Delta\alpha$ and are accordingly focused at different points in the focal plane of the telescope objective. Fermat's principle, applied to the second component λ', gives

$$FT + TW - \Delta s = (n - \Delta n)b$$

Subtracting the last two equations, we conclude

$$\Delta s = b\Delta n \tag{6-19}$$

or, introducing the dispersion,

$$\Delta s = b\left(\frac{dn}{d\lambda}\right)\Delta\lambda \tag{6-20}$$

Equation (6-20) now relates the path difference Δs to the wavelength difference $\Delta\lambda$. The angular difference $\Delta\alpha$ can also be introduced, using

$$\Delta\alpha = \frac{\Delta s}{d} = \left(\frac{b}{d}\right)\left(\frac{dn}{d\lambda}\right)\Delta\lambda \tag{6-21}$$

where d is the beam width. We appeal now to *Rayleigh's criterion*, which determines the limit of resolution of the diffraction-limited line images. This criterion is explained and used in the later treatment of diffraction, where it is shown that the minimum separation $\Delta\alpha$ of the two wavefronts, such that the images formed are just barely resolvable, is given by

$$\Delta\alpha = \frac{\lambda}{d} \tag{6-22}$$

Combining Eqs. (6-21) and (6-22), therefore,

$$\frac{\lambda}{d} = \left(\frac{b}{d}\right)\left(\frac{dn}{d\lambda}\right)\Delta\lambda$$

or the minimum wavelength separation permissible for resolvable images is

$$(\Delta\lambda)_{\min} = \frac{\lambda}{b\,(dn/d\lambda)} \tag{6-23}$$

The *resolving power* provides an alternate way of describing the resolution limit of the instrument. By definition, resolving power

$$\mathcal{R} = \frac{\lambda}{(\Delta\lambda)_{\min}} = b\,\frac{dn}{d\lambda} \tag{6-24}$$

where we have incorporated Eq. (6-23). Since dispersion is limited by the glass, prism resolving power might be improved by increasing the base b. However, this technique soon requires impractically large and heavy prisms. The dispersion $dn/d\lambda$ may be calculated, for example, from the Cauchy formula for the prism material, Eq. (6-17).

Example

Determine the resolving power and minimum resolvable wavelength difference for a prism made from flint glass with a base of 5 cm.

Solution We can calculate an approximate average value of the dispersion for $\lambda = 550$ nm by calculating

$$\frac{\Delta n}{\Delta \lambda} = \frac{n_F - n_D}{\lambda_F - \lambda_D} = \frac{1.7328 - 1.7205}{486 - 589} = -1.19 \times 10^{-4} \text{ nm}^{-1}$$

Thus the resolving power is

$$\mathcal{R} = b\left(\frac{dn}{d\lambda}\right) = (0.05 \times 10^9 \text{ nm})(1.19 \times 10^{-4} \text{ nm}^{-1}) = 5971$$

The minimun resolvable wavelength difference in the region around 550 nm is then

$$(\Delta\lambda)_{\min} = \frac{\lambda}{\mathcal{R}} = \frac{5550 \text{ Å}}{5971} \cong 1 \text{ Å}$$

Although grating spectrographs achieve higher resolving powers, they are generally more wasteful of light. Furthermore, they produce higher-order images of the same wavelength component, which can be confusing when interpreting spectral records. These instruments are discussed later.

Prisms with Special Applications. Prisms may be combined to produce *achromatic* overall behavior, that is, the net dispersion for two given wavelengths may be made zero, even though the deviation is not zero. On the other hand, the *direct vision prism* accomplishes zero deviation for a particular wavelength while at the same time providing dispersion. Schematics involving combinations of these two prism types are shown in Figure 6-12. The arrangement of prisms in Figure 6-12a, combined so that one prism cancels the dispersion of the other, can also be reversed so that the dispersion is additive, providing double dispersion.

(a) Achromatic prism   (b) Direct vision prism for wavelength λ

Figure 6-12 Nondispersive and nondeviating prisms.

A prism design useful in spectrometers is one that produces a constant deviation for all wavelengths as they are observed or detected. One example is the *Pellin-Broca prism*, illustrated in Figure 6-13. A collimated beam of light enters the prism at face AB and departs at face AD, making an angle of $90°$ with the incident direction. The dashed lines are merely added to assist in analyzing the operation of the prism, a single structure. Of the incident wavelengths, only one will refract at the precise angle that conforms to the case of minimum deviation, as shown, with the light rays parallel to the prism base AE. At face BC total internal reflection occurs to direct the light beam into the prism section ACD, where it again traverses under the condition of minimum deviation. Since the prism section BEC serves only as a mirror, the beam passes effectively with minimum deviation through sections

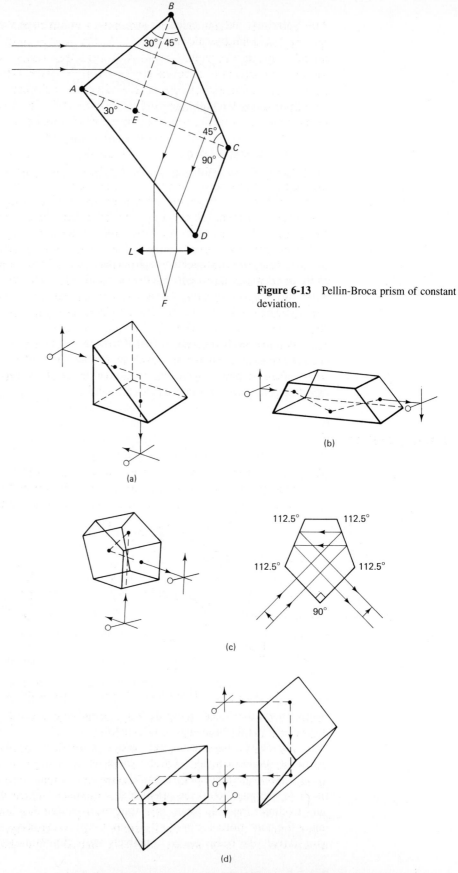

Figure 6-13 Pellin-Broca prism of constant deviation.

Figure 6-14 Image manipulation by refracting prisms. (a) Right-angle prism. (b) Dove prism. (c) Penta prism; pentagonal cross section. (d) Porro prism.

ABE and *ACD*, which together constitute a prism of 60° apex angle. In use, the spectral line is observed or recorded at *F*, the focal point of lens *L*. Thus an observing telescope may be rigidly mounted. Instead, the prism is rotated on its prism table (or about an axis normal to the page), and as it rotates, various wavelengths in the incident beam successively meet the condition of incidence angle for minimum deviation, producing the path indicated, with focus at *F*. The prism rotation may be calibrated in terms of angle, or better, in terms of wavelength.

Reflecting Prisms. Total internally reflecting prisms are frequently used in optical systems, both to alter the direction of the optical axis and to change the orientation of an image. Of course, prisms alone cannot produce images. When used in conjunction with image-forming elements, the light incident on the prism is first collimated and rendered normal to the prism face to avoid prismatic aberrations in the image. Plane mirrors may substitute for the reflecting prisms, but the prism's reflecting faces are easier to keep free of contamination, and the process of total internal reflection is capable of higher reflectivity. The stability in the angular relationship of prism faces may also be an important advantage in some applications. Some examples of reflecting prisms in use are illustrated in Figure 6-14. The *Porro prism,* Figure 6-14d, consists of two right-angle prisms, oriented in such a way that the face of one prism is partially revealed to receive the incident light and the face of the second prism is partially revealed to output the refracted light. The prism halves are separated in the figure to clarify its action. Images are inverted in both vertical and horizontal directions by the pair, so that the Porro prism is commonly used in binoculars to produce erect images.

6-3 THE CAMERA

The simplest type of camera is the pinhole camera, illustrated in Figure 6-15a. Light rays from an object are admitted into a light-tight box and onto a photographic film through a tiny pinhole, which may be provided with any simple means of shuttering,

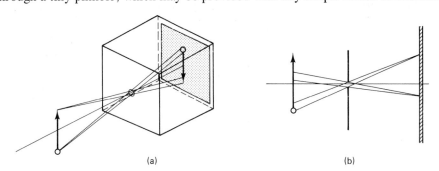

(a)   (b)

Figure 6-15   Imaging by a pinhole camera.

such as a piece of black tape. An image of the object is projected on the back wall of the box, which is lined with a piece of film.

As stated earlier, an image point is determined ideally when every ray from a corresponding object point, each processed by the optical system, intersects at the image point. A pinhole does no focusing and actually blocks out most of the rays from each object point. Because of the smallness of the pinhole, however, every point in the image is intersected only by rays that originate at *approximately* the same point of the object, as in Figure 6-15b. Alternatively, every object point sends a bundle of rays to the screen, which are limited by the small pinhole and so form a small circle of light on the screen, as in Figure 6-15a. The overlapping of these circles of light due to every object point maps out an image whose sharpness depends on the diameter of the individual circles. If they are too large, the image is blurred. Thus, as the pinhole is reduced in size, the image improves in clarity, until a certain pinhole size is reached. As the pinhole is reduced further, the images of each object point actually grow larger again due to diffraction, with consequent degradation of the image. Experimentally, one finds that the optimum pinhole size is around 0.5 mm when the pinhole-to-film distance is around 25 cm. The pinhole itself must be accurately formed in as thin an aperture as possible. A pinhole in aluminum foil, supported by a larger aperture, works well. The primary advantage of a pinhole camera (other than its elegant simplicity!) is that, since there is no focusing involved, all objects are in focus on the screen. In other words, the *depth of field* of the camera is unlimited. The primary disadvantage is that, since the pinhole admits so little of the available light, exposure times must be long. The pinhole camera is not useful in freezing the action of moving objects. The pinhole-to-film distance, while not critical, does affect the sharpness of the image and the field of view. As this distance is reduced, the angular aperture seen by the film is larger, so that more of the scene is recorded, with corresponding decrease in size of any feature of the scene. Also, the image circles decrease in size, producing a clearer image.

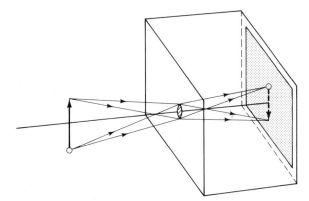

Figure 6-16 Simple camera.

If the pinhole aperture is opened sufficiently to accommodate a converging lens, we have the basic elements of the ordinary camera (Figure 6-16). The most immediate benefits of this modification are (1) an increase in the brightness of the image due to the focusing of *all* the rays of light from each object point onto its conjugate image point and (2) an increase in sharpness of the image, also due to the focusing power of the lens. The lens-to-film distance is now critical and depends on the object distance and lens focal length. For distant objects, the film must be situated in the focal plane of the lens. For closer objects, the focus falls beyond the film. Since the film plane is fixed, a focused image is procured by allowing the lens to be moved farther from the film, that is, by "focusing" the camera. The extreme possible position of the lens determines the nearest distance of objects that can be handled by the camera. "Close-ups" can be managed by changing to a lens with shorter focal length. Thus the focal length of the lens determines the subject area received by the film and the corresponding image size. In general, image size is proportional to focal length. A *wide-angle lens* is a short focal-length lens with a large field of view. A *telephoto lens* is a long focal-length lens, providing magnification at the expense of subject area. The telephoto lens avoids a correspondingly "long" camera by using a positive lens, separated from a second negative lens of shorter focal length, such that the combination remains positive.

Also important to the operation of the camera is the size of its aperture, which admits light to the film. In most cameras, the aperture is variable and is coordinated with the exposure time (shutter speed) to determine the total exposure of the film to light from the scene. The light power incident at the image plane (irradiance E_e in watts per square meter) depends directly on (1) the area of the aperture and inversely on (2) the size of the image. If, as in Figure 6-17, the aperture is circular with diameter D and the energy of the light is assumed to be distributed uniformly over a corresponding image circle of diameter d, then

$$E_e \propto \frac{\text{area of aperture}}{\text{area of image}} = \frac{D^2}{d^2} \qquad (6\text{-}25)$$

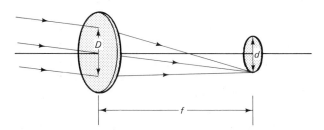

Figure 6-17 Illumination of image. The aperture (not shown) determines the useful diameter D of the lens.

The image size, as in Figure 6-17, is proportional to the focal length of the lens, so we can write

$$E_e \propto \left(\frac{D}{f}\right)^2 \qquad (6\text{-}26)$$

The quantity f/D is the *relative aperture* of the lens (also called *f*-number or f/stop), which we symbolize by the letter A,

$$A \equiv \frac{f}{D} \qquad (6\text{-}27)$$

but is, unfortunately, usually identified by the symbol f/A. For example, a lens of 4-cm focal length that is stopped down to an aperture of 0.5 cm has a relative aperture of $A = 4/0.5 = 8$. This aperture is usually referred to by photographers as $f/8$. The irradiance is now

$$E_e \propto \frac{1}{A^2} \qquad (6\text{-}28)$$

Most cameras provide selectable apertures that sequentially change the irradiance at each step by a factor of 2. The correspoinding *f*-numbers then form a geometric series with ratio $\sqrt{2}$, as in Table 6-2. Larger aperture numbers correspond to smaller exposures. Since the total *exposure* (J/m^2) of the film depends on the product of irradiance (J/m^2-s) and time (s), a desirable total exposure may be met in a variety of ways. Accordingly, if a particular film (whose *speed* is described by an ASA number) is perfectly exposed by light from a particular scene with a shutter speed of $\frac{1}{50}$ s and a relative aperture of $f/8$, it will also be perfectly exposed by any other combination that gives the same total exposure, for example, by choosing a shutter speed of $\frac{1}{100}$ s and an aperture of $f/5.6$. The change in shutter speed cuts the total exposure in half, but opening the aperture to the next f/stop doubles the exposure, leaving no change in net exposure.

The particular combination of shutter speed and relative aperture chosen for an optimum total exposure is not always arbitrary. The shutter speed must be fast, of course, to capture an action shot without blurring the image. The choice of relative

TABLE 6-2 STANDARD RELATIVE APERTURES
AND IRRADIANCE AVAILABLE ON CAMERAS

$A = f\text{-number}$	$(A = f\text{-number})^2$	E_e
1	1	E_0
1.4	2	$E_0/2$
2	4	$E_0/4$
2.8	8	$E_0/8$
4	16	$E_0/16$
5.6	32	$E_0/32$
8	64	$E_0/64$
11	128	$E_0/128$
16	256	$E_0/256$
22	512	$E_0/512$

aperture also affects another property of the image, the *depth of field*. To define this quantity precisely, we utilize Figure 6-18, which shows an axial object point at distance s_0 from a lens being imaged at distance s_0' on the other side. All objects in the object plane are precisely focused in the image plane, disregarding lens aberrations. Objects both closer to and farther from the lens, however, send bundles of rays that focus farther from and closer to the image plane, respectively. Thus a flat film, situated at distance s_0' from the lens, intercepts *circles of confusion* corresponding to such object points. If the diameters of these circles are small enough, the resultant image is still acceptable. Suppose the largest acceptable diameter is d, as shown, such that all images within a distance x of the precise image are suitably "in focus." The *depth of field* is then said to be the interval in object space conjugate to the interval $s_0' - x$ to $s_0' + x$, as shown. Notice that although the interval is symmetric about s_0' in image space, the depth of field interval is not symmetric about s_0.

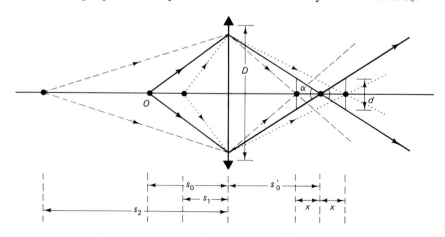

Figure 6-18 Construction illustrating depth of field. Object and image spaces are not shown to the same scale.

The near-point and far-point distances, s_1 and s_2, of the depth of field can be determined once the allowable blurring parameter d is chosen and the lens is specified by focal length and relative aperture. The angle α in Figure 6-18 may be specified in two ways,

$$\tan \alpha \cong \frac{D}{s_0'} \quad \text{and} \quad \tan \alpha \cong \frac{d}{x}$$

so that

$$x \cong \frac{ds_0'}{D} \tag{6-29}$$

It is then required to find, from the lens equation, the object distance s_1 corresponding to image distance $s_0' + x$ and the object distance s_2 corresponding to image distance $s_0' - x$. After a moderate amount of algebra, one finds

$$s_1 = \frac{s_0 f(f + Ad)}{f^2 + Ads_0} \tag{6-30}$$

$$s_2 = \frac{s_0 f(f - Ad)}{f_2 - Ads_0} \tag{6-31}$$

where the aperture is $A = f/D$. the depth of field, $s_2 - s_1$, can be expressed as

$$\text{depth of field} = \frac{2\,Ads_0(s_0 - f)f^2}{f^4 - A^2 d^2 s_0^2} \tag{6-32}$$

Acceptable values of the circle diameter d depend on the quality of the photograph desired. A slide that will be projected or a negative that will be enlarged requires better original detail and hence a smaller value for d. For most photographic work, d is of the order of thousandths of an inch. As an example, let $d = 0.04$ mm. A 5-cm focal length lens with $f/16$ aperture used to focus on an object 9 ft away will focus all objects from around 5 ft to 30 ft to an acceptable degree. Most cameras are equipped with a depth-of-field scale from which values of s_1 and s_2 can be read, once the object distance and aperture are selected. According to Eq. (6-32), depth of field is greater for smaller apertures (larger f- numbers), shorter focal lengths, and longer shooting distances.

The camera lens is called upon to perform a prodigious task. It must provide a large field of view, in the range of 35° to 65° for normal lenses and as large as 120° or more for wide-angle lenses. The image must be in focus and reasonably free from aberrations over the entire area of the film in the focal plane. The aberrations that must be reduced to an acceptable degree are, in addition to chromatic aberration, the five monochromatic aberrations: spherical aberration, coma, astigmatism, curvature of field, and distortion. Since a corrective measure for one type of aberration often causes greater degradation in the image due to another type of aberration, the optical solution represents one of many possible compromise lens designs. The labor involved in the design of a suitable lens that meets particular specifications within acceptable limits has been reduced considerably with the help of computer programming. Human ingenuity is nevertheless an essential component in the design task, since there are many more than one optical solution to a given set of specifications. The demands made upon a photographic lens cannot all be met using a single element. Various stages in solving the lens design problem are illustrated in Figure 6-19a, from the single-element meniscus lens, which may still be found in simple cameras, to the four-element Tessar lens. The use of a symmetrical placement of lenses, or groups of lenses, with respect to the aperture is often a distinctive feature of such lens designs. In such placements, one group may reverse the aberrations introduced by the other, reducing overall image degradation due to factors such as coma, distortion, and lateral chromatic aberration. The multiple-element lens in a modern 35-mm camera is shown in the cutaway photo (Figure 6-19b).

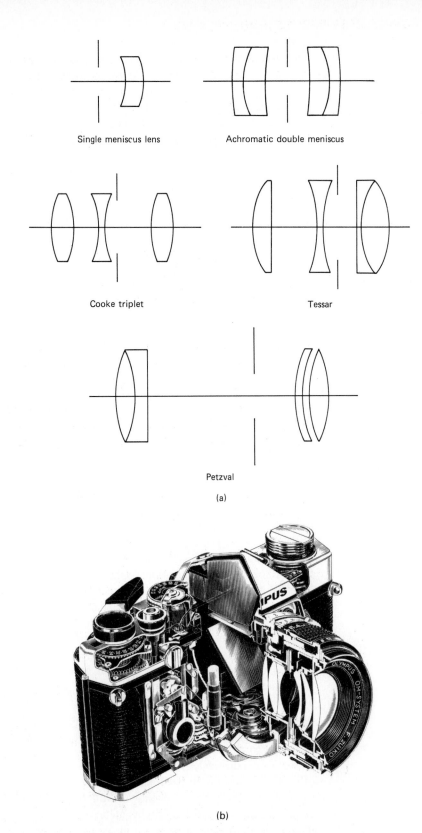

Single meniscus lens

Achromatic double meniscus

Cooke triplet

Tessar

Petzval

(a)

(b)

Figure 6-19 (a) Camera lens design. (b) Cutaway view of a modern 35-mm camera, revealing the multiple element lens. (Courtesy Olympus Corp., Woodbury, N.Y.)

The simple magnifier is essentially a positive lens used to read small print, in which case it is often called a *reading glass*, or to assist the eye in examining any small detail in a real object. It is often a simple convex lens but may be a doublet or a triplet, thereby providing for higher-quality images.

Figure 6-20 illustrates the working principle of the *simple magnifier*. A small object of dimension h, when examined by the unaided eye, is assumed to be held at

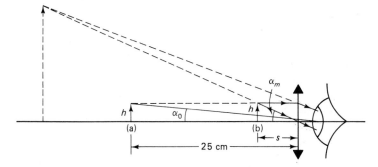

Figure 6-20 Operation of a simple magnifier.

the *near point* of the normal eye—nearest position of distinct vision—position (a), 25 cm from the eye. At this position the object subtends an angle α_0 at the eye. To project a larger image on the retina, the simple magnifier is inserted and the object is moved closer to position (b), where it is at or just inside the focal point of the lens. In this position, the lens forms a virtual image subtending a larger angle α_M at the eye. The *angular magnification*[3] of the simple magnifier is defined to be the ratio α_M/α_0. In the paraxial approximation, the angles may be represented by their tangents, giving

$$\frac{\alpha_M}{\alpha_0} = \frac{h/s}{h/25} = \frac{25}{s}$$

If the image is viewed at infinity, $s = f$ and

$$M = \frac{25}{f} \qquad \text{image at infinity} \tag{6-33}$$

At the other extreme, if the virtual image is viewed at the nearpoint of the eye, then $s' = -25$ cm, and from the thin-lens equation,

$$s = \frac{25f}{25 + f}$$

giving a magnification of

$$M = \frac{25}{f} + 1 \qquad \text{image at normal near point} \tag{6-34}$$

The actual angular magnification depends then on the particular viewer, who will move the simple magnifier until the virtual image is seen comfortably. For small focal lengths, Eqs. (6-33) and (6-34) do not differ greatly, and in citing magnifica-

---

[3] When viewing virtual images with optical instruments, the images may be at great distances, even "at infinity," when rays entering the eye are parallel. In such cases, lateral magnifications also approach infinity and are not very useful. The more convenient *angular magnification* is clearly a measure of the image size formed on the retina and is used to describe magnification when eyepieces are involved, as in microscopes and telescopes.

tions, Eq. (6-33) is most often used. Simple magnifiers may have magnifications in the range of $2\times$ to $10\times$, although the achievement of higher magnifications usually requires a lens corrected for aberrations.

In general, when magnifiers are used to aid the eye in viewing images formed by prior components of an optical system, they are called *oculars*, or *eyepieces*. The real image formed by the objective lens of a microscope, for example, serves as the object that is viewed by the eyepiece, whose angular magnification contributes to the overall magnification of the instrument. To provide quality images, the ocular is corrected to some extent for aberrations and, in particular, to reduce transverse chromatic aberration. To accomplish this improvement, two lenses are most often used. We showed earlier that the effective focal length f (measured relative to principal planes!) of two thin lenses, separated by a distance L, is given by

$$\frac{1}{f} = \frac{1}{f_1} + \frac{1}{f_2} - \frac{L}{f_1 f_2} \tag{6-35}$$

where f_1 and f_2 represent the individual focal lengths of the pair. By the lensmaker's formula, for lenses made of the same glass,

$$\frac{1}{f_1} = (n - 1)\left(\frac{1}{R_{11}} - \frac{1}{R_{12}}\right) = (n - 1)K_1 \tag{6-36}$$

and

$$\frac{1}{f_2} = (n - 1)\left(\frac{1}{R_{21}} - \frac{1}{R_{22}}\right) = (n - 1)K_2 \tag{6-37}$$

where the expressions in parentheses involving the radii of curvature of the lens surfaces are symbolized by constants K_1 and K_2, respectively. Incorporating Eqs. (6-36) and (6-37) into Eq. (6-35),

$$\frac{1}{f} = (n - 1)K_1 + (n - 1)K_2 - L(n - 1)^2 K_1 K_2 \tag{6-38}$$

To correct for transverse chromatic aberration, we require that the effective focal length of the pair remain independent of refractive index,[4] or

$$\frac{d(1/f)}{dn} = 0$$

From Eq. (6-38),

$$\frac{d(1/f)}{dn} = K_1 + K_2 - 2LK_1 K_2(n - 1) = 0$$

This condition is met, therefore, when the lenses are separated by the distance,

$$L = \frac{1}{2}\left[\frac{1}{K_1(n - 1)} + \frac{1}{K_2(n - 1)}\right]$$

or more simply, when

$$L = \tfrac{1}{2}(f_1 + f_2) \tag{6-39}$$

This condition is valid independent of the lens shapes, leaving the choice of shapes as latitude for compensating other aberrations.

---

[4] Some longitudinal chromatic aberration remains because the principal planes of the system do not coincide. Refer back to Figure 5-14 and the related discussion.

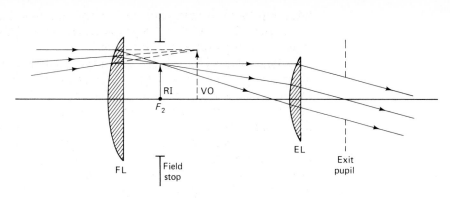

Figure 6-21 Huygens eyepiece.

Both the *Huygens* and *Ramsden eyepieces*, Figures 6-21 and 6-22, incorporate the design feature required by Eq. (6-39), that is, plano-convex lenses are separated by half the sum of their focal lengths. In the diagram of Figure 6-21, the focal length of the field lens, FL, is approximately 1.7 times the focal length of the eye lens, or ocular, EL. The primary image "observed" by the eyepiece is in this case a virtual object (VO) for the field lens, since its virtual position falls between the lenses. The field lens then forms a real image (RI) that is viewed by the eye lens. When the real image falls in the focal plane of the eye lens, the magnified image is viewed at infinity by the eye located at the exit pupil. Note that the Huygens eyepiece cannot be used as an ordinary magnifier. If cross hairs or a reticle with a scale is used with the eyepiece to make possible quantitative measurements, then to be in focus with the image, the cross hairs must be placed in the focal plane of RI, conveniently attached to the field or aperture stop placed there (Figure 6-23). The image of the cross hairs does not share in the image quality provided by the eyepiece as a whole, however, because the eye lens alone is involved in forming the image. This is not a problem in the Ramsden eyepiece, Figure 6-22, in which both the primary and intermediate images are located just in front of the field lens. In this eyepiece, the lenses have the same focal length f and, according to Eq. (6-39), are separated by f. Ideally, rays emerge from the eyepiece parallel to one another, giving a virtual magnified image at infinity, when the real object, RO, falls at the position of the first lens. A reticle is placed at this point. A disadvantage of this arrangement is that the surface of the lens is then also in focus, including dust and smudges. By using a lens separation slightly smaller than f, the reticle is in focus at a position slightly in

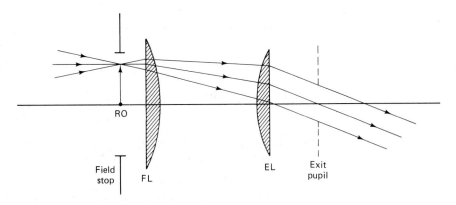

Figure 6-22 Ramsden eyepiece.

front of the lens, as shown in the ray diagram and in Figure 6-23. With a lens separation somewhat less than f, however, the requirement on L that corrects for transverse chromatic aberration is somewhat violated. A modification of the Ramsden eyepiece that almost eliminates chromatic defects is the *Kellner eyepiece*, which replaces the Ramsden eye lens with an achromatic doublet. Other eyepieces have also been designed to achieve higher magnifications and wider fields.

Figure 6-23 Construction of Huygens and Ramsden eyepieces.

Example

A Huygens eyepiece uses two lenses having focal lengths of 6.25 cm and 2.50 cm, respectively. Determine their optimum separation in reducing chromatic aberration, their equivalent focal length, and their angular magnification when viewing an image at infinity.

Solution The optimum separation is given by

$$L = \tfrac{1}{2}(f_1 + f_2) = \tfrac{1}{2}(6.25 + 2.50) = 4.375 \text{ cm}$$

The equivalent focal length is found from

$$\frac{1}{f} = \frac{1}{f_1} + \frac{1}{f_2} - \frac{L}{f_1 f_2} = \frac{1}{6.25} + \frac{1}{2.50} - \frac{4.375}{(6.25)(2.50)}$$

which gives $f = 3.57$ cm. The angular magnification is

$$M = \frac{25}{f} = \frac{25}{3.57} = 7$$

In designing eyepieces, one usually desires an exit pupil that is not much greater than the size of the pupil of the eye, so that radiance is not lost. Recall that the exit pupil is an image of the entrance pupil as formed by the ocular and that the ratio of entrance to exit pupil diameters equals the magnification. Since the entrance pupil is determined by preceding elements in the optical system (the diameter of the objective lens, in a simple telescope), this requirement places a limit on the magnifying power of the eyepiece and, thus, a lower limit on its focal length.

The important specifications of an eyepiece, assuming its aberrations are within acceptable limits for a particular application, include the following:

1. Angular magnification, given by $25/f$, where f is the focal length in centimeters. Available values are $4\times$ to $25\times$, corresponding to focal lengths of 6.25 to 1 cm.
2. Eye relief, that is, the distance from eye lens to exit pupil. Available eyepieces have eye reliefs in the range of 6 to 26 mm.
3. Field-of-view number, or size of the primary image that the eyepiece can cover, in the range of 6 to 30 mm.

6-5 MICROSCOPES

The magnification of small objects accomplished by the simple magnifier is increased further by the compound microscope. In its simplest form, the instrument consists of two positive lenses, an objective lens of small focal length that faces the object and a magnifier functioning as an eyepiece. The eyepiece "looks" at the real image formed by the objective. Referring to Figure 6-24, where the object lies outside the focal length f_o of the objective, a real image I is formed within the microscope tube. After coming to a focus, the light rays continue to the eyepiece, or ocu-

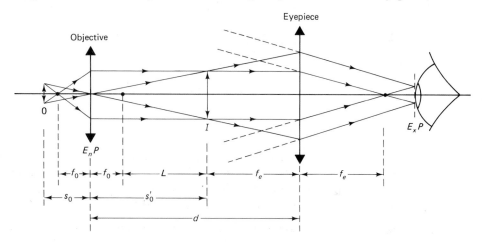

Figure 6-24 Image formation in a compound microscope.

lar lens. For visual observations, the intermediate image is made to occur at or just inside the first focal point of the eyepiece. The eye positioned near the eyepiece then sees a virtual image, inverted and magnified, as shown. The objective lens functions as the entrance pupil of the optical system. The image of the objective formed by the eyepiece is then the exit pupil, which locates the position of maximum radiant energy density and thus the optimum position for the entrance pupil of the eye. A special aperture, functioning as a field stop, is placed at the position of the intermediate image. The eye then sees both in focus together, giving the field of view a sharply defined boundary. If a camera is attached to the microscope, a real final image is required. In this case the intermediate image must be located outside the ocular focal length, f_e.

Total Magnification. When the final image is viewed by the eye, the magnification of the microscope may be defined as in the case of the simple magnifier. Thus the angular magnification for an image viewed at infinity is

$$M = \frac{25}{f_{\text{eff}}} \tag{6-40}$$

where f_{eff} (in cm) is the effective focal length of the two lenses, separated by a distance d and given by Eq. (4-32).

$$\frac{1}{f_{eff}} = \frac{1}{f_o} + \frac{1}{f_e} - \frac{d}{f_o f_e} \tag{6-41}$$

Substituting Eq. (6-41) into Eq. (6-40),

$$M = \frac{25(f_e + f_o - d)}{f_o f_e} \tag{6-42}$$

Using the thin-lens equation, however, we may express the ratio of image to object distance for the objective lens by

$$\frac{s_o'}{s_o} = \frac{d - f_e - f_o}{f_o} \tag{6-43}$$

where we have used the fact that $s_o' = d - f_e$, evident in the diagram. Incorporating Eq. (6-43) into Eq. (6-42),

$$M = -\frac{25 s_o'}{f_e s_o} \tag{6-44}$$

showing that the total magnification is just the product of the linear magnification of the objective multiplied by the angular magnification of the eyepiece when viewing the final image at infinity. The negative sign indicates an inverted image. Making use of Newton's formula for the magnification of a thin lens, Eq. (3-35),

$$m = \frac{x_o'}{f_o} = \frac{L}{f_o} \tag{6-45}$$

where L represents the distance between the objective image and its second focal point, as shown. The magnification of the microscope may then be expressed, perhaps more conveniently, as

$$M = -\left(\frac{25}{f_e}\right)\left(\frac{L}{f_o}\right) \tag{6-46}$$

In many microscopes, the length L is standardized at 16 cm. The focal lengths f_e and f_o are themselves effective focal lengths of multielement lenses, appropriately corrected for aberrations.

Example

A microscope has an objective of 3.8-cm focal length and an eyepiece of 5-cm focal length. If the distance between the lenses is 16.4 cm, find the magnification of the microscope.

Solution

$$L = d - f_o - f_e = 16.4 - 3.8 - 5 = 7.6 \text{ cm}$$

and

$$M = -\left(\frac{25}{f_e}\right)\left(\frac{L}{f_o}\right) = -\left(\frac{25}{5}\right)\left(\frac{7.6}{3.8}\right) = -10$$

Numerical Aperture. To collect more light and produce brighter images, cones of rays from the object, intercepted by the objective lens, should be as large as possible. As magnifications increase and the focal lengths and diameters of the objective lenses decrease corespondingly, the solid angle of useful rays from the ob-

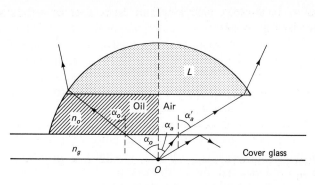

Figure 6-25 Microscope objective, illustrating the increased light-gathering power of an oil-immersion lens.

ject also decreases. In Figure 6-25, the useful light rays originating at the object point O, passing through a thin cover glass and then air to the first element of the objective lens L, make a half-angle of α_a on the right of the optical axis. Due to refraction at the glass-air interface, rays making a larger angle than α_a do not reach the lens. This limitation is somewhat relieved by using a coupling transparent fluid whose index matches as closely as possible that of the glass. On the left of the optical axis in the diagram, a layer of oil is used, and a larger half-angle α_o is possible. Typically, the cover glass index is 1.522 and the oil index is 1.516, providing an excellent match. The light-gathering capability of the objective lens is thus increased by increasing the refractive index in object space. A measure of this capability is the product of half-angle and refractive index, called the *numerical aperture*,

$$\text{N. A.} = n \sin \alpha \qquad (6\text{-}47)$$

The numerical aperture is an invariant in object space, due to Snell's law. That is, in the case of air,

$$\text{N. A.} = n_g \sin \alpha_a = \sin \alpha_a'$$

and when an *oil-immersion objective* is used,

$$\text{N. A.} = n_g \sin \alpha_o = n_o \sin \alpha_o'$$

The maximum value of the numerical aperture when air is used is unity, but when object space is filled with a fluid of index n, the maximum numerical aperture may be increased up to the value of n. In practice, the limit is around 1.6. The numerical aperture is an alternative means of defining a relative aperture or of describing how "fast" a lens is. As shown previously, image brightness is inversely proportional to the square of the f-number. Here also, image brightness is proportional to $(\text{N. A.})^2$. The numerical aperture is an important design parameter also because it limits the resolving power and the depth of focus of the lens. The resolving power is proportional to the numerical aperture, whereas the depth of focus is inversely proportional to $(\text{N. A.})^2$. Most microscopes use objectives with numerical apertures in the approximate range of 0.08 to 1.30.

Biological specimens are covered with a cover glass of 0.17 or 0.18 mm thickness. For objectives with numerical apertures over 0.30, the cover glass has increasing influence on the image quality, since it introduces a large degree of spherical aberration when oil immersion is not involved. Thus a *biological objective* compensates for the aberration introduced by a cover glass. In contrast, a *metallurgical objective* is designed without such compensation. Objectives may be classified broadly in relation to the corrections introduced into their design. For low magnifications, with focal lengths in the range of 8 to 64 mm, *achromatic objectives* are generally used. Such objectives are chromatically corrected, usually for the Fraunhofer *C* (red) and *F* (blue) wavelengths, and spherically corrected, at the Fraunhofer *D* (sodium yellow) wavelength. For higher magnifications, objective lenses with focal lengths in the range of 4 to 16 mm incorporate some fluorite elements, which together with the glass elements provide better correction over the visual spectrum. When the correction is nearly perfect throughout the visual spectrum, the objectives are said to be *apochromatic*. Since correction is more crucial at high magnifications, apochromats are usually objectives with focal lengths in the range of 1.5 to 4 mm. For even higher magnifications, the objective is usually designed as an *immersion objective*. Modern techniques and materials have also made possible *flat-field objectives* that essentially eliminate field curvature over the useful portion of the field. With ultraviolet immersion microscopes, it is customary to replace the oil with glycerine and the optical glass elements with fluorite and quartz elements because of their higher transmissivity at short wavelengths.

This discussion should make it clear that high-quality microscopes today are designed as a whole and usually for a specific use. The design of an objective or an eyepiece is directly related to the performance of other optical elements in the instrument, often including a relay lens within the body tube of the microscope as well. Thus it is generally not possible to interchange objectives and eyepieces between different model microscopes without loss or deterioration of the image.

Figure 6-26 illustrates the optical components in a standard microscope and the detailed processing of light rays through the instrument.

(a)

Figure 6-26 Standard microscope illustrating Koehler illumination. (Courtesy Carl Zeiss, Inc., Thornwood, N.Y.)

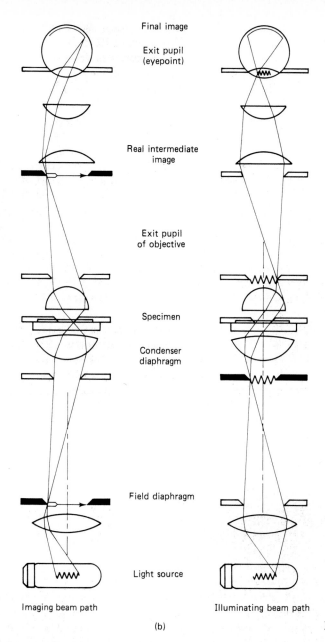

Final image

Exit pupil
(eyepoint)

Real intermediate
image

Exit pupil
of objective

Specimen

Condenser
diaphragm

Field diaphragm

Light source

Imaging beam path   Illuminating beam path

(b)

Figure 6-26   (continued)

6-6 *TELESCOPES*

Telescopes may be broadly classified as *refracting* or *reflecting*, according to whether lenses or mirrors are used to produce the image. There are, in addition, *catadioptric* systems that combine refracting and reflecting surfaces. Telescopes may also be distinguished by the erectness or inversion of the final image and by either a visual or photographic means of observation.

Refracting Telescopes.   Figures 6-27 and 6-28 show two refracting telescope types, producing, respectively, inverted and erect images. The *Keplerian telescope* in Figure 6-27 is often referred to as an *astronomical telescope* since inver-

Figure 6-27 Astronomical telescope.

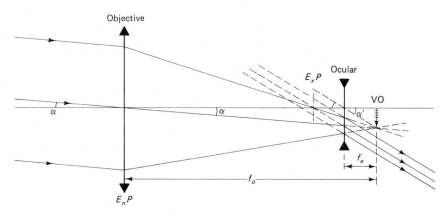

Figure 6-28 Galilean telescope.

sion of astronomical objects in the images produced creates no difficulties. The *Galilean telescope*, illustrated in Figure 6-28, produces an erect image by means of an eyepiece of negative focal length. In either case, nearly parallel rays of light from a distant object are collected by a positive objective lens, which forms a real image in its focal plane. The objective lens, being larger in diameter than the pupil of the eye, permits the collection of more light and makes visible point sources such as stars that might otherwise not be detected. The objective lens is usually a doublet, corrected for chromatic aberration. The real image formed by the objective is observed with an eyepiece, represented in the figures as a single lens. This intermediate image, located at or near the focal point of the ocular, serves as a real object (RO) for the ocular in the astronomical telescope and a virtual object (VO) in the case of the Galilean telescope. In either case, the light is refracted by the eyepiece in order to produce parallel, or nearly parallel, light rays. An eye placed near the ocular views an image at infinity but with an angular magnification given by the ratio of the angles α'/α, as shown. The object subtends the angle α at the unaided eye and the angle α' at the eyepiece. From the two right triangles formed by the intermediate image and the optical axis, it is evident that the angular magnification is

$$M = \frac{\alpha'}{\alpha} = -\frac{f_o}{f_e} \tag{6-48}$$

The negative sign is introduced, as usual, to indicate that the image is inverted in Figure 6-27, where $f_e > 0$, and is erect in Figure 6-28, where $f_e < 0$. In either case, the length L of the telescope is given by

$$L = f_o + f_e \qquad (6\text{-}49)$$

permitting a short Galilean telescope, a circumstance that makes this design convenient in the *opera glass*. The astronomical telescope may be modified to produce an erect image by the insertion of a third positive lens whose function is simply to invert the intermediate image, but this lengthens the telescope by at least four times the focal length of the additional lens. Image inversion may also be achieved without additional length, as in binoculars, through the use of inverting prisms, discussed previously.

The objective of either telescope functions as the entrance pupil, whose image in the ocular is the exit pupil, as shown. In the astronomical telescope, the exit pupil is situated just outside the eyepiece and is designed to match the size of the pupil of the eye. A telescope should produce an exit pupil at sufficient distance from the eyepiece to produce a comfortable *eye relief*. Greater ease of observation is also achieved if the exit pupil is a little larger in diameter than the eye pupil, allowing for some relative motion between eye and eyepiece. Notice that in the Galilean telescope the exit pupil falls inside the eyepiece, where it is inaccessible to the eye. This represents a disadvantage of the Galilean telescope, leading to a restriction in the field of view. Notice also that a field stop with reticle can be employed at the location of the intermediate image in the terrestrial telescope, whereas no such arrangement is possible in conjunction with the Galilean telescope. The diameter of the exit pupil D_{ex} is simply related to the diameter of the objective lens D_{obj} through the angular magnification, as follows. Since the exit pupil is the image of the entrance pupil formed by the eyepiece, we may write for the linear, transverse magnification either

$$m_e = \frac{D_{ex}}{D_{obj}} \qquad (6\text{-}50)$$

or, employing the Newtonian form of the magnification,

$$m_e = -\frac{f}{x} = -\frac{f_e}{f_0} \qquad (6\text{-}51)$$

where x is the distance of the object (objective lens) from the focal point of the eyepiece, or f_0. Combining Eqs. (6-48), (6-50), and (6-51),

$$m_e = \frac{1}{M} = \frac{D_{ex}}{D_{obj}}$$

so that

$$D_{ex} = \frac{D_{obj}}{M} \qquad (6\text{-}52)$$

Thus the diameter of the bundle of parallel rays filling the objective lens is greater by a factor of M than the diameter of the bundle of rays that pass through the exit pupil. It should be pointed out that the image is not, therefore, brighter by the same proportion, however, because the apparent size of the image increases by the same factor M. The brightness of the image cannot be greater than the brightness of the object; in fact, it is less bright due to inevitable light losses due to reflections from lens surfaces.

Binoculars (Figure 6-29) afford more comfortable telescopic viewing, allowing both eyes to remain active. In addition, the use of Porro or other prisms to pro-

Figure 6-29 Cutaway view of binoculars revealing compound objective and ocular lenses and image-inverting prism. (Courtesy Carl Zeiss, Inc., Thornwood, N.Y.)

duce erect final images also permits the distance between objective lenses to be greater than the interpupillary distance, enhancing the stereoscopic effect produced by ordinary binocular vision. The designation "6 × 30" for binoculars means that the angular magnification M produced is 6× and the diameter of the objective lens is 30 mm. Using Eq. (6-52), we conclude that the exit pupil for this pair of binoculars is 5 mm, a good match for the normal pupil diameter. For night viewing, when the pupils are somewhat larger, a rating of 7 × 50, producing an exit pupil diameter of 7 mm, would be preferable.

Example

Determine the eye relief and field of view for the 6 × 30 binoculars just described. Assume an objective focal length of 15 cm and a field lens (eyepiece) diameter of 1.50 cm.

Solution   The focal length of the ocular is found from

$$f_e = -\frac{f_o}{M} = -\frac{15}{-6} = 2.5 \text{ cm}$$

The eye relief is the distance of the next pupil from the eyepiece. Since the exit pupil is the image of the objective formed by the eyepiece, the eye relief is the image distance s', given by

$$s' = \frac{sf}{s - f} = \frac{Lf_e}{L - f_e} = \frac{(f_o + f_e)f_e}{(f_o + f_e) - f_e} = \frac{(15 + 2.5)(2.5)}{15} = 2.92 \text{ cm}$$

The angular field of view from the objective subtends both the object on one side and the field lens of the eyepiece on the other. Thus, for objects at a standard distance of 1000 yd,

$$\theta = \frac{h}{s} = \frac{D_f}{L}$$

or

$$h = s\theta = \frac{sD_f}{L} = \frac{(3000 \text{ ft})(1.50)}{15 + 2.5} = 257 \text{ ft at } 1000 \text{ yd}$$

Reflection Telescopes.   Larger-aperture objective lenses provide greater light-gathering power and resolution. Large homogeneous lenses are difficult to produce without optical defects, and their weight is difficult to support. These problems, as well as the elimination of chromatic aberrations, are solved by using curved, reflecting surfaces in place of lenses. The largest telescopes, like the Hale 200-in. reflector on Mount Palomar, use such mirrors. Such large reflecting telescopes are usually employed to examine very faint astronomical objects and use the integrating power of photographic plates, exposed over long time intervals, in observations.

Several basic designs for reflecting telescopes are shown in Figure 6-30. In the *Newtonian* design (a), a parabolic mirror is used to focus accurately all parallel rays to the same primary focal point, f_p. Before focusing, a plane mirror is used to divert the converging rays to a secondary focal point, f_s, near the body of the telescope, where an eyepiece is located to view the image. The use of a parabolic mirror avoids both chromatic and spherical aberration, but coma is present for off-axis points, severely limiting the useful field of view. In the 200-in. Hale telescope, the flat mirror can be dispensed with and the rays allowed to converge at their primary focus.

(a)

(b)

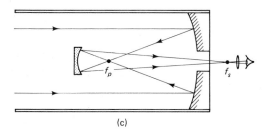

(c)

Figure 6-30   Basic designs for reflecting telescopes. (a) Newtonian telescope. (b) Cassegrain telescope. (c) Gregorian telescope.

This telescope is large enough so that the observer can be mounted on a specially built platform situated just behind the primary focus (Figure 6-31). Of course, any obstruction placed inside the telescope reduces the cross section of the incident light waves contributing to the image. In the *Cassegrain* design (Figure 6-30b), the secondary mirror is hyperboloidal convex in shape, reflecting light from the primary mirror through an aperture in the primary mirror to a secondary focus, where it is conveniently viewed or recorded. The hyperboloidal surface permits perfect imaging between the primary and secondary focal points, which function as the foci of the hyperboloid. Such accurate imaging is also possible when the secondary mirror is concave ellipsoidal, as in the *Gregorian* telescope (Figure 6-30c). The primary and secondary focal points of this telescope are now the foci of the ellipsoid.

Figure 6-31 Hale telescope (200-in.) showing observer in prime-focus cage and reflecting surface of 200-in. mirror. (California Institute of Technology.)

The Schmidt Telescope. Perhaps the most celebrated catadioptric telescope is due to a design of Bernhardt Schmidt. He sought to remove the spherical aberration of a primary spherical mirror by using a thin refracting correcting plate at the aperture of the telescope. To understand his design, refer to Figure 6-32. A concave primary reflector in (a) receives small bundles of parallel rays from various directions. Each bundle enters at the aperture, which is located at the center of curvature of the primary mirror. Since the axis of any bundle may be considered an optical axis, there are no off-axis points and thus coma and astigmatism do not enter into the aberrations of the system. When the bundles are small, each bundle consists of paraxial rays that focus at the same distance from the mirror, a distance equal to its focal length, or half the radius of curvature of the mirror. The locus of such image points is then the spherical surface indicated by the dashed line. However, when the bundles are large, as shown in (b), spherical aberration occurs, which produces a shorter focus for rays reflecting from the outer zones of the mirror relative to the optical axis of the bundle. Schmidt designed a transparent correcting plate, to be placed at the aperture, whose function was to bring the focus of all zones to the same point on the spherical focal surface, as indicated in (c). The shape suggested in the figure is designed to make the focal point of all zones agree with the focal point of a zone whose radius is 0.707 of the aperture radius, the usual choice. The result-

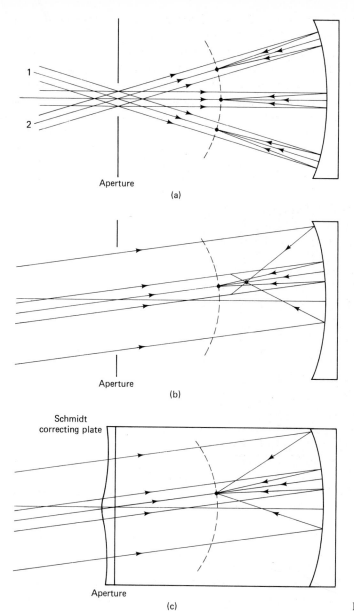

1

2

Aperture

(a)

Aperture

(b)

Schmidt
correcting plate

Aperture

(c)

Figure 6-32 The Schmidt optical system.

ing *Schmidt optical system* is therefore highly corrected for coma, astigmatism, and spherical aberration. Because the correcting plate is situated at the center of curvature of the mirror, it presents approximately the same optics to parallel beams arriving from different directions and so permits a wide field of view. Residual aberrations are due to errors in the actual fabrication of the correcting plate and because the plate does not present precisely the same cross section, and therefore the same correction, to beams entering from different directions. One disadvantage is that the focal plane is spherical, requiring a careful shaping of photographic films and plates. Notice also that with the correcting plate attached at twice the focal length of the mirror, the telescope is twice as long as the telescopes described previously, Figure 6-30. Nevertheless, the *Schmidt camera*, as it is often called, has been highly successful and has spawned a large number of variants, including designs to flatten the field near the focal plane.

6-1. An object measures 2 cm high above the axis of an optical system consisting of a 2-cm aperture stop and a thin convex lens of 5-cm focal length and 5-cm aperture. The object is 10 cm in front of the lens and the stop is 2 cm in front of the lens. Determine the position and size of the entrance and exit pupils, as well as the image. Sketch the chief ray and the two extreme rays through the optical system, from top of object to its conjugate image point.

6-2. Repeat problem 6-1 for an object 4 cm high, with a 2-cm aperture stop and a thin convex lens of 6-cm focal length and 5-cm aperture. The object is 14 cm in front of the lens and the stop is 2.50 cm behind the lens.

6-3. Repeat problem 6-1 for an object 2 cm high, with a 2-cm aperture stop and a thin convex lens of 6-cm focal length and 5-cm aperture. The object is 14 cm in front of the lens and the stop is 4 cm in front of the lens.

6-4. An optical system, centered on an optical axis, consists of (left to right)
  1. Source plane
  2. Thin lens L_1 at 40 cm from the source plane
  3. Aperture A at 20 cm farther from L_1
  4. Thin lens L_2 at 10 cm farther from A
  5. Image plane
Lens L_1 has a focal length of 40/3 cm and a diameter of 2 cm; lens L_2 has a focal length of 20/3 cm and a diameter of 2 cm; aperture A has a centered circular opening of 0.5-cm diameter.
  (a) Sketch the system.
  (b) Find the location of the image plane.
  (c) Locate the aperture stop and entrance pupil.
  (d) Locate the exit pupil.
  (e) Locate the field stop, the entrance window, and the exit window.
  (f) Determine the angular field of view.

6-5. Plot a curve of total deviation angle versus entrance angle for a prism of apex angle 60° and refractive index 1.52.

6-6. A parallel beam of white light is refracted by a 60° glass prism in a position of minimum deviation. What is the angular separation of emerging red ($n = 1.525$) and blue (1.535) light?

6-7. **(a)** Approximate the Cauchy constants A and B for crown and flint glasses, using data for the C and F Fraunhofer lines from Table 6-1. Using these constants and the Cauchy relation approximated by two terms, calculate the refractive index of the D Fraunhofer line for each case. Compare your answers with the values given in the table.
  (b) Calculate the dispersion in the vicinity of the Fraunhofer D line for each glass, using the Cauchy relation.
  (c) Calculate the chromatic resolving power of crown and flint prisms in the vicinity of the Fraunhofer D line, if each prism base is 75 mm in length. Also calculate the minimum resolvable wavelength interval in this region.

6-8. An equilateral prism of dense barium crown glass is used in a spectroscope. Its refractive index varies with wavelength, as given in the table:

nm	n
656.3	1.63461
587.6	1.63810
486.1	1.64611

  (a) Determine the minimum angle of deviation for sodium light of 589.3 nm.
  (b) Determine the dispersive power of the prism.

(c) Determine the Cauchy constants A and B in the long wavelength region; from the Cauchy relation, find the dispersion of the prism at 656.3 nm.

(d) Determine the minimum base length of the prism if it is to resolve the hydrogen doublet at 656.2716- and 656.2852-nm wavelengths. Is the project practical?

6-9. A prism of 60° refracting angle gives the following angles of minimum deviation when measured on a spectrometer: C line, 38°20′; D line, 38°33′; F line, 39°12′. Determine the dispersive power of the prism.

6-10. The refractive indices for certain crown and flint glasses are

$$\text{Crown:} \quad n_C = 1.527, \quad n_D = 1.530, \quad n_F = 1.536$$
$$\text{Flint:} \quad n_C = 1.630, \quad n_D = 1.635, \quad n_F = 1.648$$

The two glasses are to be combined in a double prism that is a direct vision prism for the D wavelength. The refracting angle of the flint prism is 5°. Determine the required angle of the crown prism and the resulting angle of dispersion between the C and the F rays. Assume that the prisms are thin and the condition of minimum deviation is satisfied.

6-11. An achromatic thin prism for the C and F Fraunhofer lines is to be made using the crown and flint glasses described in Table 6-1. If the crown glass prism has a prism angle of 15°, determine (a) the required prism angle for the flint glass and (b) the resulting "mean" deviation for the D line.

6-12. A perfectly diffuse, or *Lambertian*, surface has the form of a square, 5 cm on a side. This object radiates a total power of 25 W into the forward directions that constitute half the total solid angle of 4π. A camera with a 4-cm focal length lens and stopped down to $f/8$ is used to photograph the object when it is placed 1 m from the lens.

(a) Determine the radiant exitance, radiant intensity, and radiance of the object. (See Table 2-1.)

(b) Determine the radiant flux delivered to the film.

(c) Determine the irradiance at the film.

6-13. Investigate the behavior of Eq. (6-32), giving the dependence of the depth of field on aperture, focal length, and object distance. With the help of a calculator or computer program, generate curves showing each dependence.

6-14. A camera is used to photograph three rows of students at a distance 6 m away, focusing on the middle row. Suppose that the image defocusing or blur circles due to object points in the first and third rows is to be kept smaller than a typical silver grain of the emulsion, say 1 μm. At what object distance nearer and farther than the middle row does an unacceptable blur occur if the camera has a focal length of 50 mm and is stopped down to an $f/4$ setting?

6-15. A telephoto lens consists of a combination of two thin lenses having focal lengths of +20 cm and −8 cm, respectively. The lenses are separated by a distance of 15 cm.

(a) Determine the focal length of the combination, distance from negative lens to film plane, and image size of a distant object subtending an angle of 2° at the camera.

(b) Determine the same quantities and the position of the principal planes, as well, using the matrix approach of Chapter 4.

6-16. A 5-cm focal length camera lens with $f/4$ aperture is focused on an object 6 ft away. If the maximum diameter of the circle of confusion is taken to be 0.05 mm, determine the depth of field of the photograph.

6-17. The sun subtends an angle of 0.5° at the earth's surface, where the illuminance is about 10^5 lx at normal incidence. What is the illuminance of an image of the sun formed by a lens with diameter 5 cm and focal length 50 cm?

6-18. (a) A camera uses a convex lens of focal length 15 cm. How large an image is formed on the film of a 6-ft-tall person 100 ft away?

(b) The convex lens is replaced by a telephoto combination consisting of a 12-cm focal length convex lens and a concave lens. The concave lens is situated in the position of the original lens, and the convex lens is 8 cm in front of it. What is the required focal length of the concave lens such that distant objects form focused images on the same film plane? How much larger is the image of the person using this tele-photo lens?

6-19. The lens on a 35-mm camera is marked "50 mm, 1:1.8."

    (a) What is the maximum aperture diameter?

    (b) Starting with the maximum aperture setting, supply the next three f-numbers that would allow the irradiance to be reduced to $\frac{1}{3}$ the preceding at each successive stop.

    (c) What aperture diameters correspond to these f-numbers?

    (d) If a picture is taken at maximum aperture and at $\frac{1}{100}$ s, what exposure time at each of the other openings provides equivalent total exposures?

6-20. The magnification given by Eq. (6-33) is also valid for a double-lens eyepiece if the equivalent focal length given by Eq. (6-35) is used. Show that the magnification of a double-lens eyepiece, designed to satisfy the condition for the elimination of chromatic aberration, is, for an image at infinity,

$$M = 12.5\left(\frac{1}{f_1} + \frac{1}{f_2}\right)$$

6-21. A magnifier is made of two thin plano-convex lenses, each of 3-cm focal length and spaced 2.8 cm apart. Find (a) the equivalent focal length and (b) the magnifying power for an image formed at the near point of the eye.

6-22. The objective of a microscope has a focal length of 0.5 cm and forms the intermediate image 16 cm from its second focal point.

    (a) What is the overall magnification of the microscope when an eyepiece rated at $10\times$ is used?

    (b) At what distance from the objective is a point object viewed by the microscope?

6-23. A homemade compound microscope has, as objective and eyepiece, thin lenses of focal lengths 1 cm and 3 cm, respectively. An object is situated at a distance of 1.20 cm from the objective. If the virtual image produced by the eyepiece is 25 cm from the eye, compute (a) the magnifying power of the microscope and (b) the separation of the lenses.

6-24. Two thin convex lenses, when placed 25 cm apart, form a compound microscope whose apparent magnification is 20. If the focal length of the lens representing the eyepiece is 4 cm, determine the focal length of the other.

6-25. A level telescope contains a *graticule*—a circular glass on which a scale has been etched—in the common focal plane of objective and eyepiece, so that it is seen in focus with a distant object. If the telescope is focused on a telephone pole 30 m away, how much of the post falls between millimeter marks on the graticule? The focal length of the objective is 20 cm.

6-26. A pair of binoculars is marked "7 × 35." The focal length of the objective is 14 cm, and the diameter of the field lens of the eyepiece is 1.8 cm. Determine (a) the angular magnification of a distant object, (b) the focal length of the ocular, (c) the diameter of the exit pupil, (d) the eye relief, and (e) the field of view in terms of feet at 1000 yd.

6-27. **(a)** Show that when the final image is not viewed at infinity, the angular magnification of an astronomical telescope may be expressed by

$$M = \frac{m_{oc}f_{obj}}{s''}$$

    where m_{oc} is the linear magnification of the ocular and s'' is the distance from the ocular to the final image.

    (b) For such a telescope using two converging lenses with focal lengths of 30 cm and 4 cm, find the angular magnification when the image is viewed at infinity and when the image is viewed at a near point of 25 cm.

6-28. The moon subtends an angle of 0.5° at the objective lens of a terrestrial telescope. The focal lengths of the objective and ocular lenses are 20 cm and 5 cm, respectively. Find the diameter of the image of the moon viewed through the telescope at near point of 25 cm.

6-29. An opera glass uses an objective and eyepiece with focal lengths of $+12$ cm and -4.0 cm, respectively. Determine the length (lens separation) of the instrument and its magnifying power for a viewer whose eyes are focused (a) for infinity and (b) for a near point of 30 cm.

6-30. An astronomical telescope is used to project a real image of the moon onto a screen 25 cm from an ocular of 5-cm focal length. How far must the ocular be moved from its normal position?

6-31. (a) The Ramsden eyepiece of a telescope is made of two positive lenses of focal length 2 cm each and also separated by 2 cm. Calculate its magnifying power when viewing an image at infinity.

    (b) The objective of the telescope is a 30-cm positive lens, with a diameter of 4.50 cm. Calculate the overall magnification of the telescope.

    (c) What is the position and diameter of the exit pupil?

    (d) The diameter of the eyepiece field lens is 2 cm. Determine the angle defining the field of view of the telescope.

6-32. Show that the angular magnification of a Newtonian reflecting telescope is given by the ratio of objective to ocular focal lengths, as it is for a refracting telescope when the image is formed at infinity.

6-33. The primary mirror of a Cassegrain reflecting telescope has a focal length of 12 ft. The secondary mirror, which is convex, is 10 ft from the primary mirror along the principal axis and forms an image of a distant object at the vertex of the primary mirror. A hole in the mirror there permits viewing the image with an eyepiece of 4-in. focal length, placed just behind the mirror. Calculate the focal length of the convex mirror and the angular magnification of the instrument.

REFERENCES

[1] Horne, D. F. *Optical Instruments and Their Applications*. Bristol, England: Adam Hilger Ltd., 1980.

[2] Goodman, Douglas S. "Basic Optical Instruments." In *Geometrical and Instrumental Optics*, edited by Daniel Malacara. Boston: Academic Press, 1988.

[3] Brouwer, William. *Matrix Methods in Optical Instrument Design*. New York: W. A. Benjamin, 1964.

[4] Benford, James R., and Harold E. Rosenberger. "Microscope Objectives and Eyepieces." In *Handbook of Optics*, edited by Walter G. Driscoll and William Vaughan. New York: McGraw-Hill Book Company, 1978.

[5] Cox, A. *Photographic Optics*, 15th ed. New York: Focal Press, 1974.

[6] Horne, D. F. *Optical Production Technology*. New York: Crane, Russak and Company, 1972.

[7] Kingslake, R. *Lens Design Fundamentals*. New York: Academic Press, 1978.

[8] McLaughlin, R. B. *Special Methods in Light Microscopy*. London: Microscope Publications Ltd., 1977.

[9] Smith, W. J. *Modern Optical Engineering*. New York: McGraw-Hill Book Company, 1966.

[10] Nussbaum, Allen. *Geometric Optics: An Introduction*. Reading, Mass.: Addison-Wesley Publishing Co., 1968.

[11] Klein, Miles V. *Optics*. New York: John Wiley and Sons, 1970. Ch. 4.1: Radiometry and Photometry.

[12] Kirkpatrick, Paul. "X-ray Microscope." *Scientific American* (Mar. 1949): 44.

[13] Wald, George. "Eye and Camera." *Scientific American* (Aug. 1950): 32.

[14] Wilson, Albert G. "The Big Schmidt." *Scientific American* (Dec. 1950): 34.

[15] Muller, Erwin W. "A New Microscope." *Scientific American* (May 1952): 58.

[16] McClain, Edward, Jr. "The 600-Foot Radio Telescope." *Scientific American* (Jan. 1960): 45.

[17] Crewe, Albert. "A High-Resolution Scanning Electron Microscope." *Scientific American* (Apr. 1971): 26.

[18] Price, William H. "The Photographic Lens." *Scientific American* (Aug. 1976): 72.

[19] Quate, Calvin. "Acoustic Microscope." *Scientific American* (Oct. 1979): 62.

[20] Bahcall, J. N., and L. Spitzer, Jr. "The Space Telescope." *Scientific American* (July 1982): 40.

[21] Howells, Malcolm R., Janos Kirz, and William Sayre. "X-ray Microscopes." *Scientific American* (Feb. 1991): 88.

7

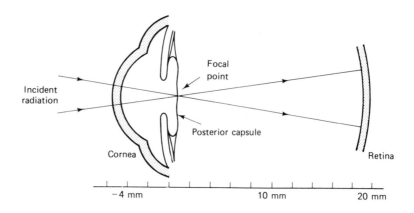

Incident radiation

Focal point

Posterior capsule

Cornea

Retina

−4 mm   10 mm   20 mm

Optics of the Eye

INTRODUCTION

In this chapter we acquaint ourselves with the optics of the eye. First we examine the structure and functions of the eye. Following this we note the *errors of refraction* in a defective eye and indicate the usual corrective optics. Finally we describe several current surgical procedures—radial keratotomy and posterior capsulotomy—wherein laser light of a specific irradiance and wavelength is used to restore visual acuity in less-than-perfect eyes.

The eyes, in conjunction with the brain, constitute a truly remarkable bio-optical system. Consider briefly the distinctive characteristics of this system. It forms images of a continuum of objects, at distances of a foot to infinity. It scans a scene as expansive as the overhead sky or focuses on detail as minute as the head of a pin. It adapts itself to an extraordinary range of intensities, from the barely visible flicker of a candle miles away on a dark night to sunlight so bright that the optical image on the retina causes serious solar burn. It distinguishes between subtle shades of color, from deep purple to deep red. Most importantly for us, functioning as a unique spatial sense organ, it localizes objects in space, accurately mapping out our three-dimensional world.

Anatomically, the eyeball is a globe, almost spherically shaped, approximately 22 mm in diameter. It lies buried in fat tissue inside the orbit, or space, in the skull surrounded by bony walls. Optically, the eyeball can be pictured as a positive lens system that refracts incident light onto its rear surface to form a real image, much as does an ordinary camera.

The basic parts of the eye are shown in Figure 7-1. Let us examine the key biological components of the eye along the optical axis, in the same order they are encountered by light rays in the usual image-forming process. Light first enters the eye through the *cornea*, a transparent tissue devoid of blood vessels but abounding in nerve cells. The cornea is roughly 12 mm in diameter and 0.6 mm thick at its center, thickening somewhat further at its edges, with a refractive index of 1.376. Upon entering the eye at the air-cornea interface, where the refractive index changes abruptly from 1.0 to 1.38, light undergoes a significant degree of bending. The corneal surface provides, in fact, about 73% of the total refractive power of the eye. Immediately behind the cornea is the *anterior chamber*, a small space filled with a watery fluid that provides nutrients for the cornea, the *aqueous humor*. This fluid has a refractive index of 1.336, almost equal to that of water (1.333). Because the refractive indices of the cornea and aqueous humor are nearly alike, little additional bending of rays occurs as light moves from the cornea into the anterior chamber. Situated in the aqueous humor is the *iris*, a diaphragm that gives the eye its characteristic color and controls the amount of light that enters. The amount and location of pigment in the iris determine whether the eye looks blue, green, gray, or brown. The adjustable hole or opening in the iris through which the light passes is called the *pupil*. The iris contains two sets of delicate muscles that change the pupil size in response to light stimulation, adjusting the diameter from a minimum of about 2 mm on a bright day to a maximum of about 8 mm under very dark conditions. While examining the inside of the eye, doctors often use drugs, such as *atropine*, to dilate or enlarge the pupil.

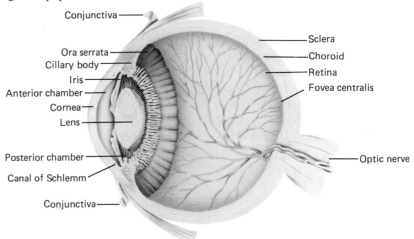

Figure 7-1 Vertical cross section of the eye. (Courtesy of Burroughs Wellcome Co.)

Immediately upon passing through the pupil, light falls on the *crystalline lens*, a transparent structure about the size and shape of a small lima bean. The lens provides the fine tuning in the final light-focusing process, changing its own shape appropriately to transform an external scene into a sharp image on the retina. The shape of the lens is controlled by the *ciliary muscle*, connected by fibers (*zonules*) to the periphery of the lens. While the muscles are relaxed, the lens assumes its flattest shape, providing the least refraction of incident light rays. In this state, the eye is focused on distant objects. When the muscles are tensed, the shape of the lens becomes increasingly curved, providing increased refraction of light. In this "strained" state, the eye is focused on nearby objects. The lens is itself a complex, onionlike layered mass of tissue, held intact by an elastic membrane. Due to the rather intricate laminar structure of fibrous tissue, the refractive index of the lens is not homogeneous. Near the center or core of the lens (on axis), the index is about 1.41; near the periphery it falls to about 1.38.

After its final refraction by the crystalline lens, light enters the *posterior chamber* or the *vitreous humor*, a transparent jellylike substance whose refractive index (1.336) is again close to that of water. The vitreous humor, essentially structureless, contains small particles of cellular debris that are referred to as *floaters*. They derive their name from the manner in which they are seen to float in one's field of view, when one looks or squints at a white ceiling, for example.

After traversing the vitreous humor, light rays reach their terminus at the inner rear layer of the eye, the *retina*, literally translated as "net." The retina, or net, is dotted with an overlapping pattern of photoreceptor cells called *rods* and *cones*. The long, thin rods, numbering over 100 million, are located more densely toward the periphery of the retina. They are exceedingly sensitive to dim light, yet they are unable to distinguish between colors. The wider cones, under 10 million in number, cluster preferentially near the center of the retina, a 3-mm diameter region called the *macula*. In sharp contrast to the response of the rods, the cones are sensitive to bright light and color but do not function well in dim light. Linked to the photoreceptor cells are three distinct types of nerve cells (amocrine, bipolar, horizontal) that transmit the visual impulse to the *optic nerve*. The optic nerve is the main trunkline that carries visual information from the retina to the brain, completing the remarkable process we call vision.

In addition to the key optical components encountered by light traveling along the axis of vision, the eye contains other components that should be mentioned. As noted in Figure 7-1, the eye is covered with a tough white coating, the *sclera*, that forms the supporting framework of the eye. Just inside the sclera lies the *choroid*, covering about four-fifths of the eye toward the back and containing most of the blood vessels that nourish the eye. The choroid, in turn, serves as the backing for the retina, the all-important net that houses the rods and cones. At the center of the macula, located somewhat above the optic nerve, is the *fovea centralis*, the region of greatest visual acuity. When it is required that one see sharp and detailed information—while removing a small splinter with a needle, for example—the eyes move continually so that light coming from the area of interest falls precisely on the fovea, a rod-free region about 200 μm in diameter. Quite by contrast, another small region in the retina, located at the point of exit of the optic nerve, is completely insensitive to light. This spot, devoid of any receptors, is appropriately called the *blind spot*.

7-2 OPTICAL REPRESENTATION OF THE EYE

As we have seen, the normal biological eye is a near spheroid, some 22 mm from cornea to retina. The optical surfaces that provide the bulk of the focusing power are essentially three: the air-cornea interface, the aqueous-lens interface, and the lens-vitreous interface. Overall, the eye can be represented, quite simply, as a thin, positive lens of focal length equal to 17 mm in the relaxed state (distant vision) or

Figure 7-2 Representation of H. V. Helmholtz's schematic eye 1, as modified by L. Laurance. For definition of symbols, refer to Table 7-1. (Adapted with permission from Mathew Alpern, "The Eyes and Vision," Section 12 in *Handbook of Optics*, New York: McGraw-Hill, 1978.)

TABLE 7-1 CONSTANTS OF A SCHEMATIC EYE

Optical surface or element	Defining symbol	Distance from corneal vertex (mm)	Radius of curvature of surface (mm)	Refractive index	Refractive power (diopters)
Cornea	S_1	—	+8[a]	—	+41.6
Lens (unit)	L	—	—	1.45	+30.5
Front surface	S_2	+3.6	+10[b]	—	+12.3
Back surface	S_3	+7.2	−6	—	+20.5
Eye (unit)	—	—	—	—	+66.6
Front focal plane	F	−13.04	—	—	—
Back focal plane	F'	+22.38	—	—	—
Front principal plane	H	+1.96	—	—	—
Back principal plane	H'	+2.38	—	—	—
Front nodal plane	N	+6.96	—	—	—
Back nodal plane	N'	+7.38	—	—	—
Anterior chamber	AC	—	—	1.333	—
Vitreous chamber	VC	—	—	1.333	—
Entrance pupil	$E_n P$	+3.04	—	—	—
Exit pupil	$E_x P$	+3.72	—	—	—

SOURCE: Adapted with permission from Mathew Alpern, "The Eyes and Vision," Table 1, Section 12, in *Handbook of Optics*, New York: McGraw-Hill Book Company, 1978.

[a] The cornea is assumed to be infinitely thin.

[b] Value given is for the relaxed eye. For the tensed or fully accommodated eye, the radius of curvature of the front surface is changed to +6 mm.

14 mm in the tensed state (near vision). In an attempt to represent the optical powers of the eye more faithfully, *schematic eyes* have been designed. While still an approximation, a schematic eye presents a fairly valid representation of the true (but complex) biological eye.

A schematic eye (after H. V. Helmholtz and L. Laurance) that represents a living, biological eye with fair accuracy is shown in Figure 7-2. Relative locations of the refracting surfaces are shown, as are the cardinal points of interest for the eye as a whole. The schematic eye shown corresponds to its relaxed state. For the fully tensed eye, the front surface of the lens sharpens its curvature from a radius of $R = 10$ mm to $R = 6$ mm. By way of summary and in conjunction with Figure 7-2, Table 7-1 lists the important optical surfaces, their distances from the corneal vertex on the optical axis, several radii of curvature, indices of refraction, and refracting powers of the optical surfaces related to the cornea and lens. Note carefully that the values for the refractive indices of various parts of the eye, as well as radii of curvature of surfaces, may not agree with values of the biological eye itself. When taken as a whole, however, the optical values that describe the schematic eye do faithfully represent the optical performance of a living, biological eye.

7-3 FUNCTIONS OF THE EYE

To operate as an effective optical system, the eye must form a retinal image of an external object or scene, either distant or nearby, in bright as well as dim light. To achieve efficient operation, the eye takes advantage of special functions. To see objects closely and far away, the eye *accommodates*. To process light signals of varying brightness, the eye *adapts*. To sense the spatial orientation of three-dimensional scenes, the eyes make use of *stereoscopic vision*. To form a faithful, detailed image of the external object, the eye relies on its *visual acuity*. In what follows, we discuss each of these visual functions in somewhat more detail.

Accommodation. Depending on the distance of the object or scene from the eye, the lens accommodates—tenses or relaxes—appropriately to fine-focus the image on the retina. For a distant object, the ciliary muscle attached to the lens relaxes and the lens assumes a flatter configuration, increasing its radii of curvature and, consequently, its focal length. As the object moves closer to the eye, the ciliary muscle tenses or contracts, squeezing or bulging the lens and resulting in decreased radii of curvature and a shorter focal length. The smaller the radii of curvature and focal length, the higher the refractive or bending power of the lens, precisely the condition needed to bring closer objects into sharp focus. In the normal eye—and before the normal aging process robs the lens of its elasticity and ability to reshape itself—accommodation produces faithful retinal images of objects from distant points (infinity) to nearby points about one foot away. The *near point* (closest point of accommodation) recedes from the eye with advancing age, starting at a position of 7 to 10 cm from the eye for a teenager, increasing to 20 to 40 cm for a middle-aged adult, and extending to as far as 200 cm in later years. For the average person, *presbyopia* (loss of accommodation) sets in during the early 40s, signaling the need for reading glasses to restore the near point to a comfortable position near 25 cm or so.

Adaptation. The ability of the eye to respond to light signals that range from very dim to very bright, a range of light intensities that differ by an astonishing factor of about 10^5, is referred to as *adaptation*. The amount of light (flux or photon number) that enters the eye is regulated first of all by the iris, with its adjustable aperture, the pupil. This adjustment of pupil diameter (from 8 mm down to 2 mm) cannot of itself account for the enormous range of intensities processed by the eye. The remarkable adaptivity of the eye is traced, in fact, to the photoreceptors in the retina, the rods and cones, and to their particular sensitivity to light. The key ingredient seems to be a pigment, called *visual pigment*, contained in both the rods and cones. The rods, stimulated by low-level light signals (*scotopic vision*), contain pigment of only one kind, called *visual purple*. The cones, sensitive to light signals of high intensity and variable color composition (*photopic vision*), each contain one of three different kinds of visual pigment. The numerous, thin rods are multiply connected to nerve fibers, making it possible for any one of a hundred rods or so to activate a single nerve fiber. The less numerous, wider cones in the macular region, by contrast, are individually connected to nerve fibers, and thus individually activated. The activation of nerve fibers—the very heart of the vision process itself—depends on chemical changes that occur in the visual pigment contained in the rods and cones. When light falls on either type of photoreceptor, the visual pigment changes from a dark state to a clearer state, undergoing a sort of bleaching process. The change in state of the visual pigment in the rods or cones is transformed into an electrical output or nerve fiber impulse. These electrical impulses are transmitted to the optic nerve and on to the brain, recording as it were the light intensity of the stimulating signal. When the visual purple is fully bleached out in the rods, the photoreceptor cells become insensitive to further light signals; a regeneration of pigment in the rods must occur before they can respond again. Apparently the single type of visual pigment in rods is much more sensitive to light than is any one of the three pigments in cones. Accordingly, rods bleach out completely at much lower light levels than do cones. A change from low-level light or scotopic vision to high-level light or photopic vision in the process of adaption consists of a rapid bleaching out of rod pigment and a resulting insensitivity of the rod receptors. The bright light is then processed efficiently by the less-sensitive cones. Conversely, adaptation from intensely bright light (handled by the cones) to very dim light involves regeneration of pigment in the rods and a restoration of "night vision." In the full process of adaptation, the scotopic response is active over light levels that range from starlight on a clear, moonless night to lunar light from a quarter-moon. The photopic response (rods completely bleached out and inactive) operates between light levels ranging roughly from twilight to bright sunlight. Between light levels of quarter-moon and twilight, rods and cones both receive light and transmit nerve impulses.

Stereoscopic Vision.   The ability to judge depth or position of objects accurately in a three-dimensional field is called *stereoscopic vision*. In humans, the optic nerves from the two eyes come together at the *optic chiasma*, near the brain. From the optic chiasma, nerve fibers originating in the right half of each eye extend to the right half of the brain. Nerve fibers originating in the left half of each eye terminate in the left half of the brain. Thus, even though each half of the brain receives an image from *both* eyes, the brain forms but a single image. The fusion by the brain of two distinct images into a single image is referred to as *binocular vision*. Nevertheless, the slight differences between the two images from the left and right eyes provide the basis for stereoscopic vision in humans. It should be noted that even monocular vision is not without some depth perception. This is due to visual clues like parallax, shadowing, and the particular perspective of familiar objects.

To have proper binocular vision without *double vision*, the images of an object must fall at corresponding points on each retina. This, of course, is what happens when the eyeballs move appropriately to focus on an object or scene, causing the image to fall on the fovea centralis of each eye. Most individuals are either right-eyed or left-eyed, indicating a dominance of one eye over the other. To determine which is your dominant eye, try the following simple test. Hold a pencil in front of you at eye level. With both eyes open, line the pencil up with the vertical edge of a picture, door, or window across the room. Holding the pencil fixed, close one eye at a time. Whichever eye is open when the pencil remains lined up with the reference object is your dominant eye. The brain records the message seen by the dominant eye, while suppressing the other.

Visual Acuity. The ability to see clearly and to perceive real differences in spatial orientation of objects is related to *visual acuity*. This ability depends directly upon the resolving power of the eye or its minimum angle of resolution of two closely spaced objects or points. Technically speaking, visual acuity is defined as the reciprocal of the minimum angle of resolution.

Operationally, assessment of resolving power or visual acuity of the eye is measured in different ways. Two-point discrimination is referred to as *minimum separable* resolution; the smallest resolvable angle subtended by a black bar on a white background is called *minimum visible*, and the smallest angle subtended by block letters that can be read (on an eye chart) is called *minimum legible*. Since most of us, at one time or another, are required to read eye charts in a vision test, we limit our discussion of visual acuity to resolving powers associated with minimum legible resolution.

The nature of the eye chart owes its existence to a Dutch ophthalmologist, Herman Snellen. According to Snellen, the letters on the eye chart are constructed so that the overall block size of a letter, from top to bottom, or side to side, subtends an angle of 5′ of arc at the test distance. The detailed lines within a letter, such as the vertical bar in the letter T or the horizontal bar in the letter H, are all constructed so that the width of each "bar" subtends an angle of 1′ of arc at the test distance. The two choices of angle grew out of the best data available to Snellen on the minimum separable resolution of the eye. For Snellen, the normal eye could just resolve a letter that subtended 5′ of arc at 20 ft, with 1′ of arc contained in the details of the letter. See Figure 7-3. In this case the eye is considered "normal," and its visual acuity is referred to as "20/20 vision."

Figure 7-3 Construction of a Snellen eye-chart letter H to measure visual acuity. The top portion of the figure shows a section of an eye chart (reduced) containing the letter H and several other letters.

To detect defects in visual acuity, Snellen letters of different sizes are also included on the eye chart. For example, a very large letter may be such that its subtense angles of 5′ and 1′ of arc hold for a test distance of, say, 300 ft. Other letters are constructed of appropriate size, subtending angles of 5′ and 1′ for other selected distances, such as 200 ft, 100 ft, 80 ft, and so on, down to 15 or even 10 ft. Then, when the letters are read by a test subject at a test distance of 20 ft, visual acuity is measured in terms of the *Snellen fraction*. The numerator of the Snellen fraction expresses the fixed testing distance, and the denominator expresses the distance at which the smallest readable letter subtends 5′ of arc overall. For example, if the large block letter E that subtends an angle of 5′ at 300 ft is just readable by a test subject seated 20 ft from the letter, visual acuity is reported as 20/300. A Snellen fraction of 20/300 means that the test subject sees poorly, reading at a distance of 20 feet what the normal eye reads as well at a distance of 300 ft. While normal vision is 20/20, visual acuity readings as good as 20/15 are not uncommon.

7-4 ERRORS OF REFRACTION AND THEIR CORRECTION

The errors of refraction of the eye lead to three well-known defects in vision: nearsightedness (*myopia*), farsightedness (*hyperopia*), and *astigmatism*. The first two are traceable, for the most part, to an abnormally shaped eyeball, axially too long or too short. Either deviation from normal length impairs the ability of the combined refracting elements, cornea and lens, to form a clear image of objects at both remote and nearby positions. The third defect, astigmatism, is due to unequal or asymmetric curvatures in the corneal surface, thereby rendering impossible the simultaneous focusing of all light incident on the eye. Whether the errors of refraction occur singly or in some combination (as they usually do), they are generally correctable with appropriately shaped external optics (eyeglasses).

As a point of reference for judging the departure of defective vision from the norm, refer to the *normal* eye depicted to the left in Figure 7-4. With accommodation, the normal eye forms a distinct image of objects located anywhere between its far point (F.P.) at infinity and its near point (N.P.), nominally a distance of 25 cm for the young adult. When the normal eye is focused at infinity (distant objects), parallel light enters the relaxed eye and forms a distinct image (Figure 7-4a). When focused at the near point, diverging light enters the tensed eye (fully accommodated) and is again brought to sharp focus on the retina (Figure 7-4b).

Myopia.   When compared with the normal eye, a myopic eye or nearsighted eye is commonly found to be longer in axial distance—from cornea to retina—than the usual, accepted span of 22 mm. As a consequence, and as illustrated schematically in Figure 7-4c, the myopic eye forms a sharp image of distant objects in *front* of the retina, and, of course, a blurred image at the retina. Distinct retinal images are not formed with the unaccommodated myopic eye until the object moves in from infinity and reaches the myopic far point, the most distant point for clear vision (Figure 7-4d). From the far point inward, with appropriate accommodation, the myopic eye sees quite clearly, even at points *closer* than the normal near point (Figure 7-4e). Since angular magnification of detail increases with proximity to the eye, the myopic eye enjoys "superior" vision of objects held close to the eye. (It can be an advantage for a watchmaker to be myopic, therefore, at least during working hours!) In short then, the nearsighted person has a contracted, drawn-in field of vision, a less-remote far point and a closer near point than a person with normal vision. While the more proximate near point might serve as an advantage, the less-remote far point is a distinct disadvantage and calls for correction.

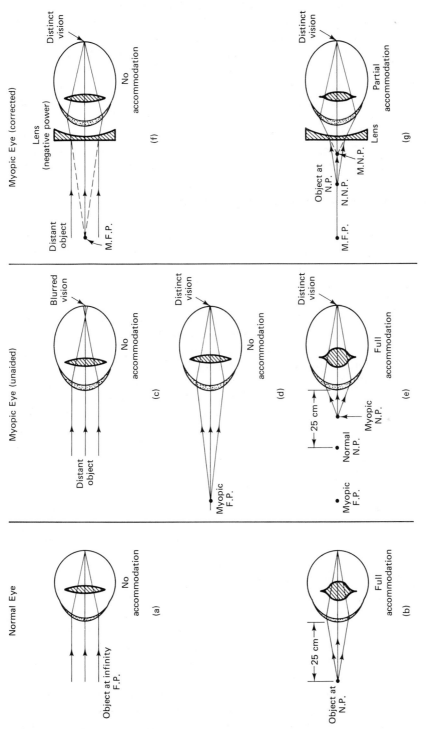

Figure 7-4 A comparison of normal and myopic vision, with optical correction. Note that refraction by the eye lens is not shown. The abbreviations read as follows: M.F.P. = myopic far point; N.P. = normal near point; N.N.P. = normal near point; M.N.P. = myopic near point.

159

Myopic vision is routinely corrected with spectacles of negative dioptic power (diverging lenses) that effectively move the myopic far point and near point outward to normal positions. Figure 7-4f shows the corrected vision for distant objects. Note that as far as the optics of the eye itself is concerned, light from distant objects appears to originate at its own myopic far point. Similarly, Figure 7-4g illustrates the situation for corrected near vision under partial accommodation. Light from an object at the normal near point appears now to originate at a point somewhat closer in than the true near point of the myopic eye.

To gain some insight into the degree of negative lens power required to correct myopic vision, consider the following example.

Example

A myopic person (without astigmatism) has a far point of 100 cm and a near point of 15 cm. (a) What correction should an optometrist prescribe to move the myopic far point out to infinity? (b) With this correction, can the myopic person also read a book held at the normal near point, 25 cm from the eye?

Solution (a) Referring to Figure 7-4f and making use of the thin-lens equation, with $s \to \infty$ and $s' = -100$ cm, one has

$$\frac{1}{s} + \frac{1}{s'} = \frac{1}{f} \quad \text{or} \quad \frac{1}{\infty} + \frac{1}{-100} = \frac{1}{f}$$

This gives $f = -100$ cm. Accordingly, the corrective spectacle lens should have a focal length of -100 cm. The optometrist would prescribe spectacles with a correction of -1.00 diopter (-1.00 D).

(b) Referring to Figure 7-4g and again using the thin-lens equation with $f = -100$ cm and $s = 25$ cm, the virtual image s' is found from

$$\frac{1}{s} + \frac{1}{s'} = \frac{1}{f} \quad \text{or} \quad \frac{1}{25} + \frac{1}{s'} = \frac{1}{-100}$$

Solving gives $s' = -20$ cm. Thus the virtual image of the print held at $s = 25$ cm is formed by the spectacles at a distance of 20 cm in front of the eye. Since the myopic person can see clearly objects brought in as close as 15 cm from the eye, the virtual image of the print formed by the lens at 20 cm is seen without difficulty. In fact, using $(1/s) + (1/s') = 1/f$, with $s' = -15$ cm (myopic near point for person) and $f = -100$ cm, solving for s, one finds that objects can be brought in as close as 17.6 cm from the eye and still be seen clearly.

Hyperopia. The farsighted, or hyperopic, eye is commonly shorter than normal. Whereas the longer-than-normal myopic eye has too much convergence in its "optical system" and requires a diverging lens to correct its over-refraction, the shorter-than-normal hyperopic eye has too little convergence and requires a converging lens to increase refraction. The drawings in Figure 7-5, in analogy with those in Figure 7-4, illustrate the defects and correction associated with the farsighted eye. In Figure 7-5a, light from a distant object enters the relaxed eye and focuses *behind* the retina, causing blurred vision. The focal point behind the retina is considered as the hyperopic far point. Figure 7-5b shows that the hyperopic eye must (and can) accommodate to see distant objects clearly. In Figure 7-5c, it is clear that for distinct vision, the hyperopic near point is farther away than the normal near point. Consequently, objects located closer than the hyperopic near point would be out of focus, even with full accommodation. The two end-point corrective meas-

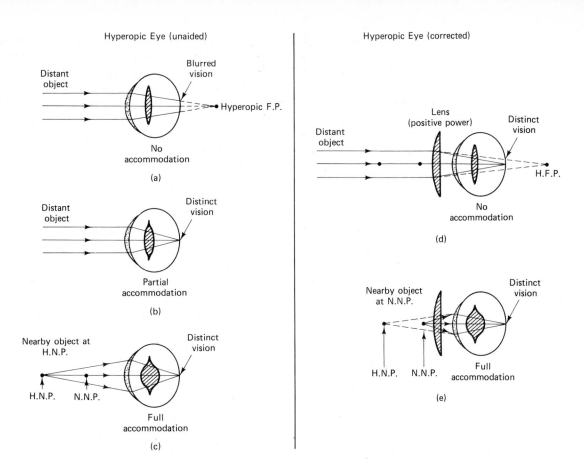

Figure 7-5 Hyperopic vision with correction. The abbreviations read as follows: H.N.P. = hyperopic near point; N.N.P. = normal near point; H.F.P. = hyperopic far point.

ures, with the appropriate positive spectacle lens in place, are indicated in Figure 7-5d and e. The corrected eye now sees distant objects clearly, without accommodation, and at the normal near point it sees objects clearly, with full accommodation.

Let's see how an optometrist might calculate the spectacle power required to correct hyperopic vision. Consider the following example.

Example

A farsighted person is diagnosed to have a near point at 150 cm. What corrective power is required for spectacles to enable this person to see objects brought in to the normal near point, 25 cm from the eye?

Solution Referring to Figure 7-5e and making use of the thin-lens equation, with $s = 25$ cm and $s' = -150$ cm, one can solve for the focal length of the required lens as follows:

$$\frac{1}{s} + \frac{1}{s'} = \frac{1}{f} \quad \text{or} \quad \frac{1}{25} + \frac{1}{-150} = \frac{1}{f}$$

Calculation gives $f = +30$ cm. Since the power in diopters is given by $1/f$ (meters), we determine a power of 3.3 D. With spectacles shaped to a power of $+3.3$ diopters, the hyperopic person can now see clearly objects brought in as close as 25 cm from the eye.

Astigmatism. The astigmatic eye suffers from uneven curvature in the surface of the refracting elements, most significantly, the cornea. Generally speaking, the radii of curvature of the corneal surface in two meridional planes (those containing the optic axis) are unequal. Such asymmetry leads to different refractive powers and, consequently, to focusing of light at different distances from the cornea, resulting, of course, in blurred vision. If the two planes are orthogonal to one another, one horizontal and the other vertical, say, the defect is referred to as *regular astigmatism*, a condition that is correctable with appropriate spectacles. If the two planes are not orthogonal, a rather rare condition called *irregular astigmatism*, the surface anomaly is not so easily corrected. Regular astigmatism is treated with cylindrical surfaces ground on the back surface of the required spectacle lens. Assume, for example, that the refractive power in the vertical meridian of the cornea is greater by 1 diopter than the power in the horizontal meridian. This situation means that the corneal surface is more sharply curved in the vertical meridian and that vertically oriented details in an object are brought to a focus nearer the cornea than are horizontally oriented details. Consider a cylindrical surface with negative power of 1 diopter in the vertical meridian. Since a cylinder has no curvature along its axis, the surface has no power in the horizontal meridian. If this surface is included in the spectacle design, it would cancel exactly the distortion introduced by the cornea and equalize powers in both meridians. As a result, vertical and horizontal details in the object scene are formed at the same distance from the cornea, and astigmatic blurring does not occur.

For most of us, blurred vision is a result of astigmatism mixed with myopia or hyperopia. If myopic astigmatism is present, for example, vision is faulty on two counts. The myopia itself causes an overall blurring of distant objects; the astigmatism compounds the problem by adding considerably more blurring in one meridian than another. Correcting for both defects is accomplished with sphero-cylindrical lenses, spherical surfaces to correct for myopia and cylindrical surfaces to correct for astigmatism.

When optometrists prescribe corrective eyeglasses for conditions of myopic or hyperopic astigmatism, they generally identify three numbers. For myopic astigmatism, the three numbers, written in prescription format, might be

$$R_x: \quad -2.00 \quad -1.00 \quad \times 180$$

For hyperopic astigmatism, the prescription might read

$$R_x: \quad +2.00 \quad -1.50 \quad \times 180$$

The first number refers to the *sphere power*, the required power in diopters of the spherical surfaces on the spectacle lens that correct for the overall myopia or hyperopia. The second number refers to the *cylinder power*, the required power of the cylindrical surface superimposed on the back surface of the spectacle lens to correct for astigmatism. The third number refers to the orientation of the *cylinder axis*, specifying whether the axis of the cylinder is to be vertical, horizontal, or somewhere in between. In optometric notation, the horizontal axis is referred to as the 180° axis, or simply "$\times$ 180," and the vertical axis as "$\times$ 90."

Figure 7-6 indicates the optical conditions associated with the corrective prescriptions just cited for both myopic and hyperopic astigmatism. For the case of myopic astigmatism, Figure 7-6a, the corneal surface is evidently less sharply curved in the horizontal meridian (power = 45.00 D) than in the vertical (power = 46.00 D). The myopic correction, always measured in the meridian of least refractive power, is found in this instance to be -2.00 D, along the horizontal meridian. The astigmatic correction, with cylinder axis horizontal ($\times$ 180), is determined to be -1.00 D.

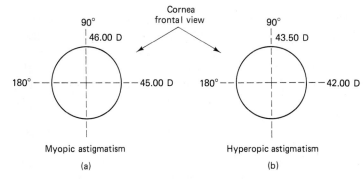

Figure 7-6 Conditions of myopic and hyperopic astigmatism with corrective spectacle prescriptions. (a) Refraction in the 180° meridian yields −2.00 D of myopia. The eyeglass prescription is R$_x$: −2.00 −1.00 ×180. (b) Refraction in the 180° meridian yields +2.00 D of hyperopia. The eyeglass prescription is R$_x$: +2.00 −1.50 ×180.

With the appropriate cylindrical surface ground on the rear of the spectacle lens, the correction of −1.00 D reduces the power in the vertical meridian from 46.00 D to 45.00 D, thereby equalizing the refracting powers in the two meridians and negating the corneal astigmatism.

Figure 7-6b shows a comparable condition and prescription for hyperopic astigmatism. Note that a sphere power correction of +2.00 D is needed to correct for the hyperopia, and a cylinder power correction of −1.50 D is needed along the vertical meridian (× 180) to equalize the refractive power in the two orthogonal meridians.

7-5 LASER THERAPY FOR OCULAR DEFECTS

Not all ocular problems can be corrected optically with appropriate eyeglasses, as are the structural defects we have just discussed. There are organic disorders that require radical treatments, often involving surgery. The laser, as discussed in Chapter 21, has emerged as a powerful tool in the operating room, used successfully in the treatment of major ocular defects.

Laser beams, both pulsed and continuous wave, are currently used to treat glaucoma, retinal bleeding, macular degeneration, retinal detachments, and opacified intraocular membranes. *Angle blockage glaucoma*, a disease of the eye characterized by increased fluid pressure within the eye and leading progressively to blindness, is treated with argon (A) and neodymium:yttrium aluminum garnet (Nd:YAG) lasers. The treatment consists of repairing structural flaws in the eye that cause increased pressure, using the laser beam to open up blocked ducts or to create new canals for better drainage between the chambers of the eye. The laser is also used effectively to treat *diabetic retinopathy*, the leading cause of blindness. To neutralize such organic disorders, the thermal energy in the laser beam is used to place thousands of tiny burns or welds at the back of the retina, thereby preventing the harmful growth or rupture of new, unwanted blood vessels (*neovascularization*). The thermal content of a laser beam, so effective where retinal coagulation is required, is also used to tack or weld retinas that fall away from the choroid at peripheral positions. This procedure, in fact, was the first successful use of lasers in a clinical environment.

Lasers with wavelength emission in the deep ultraviolet, principally the excimer lasers, offer great promise for reshaping the eyeball to correct myopic vision (*radial keratotomy*) and for removing *cataracts*, an organic disorder characterized by pockets of cloudy or opaque discoloration in the lens tissue of the eye, resulting in significantly degraded vision. Currently, the most popular use of lasers in ocular therapy involves the Nd:YAG laser in a procedure that ruptures internal ocular membranes. Following cataract surgery—removal of the cataractous lens and replacement with a plastic implant lens—certain membranes that help hold the new lens in place become opacified and effectively block light along the axis of vision. In a corrective procedure referred to as *posterior capsulotomy*, a high-power laser beam is focused to a small point near the opacified membrane. Due to the enormous power densities generated, dielectric breakdown of the optical medium, followed by acoustical shock waves, occurs. The overpressures associated with the latter rupture the membrane, allowing light to pass and vision to be restored. The procedures we have described under the headings of radial keratotomy and posterior capsulotomy are treated in greater detail in the sections that follow.

Radial Keratotomy. The eye-shaping procedure referred to as radial keratotomy introduces radial cuts in the cornea of the elongated, myopic eyeball (see Figure 7-7). After the cuts have healed, the cornea flattens, thereby reducing the axial length of the eye. As a result, normal, or near-normal, vision is restored. This radical procedure, done mostly with a surgical blade in the hands of a skilled ophthalmologist, had its beginning in the Soviet Union in 1972. As the story is told, a rather myopic Soviet lad, Boris Petrov, was engaged in a schoolyard fight in Moscow. In the usual exchange, a hard punch struck one of the thick lenses he was wearing, shattering it into multiple fragments. Some of the shards of glass, in one of those freak accidents of nature, embedded themselves in the lad's cornea in a somewhat regular, radial pattern, deep enough but still not penetrating the cornea. A Soviet ophthalmologist, Svyatoslav N. Fyodorov, who treated the youngster, did not hold out much hope for restored vision in the eye, even though the cuts were superficial. Astonishingly, though, as the corneal surface healed with all its scars, the cornea flattened out and most of the myopia disappeared. The lad saw better than ever before. Recognizing the significance of what he had witnessed, Fyodorov decided to replicate, under controlled conditions, what nature and a fist had accomplished so haphazardly. This he did, many times, in the procedure that has come to be known as radial keratotomy. This very procedure has been accompanied by unusual notoriety in the United States. It has been (and is now) performed successfully with a guarded surgical blade. Short-term results have been dramatic; nevertheless,

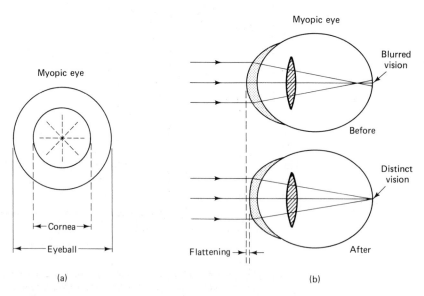

Figure 7-7 Removing myopia by reshaping the cornea with radial keratotomy. (a) Frontal view showing radial cuts. (b) Side view before and after cuts.

insufficient time has elapsed to guarantee beneficial, long-term results. Indeed, in some cases, myopia has been cured but replaced, unfortunately, with astigmatism. As a result, the procedure is currently carried out on an experimental basis.

As far as the laser is concerned, the challenge in radial keratotomy lies in using a powerful laser, sharply focused, in place of the surgical blade. Since the cornea contains abundant amounts of water, it has a very high absorption coefficient for infrared radiation at 10.6 μm (carbon dioxide CO_2 laser) and ultraviolet radiation below 400 nm. Figure 7-8 indicates the relative absorption of laser light at several wavelengths in 1 mm of both water (dashed curve) and hemoglobin (solid curve). Note that at 10.6 μm, the absorption coefficient is almost 100%, validating the claim that CO_2 laser radiation is strongly absorbed by the cornea. (Incidentally, note also that the argon ion and frequency-doubled Nd:YAG laser radiation near 500 nm is strongly absorbed in hemoglobin, making either one an excellent candidate for photo-coagulative treatments.) The powerful laser beam, almost totally absorbed by the corneal tissue, etches into the cornea, creating a neat cut or trough about the width of the laser beam. The surgical blade currently used in radial keratotomy is around 50 to 100 μm wide. CO_2 lasers can be focused down to less than 50 μm in width and excimer lasers, even narrower. Figure 7-9 shows a histologic section ($\times$ 100) of a laser cut made on the cornea of a cow with a CO_2 laser. With average power of 0.5 W focused to a beam width below 50 μm, the laser, after 20 passes, made an incision of about 60 μm wide and over 400 μm deep.

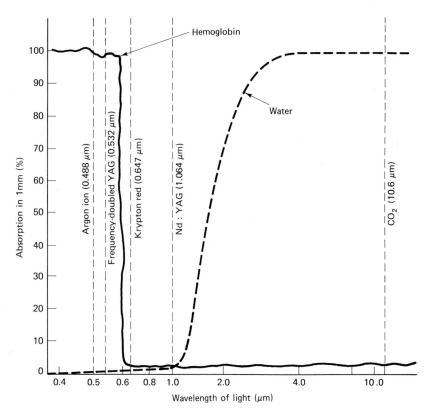

Figure 7-8 Absorption of laser light in ocular tissue for several important lasers. The percentage absorption is for 1 mm of water (dashed curve) and 1 mm of hemoglobin (solid curve). Note that radiation at 10.6 μm is almost totally absorbed in the cornea, whereas that at 1.06 μm is almost totally transmitted.

Figure 7-9 Histologic section (×100) of laser cut in the cornea of a cow. Smallest scale division is equal to 5 μm. (Reprinted from *Ophthalmic Surgery* 12, no. 2 (February 1981): 120. Copyright, 1981.)

If, then, the width of the laser beam can be adjusted and the beam power controlled, it should be possible to make clean, radial cuts of the desired width (50 μm) and depth (about 400 μm) in the corneal surface. Such an achievement, not beyond reality with microprocessor control, would produce precisely the number and nature of cuts required to remove myopia and prevent the unwanted byproduct of astigmatism. If so, individuals with myopia can look forward to correction with eyeglasses or "permanent" correction with a reshaped eyeball.

Posterior Capsulotomy. In a somewhat different procedure, identified as a posterior capsulotomy, a Nd:YAG laser is used to rupture unwanted, opacified membranes along the optical axis of the eye. The pulsed Nd:YAG laser emits a 1.06-μm laser beam that passes through the cornea, aqueous, and lens with essentially zero absorption. (See Figure 7-8 on percentage of absorption at 1.06 μm). The high-intensity laser beam penetrates the front part of the eye and comes to a sharp focus just beyond the implanted lens, some 4 mm from the front corneal surface (see Figure 7-10). Located there is the opacified membrane, the posterior segment of a weblike pillow (capsule) that previously contained the cataractous lens and now encases the implant.

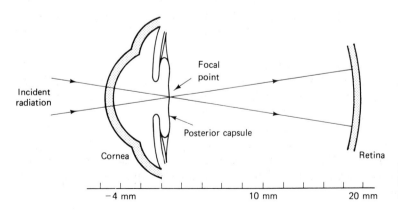

Figure 7-10 Side view of Nd:YAG laser beam focused on posterior capsule. (Reproduced with permission of Slack, Inc.)

Prior to laser surgery, the opacified membrane was removed invariably with invasive intravention by surgical instruments. In addition to the trauma involved with the operation, invasive surgery of the eyeball is always attended by possible introduction of foreign bodies and increased risk of infection. By contrast, the Nd:YAG laser surgery, performed on an outpatient basis in a matter of minutes, is neither traumatic nor infectious. Noninvasive laser intervention grew largely from the successful procedures developed by Daniele Aron-Rosa and coworkers at Trousseau Hospital in Paris and by Franz Fankhauser and colleagues at the University Eye Clinic in Bern, Switzerland. Posterior capsulotomy, as currently carried out both in the United States and abroad, focuses a short pulse of 1.06-μm laser radiation at the target. The typical pulse delivered by a Q-switched Nd:YAG laser has an energy of 1 to 4 mJ and a pulse length of several nanoseconds. With a mode-locked laser, the pulse energy is about the same, but the pulse length is much shorter, only tens of picoseconds in duration. With such short times, the laser power, even for pulse energies as small as millijoules, reaches the megawatt range and higher.

Example

Determine the irradiance (power per unit area) delivered by a Nd:YAG laser pulse of energy 4 mJ and pulse length of 1 ns when the pulse is focused on target to a tiny spot of diameter 30 μm.

Solution The irradiance E_e is calculated directly as follows:

$$E_e = \frac{P}{A} \qquad \text{where } P = \frac{4 \times 10^{-3} \text{ J}}{1 \times 10^{-9} \text{ s}} = 4 \times 10^6 \text{ W}$$

and

$$A = \frac{\pi D^2}{4} = \frac{(3.14)(30 \times 10^{-6})^2}{4} = 7.065 \times 10^{-10} \text{ m}^2$$

Thus

$$E_e = \frac{4 \times 10^6 \text{ W}}{7.065 \times 10^{-10} \text{ m}^2} = 5.7 \times 10^{15} \text{ W/m}^2$$

or

$$E_e = 5.7 \times 10^{11} \text{ W/cm}^2$$

As the example shows, when such high powers are focused sharply at the target in tiny spots ranging from 5 to 50 μm in diameter, irradiances, or power densities, of 10^{12} W/cm^2 are developed. Power densities of this magnitude are accompanied by very high electric fields that cause, first, a dielectric breakdown of optical tissue and, second, the formation of a plasma. The explosive growth of the plasma gives rise to a strong shock wave that travels radially outward, mechanically rupturing the nearby taut, opacified membrane. Neither the laser radiation, which moves on toward the retina, or the expanding shock wave causes damage elsewhere in the eye.

An expanded view of these details is shown in Figure 7-11. The laser beam incident on the focusing lens has been beam-expanded by prior optics so that its beam divergence is very small (the beam is highly collimated). Because of this and the ability to focus highly collimated, coherent laser beams onto very small target areas, the beam converges to a "point" near the opacified membrane, producing the dielectric breakdown that ultimately leads to mechanical rupture of the membrane. Figure 7-12 shows a typical Nd:YAG laser system used in membrane rupture surgery. Figure 7-13 shows the orientation of patient, ophthalmologist, and laser system during a posterior capsulotomy.

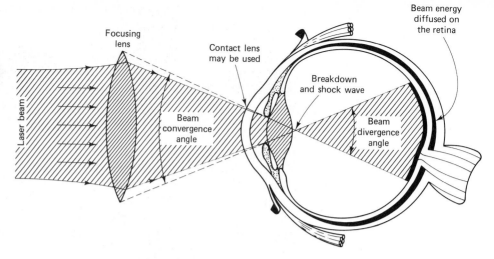

Figure 7-11 Focusing laser light to rupture membranes in posterior capsulotomies. The larger the beam convergence angle (about 17°)—limited by the pupillary aperture—the larger the beam divergence angle and the less the effect of laser radiation on the retina.

Figure 7-12 Typical Nd:YAG laser-ophthalmic slit lamp system used to perform a capsulotomy. (Reproduced by permission from *Lasers and Applications*, November 1984.)

Figure 7-13 Laser is used in conjunction with an ophthalmic slit lamp to rupture the opacified membrane. (Reproduced by permission from *Lasers and Applications*, November 1984.)

PROBLEMS

7-1. Reference to Table 7-1 indicates that the corneal radius of curvature for the unaccommodated schematic eye is 8 mm. Treating the cornea as a thin surface (whose own refraction can be neglected), bounded by air on one side and aqueous humor on the other, determine the refractive power of the corneal surface. (Note that the stated refractive power is by convention that corresponding to the shorter of its focal lengths.)

7-2. Consider the unaccommodated crystalline lens of the eye as an isolated unit, having radii of curvature and effective refractive index as given for the schematic eye in Table 7-1.

 (a) Calculate its focal length and refracting power as a thin lens in air.

 (b) Calculate its focal length and refracting power in its actual environment, surrounded on both sides with fluid of effective index 1.33. Assume a thin lens.

 (c) Calculate its focal length and refracting power again, by treating it as a thick lens, of thickness 3.6 mm. (The matrix techniques of Chapter 4 may well be applied to this problem).

7-3. Taking values for refractive indices and separation of elements from the schematic unaccommodated eye given in Table 7-1 and Figure 7-2, determine the distance behind the cornea where an image is focused for (a) an object at infinity and (b) an object at 25 cm from the eye. Use the Gaussian formula for image formation by a spherical surface in a three-step chain of calculations. In part (b), assume that the fully accommodated eye differs in the following ways: The front surface of the lens is more sharply curved, having a radius of +6 mm, but the back surface remains at −6 mm. As a result, the thickness of the lens along the axis increases to 4.0 mm, and the distance from cornea to the front surface of lens is shortened to 3.2 mm.

7-4. Use the matrix approach to find the system matrix for the unaccommodated schematic eye of Table 7-1 and Figure 7-2.

(a) Determine the four matrix elements of the system matrix, where the system extends from the first refraction at the cornea to the final refraction at the second lens surface.

(b) From the matrix elements, determine the first and second focal points and the first and second principal points, relative to the corneal surface. Compare with the distance given in Figure 7-2.

7-5. You have been asked to design a Snellen eye chart for a test distance of 5 ft. The chart is to include rows of letters to test for visual acuities of 20/300 (same as 5/75), 20/100, 20/60, 20/20, and 20/15. Determine the size of the block letter and letter detail (in inches) for each row of letters.

7-6. A hyperopic person has no astigmatism but has a near point of 125 cm. Correction with glasses requires that this person see objects at the normal near point (25 cm) clearly.

(a) What is the power of the corrective lens?

(b) Will the corrective glasses enable focusing of a distant object on the retina?

7-7. A person has a far point of 50 cm and a near point of 15 cm. What power eyeglasses is needed to correct the far point? Using the eyeglasses, what is the person's new near point?

7-8. From an examination of Figure 7-8, determine which lasers would be suitable for (a) photocoagulation of bleeding vessels on the retina; (b) thermal cutting of corneal layers; (c) focusing of light energy in the vitreous chamber without absorption during passage through the cornea, aqueous humor, lens, and associated blood vessels.

7-9. Consider each of the following spectacle prescriptions and describe the refractive errors that are involved:

(a) −1.50, −1.50, axis 180

(b) −2.00

(c) +2.00

(d) +2.00, −1.50, axis 180

7-10. The CO_2 laser used to make the corneal incision shown in Figure 7-9 has an average power of 5 W and a beam divergence of 2.2 mrad. After emerging from the laser, it is sent through a 5× beam expander and then focused onto the cornea by a 3.3-cm focal length germanium lens.

(a) Why is germanium (and not glass) used as a lens material?

(b) In Section 21-4, where *laser focusability* is discussed, it is shown that the angular beam spread (divergence) of an expanded laser beam is equal to the divergence of the incident beam divided by the beam expansion factor. What then is the beam divergence of this CO_2 laser beam after passing through the 5× beam expander?

(c) Using the approximate formula, $D = f\phi$, where f is the focal length and ϕ is the beam divergence of the expanded beam, determine the diameter D of the focal spot on the cornea. (Refer to Section 21-4 for a discussion of this equation.)

(d) What is the power density (irradiance) of the focused CO_2 laser beam on the cornea?

7-11. For the posterior capsulotomy surgery described in this chapter, the following data are typical:

      Laser: Nd:YAG
      Wavelength: 1.06 μm
      Pulsewidth: 10 ns
      Energy per pulse: 10 mJ
      Beam divergence at focusing lens: 0.1 mrad
      Power of focusing lens: 20 D

(a) What is the average power per pulse?

(b) What is the spot size of the focused Nd:YAG laser beam at the opaque membrane in the interior of the eye? (See problem 7-10c.)

(c) Assuming that none of the incident power is lost, what is the beam irradiance on target?

7-12. When working with lasers, one is admonished to "never stare into the beam." With low-power helium-neon (He-Ne) lasers (milliwatt level), one might be tempted to consider this caution somewhat loosely. Even though the power in a laser beam may be fairly small, the focusing action of the eye can concentrate the laser power onto the retinal surface to create a large irradiance level. To illustrate this effect, consider a 4-mW He-Ne laser that emits a collimated beam of 7 mm diameter at 632.8 nm, with a beam divergence of 1.5 mrad.

(a) Calculate the irradiance of the beam, in W/cm^2, just outside the laser.

(b) Assume that the entire beam enters a dark-adapted eye (pupil diameter of 7 mm) that is staring at the beam and focused at infinity. Treating the eye as a simple thin lens of 17 mm focal length, determine the irradiance of the focused spot on the retinal surface. (Refer to problem 7-10c for help in calculating the size of the focused spot.)

(c) By approximately what factor does the focusing action of the eye increase the irradiance? What do you conclude about the seriousness of the admonition?

7-13. Laser protective eyewear filters are used by personnel who operate laser beams. The optical density (OD) for a specific filter is given by the formula OD = $\log_{10}(E_p/\text{MPE})$, where E_p is the energy density (J/cm^2) in the laser pulse and MPE is the "maximum permissible exposure" to the eye, also in units of energy density. Suppose you are to determine an appropriate filter for use with a Nd:YAG laser (1.06 μm) that emits single laser pulses of 80 mJ. For safety reasons, this laser is rated as a Class IV laser with an MPE of 5.0×10^{-6} J/cm^2.

(a) Determine the energy density in J/cm^2 of the 80-mJ laser emitting a 7-mm diameter laser pulse.

(b) What should be the OD of the protective filter that absorbs 1.06 μm radiation?

REFERENCES

[1] Alpern, Mathew. "The Eyes and Vision." In *Handbook of Optics*, edited by Walter G. Driscoll and William Vaughan. New York: McGraw-Hill Book Company, 1978.

[2] Duke-Elder, S., and D. Abrams. *Ophthalmic Optics and Refraction*. Vol. 5 of *Systems of Ophthalmology*, edited by S. Duke-Elder. St. Louis: C. V. Mosby Company, 1970.

[3] Michaels, D. D. *Visual Optics and Refraction*, 2d ed. St. Louis: C. V. Mosby Company, 1980.

[4] Rubin, M. L. *Optics for Clinicians*, 2d ed. Gainsville, Fla.: Triad Scientific Publishers, 1974.

[5] Fincham, W. H. A., and M. H. Freeman. *Optics*, 9th ed. Boston: Butterworth Publishers, 1980. Ch. 20.

[6] Feynman, Richard P., Robert B. Leighton, and Matthew Sands. *The Feynman Lectures in Physics*, vol. 1. Reading, Mass.: Addison-Wesley Publishing Company, 1975. Ch. 35, 36.

[7] Sperry, R. W. "The Eye and the Brain." *Scientific American* (May 1956): 48.

[8] Milne, L. J., and M. J. Milne. "Electrical Events in Vision." *Scientific American* (Dec. 1956): 113.

[9] Michael, Charles R. "Retinal Processing of Visual Images." *Scientific American* (May 1969): 104.

[10] Sliney, D. H., and M. L. Wolbarsht. *Safety with Lasers and Other Optical Sources: A Comprehensive Handbook*. New York: Plenum Press, 1980.

8

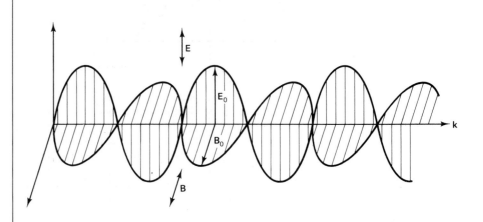

Wave Equations

INTRODUCTION

In this chapter we develop mathematical expressions for wave motion in general but concentrate on the most useful special case, the harmonic wave. Harmonic wave functions are then specified further to represent electromagnetic waves, which include light waves. Results from electromagnetism describing the physics of electromagnetic waves are borrowed to enable a determination of the energy delivered by such waves.

8-1 ONE-DIMENSIONAL WAVE EQUATION

The most general form of a traveling wave, and the differential equation it satisfies, can be determined in the following way. Consider first a one-dimensional wave pulse of arbitrary shape, described by $y' = f(x')$, fixed to a coordinate system $O'(x', y')$, as in Figure 8-1a. Consider next that the O' system, together with the pulse, moves to the right along the x-axis at uniform speed v relative to a fixed coordinate system, $O(x, y)$, as in Figure 8-1b. As it moves, the pulse is assumed to maintain its shape. Any point on the pulse, such as P, can be described by either of two coordinates, x or x', where $x' = x - vt$. The y-coordinate is identical in either

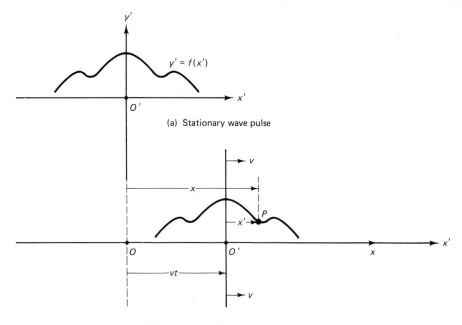

(a) Stationary wave pulse

(b) Wave pulse translating at constant speed

Figure 8-1   Translating wave pulses.

system. From the point of view of the stationary coordinate system, then, the moving pulse has the mathematical form

$$y = y' = f(x') = f(x - vt)$$

If the pulse moves to the left, the sign of v must be reversed, so that in general we may write

$$y = f(x \pm vt) \qquad (8\text{-}1)$$

as the general form of a traveling wave. Notice that we have assumed $x = x'$ at $t = 0$. The original shape of the pulse, $y' = f(x')$, does not vary but is found simply translated along the x-direction by the amount vt at time t. The function f is any function whatsoever, so that for example,

$$y = A \sin (x - vt)$$
$$y = A (x + vt)^2$$
$$y = e^{(x-vt)}$$

all represent traveling waves. Only the first, however, represents the important case of a periodic wave.

We wish to find next the partial differential equation that is satisfied by all such periodic waves, regardless of the particular function f. Since y is a function of two variables, x and t, we use the chain rule of partial differentiation and write

$$y = f(x')$$

where

$$x' = x \pm vt$$

so that

$$\partial x'/\partial x = 1 \quad \text{and} \quad \partial x'/\partial t = \pm v$$

Employing the chain rule, the space derivative is

$$\frac{\partial y}{\partial x} = \frac{\partial f}{\partial x'} \frac{\partial x'}{\partial x} = \frac{\partial f}{\partial x'}$$

Repeating the procedure to find the second derivative,

$$\frac{\partial^2 y}{\partial x^2} = \frac{\partial}{\partial x}\left(\frac{\partial y}{\partial x}\right) = \frac{\partial(\partial y/\partial x)}{\partial x'}\frac{\partial x'}{\partial x} = \frac{\partial}{\partial x'}\left(\frac{\partial f}{\partial x'}\right) = \frac{\partial^2 f}{\partial x'^2}$$

Similarly, the time derivatives are found:

$$\frac{\partial y}{\partial t} = \frac{\partial f}{\partial x'}\frac{\partial x'}{\partial t} = \pm v \frac{\partial f}{\partial x'}$$

$$\frac{\partial^2 y}{\partial t^2} = \frac{\partial}{\partial t}\left(\frac{\partial y}{\partial t}\right) = \frac{\partial(\partial y/\partial t)}{\partial x'}\frac{\partial x'}{\partial t} = \frac{\partial}{\partial x'}\left(\pm v\frac{\partial f}{\partial x'}\right)(\pm v) = v^2 \frac{\partial^2 f}{\partial x'^2}$$

Combining the results for the two second derivatives, we arrive at the one-dimensional differential wave equation,

$$\frac{\partial^2 y}{\partial x^2} = \frac{1}{v^2}\frac{\partial^2 y}{\partial t^2} \tag{8-2}$$

Any wave of the form of Eq. (8-1) must satisfy the wave Eq. (8-2), regardless of the physical nature of the wave itself. Thus, to determine whether a given function of x and t represents a traveling wave, it is sufficient to show either that it is of the general form of Eq. (8-1) or that it satisfies the wave Eq. (8.2).

8-2 HARMONIC WAVES

Of special importance are *harmonic* waves that involve the sine or cosine functions,

$$y = A \, {}^{\sin}_{\cos}[k(x \pm vt)] \tag{8-3}$$

where A and k are constants that can be varied without changing the harmonic character of the wave. These are periodic waves, representing smooth pulses that repeat themselves endlessly. Such waves are often generated by undamped oscillators undergoing simple harmonic motion. More important, the sine and cosine functions together form a *complete set* of functions; that is, a linear combination of terms like those in Eq. (8-3) can be found to represent any actual periodic wave form. Such a series of terms is called a *Fourier series* and is treated further in Section 12-1. Thus combinations of harmonic waves are potentially capable of representing more complicated wave forms, even a series of rectangular pulses or square waves.

Since $\sin x = \cos(x - \pi/2)$, the only difference between the sine and cosine functions is a relative translation of $\pi/2$ radians. It is sufficient in what follows, therefore, to treat only one of these functions. Accordingly, a sine wave is pictured in Figure 8-2. In Figure 8-2a, a section of the wave with *amplitude A* is shown at a fixed time, as in a snapshot; in Figure 8-2b the time variations of the wave are pictured at a fixed point x along the wave. In Figure 8-2a the repetitive spatial unit of the wave is shown as the *wavelength* λ. Because of this periodicity, increasing all x by λ should reproduce the same wave. Mathematically, the wave is reproduced be-

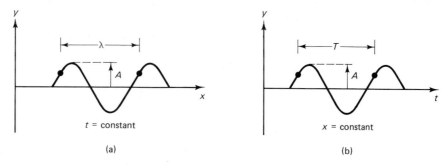

Figure 8-2 Extension of a sine wave in space and time. (a) Sine wave at a fixed time. (b) Sine wave at a fixed point.

cause the argument of the sine function is advanced by 2π. Symbolically,

$$A \sin k[(x + \lambda) + vt] = A \sin [k(x + vt) + 2\pi]$$

or

$$A \sin (kx + k\lambda + kvt) = A \sin (kx + kvt + 2\pi)$$

It follows that $k\lambda = 2\pi$, so that the *propagation constant k* contains information regarding the wavelength.

$$k = \frac{2\pi}{\lambda} \tag{8-4}$$

Alternatively, if the wave is viewed from a fixed position, as in Figure 8-2b, it is periodic in time with a repetitive temporal unit called the *period T*. Increasing all t by T, the wave form is exactly reproduced, so that

$$A \sin k[x + v(t + T)] = A \sin [k(x + vt) + 2\pi]$$

or

$$A \sin (kx + kvt + kvT) = A \sin (kx + kvt + 2\pi)$$

Clearly, $kvT = 2\pi$, and we have an expression that relates the period T to the propagation constant k and wave velocity v. The same information is included in the relation

$$v = \nu\lambda \tag{8-5}$$

where we have used Eq. (8-4) together with the reciprocal relation between period T and frequency ν,

$$\nu = \frac{1}{T} \tag{8-6}$$

Related descriptions of wave parameters are often used. The combination $\omega = 2\pi\nu$ is called the *angular frequency*, and the reciprocal of the wavelength $\kappa = 1/\lambda$ is called *wave number*. With these relationships it is easy to show the equivalence of the following common forms for harmonic waves:

$$y = A \, {}^{\sin}_{\cos}[k(x \pm vt)] \tag{8-7}$$

$$y = A \, {}^{\sin}_{\cos}\left[2\pi\left(\frac{x}{\lambda} \pm \frac{t}{T}\right)\right] \tag{8-8}$$

$$y = A \, {}^{\sin}_{\cos}[(kx \pm \omega t)] \tag{8-9}$$

In any case, the argument of the sine or cosine, which is an angle that depends on space and time, is called the *phase*, φ. For example, in Eq. (8-7),

$$\varphi = k(x \pm vt) \qquad\qquad (8\text{-}10)$$

When x and t change together in such a way that φ is constant, the displacement $y = A \sin \varphi$ is also constant. The condition of constant phase evidently describes the motion of a fixed point on the wave form, which moves with the velocity of the wave. Thus if φ is constant,

$$d\varphi = 0 = k(dx \pm v\, dt)$$

and

$$\frac{dx}{dt} = \mp v$$

confirming that v represents the wave velocity, which is negative when $\varphi = k(x + vt)$ and positive when $\varphi = k(x - vt)$.

In any of the wave equations, Eqs. (8-7) to (8-9), notice that under initial conditions $x = 0$ and $t = 0$, $y = 0$ if the sine function is used and $y = A$ if the cosine function is used. As pointed out previously, both situations could be handled by either the sine or cosine function if an angle of 90° is added to the phase. In general, to accommodate any arbitrary initial displacement, some angle φ_0 must be added to the phase. For example, Eq. (8-7) with the sine function becomes

$$y = A \sin [k(x \pm vt) + \varphi_0]$$

Now suppose our initial boundary conditions are such that $y = y_0$ when $x = 0$ and $t = 0$. Then

$$y = A \sin \varphi_0 = y_0$$

from which the required *initial phase angle* φ_0 can be calculated as

$$\varphi_0 = \sin^{-1}\left(\frac{y_0}{A}\right)$$

The wave Eqs. (8-7) to (8-9) can be generalized further to yield any initial displacement, therefore, by the addition of an initial phase angle φ_0 to the phase. In many cases, the precise phase of the wave is not of interest. Then φ_0 is set equal to zero for simplicity.

Example

A traveling wave propagates according to the expression

$$y(x, t) = 0.35 \sin \left(3\pi x - 10\,\pi t + \frac{\pi}{4}\right)$$

where x is in meters and t is in seconds. Determine the wavelength, frequency, velocity, and initial phase angle. Also find the displacement at $x = 10$ cm and $t = 0$.

Solution By comparison with Eq. (8-9), $k = 3\pi$ and $\omega = 10\pi$. Thus

$$\lambda = \frac{2\pi}{k} = \frac{2}{3}\,\text{m} \quad \text{and} \quad f = \frac{\omega}{2\pi} = 5\,\text{Hz}$$

The initial phase ($x = 0$, $t = 0$) is $\pi/4$. The velocity of the wave may be found from $v = \lambda f = (2/3)\,5 = 3.33$ m/s in the positive x-direction (due to the negative sign in the phase). One can also set the phase $\varphi = 3\pi x - 10\pi t + \pi/4$ equal to a constant, so that

$$d\varphi = 3\pi \, dx - 10\pi \, dt = 0$$

or $v = dx/dt = 10\pi/3\pi = +3.33$ m/s. Furthermore, the displacement

$$y(0.1, 0) = 0.35 \sin\left(0.3\pi + \frac{\pi}{4}\right) = +0.346 \text{ m}$$

8-3 COMPLEX NUMBERS

In many situations it proves to be convenient to represent harmonic waves in complex-number notation. To this end, we first review briefly the forms in which we may write a complex number and their most useful relationships.

A complex number $\tilde{z}$ is expressed as the sum of its *real* and *imaginary* parts,

$$\tilde{z} = a + ib \qquad (8\text{-}11)$$

where

$$a = \text{Re} (\tilde{z}) \quad \text{and} \quad b = \text{Im} (\tilde{z})$$

are real numbers and $i = \sqrt{-1}$. The form of the complex number given by Eq. (8-11) can also be cast into polar form. Referring to Figure 8-3, the complex number $\tilde{z}$ is represented in terms of its real and imaginary parts along the corresponding axes. The magnitude of $\tilde{z}$, symbolized by $|\tilde{z}|$, also called its *absolute value* or *modulus*, is given by the Pythagorean theorem as

$$|\tilde{z}|^2 = a^2 + b^2 \qquad (8\text{-}12)$$

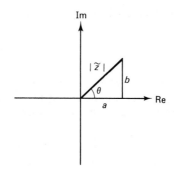

Figure 8-3 Graphical representation of a complex number along real (Re) and imaginary (Im) axes.

Since from Figure 8-3, $a = |\tilde{z}| \cos \theta$ and $b = |\tilde{z}| \sin \theta$, it is also possible to express $\tilde{z}$ by

$$\tilde{z} = |\tilde{z}|(\cos \theta + i \sin \theta)$$

The expression in parentheses is, by Euler's formula,

$$e^{i\theta} = \cos \theta + i \sin \theta \qquad (8\text{-}13)$$

so that

$$\tilde{z} = |\tilde{z}|e^{i\theta} \qquad (8\text{-}14)$$

where

$$\theta = \tan^{-1}\left(\frac{b}{a}\right) \qquad (8\text{-}15)$$

θ	$e^{i\theta}$
$0°$	1
$90°$	i
$180°$	-1
$270°$	$-i$

Figure 8-4 Frequently used values of $e^{i\theta}$.

The *complex conjugate* $\tilde{z}*$ is simply the complex number $\tilde{z}$ with i replaced by $-i$. Thus if $\tilde{z} = a + ib$,

$$\tilde{z}* = a - ib \quad \text{or} \quad \tilde{z}* = |\tilde{z}| e^{-i\theta} \tag{8-16}$$

where the asterisk is used to denote the complex conjugate. A very useful minitheorem is that the product of a complex number with its complex conjugate equals the square of its absolute value. Using the polar form,

$$\tilde{z}\tilde{z}* = (|\tilde{z}|e^{i\theta})(|\tilde{z}|e^{-i\theta}) = |\tilde{z}|^2 \tag{8-17}$$

Finally, it will be helpful to list the values of $e^{i\theta}$, using Euler's formula, Eq. (8-13), for frequently occurring special cases. These are given in Figure 8-4, together with a mnemonic device to assist in recalling them quickly.

8-4 HARMONIC WAVES AS COMPLEX NUMBERS

Using Euler's formula, it is possible to express a harmonic wave by

$$\tilde{y} = Ae^{i(kx-\omega t)} \tag{8-18}$$

where

$$\text{Re}\,(\tilde{y}) = A\cos(kx - \omega t) \tag{8-19}$$

and

$$\text{Im}\,(\tilde{y}) = A\sin(kx - \omega t) \tag{8-20}$$

Expressed in the form of Eq. (8-18), the harmonic wave function thus includes both sine and cosine waves as its real and imaginary parts. Calculations employing the complex form implictly carry correct results for both sine and cosine waves. At any point in such calculations, appropriate expressions for either form can be extracted by taking the real or the imaginary parts of both sides of the equation. Because the mathematics with exponential functions is usually simpler than with trigonometric functions, it is often convenient to deal with harmonic waves written in the form of Eq. (8-18).

8-5 PLANE WAVES

We wish now to generalize the harmonic wave equation further so that it can represent a propagation along any direction in space. Since an arbitrary direction involves the three spatial coordinates x, y, and z, we represent the wave displacement by ψ rather than y; for example,

$$\psi = A\sin(kx - \omega t) \tag{8-21}$$

Equation (8-21) represents a traveling wave moving along the $+x$-direction. At fixed time (for simplicity we take $t = 0$), the spatial extent of this wave is

$$\psi = A\sin kx \tag{8-22}$$

When x = constant, the phase $\varphi = kx$ = constant. Thus the surfaces of constant phase are the family of planes given in Figure 8-5. The surfaces of constant phase constitute the *wavefronts* of the disturbance. Evidently, then, the wave displacement given by ψ is the same for all points of a wavefront. The wave disturbance at an arbitrary point in space, defined by the vector **r** in Figure 8-6a, is therefore the same as for the point x along the x-axis, where $x = r \cos \theta$. Eq. (8-22) may then be written as

$$\psi = A \sin (kr \cos \theta)$$

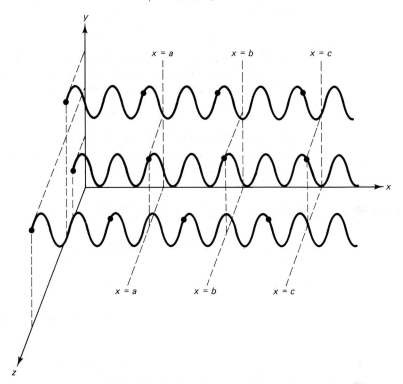

Figure 8-5 Plane waves along x-axis. Surfaces of constant phase are the planes x = constant. The waves penetrate the planes $x = a$, $x = b$, and $x = c$ at the points shown.

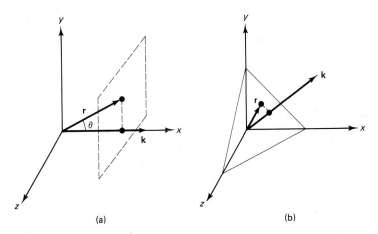

Figure 8-6 Generalization of the plane wave to an arbitrary direction. The wave direction is given by the vector **k** along the x-axis in (a) and an arbitrary direction in (b).

Some simplification results if the propagation constant, whose magnitude $2\pi/\lambda$ has already been determined in Eq. (8-4), is now considered to be a vector quantity, pointing in the direction of propagation. Then $kr \cos \theta = \mathbf{k} \cdot \mathbf{r}$, and the harmonic wave of Eq. (8-21) becomes

$$\psi = A \sin (\mathbf{k} \cdot \mathbf{r} - \omega t) \tag{8-23}$$

In this form, Eq. (8-23) can represent plane waves propagating in any arbitrary direction given by $\mathbf{k}$, as shown in Figure 8-6b. In the general case,

$$\mathbf{k} \cdot \mathbf{r} = xk_x + yk_y + zk_z$$

where (k_x, k_y, k_z) are the components of the propagation direction and (x, y, z) are the components of the point in space where the displacement ψ is evaluated.

The harmonic wave equation is now a three-dimensional wave equation that might also be expressed in complex form as

$$\psi = A e^{i(\mathbf{k} \cdot \mathbf{r} - \omega t)} \tag{8-24}$$

The partial differential equation satisfied by such three-dimensional waves is a generalization of Eq. (8-2) in the form

$$\frac{\partial^2 \psi}{\partial x^2} + \frac{\partial^2 \psi}{\partial y^2} + \frac{\partial^2 \psi}{\partial z^2} = \frac{1}{v^2} \frac{\partial^2 \psi}{\partial t^2} \tag{8-25}$$

as can easily be verified by computing the second partial derivatives of ψ from Eq. (8-24). The wave Eq. (8-25) is often written more compactly by separating the second spatial derivatives from the wave function ψ, treating them as operators,

$$\left(\frac{\partial^2}{\partial x^2} + \frac{\partial^2}{\partial y^2} + \frac{\partial^2}{\partial z^2} \right) \psi = \frac{1}{v^2} \frac{\partial^2 \psi}{\partial t^2}$$

and defining the entire operator in parentheses as the *Laplacian operator*,

$$\nabla^2 \equiv \frac{\partial^2}{\partial x^2} + \frac{\partial^2}{\partial y^2} + \frac{\partial^2}{\partial z^2}$$

so that Eq. (8-25) becomes simply

$$\nabla^2 \psi = \frac{1}{v^2} \frac{\partial^2 \psi}{\partial t^2} \tag{8-26}$$

8-6 SPHERICAL WAVES

Harmonic wave disturbances emanating from a point source in a homogeneous medium travel at equal rates in all directions. Surfaces of constant phase, that is, wavefronts, are then spherical surfaces centered at the source. Such waves may also be represented by the harmonic wave equations developed for plane waves, with one modification: The amplitude must be divided by the distance r to give

$$\psi = \left(\frac{A}{r} \right) e^{i(\mathbf{k} \cdot \mathbf{r} - \omega t)} \tag{8-27}$$

The spherical wave, as it propagates further from the source, decreases in amplitude, in contrast to a plane wave for which amplitude is constant. If the amplitude at distance r from the source is A/r, then the irradiance (W/m²) of the wave there is proportional to $(A/r)^2$, and we see that we are simply describing the familiar inverse square law of propagation for spherical wave disturbances. Notice that in this case the meaning of A must be carefully described. Clearly, we cannot allow the amplitude to become infinite at the source point, as r approaches zero. The value of A must correspond to the amplitude of the wave at unit distance ($r = 1$) from the source.

8-7 ELECTROMAGNETIC WAVES

The harmonic wave equations discussed so far can represent any type of wave disturbance that varies in a sinusoidal manner. This includes, for example, waves on a string, water waves, and sound waves. The equations apply to a specific situation as soon as the physical significance of the displacement ψ is identified. The quantity ψ may refer to vertical displacements of a string or pressure variations due to a sound wave propagating in a gas. For electromagnetic waves that can represent the propagation of light, ψ stands for either of the varying electric or magnetic fields that together constitute the wave. Figure 8-7 depicts a plane electromagnetic wave traveling in some arbitrary direction. From Maxwell's equations, which describe such waves, we know that the harmonic variations of the electric and magnetic fields are always perpendicular to one another and to the direction of propagation given by $\mathbf{k}$, as suggested by the orthogonal set of axes in Figure 8-7. These variations may be described by the harmonic wave equation in the form

$$\mathbf{E} = \mathbf{E}_0 e^{i(\mathbf{k}\cdot\mathbf{r} - \omega t)} \tag{8-28}$$

$$\mathbf{B} = \mathbf{B}_0 e^{i(\mathbf{k}\cdot\mathbf{r} - \omega t)} \tag{8-29}$$

where $\mathbf{E}$ and $\mathbf{B}$ represent the electric and magnetic fields, respectively, and $\mathbf{E}_0$ and $\mathbf{B}_0$ are their amplitudes. Both components of the wave travel with the same propagation vector $\mathbf{k}$ and frequency ω and thus with the same wavelength and speed. Furthermore, electromagnetic theory tells us that the field amplitudes are related by $E_0 = cB_0$, where c is the speed of the wave. At any specified time and place,

$$E = cB \tag{8-30}$$

In free space, the velocity c is given by

$$c = \frac{1}{\sqrt{\epsilon_0 \mu_0}} \tag{8-31}$$

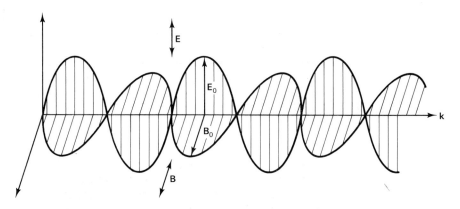

Figure 8-7 Plane electromagnetic wave. The electric field $\mathbf{E}$, magnetic field $\mathbf{B}$, and propagation vector $\mathbf{k}$ are everywhere mutually perpendicular.

where the constants ϵ_0 and μ_0 are, respectively, the permittivity and permeability of vacuum. Measured values for these constants, $\epsilon_0 = 8.8542 \times 10^{-12}$ $(C\text{-}s)^2/\text{kg-m}^3$ and $\mu_0 = 4\pi \times 10^{-7}$ kg-m/(A-s)^2, provide an indirect method of determining the speed of light in free space and yield a value of $c = 2.998 \times 10^8$ m/s.

Such a wave, of course, represents the transmission of energy. The energy density (J/m^3) associated with the electric field in free space is

$$u_E = \tfrac{1}{2} \epsilon_0 E^2 \qquad (8\text{-}32)$$

and the energy density associated with the magnetic field in free space is

$$u_B = \frac{1}{2\mu_0} B^2 \qquad (8\text{-}33)$$

These expressions, easily derived for the static electric field of an ideal capacitor and the static magnetic field of an ideal solenoid, are generally valid. Incorporating Eqs. (8-30) and (8-31) into either of the Eqs. (8-32) or (8-33), u_E and u_B are shown to be equal. For example, starting with Eq. (8-33),

$$u_B = \frac{1}{2\mu_0} \left(\frac{E}{c}\right)^2 = \left(\frac{\epsilon_0 \mu_0}{2\mu_0}\right) E^2 = u_E \qquad (8\text{-}34)$$

The energy of an electromagnetic wave is therefore divided equally between its constituent electric and magnetic fields. The total energy density is the sum

$$u = u_E + u_B = 2u_E = 2u_B$$

or

$$u = \epsilon_0 E^2 = \left(\frac{1}{\mu_0}\right) B^2 \qquad (8\text{-}35)$$

Consider next the rate at which energy is transported by the electromagnetic wave, or its *power*. In a time Δt, the energy transported through a cross section of area A (Figure 8-8) is the energy associated with the volume ΔV of a rectangular volume of length $c \Delta t$. Thus

$$\text{power} = \frac{\text{energy}}{\Delta t} = \frac{u \, \Delta V}{\Delta t} = \frac{u(Ac \, \Delta t)}{\Delta t} = ucA \qquad (8\text{-}36)$$

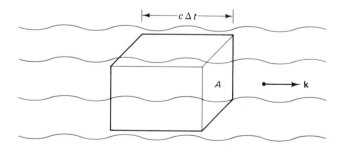

Figure 8-8 Energy flow of an electromagnetic wave. In time Δt, the energy enclosed in the rectangular volume flows across the surface A.

or the power transferred per unit area, S, is

$$S = uc \qquad (8\text{-}37)$$

We now express the energy density u in terms of E and B, as follows, making use of Eqs. (8-31) and (8-35):

$$u = \sqrt{u} \, \sqrt{u} = (\sqrt{\epsilon_0} E)\left(\frac{B}{\sqrt{\mu_0}}\right) = \frac{\epsilon_0}{\sqrt{\epsilon_0 \mu_0}} EB = \epsilon_0 c EB \qquad (8\text{-}38)$$

Inserting this result into Eq. (8-37),

$$S = \epsilon_0 c^2 E B \tag{8-39}$$

The power per unit area, S, when assigned the direction of propagation, is called the *Poynting vector*. Since this direction is the same as that of the cross product of the orthogonal vectors, **E** and **B**, we can write, finally

$$\mathbf{S} = \epsilon_0 c^2 \mathbf{E} \times \mathbf{B} \tag{8-40}$$

Because of the rapid variation of the electric and magnetic fields, whose frequencies are 10^{14} to 10^{15} Hz in the visible spectrum, the magnitude of the Poynting vector in Eq. (8-39) is also a rapidly varying function of time. In most cases a time average of the power delivered per unit area is all that is required. This quantity is called the *irradiance*, E_e.

$$E_e = \langle |\mathbf{S}| \rangle = \epsilon_0 c^2 \langle E_0 B_0 \sin^2 (\mathbf{k} \cdot \mathbf{r} \pm \omega t) \rangle \tag{8-41}$$

where the angle brackets denote a time average and we have expressed the fields as sine functions of the phase. The average of the functions $\sin^2 \theta$ or $\cos^2 \theta$ over a period is easily shown to be exactly $\frac{1}{2}$, so that

$$E_e = \tfrac{1}{2}\epsilon_0 c^2 \, E_0 B_0$$

$$E_e = \tfrac{1}{2} \epsilon_0 c E_0^2 \tag{8-42}$$

$$E_e = \tfrac{1}{2}\left(\frac{c}{\mu_0}\right) B_0^2$$

The alternative forms of Eq. (8-42) are expressed for the case of free space. They apply also to a medium of refractive index n if ϵ_0 is replaced by $n^2 \epsilon_0$ and c is replaced by the velocity c/n. Notice that these changes leave the first of the alternate forms invariant.

Example

A laser beam of radius 1 mm carries a power of 6 kW. Determine its average irradiance and the amplitude of its E and B fields.

Solution   The average irradiance

$$E_e = \frac{\text{power}}{\text{area}} = \frac{6000}{\pi (10^{-3})^2} = 1.91 \times 10^9 \text{ W/m}^2$$

From Eq. (8-42),

$$E_0 = \left(\frac{2E_e}{\epsilon_0 c}\right)^{1/2} = \left[\frac{2(1.91 \times 10^9)}{\epsilon_0 c}\right]^{1/2} = 1.20 \times 10^6 \text{ V/m}$$

and, from Eq. (8-30),

$$B_0 = \frac{E_0}{c} = \frac{1.20 \times 10^6}{c} = 4.00 \times 10^{-3} \text{ T}$$

8-8 DOPPLER EFFECT

The familiar *Doppler effect* for sound waves has its counterpart in light waves, but with an important difference. Recall that when dealing with sound waves, the apparent frequency of a source increases or decreases depending on the motion of both source and observer along the line joining them. The frequency shift due to a moving *source* is based physically on a change in transmitted wavelength. The frequency shift due to a moving *observer* is based physically on the change in speed of the sound waves relative to the observer. The two effects are physically distinct and described by different equations. They are also essentially different from the case of light waves. The difference between the Doppler effect in sound and light waves is more than the difference in wave speeds. Whereas sound waves propagate through a material medium, light waves propagate in vacuum. As soon as the medium of propagation is removed, there is no longer a physical basis for the distinction between moving observer and moving source. There is *one* relative motion between them that determines the frequency shift in the Doppler effect for light. The derivation of the Doppler effect for light requires the theory of special relativity and so is not carried out here. The result [1] is expressed by

$$\frac{\lambda'}{\lambda} = \sqrt{\frac{1 - \dfrac{v}{c}}{1 + \dfrac{v}{c}}} \tag{8-43}$$

where λ' is the Doppler-shifted wavelength and v is the relative velocity between source and observer. The sign of v is positive when they are approaching one another. When $v \ll c$, this equation is approximated by

$$\frac{\lambda'}{\lambda} = 1 - \frac{v}{c} \tag{8-44}$$

The Doppler effect is especially important when used to determine the speed of astronomical sources emitting electromagnetic radiation. The *red shift* is the shift in wavelength of such radiation toward longer wavelengths, due to a relative speed of the source *away* from us. *Doppler broadening* of spectral lines represents another important application in which fast-moving atoms of a gas radiate light with both increases and decreases in frequency due to their random motion toward and away from the observer making spectroscopic measurements.

Example

Light from a distant galaxy shows the characteristic lines of the oxygen spectrum, except that the wavelengths are shifted from their values as measured using laboratory sources. In particular, the line expected at 513 nm shows up at 525 nm. What is the speed of the galaxy relative to the earth?

Solution   Here, $\lambda = 513$ nm and $\lambda' = 525$ nm. Thus, using Eq. (8-44),

$$\frac{525}{513} = 1 - \frac{v}{c}$$
$$v = -0.0234c = -7020 \text{ km/s}$$

Since the apparent λ is larger (the frequency less), the galaxy is moving away from the earth with a speed of approximately 7020 km/s.

PROBLEMS

8-1. A pulse of the form $y = ae^{-bx^2}$ is formed in a rope, where a and b are constants and x is in centimeters. Sketch this pulse. Then write an equation that represents the pulse moving in the negative direction at 10 cm/s.

8-2. A transverse wave pulse, described by

$$y = \frac{4}{x^2 + 2}$$

is initiated at $t = 0$ in a stretched string.

(a) Write an equation for the traveling pulse if it moves with a speed of 2.5 m/s in the negative x-direction.

(b) Plot the pulse at $t = 0$, $t = 2$, and $t = 5$ seconds.

8-3. Consider the following mathematical expressions, where distances are in meters:

1. $y(z, t) = A \sin^2 [4\pi (t + z)]$
2. $y(x, t) = A(x - t)^2$
3. $y(x, t) = A/(Bx^2 - t)$

(a) Which qualify as traveling waves? Prove your conclusion.

(b) If they qualify, give the magnitude and direction of the wave velocity.

8-4. If the following represents a traveling wave, determine its velocity (magnitude and direction), where distances are in meters.

$$y = \frac{100e^{x^2 - 20xt + 100t^2}}{x - 10t}$$

8-5. A harmonic traveling wave is moving in the negative z-direction with an amplitude (arbitrary units) of 2, a wavelength of 5 m, and a period of 3 s. Its displacement at the origin is zero at time zero. Write a wave equation for this wave (a) that exhibits directly both wavelength and period; (b) that exhibits directly both propagation constant and velocity; (c) in complex form.

8-6. **(a)** Write the equation of a harmonic wave traveling along the x-direction at $t = 0$ if it is known to have an amplitude of 5 m and a wavelength of 50 m.

(b) Write an expression for the disturbances at $t = 4s$ if it is moving in the negative x-direction at 2 m/s.

8-7. For a harmonic wave given by $y = 10 \sin (628.3x - 6283t)$, with x and y in centimeters and t in seconds, determine (a) wavelength; (b) frequency; (c) propagation constant; (d) angular frequency; (e) period; (f) velocity; (g) amplitude.

8-8. Use the constant phase condition to determine the velocity of each of the following waves in terms of the constants A, B, C, and D. Distances are in meters and time in seconds. Verify your results dimensionally.

(a) $f(y, t) = A(y - t)^2$

(b) $f(x, t) = A(Bx + Ct + D)^2$

(c) $f(z, t) = A \exp (Bz^2 + BC^2t^2 - 2BCzt)$

8-9. A harmonic wave traveling in the $+x$-direction has, at $t = 0$, a displacement of 13 units at $x = 0$ and a displacement of -7.5 units at $x = 3\lambda/4$. Write the equation for the wave at $t = 0$.

8-10. **(a)** Show that if the maximum positive displacement of a sinusoidal wave occurs at distance x_0 centimeters from the origin when $t = 0$, its initial phase angle φ_0 is given by

$$\varphi_0 = \frac{\pi}{2} - \left(\frac{2\pi}{\lambda}\right) x_0$$

where the wavelength λ is in centimeters.

(b) Determine the initial phase and sketch the wave when $\lambda = 10$ cm and $x_0 = 0, \frac{5}{6}, \frac{5}{2}, 5$, and $-\frac{1}{2}$ cm.

(c) What are the appropriate initial phase angles for (b) when a cosine function is used instead?

8-11. By finding appropriate expressions for $\mathbf{k} \cdot \mathbf{r}$, write equations describing a sinusoidal plane wave in three dimensions, displaying wavelength and velocity, if propagation is

(a) along the $+z$-axis;

(b) along the line $x = y$, $z = 0$;

(c) perpendicular to the planes $x + y + z$ = constant.

8-12. Show that if $\bar{z}$ is a complex number, **(a)** Re $(\bar{z}) = (\bar{z} + \bar{z}*)/2$; **(b)** Im $(\bar{z}) = (\bar{z} - \bar{z}*)/2i$; **(c)** $\cos \theta = (e^{i\theta} + e^{-i\theta})/2$; **(d)** $\sin \theta = (e^{i\theta} - e^{-i\theta})/2i$.

8-13. Show that a wave function, expressed in complex form, is shifted in phase **(a)** by $\pi/2$ when multiplied by i and **(b)** by π when multiplied by -1.

8-14. Two waves of the same amplitude, speed, and frequency travel together in the same region of space. The resultant wave may be written as a sum of the individual waves,

$$\psi(y, t) = A \sin(ky + \omega t) + A \sin(ky - \omega t + \pi)$$

With the help of complex exponentials, show that

$$\psi(y, t) = 2A \cos(ky) \sin(\omega t)$$

8-15. The energy flow to the earth associated with sunlight is about 1.4 kW/m². Find the maximum values of E and B for a wave of this power density.

8-16. A light wave is traveling in glass of index 1.50. If the electric field amplitude of the wave is known to be 100 V/m, find **(a)** the amplitude of the magnetic field and **(b)** the average magnitude of the Poynting vector.

8-17. The solar constant is the radiant flux density (irradiance) from the sun at the earth's surface and is about 0.135 W/cm². Assume an average wavelength of 700 nm for the sun's radiation which reaches the earth. Find **(a)** the amplitude of the **E**- and **B**-fields; **(b)** the number of photons that arrive each second on each square meter of a solar panel; **(c)** a harmonic wave equation for the **E**-field of the solar radiation, inserting all constants numerically.

8-18. **(a)** The light from a 220-W lamp spreads uniformly in all directions. Find the irradiance of these optical electromagnetic waves and the amplitude of their **E**-field at a distance of 10 m from the lamp. Assume that 5% of the lamp energy is converted to light.

(b) Suppose a 2000-W laser beam is concentrated by a lens into a cross-sectional area of about 1×10^{-6} cm². Find the corresponding irradiance and amplitudes of the **E**- and **B**-fields there.

8-19. Show that the amplitude of a cylindrical wave must vary inversely with $\sqrt{r}$.

8-20. Show that Eq. (8-44) for the Doppler effect follows from Eq. (8-43) when $v \ll c$.

8-21. How fast does one have to approach a red traffic light to see a green signal? So that we all get the same answer, say that a good red is 640 nm and a good green is 540 nm.

8-22. A quasar near the limits of the observed universe to date shows a wavelength that is 4.80 times the wavelength emitted by the same molecules on the earth. If the Doppler effect is responsible for this shift, what velocity does it determine for the quasar?

8-23. Estimate the Doppler broadening of the 706.52-nm line of helium when the gas is at 1000 K. Use the root-mean-square velocity of a gas molecule given by

$$v_{rms} = \sqrt{\frac{3RT}{M}}$$

where R is the gas constant, T the Kelvin temperature, and M the molecular weight.

REFERENCES

[1] Resnick, Robert. *Basic Concepts in Relativity and Early Quantum Mechanics*. New York: John Wiley and Sons, 1972. Ch. 2.

[2] Hecht, Eugene, and Alfred Zajac. *Optics*. Reading, Mass.: Addison-Wesley Publishing Company, 1974. Ch. 2.

[3] Ghatak, Ajoy K. *An Introduction to Modern Optics*. New York: McGraw-Hill Book Company, 1972. Ch. 1.

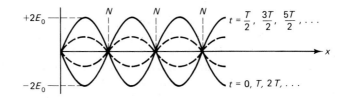

Superposition of Waves

INTRODUCTION

In Chapter 8 equations describing waves of a given amplitude, wavelength, and frequency were developed. Quite commonly, it is necessary to deal with situations in which two or more such waves arrive at the same point in space or exist together along the same direction. Several important cases of the combined effects of two or more harmonic waves are treated in this chapter. The first case deals with the superposition of harmonic waves of differing amplitudes and phases but with the same frequency. The analysis shows that the resultant is just another harmonic wave having the same frequency. This leads to an important difference between the irradiance attainable from randomly phased and coherent harmonic waves. The chapter next treats standing waves that result from the superposition of a harmonic wave with its reflected counterpart. We end with a discussion of the speed of the modulating envelope of a group of harmonic waves differing in frequency, an application that applies immediately to the important case of dispersion.

9-1 SUPERPOSITION PRINCIPLE

To explain the combined effects of waves successfully one must ask specifically: What is the net displacement ψ at a point in space where waves with independent displacements ψ_1 and ψ_2 exist together? In most cases of interest, the correct answer is given by the *superposition principle*: The resultant displacement is the sum of the separate displacements of the constituent waves:

$$\psi = \psi_1 + \psi_2 \qquad (9\text{-}1)$$

Using this principle, the resultant wave amplitude and power density (W/m^2) can be calculated and verified by measurement. In this way, the superposition principle has been determined to be valid for all kinds of waves.

The same principle can be stated more formally as follows. If ψ_1 and ψ_2 are independently solutions of the wave equation,

$$\nabla^2 \psi = \frac{1}{v^2} \frac{\partial^2 \psi}{\partial t^2}$$

then the *linear combination*,

$$\psi = a\psi_1 + b\psi_2$$

where a and b are constants, is also a solution.

The superposition of electromagnetic (em) waves may be expressed in terms of their electric or magnetic fields by the vector equations,

$$\mathbf{E} = \mathbf{E}_1 + \mathbf{E}_2 \quad \text{and} \quad \mathbf{B} = \mathbf{B}_1 + \mathbf{B}_2$$

In general, the orientation of the electric or magnetic fields must be taken into account. The superposition of waves at a point where their electric fields are othogonal, for example, does not yield the same result as the case in which they are parallel. A more formal accounting of the vector nature of $\mathbf{E}$ in the superposition of two em waves will be taken in the next chapter. The case of orthogonal $\mathbf{E}$ waves is considered in detail in the discussion on the polarization of light. For the present, we treat electric fields as scalar quantities. This treatment is strictly valid for cases where the individual $\mathbf{E}$ vectors are parallel; it is often applied in cases where they are nearly parallel. The treatment is valid also for cases of unpolarized light, in which the $\mathbf{E}$ field can be represented by two orthogonal components. The scalar theory applies to each component and its parallel counterpart in the superposing waves, and thus to the entire wave.

Nonlinear effects for which the superposition principle does not predict all the observed results can occur when light of very large amplitude interacts with matter. The possibility of producing high-energy densities, using laser light, has facilitated the study and use of such effects, making *nonlinear optics* an important branch of modern optics. An introduction to this subject is taken up in Chapter 26.

9-2 SUPERPOSITION OF WAVES OF THE SAME FREQUENCY

The first case of superposition to be considered is the situation in which two harmonic waves of the same frequency combine to form a resultant wave disturbance. We permit the two waves to differ in amplitude and phase. Beginning with a wave in the form

$$E = E_0 \sin (\mathbf{k} \cdot \mathbf{r} + \omega t + \varphi_0)$$

where an initial phase angle φ_0 is added for generality, we set $\mathbf{k} \cdot \mathbf{r}$ equal to a constant because we wish to examine waves at a fixed point in space. Thus

$$E = E_0 \sin (\omega t + \alpha) \qquad (9\text{-}2)$$

where the constant phase angle

$$\alpha = \mathbf{k} \cdot \mathbf{r} + \varphi_0 \qquad (9\text{-}3)$$

Two such waves, intersecting at a fixed point, may differ in phase by

$$\alpha_2 - \alpha_1 = \mathbf{k} \cdot (\mathbf{r}_2 - \mathbf{r}_1) + (\varphi_{02} - \varphi_{01})$$

due to a path difference (given by the first term) and an initial phase difference (given by the second term). The time variations of the em waves at the given point can be expressed by

$$E_1 = E_{01} \sin (\omega t + \alpha_1) \tag{9-4}$$

$$E_2 = E_{02} \sin (\omega t + \alpha_2) \tag{9-5}$$

By the superposition principle, the resultant electric field E_R at the point is

$$E_R = E_1 + E_2 = E_{01} \sin (\omega t + \alpha_1) + E_{02} \sin (\omega t + \alpha_2)$$

Using the trigonometric identity for the sum of two angles,

$$\sin (A + B) \equiv \sin A \cos B + \cos A \sin B$$

and recombining terms,

$$E_R = (E_{01} \cos \alpha_1 + E_{02} \cos \alpha_2) \sin \omega t + (E_{01} \sin \alpha_1 + E_{02} \sin \alpha_2) \cos \omega t \tag{9-6}$$

Leaving this result for a moment, notice that if we picture each of the component waves, Eqs. (9-4) and (9-5), graphically as *phasors* by plotting magnitude and phase angle and add them (Figure 9-1a) as if they were vectors, a resultant, or sum, is found with magnitude E_0 and phase α. From Figure 9-1b, the components of the resultant are

$$E_0 \cos \alpha = E_{01} \cos \alpha_1 + E_{02} \cos \alpha_2$$

and

$$E_0 \sin \alpha = E_{01} \sin \alpha_1 + E_{02} \sin \alpha_2$$

In terms of the quantities E_0 and α defined by this graphical technique, Eq. (9-6) becomes

$$E_R = E_0 \cos \alpha \sin \omega t + E_0 \sin \alpha \cos \omega t$$

or

$$E_R = E_0 \sin (\omega t + \alpha) \tag{9-7}$$

We conclude that the resultant wave E_R is another harmonic wave of the same frequency ω, with amplitude E_0 and phase α, related to the constituent harmonic waves

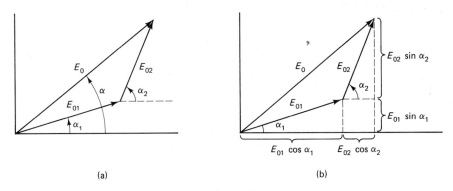

(a)   (b)

Figure 9-1 Phasor diagrams for the superposition of two harmonic waves. (a) Adding two harmonic waves. (b) Phasor components.

by the phasor diagram, Figure 9-1. The cosine law may be applied to Figure 9-1a, yielding an expression for E_0,

$$E_0^2 = E_{01}^2 + E_{02}^2 + 2E_{01}E_{02}\cos(\alpha_2 - \alpha_1) \qquad (9\text{-}8)$$

and from Figure 9-1b, the phase angle is clearly given by

$$\tan \alpha = \frac{E_{01}\sin\alpha_1 + E_{02}\sin\alpha_2}{E_{01}\cos\alpha_1 + E_{02}\cos\alpha_2} \qquad (9\text{-}9)$$

As with vectors, the graphical procedure could be extended to accommodate any number of component waves of the same frequency, as shown in Figure 9-2 for four such waves. The diagram makes apparent the proper generalization of Eqs. (9-8) and (9-9) for N such harmonic waves:

$$\tan \alpha = \frac{\displaystyle\sum_{i=1}^{N} E_{0i}\sin\alpha_i}{\displaystyle\sum_{i=1}^{N} E_{0i}\cos\alpha_i} \qquad (9\text{-}10)$$

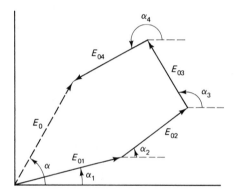

Figure 9-2 Phasor diagram for four harmonic waves of the same frequency. Superposition produces a resultant wave of the same frequency, with amplitude E_0 and phase α.

and by the Pythagorean theorem,

$$E_0^2 = \left(\sum_{i=1}^{N} E_{0i}\sin\alpha_i\right)^2 + \left(\sum_{i=1}^{N} E_{0i}\cos\alpha_i\right)^2 \qquad (9\text{-}11)$$

Eq. (9-11) may profitably be cast into a form that looks more like a generalization of the cosine law in Eq. (9-8). Expanding each term,

$$\left(\sum_{i=1}^{N} E_{0i}\sin\alpha_i\right)^2 = \sum_{i=1}^{N} E_{0i}^2\sin^2\alpha_i + 2\sum_{j>i}^{N}\sum_{i=1}^{N} E_{0i}E_{0j}\sin\alpha_i\sin\alpha_j \qquad (9\text{-}12)$$

$$\left(\sum_{i=1}^{N} E_{0i}\cos\alpha_i\right)^2 = \sum_{i=1}^{N} E_{0i}^2\cos^2\alpha_i + 2\sum_{j>i}^{N}\sum_{i=1}^{N} E_{0i}E_{0j}\cos\alpha_i\cos\alpha_j \qquad (9\text{-}13)$$

The first term of the right members is the sum of the squares of the individual terms of the series in the left members. The double sums represent all the cross products, excluding—by the use of notation $j > i$—the self-products already accounted for in the first term and also avoiding a duplication of products already tallied by the factor 2. Adding Eqs. (9-12) and (9-13),

$$E_0^2 = \sum_{i=1}^{N} E_{0i}^2(\sin^2\alpha_i + \cos^2\alpha_i) + 2\sum_{j>i}^{N}\sum_{i=1}^{N} E_{0i}E_{0j}(\cos\alpha_i\cos\alpha_j + \sin\alpha_i\sin\alpha_j)$$

The expressions in parentheses are equivalent to unity in the first term and are equivalent to $\cos(\alpha_i - \alpha_j)$ in the second, so that, finally

$$E_0^2 = \sum_{i=1}^{N} E_{0i}^2 + 2 \sum_{j>i}^{N} \sum_{i=1}^{N} E_{0i}E_{0j} \cos(\alpha_j - \alpha_i) \qquad (9\text{-}14)$$

Summarizing, the sum of N harmonic waves of identical frequency is again a harmonic wave of the same frequency, with amplitude given by Eq. (9-11) or (9-14) and phase given by Eq. (9-10).

Example

Determine the result of the superposition of the following harmonic waves: $E_1 = 7 \sin(\omega t + \pi/3)$, $E_2 = 12 \cos(\omega t + \pi/4)$, and $E_3 = 20 \sin(\omega t + \pi/5)$.

Solution To make all phase angles consistent, first change the cosine wave to a sine wave: $E_2 = 12 \sin(\omega t + \pi/4 + \pi/2) = 12 \sin(\omega t + 3\pi/4)$. Then, using Eq. (9-11),

$$E_0^2 = \left[7 \sin\left(\frac{\pi}{3}\right) + 12 \sin\left(\frac{3\pi}{4}\right) + 20 \sin\left(\frac{\pi}{5}\right) \right]^2$$
$$+ \left[7 \cos\left(\frac{\pi}{3}\right) + 12 \cos\left(\frac{3\pi}{4}\right) + 20 \cos\left(\frac{\pi}{5}\right) \right]^2$$

or $E_0^2 = 26.303^2 + 11.195^2$ and $E_0 = 28.6$. The same result can be found using Eq. (9-14), which would take the form

$$E_0^2 = 7^2 + 12^2 + 20^2 + 2\left[7 \times 12 \cos\left(\frac{3\pi}{4} - \frac{\pi}{3}\right) \right.$$
$$\left. + 7 \times 20 \cos\left(\frac{\pi}{5} - \frac{\pi}{3}\right) + 12 \times 20 \cos\left(\frac{\pi}{5} - \frac{3\pi}{4}\right) \right]$$

The phase angle of the resulting harmonic wave is found using Eq. (9-10). Since the sums forming the numerator and denominator have already been calculated in the first part, we have

$$\tan \alpha = \frac{26.303}{11.195} \quad \text{and} \quad \alpha = 1.17 \text{ (radians)}$$

Thus the resulting harmonic wave can be expressed as

$$E_0 = 28.6 \sin(\omega t + 1.17) \quad \text{or} \quad 28.6 \sin(\omega t + 0.372\pi)$$

9-3 RANDOM AND COHERENT SOURCES

The effort expended in achieving the form of Eq. (9-14) pays immediate dividends in enabling us to distinguish rather neatly two important cases of superposition: (1) the case of N randomly phased sources of equal amplitude and frequency, where N is a large number, and (2) the case of N coherent sources of the same type. In the first instance, if phases are random, the phase differences $(\alpha_i - \alpha_j)$ are also random. The sum of cosine terms in Eq. (9-14) then approaches zero as N increases, because terms are equally divided between positive and negative fractions ranging from -1 to $+1$. This leaves

$$E_0^2 = \sum_{i=1}^{N} E_{0i}^2 = NE_{01}^2 \qquad (9\text{-}15)$$

because there are N sources of equal amplitude. Thus for N randomly phased sources, the squares of the individual amplitudes add up to produce the square of the resultant amplitude. Recalling that the irradiance (W/m²) is proportional to the square of the amplitude of the electric field, we can say that *the resultant irradiance of N identical but randomly phased sources is the sum of the individual irradiances.* On the other hand, if the N sources are *coherent, and in phase,* so that all α are equal, then Eq. (9-14) becomes

$$E_0^2 = \sum_{i=1}^{N} E_{0i}^2 + 2 \sum_{j>i}^{N} \sum_{i=1}^{N} E_{0i}E_{0j}$$

since all cosine factors are unity. The right side should be recognizable as the square of the sum of the individual amplitudes. Then, more simply,

$$E_0^2 = \left(\sum_{i=1}^{N} E_{0i} \right)^2 = (NE_{01})^2 = N^2 E_{01}^2 \tag{9-16}$$

Here the individual amplitudes simply add to produce a resultant $E_0 = NE_{01}$ rather than $\sqrt{N}E_{01}$, as before. Thus *the resultant irradiance of N identical coherent sources, radiating in phase with each other, is N^2 times the irradiance of the individual sources.* Notice that in this case the result does not require that N be a large number. We conclude that the irradiance of 100 coherent in-phase sources, for example, is 100 times greater than the more usual case of 100 random sources. If E is interpreted as the amplitude of a compressional wave, the result holds for sound intensities as well.

9-4 STANDING WAVES

Another important case of superposition arises when a given wave exists in both forward and reverse directions along the same medium. This condition occurs most frequently when the forward wave experiences a reflection at some point along its path, as in Figure 9-3a. Let us assume for the moment an ideal situation in which none of the energy is lost on reflection nor absorbed by the transmitting medium. This permits us to write both waves with the same amplitude. Forward and reverse waves are, respectively,

$$E_1 = E_0 \sin (kx - \omega t) \tag{9-17}$$

$$E_2 = E_0 \sin (kx + \omega t) \tag{9-18}$$

(a)

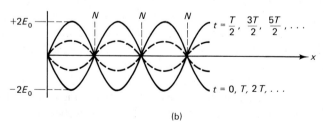

(b)

Figure 9-3 Standing waves. (a) A typical standing wave situation occurs when a wave and its reflection exist along the same medium. A phase shift (not shown) generally occurs on reflection. (b) Resultant displacement of a standing wave, shown at various instants. The solid lines represent the maximum displacement of the wave. The displacement at the nodes (N) is always zero.

The resultant wave in the medium, by the principle of superposition, is

$$E_R = E_1 + E_2 \doteq E_0 [\sin (kx + \omega t) + \sin (kx - \omega t)] \qquad (9\text{-}19)$$

It is expedient in this case to define

$$\alpha = kx + \omega t \quad \text{and} \quad \beta = kx - \omega t$$

and employ the trigonometric identity

$$\sin \alpha + \sin \beta \equiv 2 \sin \tfrac{1}{2}(\alpha + \beta) \cos \tfrac{1}{2}(\alpha - \beta)$$

Applied to Eq. (9-19), this leads immediately to the result

$$E_R = (2E_0 \sin kx) \cos \omega t \qquad (9\text{-}20)$$

which represents a standing wave, plotted in Figure 9-3b. Interpretation is facilitated by regarding the quantity in parentheses as a space-dependent amplitude. At any point x along the medium, the oscillations are given by

$$E_R = A(x) \cos \omega t$$

where $A(x) = 2E_0 \sin kx$. There exist values of x for which $A(x) = 0$, and thus $E_R = 0$ for all t. These values occur whenever

$$\sin kx = 0, \quad \text{or} \quad kx = \frac{2\pi x}{\lambda} = m\pi, \qquad m = 0, \pm 1, \pm 2, \ldots$$

or

$$x = m\left(\frac{\lambda}{2}\right) = 0, \frac{\lambda}{2}, \lambda, \frac{3\lambda}{2}, \ldots \qquad (9\text{-}21)$$

Such points are called the *nodes* of the standing wave and are separated by half a wavelength. At various times, the standing wave will appear as sine waves, like those shown in Figure 9-3b. Although their amplitudes vary with time, all pass through zero at the fixed nodal points. E_R has its maximum value at all points when $\cos \omega t = \pm 1$, or when

$$\omega t = 2\pi f t = \left(\frac{2\pi}{T}\right) t = m\pi$$

Thus the outer envelope of the standing wave occurs at times

$$t = m\left(\frac{T}{2}\right) = 0, \frac{T}{2}, T, \frac{3T}{2}, \ldots$$

where T is the period. There are also periodic times when the standing wave is everywhere zero, since $\cos \omega t = 0$ for $t = T/4, 3T/4, \ldots$.

Unlike traveling waves, standing waves transmit no energy. All the energy in the wave goes into sustaining the oscillations between nodes, at which points forward and reverse waves cancel. However, since mirrors are not perfect reflectors and the transmitting medium generally absorbs some of the wave energy, wave amplitude decreases with x. Unless the source continues to replace lost energy, the amplitude also decreases with time. In this case, the two waves do not cancel exactly at the nodes nor do they add to the maximum of $2E_0$ at the antinodes, points halfway between the nodes. The resultant wave will then be found to include a traveling wave component that carries energy to the mirror and back.

Introduction of a relative phase between the waves of Eqs. (9-17) and (9-18), such as would be expected on reflection, leads to a phase angle component in the sine and cosine factors of Eq. (9-20). Nodes will then be displaced from the positions shown in Figure 9-3, but their separation remains $\lambda/2$. Times at which the form is everywhere zero or everywhere at its maximum displacement also change. The principal features of the standing wave, however, remain unaffected.

9-5 PHASE AND GROUP VELOCITIES

Yet another case of superposition, with important applications in optics, is that of waves of the same or comparable amplitude but differing in frequency. Differences in frequency imply differences in wavelength and velocity. The superposition of several such waves, with wave crests moving at different speeds, exhibits periodically large and small resultant amplitudes. A point where individual crests are coincident, yielding the maximum net amplitude, is itself a location changing with time and therefore possessing its own characteristic speed. Between such points of maximum response, at any time, there appear locations of minimum amplitude due to the juxtaposition of constituent waves more or less out of phase. These features in the resultant wave are manifest even in the case of two component waves, as will be made clear in what follows.

Let the two waves differing in frequency and wave number be represented by

$$E_1 = E_0 \cos (k_1 x - \omega_1 t) \tag{9-22}$$

$$E_2 = E_0 \cos (k_2 x - \omega_2 t) \tag{9-23}$$

The superposition of these waves, which are traveling together in a given medium, is

$$E_R = E_1 + E_2 = E_0[\cos (k_1 x - \omega_1 t) + \cos (k_2 x - \omega_2 t)]$$

Making use of the trigonometric identity

$$\cos \alpha + \cos \beta \equiv 2 \cos \tfrac{1}{2}(\alpha + \beta) \cos \tfrac{1}{2}(\alpha - \beta) \tag{9-24}$$

and identifying

$$\alpha = k_1 x - \omega_1 t$$

$$\beta = k_2 x - \omega_2 t$$

we have

$$E_R = 2E_0 \cos \left[\frac{(k_1 + k_2)}{2} x - \frac{(\omega_1 + \omega_2)}{2} t \right] \cos \left[\frac{(k_1 - k_2)}{2} x - \frac{(\omega_1 - \omega_2)}{2} t \right] \tag{9-25}$$

Now let

$$\omega_p = \frac{\omega_1 + \omega_2}{2}, \qquad k_p = \frac{k_1 + k_2}{2} \tag{9-26}$$

and

$$\omega_g = \frac{\omega_1 - \omega_2}{2}, \qquad k_g = \frac{k_1 - k_2}{2} \tag{9-27}$$

Then

$$E_R = 2E_0 \cos (k_p x - \omega_p t) \cos (k_g x - \omega_g t) \tag{9-28}$$

Equation (9-28) represents a product of two cosine waves. The first possesses a frequency ω_p and propagation constant k_p that are, respectively, the averages of the frequencies and propagation constants of the component waves. The second cosine factor represents a wave with frequency ω_g and propagation constant k_g that are much smaller by comparison, since differences of the original values are taken in Eq. (9-27). With $\omega_p \gg \omega_g$, plots of the cosine functions may appear like those of Figure 9-4a, calculated at the same point x_0. The slowly varying cosine function may be considered as a fraction that ranges between $+1$ and -1 for various t. Such a fraction multiplying the rapidly varying function reduces its displacement proportionately.

(a)

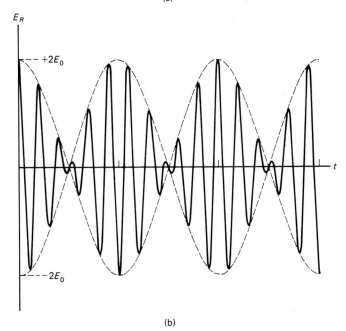

(b)

Figure 9-4 (a) Separate plots of the cosine factors of Eq. (9-28) at $x = x_0$, where $\omega_p \gg \omega_g$. (b) Modulated wave representing Eq. (9-28) at $x = x_0$.

The overall effect is that the low-frequency wave serves as an envelope modulating the high-frequency wave, as shown in Figure 9-4b. The dashed lines depict the envelope of the resulting wave disturbance. Such a wave disturbance exhibits the phenomenon of *beats*. Because the square of the displacement of the wave at any time is a measure of its radiant flux density, the energy delivered by the traveling sequence of pulses in Figure 9-4b is itself pulsating at a *beat frequency*, ω_b. The figure shows that the beat frequency is twice the frequency of the modulating envelope, or

$$\omega_b = 2\omega_g = 2\left(\frac{\omega_1 - \omega_2}{2}\right) = \omega_1 - \omega_2 \tag{9-29}$$

From Eq. (9-29) we see that the beat frequency is simply the difference frequency for the two waves. In the case of sound, this is the usual beat frequency heard when two tuning forks are made to vibrate simultaneously, equal to the difference in fork frequencies.

The preceding discussion has immediate application to optics in the phenomenon of dispersion. Due to dispersion, light components of different wavelengths travel with different speeds through a refractive medium. Even so-called monochromatic light possesses a spread of wavelengths, however narrow, about the average. Any two wavelength components of such a light beam, moving through a dispersive medium, can be represented by Eqs. (9-22) and (9-23) and thus produce a resultant like the one pictured in Figure 9-4b. The velocity of the higher-frequency wave as well as that of the lower-frequency envelope can be found from the general relation for velocity,

$$v = \nu\lambda = \frac{\omega}{k} \tag{9-30}$$

The velocity of the higher-frequency wave, from Eq. (9-26), is then the *phase velocity*,

$$v_p = \frac{\omega_p}{k_p} = \frac{\omega_1 + \omega_2}{k_1 + k_2} \cong \frac{\omega}{k} \tag{9-31}$$

where the final member is an approximation in the case $\omega_1 \cong \omega_2 = \omega$ and $k_1 \cong k_2 = k$ for neighboring frequency and wavelength components in a continuum. On the other hand, the velocity of the envelope, called the *group velocity*, is

$$v_g = \frac{\omega_g}{k_g} = \frac{\omega_1 - \omega_2}{k_1 - k_2} \cong \frac{d\omega}{dk} \tag{9-32}$$

assuming again that the differences between frequencies and propagation constants are small. Now group velocity $v_g = d\omega/dk$ and phase velocity $v_p = \omega/k$ need not be the same. If $v_p > v_g$, the high-frequency waves would appear to have a velocity to the right relative to the envelope, also in motion. These waves, which can be produced by an oscilloscope, would seem to disappear at the right node and be generated at the left node of each pulse. If $v_p < v_g$, their relative motion would, of course, be reversed. When $v_p = v_g$, the high-frequency waves and envelope would move together at the same rate, without relative motion. The relation between group and phase velocities can be found as follows. Substituting Eq. (9-31) into Eq. (9-32) and performing the differentiation of a product,

$$v_g = \frac{d\omega}{dk} = \frac{d}{dk}(kv_p)$$

$$\tag{9-33}$$

$$v_g = v_p + k\left(\frac{dv_p}{dk}\right)$$

When the velocity of a wave does not depend on wavelength, that is, in a nondispersive medium, $dv_p/dk = 0$, and phase and group velocities are equal. This is the case of light propagating in a vacuum, where $v_p = v_g = c$. In dispersive media, however, $v_p = c/n$, where the refractive index n is a function of λ or k. Then $n = n(k)$, and

$$\frac{dv_p}{dk} = \frac{d}{dk}\left(\frac{c}{n}\right) = \frac{-c}{n^2}\left(\frac{dn}{dk}\right)$$

When incorporated into Eq. (9-33), we have an alternate relation between phase and group velocities,

$$v_g = v_p\left[1 - \frac{k}{n}\left(\frac{dn}{dk}\right)\right] \qquad (9\text{-}34)$$

Again, using $k = 2\pi/\lambda$ and $dk = -(2\pi/\lambda^2)\,d\lambda$, Eq. (9-34) can be reformulated as

$$v_g = v_p\left[1 + \frac{\lambda}{n}\left(\frac{dn}{d\lambda}\right)\right] \qquad (9\text{-}35)$$

In regions of normal dispersion, $dn/d\lambda < 0$ and $v_g < v_p$.

These results, derived here for the case of two wave components, holds in general for a number of waves with a narrow range of frequencies.[1] Their sum can be characterized both by a phase velocity, the average velocity of the individual waves, and by the group velocity, the velocity of the modulating wave form itself. Since the latter determines the speed with which energy is transmitted, it is the directly measurable speed of the waves. When carrier waves are modulated to contain information, as in amplitude modulation (AM) of radio waves, we may speak of the group velocity as the signal velocity, which is usually less than the phase velocity of the carrier waves. When light pulses, consisting of a number of harmonic waves extending over a range of frequencies, are transmitted through a dispersive medium, the velocity of the group is the velocity of the pulses and is different from the velocity of the individual harmonic waves. In wave mechanics the electron itself is represented by a localized wave packet that can be decomposed into a number of harmonic waves with a range of wavelengths. The measured velocity of the electron is the velocity of the wave packet, that is, the group velocity of the constituent waves.

PROBLEMS

9-1. Two plane waves are given by

$$E_1 = \frac{5E_0}{(3x - 4t)^2 + 2} \quad \text{and} \quad E_2 = \frac{-5E_0}{(3x + 4t - 6)^2 + 2}$$

with x in meters and t in seconds.
(a) Describe the motion of the two waves.
(b) At what instant is their superposition everywhere zero?
(c) At what point is their superposition always zero?

9-2. (a) Show in a phasor diagram the following two harmonic waves:

$$E_1 = 2 \sin \omega t \quad \text{and} \quad E_2 = 7 \sin\left(\omega t + \frac{\pi}{4}\right)$$

(b) Determine the mathematical expression for the resultant wave.

---

[1] *Fourier analysis* provides techniques for dealing with the superposition of many harmonic components. This subject is introduced in Chapter 12.

9-3. Find the resultant of the superposition of two harmonic waves in the form

$$E = E_0 \sin (\omega t + \alpha)$$

with amplitudes of 3 and 4 and phases of 30° and 90°, respectively. Both waves have a period of 1 s.

9-4. Two waves traveling together along the same line are given by

$$y_1 = 5 \sin \left[\omega t + \frac{\pi}{2} \right]$$

$$y_2 = 7 \sin \left[\omega t + \frac{\pi}{3} \right]$$

Write the resultant wave equation.

9-5. Plot and write the equation of the superposition of the following harmonic waves: $E_1 = \sin (\omega t - 10°)$, $E_2 = 3 \cos (\omega t + 100°)$, and $E_3 = 2 \sin (\omega t - 30°)$, where the period of each is 2 s.

9-6. One hundred antennas are putting out identical waves, given by

$$E = 0.02 \sin (\omega t + \epsilon) \text{ V/m}$$

The waves are brought together at a point. What is the amplitude of the resultant when (a) all waves are in phase (coherent sources) and (b) the waves have random phase differences.

9-7. Two plane waves of the same frequency and with vibrations in the z-direction are given by

$$\psi (x, t) = 4 \sin \left[20t + \left(\frac{\pi}{3} \right) x + \pi \right]$$

$$\psi (y, t) = 2 \sin \left[20t + \left(\frac{\pi}{4} \right) y + \pi \right]$$

Write the resultant wave equation expressing their superposition at the point $x = 5$ and $y = 2$.

9-8. Beginning with the relation between group velocity and phase velocity in the form

$$v_g = v_p - \lambda (dv/d\lambda)$$

(a) express the relation in terms of n and ω and (b) determine whether the group velocity is greater or less than the phase velocity in a medium having a normal dispersion.

9-9. The *dispersive power* of glass is defined as the ratio $(n_F - n_C)/(n_D - 1)$, where C, D, and F refer to the Fraunhofer wavelengths, $\lambda_c = 6563$ Å, $\lambda_D = 5890$ Å, and $\lambda_F = 4861$ Å. Find the approximate group velocity in glass whose dispersive power is $\frac{1}{30}$ and for which $n_D = 1.50$.

9-10. The dispersion curve of glass can be represented approximately by Cauchy's empirical equation, $n = A + B/\lambda^2$. Find the phase and group velocities for light of 500-nm wavelength in a particular glass for which $A = 1.40$ and $B = 2.5 \times 10^6$ Å².

9-11. The dielectric constant K of a gas is related to its index of refraction by the relation $K = n^2$.

(a) Show that the group velocity for waves traveling in the gas may be expressed in terms of the dielectric constant by

$$v_g = \frac{c}{\sqrt{K}} \left[1 - \frac{\omega}{2K} \frac{dK}{d\omega} \right]$$

where c is the speed of light in vacuum.

(b) An empirical relation giving the variation of K with frequency is

$$K = 1 + [A/(\omega_0^2 - \omega^2)]$$

where A and ω_0 are constants for the gas. If the second term is very small compared to the first, show that

$$v_g \cong c\left[1 - \frac{\omega^2 A}{(\omega_0^2 - \omega^2)^2}\right]$$

9-12. **(a)** Show that group velocity can be expressed as

$$v_g = v_p - \lambda\left(\frac{dv_p}{d\lambda}\right)$$

(b) Find the group velocity for plane waves in a dispersive medium, for which $v_p = A + B\lambda$, where A and B are constants. Interpret the result.

9-13. Waves on the ocean have different velocities, depending on their depth. Long wavelength waves, traveling deep in the ocean, have a speed given approximately by

$$v = \left(\frac{g\lambda}{2\pi}\right)^{1/2}$$

where g is the acceleration of gravity. Short wavelength waves, corresponding to surface ripples, have a velocity given approximately by

$$v = \left(\frac{2\pi T}{\lambda\rho}\right)^{1/2}$$

where ρ is the density and T is the surface tension. Show that the group velocity for long wavelength waves is $\frac{1}{2}$ their phase velocity and the group velocity for short wavelength waves is $\frac{3}{2}$ their phase velocity.

9-14. A laser emits a monochromatic beam of wavelength λ, which is reflected normally from a plane mirror, receding at a speed v. What is the beat frequency between the incident and reflected light?

9-15. Standing waves are produced by the superposition of the wave

$$y = 7 \sin 2\pi\left[\frac{t}{T} - \frac{2x}{\pi}\right] \qquad (x, y \text{ in cm}; t \text{ in s})$$

and its reflection in a medium whose absorption is negligible. For the resultant wave, find the amplitude, wavelength, length of one loop, velocity, and period.

9-16. A medium is disturbed by an oscillation described by

$$y = 3 \sin\left(\frac{\pi x}{10}\right) \cos(50\pi t) \qquad (x, y \text{ in cm}; t \text{ in s})$$

(a) Determine the amplitude, frequency, wavelength, speed, and direction of the component waves whose superposition produces this result.
(b) What is the internodal distance?
(c) What are the displacement, velocity, and acceleration of a particle in the medium at $x = 5$ cm and $t = 0.22$ s?

9-17. Express the plane waves of Eqs. (9-17) and (9-18) in the complex representation. In this form, show that the superposition of the waves is the standing wave given by Eq. (9-20).

REFERENCES

[1] Hecht, Eugene, and Alfred Zajac. *Optics.* Reading, Mass.: Addison-Wesley Publishing Company, 1974. Ch. 7.

[2] Ghatak, Ajoy K. *An Introduction to Modern Optics.* New York: McGraw-Hill Book Company, 1972. Ch. 1.

10

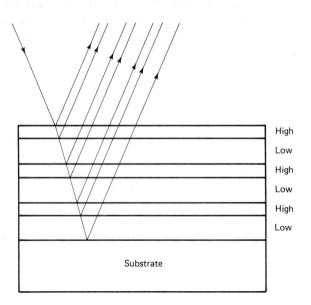

High
Low
High
Low
High
Low

Substrate

Interference of Light

INTRODUCTION

Like standing waves and beats, treated in the preceding chapter, the phenomenon of interference depends on the superposition of two or more individual waves under rather strict conditions that will soon be clarified. When interest lies primarily in the effects of enhancement or diminution of light waves, due precisely to their superposition, these effects are usually said to be due to the interference of light. When enhancement, or *constructive interference*, and diminution, or *destructive interference*, conditions alternate in a spatial display, the interference is said to produce a pattern of *fringes*, as in the double-slit interference pattern. The same conditions may lead to the enhancement of one visible wavelength interval or color at the expense of the others, in which case interference colors are produced, as in oil slicks and soap films. The simplest explanation of these phenomena can be successfully undertaken by treating light as a wave motion. In this and following chapters, several such applications, considered under the general heading of interference, are presented.

10-1 TWO-BEAM INTERFERENCE

We consider first the interference of two waves, represented by $\mathbf{E}_1$ and $\mathbf{E}_2$, where we take into account the vector property of the electric fields. In cases of interference, both waves typically originate from a single source and reunite after traveling along different paths. The direction of travel of the waves need not be the same when they come together, however, so that whereas they maintain the same frequency, they generally do not have the same propagation vector $\mathbf{k}$. Accordingly, we may express the wave equations by

$$\mathbf{E}_1 = \mathbf{E}_{01} \cos (\mathbf{k}_1 \cdot \mathbf{r} - \omega t + \epsilon_1) \tag{10-1}$$

$$\mathbf{E}_2 = \mathbf{E}_{02} \cos (\mathbf{k}_2 \cdot \mathbf{r} - \omega t + \epsilon_2) \tag{10-2}$$

At some general point P, defined by *position vector* $\mathbf{r}$, the waves intersect to produce a disturbance whose electric field $\mathbf{E}_p$ is given by the principle of superposition,

$$\mathbf{E}_p = \mathbf{E}_1 + \mathbf{E}_2$$

Now $\mathbf{E}_1$ and $\mathbf{E}_2$ are rapidly varying functions with optical frequencies of the order of 10^{14} to 10^{15} Hz for visible light. Thus both $\mathbf{E}_1$ and $\mathbf{E}_2$ average to zero over very short time intervals. Measurement of the waves by their effect on the eye or some other light detector depends on the energy of the light beam. The radiant power density, or *irradiance*, E_e (W/m²), measures the time average of the square of the wave amplitude. Unfortunately, the standard symbol for irradiance, except for subscript, is the same as that for the electric field. To avoid confusion we use, temporarily, the symbol I for irradiance.

$$I = \epsilon_0 c \langle \mathbf{E}^2 \rangle \tag{10-3}$$

Thus the resulting irradiance at P is given by

$$I = \epsilon_0 c \langle \mathbf{E}_p^2 \rangle = \epsilon_0 c \langle \mathbf{E}_p \cdot \mathbf{E}_p \rangle = \epsilon_0 c \langle (\mathbf{E}_1 + \mathbf{E}_2) \cdot (\mathbf{E}_1 + \mathbf{E}_2) \rangle$$

or

$$I = \epsilon_0 c \langle \mathbf{E}_1^2 + \mathbf{E}_2^2 + 2\mathbf{E}_1 \cdot \mathbf{E}_2 \rangle \tag{10-4}$$

By Eq. (10-3), the first two terms correspond to the irradiances of the individual waves, I_1 and I_2. The last term depends on an interaction of the waves and is called the *interference term*, I_{12}. We may then write

$$I = I_1 + I_2 + I_{12} \tag{10-5}$$

If light behaved without interference, like classical particles, we would then expect $I = I_1 + I_2$. The presence of the third term I_{12} is indicative of the wave nature of light, which can produce enhancement or diminution of the irradiance through interference. Notice that when $\mathbf{E}_1$ and $\mathbf{E}_2$ are orthogonal, so that their dot product vanishes, no interference results. When the electric fields are parallel, on the other hand, the interference term makes its maximum contribution. Two beams of unpolarized light produce interference because each can be resolved into orthogonal components of $\mathbf{E}$ that can then be paired off with similar components of the other beam. Each component produces an interference term with $\mathbf{E}_1 \parallel \mathbf{E}_2$ ($\mathbf{E}_1$ parallel to $\mathbf{E}_2$).

Consider the interference term,

$$I_{12} = 2\epsilon_0 c \langle \mathbf{E}_1 \cdot \mathbf{E}_2 \rangle \tag{10-6}$$

where $\mathbf{E}_1$ and $\mathbf{E}_2$ are given by Eqs. (10-1) and (10-2). Their dot product,

$$\mathbf{E}_1 \cdot \mathbf{E}_2 = \mathbf{E}_{01} \cdot \mathbf{E}_{02} \cos (\mathbf{k}_1 \cdot \mathbf{r} - \omega t + \epsilon_1) \cos (\mathbf{k}_2 \cdot \mathbf{r} - \omega t + \epsilon_2)$$

can be simplified by first expanding the cosine factors, interpreted as the difference of two angles. To this end, let us define

$$\alpha \equiv \mathbf{k}_1 \cdot \mathbf{r} + \epsilon_1 \quad \text{and} \quad \beta \equiv \mathbf{k}_2 \cdot \mathbf{r} + \epsilon_2$$

so that

$$\mathbf{E}_1 \cdot \mathbf{E}_2 = \mathbf{E}_{01} \cdot \mathbf{E}_{02} \cos (\alpha - \omega t) \cos (\beta - \omega t)$$

Expanding and multiplying the cosine factors, we arrive at

$$\langle \mathbf{E}_1 \cdot \mathbf{E}_2 \rangle = \mathbf{E}_{01} \cdot \mathbf{E}_{02}[\cos \alpha \cos \beta \langle \cos^2 \omega t \rangle + \sin \alpha \sin \beta \langle \sin^2 \omega t \rangle$$
$$+ (\cos \alpha \sin \beta + \sin \alpha \cos \beta)\langle \sin \omega t \cos \omega t \rangle]$$

where time averages are indicated for each time-dependent factor. Over any number of complete cycles, one can easily show that

$$\langle \cos^2 \omega t \rangle = \tfrac{1}{2}, \qquad \langle \sin^2 \omega t \rangle = \tfrac{1}{2}$$

and

$$\langle \sin \omega t \cos \omega t \rangle = 0$$

Thus

$$\langle \mathbf{E}_1 \cdot \mathbf{E}_2 \rangle = \tfrac{1}{2}\mathbf{E}_{01} \cdot \mathbf{E}_{02} \cos (\alpha - \beta)$$

or

$$\langle \mathbf{E}_1 \cdot \mathbf{E}_2 \rangle = \tfrac{1}{2}\mathbf{E}_{01} \cdot \mathbf{E}_{02} \cos [(\mathbf{k}_1 - \mathbf{k}_2) \cdot \mathbf{r} + (\epsilon_1 - \epsilon_2)] \qquad (10\text{-}7)$$

where the expression in brackets is the phase difference between $\mathbf{E}_1$ and $\mathbf{E}_2$, as given in Eqs. (10-1) and (10-2):

$$\delta = (\mathbf{k}_1 - \mathbf{k}_2) \cdot \mathbf{r} + (\epsilon_1 - \epsilon_2) \qquad (10\text{-}8)$$

Combining Eqs. (10-6), (10-7), and (10-8),

$$I_{12} = \epsilon_0 c \mathbf{E}_{01} \cdot \mathbf{E}_{02} \cos \delta \qquad (10\text{-}9)$$

Similarly, the irradiance terms I_1 and I_2 of Eq. (10-5) can be shown to yield

$$I_1 = \epsilon_0 c \langle \mathbf{E}_1^2 \rangle = \tfrac{1}{2}\epsilon_0 c E_{01}^2 \qquad (10\text{-}10)$$

and

$$I_2 = \epsilon_0 c \langle \mathbf{E}_2^2 \rangle = \tfrac{1}{2}\epsilon_0 c E_{02}^2 \qquad (10\text{-}11)$$

In the case $\mathbf{E}_{01} \parallel \mathbf{E}_{02}$, their dot product in Eq. (10-9) is identical with the product of their magnitudes. These may be expressed in terms of I_1 and I_2 by the use of Eqs. (10-10) and (10-11), resulting in

$$I_{12} = 2\sqrt{I_1 I_2} \cos \delta \qquad (10\text{-}12)$$

so that we may write, finally,

$$I = I_1 + I_2 + 2\sqrt{I_1 I_2} \cos \delta \qquad (10\text{-}13)$$

Notice that once we have made the assumption that the $\mathbf{E}$-fields are parallel, the treatment becomes the same as the scalar theory of the preceding chapter. In particular, Eq. (10-13) follows directly from Eq. (9-14) for $N = 2$ and when Eqs. (10-10) and (10-11) are used to introduce irradiances in place of amplitudes.

Depending on whether $\cos \delta > 0$ or $\cos \delta < 0$ in Eq. (10-13), the interference term either augments or diminishes the sum of the individual irradiances I_1 and I_2, leading to constructive or destructive interference, respectively. On the other hand, if the initial phase difference $(\epsilon_1 - \epsilon_2)$ in Eq. (10-8) varies randomly, the waves are said to be mutually incoherent, and $\cos \delta$ becomes a time-dependent factor whose average is zero. Even though interference is always occurring, no pattern can be sustained long enough to be detected. Thus some degree of coherence, that is, $\langle \cos \delta \rangle \neq 0$, is necessary to observe interference. In particular, if the two waves originate from independent sources, such as incandescent bulbs or gas-discharge lamps, the waves will be mutually incoherent. Laser sources, even though independent, can possess sufficient mutual coherence for interference to be observed over short periods of time. The other term in $\cos \delta$ is, from Eq. (10-8), $(\mathbf{k_1} - \mathbf{k_2}) \cdot \mathbf{r}$. As the point of observation given by $\mathbf{r}$ varies, $\cos \delta$ takes on alternating maximum and minimum values and interference fringes occur, spatially separated.

To be more specific, when $\cos \delta = +1$, constructive interference yields the maximum irradiance

$$I_{max} = I_1 + I_2 + 2\sqrt{I_1 I_2} \qquad (10\text{-}14)$$

This condition occurs whenever the phase difference $\delta = 2m\pi$, where m is any integer or zero. On the other hand, when $\cos \delta = -1$, destructive interference yields the minimum, or background, irradiance

$$I_{min} = I_1 + I_2 - 2\sqrt{I_1 I_2} \qquad (10\text{-}15)$$

a condition that occurs whenever $\delta = (2m + 1)\pi$. A plot of irradiance I versus phase δ, in Figure 10-1a, exhibits periodic fringes. Destructive interference is complete, that is, cancellation is complete, when $I_1 = I_2 = I_0$. Then, Eqs. (10-14) and (10-15) give

$$I_{max} = 4I_0 \quad \text{and} \quad I_{min} = 0$$

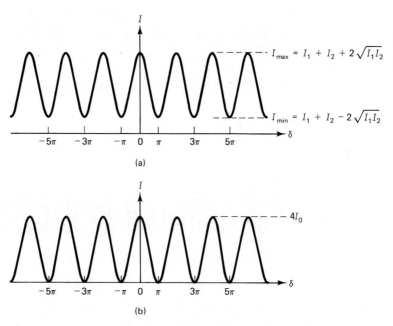

(a)

(b)

Figure 10-1 Irradiance of interference fringes as a function of phase. Fringe contrast is enhanced in (b), where the background irradiance $I_{min} = 0$ when $I_1 = I_2$.

Resulting fringes, shown in Figure 10-1b, now exhibit better contrast. A measure of *fringe contrast*, also called *visibility*, with values between 0 and 1, is given by the quantity

$$\text{fringe contrast} = \frac{I_{max} - I_{min}}{I_{max} + I_{min}}$$

In the experimental utilization of fringe patterns, it is therefore usually desirable to arrange that the interfering beams have the same amplitudes.

Another useful form of Eq. (10-13), for the case of interfering beams of equal amplitude, is found by writing

$$I = I_0 + I_0 + 2\sqrt{I_0^2} \cos \delta = 2I_0(1 + \cos \delta)$$

and then making use of the trigonometric identity

$$1 + \cos \delta \equiv 2 \cos^2 \left(\frac{\delta}{2} \right)$$

The irradiance for two equal interfering beams is then

$$I = 4I_0 \cos^2 \left(\frac{\delta}{2} \right) \tag{10-16}$$

Notice that energy is not conserved at each point of the superposition, that is, $I \neq 2I_0$, but that over at least one spatial period of the fringe pattern $I_{av} = 2I_0$. This situation is typical of interference and diffraction phenomena: If the power density falls below the average at some points, it rises above the average at other points in such a way that the total pattern satisfies the principle of energy conservation.

Example

Two interfering beams with parallel electric fields are given by

$$E_1 = 2 \cos \left(\mathbf{k}_1 \cdot \mathbf{r} - \omega t + \frac{\pi}{3} \right) \quad \text{(kV/m)}$$

$$E_2 = 5 \cos \left(\mathbf{k}_2 \cdot \mathbf{r} - \omega t + \frac{\pi}{4} \right) \quad \text{(kV/m)}$$

Let us determine the irradiance contributed by each beam acting alone and that due to their mutual interference at a point where their path difference is zero. We have

$$I_1 = \tfrac{1}{2} \epsilon_0 c E_{01}^2 = \tfrac{1}{2} \epsilon_0 c (2000)^2 = 5309 \text{ W/m}^2$$

$$I_2 = \tfrac{1}{2} \epsilon_0 c E_{02}^2 = \tfrac{1}{2} \epsilon_0 c (5000)^2 = 33{,}180 \text{ W/m}^2$$

$$I_{12} = 2\sqrt{I_1 I_2} \cos \delta = 2\sqrt{(5309 \times 33180)} \cos \left(\frac{\pi}{3} - \frac{\pi}{4} \right) = 25{,}640 \text{ W/m}^2$$

To find the fringe contrast near the region of superposition we must calculate

$$I_{max} = I_1 + I_2 + 2\sqrt{I_1 I_2} = 5309 + 33180 + 2\sqrt{(5309 \times 33180)}$$

$$= 65{,}034 \text{ W/m}^2$$

$$I_{min} = I_1 + I_2 - 2\sqrt{I_1 I_2} = 5309 + 33180 - 2\sqrt{(5309 \times 33180)}$$

$$= 11{,}945 \text{ W/m}^2$$

The contrast is then given by $(I_{max} - I_{min})/(I_{max} + I_{min})$, or

$$\text{fringe contrast} = \frac{65{,}034 - 11{,}945}{65{,}034 + 11{,}945} = 0.690$$

If the amplitudes of the two waves were equal, then $I_{max} = 4I_0$, $I_{min} = 0$, and the fringe contrast would be 1.

The decisive experiment performed by Thomas Young in 1802 is shown schematically in Figure 10-2. Monochromatic light is first allowed to pass through a single small hole in an aperture in order to approximate a single point source S. The light spreads out in spherical waves from the source according to Huygens' principle and is allowed to fall on the two closely spaced holes, S_1 and S_2, in an aperture. The holes thus become two coherent sources of light, whose interference can be observed on a screen some distance away. If the two holes are equal in size, light emanating from the holes have comparable amplitudes, and the irradiance at any point of superposition is given by Eq. (10-16). For observation points, such as P, on the screen a distance s from the aperture, the phase difference δ between the two waves arriving must be determined to calculate the resultant irradiance there. Clearly, if $S_2 P - S_1 P = m\lambda$, the waves will arrive in phase, and maximum irradiance or brightness results. If $S_2 P - S_1 P = (m + \frac{1}{2})\lambda$, the requisite condition for destructive interference or darkness is met. Practically speaking, the hole separation a is much smaller than the screen distance s, allowing a simple expression for the path distance, $S_2 P - S_1 P$. Using P as a center, let an arc $S_1 Q$ be drawn of radius $S_1 P$, so that it intersects the line $S_2 P$ at Q. Then $S_2 P - S_1 P$ is equal to the segment Δ, as shown. The first approximation is to regard arc $S_1 Q$ as a straight-line segment that forms one leg of the right triangle, $S_1 S_2 Q$. If θ is the angle between the aperture and $S_1 Q$, $\Delta = a \sin \theta$. The second approximation identifies the angle θ with the angle between the optical axis OX and the line drawn from the midpoint O between holes to the point P at the screen. Observe that the corresponding sides of the two angles θ are related such that $OX \perp S_1 S_2$, and OP is almost exactly perpendicular to $S_1 Q$. The condition for constructive interference at a point P on the screen is then, to a very good approximation,

$$S_2 P - S_1 P = \Delta = m\lambda \cong a \sin \theta \qquad (10\text{-}17)$$

whereas for destructive interference,

$$\Delta = (m + \tfrac{1}{2})\lambda = a \sin \theta \qquad (10\text{-}18)$$

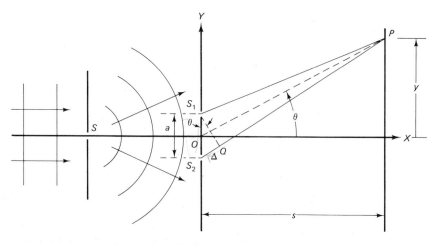

Figure 10-2   Schematic for Young's double-slit experiment. The holes S_1 and S_2 are usually slits, with the long dimensions extending into the page.

where m is zero or of integral value. The irradiance on the screen, at a point determined by the angle θ, is found using Eq. (10-16) and the relationship between path difference Δ and phase difference δ,

$$\delta = \left(\frac{2\pi}{\lambda}\right)\Delta$$

The result is

$$I = 4I_0 \cos^2\left(\frac{\pi\Delta}{\lambda}\right) = 4I_0 \cos^2\left(\frac{\pi a \sin\theta}{\lambda}\right).$$

For points P near the optical axis, where $y \ll s$, we may approximate further: $\sin\theta \cong \tan\theta \cong y/s$, so that

$$I = 4I_0 \cos^2\left(\frac{\pi a y}{\lambda s}\right) \tag{10-19}$$

By allowing the cosine function in Eq. (10-19) to become alternately ± 1 and 0, the conditions expressed by Eqs. (10-17) and (10-18) for constructive and destructive interference are reproduced.

Arguing now from Eq. (10-17) for bright fringe positions in the form

$$y_m = \frac{m\lambda s}{a}, \qquad m = 0, 1, 2, \dots \tag{10-20}$$

we find a constant separation between irradiance maxima, corresponding to successive values of m, given by

$$\Delta y = \frac{\lambda s}{a} \tag{10-21}$$

with minima situated midway between them. Thus fringe separation is proportional both to wavelength and screen distance and inversely proportional to the hole spac-

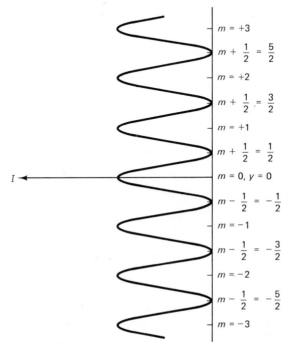

Figure 10-3 Irradiance versus distance from optical axis for double-slit fringe pattern. The *order* of the interference pattern is indicated by m, with integral values of m determining positions of fringe maxima.

ing. Reducing the hole spacing expands the fringe pattern formed by each color. Measurement of the fringe separation provides a means of determining the wavelength of the light. The single hole, used to secure a degree of spatial coherence, may be eliminated if laser light, both highly monochromatic and spatially coherent, is used to illuminate the double slit. In the observational arrangement just described, fringes are observed on a screen placed perpendicular to the optical axis at some distance from the aperture, as indicated in Figure 10-3. Fringe maxima coincide with integral orders of m, and fringe minima fall halfway between maxima.

Example

Light from a narrow slit passes through two identical and parallel slits, 0.2 mm apart. Interference fringes are seen on a screen 1 m away, with a separation of 3.29 mm. How does the irradiance at the screen vary, if the contribution of one slit alone is I_0? What is the wavelength of the light?

Solution   According to Eq. (10-19), $I = 4I_0 \cos^2 [\pi ay/\lambda s]$. In this case,

$$I = 4I_0 \cos^2 [\pi (0.0002)y/(658 \times 10^{-9})(1)] = 4I_0 \cos^2 (955y)$$

From Eq. (10-21),

$$\lambda = a\,\Delta y/s = (0.0002)(3.29 \times 10^{-3})/1 = 658 \text{ nm}$$

An alternative way to view the formation of bright (B) positions of constructive interference and dark (D) positions of destructive interference is shown in Figure 10-4. The crests and valleys of spherical waves from S_1 and S_2 are shown approaching the screen. Along directions marked B, wave crests (or wave valleys) from both slits coincide, producing maximum irradiance. Along directions marked D, on the other hand, the waves are seen to be out of step by half a wavelength, and destructive interference results.

Obviously, fringes should be present in all the space surrounding the holes, where light from the holes is allowed to interfere, though the irradiance is greatest in

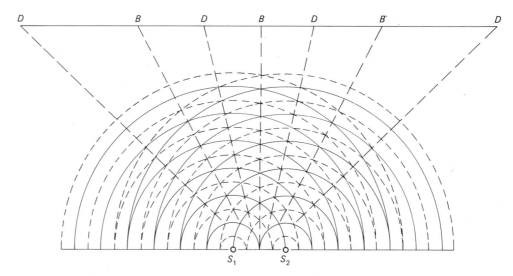

Figure 10-4 Alternating bright and dark interference fringes are produced by light from two coherent sources. Along directions where crests (solid circles) from S_1 intersect crests from S_2, brightness (B) results. Along directions where crests meet valleys (dashed circles), darkness (D) results.

the forward direction. If we imagine two coherent point sources of light radiating in all directions, then the condition given by Eq. (10-17) for bright fringes,

$$S_2P - S_1P = m\lambda \qquad (10\text{-}22)$$

defines a family of bright fringe surfaces in the space surrounding the holes. To visualize this set of surfaces, we may take advantage of the inherent symmetry in the arrangement. In Figure 10-5, the intersection of several bright fringe surfaces with a plane that includes the two sources is shown, each surface corresponding to an integral value of order m. The surfaces are hyperbolic, since Eq. (10-22) is precisely the condition for a family of hyperbolic curves with parameter m. Inasmuch as the y-axis is an axis of symmetry, the corresponding bright fringe surfaces are generated by rotating the entire pattern about the y-axis. One should then be able to visualize the intercept of these surfaces with the plane of an observational screen placed anywhere in the vicinity. In particular, a screen placed perpendicular to the OX axis, as in Figure 10-2, intercepts hyperbolic arcs that appear as straight-line fringes near the axis, whereas a screen placed perpendicular to the OY axis shows concentric circular fringes centered on the axis. Because the fringe system extends throughout the space surrounding the two sources, the fringes are said to be *nonlocalized*.

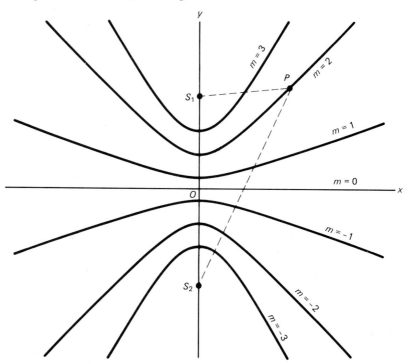

Figure 10-5 Bright fringe surfaces for two coherent point sources. The distances from S_1 and S_2 to any fringe point P differ by an integral number of wavelengths. The surfaces are generated by rotating the pattern about the y-axis.

The holes S, S_1, and S_2 of Figure 10-2 are usually replaced by parallel, narrow slits (oriented with their long sides perpendicular to the page in Figure 10-2) to illuminate more fully the interference pattern. The effect of the array of point sources along the slits, each set producing its own fringe system as just described, is simply to elongate the pattern parallel to the fringes, without changing their geometrical relationships. This is true even when two points along a source slit are not mutually coherent.

Interference fringes may sometimes appear in arrangements when only one light source is present. It is possible, through reflection or refraction, to produce virtual images that, acting together or with the actual source, behave as two coherent sources that can produce an interference pattern. Figures 10-6 to 10-8 illustrate three such examples. These examples are not only of some historic importance; they also serve to impress us with the variety of ways unexpected fringe patterns may appear in optical experiments, especially when the extremely coherent light of a laser is being used. In Figure 10-6, interference fringes are produced due to the superposition of light at the screen that originates at the actual source S and, by reflection, also originates effectively from its virtual source S' below the surface of the plane mirror MM'. Where the direct and reflected beams strike the screen, fringes will appear. The position of bright fringes is given by the double slit Eq. (10-20), where a is twice the distance of source S above the mirror plane. The arrangement is known as *Lloyd's mirror*. If the screen contacts the mirror at M', the fringe at their intersection is found to be dark. Since at this point the optical-path difference between the two interfering beams vanishes, one should expect a bright fringe. The

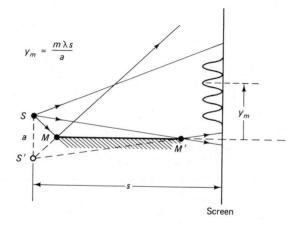

$$y_m = \frac{m\lambda s}{a}$$

Figure 10-6 Interference with Lloyd's mirror. Coherent sources are the point source S and its virtual image, S'.

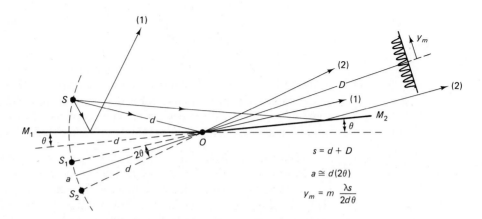

$$s = d + D$$

$$a \cong d(2\theta)$$

$$y_m = m\frac{\lambda s}{2d\theta}$$

Figure 10-7 Interference with Fresnel's mirrors. Coherent sources are the two virtual images of point source S, formed in the two plane mirrors M_1 and M_2. Direct light from S is not allowed to reach the screen.

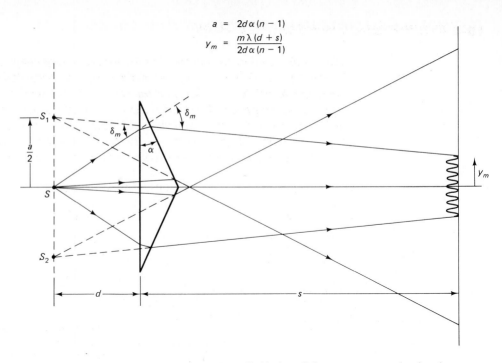

$$a = 2d\alpha(n-1)$$
$$y_m = \frac{m\lambda(d+s)}{2d\alpha(n-1)}$$

Figure 10-8 Interference with Fresnel's biprism. Coherent sources are the virtual images S_1 and S_2 of source S, formed by refraction in the two halves of the prism.

contrary experimental result is explained by requiring a phase shift of π for the air-glass reflection.[1]

Another closely related arrangement is *Fresnel's mirrors*, Figure 10-7. Interference occurs between the light reflected from each of two mirrors M_1 and M_2, inclined at a small relative angle θ. Two rays reflected from each are shown labeled as (1) from M_1 and (2) from M_2. Interference fringes appear in the region of overlap. Interference effectively occurs between the two coherent virtual images S_1 and S_2, acting as sources. Once the virtual image separation a is related to the tilt angle θ and to the distance d from actual source to the intersection of the mirrors at O, the fringe pattern may again be described by Eq. (10-20). The screen is shown at distance D from point O.

Figure 10-8 shows *Fresnel's biprism*, which refracts light from a small source S in such a way that it appears to come from two coherent, virtual sources, S_1 and S_2. Extreme rays for refraction at the top and bottom halves are shown. Interference fringes are seen in the overlap region on the screen. In practice, the prism angle α is very small, of the order of a degree. One of the rays (shown) passes through the wedge in a symmetrical fashion, making equal entrance and exit angles with the two sides and satisfying the condition for minimum deviation. For this ray the deviation angle δ_m is given by $\delta_m = \alpha(n-1)$. The geometry of this particular ray provides a means of approximately determining the virtual source separation a in terms of prism index n and angle α:

$$a = 2d\delta_m = 2d\alpha(n-1) \tag{10-23}$$

Interference fringes are then described by Eq. (10-20), as usual.

---

[1] A theoretical explanation for phase changes on reflection results from an analysis based on Maxwell's equations and requires identification of the state of polarization of the light. The subject is treated in Chapter 20.

The familiar appearance of colors on the surface of oily water and soap films and the beautiful iridescence often seen in mother-of-pearl, peacock feathers, and butterfly wings are associated with the interference of light in single or multiple thin surface layers of transparent material. There exists a variety of situations in which such interference can take place, affecting the nature of the interference pattern and the conditions under which it can be observed. Variables in the situation include the size and spectral width of the source and the shape and reflectance of the film.

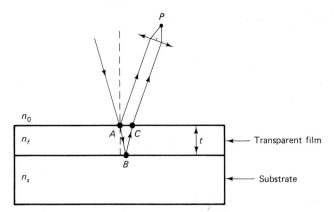

Figure 10-9 Double-beam interference from a film. Rays reflected from the top and bottom plane surfaces of the film are brought together at P by a lens.

Consider the case of a film of transparent material bounded by parallel planes, such as might be formed by an oil slick, a metal oxide layer, or an evaporated coating on a flat, glass substrate (Figure 10-9). A beam of light incident on the film surface at A divides into reflected and refracted portions. This separation of the original light into two parts, preliminary to recombination and interference, is usually referred to as *amplitude division*, in contrast to a situation like Young's double slit, in which separation is said to occur by *wavefront division*. The refracted beam reflects again at the film-substrate interface B and leaves the film at C, in the same direction as the beam reflected at A. Part of the beam may reflect internally again at C and continue to experience multiple reflections within the film layer until it has lost its intensity. There will thus exist multiple parallel beams emerging from the top surface, although with rapidly diminishing amplitudes. Unless the reflectance of the film is large, a good approximation to the more complex situation of multiple reflection (Section 11-5) is to consider only the first two emerging beams. The two parallel beams leaving the film at A and C can be brought together by a converging lens, the eye, for example. The two beams intersecting at P superpose and interfere. Since the two beams travel different paths from point A onward, a relative phase difference develops that can produce constructive or destructive interference at P. The optical path difference Δ, *in the case of normal incidence*, is the additional path length ABC traveled by the refracted ray times the refractive index of the film. Thus

$$\Delta = n(AB + BC) = n(2t) \qquad (10\text{-}24)$$

where t is the film thickness. For example, if $2nt = \lambda_0$, the wavelength of the light in vacuum, the two interfering beams, on the basis of optical path difference alone, would be in phase and produce constructive interference. However, an additional phase difference, due to the phenomenon of phase changes on reflection, must be considered. Suppose that $n_f > n_0$ and $n_f > n_s$. In fact, often $n_0 = n_s$ because the media bounding the film are identical, as in the case of a water film (soap bubble) in air. Then the reflection at A occurs with light going from a lower index n_0 toward a higher index n_f, a condition usually called *external reflection*. The reflection at B, on the other hand, occurs for light going from a higher index n_f toward a lower index n_s, the condition of *internal reflection*. A relative phase shift of π occurs between the externally and internally reflected beams, so that equivalently, an additional path difference of $\lambda/2$ is introduced between the two beams. The net optical path difference between the beams is then $\lambda + \lambda/2$, which puts them precisely out of phase, and destructive interference results at P. If, instead, both reflections are external $(n_0 < n_f < n_s)$ or if both reflections are internal $(n_0 > n_f > n_s)$, no relative phase difference due to reflection needs to be taken into account. In that case, constructive interference occurs at P.

A frequent use of such single-layer films is in the production of *antireflecting coatings* on optical surfaces. In most cases, the light enters the film from air, so that $n_0 = 1$. Furthermore, if $n_s > n_f$, no relative phase shift between the two reflected beams occurs, and the optical-path difference alone determines the type of interference to be expected. If the film thickness is $\lambda_f/4$, where λ_f is the wavelength of the light in the film, then $2t = \lambda_f/2$ and the optical path difference $2n_f t = \lambda_0/2$, since $\lambda_0 = n_f \lambda_f$. Destructive interference occurs at this wavelength and to some extent at neighboring wavelengths, which means that the light reflected from such a film is the incident spectrum minus the wavelength region around λ_0. If the incident light is white and λ_0 is in the visible region, the reflected light is colored. Extinction of a region of the spectrum by nonreflecting films of $\lambda/4$ thickness is, of course, more effective if the amplitudes of the two reflected beams are equal. In general, all one can say is that for constructive interference the two amplitudes add (being in phase), and for destructive interference the amplitudes subtract (being exactly out of phase). For the difference to be zero, that is, for destructive interference to be complete, the amplitudes must be equal. We show later (Chapter 19) that, in the case of normal incidence, the *reflection coefficient* (or ratio of reflected to incident electric field amplitudes) is given by

$$r = \frac{1 - n}{1 + n} \tag{10-25}$$

where the *relative index* $n = n_2/n_1$. The amplitudes of the electric field reflected internally and externally from the film of Figure 10-9 are then equal, assuming a nonabsorbing film, if the relative indices are equivalent for these cases, that is, if

$$\frac{n_f}{n_0} = \frac{n_s}{n_f} \quad \text{or} \quad n_f = \sqrt{n_0 n_s} \tag{10-26}$$

Since usually $n_0 = 1$, the requirement that reflected beams be of equal amplitude is met by choosing a film whose refractive index is the square root of the substrate's refractive index. A suitable film material for the application may or may not exist, and some compromise is made. For example, to reduce the reflectance of lenses employed in optical instruments handling white light, the film thickness of $\lambda/4$ is determined with a λ in the center of the visible spectrum or wherever the detection system is most sensitive. In the case of the eye, this is the yellow-green portion near 550 nm. Assuming $n = 1.50$ for the glass lens, ideally $n_f = \sqrt{1.50} = 1.22$. The nearest practical film material with a matching index is MgF_2, with $n = 1.38$. The beneficial loss of reflected light near the middle of the spectrum results in a predominance of the blue and red ends of the spectrum, so that the coatings appear purple in reflected light.

As another example, consider a multilayer stack of alternating high-low index dielectric films (Figure 10-10). If each film has an optical thickness of $\lambda_f/4$, a little analysis shows that in this case all emerging beams are in phase. Multiple reflections

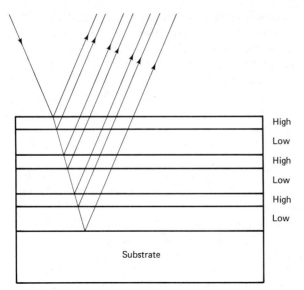

Figure 10-10 Multilayer dielectric mirror of alternating high and low index. Each film is $\lambda_f/4$ in optical thickness.

in the region of λ_0 increase the total reflected intensity and the quarter-wave stack performs as an efficient mirror. Such multilayer stacks can be designed to satisfy extinction or enhancement of reflected light over a greater portion of the spectrum than the single layer film. They are treated in greater detail in Chapter 19.

Returning now to the single-layer film, we want first to generalize the conditions for constructive and destructive interference by calculating the optical path difference in the case incident rays are not normal. Figure 10-11 illustrates a ray incident on a film at an angle θ_i. The phase difference at points C and D between emerging beams is due to the optical path difference between paths AD and ABC. After points C and D are reached, the respective beams are parallel and in the same medium, so that no further phase difference occurs. To assist in the calculation, point G is shown midway between A and C at the foot of the altitude BG in the isosceles triangle ABC. Points E and F are determined by constructing the perpendiculars GE and GF to the ray paths AB and BC, respectively. The optical path difference between the emerging beams is then

$$\Delta = n_f(AB + BC) - n_0(AD)$$

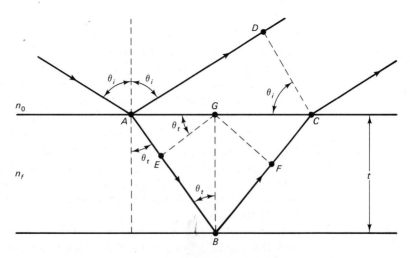

Figure 10-11 Single-film interference with light incident at arbitrary angle θ_i.

where n_f and n_0 are the refractive indices of film and external medium, as shown. It is helpful to break the distances AB and BC into parts, resulting in

$$\Delta = [n_f(AE + FC) - n_0 AD] + n_f(EB + BF) \qquad (10\text{-}27)$$

The quantity in square brackets vanishes, as we now show. By Snell's law,

$$n_0 \sin \theta_i = n_f \sin \theta_t \qquad (10\text{-}28)$$

In addition, by inspection,

$$AE = AG \sin \theta_t = \left(\frac{AC}{2}\right) \sin \theta_t \qquad (10\text{-}29)$$

and

$$AD = AC \sin \theta_i \qquad (10\text{-}30)$$

From Eq. (10-29) and incorporating, in turn, Eqs. (10-30) and (10-28),

$$2AE = AC \sin \theta_t = AD \left(\frac{\sin \theta_t}{\sin \theta_i}\right) = AD \left(\frac{n_0}{n_f}\right)$$

so that

$$n_0 AD = 2n_f AE = n_f(AE + FC) \qquad (10\text{-}31)$$

which was to be proved. There remains then, from Eq. (10-27),

$$\Delta = n_f(EB + BF) = 2n_f EB \qquad (10\text{-}32)$$

The length EB is related to the film thickness t by $EB = t \cos \theta_t$, so we have, finally,

$$\Delta = 2n_f t \cos \theta_t \qquad (10\text{-}33)$$

The optical path difference Δ is economically expressed by Eq. (10-33) in terms of the angle of refraction, not the angle of incidence, which of course can be recovered through Snell's law, Eq. (10-28). Notice that for normal incidence, $\theta_i = \theta_t = 0$ and $\Delta = 2n_f t$, as expected. The corresponding phase difference is $\delta = k\Delta = (2\pi/\lambda_0) \Delta$. The net phase difference must also take into account possible phase differences that arise on reflection, as discussed previously. Nevertheless, if we call Δ_p the optical path difference given by Eq. (10-33) and Δ_r the equivalent path difference arising from phase change on reflection, we can state quite generally the conditions for

$$\text{constructive interference:} \quad \Delta_p + \Delta_r = m\lambda \qquad (10\text{-}34)$$

and

$$\text{destructive interference:} \quad \Delta_p + \Delta_r = (m + \tfrac{1}{2})\lambda \qquad (10\text{-}35)$$

where $m = 0, 1, 2, \ldots$.

If, for example, constructive interference results between the two parts of a single beam incident at angle θ_i, the same condition will hold for all beams incident at the same angle. This is possible if the source is an extended source, as in Figure 10-12. Independent point sources S_1, S_2, and S_3 are shown, all contributing to the intensity of the light at P. Since these sources are noncoherent, interference is sustained only between pairs of reflected rays originating from the same source. If the lens aperture becomes too small to admit two such beams, such as (a) and (b) from S_1, no interference is detected. This may happen, for example, if the film thickness and, therefore, the spatial separation of two interfering beams—such as (a) and

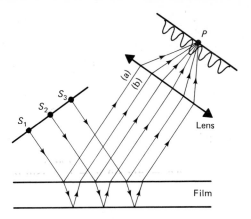

Figure 10-12 Interference by a dielectric film with an extended source. Fringes of equal inclination are focused by a lens.

(b)—are increased, while the pupil of the eye viewing the reflected light is limited in size. Without a focusing device, these *virtual fringes* do not appear. They are called *localized fringes* because they are, so to speak, localized at infinity. Recall that *nonlocalized fringes* (Figure 10-5) are, in contrast, formed everywhere. Fringes formed as in Figure 10-12 are also referred to as *Haidinger fringes*, or *fringes of equal inclination*, since they are formed by parallel incident beams from an extended source. If a different inclination is chosen, parallel rays from the various source points are incident on the film at a different angle, reflect as parallel rays from the film at a different angle, and all focus at some other point where they interfere, according to the conditions expressed by Eqs. (10-34) and (10-35).

The fringes of equal inclination just described are not possible if the source is a point or is very small, since every ray of light from the source to the film must, in that case, arrive at a different angle of incidence (Figure 10-13). Fringes of a different kind are nonetheless formed. Since rays are reflected to any point P from the two film surfaces as if they originated at the virtual sources S_1 and S_2, this may be considered an instance of the two-point source pattern already discussed in connection with Figure 10-5. Real, nonlocalized fringes are formed in the space above the film. If the source of light is a laser, the fringe pattern is clearly visible on a screen placed anywhere in the vicinity of the film. The condition for interference is just that of the two-source interference pattern, where the slit separation is the distance between virtual sources S_1 and S_2. In Figure 10-13, S_1 and S_2 are located approximately by ignoring refraction in the film.

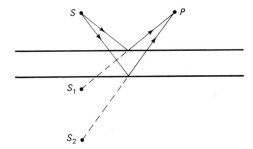

Figure 10-13 Interference by a dielectric film with a point source. Real, nonlocalized fringes appear as in the two-point source pattern of Figure 10-5. Refraction has been ignored.

10-5 FRINGES OF EQUAL THICKNESS

If the film is of varying thickness t, the optical path difference $\Delta = 2 n_f t \cos \theta_t$ varies even without variation in the angle of incidence. Thus, if the direction of the incident light is fixed, say at normal incidence, a bright or dark fringe will be associated with a particular thickness for which Δ satisfies the condition for constructive or destructive interference, respectively. For this reason, fringes produced by a variable-thickness film are called *fringes of equal thickness*. A typical arrangement for viewing these fringes is shown in Figure 10-14a. An extended source is used in conjunction with a beam splitter set at an angle of 45° to the incident light. The beam splitter in this position enables light to strike the film at normal incidence, while at

(a)

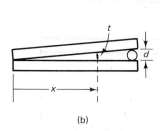

(b)

Figure 10-14 Interference from a wedge-shaped film, producing localized fringes of equal thickness. (a) Viewing assembly. (b) Air wedge formed with two microscope slides.

the same time providing for the transmission of part of the reflected light into the detector (eye). Fringes, often called *Fizeau fringes*, are seen localized at the film, from which the interfering rays diverge. At normal incidence, $\cos \theta_t = 1$ and $\Delta = 2n_f t$. Thus the condition for bright and dark fringes, Eqs. (10-34) and (10-35), is

$$2n_f t + \Delta_r = \begin{cases} m\lambda, & \text{bright} \\ (m + \frac{1}{2})\lambda, & \text{dark} \end{cases} \tag{10-36}$$

where Δ_r is either $\lambda/2$ or 0, depending on whether there is or is not a relative phase shift of π between the rays reflected from the top and bottom surfaces of the film. One way of forming a suitable wedge for experimentation is to use two clean, glass microscope slides, wedged apart at one end by a thin spacer, perhaps a hair, as in Figure 10-14b. The resulting air layer between the slides shows Fizeau fringes when the slides are illuminated by monochromatic light. For this film, the two reflections are from glass to air (internal reflection) and from air to glass (external reflection), so that Δ_r in Eq. (10-36) is $\lambda/2$. As t increases in a linear fashion along the length of the slides from $t = 0$ to $t = d$, Eq. (10-36) is satisfied for consecutive orders of m, and a series of equally spaced, alternating bright and dark fringes will be seen by reflected light. These fringes are virtual, localized fringes and cannot be projected onto a screen.

If the extended source of Figure 10-14a is the sky and white light is incident at some angle on a film of variable thickness, as in Figure 10-15, the film may appear in a variety of colors, like an oil slick after a rain. Suppose that in a small region of the film the thickness is such as to produce constructive interference for wavelengths

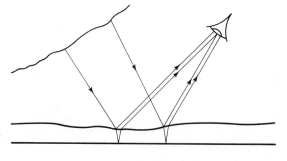

Figure 10-15 Interference by an irregular film illuminated by an extended source. Variations in film thickness, as well as angle of incidence, determine the wavelength region reinforced by interference.

in the red portion of the spectrum at some order m. If the wavelengths at which constructive interference occurs again for orders $m + 1$ and $m - 1$ are outside the visible spectrum, the reflected light appears red. This can occur readily for low orders and therefore for thin films.

10-6 NEWTON'S RINGS

Since Fizeau fringes are fringes of equal thickness, their contours directly reveal any nonuniformities in the thickness of the film. Figure 10-16a shows how this circumstance can be put to practical use in determining the quality of the spherical surface of a lens, for example, in an arrangement in which the Fizeau fringes have come to be referred to as *Newton's rings*. An air wedge, formed between the spherical surface and an optically flat surface, is illuminated with normally incident monochromatic light, such as a sodium lamp or a mercury lamp with a filter, to isolate one of its spectral lines. Equal-thickness contours for a perfectly spherical surface, and therefore the fringes viewed, are concentric circles around the point of contact with the optical flat. At that point, $t = 0$ and the path difference between reflected rays is $\lambda/2$, as a result of reflection. The center of the fringe pattern thus appears dark, and Eq. (10-36) gives $m = 0$ for the order of the destructive interference. Irregularities in the surface of the lens show up as distortions in the concentric ring pattern. This arrangement can also be used as an optical means of measuring the radius of curvature of the lens surface. A geometrical relation exists between the radius r_m of the mth-order dark fringe, the corresponding air-film thickness t_m, and the radius of curvature R of the air film or the lens surface. Referring to Figure 10-16b and making use of the Pythagorean theorem, we have

$$R^2 = r_m^2 + (R - t_m)^2$$

or

$$R = \frac{r_m^2 + t_m^2}{2t_m} \qquad (10\text{-}37)$$

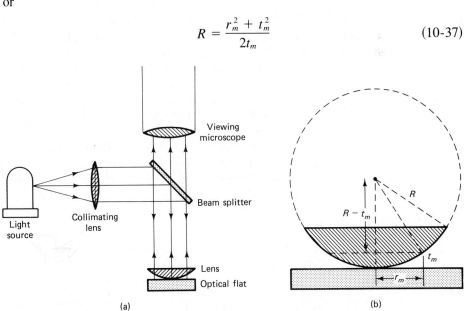

(a) (b)

Figure 10-16 (a) Newton's rings apparatus. Interference fringes of equal thickness are produced by the air wedge between lens and optical flat. (b) Essential geometry for production of Newton's rings.

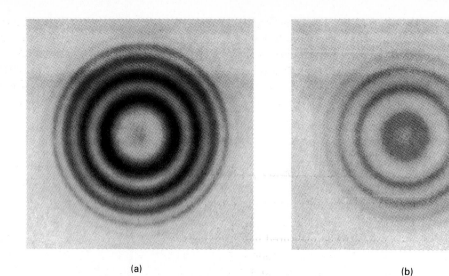

(a)   (b)

Figure 10-17 Newton's rings in (a) reflected light and (b) transmitted light are complementary. (From M. Cagnet, M. Francon, and J. C. Thrierr, *Atlas of Optical Phenomenon*, Plate 9, Berlin: Springer-Verlag, 1962.)

The radius of the mth dark ring is measured and the corresponding thickness of air is determined from the interference condition of Eq. (10-36). Thus R can be found. A little thought should convince one that light transmitted through the optical flat will also show circular interference fringes. As shown in Figure 10-17, the pattern differs in two important respects from the fringes by reflected light. First, the fringes show poor contrast, because the two transmitted beams with largest amplitudes have quite different values and result in incomplete cancellation. Second, the center of the fringe pattern is bright rather than dark, and the entire fringe system is complementary to the system by reflection.

Example

A plano-convex lens ($n = 1.523$) of $\frac{1}{8}$ diopter power is placed, convex surface down, on an optically flat surface. Using a traveling microscope and sodium light ($\lambda = 589.3$ nm), interference fringes are observed. Determine the radii of the first and tenth dark rings.

Solution In this case, $\Delta_r = \lambda/2$, so that Eq. (10-36) leads to an air-film thickness at the mth dark ring given by $t_m = m\lambda/2n_f$. Since the film is air, $n_f = 1$ and $t_m = m\lambda/2$. The ring radii are given by Eq. (10-37). On neglecting the very small term in t_m^2, this is $r_m^2 = 2Rt_m$. The radius of curvature of the convex surface of the lens is found from the lensmaker's equation:

$$\frac{1}{f} = (n - 1)\left(\frac{1}{R_1} - \frac{1}{R_2}\right)$$

With $f = 8$ m, $n = 1.523$, and $R_2 \rightarrow \infty$, this gives $R = 4.184$ m. Then

$$r_m^2 = 2Rt_m = 2R\left(\frac{m\lambda}{2}\right) = mR\lambda$$

$$r_1^2 = (1)(4.184)(589.3 \times 10^{-9}) = 2.466 \times 10^{-6} \text{ m}^2$$

$$r_{10}^2 = (10)(4.184)(589.3 \times 10^{-9}) = 24.66 \times 10^{-6} \text{ m}^2$$

or $r_1 = 1.57$ mm and $r_{10} = 4.97$ mm.

It is ironic that the phenomenon we have been describing, involving so intimately the wave nature of light, should be known as Newton's rings after one who championed the corpuscular theory of light. Probably the first measurement of the wavelength of light was made by Newton, using this technique. Consistent with his corpuscular theory, however, Newton interpreted this quantity as a measurement of the distance between the "easy fits of reflection" of light corpuscles.

10-7 FILM-THICKNESS MEASUREMENT BY INTERFERENCE

Fringes of equal thickness provide a sensitive optical means for measuring thin films. A sketch of one possible arrangement is shown in Figure 10-18. Suppose the film F to be measured has a thickness d. The film has been deposited on some substrate S. Monochromatic light is channeled from a light source LS through a fiber-optics light pipe LP to a right-angle beam-splitting prism BS, which transmits one beam to a flat mirror M and the other to the film surface. After reflection, each is transmitted by the beam splitter into a microscope MS, where they are allowed to interfere. Equivalently, the beam reflected from the mirror M can be considered to arise from its virtual image M'. The virtual mirror M' is constructed by imaging M through the beam-splitter reflecting plane. This construction makes it clear that the interference pattern results from interference due to the air film between the reflecting plane at M' and the film F. In practice, mirror M can be moved toward or away from the beam splitter to equalize optical path lengths and can be tilted to make M' more or less parallel to the film surface. Furthermore, the beam splitter and mirror assembly form one unit that can be attached to the microscope in place of its objective lens. When M' and the film surface are not precisely parallel, the usual Fizeau fringes due to a wedge will be seen through the microscope, which has been prefocused on the film. The light beam striking the film is allowed to cover the edge of the film F, so that two fringe systems are seen side by side, corresponding to air films that differ by the required thickness at their juncture. Figure 10-19a shows a typical photograph of the fringe systems, made through a microscope. The transla-

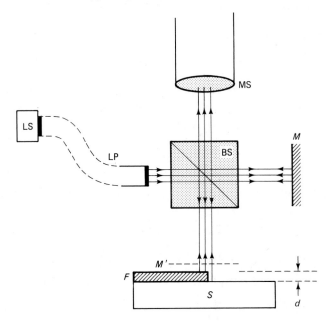

Figure 10-18 Film-thickness measurement. Interference fringes produced by light reflected from the film surface and substrate allow a determination of the film thickness d.

(a)

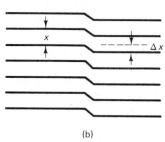

(b)

Figure 10-19 (a) Photograph of interference fringes produced by the arrangement shown in Figure 10-18. The troughlike depression evident in the interference pattern was made by evaporating the film over a thin, straight wire. (b) Sketch of one side of the trough shown in the photo. The fringe pattern shifts by an amount Δx at the film edge. (Photo by J. Feldott.)

tion of one fringe system relative to the other provides a means of determining d, as follows. For normal incidence, bright fringes satisfy Eq. (10-34),

$$\Delta_p + \Delta_r = 2nt + \Delta_r = m\lambda$$

where t represents the thickness of the air film at some point. If the air-film thickness now changes by an amount $\Delta t = d$, the order of interference m changes accordingly, and we have

$$2n\,\Delta t = 2d = (\Delta m)\lambda$$

where we have set $n = 1$ for an air film. Increasing the thickness t by $\lambda/2$, for example, changes the order of any fringe by $\Delta m = 1$, that is, the fringe pattern translates by one whole fringe. For a shift of fringes of magnitude Δx (Figure 10-19b) the change in m is given by $\Delta m = \Delta x/x$, resulting in

$$d = (\Delta x/x)(\lambda/2) \qquad (10\text{-}38)$$

Since both fringe spacing x and fringe shift Δx can be measured with a stable microscope—or from a photograph like that of Figure 10-19—the film thickness d is determined. When using monochromatic light, the net shift of fringe systems is ambiguous because a shift $\Delta x = 0.5x$, for example, will look exactly like a shift $\Delta x = 1.5x$. This ambiguity may be removed in one of two ways. If the shift is more than one fringe width, this situation is apparent when viewing white-light fringes, formed in the same way. The superposition of colors that form the white-light fringes creates a pattern whose center at $m = 0$ is unique, serving as an unambiguous index of fringe location. The integral shift of fringe patterns is then easily seen and can be combined with the monochromatic measurement of Δx described previously. A second method is to prepare the film so that its edge is not sharp but tails off gradually. In this case each fringe of one set can be followed down the film edge into the corresponding fringe of the second set, as in Figure 10-19. If the film cannot be provided with a gradually tailing edge, a thin film of silver, for example, can be evaporated over both the film and substrate. The step in the metal film will usually be somewhat sloped, but the total step will be the same as the thickness of the film to be measured. A one-to-one correspondence between individual fringes of each set can then be made visually.

PROBLEMS

10-1. Two beams having parallel electric fields are described by

$$E_1 = 3 \sin \left(\mathbf{k}_1 \cdot \mathbf{r} - \omega t + \frac{\pi}{5} \right)$$

$$E_2 = 4 \sin \left(\mathbf{k}_2 \cdot \mathbf{r} - \omega t + \frac{\pi}{6} \right)$$

with amplitudes in kV/m. The beams interfere at a point where the phase difference due to path is $\pi/3$ (the first beam having the longer path). At the point of superposition, calculate (a) the irradiances I_1 and I_2 of the individual beams; (b) the irradiance I_{12} due to their interference; (c) the net irradiance; (d) the fringe visibility.

10-2. Two harmonic light waves with amplitudes of 1.6 and 2.8 interfere at some point on a screen. What fringe contrast or visibility results there if their electric field vectors are parallel and if they are perpendicular?

10-3. The ratio of the amplitudes of two beams forming an interference fringe pattern is 2/1. What is the fringe contrast? What ratio of amplitudes produces a fringe contrast of 0.5?

10-4. (a) Show that if one beam of a two-beam interference setup has an irradiance of N times that of the other beam, the fringe visibility is given by

$$V = \frac{2\sqrt{N}}{N + 1}$$

(b) Determine the beam irradiance ratios for visibilities of 0.96, 0.9, 0.8, and 0.5.

10-5. A mercury source of light is positioned behind a glass filter, which allows transmission of the 546.1-nm green light from the source. The light is allowed to pass through a narrow, horizontal slit positioned 1 mm above a flat mirror surface. Describe both qualitatively and quantitatively what appears on a screen 1 m away from the slit.

10-6. Two slits are illuminated by light that consists of two wavelengths. One wavelength is known to be 436 nm. On a screen, the fourth minimum of the 436-nm light coincides with the third maximum of the other light. What is the wavelength of the unknown light?

10-7. In a Young's experiment, narrow double slits 0.2 mm apart diffract monochromatic light onto a screen 1.5 m away. The distance between the fifth minima on either side of the zeroth-order maximum is measured to be 34.73 mm. Determine the wavelength of the light.

10-8. A quasi-monochromatic beam of light illuminates Young's double-slit setup, generating a fringe pattern having a 5.6-mm separation between consecutive dark bands. The distance between the plane containing the apertures and the plane of observation is 10 m, and the two slits are separated by 1.0 mm. Sketch the experimental arrangement. Why is an initial single slit necessary? What is the wavelength of the light?

10-9. In an interference experiment of the Young type, the distance between slits is 0.5 mm, and the wavelength of the light is 600 nm.

  (a) If it is desired to have a fringe spacing of 1 mm at the screen, what is the proper screen distance?

  (b) If a thin plate of glass ($n = 1.50$) of thickness 100 microns is placed over one of the slits, what is the lateral fringe displacement at the screen?

  (c) What path difference corresponds to a shift in the fringe pattern from a peak maximum to the (same) peak half-maximum?

10-10. White light (400 to 700 nm) is used to illuminate a double slit with a spacing of 1.25 mm. An interference pattern falls on a screen 1.5 m away. A pinhole in the screen allows some light to enter a spectrograph of high resolution. If the pinhole in the screen is 3 mm from the central white fringe, where would one expect dark lines to show up in the spectrum of the pinhole source?

10-11. Sodium light (589.3 nm) from a narrow slit illuminates a Fresnel biprism made of glass of index 1.50. The biprism is twice as far from a screen on which fringes are observed as it is from the slit. The fringes are observed to be separated by 0.03 cm. What is the biprism angle?

10-12. The small angle between two-plane, adjacent reflecting surfaces is determined by examining the interference fringes produced in a Fresnel mirror experiment. A source slit is parallel to the intersection between the mirrors and 50 cm away. The screen is 1 m from the same intersection, measured along the normal to the screen. When illuminated with sodium light (589.3 nm), fringes appear on the screen with a spacing of 0.5 mm. What is the angle?

10-13. The prism angle of a very thin prism is measured by observing interference fringes as in the Fresnel biprism technique. The distances from slit to prism and from prism to eye are in the ratio of 1 : 4. Twenty dark fringes are found to span a distance of 0.5 cm when green mercury light is used. If the refractive index of the prism is 1.50, determine the prism angle.

10-14. Light of continuously variable wavelength illuminates normally a thin oil (index of 1.30) film on a glass surface. Extinction of the reflected light is observed to occur at wavelengths of 525 and 675 nm in the visible spectrum. Determine the thickness of the oil film and the orders of the interference.

10-15. A thin film of MgF_2 ($n = 1.38$) is deposited on glass so that it is antireflecting at a wavelength of 580 nm under normal incidence. What wavelength is minimally reflected when the light is incident instead at 45°?

10-16. A nonreflecting, single layer of a lens coating is to be deposited on a lens of refractive index $n = 1.78$. Determine the refractive index of a coating material and the thickness required to produce zero reflection for light of wavelength 550 nm.

10-17. Remember that the energy of a light beam is proportional to the square of its amplitude.

  (a) determine the percentage of light energy reflected in air from a single surface separating a material of index 1.40 for light of $\lambda = 500$ nm.

  (b) When deposited on glass of index 1.60, how thick should a film of this material be in order to reduce the reflected energy by destructive interference?

  (c) What is then the effective percent reflection from the film layer?

10-18. A soap film is formed using a rectangular wire frame and held in a vertical plane. When illuminated normally by laser light at 632.8 nm, one sees a series of localized interference fringes that measure 15 per cm. Explain their formation.

10-19. A beam of white light (a continuous spectrum from 400 to 700 nm, let us say) is incident at an angle of 45° on two parallel glass plates separated by an air film 0.001 cm thick. The reflected light is admitted into a prism spectroscope. How many dark "lines" are seen across the entire spectrum?

10-20. Two microscope slides are placed together but held apart at one end by a thin piece of tin foil. Under sodium light (589 nm) normally incident on the air film formed between the slides, one observes exactly 40 bright fringes from the edges in contact to the edge of the tin foil. Determine the thickness of the foil.

10-21. Plane plates of glass are in contact along one side and held apart by a wire 0.05 mm in diameter, parallel to the edge in contact and 20 cm distant. Using filtered green mercury light, directed normally on the air film between plates, interference fringes are seen. Calculate the separation of the dark fringes. How many dark fringes appear between the edge and the wire?

10-22. Show that the separation of the virtual sources producing interference from a film of index n and uniform thickness t, when illuminated by a point source, is $2t/n$. Assume the film is in air and light is incident at near-normal incidence.

10-23. Newton's rings are formed between a spherical lens surface and an optical flat. If the tenth bright ring of green light (546.1 nm) is 7.89 mm in diameter, what is the radius of curvature of the lens surface?

10-24. Newton's rings are viewed both with the space between lens and optical flat empty and filled with a liquid. Show that the ratio of the radii observed for a particular order fringe is very nearly the square root of the liquid's refractive index.

10-25. A Newton's ring apparatus is illuminated by light with two wavelength components. One of the wavelengths is 546 nm. If the eleventh bright ring of the 546-nm fringe system coincides with the tenth ring of the other, what is the second wavelength? What is the radius at which overlap takes place and the thickness of the air film there? The spherical surface has a radius of 1 m.

10-26. A fringe pattern found using an interference microscope objective is observed to have a regular spacing of 1 mm. At a certain point in the pattern, the fringes are observed to shift laterally by 3.4 mm. If the illumination is green light of 546.1 nm, what is the dimension of the "step" in the film that caused the shift?

REFERENCES

[1] Baumeister, Philip, and Gerald Pincus. "Optical Interference Coatings." *Scientific American* (Dec. 1970): 58.

[2] Fincham, W. H. A., and M. H. Freeman. *Optics*, 9th ed. London: Butterworths, 1980. Ch. 14.

[3] Klein, Miles V. *Optics*. New York: John Wiley and Sons, 1970. Ch. 5.

[4] Hecht, Eugene, and Alfred Zajac. *Optics*. Reading, Mass.: Addison-Wesley Publishing Company, 1974. Ch. 9.

[5] Ghatak, Ajoy K. *An Introduction to Modern Optics*. New York: McGraw-Hill Book Company, 1972. Ch. 4.

[6] Feynman, Richard P., Robert B. Leighton, and Matthew Sands. *The Feynman Lectures in Physics*, vol. 1. Reading, Mass.: Addison-Wesley Publishing Company, 1975. Ch. 28, 29.

[7] Longhurst, R. S. *Geometrical and Physical Optics*, 2d ed. New York: John Wiley and Sons, 1967. Ch. 7, 8.

11

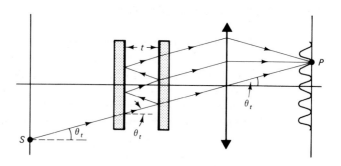

Optical Interferometry

INTRODUCTION

An instrument designed to exploit the interference of light and the fringe patterns that result from optical path differences, in any of a variety of ways, is called an *optical interferometer*. This general description of the instrument should reflect the wide variety of designs and uses of interferometers. Applications extend also to acoustic and radio waves, but here we are interested in the optical interferometer. In this chapter we discuss chiefly the Michelson and the Fabry-Perot interferometers and suggest only a few of their many applications.

To achieve interference between two coherent beams of light, an interferometer divides an initial beam into two or more parts that travel diverse optical paths and then reunite to produce an interference pattern. One criterion for broadly classifying interferometers distinguishes the manner in which the initial beam is separated. *Wavefront division interferometers* sample portions of the same wavefront of a coherent beam of light, as in the case of Young's double slit, or adaptations like those using Lloyd's mirror or Fresnel's biprism. *Amplitude-division interferometers* instead use some type of *beam splitter* that divides the initial beam into two parts. The Michelson interferometer is of this type. Usually the beam splitting is managed by a semireflecting metallic or dielectric film; it can also occur by frustrated total internal reflection at the interface of two prisms forming a cube, or by means of double refraction or diffraction. Another means of classification distinguishes between those interferometers that make use of the interference of two beams, as in the case of the Michelson interferometer, and those that operate with multiple beams, as in the Fabry-Perot interferometer.

11-1 THE MICHELSON INTERFEROMETER

The Michelson interferometer, first introduced by Albert Michelson in 1881, has played a vital role in the development of modern physics. This simple and versatile instrument was used, for example, to establish experimental evidence for the validity of the special theory of relativity, to detect and measure hyperfine structure in line spectra, to measure the tidal effect of the moon on the earth, and to provide a substitute standard for the meter in terms of wavelengths of light. Michelson himself pioneered much of this work.

A schematic of the Michelson interferometer is shown in Figure 11-1a. From an extended source of light S, a beam 1 of light is split by beam splitter (BS) by means of a thin, semitransparent front surface metallic or dielectric film, deposited on glass. The interferometer is therefore of the amplitude-splitting type. Reflected beam 2 and transmitted beam 3, of roughly equal amplitudes, continue to fully reflecting mirrors $M2$ and $M1$, respectively, where their directions are reversed. On returning to the beam splitter, beam 2 is now transmitted and beam 3 is reflected by the semitransparent film so that they come together again and leave the interferometer as beam 4. The useful aperture of this double-beam interferometer is such that all rays striking $M1$ and $M2$ will be normal or nearly so. Thus beam 4 includes rays that have traveled different optical paths and will demonstrate interference. At least one of the mirrors is equipped with tilting adjustment screws that allow the surface of $M1$ to be made perpendicular to that of $M2$. One of the mirrors is also movable

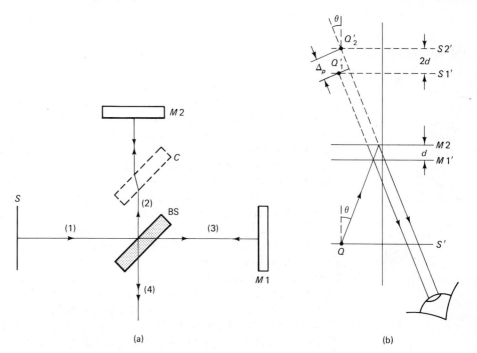

Figure 11-1 (a) The Michelson interferometer. (b) Equivalent optics for the Michelson interferometer.

along the direction of the beam by means of an accurate track and micrometer screw. In this way the difference between the optical paths of beams 2 and 3 can be gradually varied. Notice that beam 3 traverses the beam splitter three times, whereas beam 2 traverses it only once. In some applications, where white light is used, it is essential that the optical paths of the two beams be made precisely equal. Although this can be accomplished at one wavelength by appropriately increasing the distance of $M2$ from BS, the correction would not suffice at another wavelength because of the dispersion of the glass. To compensate for all wavelengths at once, a plate C made of the same material and dimensions as BS is inserted parallel to BS in the path of beam 2. Any small, remaining inequalities in optical paths can be removed by allowing the compensator to rotate, thus varying the optical path through the thickness of its glass plate.

The actual interferometer in Figure 11-1a possesses two optical axes at right angles to one another. A simpler but equivalent optical system, having a single optical axis, can be displayed by working with virtual images of source S and mirror $M1$ via reflection in the BS mirror. These positions are most simply found by regarding the assembly including S, $M1$, and beams 1 and 3 of Figure 11-1a as rotated counterclockwise by 90° about the point of intersection of the beams with the BS mirror. The resulting geometry is shown in Figure 11-1b. The new position of the source plane is S', and the new position of the mirror $M1$ is $M1'$. Light from a point Q on the source plane S' then effectively reflects from both mirrors $M2$ and $M1'$, shown parallel and with an optical path difference of d. The two reflected beams appear to come from the two virtual images, Q_1' and Q_2', of object point Q. Since the images $S1'$ and $S2'$ of the source plane in the mirrors must be separated by twice the mirror separation, the distance between Q_1' and Q_2' is $2d$, and the optical path difference between the two beams emerging from the interferometer is

$$\Delta_p = 2d \cos \theta \qquad (11\text{-}1)$$

where the angle θ measures the inclination of the beams relative to the optical axis. For a normal beam, $\theta = 0$ and $\Delta_p = 2d$. We expect this result, since, if one mirror is farther from BS than the other by a distance d, the extra distance traversed by the beam taking the longer route includes distance d twice, once before and once after reflection. If, in addition, $\Delta = m\lambda$, so that the two beams interfere constructively, it follows that they will do so repeatedly for every $\lambda/2$ translation of one of the mirrors.

The optical system of Figure 11-1b is now equivalent to the case of interference due to a plane parallel air film, illuminated by an extended source. Virtual fringes of equal inclination may be seen by looking into the beam splitter along ray 4, with the eye or a telescope focused at infinity. Assuming that the two interfering beams are of equal amplitude, the irradiance of the fringe system of circles concentric with the optical axis is given, as in Eq. (10-16), by

$$I = 4I_0 \cos^2 \left(\frac{\delta}{2} \right) \qquad (11\text{-}2)$$

where the phase difference is

$$\delta = k\Delta = \left(\frac{2\pi}{\lambda} \right) \Delta \qquad (11\text{-}3)$$

The net optical path difference is $\Delta = \Delta_p + \Delta_r$, as usual. A relative π phase shift between the two beams occurs because beam 2 experiences two external reflections but beam 3 experiences only one.[1] For dark fringes, then,

$$\Delta_p + \Delta_r = 2d \cos \theta + \frac{\lambda}{2} = \left(m + \frac{1}{2}\right)\lambda$$

or, more simply,

$$2d \cos \theta = m\lambda, \qquad m = 0, 1, 2, \ldots \qquad \text{dark fringes} \qquad (11\text{-}4)$$

If d is of such magnitude that the normal rays forming the center of the fringe system satisfy Eq. (11-4), that is, the center fringe is dark, then its order, given by

$$m_{max} = \frac{2d}{\lambda} \qquad (11\text{-}5)$$

is a large integer. Neighboring dark fringes decrease in order outwards from the center of the pattern, as $\cos \theta$ decreases from its maximum value of 1. This ordering of fringes may be inverted for convenience by associating another integer p with each fringe of order m, where

$$p = m_{max} - m = \frac{2d}{\lambda} - m \qquad (11\text{-}6)$$

Using Eq. (11-6) to replace m in Eq. (11-4), we arrive at

$$p\lambda = 2d(1 - \cos \theta), \qquad p = 0, 1, 2, \ldots \qquad \text{dark fringes} \qquad (11\text{-}7)$$

where now the central fringe is of order zero and the neighboring fringes increase in order, outward from the center. Figure 11-2 illustrates the relationship between orders m and p for the arbitrary case where $m_{max} = 100$. Equation (11-4) or (11-7) indicates that, as d is varied, a particular point in the fringe pattern ($\theta = $ constant) will correspond to gradually changing values of order m or p. Integral values occur whenever the point coincides with a dark fringe. Equivalently, this means that as d is varied, fringes of the pattern appear to shrink toward the center, where they disappear, or else expand outward from the center, where they seem to originate, depending on whether the optical path difference is decreasing or increasing. The motion of the fringe pattern thus reverses as one of the mirrors is moved continually through the point of zero path difference. Viewed in another way, Eq. (11-4) re-

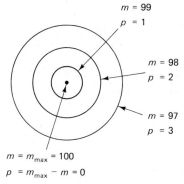

$m = 99$

$p = 1$

$m = 98$

$p = 2$

$m = 97$

$p = 3$

$m = m_{max} = 100$

$p = m_{max} - m = 0$

Figure 11-2 Alternate orderings of fringes.

[1] This conclusion assumes dielectric coatings of index less than that of glass. The assumption is made to allow discussion of the fringe pattern in a concrete (and common) situation. It does not affect the validity of results like Eq. (11-8), for example, because measurements depend on the net *motion* of the fringe pattern, not on precisely where it is dark and where it is bright.

quires an increase in the angular separation $\Delta\theta$ of a given small fringe interval Δm as the mirror spacing d becomes smaller, since

$$|\Delta\theta| = \frac{\lambda\,\Delta m}{2d\sin\theta}$$

This means that the fringes are more widely separated when optical path differences are small. In fact, if $d = \lambda/2$, then from Eq. (11-4), $m = \cos\theta$, and the entire field of view encompasses no more than one fringe! For a mirror translation Δd, the number Δm of fringes passing a point at or near the center of the pattern is, according to Eq. (11-4),

$$\Delta m = \frac{2\,\Delta d}{\lambda} \tag{11-8}$$

Equation (11-8) suggests an experimental way of either measuring λ when Δd is known or calibrating the micrometer translation screw when λ is known.

Example

Fringes are observed due to monochromatic light in a Michelson interferometer. When the movable mirror is translated by 0.73 mm, a shift of 300 fringes is observed. What is the wavelength of the light? What displacement of the fringe system takes place when a flake of glass of index 1.51 and 0.005 mm thickness is placed in one arm of the interferometer? (Assume that the light beam is normal to the glass surface.)

Solution Using Eq. (11-8),

$$\lambda = \frac{2\,\Delta d}{\Delta m} = \frac{(2)(0.073)}{300} = 4.87 \times 10^{-4}\ \text{mm} = 487\ \text{nm}$$

With the glass inserted, one arm is effectively lengthened by a path difference of $\Delta d = n_g t - n_{\text{air}} t$, so that

$$\Delta m = \frac{2\,\Delta d}{\lambda} = \frac{2(n_g - 1)t}{\lambda} = \frac{2(0.51)(0.005 \times 10^{-3})}{487 \times 10^{-9}}$$

or $\Delta m = 10.5$ fringes.

11-2 APPLICATIONS OF THE MICHELSON INTERFEROMETER

The Michelson interferometer is easily adaptable to the measurement of thin films, a technique essentially the same as that described in the preceding chapter. It is also easily adaptable to the determination of the index of refraction of a gas. An evacuable cell with plane, parallel windows is interposed in the path of beam 3, Figure 11-1a, and is filled with a gas at a pressure and temperature for which its index of refraction is desired. The fringe system established under these conditions is monitored as the gas is gradually pumped out of the cell. A count Δm of the net fringe shift is related to the change in optical path during the replacement of the gas by vacuum. If the actual length of the cell is accurately known to be L, the change in optical path is given by

$$\Delta d = nL - L = L(n - 1) \tag{11-9}$$

and using Eq. (11-8), it follows that the index can be determined from

$$n - 1 = \left(\frac{\lambda}{2L}\right) \Delta m \qquad (11\text{-}10)$$

Consider another direct application of the Michelson interferometer, the determination of wavelength difference between two closely spaced components of a spectral "line," λ and λ'. Each wavelength forms its own system of circular fringes according to Eq. (11-4). Suppose we view the circular systems near their center, so that $\cos \theta \cong 1$. Then for a given path difference d of the interferometer, the product $m\lambda$ is fixed, that is, $m\lambda = m'\lambda'$. When the fringe systems coincide, the pattern appears sharp, whereas when the fringes of one system in the region of observation lie midway between the fringes of the second system, the pattern appears rather uniform in brightness, or "washed out." The mirror movement Δd required between consecutive coincidences is related to the wavelength difference $\Delta \lambda$ as follows. At one coincidence, when fringes are "in step," the orders of the two systems corresponding to λ and λ' must be related by

$$m = m' + N$$

where N is an integer. If the optical path difference at this time is d_1, then from Eq. (11-4),

$$\frac{2d_1}{\lambda} = \frac{2d_1}{\lambda'} + N \qquad (11\text{-}11)$$

Let the optical path difference be increased to d_2, when the next coincidence is found. Then

$$m = m' + (N + 1)$$

or

$$\frac{2d_2}{\lambda} = \frac{2d_2}{\lambda'} + N + 1 \qquad (11\text{-}12)$$

By subtracting Eq. (11-11) from Eq. (11-12) and by writing the mirror movement $\Delta d = d_2 - d_1$, we find

$$\lambda' - \lambda = \frac{\lambda\lambda'}{2\,\Delta d} \qquad (11\text{-}13)$$

Now since λ and λ' are very close, the wavelength difference of the two unresolved components can be approximated by

$$\Delta\lambda = \frac{\lambda^2}{2\,\Delta d} \qquad (11\text{-}14)$$

This technique is often employed in an optics laboratory course to measure the wavelength difference of 6 Å between the two components of the yellow "line" of sodium.

All the preceding discussion of the fringes from a Michelson interferometer has been in terms of virtual fringes of equal inclination. We have assumed that mirrors $M1$ and $M2$ are precisely perpendicular, or what amounts to the same thing, precisely parallel in the equivalent optical system of Figure 11-1b. If the alignment is such that the air space between $M1'$ and $M2$ in Figure 11-1b is a wedge, fringes of equal thickness may be seen localized at the mirrors. These fringes will be straight, oriented parallel to the line that represents the intersection of $M1'$ and $M2$. If the wedge is of large angle, they will be curved in a way that can be shown to be hyperbolic arcs. Again, if the source is small, then real, nonlocalized fringes appear in the light emerging from the interferometer, as if formed by the two virtual images of the source in $M1'$ and $M2$. These fringes appear without effort when the intense, coherent light of a laser is used. These possibilities have already been discussed in the previous chapter, where we treated the various interference fringes that can be formed by illumination of a film. Figure 11-3 is a photograph showing the distortion of fringes of equal thickness produced by a candle flame when situated in one arm of a Michelson interferometer. Variations in temperature produce variations in optical path length by changing the refractive index of the air.

Figure 11-3 Deformation of fringes of equal thickness in the neighborhood of a candle flame. (From M. Cagnet, M. Francon, and J. C. Thrierr, *Atlas of Optical Phenomenon*, Plate 12, Berlin: Springer-Verlag, 1962.)

11-3 VARIATIONS OF THE MICHELSON INTERFEROMETER

Although there are many ways in which a beam of light may be split into two parts and reunited after traversing diverse paths, we examine briefly two variations that can be considered adaptations of the Michelson interferometer. A slight modification by Twyman and Green is shown in Figure 11-4a. Instead of using an extended source, this interferometer uses a point source together with a collimating lens $L1$, so that all rays enter the interferometer parallel to the optical axis, or $\cos\theta = 1$. The parallel rays emerging from the interferometer are brought to a focus by lens $L2$ at P, where the eye is placed. The circular fringes of equal inclination no longer appear; in their place are seen fringes of equal thickness. These fringes reveal imperfections in the optical system that cause variations in optical path length. When no distortions appear in the plane wavefronts through the interferometer, uniform illumination is seen near P. If the interferometer components are of high quality, this system can be used to test the optical quality of another optical component, such as

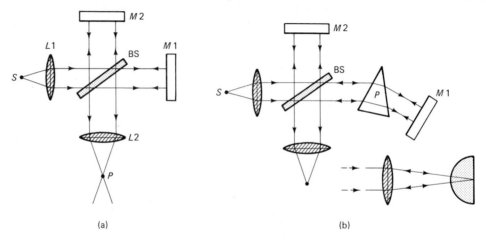

Figure 11-4 (a) Twyman-Green interferometer. (b) Twyman-Green interferometer used in the testing of a prism and a lens (inset).

a prism, situated as shown in Figure 11-4b. Any surface imperfection or internal variation in refractive index shows up as a distortion of the fringe pattern. Lenses are tested for aberrations in the same way, once plane mirror $M1$ is replaced by a convex spherical surface that can reflect the refracted rays back along themselves, as suggested in the inset of Figure 11-4b.

A more radical variation, sketched in Figure 11-5, is the Mach-Zehnder interferometer. The incident beam of roughly parallel light is divided into two beams at beam splitter BS. Each beam is again totally reflected by mirrors $M1$ and $M2$, and the beams are made coincident again by the semitransparent mirror $M3$. Path lengths of beams 1 and 2 around the rectangular system and through the glass of the beam splitters are identical. This interferometer has been used, for example, in aerodynamic research, where the geometry of air flow around an object in a wind tunnel is revealed through local variations of pressure and refractive index. A windowed test chamber, into which the model and a streamline flow of air is introduced, is placed in path 1. An identical chamber is placed in path 2 to maintain equality of optical paths. The air-flow pattern is revealed by the fringe pattern. For such applications the interferometer must be constructed on a rather large scale. An advantage of the Mach-Zehnder over the Michelson interferometer is that, by appropriate small rotations of the mirrors, the fringes may be made to appear at the object being tested, so that both can be viewed or photographed together. In the Michelson interferometer, fringes appear localized on the mirror and so cannot be seen in sharp focus at the same time as a test object placed in one of its arms.

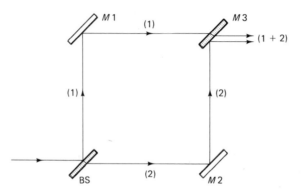

Figure 11-5 Mach-Zehnder interferometer.

The Michelson, Twyman-Green, and Mach-Zehnder interferometers are all two-beam interference instruments that operate by division of amplitude. We turn now to an important case of a multiple-beam instrument, the Fabry-Perot interferometer. Before discussing the instrument itself, however, it will be necessary to examine the phenomenon of multiple reflections from a parallel transparent plate.

11-4 STOKES RELATIONS

We begin with an argument due to Sir George Stokes, which yields information concerning the amplitudes of reflected and transmitted portions of a plane wavefront incident on a plane refracting surface, as in Figure 11-6a. Let E_i represent the amplitude of the incident light. We define reflection and transmission coefficients[2] by

$$r = \frac{E_r}{E_i}, \qquad t = \frac{E_t}{E_i} \tag{11-15}$$

so that at the interface, E_i is divided into a reflected part, $E_r = rE_i$, and a transmitted part, $E_t = tE_i$ as shown. For a ray incident from the second medium, we define similar quantities, which we distinguish with prime notation, r' and t'. According to the principle of ray reversibility, the situation shown in Figure 11-6b must also be valid. In general, however, two rays incident at the interface, as in Figure 11-6b, each result in a reflected and a transmitted ray, all of which are shown, with appropriate amplitudes, in Figure 11-6c. We conclude that the situations depicted in Figure 11-6b and c must be physically equivalent, so that we can write

$$E_i = (r^2 + t't)E_i$$

and

$$0 = (r't + tr)E_i$$

or

$$tt' = 1 - r^2 \tag{11-16}$$

$$r = -r' \tag{11-17}$$

Equations (11-16) and (11-17) are the *Stokes relations* between amplitude coefficients for angles of incidence related through Snell's law. Equation (11-17) states that the amplitudes of reflected beams for rays incident from either direction

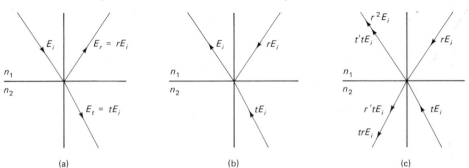

Figure 11-6 Figures used in deriving Stokes relations.

---

[2] We will have occasion later to also use *reflectance* (R) and *transmittance* (T), defined as the ratio of the corresponding irradiances. Although $R = r^2$, $T \neq t^2$, as explained in Section 20-4.

are the same in magnitude but differ by a π phase shift. This becomes clearer if Eq. (11-17) is written in the equivalent form, $r = e^{i\pi}r'$. This result agrees with the predictions of the more complete *Fresnel equations*, treated in Chapter 20. Both the Fresnel theory and experiments, such as Lloyd's mirror, establish the fact that the phase shift occurs for the ray incident on the interface from the side of higher velocity or lower index. This wave phenomenon has its analogy in the reflection of waves from the fixed end of a rope. Both of the Stokes relations will be needed in the discussion that follows.

11-5 MULTIPLE-BEAM INTERFERENCE IN A PARALLEL PLATE

We return now to the problem of reflections from a parallel plate, already considered in a two-beam approximation. Consider the multiple reflections of the narrow beam of light of amplitude E_0 and angle of incidence θ_i, as shown in Figure 11-7. The reflection and transmission amplitude coefficients are r and t at an external reflection and r' and t' at an internal reflection. The amplitude of each segment of the beam can be assigned by multiplying the previous amplitude by the appropriate reflection or transmission coefficient, beginning with the incident wave of amplitude E_0 and working progressively through the train of reflections. Multiple parallel beams emerge from the top and from the bottom of the plate. Multiple-beam interference takes place when either set is focused to a point by a converging lens, as shown for the transmitted beam. Having originated from a single beam, the multiple beams are coherent. Further, if the incident beam is near normal, the beams are brought together with their E vibrations nearly parallel.

We consider the superposition of the reflected beams from the top of the plate. According to Eq. (10-33), the phase difference between successive reflected beams is given by

$$\delta = k\Delta, \quad \text{where} \quad \Delta = 2n_f t \cos \theta_t \qquad (11\text{-}18)$$

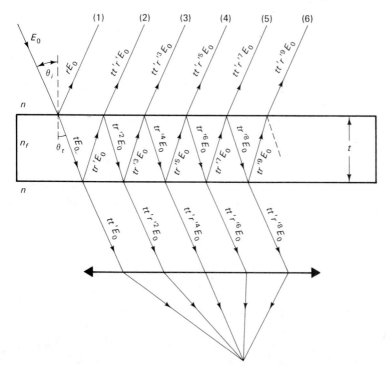

Figure 11-7 Multiply reflected and transmitted beams in a parallel plate.

Here n_f is the refractive index of the plate and t is its thickness. If the incident ray is expressed as $E_0 e^{i\omega t}$, the successive reflected rays can be expressed by appropriately modifying both the amplitude and phase of the initial wave. Referring to Figure 11-7, these are

$$E_1 = (rE_0)e^{i\omega t}$$

$$E_2 = (tt'r'E_0)e^{i(\omega t - \delta)}$$

$$E_3 = (tt'r'^3 E_0)e^{i(\omega t - 2\delta)}$$

$$E_4 = (tt'r'^5 E_0)e^{i(\omega t - 3\delta)}$$

and so on. A little inspection of these equations shows that the Nth such reflected wave can be written as

$$E_N = (tt'r'^{(2N-3)} E_0)e^{i[\omega t - (N-1)\delta]} \tag{11-19}$$

a form that holds good for all but E_1, which never traverses the plate. When these waves are superposed, therefore, the resultant E_R may be written as

$$E_R = \sum_{N=1}^{\infty} E_N = rE_0 e^{i\omega t} + \sum_{N=2}^{\infty} tt'E_0 r'^{(2N-3)} e^{i[\omega t - (N-1)\delta]}$$

Factoring a bit, we have

$$E_R = E_0 e^{i\omega t}\left[r + tt'r'e^{-i\delta}\sum_{N=2}^{\infty} r'^{(2N-4)} e^{-i(N-2)\delta}\right]$$

The summation is now in the form of a geometric series,

$$\sum_{N=2}^{\infty} x^{N-2} = 1 + x + x^2 + \ldots$$

where

$$x = r'^2 e^{-i\delta}$$

Since $|x| < 1$, the series converges to the sum $S = 1/(1 - x)$. Thus

$$E_R = E_0 e^{i\omega t}\left(r + \frac{tt'r'e^{-i\delta}}{1 - r'^2 e^{-i\delta}}\right)$$

Making use next of the Stokes relations, Eqs. (11-16) and (11-17),

$$E_R = E_0 e^{i\omega t}\left[r - \frac{(1 - r^2)re^{-i\delta}}{1 - r^2 e^{-i\delta}}\right]$$

After simplifying,

$$E_R = E_0 e^{i\omega t}\left[\frac{r(1 - e^{-i\delta})}{1 - r^2 e^{-i\delta}}\right]$$

The irradiance, I_R, of the resultant beam is proportional to the square of the amplitude, E_R, which is itself complex, so we calculate $|E_R|^2 = E_R E_R^*$, or

$$|E_R|^2 = E_0^2 r^2\left[\frac{e^{i\omega t}(1 - e^{-i\delta})}{1 - r^2 e^{-i\delta}}\right]\left[\frac{e^{-i\omega t}(1 - e^{i\delta})}{1 - r^2 e^{i\delta}}\right]$$

After processing the product of the bracketed terms and making use of the identity,

$$2 \cos \delta \equiv (e^{i\delta} + e^{-i\delta})$$

there results

$$|E_R|^2 = E_0^2 r^2 \left(\frac{1 - \cos \delta}{1 + r^4 - 2r^2 \cos \delta} \right) \qquad (11\text{-}20)$$

or, in terms of irradiance,

$$I_R = \left[\frac{2r^2(1 - \cos \delta)}{1 + r^4 - 2r^2 \cos \delta} \right] I_i \qquad (11\text{-}21)$$

where I_i represents the irradiance of the incident beam, and we have used the proportionality

$$\frac{I_R}{I_i} = \frac{|E_R|^2}{|E_0|^2} \qquad (11\text{-}22)$$

A similar treatment of the transmitted beams leads to the resultant transmitted irradiance,

$$I_T = \left[\frac{(1 - r^2)^2}{1 + r^4 - 2r^2 \cos \delta} \right] I_i \qquad (11\text{-}23)$$

Equation (11-23) also follows more directly by combining Eq. (11-21) with the relation $I_R + I_T = I_i$, required by the conservation of energy for nonabsorbing films.

A minimum in reflected irradiance occurs, according to Eq. (11-21), when $\cos \delta = 1$, or when

$$\delta = 2\pi m \quad \text{and} \quad \Delta = 2n_f t \cos \theta_t = m\lambda \qquad (11\text{-}24)$$

Necessarily, this must also be the condition for a transmission maximum. Equation (11-23) gives $I_T = I_i$, as expected. A study of Figure 11-7, or the equations describing the set of reflected beams, shows that in the case of a reflection minimum, the second reflected beam and all subsequent beams are in phase with one another but exactly out of phase with the first reflected beam. Since the net reflected irradiance vanishes, there is a perfect cancellation of the first beam with the sum of all the remaining beams. The two-beam approximation works well, then, if the amplitude of the second beam is close to the amplitude of the first beam. Our equations show that their ratio is

$$\left| \frac{E_2}{E_1} \right| = \left| \frac{tt'r'E_0}{rE_0} \right| = 1 - r^2$$

which is close to unity when r^2 is small. For normal incidence on glass of index $n = 1.5$, $r^2 = 0.04$. Thus 96% of the cancellation occurs between the first two reflected beams alone, and the two-beam treatment is well justified.

Reflection maxima occur, in the other extreme, when $\cos \delta = -1$, or when

$$\delta = \pi, 3\pi, \ldots = (m + \tfrac{1}{2})2\pi$$

and

$$\Delta = 2n_f t \cos \theta_t = (m + \tfrac{1}{2})\lambda \qquad (11\text{-}25)$$

In this case, Eqs. (11-21) and (11-23) yield

$$I_R = \left[\frac{4r^2}{(1 + r^2)^2} \right] I_i \qquad (11\text{-}26)$$

$$I_T = \left[\frac{(1 - r^2)}{(1 + r^2)} \right]^2 I_i \qquad (11\text{-}27)$$

It is easily verified that $I_R + I_T = I_i$. Also, inspection of Eq. (11-23) shows that the condition for a transmission minimum occurs when cos $\delta = -1$, so that Eq. (11-26) does indeed give the maximum reflected intensity.

11-6 FABRY-PEROT INTERFEROMETER

The Fabry-Perot interferometer makes use of the plane parallel plate we have been discussing to produce an interference pattern by the multiple beams of the transmitted light. This instrument, probably the most adaptable of all interferometers, has been used, for example, in precision wavelength measurements, analysis of hyperfine spectral line structure, determination of refractive indices of gases, and the calibration of the standard meter in terms of wavelengths. Although simple in structure, it is a high-resolution instrument that has proved to be a powerful tool in a wide variety of applications.

A typical arrangement is shown in Figure 11-8. Two thick glass or quartz plates are used to enclose a plane parallel "plate" of air between them, which forms the medium within which the beams are multiply reflected. The important surfaces of the glass plates are therefore the inner ones. Their surfaces are generally polished to a flatness of better than $\lambda/50$ and coated with a highly reflective layer of silver or aluminum. Silver films are most useful in the visible region of the spectrum, but their reflectivity drops off sharply around 400 nm, so that for applications below 400 nm, aluminum is usually used. Of course the films must be thin enough to be partially transmitting. Optimum thicknesses for silver coatings are around 50 nm. The outer surfaces of the glass plate are purposely formed at a small angle relative to the inner faces (several minutes of arc are sufficient) to eliminate spurious fringe patterns that can arise from the glass itself acting as a parallel plate. The spacing, or thickness, t of the air layer, is an important performance parameter of the interferometer, as we shall see. When the spacing is fixed, the instrument is often referred to as an *etalon*.

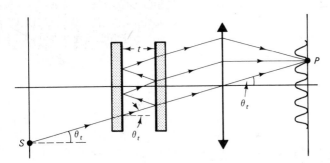

Figure 11-8 Fabry-Perot interferometer.

Consider a narrow, monochromatic beam from an extended source point S making an angle (in air) of θ_t with respect to the optical axis of the system, as in Figure 11-8. The single beam produces multiple coherent beams in the interferometer, and the emerging set of parallel rays are brought together at a point P in the focal plane of the converging lens L. The nature of the superposition at P is determined by the path difference between successive parallel beams, $\Delta = 2n_f t \cos \theta_t$. Using $n_f = 1$ for air, the condition for brightness is

$$2t \cos \theta_t = m\lambda \qquad (11\text{-}28)$$

Other beams from different points of the source but in the same plane and making the same angle θ_t with the axis satisfy the same path difference and also arrive at P.

With t fixed, Eq. (11-28) is satisfied for certain angles θ_t, and the fringe system is the familiar concentric rings due to the focusing of fringes of equal inclination. When a collimating lens is used between source and interferometer, as shown in Figure 11-9a, every set of parallel beams entering the etalon must arise from the same source point. A one-to-one correspondence then exists between source and screen points. The screen may be the retina or a photographic plate. Figure 11-9b illustrates another arrangement, in which the source is small. Collimated light in this instance reaches the plates at a fixed angle θ_t ($\theta_t = 0$ is shown) and comes to a focus at a light detector. As the spacing t is varied, the detector records the interference pattern as a function of time in an *interferogram*. If the souce light consists of two wavelength components, for example, the output of the two systems is either a double set of circular fringes on a photographic plate or a plot of resultant irradiance I versus the plate spacing or time, as suggested in Figure 11-9b.

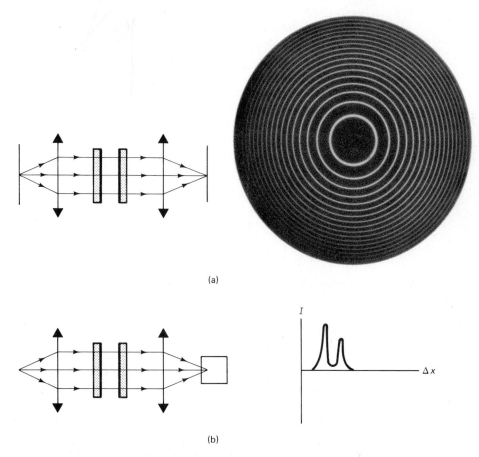

(a)

(b)

Figure 11-9 (a) Fabry-Perot interferometer, used with an extended source and a fixed plate spacing. A circular fringe pattern like the one shown may be photographed at the screen. (Photo from M. Cagnet, M. Francon, and J. C. Thrierr, *Atlas of Optical Phenomenon*, Plate 10, Berlin: Springer-Verlag, 1962.) (b) Fabry-Perot interferometer, used with a point source and a variable plate spacing. A detector at the focal point of the second lens is connected to a plotter to give an output like the one shown, for a translation through one order of interference.

11-7 FRINGE PROFILES: THE AIRY FUNCTION

The variation of the irradiance in the fringe pattern of the Fabry-Perot as a function of the phase or path difference is called the *fringe profile*. The sharpness of the fringes is, of course, important to the ultimate resolving power of the instrument. The irradiance provided by the resultant of the transmitted beams has already been found in Eq. (11-23) and is repeated here:

$$I_T = \left[\frac{(1 - r^2)^2}{1 + r^4 - 2r^2 \cos \delta} \right] I_i$$

Using the trigonometric identity

$$\cos \delta \equiv 1 - 2 \sin^2 \left(\frac{\delta}{2} \right)$$

and simplifying a bit, the *transmittance T*, or *Airy function*, can be expressed as

$$T = \frac{I_T}{I_i} = \frac{1}{1 + [4r^2/(1 - r^2)^2] \sin^2 (\delta/2)} \qquad (11\text{-}29)$$

The square-bracketed factor, which is a function of the reflection coefficient, was called the *coefficient of finesse* by Fabry:

$$F \equiv \frac{4r^2}{(1 - r^2)^2} \qquad (11\text{-}30)$$

Equation (11-29), known as *Airy's formula* for the transmitted irradiance, can then be expressed more compactly as

$$T = \frac{1}{1 + F \sin^2 (\delta/2)} \qquad (11\text{-}31)$$

The quantity F is a sensitive function of the reflection coefficient since, as r varies from 0 to 1, F varies from 0 to infinity. We show that F also represents a certain measure of fringe contrast, written as the ratio

$$\frac{(I_T)_{\max} - (I_T)_{\min}}{(I_T)_{\min}} = \frac{T_{\max} - T_{\min}}{T_{\min}} \qquad (11\text{-}32)$$

From the Airy formula, Eq. (11-31), $T_{\max} = 1$ when $\sin \delta/2 = 0$, and $T_{\min} = 1/(1 + F)$ when $\sin \delta/2 = \pm 1$. Thus

$$\frac{1 - 1/(1 + F)}{1/(1 + F)} = F \qquad (11\text{-}33)$$

The fringe profile may be plotted once a value of r is chosen. Such a plot, for several choices of r, is given in Figure 11-10. For each curve, $T = T_{\max} = 1$ at $\delta = m(2\pi)$, and $T = T_{\min} = 1/(1 + F)$ at $\delta = (m + \frac{1}{2})2\pi$. Notice that $T_{\max} = 1$ regardless of r and that $T_{\min}$ is never zero but approaches this value as r approaches 1. The transmittance peaks sharply at higher values of r as the phase difference approaches integral multiples of 2π, remaining near zero for most of the region between fringes. As r increases even more to an attainable value of 0.97, for example, F increases to 1078 and the fringe widths are less than a third of their values at half-maximum for $r = 0.9$. The sharpness of these fringes is to be compared with the broader fringes from a Michelson interferometer, which have a simple $\cos^2 (\delta/2)$ dependence on the phase (Eq. (11-2)). These are also shown in Figure 11-10 by the dashed lines, normalized to the same maximum value.

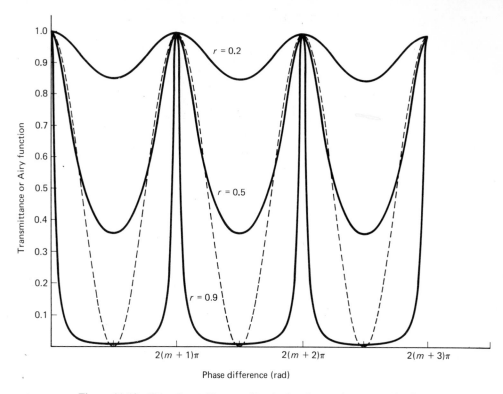

Figure 11-10 Fabry-Perot fringe profile. A plot of transmittance or Airy function versus phase difference for selected values of reflection coefficient. Dashed lines represent comparable fringes from a Michelson interferometer.

11-8 RESOLVING POWER

When two wavelength components are present in the incident light, the Fabry-Perot interferometer gives a double set of circular fringes, each set belonging to one of the wavelengths. A detector scanning across the width of two closely spaced rings, in a particular order m of the interference, might give a plot like that of Figure 11-9b. Such a plot for two wavelength components of comparable irradiance is shown in Figure 11-11a. Although the two peaks are shown separately, only the sum of the two, which follows the dashed line between peaks, is measured. Clearly, if the wavelengths are very close the fringes are very close, and it may not be possible to distinguish two separate peaks in the measured irradiance. The minimum wavelength separation $(\Delta\lambda)_{min}$ that can be resolved by the instrument depends on one's ability to detect the dip in the measured pattern between peaks. According to a resolution criterion due to Rayleigh, the dip caused by identical peaks may not be less than about 20% of the maximum irradiance. Rayleigh's criterion, as it applies to the profile of diffraction images with secondary maxima (Chap. 16), cannot strictly be applied here. It will be sufficient for our purposes to approximate that criterion by requiring that the crossover point be not more than half the maximum irradiance of either individual peak. The phase interval δ_c (Figure 11-11b) for the fringe between maximum and half-maximum values of T can be found from Eq. (11-31) solved for $\sin \delta_c/2$,

$$\sin \frac{\delta_c}{2} = \frac{1}{\sqrt{F}}$$

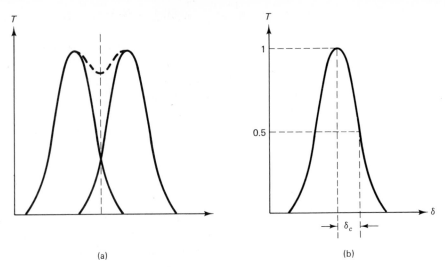

(a)   (b)

Figure 11-11 (a) Scan of two wavelength components of comparable strength in a Fabry-Perot fringe pattern. (b) Application of resolution criterion. The half-width of the peak at half maximum corresponds to the phase interval δ_c.

Since δ_c will be small,

$$\delta_c \cong \frac{2}{\sqrt{F}} \tag{11-34}$$

The phase difference between the two fringe maxima is twice this value, or

$$(\Delta\delta)_{\min} = \frac{4}{\sqrt{F}} \tag{11-35}$$

The corresponding minimum resolvable wavelength difference may be found as follows. The phase difference, Eq. (11-18),

$$\delta = 4\pi t \cos \theta_t \left(\frac{1}{\lambda}\right)$$

For small wavelength intervals, the magnitude of $\Delta\delta$ is given by

$$\Delta\delta = (4\pi t \cos \theta_t) \Delta \left(\frac{1}{\lambda}\right) = \left(\frac{4\pi t \cos \theta_t}{\lambda^2}\right) \Delta\lambda$$

Combining with Eq. (11-35),

$$(\Delta\lambda)_{\min} = \frac{\lambda^2}{\pi \sqrt{F} t \cos \theta_t}$$

Since at the fringe maxima,

$$2t \cos \theta_t = m\lambda \tag{11-36}$$

we may write, more simply,

$$(\Delta\lambda)_{\min} = \frac{2\lambda}{m\pi \sqrt{F}} \tag{11-37}$$

Here λ may be considered to be either of the two wavelengths or their average, since they are close in value. The resolving power $\mathcal{R}$ is defined in general as

$$\mathcal{R} \equiv \frac{\lambda}{(\Delta\lambda)_{\min}} \tag{11-38}$$

When Eq. (11-38) is applied to the Fabry-Perot interferometer,

$$\mathscr{R} = \left(\frac{\pi}{2}\right) m \sqrt{F} \tag{11-39}$$

Large resolving powers are, of course, desirable. For the Fabry-Perot, we see that large values occur when the order is large, near the center of the fringe pattern, and for large coefficients of finesse, which correspond to high reflectance. Notice that to maximize m at the pattern center, Eq. (11-36) requires that the plate separation t be as large as possible, giving

$$m_{\text{max}} = \frac{2t}{\lambda} \tag{11-40}$$

Example

A Fabry-Perot interferometer has a 1-cm spacing and a reflection coefficient of $r = 0.95$. For a wavelength around 500 nm, determine its maximum order of interference, its coefficient of finesse, its minimum resolvable wavelength interval, and its resolving power.

Solution   Using Eq. (11-30) and Eqs. (11-37) to (11-40), we find

$$m_{\text{max}} = \frac{2t}{\lambda} = \frac{2(1 \times 10^{-2})}{500 \times 10^{-9}} = 40{,}000$$

$$F = \frac{4r^2}{(1 - r^2)^2} = \frac{4(0.95)^2}{(1 - 0.95^2)^2} = 380$$

$$(\Delta\lambda)_{\text{min}} = \frac{2\lambda}{m\pi\sqrt{F}} = \frac{2(5000\ \text{Å})}{(40{,}000)\,\pi\,(\sqrt{380})} = 0.004\text{Å}$$

$$\mathscr{R} = \frac{\lambda}{(\Delta\lambda)_{\text{min}}} = \frac{5000}{0.004} = 1.2 \times 10^6$$

Good Fabry-Perot interferometers may be expected to have resolving powers of a million or so, and values around 20 million have been used. This represents one to two orders of improvement over the performance of comparable prism and grating instruments. The high-resolution performance of a Fabry-Perot instrument is illustrated by the photograph of the ring pattern of the mercury green line, revealing its fine structure, in Figure 11-12.

Figure 11-12   Fabry-Perot rings obtained with the mercury green line, revealing fine structure. (From M. Cagnet, M. Francon, and J. C. Thrierr, *Atlas of Optical Phenomenon*, Plate 10, Berlin: Springer-Verlag, 1962.)

Individual sets of circular fringes, one set produced by each wavelength component, appear simultaneously in a Fabry-Perot interference pattern. Interpretation becomes complicated unless some means are found to limit the range of wavelengths analyzed by the interferometer. For example, consider two wavelengths λ_1 and λ_2, where $\lambda_2 = \lambda_1 + \Delta\lambda$. For small values of $\Delta\lambda$ the two sets of circular fringes will be close together in each order of interference. As $\Delta\lambda$ increases, however, the fringes separate. When the separation becomes equal to the distance between consecutive orders, confusion of orders results. Let us calculate the wavelength difference $\Delta\lambda$ such that the mth order of λ_2 falls on the $(m + 1)$ order of λ_1. This is called the *free spectral range* (fsr) of the interferometer. Since each product of m and λ corresponds to the same t and θ_t in Eq. (11-36), we can write

$$m\lambda_2 = (m + 1)\lambda_1$$

Then, with $\lambda_2 = \lambda_1 + \Delta\lambda$, we find

$$(\Delta\lambda)_{fsr} = \frac{\lambda_1}{m} \qquad (11\text{-}41)$$

The fsr is the change in λ_1 necessary to shift its circular fringe pattern by the distance of consecutive orders. Incorporating Eq. (11-36), with $\cos \theta_t = 1$ near the center of the fringe pattern where resolution is best, we may also write

$$(\Delta\lambda)_{fsr} \cong \frac{\lambda^2}{2t} \qquad (11\text{-}42)$$

To avoid associating fringes of one order with those of the next, then, one should have

$$\Delta\lambda < \frac{(\Delta\lambda)_{fsr}}{2}$$

Notice that a large order m is detrimental to a large fsr by Eq. (11-41), whereas it is favorable to good resolution by Eq. (11-37). For example, in the preceding high-resolution numerical example, the fsr is only 0.125 Å. One would like to maximize the quantity

$$\frac{(\Delta\lambda)_{fsr}}{(\Delta\lambda)_{min}} = \frac{\pi\sqrt{F}}{2} \qquad (11\text{-}43)$$

The ratio expressed by Eq. (11-43) thus represents a figure of merit for the Fabry-Perot interferometer and is called, simply, its *finesse*, $\mathscr{F}$, not to be confused with the coefficient of finesse F, on which it depends. The finesse is usually defined as the ratio of separation of adjacent maxima to the half-width of the individual fringes, as illustrated in Figure 11-13. Let us demonstrate the equivalence of this definition and that of Eq. (11-43). The phase difference between fringes is 2π. The phase width of a fringe at half its maximum irradiance is twice δ_c, as calculated in Eq. (11-34):

$$\delta_{1/2} = \frac{4}{\sqrt{F}}$$

Thus

$$\mathscr{F} = \frac{2\pi}{(4/\sqrt{F})} = \frac{\pi\sqrt{F}}{2} \qquad (11\text{-}44)$$

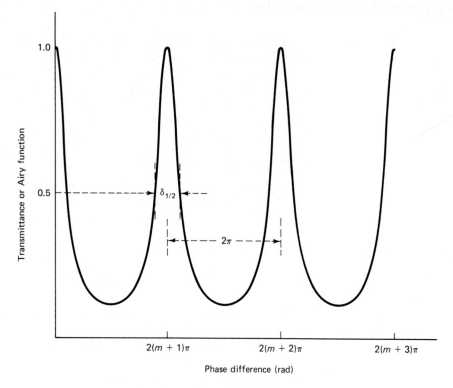

Figure 11-13 The *finesse* $\mathcal{F}$ is the ratio of the separation of adjacent fringe maxima to their individual width at half-maximum.

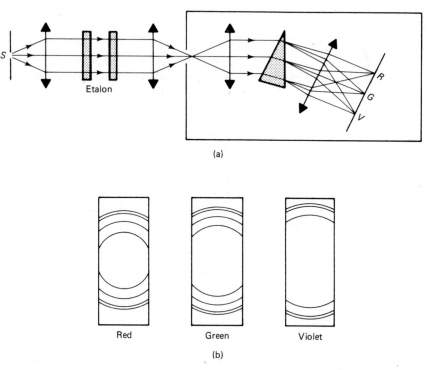

(a)

Red   Green   Violet

(b)

Figure 11-14 (a) Use of a Fabry-Perot etalon in tandem with a prism spectrograph. (b) Fringed spectral lines formed by the system in (a).

We conclude that the largest finesse of the interferometer represents the best compromise between the demands of resolution and free spectral range.

The limitations of this compromise may be overcome in several ways. One is to use two etalons in tandem, one of high resolution and the other of large fsr. In this way, it turns out that one can combine both capabilities. Another solution is to follow the etalon with a spectrograph, as shown in Figure 11-14a. Suppose that the light source has several well-separated spectral components, each with its detailed structure. The circular fringe patterns produced by the etalon alone would consist of a confusing superposition of fringes due to each of the constituent wavelengths. If the slit of the spectrograph, opened rather wide, intercepts a wide band through the center of the circular fringe pattern, the prism will perform its own spatial separation of wavelengths. Each wavelength interval of the source then appears as a broad image of the slit but with fringe patterns corresponding to each wavelength and its fine structure components, as suggested in Figure 11-14b.

PROBLEMS

11-1. When one mirror of a Michelson interferometer is translated by 0.0114 cm, 523 fringes are observed to pass the cross hairs of the viewing telescope. Calculate the wavelength of the light.

11-2. When looking into a Michelson interferometer illuminated by the 546.1-nm light of mercury, one sees a series of straight-line fringes that number 12 per centimeter. Explain their occurrence.

11-3. A thin sheet of fluorite of index 1.434 is inserted normally into one beam of a Michelson interferometer. Using light of wavelength 589 nm, the fringe pattern is found to shift by 35 fringes. What is the thickness of the sheet?

11-4. Looking into a Michelson interferometer, one sees a dark central disk surrounded by concentric bright and dark rings. One arm of the device is 2 cm longer than the other, and the wavelength of the light is 500 nm. Determine (a) the order of the central disc and (b) the order of the sixth dark ring from the center.

11-5. A Michelson interferometer is used to measure the refractive index of a gas. The gas is allowed to flow into an evacuated glass cell of length L placed in one arm of the interferometer. The wavelength is λ.
   (a) If N fringes are counted as the pressure in the cell changes from vacuum to atmospheric pressure, what is the index of refraction n in terms of N, λ, and L?
   (b) How many fringes would be counted if the gas were carbon dioxide ($n = 1.00045$) for a 10-cm cell length, using sodium light at 589 nm?

11-6. A Michelson interferometer is used with red light of wavelength 632.8 nm and is adjusted for a path difference of 20 μm. Determine the angular radius of the (a) first (smallest diameter) ring observed and (b) tenth ring observed.

11-7. A polished surface is examined using a Michelson interferometer with the polished surface replacing one of the mirrors. A fringe pattern characterizing the surface contour is observed using He-Ne light of wavelength 632.8 nm. Fringe distortion over the surface is found to be less than $\frac{1}{4}$ the fringe separation at any point. What is the maximum depth of polishing defects on the surface?

11-8. A laser beam from a 1-mW He-Ne laser (632.8 nm) is directed onto a parallel film with an incident angle of 45°. Assume a beam diameter of 1 mm and a film index of 1.414. Determine (a) the amplitude of the E-vector of the incident beam; (b) the angle of refraction of the laser beam into the film; (c) the magnitudes of r' and tt', using the Stokes relations and a reflection coefficient, $r = 0.280$; (d) the independent amplitudes of the first three reflected beams and, by comparison with the incident beam, the percentage of radiant power density reflected in each; (e) the same information as in (d) for the first two transmitted beams; (f) the minimum thickness of film that would lead to total cancellation of the reflected beams when they are brought together at a point by a lens.

11-9. (a) Using Eq. (10-25), show that amplitudes of the first three reflected and first three transmitted beams from a parallel, nonabsorbing glass ($n = 1.52$) plate, when the incident beam is near normal and of unit amplitude, are given by

	(1)	(2)	(3)
reflected	0.206	0.198	0.0084
transmitted	0.957	0.041	0.0017

(b) Show as a result that the first two reflected rays produce a fringe contrast or visibility of 0.999, whereas the first two transmitted rays produce a fringe contrast of only 0.085.

11-10. The plates of a Fabry-Perot interferometer have a reflectance coefficient of $r = 0.99$. Calculate the minimum (a) resolving power and (b) plate separation that will accomplish the resolution of the two components of the H-alpha doublet of the hydrogen spectrum, whose separation is 0.1360 Å at 6563 Å.

11-11. A Fabry-Perot interferometer is to be used to resolve the mode structure of a He-Ne laser operating at 632.8 nm. The frequency separation between the modes is 150 MHz. The plates are separated by an air gap and have a reflectance (r^2) of 0.999.
(a) What is the coefficient of finesse of the instrument?
(b) What is the resolution required?
(c) What plate spacing is required?
(d) What is the free spectral range of the instrument under these conditions?
(e) What is the minimum resolvable wavelength interval under these conditions?

11-12. A Fabry-Perot etalon is fashioned from a single slab of transparent material having a high refractive index ($n = 4.5$) and a thickness of 2 cm. The uncoated surfaces of the slab have a reflectance (r^2) of 0.90. If the etalon is used in the vicinity of wavelength 546 nm, determine (a) the highest-order fringe in the interference pattern; (b) the ratio T_{max}/T_{min}; (c) the resolving power.

11-13. The separation of a certain doublet is 0.0055 nm at a wavelength of 490 nm. A variable-spaced Fabry-Perot interferometer is used to examine the doublet. At what spacing does the mth order of one component coincide with the $(m + 1)$th order of the other?

11-14. The *reflectance* r^2 of a Fabry-Perot etalon is 60%. Determine the ratio of transmittance of the etalon at maximum to the trasmittance at halfway between maxima.

11-15. White light is passed through a Fabry-Perot interferometer in the arrangement shown in Figure 11-9b, where the detector is a spectroscope. A series of bright bands appear. When mercury light is simultaneously admitted into the spectroscope slit, 150 of the bright bands are seen to fall between the violet and green lines of mercury at 435.8 nm and 546.1 nm, respectively. What is the thickness of the etalon?

11-16. Apply the reasoning used to calculate the finesse of a Fabry-Perot interferometer to the Michelson interferometer: Using the irradiance of Michelson fringes as a function of phase, calculate (a) the fringe separation; (b) the fringe width at half-maximum; (c) their ratio, the finesse.

11-17. Assume that in a Mach-Zehnder interferometer (Figure 11-5), the beam splitter and mirror $M3$ each transmit 80% and reflect 20% of the incident light. Compare the fringe contrast when observing the interference of the two emerging beams (shown) with the fringe contrast that results from the two beams emerging from $M3$ along a direction at 90° relative to the first (not shown). For the second case, beam (1) is reflected and beam (2) is transmitted at $M3$.

REFERENCES

[1] Hariharan, P. *Optical Interferometry*. Orlando, Fla.: Academic Press, 1985.

[2] Connes, Pierre. "How Light Is Analyzed." *Scientific American* (Sept. 1968): 72.

[3] Smith, F. Dow. "How Images Are Formed." *Scientific American* (Sept. 1968): 96.

[4] Francon, Maurice. *Optical Interferometry*. New York: Academic Press, 1966.

[5] Tolansky, Samuel. *An Introduction to Interferometry*. New York: John Wiley and Sons, 1973.

[6] Hernandez, G. *Fabry-Perot Interferometers*. New York: Cambridge University Press, 1986.

[7] James, J. F., and R. S. Sternberg. *The Design of Optical Spectrometers*. London: Chapman and Hall Ltd., 1969. Ch. 7.

[8] Robinson, Glen M., David M. Perry, and Richard W. Peterson. "Optical Interferometry of Surfaces." *Scientific American* (July 1991): 66.

[9] Reynolds, George O., John B. DeVelis, George B. Parrent, Jr., and Brian J. Thompson. *Physical Optics Notebook: Tutorials in Fourier Optics*. Bellingham, Wash.: SPIE Optical Engineering Press, 1989. Ch. 22–24.

12

Coherence

INTRODUCTION

The term *coherence* is used to describe the correlation between phases of monochromatic radiations. Beams with random phase relationships are, generally speaking, incoherent beams, whereas beams with a constant phase relationship are coherent beams. The requirement of coherence between interfering beams of light, if they are to produce observable fringe patterns, was mentioned in connection with the interference mechanisms discussed in Chapter 10. We also discussed the relationship between coherence and the net irradiance of interfering beams in Chapter 9. There we concluded that in the superposition of in-phase coherent beams, individual amplitudes add together, whereas in the superposition of incoherent beams, individual irradiances add together. In this chapter, we examine the property of coherence in greater detail, distinguishing between *longitudinal coherence*, which is related to the spectral purity of the source, and *lateral* or *spatial coherence*, which is related to the size of the source. We also describe a quantitative measure of *partial coherence*, the condition under which most experimental measurements of interference take place. We begin our treatment with a brief description of Fourier analysis, which we will need in this chapter.

When a number of harmonic waves of the same frequency are added together, even though they differ in amplitude and phase, the result is again a harmonic wave of the given frequency, as shown in Chapter 9. If the superposed waves differ in frequency as well, the result is periodic but anharmonic and may assume an arbitrary shape, such as that shown in Figure 12-1. An infinite variety of shapes may be synthesized in this way. The inverse process of decomposition of a given waveform into its harmonic components is called *Fourier analysis*.

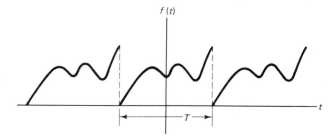

Figure 12-1 Anharmonic function of time with period T.

The successful decomposition of a waveform into a series of harmonic waves is insured by the *theorem of Dirichlet*:

If $f(t)$ is a bounded function of period T with at most a finite number of maxima or minima or discontinuities in a period, then the *Fourier series*,

$$f(t) = \frac{a_0}{2} + \sum_{m=1}^{\infty} a_m \cos m\omega t + \sum_{m=1}^{\infty} b_m \sin m\omega t \qquad (12\text{-}1)$$

converges to $f(t)$ at all points where $f(t)$ is continuous and to the average of the right and left limits at each point where $f(t)$ is discontinuous.

In Eq. (12-1), m takes on integral values and $\omega = 2\pi f = 2\pi/T$, where T is the period of the arbitrary $f(t)$. The sine and cosine terms can be interpreted as harmonic waves with amplitudes of b_m and a_m, respectively, and frequencies of $m\omega$. The magnitudes of the coefficients or amplitudes determine the contribution each harmonic wave makes to the resultant anharmonic waveform. If Eq. (12-1) is multiplied by dt and integrated over one period T, the sine and cosine integrals vanish, and the result is

$$a_0 = \frac{2}{T} \int_{t_0}^{t_0+T} f(t)\, dt \qquad (12\text{-}2)$$

If Eq. (12-1) is multiplied throughout instead by $\cos n\omega t\, dt$, where n is any integer, and then integrated over a period, the only nonvanishing integral on the right side is the one including the coefficient a_m, and one finds

$$a_m = \frac{2}{T} \int_{t_0}^{t_0+T} f(t) \cos m\omega t\, dt \qquad (12\text{-}3)$$

Similarly, multiplying Eq. (12-1) by $\sin n\omega t\, dt$ and integrating gives

$$b_m = \frac{2}{T} \int_{t_0}^{t_0+T} f(t) \sin m\omega t\, dt \qquad (12\text{-}4)$$

Thus, once $f(t)$ is specified, each of the coefficients a_0, a_m, and b_m can be calculated, and the analysis is complete.

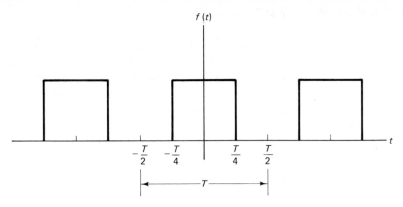

$f(t)$

$$-\frac{T}{2} \quad -\frac{T}{4} \qquad \frac{T}{4} \quad \frac{T}{2}$$

t

T

Figure 12-2 Square wave.

As an example, consider the Fourier analysis of the square wave shown in Figure 12-2 and represented over a period symmetric with the origin by

$$f(t) = \begin{cases} 0, & -T/2 < t < -T/4 \\ 1, & -T/4 < t < T/4 \\ 0, & T/4 < t < T/2 \end{cases}$$

Since the function is even in t, the coefficients b_m are found to vanish, and only cosine terms (also even functions of t) remain. From Eqs. (12-2) and (12-3), we find

$$a_0 = 1$$

$$a_m = \left(\frac{2}{m\pi}\right) \sin\left(\frac{m\pi}{2}\right)$$

so that the Fourier series that converges to the square wave of Figure 12-2 as more terms are included in the summation is

$$f(t) = \frac{1}{2} + \sum_{m=1}^{\infty} \left[\left(\frac{2}{m\pi}\right) \sin\left(\frac{m\pi}{2}\right) \right] \cos m\omega t$$

Writing out the first few terms explicitly,

$$f(t) = \frac{1}{2} + \frac{2}{\pi}\left(\cos \omega t - \frac{1}{3} \cos 3\omega t + \frac{1}{5} \cos 5\omega t + \cdots \right)$$

Notice that the contribution of each successive term decreases because its amplitude decreases. Thus a finite number of terms may represent the function rather well. The more rapidly the series converges, the fewer are the terms needed for an adequate fit. Notice also that some amplitudes may be negative, that is, some harmonic waves must be subtracted from the sum to accomplish the convergence. Quite reasonably, fine features in the given $f(t)$, such as the corners of the square waves, require waves of smaller wavelength, or higher frequency components, to represent them. Accordingly, if the widths of the square waves were allowed to diminish, so that the individual squares approached spikes, one would expect a greater contribution from the high-frequency components for an adequate synthesis of the function.

With the help of Euler's equation, the Fourier series given in general by Eq. (12-1), involving as it does both sine and cosine terms, can be expressed in complex notation using exponential functions. The result is

$$f(t) = \sum_{n=-\infty}^{+\infty} c_n e^{-in\omega t} \qquad (12\text{-}5)$$

where now the coefficients

$$c_n = \frac{1}{T} \int_{t_0}^{t_0+T} f(t)e^{in\omega t}\, dt \tag{12-6}$$

In cases where we wish instead to represent a nonperiodic function (cleverly interpreted mathematically as a periodic function whose period T approaches infinity), it is possible to generalize the Fourier series to a *Fourier integral*. For example, a single pulse is a nonperiodic function but can be interpreted as a periodic function whose period extends from $t = -\infty$ to $t = +\infty$. It can be shown that the discrete Fourier series now becomes an integral given by

$$f(t) = \int_{-\infty}^{+\infty} g(\omega)e^{-i\omega t}\, d\omega \tag{12-7}$$

where the coefficient

$$g(\omega) = \frac{1}{2\pi} \int_{-\infty}^{+\infty} f(t)e^{i\omega t}\, dt \tag{12-8}$$

The Fourier integral, Eq. (12-7), and the expression for its associated coefficient, Eq. (12-8), have a certain degree of mathematical symmetry and are together referred to as a *Fourier-transform pair*. Instead of a discrete spectrum of frequencies given by the Fourier series, Eq. (12-6), we are led to a continuous spectrum, as given by Eq. (12-8). In Figure 12-3, a sample discrete set of coefficients, as might be calculated from Eq. (12-6), is shown together with a continuous distribution approximated by the coefficients, such as might result from Eq. (12-8).

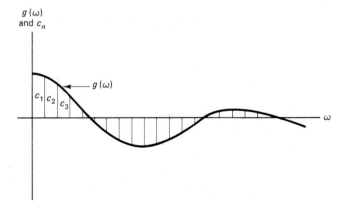

Figure 12-3 Fourier coefficients of a periodic function specify discrete harmonic components of amplitude c_n at frequency ω_n. The Fourier transform of a nonperiodic function requires instead a continuous frequency spectrum $g(\omega)$.

It should be pointed out that if the function to be represented is a function of spatial position x with period L, say, rather than of time t with period T, then in Eqs. (12-1) through (12-8) T should be replaced by L and the temporal frequency $\omega = 2\pi/T$ should be replaced by the *spatial frequency*, $k = 2\pi/L$. For example, the Fourier transforms in Eqs. (12-7) and (12-8) become

$$f(x) = \int_{-\infty}^{+\infty} g(k)e^{-ikx}\, dk \tag{12-9}$$

$$g(k) = \frac{1}{2\pi} \int_{-\infty}^{+\infty} f(x)e^{ikx}\, dx \tag{12-10}$$

The spectral resolution of an infinitely long sinusoidal wave is extremely simple: It is one term of the Fourier series, the term corresponding to the actual frequency of the wave. In this case, all other coefficients vanish. Sinusoidal waves without a beginning or an end are, however, mathematical idealizations. In practice, the wave is turned on and off at finite times. The result is a wave train of finite length, such as the one pictured in Figure 12-4. Fourier analysis of such a wave train must regard it as a nonperiodic function. Clearly, it cannot be represented by a single sine wave that has no beginning or end. Rather, the various harmonic waves that combine to produce the wave train must be numerous and so selected that they produce exactly the wave train during the time interval it exists and cancel exactly everywhere outside that interval. Evidently, the turning "on" and "off" of the wave adds many other spectral components to that of the temporary wave train itself. The use of the Fourier-transform integrals leads, in fact, to a *continuous* distribution of frequency components. What we have said here of a finite wave train is also true of any isolated pulse, regardless of its shape. We consider for simplicity the spectral resolution of a pulse that is, while it exists at some point, a harmonic wave. The problem must be handled, as suggested, by the Fourier integral transforms, Eqs. (12-7) and (12-8). We have placed the origin of the time frame, Figure 12-4, so that the wave train is symmetrical about it. The wave train has a lifetime of τ_0 and a frequency of ω_0. Thus it may be represented by

$$f(t) = \begin{cases} e^{-i\omega_0 t}, & -\dfrac{\tau_0}{2} < t < \dfrac{\tau_0}{2} \\ 0, & \text{elsewhere} \end{cases} \tag{12-11}$$

The frequency spectrum $g(\omega)$ is calculated from Eq. (12-8), with the specific function $f(t)$ of Eq. (12-11),

$$g(\omega) = \frac{1}{2\pi} \int_{-\infty}^{+\infty} f(t)e^{i\omega t}\,dt = \frac{1}{2\pi} \int_{-\tau_0/2}^{+\tau_0/2} e^{i(\omega-\omega_0)t}\,dt$$

Integrating, we have

$$g(\omega) = \left[\frac{e^{i(\omega-\omega_0)t}}{2\pi i(\omega - \omega_0)}\right]_{-\tau_0/2}^{+\tau_0/2}$$

$$g(\omega) = \frac{1}{\pi(\omega - \omega_0)}\left[\frac{e^{i(\omega-\omega_0)\tau_0/2} - e^{-i(\omega-\omega_0)\tau_0/2}}{2i}\right]$$

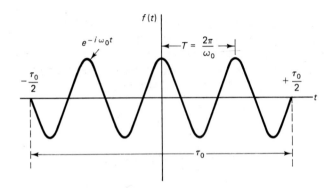

Figure 12-4 Finite harmonic wave train of lifetime τ_0 and period $2\pi/\omega_0$. The spatial extension of the pulse is $\ell_0 = c\tau_0$.

or, after using the identity,

$$e^{ix} - e^{-ix} \equiv 2i \sin x$$

$$g(\omega) = \frac{\sin\left[(\tau_0/2)(\omega - \omega_0)\right]}{\pi(\omega - \omega_0)} = \frac{\tau_0}{2\pi}\left\{\frac{\sin\left[(\tau_0/2)(\omega - \omega_0)\right]}{\left[(\tau_0/2)(\omega - \omega_0)\right]}\right\} \qquad (12\text{-}12)$$

Calling $u = (\tau_0/2)(\omega - \omega_0)$, we then have $g(\omega) = (\tau_0/2\pi)[(\sin u)/u]$. The function $(\sin u)/u$, often called simply $sinc\,(u)$, shows up frequently. It has the property that as u approaches 0, the function approaches a value of 1. Thus, from Eq. (12-12), we conclude that

$$\lim_{\omega \to \omega_0} g(\omega) = \frac{\tau_0}{2\pi} \qquad (12\text{-}13)$$

Furthermore, the sinc function $(\sin u)/u$ vanishes whenever $\sin u = 0$, except at $u = 0$, the case already described by Eq. (12-13). In every other case, $\sin u = 0$ for $u = n\pi$, $n = \pm 1, \pm 2, \ldots$, and so

$$g(\omega) = 0 \quad \text{when} \quad \omega = \omega_0 \pm \frac{2n\pi}{\tau_0} \qquad (12\text{-}14)$$

As ω increases (or decreases) from ω_0 then, $g(\omega)$ passes periodically through zero. The accompanying increase in the magnitude of u, or of the denominator of Eq. (12-12), gradually decreases the amplitude of an otherwise harmonic variation. These results are all displayed in Figure 12-5, where the origin of the frequency spectrum is chosen at its point of symmetry, $\omega = \omega_0$. When the amplitude $g(\omega)$ is squared, the resulting curve is the *power spectrum*, shown as the solid curve in Figure 12-5. Although frequencies far from ω_0 contribute to the power spectrum, the bulk of the energy of the wave train is clearly carried by the frequencies present in the central maximum, of width $4\pi/\tau_0$. Notice that the shorter the wave train of Figure 12-4, that is, the smaller the lifetime τ_0, the wider is the central maximum of Figure 12-5. This means that the harmonic waves making important contributions to the actual wave train span a greater frequency interval. We take the half-width of the central maximum, or $2\pi/\tau_0$, to indicate in a rough way the range of dominant frequencies required. This criterion at least preserves the important inverse relation-

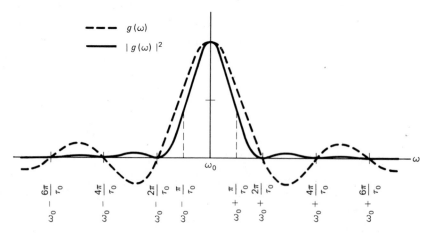

Figure 12-5 Fourier transform of the finite harmonic wave train of Figure 12-4. The dashed line gives the amplitude of the frequency spectrum and the solid line gives its square, the power spectrum. The curves have been normalized to the same maximum amplitude.

ship with τ_0. Accordingly, we write, as a measure of the frequency band centered around ω_0 required to represent the harmonic wave train of frequency ω_0 and life-time τ_0,

$$\Delta\omega = \frac{2\pi}{\tau_0} \quad \text{or} \quad \Delta f = \frac{1}{\tau_0} \tag{12-15}$$

Equation (12-15) shows that if $\tau_0 \to \infty$, corresponding to a wave train of infinite length, $\Delta\omega \to 0$, and a single frequency ω_0 or wavelength λ_0 suffices to represent the wave train. In this idealized case we have a perfectly monochromatic beam, as considered previously. On the other hand, as $\tau_0 \to 0$, approximating a harmonic "spike," $\Delta\omega \to \infty$. Thus the sharper or narrower the pulse, the greater is the number of frequencies required to represent it, and so the greater the *line width*, or $\Delta\lambda$, of the harmonic wave package.

12-3 TEMPORAL COHERENCE AND NATURAL LINE WIDTH

Clearly, there are no perfectly monochromatic sources. Sources we call "monochromatic" emit light that can be represented as a sequence of harmonic wave trains of finite length, as suggested in Figure 12-6, each separated from the others by a discontinuous change in phase. These phase changes reflect the erratic process by which excited atoms in a light source undergo transitions between energy levels, producing brief and random radiation wave trains. A given source can be character-ized by an average wave train lifetime τ_0, called its *coherence time*. Thus the physi-cal implications of Eq. (12-15) may be summarized as follows: The natural width of a spectral line is inversely proportional to the coherence time of the source. The greater its coherence time, the more monochromatic the source. The *coherence length* l_t of a wavetrain is the length of its coherent pulse, or

$$l_t = c\tau_0 \tag{12-16}$$

Combining Eqs. (12-15) and (12-16), the coherence length is

$$l_t = \frac{c}{\Delta f}$$

Then, approximating Δf by the magnitude of its differential from the expression $f = c/\lambda$, we may also write

$$l_t \cong \frac{\lambda^2}{\Delta\lambda} \tag{12-17}$$

Thus the natural line width is

$$\Delta\lambda \cong \frac{\lambda^2}{l_t} \tag{12-18}$$

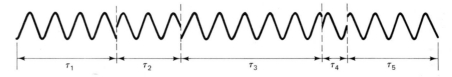

Figure 12-6 Sequence of harmonic wavetrains of varying finite lengths or life-times τ. The wavetrain may be characterized by an average lifetime, the coherence time τ_0.

To digress briefly, it is interesting to note that Eq. (12-18) is an expression of the uncertainty principle of quantum mechanics, where a wave pulse is used to represent, say, an electron. If the coherence length l_t is interpreted as the interval Δx within which the particle is to be found—that is, its uncertainty in location—and the uncertainty in momentum Δp is expressed by the differential of the deBroglie wavelength in the equation $p = h/\lambda$, the result is $\Delta x\,\Delta p = h$. The inequality associated with the Heisenberg uncertainty relation arises from the inequality inherent in Eq. (12-15).

Since the line width of spectral sources can be measured, average coherence times and coherent lengths my be surmised. White light, for example, has a "line width" of around 300 nm, extending roughly from 400 to 700 nm. Taking the average wavelength at 550 nm, Eq. (12-17) gives

$$l_t = \frac{550^2}{300} \cong 1000 \text{ nm} \cong 2\lambda_{\text{av}}$$

a very small coherence length indeed, of around a millionth of a centimeter or two "wavelengths" of white light. Understandably, interference fringes by white light are difficult to obtain since the difference in the path lengths of the interfering beams should not be greater than the coherence length for the light. Sodium or mercury gas-discharge lamp sources are far more monochromatic and coherent. For example, the green line of mercury at 546 nm may have a line width of around 0.025 nm, giving a coherence length of 1.2 cm. One of the most monochromatic gas-discharge sources is a gas of the krypton 86 isotope, whose orange emission line at 606 nm has a line width of only 0.00047 nm. The coherence length of this radiation, by Eq. (12-17), is 78 cm! Laser radiation has far surpassed even the coherence of this gas-discharge source. The short-term stability of commercially available CO_2 lasers, for example, is such that line widths of around 1×10^{-5} nm are attainable at the infrared emission wavelength of 10.6 μm. These numbers give a coherence length of around 11 km! Under carefully controlled conditions, He-Ne lasers can improve this figure by another order of magnitude. Somewhat discouragingly, the common He-Ne laser used in instructional laboratories may not have coherence lengths much greater than its cavity length, due to random temperature fluctuations and mirror vibrations. These spurious effects change the cavity length, lead to multimode oscillations, and adversely affect the coherence length of the laser. Hence the use of these lasers, in holography experiments, for example, still requires some care in equalizing optical-path lengths.

12-4 PARTIAL COHERENCE

As pointed out previously, when the phase difference between two waves is constant, they are mutually coherent waves. In practice, this condition is only approximately met, and we speak of *partial coherence*. The concept is defined more precisely in what follows. Consider, as in Figure 12-7, a general situation in which

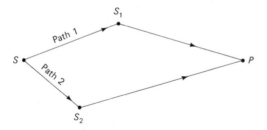

Figure 12-7 Interference at P due to waves from S traveling different paths. The waves are redirected at S_1 and S_2 by various means, including reflection, refraction, and diffraction.

interference is produced at P between two beams that originate from a single coherent source S after traveling different paths. Let the two beams be represented in general by

$$\mathbf{E}_1(\mathbf{r}, t) = \mathbf{E}_{01} e^{i(\mathbf{k}_1 \cdot \mathbf{r} - \omega t + \epsilon)} \tag{12-19}$$

$$\mathbf{E}_2(\mathbf{r}, t) = \mathbf{E}_{02} e^{i(\mathbf{k}_2 \cdot \mathbf{r} - \omega t + \epsilon)} \tag{12-20}$$

At point P, each field varies only with time, and if the lower path is longer, requiring an extra time interval τ, we may simply write at P:

$$\mathbf{E}_p(t) = \mathbf{E}_1(t) + \mathbf{E}_2(t + \tau) = \mathbf{E}_{01} e^{-i\omega t} + \mathbf{E}_{02} e^{-i\omega(t + \tau)}$$

where we have set $r = 0$ and $\epsilon = 0$, for convenience. Except for a multiplicative constant, the irradiance at P is given by

$$I_p = \langle \mathbf{E}_p \cdot \mathbf{E}_p^* \rangle = \langle (\mathbf{E}_1 + \mathbf{E}_2) \cdot (\mathbf{E}_1^* + \mathbf{E}_2^*) \rangle$$

where the angle brackets represent time averages. Expanding,

$$I_p = \langle |E_1|^2 + |E_2|^2 + (\mathbf{E}_1 \cdot \mathbf{E}_2^* + \mathbf{E}_2 \cdot \mathbf{E}_1^*) \rangle$$

The quantity in parentheses is just the sum of a complex number and its complex conjugate, which is always equal to twice its real part. Thus

$$I_p = I_1 + I_2 + 2 \, \mathrm{Re} \, \langle E_1 E_2^* \rangle \tag{12-21}$$

where I_1 and I_2 represent the irradiances of the individual beams and the third term represents interference between them. In Eq. (12-21), we have also assumed that the beams have the same polarization and so have suppressed the dot product. The interference term can add to or subtract from the simple sum of the two beam irradiances, depending on the correlation in phase between the two fields at P. We define, accordingly, a *correlation function*,

$$\Gamma_{12}(\tau) \equiv \langle E_1(t) E_2^*(t + \tau) \rangle \tag{12-22}$$

and, dividing by the amplitudes of the fields, a *normalized correlation function*,

$$\gamma_{12}(\tau) \equiv \frac{\Gamma_{12}(\tau)}{\sqrt{I_1 I_2}} \tag{12-23}$$

The irradiance at P may then be expressed by

$$I_p = I_1 + I_2 + 2\sqrt{I_1 I_2} \, \mathrm{Re} \, [\gamma_{12}(\tau)] \tag{12-24}$$

The function $\gamma_{12}(\tau)$, now the heart of the interference term, is a function of τ and therefore of the location of point P. We know that the time difference between paths, relative to the average coherence time of the source, is crucial to the degree of coherence achieved. We expect that for $\tau > \tau_0$, some coherence between the two beams will be lost. The dependence of $\gamma_{12}(\tau)$ on τ_0 is now derived, under the assumption that τ_0 represents a constant coherence time rather than an average. Such a wave train is shown at the top of Figure 12-8a, with regular discontinuities in phase, separated by the time interval τ_0. The phase $\varphi(t)$ of the harmonic wave is changed in an arbitrary fashion and is plotted by the solid line below. The wave with harmonic frequency ω can then be represented by

$$E = E_0 e^{-i[\omega t - \varphi(t)]} \tag{12-25}$$

The phase changes $\varphi(t + \tau)$ of the second wave, arriving τ earlier, are also shown (dashed line) as the same function but displaced by the time τ. The normalized cor-

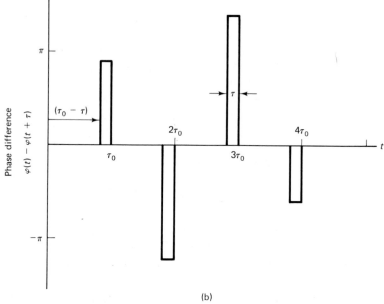

Figure 12-8 (a) Random phase fluctuations $\varphi(t)$ every τ_0 of a wave (solid line) and the same phase fluctuations $\varphi(t + \tau)$ of the wave (dashed line) at a time τ earlier. (b) Difference in the phase between the two waves described in (a).

relation function, often called the *degree of coherence* between two waves of equal amplitude E_0 and out of step by time τ, is, by Eq. (12-23),

$$\gamma_{12}(\tau) = \frac{\Gamma_{12}(\tau)}{I} = \frac{\langle E(t)E^*(t + \tau)\rangle}{|E_0|^2}$$

Writing out the product of the fields in terms of Eq. (12-25), we have

$$E(t)E^*(t + \tau) = E_0 e^{-i\omega t} e^{i\varphi(t)} E_0 e^{i\omega(t+\tau)} e^{-i\varphi(t+\tau)}$$

or

$$E(t)E^*(t + \tau) = E_0^2 e^{i\omega\tau} e^{i[\varphi(t) - \varphi(t+\tau)]}$$

so that

$$\gamma_{12}(\tau) = e^{i\omega\tau} \langle e^{i[\varphi(t) - \varphi(t+\tau)]} \rangle$$

The time average expressed in this equation may be calculated from

$$\langle e^{i[\varphi(t) - \varphi(t + \tau)]} \rangle = \frac{1}{T} \int_0^T e^{i[\varphi(t) - \varphi(t+\tau)]} \, dt \qquad (12\text{-}26)$$

where T is a sufficiently long time. The function $\varphi(t) - \varphi(t + \tau)$ in the exponent is pictured in Figure 12-8b and is seen to be a regularly spaced series of rectangular pulses with random magnitude falling between -2π and $+2\pi$. Consider the first coherence time interval τ_0, in which the pulse function may be expressed by

$$\varphi(t) - \varphi(t + \tau) = \begin{cases} 0, & 0 < t < (\tau_0 - \tau) \\ H, & (\tau_0 - \tau) < t < \tau_0 \end{cases}$$

In successive intervals, the expression is the same, except for the value of H. We may then write the normalized coherence function γ_{12} for a large number N of intervals as

$$\gamma_{12} = e^{i\omega\tau} \frac{1}{N\tau_0} \left[\underbrace{\int_0^{\tau_0 - \tau} e^{i(0)} \, dt + \int_{\tau_0 - \tau}^{\tau_0} e^{iH_1} \, dt}_{\text{interval } N = 1} + \begin{array}{l} \text{similar terms for } (N - 1) \\ \text{successive intervals} \end{array} \right]$$

Integrating over N terms,

$$\gamma_{12} = \left(\frac{e^{i\omega\tau}}{N\tau_0} \right) [(\tau_0 - \tau + \tau e^{iH_1}) + (\tau_0 - \tau + \tau e^{iH_2}) + \cdots]$$

Combining the first terms of each interval and summing the rest,

$$\gamma_{12} = \left(\frac{e^{i\omega\tau}}{N\tau_0} \right) \left[N(\tau_0 - \tau) + \tau \sum_{j=1}^N e^{iH_j} \right]$$

Because of the random nature of H_j, the terms in the summation average to zero for N sufficiently large. Thus only those times during which the waves coincide—when $\varphi(t) = \varphi(t + \tau)$—contribute to the integral, and we are left with

$$\gamma_{12}(\tau) = \left(1 - \frac{\tau}{\tau_0} \right) e^{i\omega\tau} \qquad (12\text{-}27)$$

The real part of γ_{12}, required in Eq. (12-24), is given by

$$\text{Re} \, [\gamma_{12}(\tau)] = \left(1 - \frac{\tau}{\tau_0} \right) \cos \omega\tau \qquad (12\text{-}28)$$

and so takes on a maximum value of 1 when $\tau = 0$ (equal path lengths), a value of 0 when $\tau = \tau_0$ (path difference equals coherence length), and any value between. The amplitude of the cosine term in Eq. (12-28) is just the magnitude of the degree of coherence γ_{12}, that is,

$$|\gamma_{12}(\tau)| = 1 - \frac{\tau}{\tau_0} \qquad (12\text{-}29)$$

This quantity sets the limits of the variations in the interference term and thus controls the contrast or visibility of the fringes as a function of τ. This amplitude, $|\gamma_{12}(\tau)|$, is plotted in Figure 12-9. Combining the last three equations,

$$\gamma_{12}(\tau) = |\gamma_{12}| e^{i\omega\tau} \tag{12-30}$$

$$\mathrm{Re}\ \gamma_{12}(\tau) = |\gamma_{12}| \cos \omega\tau \tag{12-31}$$

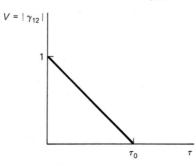

Figure 12-9 Fringe visibility or degree of coherence as a function of the difference in arrival times of two waves with coherence time τ_0.

Recalling the empirical expression for fringe contrast or visibility introduced earlier,

$$V = \frac{I_{\max} - I_{\min}}{I_{\max} + I_{\min}} \tag{12-32}$$

we may now delineate the following special cases:

1. *Complete incoherence:* $\tau \rightarrow \tau_0$ and $|\gamma_{12}| = 0$

$$I_p = I_1 + I_2$$

$$I_p = 2I_0, \quad \text{for equal beams}$$

$$V = \frac{2I_0 - 2I_0}{4I_0} = 0$$

2. *Complete coherence:* $\tau = 0$ and $|\gamma_{12}| = 1$

$$I_p = I_1 + I_2 + 2\sqrt{I_1 I_2} \cos \omega\tau$$

$$I_{\max} = I_1 + I_2 + 2\sqrt{I_1 I_2} = 4I_0, \quad \text{for equal beams}$$

$$I_{\min} = I_1 + I_2 - 2\sqrt{I_1 I_2} = 0, \quad \text{for equal beams}$$

$$V = \frac{4I_0}{4I_0} = 1$$

3. *Partial coherence:* $0 < \tau < \tau_0$ and $1 > |\gamma_{12}| > 0$

$$I_p = I_1 + I_2 + 2\sqrt{I_1 I_2}\, \mathrm{Re}\,(\gamma_{12})$$

$$I_p = 2I_0[1 + \mathrm{Re}\,(\gamma_{12})], \quad \text{for equal beams}$$

$$I_{\max} = 2I_0(1 + |\gamma_{12}|) \quad \text{and} \quad I_{\min} = 2I_0(1 - |\gamma_{12}|)$$

$$V = \frac{4I_0 |\gamma_{12}|}{4I_0} = |\gamma_{12}|$$

In all cases of equal beams, therefore, the fringe visibility V is equal to the magnitude of the correlation function $|\gamma_{12}|$, and either one is a measure of the degree of coherence. Fringes corresponding to cases 3 and 2 were depicted in Figure 10-1a and b, respectively, in the earlier discussion of interference.

Example

In an interference experiment, a light beam is split into two equal-amplitude parts. The two parts are superimposed again after traveling diverse paths. The light is of wavelength 541 nm with a line width of 1 Å, and the path difference is 1.50 mm. Determine the visibility of the interference fringes. How is the visibility modified if the path difference is doubled?

Solution   The visibility is given by

$$V = 1 - \frac{\tau}{\tau_0} = 1 - \frac{\Delta}{\ell_t}$$

where the ratio of time delay to coherence time is replaced by the corresponding ratio of path difference to coherence length. In this case, $\ell_t = \lambda^2/\Delta\lambda = (5410 \text{ Å})^2/(1 \text{ Å}) = 2.93$ mm. Thus,

$$V = 1 - \frac{1.5}{2.93} = 0.49$$

When the path difference is doubled, $\Delta > \ell_t$ and $\tau > \tau_0$, so that the beams are incoherent and $V = 0$.

12-5 SPATIAL COHERENCE

In speaking of temporal coherence, we have been considering the correlation in phase between temporally distinct points of the radiation field of a source along its line of propagation. For this reason, temporal coherence is also called *longitudinal coherence*. The degree of coherence can be observed by examining the interference fringe contrast in an amplitude-splitting instrument, such as the Michelson interferometer. As we have seen, temporal coherence is a measure of the average length of the constituent harmonic waves, which depends on the radiation properties of the source. In contrast, we now turn our attention to what is referred to as *spatial*, or *lateral*, *coherence*, the correlation in phase between spatially distinct points of the radiation field. This type of coherence is important when using a wavefront-splitting interferometer, such as the double slit. The quality of the interference pattern in the double-slit experiment depends on the degree of coherence between distinct regions of the wave field at the two slits.

To sharpen our understanding of the coherence of a wavefield radiating from a source, consider the situation depicted in Figure 12-10. Light from a source S passes through a double slit and is also sampled by a Michelson interferometer located nearby. Spatial coherence between wavefront points A and B at the slits is insured as long as the source S is a true point source. In that case all rays emanating from S are associated with a single set of spherical waves that have the same phase on any given wavefront. Are clear distinguishable fringes then formed on a screen near point P_1? The answer, of course, depends on whether the light from S, traveling along the two distinct paths SAP_1 and SBP_1, is temporally as well as spatially coherent. The matter of temporal coherence requires a comparison between the path difference $\Delta = SAP_1 - SBP_1$ and the coherence length of the radiation. This is equivalent to a comparison of coherence along any radial direction of light propagation from the source at two wavefronts separated by the same path difference. It is this property of temporal coherence that is measured by the Michelson interferometer. If the path difference Δ is much less than the coherence length ($\Delta \ll l_t$), clean interference fringes are formed at P_1; if the path difference is equal to or greater than the coherence

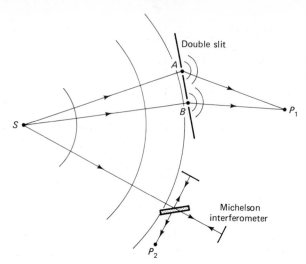

Double slit

S

A

B

P_1

Michelson
interferometer

P_2

Figure 12-10 Wavefront and amplitude division of radiation from source S, illustrating the practical requirements of spatial and temporal coherence.

length ($\Delta \gtrsim l_t$), interference fringes are poorly defined or absent altogether. In practice, of course, S is always an extended source, so that rays reach A and B from many points of the source. In ordinary (nonlaser) sources, light emitted by different points of a source, well over a wavelength in separation, is not correlated in phase and so lacks coherence. Thus the spatial coherence of light at the slits depends on how closely the source resembles a point source of light, either in extension or in its actual coherence properties.

We wish to show now that if two source points S_1 and S_2, as in Figure 12-11, are separated by a distance s and if light of wavelength λ from these sources is observed at a distance r away, there will be a region of high spatial coherence of dimension l_s, given by

$$l_s < \frac{\lambda}{\theta} \tag{12-33}$$

where θ is the angle subtended by the point sources at the observation point P. Accepting this result for the moment and combining it with the temporal or longitudinal coherence length l_t, we conclude that there exists at any point in the radiation field of a real light source a region of space in which the light is coherent. This region has lateral dimensions of l_s and longitudinal dimensions of l_t relative to the source and thus occupies a volume or roughly $l_s^2 l_t$ around the point P. It is from this volume that any interferometer must accept radiation if it is to produce observable interference fringes.

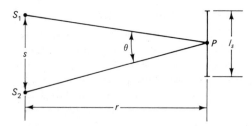

S_1

θ

s

P l_s

r

S_2

Figure 12-11 Lateral region of coherence l_s, due to two independent point sources.

12-6 SPATIAL COHERENCE WIDTH

Consider now the spatial coherence at points P_1 and P_2 in the radiation field of a quasi-monochromatic extended source, simply represented by two mutually incoherent emitting points A and B at the edges of the source (Figure 12-12). We may think

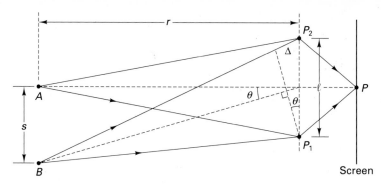

Figure 12-12 Light from each of two point sources A and B reach points P_1 and P_2 in the radiation field and are allowed to interfere at the screen. In practice, $s \ll \ell$ and angles θ are approximately equal.

of P_1 and P_2 as two slits that propagate light to a screen, where interference fringes may be viewed. Each point source, acting alone, then produces a set of double-slit interference fringes on the screen. When both sources act together, however, the fringe systems overlap. If the fringe systems overlap with their maxima and minima falling together, the resulting fringe pattern is highly visible, and the radiation from the two incoherent sources is considered highly coherent! When the fringe systems are relatively displaced, however, so that maxima of one fall on the minima of the other, the composite pattern is not visible and the radiation is considered incoherent. Suppose that source B is at the position of source A, or that the distance s in Figure 12-12 is zero. The fringe systems at the screen then coincide and correspond to the fringes of a single point source. A maximum in the interference pattern occurs at P if P lies on the perpendicular bisector of the two slits. In this condition,

$$BP_2 - BP_1 = AP_2 - AP_1 = 0$$

If source B is moved below A, the fringe systems separate until, at a certain distance s, where

$$BP_2 - BP_1 = \Delta = \frac{\lambda}{2}$$

the maximum in the fringe system at P due to source B is replaced by a minimum, and the composite fringe pattern disappears.

If the angle θ represents the angular separation of the sources from the plane of the slits, then from the diagram, $\Delta \cong \ell\theta$, where ℓ is the distance between slits, and $\theta \cong s/r$, where r is the distance to the sources. It follows that

$$\Delta = \frac{\lambda}{2} = \frac{s\ell}{r} \quad \text{or} \quad s = \frac{r\lambda}{2\ell} \tag{12-34}$$

When the distance AB is considered instead to be a continuous array of point sources, the individual fringe systems do not give complete cancellation until the spatial extent AB of the source reaches twice the value s in Eq. (12-34). If extreme points are separated by an amount $s < r\lambda/\ell$, then fringe definition is assured. Regarding this result as describing instead the maximum slit separation ℓ, given a source dimension s, we have for the *spatial coherence width* ℓ_s,

$$\ell_s < \frac{r\lambda}{s} \cong \frac{\lambda}{\theta} \tag{12-35}$$

As ℓ_s is restricted to smaller fractions of this value, the fringe contrast is correspondingly improved.

According to this argument, moving the source B even farther should bring the fringe system into coincidence again, so that the degree of coherence $|\gamma_{12}|$ between P_1 and P_2 is a periodic function. In a more complete mathematical argument, the extended source is represented by a continuous array of elemental emitting areas rather than by two point sources. Results show that outside the coherence width given by Eq. (12-35), the fringe visibility, while oscillatory, is negligible. According to a general theorem, known as the *Van Cittert-Zernike theorem* [1], a plot of the degree of coherence versus spatial separation ℓ of points P_1 and P_2 is the same as a plot of the diffraction pattern due to an aperture of the same size and shape as the extended source. Such patterns for rectangular and circular sources are discussed in Chapter 16.

The significance of Eq. (12-35) is apparent in the case of Young's double-slit experiment, where an extended source is used together with a single slit to render the light striking the double slit reasonably coherent, as in Figure 12-13. We may now use Eq. (12-35) to determine how small the single slit must be to ensure coherence and the production of fringes at the screen. The two slits S_1 and S_2 must fall within the coherence width l_s due to the primary slit of width s.

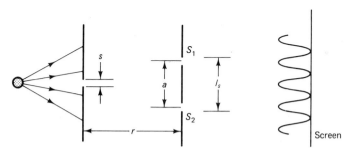

Figure 12-13 Young's double-slit setup. Slits S_1 and S_2 must fall within the lateral coherence width l_s due to the single-slit source.

Example

Let the source-to-slit distance be 20 cm, the slit separation 0.1 mm, and the wavelength 546 nm. Determine the maximum width of the primary or single slit.

Solution   Using Eq. (12-35),

$$s < \frac{r\lambda}{l_s} = \frac{(0.2)(546 \times 10^{-9})}{1 \times 10^{-4}} = 1.1 \text{ mm}$$

Now suppose that the source slit in the example is made exactly 1.1 mm in width and that the separation between slits S_1 and S_2 is adjustable. When the slits are very close together ($a \ll l_s$), they fall within a high coherence region and the fringes in the interference pattern appear sharply defined. As the slits are moved farther apart, the degree of coherence $|\gamma_{12}|$ decreases and the fringe contrast begins to degrade. When the slit separation a reaches a value of 0.1 mm, $|\gamma_{12}| = 0$ and the fringes disappear. Evidently an experimental determination of this slit separation could be used to deduce the size s of the extended source. This technique was employed by Michelson to measure the angular diameter of stars. Stars are so distant that imaging techniques are unable to resolve their diameters. If a star is regarded as an extended, incoherent source with light emanating from a continuous array of points extending across a diameter s of the star (see Figure 12-14b), then the spatial coherent width l_s in Eq. (12-35) becomes

$$l_s < \frac{1.22\lambda}{\theta} \tag{12-36}$$

Here the factor 1.22 arises from the circular shape of the source, as it does in the Fraunhofer diffraction of a circular aperture. Since the angular diameter θ of a star is

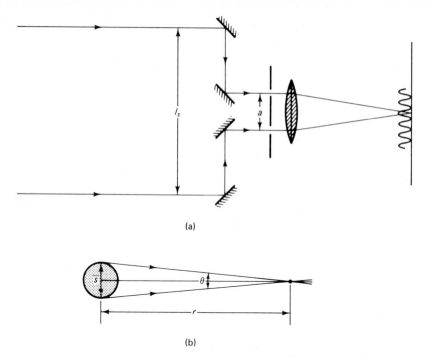

(a)

(b)

Figure 12-14 Michelson stellar interferometer (a) used to determine a stellar diameter (b).

extremely small, l_s will be correspondingly large. The movable slits were therefore arranged as in Figure 12-14a, using mirrors that direct widely separated portions of the radiation wavefront into a double-slit-telescope instrument. The spacing of the interference fringes depend on the double slit separation a, whereas their visibility depends on the separation l_s. As l_s is increased, the fringes disappear when equality in Eq. (12-36) is satisfied.

Example

When Michelson used this technique on the star Betelgeuse in the constellation Orion, he found a first minimum in the fringes at $l_s = 308$ cm. Using an average wavelength of 570 nm, what is the angular diameter of the star?

Solution   Taking Eq. (12-36) as an equality,

$$\theta = \frac{1.22\lambda}{l_s} = \frac{1.22(570 \times 10^{-9})}{3.08} = 2.26 \times 10^{-7} \text{ rad}$$

Since Orion is known to be about 1×10^{15} mi away, the stellar diameter is $s = r\theta = 2.26 \times 10^{8}$ mi, or about 260 solar diameters.

PROBLEMS

12-1. Determine the Fourier series for the square wave of spatial period L given by

$$f(x) = \begin{cases} -1, & \dfrac{-L}{2} < x < 0 \\ +1, & 0 < x < \dfrac{+L}{2} \end{cases}$$

12-2. A half-wave rectifier removes the negative half-cycles of a sinusoidal waveform, given by $E = E_0 \cos \omega t$. Find the Fourier series of the resulting wave.

12-3. Find the Fourier transform of the Gaussian function given by

$$f(t) = he^{-t^2/2\sigma^2}$$

where h is the height and σ the "width." (*Hint*: Remember how to complete a square? You will also need the definite integral

$$\int_{-\infty}^{+\infty} e^{-x^2} \, dx = \sqrt{\pi}$$

in your calculations.) Does the transform, interpreted as the frequency spectrum, show the proper relationship to the original "pulse" width?

12-4. Using the Fourier transform, determine the power spectrum of a single square pulse of amplitude A and duration τ_0. Sketch the power spectrum, locating its zeros, and show that the frequency bandwidth for the pulse is inversely proportional to its duration.

12-5. Two light filters are used to transmit yellow light centered around a wavelength of 590 nm. One filter has a "broad" transmission width of 100 nm, whereas the other has a "narrow" pass band of 10 nm. Which filter would be better to use for an interference experiment? Compare the coherence lengths of the light from each.

12-6. A continuous He-Ne laser beam (632.8 nm) is "chopped," using a spinning aperture, into 0.1-ns pulses. Compute the resultant line width $\Delta\lambda$, bandwidth Δf, and coherence length.

12-7. The angular diameter of the sun viewed from the earth is approximately 0.5 degree. Determine the spatial coherence length for "good" coherence, neglecting any variations in brightness across the surface. Let us consider, somewhat arbitrarily, that "good" coherence will exist over an area that is 10% of the maximum area of coherence.

12-8. Michelson found that the cadmium red line (643.8 nm) is one of the most ideal monochromatic sources available, allowing fringes to be discerned up to a path difference of 30 cm in a beam-splitting interference experiment, such as with a Michelson interferometer. Calculate (a) the wavelength spread of the line and (b) the coherence time of the source.

12-9. A narrow band-pass filter transmits wavelengths in the range 5000 ± 0.5 Å. If this filter is placed in front of a source of white light, what is the coherence length of the transmitted light?

12-10. Let a collimated beam of white light fall on one refracting face of a prism and let the light emerging from the second face be focused by a lens onto a screen. Suppose that the linear dispersion at the screen is 20 Å/mm. By introducing a narrow "exit slit" in the screen, one has a type of monochromator that provides a nearly monochromatic beam of light. Sketch the setup. For an exit slit of 0.02 cm, what is the coherence time and coherence length of the light of mean wavelength 5000 Å?

12-11. A pinhole of diameter 0.5 mm is used in front of a sodium lamp (5890 Å) as a source in a Young interference experiment. The distance from pinhole to slits is 1 m. What is the maximum slit space insuring interference fringes that are just visible?

12-12. Determine the linewidth in angstroms and hertz for laser light whose coherence length is 10 km. The mean wavelength is 6328 Å.

12-13. (a) A monochromator is used to obtain quasi-monochromatic light from a tungsten lamp. The linear dispersion of the instrument is 20 Å/mm and an exit slit of 200 μm is used. What is the coherence time and length of the light from the monochromator when set to give light of mean wavelength 500 nm?

(b) This light is used to form fringes in an interference experiment in which the light is first amplitude-split into two equal parts and then brought together again. If the optical path difference between the two paths is 0.400 mm, calculate the magnitude of the normalized correlation function and the visibility of the resulting fringes.

(c) If the maximum irradiance produced by the fringes is 100 on an arbitrary scale, what is the difference between maximum irradiance and background irradiance on this scale?

12-14. Determine the length and base area of the cylindrical volume within which light received from the sun is coherent. For this purpose, let us assume "good" spatial coherence occurs within a length that is 25% of the maximum value given by Eq. (12-36). The sun subtends an angle of 0.5° at the earth's surface. The mean value of the visible spectrum may be taken at 550 nm. Express the coherence volume also in terms of number of wavelengths across cylindrical length and diameter.

12-15. (a) Show that the fringe visibility may be expressed by

$$V = \frac{2\sqrt{I_1 I_2}\,|\gamma_{12}(\tau)|}{(I_1 + I_2)}$$

(b) What irradiance ratio of the interfering beams reduces the fringe visibility by 10% of that for equal-amplitude beams?

12-16. Show that the visibility of double-slit fringes in the mth order is given by

$$V = 1 - \left(m\frac{\Delta\lambda}{\lambda}\right)$$

where λ is the average wavelength of the light and $\Delta\lambda$ is its linewidth.

12-17. A filtered mercury lamp produces green light at 546.1 nm with a linewidth of 0.05 nm. The light illuminates a double slit of spacing 0.1 mm. Determine the visibility of the fringes on a screen 1 m away, in the vicinity of the $m = 20$ order fringe. (See problem 12-16.) If the discharge lamp is replaced with a white light source and a filter of bandwidth 10 nm at 546 nm, how does the visibility change?

12-18. A Michelson interferometer forms fringes with cadmium red light of 643.847 nm and linewidth of 0.0013 nm. What is the visibility of the fringes when one mirror is moved 1 cm from the position of zero path difference between arms? How does this change when the distance moved is 5 cm? At what instance does the visibility go to zero?

12-19. (a) Repeat problem 12-18 when the light is the green mercury line of 546.1 nm with a linewidth of 0.025 nm.

(b) How far can the mirror be moved from zero path difference so that fringe visibility is at least 0.85?

REFERENCES

[1] Born, M., and E. Wolf. *Principles of Optics*. New York: Pergamon Press, 1959.

[2] Fowles, Grant R. *Introduction to Modern Optics*. New York: Holt, Rinehart and Winston, 1968.

[3] Parrent, Mark J., and George B. Parrent, Jr. *Theory of Partial Coherence*. Englewood Cliffs, N.J.: Prentice-Hall, 1964.

[4] Feynman, Richard P., Robert B. Leighton, and Matthew Sands. *The Feynman Lectures in Physics*, vol. 1. Reading, Mass.: Addison-Wesley Publishing Company, 1975. Ch. 50.

[5] Reynolds, George O., John B. DeVelis, George B. Parrent, Jr., and Brian J. Thompson. *Physical Optics Notebook: Tutorials in Fourier Optics*. Bellingham, Wash.: SPIE Optical Engineering Press, 1989. Ch. 11, 17, 18.

[6] Peřina, Jan. *Coherence of Light*. New York: Van Nostrand Reinhold Co., 1971.

13

Holography

INTRODUCTION

Holography is one of the many flourishing fields that owes its success to the laser. Although the technique was invented in 1948 before the advent of coherent laser light by the British scientist Dennis Gabor, the assurance of success was made possible by the laser. Emmett Leith and Juris Upatnieks at the University of Michigan first applied laser light to holography in 1962 and also introduced an important off-axis technique of illumination that we explain presently.

The spectacular improvement in three-dimensional photography made possible by the hologram has aroused unusual interest in nonscientific circles as well, so that the fast-multiplying applications of holography today also include its use in art and advertising.

13-1 CONVENTIONAL VERSUS HOLOGRAPHIC PHOTOGRAPHY

We are aware that a conventional photograph is a two-dimensional version of a three-dimensional scene, bringing into focus every part of the scene that falls within the depth of field of the lens. As a result the photograph lacks the perception of depth or the parallax with which we view a real-life scene. In contrast, the hologram provides a record of the scene that preserves these qualities. The hologram succeeds in effectively "freezing" and preserving for later observation the intricate wavefront of light that carries all the visual information of the scene. In viewing a hologram, this wavefront is reconstructed or released, and we view what we would have seen if present at the original scene through the "window" defined by the hologram. The reconstructed wavefront provides depth perception and parallax, allowing us to look around the edge of an object to see what is behind. It may be manipulated by a lens, for example, in the same way as the original wavefront. Thus a "hologram," as its etymology suggests, includes the "whole message."

The real-life qualities of the image provided by a hologram stem from the preservation of information relating to the phase of the wavefront in addition to its amplitude or irradiance. Recording devices like ordinary photographic film and photomultipliers are sensitive only to the radiant energy received. In a developed photograph, for example, the *optical density* of the emulsion at each point is a function of the optical energy received there due to the light-sensitive chemical reaction that reduces silver to its metallic form. When energy alone is recorded, the phase relationships of waves arriving from different directions and distances, and hence the visual lifelikeness of the scene, are lost. To record these phase relationships as well, it is necessary to convert phase information into amplitude information. The interference of light waves provides the requisite means. Recall that when waves interfere to produce a large *amplitude*, they must be in *phase*, and when the amplitude is a minimum, the waves are out of phase, so that various contributions effectively cancel one another. If the wavefront of light from a scene is made to interfere with a coherent reference wavefront, then, the resultant interference pattern includes information regarding the phase relationships of each part of the wavefront with the reference wave and, therefore, with every other part. The situation is sometimes described by referring to the reference wave as a *carrier wave* that is *modulated* by the *signal wave* from the scene. This language provides a fruitful comparison with the techniques of radio wave communication.

In conventional photography a lens is used to focus the scene onto a film. All the light originating from a single point of the scene and collected by the lens is focused to a single conjugate point in the image. We can say that a one-to-one relationship exists between object and image points. By contrast, a hologram is made, as we shall see, without use of a lens or any other focusing device. The hologram is a complex interference pattern of microscopically spaced fringes, not an image of the scene. Each point of the hologram receives light from every point of the scene or, to put it another way, every object point illuminates the entire hologram. There is no one-to-one correspondence between object points and points in the wavefront before reconstruction occurs. The hologram is a record of this wavefront.

13-2 HOLOGRAM OF A POINT SOURCE

To see how the process is realized in practice, both making the hologram and using the hologram to reconstruct the original scene, we begin with a very basic example, the hologram of a point source. In Figure 13-1a, plane wavefronts of coherent, monochromatic radiation illuminate a photographic plate. In addition, spherical wavefronts reach the plate after scattering from object point O. The plate, when developed, then shows a series of concentric interference rings about X as a center. Point P falls on such a ring, for example, if the optical path difference $OP - OX$ is an integral number of wavelengths, ensuring that the *reference beam* of plane wavefront light arrives at P in step with the scattered *subject beam* of light. The developed plate is called a *Gabor zone plate*—or zone lens—with circular transmitting zones, whose transmittance is a gradually varying function of radius. The Gabor

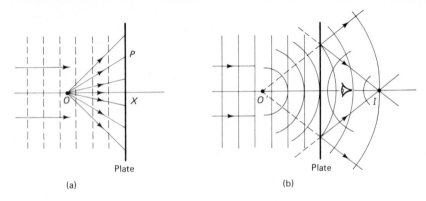

Figure 13-1 Hologram of a point source O is constructed in (a) and used in (b) to reconstruct the wavefront. Two images are formed in reconstruction.

zone plate is called a "sinusoidal" circular grating because the optical density, and therefore the transmittance of the grating, varies as $\cos^2 r^2$ along the radius of the zone pattern.[1] This sinusoidal plate is, in fact, a hologram of the point O. The hologram itself is a series of circular interference fringes that do not resemble the object, but the object may be reconstructed, as in Figure 13-1b, by placing the hologram back into the reference beam without the presence of the object O. Just as light directed from O originally interferes with the reference beam to produce the zone rings, so the same reference beam is now reinforced in diffraction from the rings along directions that diverge from the equivalent point O'. The point O' thus locates a *virtual image* of the original object point O, seen on reconstruction by looking into the hologram. The condition for reinforcement must also be satisfied by a second point on the exit side of the hologram, a point I that is symmetrically placed with O' relative to the film. Clearly the set of distances from I to the consecutive zones satisfies the same geometric relationships as the corresponding distances from O' to the zones. Thus the diffracted light also converges to point I, a *real image* of the original object point O that can be focused onto a screen. If, in the making of this hologram, the object point O is moved farther away, the radius of each zone increases. For an off-axis object at infinity, the zones are straight, parallel interference fringes. The hologram is then a *grating hologram*, formed by the intersection of two plane wavefronts of light arriving at the plate along different directions. The grating hologram is discussed further in Chapter 17. As explained there, the greater the angle between these wavefronts, the finer the spacing of the interference fringes. The family of circular and straight, parallel fringes we have been discussing can be seen as special cases of two point-source interference, observed in planes perpendicular and parallel, respectively, to the axis joining the points. (See the discussion relating to Figure 10-5.) When object point O is replaced by an extended object or three-dimensional scene, each point of the scene produces its own Gabor zone pattern on the film. The hologram is now a complex montage of zones in which is coded all the information of the wavefront from the scene. On reconstruction, each set of zones produces its own real and virtual images, and the original scene is reproduced. One usually views the virtual image by looking into the hologram. Figure 13-1b shows that when viewing the virtual image in this way, undesirable light forming the real image is also intercepted. Leith and Upatnieks introduced an off-axis technique, using one or more mirrors to bring in the reference beam from a different angle, so that the directions of the reconstructed real and virtual wavefronts are separated.

---

[1] More precisely, the transmittance can be expressed as $A + B \cos^2 (ar^2)$, where A, B, and a are constants. See problem 13-12.

The two basic types of holograms discussed in the preceding paragraph are the Gabor zone plate and holographic grating, corresponding to point objects at a finite distance and at an infinite distance from the plate, respectively. If the zone plate or grating provides a square wave type of transmittance, alternating between minimum and maximum, then multiple diffracted images are possible. The familiar diffraction grating of this type is known to produce orders of diffraction with $m = 0$, ± 1, ± 2, . . . , limited by the maximum diffraction angle. The zone plate with such transmittance properties is the *Fresnel zone plate*, which produces multiple focal points along its axis beyond those discussed here for the Gabor zone plate. It can be shown, however, that when the transmittance profile of the grooves or zones is not sharp but varies continuously, these general remarks concerning orders have to be modified. In particular, when the grating or circular zones are "sinusoidal" in character, that is, their transmittance profiles follow a $\cos^2 x$ (grating) or $\cos^2 r^2$ (circular zone plate) irradiance, only first-order images appear, in addition to the zeroth order, on reconstruction. For the circular zones, the two first-order images are the real and virtual images discussed.

In the formation of holograms as shown in Figure 13-1a, the sinusoidal irradiance at the film can fall to zero at points of destructive interference when the signal and reference beams are equal in amplitude. The emulsion, however, is incapable of responding linearly to all irradiances, varying from zero to maximum, so that the developed film will show a distorted $\cos^2 r^2$ transmittance, and higher-order diffractions will not be suppressed. By making the reference beam stronger than the signal beam, the minimum irradiance on the emulsion can be raised to the level of its linear response characteristics. A variation in transmittance of the type

$$T = T_0 + T_m \cos^2 r^2$$

is produced, and higher-order images are eliminated. The compromise is that the $\cos^2 r^2$ transmittance is now superimposed over a nonzero minimum transmittance T_0, and fringe contrast is somewhat reduced.

As we have just pointed out, the amplitude of the reference beam is made somewhat greater than the average amplitude of the signal or object beam, so that the reference wave is modulated by the signal. Even when the signal is zero, the reference beam is of sufficient strength to stimulate the emulsion within its region of linear response to radiant energy. The effect of variations in signal *strength* is then to produce variations in the *contrast* of the interference fringes, whereas variations in *phase* (or *direction*) of the signal waves produce variations in *spacing* of the fringes. Thus it is in the local variations of fringe contrast and spacing across the hologram that the corresponding variations in amplitude and phase of the object waves are encoded. High-resolution film is used to record this information faithfully.

13-3 HOLOGRAM OF AN EXTENDED OBJECT

One of many holographic techniques for producing an off-axis reference beam in conjunction with the beam of diffusely reflected light from a three-dimensional scene is shown in Figure 13-2a. A combination of pinhole and lens is used to expand the beam from a laser. The expanded beam is then split by a semireflecting plate BS to produce two coherent beams. One beam, the *reference beam E_R*, is directed by two plane mirrors $M1$ and $M2$ onto the photographic plate, as shown. The other beam, E_S, reflects diffusely from the subject and some of this beam, which we call the *subject beam*, also strikes the film, where it interferes with the reference beam and produces the hologram.

(a)

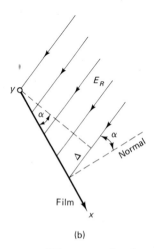

(b)

Figure 13-2 (a) Off-axis holographic system. (b) Orientation of film with reference beam in (a).

We now make the previous qualitative explanation somewhat quantitative. Let the reference beam be represented by the field

$$E_R = re^{i(\omega t + \varphi)} \qquad (13\text{-}1)$$

at the plane of the film. The amplitude $r = r(x, y)$ of the reference beam can be assumed constant over the essentially plane wavefront. The phase angle φ arises from the angle α between the film plane and the wavefront of the reference beam, as indicated in Figure 13-2b. If the top edge of the beam strikes the film at $x = 0$, then φ is a linear function of distance x along the film plane, since

$$\varphi = \left(\frac{2\pi}{\lambda}\right)\Delta = \left(\frac{2\pi}{\lambda}\right) x \sin \alpha \qquad (13\text{-}2)$$

Thus the phase angle φ relates only to the tilt of the film plane relative to the reference beam and appears as an exponential factor in Eq. (13-1):

$$E_R = re^{i\omega t}e^{i\varphi} \qquad (13\text{-}3)$$

If the reference beam were not present, the film would be illuminated only by the subject beam,

$$E_S = se^{i(\omega t + \theta)} \qquad (13\text{-}4)$$

where $s(x, y)$ is the amplitude of the reflected light at different points of the film and $\theta = \theta(x, y)$ is a complicated function due to the variations in phase of the light reaching the film from different parts of the subject. If the subject beam alone were present, the film would be darkened in proportion to the irradiance of the subject beam. Omitting for simplicity the constant factors between irradiance and square of amplitude, we write

$$I_S = |E_S|^2 = E_S^* E_S = [s(x, y)]^2 \qquad (13\text{-}5)$$

The irradiance function thus includes no information regarding phase of the subject beam. With the reference beam also present, however, the resultant amplitude E_F at each point of the film—subject to the scalar approximation—is given by

$$E_F = E_R + E_S$$

so that

$$I_F = |E_F|^2 = (E_R + E_S)(E_R^* + E_S^*)$$

Multiplying the binomials,

$$
\begin{aligned}
I_F &= E_R E_R^* + E_S E_S^* + E_R E_S^* + E_S E_R^* \\
I_F &= r^2 + s^2 + E_R E_S^* + E_S E_R^*
\end{aligned}
\qquad (13\text{-}6)
$$

The right side of Eq. (13-6) is a function of x and y and so varies from point to point on the film plane. The last two terms now incorporate the important function $\theta(x, y)$. Explicitly,

$$
\begin{aligned}
I_F &= r^2 + s^2 + rse^{i(\omega t+\theta)} e^{-i(\omega t+\varphi)} + rse^{i(\omega t+\varphi)} e^{-i(\omega t+\theta)} \\
I_F &= r^2 + s^2 + rse^{i(\theta-\varphi)} + rse^{-i(\theta-\varphi)}
\end{aligned}
\qquad (13\text{-}7)
$$

This irradiance function describes the hologram. When the film is developed, its transmittance is determined by I_F.

To reconstruct the image of the scene, the hologram is situated in the reference beam again, as in the formation of the hologram (Figure 13-2b). Of course, the subject is now absent. When illuminated by the reference beam, the hologram, due to its transmittance function, modulates both the amplitude and the phase of the beam. As before,

$$E_R = re^{i(\omega t+\varphi)} \qquad (13\text{-}8)$$

The resulting emergent beam can then be expressed, except for constants, in terms of the field E_H by

$$E_H \propto I_F E_R = (r^2 + s^2)E_R + r^2 se^{i(\omega t+\theta)} + r^2 e^{i(2\varphi)} se^{i(\omega t-\theta)} \qquad (13\text{-}9)$$

where we have multiplied together Eqs. (13-7) and (13-8). We now interpret the three terms in Eq. (13-9) as the reconstruction of three distinct beams from the hologram. Each beam is also illustrated in Figure 13-3. The first term,

$$E_{H1} = (r^2 + s^2)E_R = (r^2 + s^2)re^{i(\omega t+\varphi)} \qquad (13\text{-}10)$$

represents a reference beam modulated in amplitude but not in phase. It therefore appears like the incident beam and passes through the hologram without deviation. In analogy with the holographic grating, it corresponds to zeroth-order diffraction. The second term is

$$E_{H2} = r^2 se^{i(\omega t+\theta)} \qquad (13\text{-}11)$$

which describes the subject beam, amplitude-modulated by the factor r^2. Thus the beam represents a *reconstructed wavefront* from the subject, making the same angle α relative to the reference beam. Since this beam is essentially the subject beam, it appears to come from the subject. Hence it diverges on emerging from the holo-

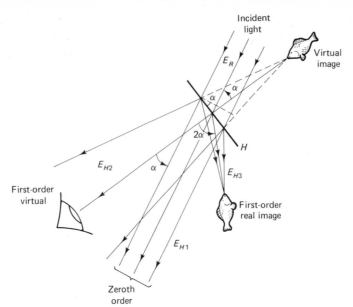

Figure 13-3 Reconstruction of hologram formed in Figure 13-2a.

gram, as if coming from a *virtual image* behind the hologram. This virtual image is what we customarily view.

The third term is given by

$$E_{H3} = r^2 e^{i(2\varphi)} \, s e^{i(\omega t - \theta)} \tag{13-12}$$

and represents the subject beam, modulated in both amplitude and phase. This beam reconstructs the subject beam of Eq. (13-4) but with *phase reversal*. Every delay in phase in E_S now shows up as a phase advance. The image is turned inside out. Because of phase reversal, originally diverging rays become converging and focus as a *real image* on the viewing side of the hologram. The factor $e^{i(2\varphi)}$, when compared with the phase term in Eq. (13-3), indicates an angular displacement of the image direction by 2α relative to the normal to the film plane. Notice that the off-axis system illustrated in Figure 13-2a produces a hologram in which the two first-order beams are separate in direction from each other and the zeroth-order beam. The virtual image can be observed clearly, without confusion from the other beams.

The hologram made of an extended object shows the same essential features as the hologram of the point object. Photography by holography is a two-step process. Recall that in the making of a hologram, no lens is used, and the presence of the reference beam is essential. The light must have sufficient temporal coherence so that path differences between the two beams do not exceed the coherence length of the light; it must also possess sufficient spatial coherence so that the beam is coherent across that portion of the wavefront needed to encompass the scene. Of course, the holographic system must be vibration-free to within a fraction of the wavelength of the light during the exposure, a condition that is easily satisfied when high-power laser pulses of very short duration are used to freeze undesirable motion.

A three-dimensional view of the object from all sides can be produced on a holographic film that is wrapped around the object on a cylindrical form, as shown

Figure 13-4 Cylindrical film surrounding the subject records a 360° hologram.

in Figure 13-4. Light reaches the film both directly and with the help of a mirror at the end of the cylinder (the reference beam) and by light scattered from the object. When viewed under the same conditions, the 360° hologram produces a view of the fish from all sides.

13-4 HOLOGRAM PROPERTIES

As stated earlier, the entire hologram receives light from each object point in the scene. As a result, any portion of the hologram contains information of the whole scene. If a hologram is cut up into small squares, each square is a hologram of the whole scene, although the reduction in aperture degrades the resolution of the image. The situation is much the same as when looking through a small, square aperture placed in front of a window. The same scene is viewed, though with slightly varying perspective, as the opening is moved to different parts of the window. Each view is complete, exhibiting both depth and parallax. Another interesting property of a hologram is that a contact print of the hologram, which interchanges the optically dense and transparent regions, has the same properties in use. The "negative" of a hologram alters neither fringe contrast nor spacing, and hence does not modify the stored information. Furthermore, the hologram may contain a number of separate exposures, each taken with the film at a different angle relative to the reference beam and with different wavelengths of light. On reconstruction, each scene appears in its own light when viewed along the direction of the original scene, without mutual interference.

13-5 WHITE-LIGHT HOLOGRAMS

If the hologram of Figure 13-3 is viewed in a reference beam of different color than that used in its construction, it can be shown that the image of the fish will appear at a different angle. The hologram, like the holographic grating, operates as a dispersing element. If the reference beam is white light, therefore, the continuously displaced images due to different spectral regions of the light overlap and produce a colored blur. By producing a hologram that restricts the possible angular views of the subject to one through a horizontal slit, the confusion of images is reduced. In reconstruction the hologram creates an improved image in white light. The virtual image now appears colored. The particular color seen depends on the direction along which the hologram is viewed as the head is moved along a vertical line. Such a hologram is called a *rainbow hologram*. Since the view is now restricted to what one would see by viewing the subject through a horizontal slit placed in front of it, the rainbow hologram reproduces horizontal parallax but suffers from a loss of vertical parallax. The hologram may be viewed in reflected light by coating the back side of the hologram with a thin layer of aluminum, which then serves as a mirror to redirect white light back through the hologram.

When the thickness of the emulsion is large compared with the fringe spacing, the hologram may be considered a three-dimensional, or *volume*, *hologram*. The interference fringes are now interference surfaces within the emulsion that behaves as crystalline planes of atoms in diffracting light, that is, like a three-dimensional grating. Unlike two-dimensional holograms, volume holograms can reproduce images in their original colors when illuminated with white light. To see how this comes about, consider the formation of closely spaced interference surfaces within a thick emulsion by using coherent subject and reference beams with the largest angular

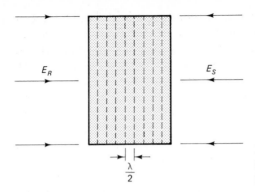

separation possible, 180°, as in Figure 13-5. If the two monochromatic beams have undistorted plane wavefronts, for example, the standing-wave pattern produces antinodal planes perpendicular to the beam directions and spaced $\lambda/2$ apart, as shown. The maximum irradiance in these planes produces, after film development, planes consisting of excess free silver, which function as partially reflecting planes. Of course the emulsion must itself possess high-resolution potential to record faithfully such detail. When illuminated from the reference beam direction with white light, for instance, the developed hologram partially reflects light from each silver layer, but only light of the wavelength used in making the hologram is reinforced by such multiple reflections. The physics of the process is, of course, the same as that for X-ray diffraction from crystalline planes, governed by the Bragg equation,

$$m\lambda = 2d \sin \theta$$

and illustrated in Figure 13-6. Thus if a volume hologram is illuminated at a given angle θ, only the one wavelength that satisfies the Bragg equation locally, where planar spacing is d, is reinforced and appears as a brightly reflected beam. The thicker the emulsion and the greater the number of contributing reflecting planes, the more selective the hologram will be in reinforcing the correct wavelength. If a volume hologram is made by multiple exposures of a scene in each of three primary colors, the reconstruction process with white-light illumination can produce a three-dimensional image in full color.

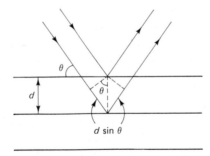

Figure 13-6 Constructive interference of reflected waves from planes of separation d is governed by the Bragg equation $m\lambda = 2d \sin \theta$.

13-6 OTHER APPLICATIONS OF HOLOGRAPHY

Holography offers a wide variety of fascinating applications, of which we briefly describe only a few. The hologram, itself a product of the interference of light, has been used as an alternate technique in *interferometry*, the science of using the wavelength of light and interference to measure very small optical path lengths with precision. Suppose that the hologram of the fish in Figure 13-3 and the fish itself are returned exactly to their original positions, and suppose that the same reference beam illuminates the scene. In looking through the hologram, one now sees the virtual image superimposed over the object itself. Both are viewed with the same coherent light. If no change has occurred since recording the hologram, the view appears as if the subject or the hologram alone were in place. Suppose, however, that the model of the fish has undergone some small changes in shape, by thermal expansion, for example. Now the direct image of the object and the holographic image are slightly different, and the light forming the two images interferes, producing fringes that measure the extent of the change at specific locations, as in the case of Newton's rings. This techniqe is often applied to determine maximum stress points on the subject as pressure is applied, as in the case of an automobile tire, for example. The sensitivity of this technique has been dramatically demonstrated in holographic recordings of convection currents around a hot filament, compressional waves surrounding a speeding bullet, and the wings of a fruit fly in motion. Changes that occur over a period of time can be monitored in the same way, by returning the model to the holographic system.

Another useful application of holography is in *microscopy*. When specimens of cells or microscopic particles are viewed conventionally under high magnification, the depth of field is correspondingly small. A photograph that freezes motion of the specimen captures in a focused image a very limited depth of field within the specimen. The disadvantages of this restriction can be overcome if the photograph is a hologram, which in a single snapshot contains potentially all the ordinary photographs that could be made after successive refocusings throughout the depth of the living specimen. The image provided by the hologram may be viewed by focusing at leisure on any depth of an unchanging field. In making a hologram with a microscope, the specimen is illuminated by laser light, part of which is first split off outside the microscope and routed independently to the photographic plate, where it rejoins the subject beam processed by the microscope optics. Furthermore it can be shown that, if the reconstructing light of wavelength λ_r is longer than the wavelength λ_s used in "holographing" the subject, a magnification given by

$$M = \left(\frac{q}{p}\right) \left(\frac{\lambda_r}{\lambda_s}\right) \tag{13-13}$$

results, where p is the object distance (subject from film) and q is the corresponding image distance (image from hologram). Object and image distances are equal when the reference and reconstructing wavefronts are both plane waves. The content of Eq. (13-13) implies, for example, that if the hologram were made with laser X-radiation and viewed with visible light, magnifications as large as 10^6 could be achieved without deterioration in resolution. This prospect has contributed to the interest in developing X-ray lasers. X-ray holograms could provide strikingly detailed three-dimensional images of microscopic objects as small as viruses and DNA molecules. The ability to view a hologram with radiation of a different wavelength than that used in making the hologram offers other interesting possibilities, the use of an ultrasonic wave hologram to replace medical X-rays, for example, or the reading of a radar hologram with visible wavelengths. In fact, in his original work, Gabor proposed reconstruction of an electron wave hologram with optical wavelengths in an effort to improve the resolution of electron microscopes.

The mention of *ultrasonic holograms* above implies that the waves producing a hologram need not be electromagnetic in nature. Indeed, the principles of holography do not depend on the transverse character of the radiation. Because of the ability of ultrasonic waves to penetrate objects opaque to visible light, holograms formed with such waves can be very useful. Opaque bodies that are promising candidates range from the human body to archeological tombs. Structures and cavities inside can be revealed in three-dimensional images formed by ultrasonic holography. Figure 13-7 illustrates another application of ultrasonic holography to reveal objects under the surface of the ocean. $G1$ and $G2$ represent two phase-coupled generators radiating coherent ultrasonic waves. The wavefront from $G2$ is deformed by an underwater object and interferes with the undeformed reference beam from $G1$. The deformations of the water surface represent an acoustic hologram. If this region is illuminated with monochromatic light, the light diffracted from the deformations can be photographed and converted into a visual image of the underwater object. The potential offered for submarine detection is an obvious military application.

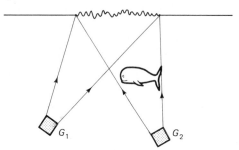

Figure 13-7 Deformations in the surface of the water due to two coherent ultrasonic waves.

Holographic data storage also offers tremendous potential. Because data can be reduced by the holographic technique to dimensions of the order of the wavelength of light, volume holograms can be used to record vast quantities of information. As the hologram is rotated, new exposures can be made. Photosensitive crystals, such as potassium bromide crystals with color centers or the lithium niobate crystal, can be used in place of thick-layered photoemulsions. Because information can be reduced to such tiny dimensions and the crystal can be repeatedly exposed after small rotations that take the place of turning pages, it is said that all the information in the Library of Congress could theoretically be recorded on a crystal the size of a sugar cube! Information may, of course, be recorded in digital form and thus read by a computer, so that holographic storage offers a means of providing computer storage. In conjunction with the optical transport of computer information through optical fibers, information handling, storage, and retrieval can all be done using light. A fascinating aspect of holographic data storage lies in its reliability. Since every data unit is recorded throughout the volume of the hologram, in unique holographic fashion, damage to a portion of the hologram, while affecting the signal-to-noise level of the reconstructed image, does not affect its reliability. Information is not lost, as would be the case in other memory devices, where every bit of information has unique storage coordinates.

In a reciprocal sense, computers are used to advantage in the science of holography by making possible the construction of *synthetic holograms* that faithfully represent three-dimensional objects. The object is first defined mathematically by specifying its coordinates and the intensity of all its points. The computer calculates the complex amplitude that is the sum of radiation due to the object and the reference wave and then directs the drawing of the hologram, which can be photographed and reduced to the appropriate fringe spacings required. For example, an ideal aspheric wavefront can be created synthetically to serve as a model against which a mirror may be shaped, using interference between the two surfaces as a guide to making appropriate corrections.

Another area in which holograms may be very useful is in *pattern recognition*. Briefly, the procedure is as follows. A text is scanned, for example, for the presence of a particular letter or word. Light from the text to be searched is passed through a hologram of the letter or word to be identified in an appropriate optical system. The presence of the letter is indicated by the formation of a bright spot in a location that indicates the position of the letter in the text. The hologram acts as a matched filter, recognizing and transmitting only that spatial spectrum similar to the one recorded on it. The technique can be applied to holographic reading of microfilms, for example. Military applications include the use of a memory bank of holograms of particular objects or targets constructed from aerial photographs. Weapons could, by pattern recognition, select proper targets. It has also been suggested that robots could identify and be directed toward appropriate objects in the same way. Pattern recognition is discussed further in Chapter 25.

Holograms that simply redirect light may be used as inexpensive optical elements, serving in the place of lenses and mirrors. To cite one popular application, laser readers of the universal product code on groceries use a spinning disc outfitted with a number of holographic lenses. By continuously providing many angles of laser scanning, the product code can be identified even when the item is passed casually over the scanner.

PROBLEMS

13-1. Use Eq. (10-13) for the superposition of two unequal beams to show that the irradiance pattern of a Gabor zone plate (the hologram of a point source) is given approximately by

$$I = A + B \cos^2 (ar^2)$$

where $A = I_1 + I_2 - 2\sqrt{I_1 I_2}$, $B = 4\sqrt{I_1 I_2}$, and $a = \pi/2s\lambda$. Here I_1 and I_2 are the irradiances due to the reference and signal beams, respectively, s is the distance of the object point from the film, and λ is the wavelength of the light. For the approximation, assume the path difference between the two beams is much smaller than s, so that we are looking at the inner zones of the hologram.

13-2. (a) Show that if the local ratio of reference to subject beam irradiances is a factor N at some region of a hologram, then the visibility of the resulting fringes is $2\sqrt{N}/(N + 1)$.

(b) What is the fringe visibility in a region where the irradiance of the reference beam is three times that of the subject beam?

13-3. Show that the separation d of fringes in the formation of a holographic grating, as in Figure 13-5, is given by $\lambda/(2 \sin \theta)$, where 2θ is the angle between the coherent beams in the film and λ is the wavelength of the beams in the film. Assume the beams are incident symmetrically on the film's surface. If the beams are argon laser beams of 488-nm wavelength and the angle between beams is 120°, how many grooves per millimeter are formed in a plane emulsion oriented perpendicular to the fringes? Assume $n = 1$.

13-4. The angle between the signal and reference beams during construction of a hologram is 20°. If the light is from a He-Ne laser at 633 nm, what is the fringe spacing? Assume a refractive index of 1 for the emulsion. (See problem 13-3).

13-5. Suppose a hologram is to be made of moving object using a 1-nanosecond laser pulse at a wavelength of 633 nm. What is the permissible speed of the motion such that the object does not move more than $\lambda/10$ during the exposure?

13-6. During the construction of a hologram, a beam splitter is selected that makes the amplitude of the reference beam eight times that of the signal beam at the emulsion. What is the maximum ratio of beam irradiances there?

13-7. Let us suppose that as a theoretical limit, one bit of information can be stored in each λ^3 of hologram volume. At a wavelength of 492 nm and a refractive index of 1.30, determine the storage capacity of 1 mm³ of hologram volume.

13-8. A volume hologram is made using oppositely directed monochromatic beams of coherent, collimated laser light at 500 nm, as in Figure 13-5.
   (a) Determine the spacing of the developed silver planes within the emulsion.
   (b) What wavelength is reinforced in reflected light when white light is incident normally on the hologram?
   (c) Repeat (b) when the angle of incidence (relative to the normal) is 30°. Assume a film refractive index of 1.

13-9. Two beams of planar wavefront, 633-nm coherent light, whose directions are 120° apart, strike a photographic emulsion.
   (a) Sketch the arrangement, showing the orientation of the planes of constructive interference within the emulsion.
   (b) Determine the planar spacing of the developed volume hologram.
   (c) At what angle of incidence relative to the silver planes is a wavelength of 450 nm reinforced? Assume $n = 1$ in the emulsion.

13-10. Suppose that the blue component of a white-light hologram is formed as in Figure 13-5, using light of 430-nm wavelength. If emulsion shrinkage is 15% during processing, what wavelength is reinforced by the blue-light fringes on reconstruction? How does this affect the holographic image under white-light viewing?

13-11. A hologram is constructed with ultraviolet laser light of 337 nm and viewed in red laser light at 633 nm.
   (a) If the original reference beam and the reconstructing beam are both collimated, what is the magnification of the holographic image, compared with the original subject?
   (b) What magnification would result if coherent X-radiation of 1 Å wavelength were available to construct the hologram?

13-12. (a) Verify that the reconstructed wavefront from the hologram of a point source produces both the real and virtual images shown in Figure 13-1b. First, find the irradiance at the film due to the superposition of a plane and a spherical wave. Then, find the amplitude of the light transmitted by the developed film when irradiated by the reference beam. Interpret the terms as done in the discussion of a hologram of a three-dimensional subject.
   (b) Show that the phase delay of the diverging subject beam, at a point on the film a distance y from the axis, is given by $\pi y^2 / \lambda d$, where d is the distance of the point source from the film. This result follows when $y \ll d$. Show also that reversal of the phase angle produces a converging spherical wavefront associated with the real image on reconstruction.

REFERENCES

[1] Smith, Howard Michael. *Principles of Holography*. New York: John Wiley and Sons, 1975.

[2] Francon, M. *Holography*. New York: Academic Press, 1974.

[3] Stroke, George W. *An Introduction to Coherent Optics and Holography*, 2d ed. New York: Academic Press, 1969.

[4] Vest, C. M. *Holographic Interferometry*. New York: John Wiley and Sons, 1979.

[5] Caulfield, H. John, ed. *Handbook of Optical Holography*. New York: Academic Press, 1979.

[6] Caulfield, H. John. "The Wonder of Holography." *National Geographic* 165 (March 1984): 364.

[7] Leith, Emmett N., and Juris Upatnieks. "Photography by Laser." *Scientific American* (June 1965): 24.

[8] Pennington, Keith S. "Advances in Holography." *Scientific American* (Feb. 1968): 40.

[9] Metherell, Alexander F. "Acoustical Holography." *Scientific American* (Oct. 1969): 36.

[10] Leith, Emmett N. "White-Light Holograms." *Scientific American* (Oct. 1976): 80.

[11] Reynolds, George O., John B. DeVelis, George B. Parrent, Jr., and Brian J. Thompson. *Physical Optics Notebook: Tutorials in Fourier Optics*. Bellingham, Wash.: SPIE Optical Engineering Press, 1989. Ch. 25–27.

14

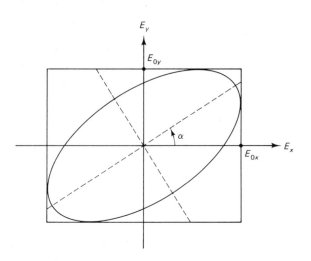

Matrix Treatment of Polarization

INTRODUCTION

The representation of a plane wave of electromagnetic radiation by a drawing such as Figure 8-7 is not applicable to ordinary light. In a plane wave, the electric field vector always oscillates parallel to a fixed direction in space. Light of such character is said to be *linearly polarized*. The same can be said of the magnetic field vector, which maintains an orientation perpendicular to the electric field vector such that the direction of $\mathbf{E} \times \mathbf{B}$ is everywhere the direction of wave propagation. Such a wave might be produced by a distant single-dipole oscillator or by a collection of dipole oscillators radiating in synchronization. Ordinary light, however, is produced by a number of independent atomic sources whose radiation is not synchronized. Consider a beam of ordinary light, such as that produced by a hot filament. The resultant **E**-field vector, generated from a collection of radiating atoms, does not maintain a constant direction of oscillation, nor does it vary spatially in a regular manner, producing *elliptically polarized* or *circularly polarized* light. Such ordinary light is simply said to be *unpolarized*. Of course a beam of light can consist of a mixture of polarized and unpolarized light, in which case it is said to be *partially polarized*.

The possibility of polarizing light is essentially related to its transverse character. If light were a longitudinal wave, the production of polarized light in the ways to be described would simply not be possible. Thus the polarization of light constitutes experimental proof of its transverse character.

In our mathematical description of polarization, we shall employ a matrix technique developed by R. Clark Jones [1]. First we develop two-element column matrices or vectors to represent light in various modes of polarization. Then we examine the physical elements that produce polarized light and discover corresponding 2×2 matrices that function as mathematical operators on the Jones vectors. In the following chapter, we examine in more detail the physical processes that are responsible for producing polarized light.

14-1 MATHEMATICAL REPRESENTATION OF POLARIZED LIGHT: JONES VECTORS

Consider a ray of light directed perpendicularly out of the page, situated at the origin of the axis system in Figure 14-1. Let the **E**-field of the light be represented by the vector shown. Since the **E**-field varies continuously in magnitude and changes direction every half-period, the figure shows the magnitude and direction of **E** at a particular instant. Let the components of **E** along the x-and y-axes be E_x and E_y, respectively. Then, in terms of the unit vectors **i** and **j**,

$$\mathbf{E} = \mathbf{i}E_x + \mathbf{j}E_y \tag{14-1}$$

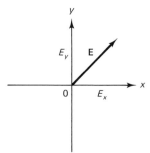

Figure 14-1 Representation of the instantaneous **E**-vector of a light ray traveling in the $+z$-direction. Oscillations of the **E**-vector are equivalent to oscillations of the two orthogonal components, E_x and E_y.

Introducing the space and time dependence of the component vibrations,

$$E_x = E_{0x}e^{i(kz - \omega t + \varphi_x)} \tag{14-2}$$

and

$$E_y = E_{0y}e^{i(kz - \omega t + \varphi_y)} \tag{14-3}$$

for component waves traveling in the $+z$-direction with amplitudes E_{0x} and E_{0y} and phases φ_x and φ_y. Combining with Eq. (14-1),

$$\mathbf{E} = \mathbf{i}E_{0x}e^{i(kz - \omega t + \varphi_x)} + \mathbf{j}E_{0y}e^{i(kz - \omega t + \varphi_y)}$$

which may also be written

$$\mathbf{E} = [\mathbf{i}E_{0x}e^{i\varphi_x} + \mathbf{j}E_{0y}e^{i\varphi_y}]\, e^{i(kz - \omega t)} = \tilde{\mathbf{E}}_0 e^{i(kz - \omega t)} \tag{14-4}$$

The bracketed quantity, separated into x-and y-components, is now recognized as the complex amplitude $\tilde{\mathbf{E}}_0$ for the polarized wave. Since the state of polarization of the light is completely determined by the relative amplitudes and phases of these components, we need concentrate only on the complex amplitude, written as a two-element matrix, or *Jones vector*,

$$\tilde{\mathbf{E}}_0 = \begin{bmatrix} \tilde{E}_{0x} \\ \tilde{E}_{0y} \end{bmatrix} = \begin{bmatrix} E_{0x}e^{i\varphi_x} \\ E_{0y}e^{i\varphi_y} \end{bmatrix} \tag{14-5}$$

Figure 14-2 Representation of E-vectors of linearly polarized light with various special orientations. The direction of the light is along the z-axis.

(a)   (b)   (c)

Let us determine the particular forms for Jones vectors that describe linear, circular, and elliptical polarization. In Figure 14-2a, vertically polarized light travels in the $+z$-direction with its **E**-oscillations along the y-axis. Since **E** actually has a sinusoidally varying magnitude as it progresses, only the amplitude of the electric field is symbolized in both the positive y and negative y directions. In this case we set $E_{0x} = 0$ and $E_{0y} = A$, say. In the absence of an E_x-component, the phase φ_y may be set equal to zero for convenience. Then, by Eq. (14-5), the corresponding Jones vector is

$$E_0 = \begin{bmatrix} 0 \\ A \end{bmatrix} = A \begin{bmatrix} 0 \\ 1 \end{bmatrix}$$

Furthermore, when only the mode of polarization is of interest, the amplitude A may be set equal to one. The Jones vector for vertically linearly polarized light is then simply $\begin{bmatrix} 0 \\ 1 \end{bmatrix}$. This simplified form is the *normalized* form of the vector. In general, a vector $\begin{bmatrix} a \\ b \end{bmatrix}$ is expressed in normalized form when

$$|a|^2 + |b|^2 = 1$$

Similarly, Figure 14-2b represents horizontally polarized light, for which, letting $E_{0y} = 0$, $\varphi_x = 0$, and $E_{0x} = A$,

$$E_0 = \begin{bmatrix} A \\ 0 \end{bmatrix} = A \begin{bmatrix} 1 \\ 0 \end{bmatrix}$$

On the other hand, Figure 14-2c represents linearly polarized light whose vibrations occur along a line making an angle α with respect to the x-axis. Both x- and y-components of E are simultaneously present. Evidently this is the general case that reduces to the vertically polarized mode when $\alpha = 90°$ and to the horizontally polarized mode when $\alpha = 0°$. Notice that to produce the resultant shown, the two perpendicular vibrations must be in phase, that is, they must pass through the origin together, increase along their respective positive axes together, reach their maximum values together, and then return together to continue the cycle. Figure 14-3a makes this sequence clear. Accordingly, since we require merely a *relative* phase of zero, we set $\varphi_x = \varphi_y = 0$. For a resultant with amplitude A, the perpendicular component amplitudes are $E_{0x} = A \cos \alpha$ and $E_{0y} = A \sin \alpha$. The Jones vector takes the form

$$\tilde{E}_0 = \begin{bmatrix} E_{0x} e^{i\varphi_x} \\ E_{0y} e^{i\varphi_y} \end{bmatrix} = \begin{bmatrix} A \cos \alpha \\ A \sin \alpha \end{bmatrix} = A \begin{bmatrix} \cos \alpha \\ \sin \alpha \end{bmatrix} \tag{14-6}$$

For the normalized form of the vector, we set $A = 1$, since $\cos^2 \alpha + \sin^2 \alpha = 1$. Notice that this general form does indeed reduce to the Jones vectors found for the case $\alpha = 0°$ and $\alpha = 90°$. For other orientations, for example, $\alpha = 60°$,

$$E_0 = \begin{bmatrix} \cos 60 \\ \sin 60 \end{bmatrix} = \begin{bmatrix} \dfrac{1}{2} \\ \dfrac{\sqrt{3}}{2} \end{bmatrix} = \dfrac{1}{2} \begin{bmatrix} 1 \\ \sqrt{3} \end{bmatrix}$$

(a)

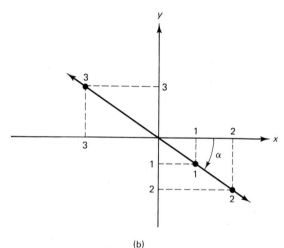

(b)

Figure 14-3 (a) Perpendicular vibrations in phase produce linearly polarized light with **E**-vectors lying in the first and third quadrants. (b) Perpendicular vibrations 180° out of phase produce linearly polarized light with **E**-vectors lying in the second and fourth quadrants.

Alternatively, given a vector $[\begin{smallmatrix} a \\ b \end{smallmatrix}]$, where a and b are real numbers, the inclination of the corresponding linearly polarized light is given by

$$\alpha = \tan^{-1}\left(\frac{E_{0y}}{E_{0x}}\right) = \tan^{-1}\left(\frac{b}{a}\right) \qquad (14\text{-}7)$$

Generalizing a bit, suppose α were a negative angle, as in Figure 14-3b. In this case, the E_{0y}-element of the vector is a negative number, since the sine is an odd function, whereas the E_{0x} element remains positive. The negative sign ensures that the two vibrations are 180° out of phase, as needed to produce linearly polarized light with **E**-vectors lying in the second and fourth quadrants. Referring to Figure 14-3b again, this means that if the x-vibration is increasing from the origin along its positive direction, the y-vibration must be increasing from the origin along its negative direction. The resultant vibration takes place along a line with negative slope.

Summarizing, a Jones vector $[\begin{smallmatrix} a \\ b \end{smallmatrix}]$ with both a and b real numbers, not both zero, represents linearly polarized light at inclination angle $\alpha = \tan^{-1}(b/a)$.

By now it is probably apparent that, in determining the resultant vibration due to two perpendicular components, we are in fact determining the appropriate *Lissajous figure*. If the phase difference between the vibrations is other than 0° or 180°, the resultant **E**-vector traces out an ellipse rather than a straight line. Of course, the straight line can be considered a special case of the ellipse, as can the circle. Figure 14-4 summarizes the sequence of Lissajous figures as a function of relative phase $\Delta\varphi = \varphi_y - \varphi_x$ for the general case $E_{0x} \neq E_{0y}$. Notice the sense of rotation of the tip of the **E**-vector around the ellipses shown in Figure 14-4, which makes the case $\Delta\varphi = 45°$, for example, different from the case $\Delta\varphi = 315°$. When $E_{0x} = E_{0y}$, the ellipses corresponding to $\Delta\varphi = 90°$ or $270°$ reduce to circles.

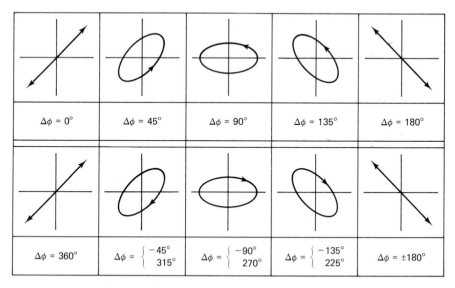

$\Delta\phi = 0°$	$\Delta\phi = 45°$	$\Delta\phi = 90°$	$\Delta\phi = 135°$	$\Delta\phi = 180°$

| $\Delta\phi = 360°$ | $\Delta\phi = \begin{cases} -45° \\ 315° \end{cases}$ | $\Delta\phi = \begin{cases} -90° \\ 270° \end{cases}$ | $\Delta\phi = \begin{cases} -135° \\ 225° \end{cases}$ | $\Delta\phi = \pm180°$ |

Figure 14-4 Lissajous figures as a function of relative phase for orthogonal vibrations of unequal amplitude. An angle lead greater than 180° may also be represented as an angle lag of less than 180°. For all figures we have adopted the phase lag convention $\Delta\varphi = \varphi_y - \varphi_x$.

Now suppose $E_{0x} = E_{0y} = A$ and E_x leads E_y by 90°. Then at the instant E_x has reached its maximum displacement $+A$, for example, E_y is zero. A fourth of a period later, E_x is zero and $E_y = +A$, and so on. Figure 14-5 shows a few samples in the process of forming the resultant vibration. For the cases illustrated there, where the x-vibration leads the y-vibration, it is necessary to make $\varphi_y > \varphi_x$. This apparent contradiction results from our choice of phase in the formulation of the **E**-field in Eqs. (14-2) and (14-3), where the time-dependent term in the exponent is negative. To show this, let us observe the wave at $z = 0$ and choose $\varphi_x = 0$ and $\varphi_y = \epsilon$, so that $\varphi_y > \varphi_x$. Equations (14-2) and (14-3) then become

$$\tilde{E}_x = E_{0x}e^{-i\omega t}$$

$$\tilde{E}_y = E_{0y}e^{-i(\omega t - \epsilon)}$$

The negative sign before ϵ indicates a lag ϵ in the y-vibration relative to the x-vibration. To see that these equations represent the sequence in Figure 14-5, we take their real parts and set $E_{0x} = E_{0y} = A$ and $\epsilon = \pi/2$, giving

$$E_x = A \cos \omega t$$

$$E_y = A \cos \left(\omega t - \frac{\pi}{2} \right) = A \sin \omega t$$

Recalling that $\omega = 2\pi f = 2\pi/T$, each of the cases in Figure 14-5 can be easily verified. Also, since

$$E^2 = E_x^2 + E_y^2 = A^2 (\cos^2 \omega t + \sin^2 \omega t) = A^2$$

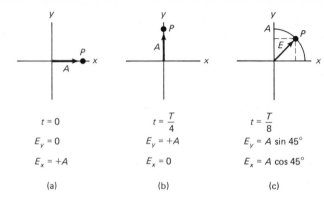

$t = 0$

$E_y = 0$

$E_x = +A$

(a)

$t = \dfrac{T}{4}$

$E_y = +A$

$E_x = 0$

(b)

$t = \dfrac{T}{8}$

$E_y = A \sin 45°$

$E_x = A \cos 45°$

(c)

Figure 14-5 Resultant **E**-vibration due to orthogonal component vibrations of equal magnitude and phase difference of 90°, shown at three different times. The points P represent the position of the resultant. In (c) a sketch of the circular path traced by E is also shown. Notice that the E-vector rotates counterclockwise in this case.

the tip of the resultant vector traces out a circle of radius A.

We now deduce the Jones vector for this case—where E_x leads E_y—taking $E_{0x} = E_{0y} = A$, $\varphi_x = 0$, and $\varphi_y = \pi/2$. Then

$$\tilde{E}_0 = \begin{bmatrix} E_{0x} e^{i\varphi_x} \\ E_{0y} e^{i\varphi_y} \end{bmatrix} = \begin{bmatrix} A \\ A e^{i\pi/2} \end{bmatrix} = A \begin{bmatrix} 1 \\ i \end{bmatrix} \qquad (14\text{-}8)$$

To determine the normalized form of the vector, notice that $1^2 + |i|^2 = 1 + 1 = 2$, so that each element must be divided by $\sqrt{2}$ to produce unity. Thus the Jones vector $(1/\sqrt{2})\begin{bmatrix} 1 \\ i \end{bmatrix}$ represents circularly polarized light when **E** rotates counterclockwise, viewed head-on. This mode is called *left-circularly polarized* light.

Similarly, if E_y leads E_x by $\pi/2$, the result will again be circularly polarized light with clockwise rotation, or *right-circularly polarized* light. Replacing $\pi/2$ by $(-\pi/2)$ in Eq. (14-8) gives the normalized Jones vector,

$$E_0 = \frac{1}{\sqrt{2}} \begin{bmatrix} 1 \\ -i \end{bmatrix}$$

Notice that one of the elements in the Jones vector for circularly polarized light is now pure imaginary, and the magnitudes of the elements are the same. Due to the mathematical form of the vector, the actual character of the light may not always be immediately apparent. For example, the Jones vector $\begin{bmatrix} 2i \\ 2 \end{bmatrix}$ represents right-circularly polarized light since

$$\begin{bmatrix} 2i \\ 2 \end{bmatrix} = 2 \begin{bmatrix} i \\ 1 \end{bmatrix} = 2i \begin{bmatrix} 1 \\ -i \end{bmatrix}$$

The prefactor of the Jones vector may affect the amplitude and, hence, the irradiance of the light but not the polarization mode. Prefactors such as 2 and 2i may therefore be ignored unless information regarding energy is required.

Next suppose that, even though the phase difference between the component orthogonal vibrations is still 90°, the vibrations are of unequal amplitude. If $E_{0x} = A$ and $E_{0y} = B$, say, Eq. (14-8) is modified to give

$$\begin{bmatrix} A \\ iB \end{bmatrix} \quad \text{counterclockwise} \qquad \text{and} \qquad \begin{bmatrix} A \\ -iB \end{bmatrix} \quad \text{clockwise}$$
$$\phantom{\begin{bmatrix} A \\ iB \end{bmatrix}} \quad \text{rotation} \qquad \phantom{\text{and}} \qquad \phantom{\begin{bmatrix} A \\ -iB \end{bmatrix}} \quad \text{rotation}$$

These instances of elliptical polarization are illustrated in Figure 14-4 for $\Delta\varphi = 90°$ and $\Delta\varphi = 270°$. Notice that a lag of 90° is equivalent to a lead of 270°. The ellipse is oriented with its major axis along the x- or y-axis, as in Figure 14-6, depending on the relative magnitudes of E_{0x} and E_{0y}. In addition, either case may produce clockwise rotation of **E** around the ellipse (when E_y leads E_x) or counterclockwise rotation (when E_x leads E_y). Based on these observations, we conclude that a Jones vector with elements of unequal magnitude, one of which is pure imaginary, repre-

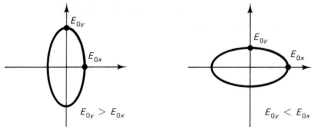

Figure 14-6 Elliptically polarized light for the case $\Delta\varphi = 90°$.

sents elliptically polarized light oriented along the xy-axes. The normalized forms of the Jones vectors now must include a prefactor of $1/\sqrt{A^2 + B^2}$.

It is also possible to produce elliptically polarized light with principal axes inclined to the xy-axes, as evident in Figure 14-4. This situation occurs when the phase difference between component vibrations is some angle other than $m\pi$ (linear polarization) or $(m + \frac{1}{2})\pi$ (circular or elliptical polarization oriented symmetrically about the xy-axes). Here $m = 0, 1, 2, \ldots$. For example, consider the case where E_x leads E_y by some angle ϵ, that is, $\varphi_y - \varphi_x = \epsilon$. Taking $\varphi_x = 0$, $\varphi_y = \epsilon$, $E_{0x} = A$, and $E_{0y} = b$, the Jones vector is

$$\tilde{E}_0 = \begin{bmatrix} E_{0x}e^{i\varphi_x} \\ E_{0y}e^{i\varphi_y} \end{bmatrix} = \begin{bmatrix} A \\ be^{i\epsilon} \end{bmatrix}$$

Using Euler's theorem we write

$$be^{i\epsilon} = b(\cos \epsilon + i \sin \epsilon) = B + iC$$

The Jones vector for this general case is then

$$\tilde{E}_0 = \begin{bmatrix} A \\ B + iC \end{bmatrix} \tag{14-9}$$

where one of the elements is now a complex number having both real and imaginary parts. The normalized form must be divided by $\sqrt{A^2 + B^2 + C^2}$. This form of the Jones vector is the most general, including all those discussed previously as special cases. With the help of analytical geometry, it is possible to show that the ellipse whose Jones vector is given by Eq. (14-9) is inclined at an angle α with respect to the x-axis, as shown in Figure 14-7. The angle of inclination is determined from

$$\tan 2\alpha = \frac{2E_{0x}E_{0y} \cos \epsilon}{E_{0x}^2 - E_{0y}^2} \tag{14-10}$$

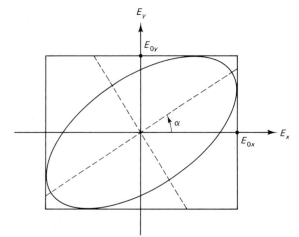

Figure 14-7 Elliptically polarized light oriented at an angle relative to the x-axis.

The ellipse is situated in a rectangle of sides $2E_{0x}$ and $2E_{0y}$. In terms of the parameters A, B, and C, the derivation of Eq. (14-9) makes clear that

$$E_{0x} = A, \qquad E_{0y} = \sqrt{B^2 + C^2}, \qquad \text{and} \quad \epsilon = \tan^{-1}\left(\frac{C}{B}\right) \qquad (14\text{-}11)$$

Example

Analyze the Jones vector given by

$$\begin{bmatrix} 3 \\ 2 + i \end{bmatrix}$$

to show that it represents elliptically polarized light.

Solution The light has relative phase between component vibrations of $\varphi_y - \varphi_x = \epsilon = \tan^{-1}(\frac{1}{2}) = 26.6°$. Since $E_{0x} = 3$ and $E_{0y} = \sqrt{2^2 + 1^2} = \sqrt{5}$, the inclination angle of the axis is given by

$$\alpha = \frac{1}{2} \tan^{-1} \frac{(2)(3)(\sqrt{5}) \cos{(26.6°)}}{9 - 5} = 35.8°$$

With this data the ellipse can be sketched as indicated in Figure 14-7. More precisely, the equation of the ellipse is given by

$$\left(\frac{E_x}{E_{0x}}\right)^2 + \left(\frac{E_y}{E_{0y}}\right)^2 - 2 \left(\frac{E_x}{E_{0x}}\right)\left(\frac{E_y}{E_{0y}}\right) \cos \epsilon = \sin^2 \epsilon \qquad (14\text{-}12)$$

For this example, the equation of the ellipse is

$$\frac{E_x^2}{9} + \frac{E_y^2}{5} - 0.267 E_x E_y = 0.2$$

When E_x lags E_y, the phase angle ϵ becomes negative and leads to the Jones vector representing a clockwise rotation instead,

$$E_0 = \begin{bmatrix} A \\ B - iC \end{bmatrix}$$

Table 14-1 provides a convenient summary of the most common Jones vectors in their normalized forms. It must be emphasized that the forms given in Table 14-1 are not unique. First, any Jones vector may be multiplied by a real constant, changing amplitude but not polarization mode. Vectors in Table 14-1 have all been multiplied by prefactors, when necessary, to put them in normalized form. Thus, for example, the vector $\begin{bmatrix} 2 \\ 2 \end{bmatrix} = 2\begin{bmatrix} 1 \\ 1 \end{bmatrix}$ and so represents linearly polarized light making an angle of 45° with the x-axis and with amplitude of $2\sqrt{2}$. Second, each of the vectors in Table 14-1 can be multiplied by a factor of the form $e^{i\varphi}$, which has the effect of promoting the phase of each element by φ, that is, $\varphi_x \rightarrow \varphi_x + \varphi$ and $\varphi_y \rightarrow \varphi_y + \varphi$. Since the phase difference is unchanged in this process, the new vector represents the same polarization mode. Recall that the vectors in Table 14-1 were formulated by choosing, somewhat arbitrarily, $\varphi_x = 0$. Thus, for example, multiplying the vector representing left-circularly polarized light by $e^{i\pi/2} = i$,

$$i\begin{bmatrix} 1 \\ i \end{bmatrix} = \begin{bmatrix} i \\ -1 \end{bmatrix}$$

produces an alternate form of the vector. Clearly, given the second form, one could deduce the standard form in Table 14-1 by extracting the factor i.

TABLE 14-1 SUMMARY OF JONES VECTORS $E_0 = \begin{bmatrix} E_{0x}e^{i\varphi_x} \\ E_{0y}e^{i\varphi_y} \end{bmatrix}$

I. Linear Polarization ($\Delta\varphi = m\pi$)

General:
$$E_0 = \begin{bmatrix} \cos\alpha \\ \sin\alpha \end{bmatrix}$$

Vertical: $E_0 = \begin{bmatrix} 0 \\ 1 \end{bmatrix}$

Horizontal: $E_0 = \begin{bmatrix} 1 \\ 0 \end{bmatrix}$

At $+45°$: $E_0 = \dfrac{1}{\sqrt{2}}\begin{bmatrix} 1 \\ 1 \end{bmatrix}$

At $-45°$: $E_0 = \dfrac{1}{\sqrt{2}}\begin{bmatrix} 1 \\ -1 \end{bmatrix}$

II. Circular Polarization $\left(\Delta\phi = \dfrac{\pi}{2}\right)$

Left:
$$E_0 = \dfrac{1}{\sqrt{2}}\begin{bmatrix} 1 \\ i \end{bmatrix}$$

Right:
$$E_0 = \dfrac{1}{\sqrt{2}}\begin{bmatrix} 1 \\ -i \end{bmatrix}$$

III. Elliptical Polarization

Left:
$$E_0 = \dfrac{1}{\sqrt{A^2 + B^2}}\begin{bmatrix} A \\ iB \end{bmatrix}$$

$\left(\Delta\phi = (m + \tfrac{1}{2})\,\pi\right)$

Right: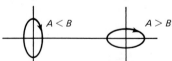
$$E_0 = \dfrac{1}{\sqrt{A^2 + B^2}}\begin{bmatrix} A \\ -iB \end{bmatrix}$$

Left:
$$E_0 = \dfrac{1}{\sqrt{A^2 + B^2 + C^2}}\begin{bmatrix} A \\ B + iC \end{bmatrix}$$

$\left(\Delta\phi \neq \begin{cases} m\pi \\ (m + \tfrac{1}{2})\pi \end{cases}\right)$

Right:
$$E_0 = \dfrac{1}{\sqrt{A^2 + B^2 + C^2}}\begin{bmatrix} A \\ B - iC \end{bmatrix}$$

The usefulness of these Jones vectors will be demonstrated after Jones matrices representing polarizing elements are also developed. However, at this point it is already possible to calculate the result of the superposition of two or more polarized modes by adding their Jones vectors. The addition of left- and right-circularly polarized light, for example, gives

$$\begin{bmatrix} 1 \\ i \end{bmatrix} + \begin{bmatrix} l \\ -i \end{bmatrix} = \begin{bmatrix} 1 + 1 \\ i - i \end{bmatrix} = \begin{bmatrix} 2 \\ 0 \end{bmatrix}$$

or linearly polarized light of twice the amplitude. We conclude that linearly polarized light can be regarded as being made up of left- and right-circularly polarized light in equal proportions. As another example, consider the superposition of vertically and horizontally linearly polarized light in phase:

$$\begin{bmatrix} 0 \\ 1 \end{bmatrix} + \begin{bmatrix} 1 \\ 0 \end{bmatrix} = \begin{bmatrix} 1 \\ 1 \end{bmatrix}$$

The result is linearly polarized light at 45° inclination. Notice that the addition of orthogonal components of linearly polarized light is *not* unpolarized light, even though unpolarized light is often symbolized by such components. There is no Jones vector representing unpolarized or partially polarized light.[1]

14-2 MATHEMATICAL REPRESENTATION OF POLARIZERS: JONES MATRICES

Various devices can serve as optical elements that transmit light but modify the state of polarization. The physical mechanisms underlying their operation will be discussed in the next chapter. Here it will be sufficient to categorize such polarizers in terms of their effects, which are basically three in number.

Linear Polarizer. The linear polarizer selectively removes all or most of the **E**-vibrations in a given direction, while allowing vibrations in the perpendicular direction to be transmitted. In most cases, the selectivity is not 100% efficient, so that the transmitted light is partially polarized. Figure 14-8 illustrates the operation schematically. Unpolarized light traveling in the $+z$-direction passes through a plane polarizer, whose preferential axis of transmission, or transmission axis (TA), is ver-

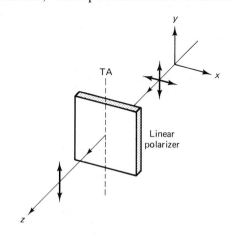

Figure 14-8 Operation of a linear polarizer.

---

[1] A matrix approach that handles partially polarized light, using 1×4 *Stokes vectors* and 4×4 *Mueller matrices* can be found in [2] and [3].

tical. The unpolarized light is represented by two perpendicular (x and y) vibrations, since any direction of vibration present can be resolved into components along these directions. The light transmitted includes components only along the TA direction and is therefore linearly polarized in the vertical, or y-direction. The horizontal components of the original light have been removed by absorption. In the figure, the process is assumed to be 100% efficient.

Phase Retarder. The phase retarder does not remove either of the component orthogonal **E**-vibrations but introduces a phase difference between them. If light corresponding to each vibration travels with different speeds through such a *retardation plate*, there will be a cumulative phase difference $\Delta\varphi$ between the two waves as they emerge.

Symbolically, Figure 14-9 shows the effect of a retardation plate on unpolarized light in a case where the vertical component travels through the plate faster than the horizontal component. This is suggested by the schematic separation of the two components on the optical axis, although of course both waves are simultaneously present at each point along the axis. The fast axis (FA) and slow axis (SA) directions of the plate are also indicated. When the net phase difference $\Delta\varphi = 90°$, the retardation plate is called a *quarter-wave plate*; when it is 180°, it is called a *half-wave plate*.

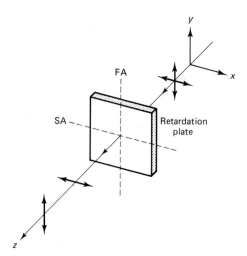

Figure 14-9 Operation of a phase retarder.

Rotator. The rotator has the effect of rotating the direction of linearly polarized light incident on it by some particular angle. Vertical linearly polarized light is shown incident on a rotator in Figure 14-10. The effect of the rotator element is to transmit linearly polarized light whose direction of vibration has, in this case, rotated counterclockwise by an angle θ.

We desire now to create a set of matrices corresponding to these three types of polarizers so that, just as the optical element alters the polarization mode of the actual light beam, an element matrix operating on a Jones vector will produce the same result mathematically. We adopt a pragmatic point of view in formulating appropriate matrices. For example, consider a linear polarizer with a transmission axis along the vertical, as in Figure 14-8. Let a 2×2 matrix representing the polarizer operate on vertically polarized light, and let the elements of the matrix to be determined

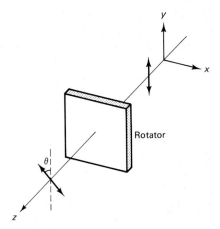

Figure 14-10 Operation of a rotator.

be represented by letters a, b, c, and d. The resultant transmitted or product light in this case must again be vertically linearly polarized light. Symbolically,

$$\begin{bmatrix} a & b \\ c & d \end{bmatrix} \begin{bmatrix} 0 \\ 1 \end{bmatrix} = \begin{bmatrix} 0 \\ 1 \end{bmatrix}$$

This matrix equation is equivalent to the algebraic equations

$$a(0) + b(1) = 0$$

$$c(0) + d(1) = 1$$

from which we conclude $b = 0$ and $d = 1$. To determine elements a and c, let the same polarizer operate on horizontally polarized light. In this case no light is transmitted, or

$$\begin{bmatrix} a & b \\ c & d \end{bmatrix} \begin{bmatrix} 1 \\ 0 \end{bmatrix} = \begin{bmatrix} 0 \\ 0 \end{bmatrix}$$

The corresponding algebraic equations are now

$$a(1) + b(0) = 0$$

$$c(1) + d(0) = 0$$

from which $a = 0$ and $c = 0$. We conclude here without further proof that the appropriate matrix is

$$M = \begin{bmatrix} 0 & 0 \\ 0 & 1 \end{bmatrix} \qquad \text{linear polarizer, TA vertical} \qquad (14\text{-}13)$$

The matrix for a linear polarizer, TA horizontal, can be obtained in a similar manner and is included in Table 14-2, near the end of this chapter. Suppose next that the linear polarizer has a TA inclined at 45° to the x-axis. To keep matters as simple as possible we allow light linearly polarized in the same direction as—and perpendicular to—the TA to pass in turn through the polarizer. Following the approach used earlier,

$$\begin{bmatrix} a & b \\ c & d \end{bmatrix} \begin{bmatrix} 1 \\ 1 \end{bmatrix} = \begin{bmatrix} 1 \\ 1 \end{bmatrix} \qquad \text{and} \qquad \begin{bmatrix} a & b \\ c & d \end{bmatrix} \begin{bmatrix} 1 \\ -1 \end{bmatrix} = \begin{bmatrix} 0 \\ 0 \end{bmatrix}$$

Equivalently,

$$a + b = 1$$
$$c + d = 1$$
$$a - b = 0$$
$$c - d = 0$$

or $a = b = c = d = \frac{1}{2}$. Thus the correct matrix is

$$M = \frac{1}{2}\begin{bmatrix} 1 & 1 \\ 1 & 1 \end{bmatrix} \qquad \text{linear polarizer, TA at } 45° \qquad (14\text{-}14)$$

In the same way, a general matrix representing a linear polarizer with TA at angle θ can be determined. This is left as an exercise for the student. The result is

$$M = \begin{bmatrix} \cos^2 \theta & \sin \theta \cos \theta \\ \sin \theta \cos \theta & \sin^2 \theta \end{bmatrix} \qquad (14\text{-}15)$$

which includes Eqs. (14-13) and (14-14) as special cases, with $\theta = 90°$ and $\theta = 45°$, respectively.

Proceeding to the case of a phase retarder, we desire a matrix that will transform the elements

$$E_{0x}e^{i\varphi_x} \quad \text{into} \quad E_{0x}e^{i(\varphi_x + \epsilon_x)}$$

and

$$E_{0y}e^{i\varphi_y} \quad \text{into} \quad E_{0y}e^{i(\varphi_y + \epsilon_y)}$$

Inspection is sufficient to show that this is accomplished by the matrix operation

$$\begin{bmatrix} e^{i\epsilon_x} & 0 \\ 0 & e^{i\epsilon_y} \end{bmatrix} \begin{bmatrix} E_{0x}e^{i\varphi_x} \\ E_{0y}e^{i\varphi_y} \end{bmatrix} = \begin{bmatrix} E_{0x}e^{i(\varphi_x+\epsilon_x)} \\ E_{0y}e^{i(\varphi_y + \epsilon_y)} \end{bmatrix}$$

Thus the general form of a matrix representing a phase retarder is

$$M = \begin{bmatrix} e^{i\epsilon_x} & 0 \\ 0 & e^{i\epsilon_y} \end{bmatrix} \qquad \text{phase retarder} \qquad (14\text{-}16)$$

where ϵ_x and ϵ_y represent the advance in phase of the E_x- and E_y-components of the incident light. Of course, ϵ_x and ϵ_y may be negative quantities. As a special case, consider a quarter-wave plate (QWP) for which $|\Delta\epsilon| = \pi/2$. We distinguish the case for which $\epsilon_y - \epsilon_x = \pi/2$ (SA vertical) from the case for which $\epsilon_x - \epsilon_y = \pi/2$ (SA horizontal). In the former case, then, let $\epsilon_x = -\pi/4$ and $\epsilon_y = +\pi/4$. Obviously, other choices—an infinite number of them—are possible, so that Jones matrices, like Jones vectors, are not unique. This particular choice, however, leads to a common form of the matrix, due to its symmetrical form:

$$M = \begin{bmatrix} e^{-i\pi/4} & 0 \\ 0 & e^{i\pi/4} \end{bmatrix} = e^{-i\pi/4}\begin{bmatrix} 1 & 0 \\ 0 & i \end{bmatrix} \qquad \text{QWP, SA vertical} \qquad (14\text{-}17)$$

Similarly, when $\epsilon_x > \epsilon_y$,

$$M = e^{i\pi/4}\begin{bmatrix} 1 & 0 \\ 0 & -i \end{bmatrix} \qquad \text{QWP, SA horizontal} \qquad (14\text{-}18)$$

Corresponding matrices for half-wave plates (HWP), where $|\Delta\epsilon| = \pi$, are given by

$$M = \begin{bmatrix} e^{-i\pi/2} & 0 \\ 0 & e^{i\pi/2} \end{bmatrix} = e^{-i\pi/2}\begin{bmatrix} 1 & 0 \\ 0 & -1 \end{bmatrix} \qquad \text{HWP, SA vertical} \qquad (14\text{-}19)$$

$$M = \begin{bmatrix} e^{i\pi/2} & 0 \\ 0 & e^{-i\pi/2} \end{bmatrix} = e^{i\pi/2}\begin{bmatrix} 1 & 0 \\ 0 & -1 \end{bmatrix} \qquad \text{HWP, SA horizontal} \qquad (14\text{-}20)$$

The elements of the matrices are identical in this case, since advancement of phase by π is physically equivalent to retardation by π. The only difference lies in the prefactors that modify the phases of all the elements of the Jones vector in the same way and hence do not affect interpretation of the results.

The requirement for a rotator of angle β is that an **E**-vector oscillating linearly at angle θ be converted to one that oscillates linearly at angle $(\theta + \beta)$. Thus the matrix elements must satisfy

$$\begin{bmatrix} a & b \\ c & d \end{bmatrix} \begin{bmatrix} \cos\theta \\ \sin\theta \end{bmatrix} = \begin{bmatrix} \cos(\theta + \beta) \\ \sin(\theta + \beta) \end{bmatrix}$$

or

$$a\cos\theta + b\sin\theta = \cos(\theta + \beta)$$

$$c\cos\theta + d\sin\theta = \sin(\theta + \beta)$$

From the trigonometric identities for the sine and cosine of the sum of two angles,

$$\cos(\theta + \beta) = \cos\theta\cos\beta - \sin\theta\sin\beta$$

$$\sin(\theta + \beta) = \sin\theta\cos\beta + \cos\theta\sin\beta$$

it follows that

$$a = \cos\beta \qquad c = \sin\beta$$

$$b = -\sin\beta \qquad d = \cos\beta$$

so that the desired matrix is

$$M = \begin{bmatrix} \cos\beta & -\sin\beta \\ \sin\beta & \cos\beta \end{bmatrix} \qquad \text{rotator through angle } +\beta \qquad (14\text{-}21)$$

The Jones matrices derived in this chapter are summarized in Table 14-2. As an important example, consider the production of circularly polarized light by

TABLE 14-2 SUMMARY OF JONES MATRICES

I. Linear polarizers

TA horizontal $\begin{bmatrix} 1 & 0 \\ 0 & 0 \end{bmatrix}$ TA vertical $\begin{bmatrix} 0 & 0 \\ 0 & 1 \end{bmatrix}$ TA at 45° to horizontal $\frac{1}{2}\begin{bmatrix} 1 & 1 \\ 1 & 1 \end{bmatrix}$

II. Phase retarders

General $\begin{bmatrix} e^{i\epsilon_x} & 0 \\ 0 & e^{i\epsilon_y} \end{bmatrix}$

QWP, SA vertical $\quad e^{-i\pi/4}\begin{bmatrix} 1 & 0 \\ 0 & i \end{bmatrix}$ QWP, SA horizontal $\quad e^{i\pi/4}\begin{bmatrix} 1 & 0 \\ 0 & -i \end{bmatrix}$

HWP, SA vertical $\quad e^{-i\pi/2}\begin{bmatrix} 1 & 0 \\ 0 & -1 \end{bmatrix}$ HWP, SA horizontal $\quad e^{i\pi/2}\begin{bmatrix} 1 & 0 \\ 0 & -1 \end{bmatrix}$

III. Rotator

Rotator $\qquad (\theta \to \theta + \beta) \qquad \begin{bmatrix} \cos\beta & -\sin\beta \\ \sin\beta & \cos\beta \end{bmatrix}$

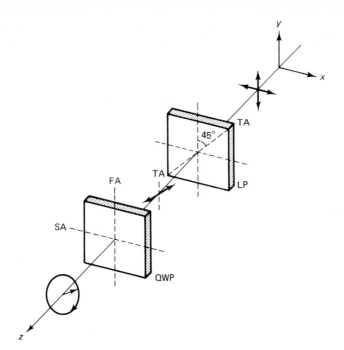

Figure 14-11 Production of circularly polarized light.

combining a linear polarizer with a QWP. Let the linear polarizer (LP) produce light vibrating at an angle of 45°, as in Figure 14-11, which is then transmitted by the QWP. In this arrangement the light incident on the QWP is divided equally between fast and slow axes. On emerging, a phase difference of 90° results in circularly polarized light. With the Jones calculus, this process is equivalent to allowing the QWP matrix to operate on the Jones vector for the linearly polarized light,

$$e^{i\pi/4}\begin{bmatrix} 1 & 0 \\ 0 & -i \end{bmatrix} \frac{1}{\sqrt{2}} \begin{bmatrix} 1 \\ 1 \end{bmatrix} = \left(\frac{1}{\sqrt{2}}\right) e^{i\pi/4} \begin{bmatrix} 1 \\ -i \end{bmatrix}$$

giving right-circularly polarized light with amplitude $1/\sqrt{2}$ times the amplitude of the original linearly polarized light. If the fast and slow axes of the QWP are interchanged, a similar calculation shows that the result is left-circularly polarized instead.

Example

Consider the result of allowing left-circularly polarized light to pass through an eighth-wave plate.

Solution We first need a matrix that can represent the eighth-wave plate, a phase retarder that introduces a relative phase of $2\pi/8 = \pi/4$, or 45°. Thus, letting $\epsilon_x = 0$,

$$M = \begin{bmatrix} e^{i\epsilon_x} & 0 \\ 0 & e^{i\epsilon_y} \end{bmatrix} = \begin{bmatrix} 1 & 0 \\ 0 & e^{i\pi/4} \end{bmatrix}$$

This matrix is then allowed to operate on the Jones vector representing the left-circularly polarized light:

$$\begin{bmatrix} 1 & 0 \\ 0 & e^{i\pi/4} \end{bmatrix} \begin{bmatrix} 1 \\ i \end{bmatrix} = \begin{bmatrix} 1 \\ ie^{i\pi/4} \end{bmatrix} = \begin{bmatrix} 1 \\ e^{i3\pi/4} \end{bmatrix}$$

The resultant Jones vector indicates that the light is elliptically polarized, and the components are out of phase by 135°. Using Euler's equation to expand $e^{i3\pi/4}$, we obtain

$$e^{i3\pi/4} = -\frac{1}{\sqrt{2}} + i\left(\frac{1}{\sqrt{2}}\right)$$

and using our standard notation for this case, we have

$$M = \begin{bmatrix} A \\ B + iC \end{bmatrix}, \qquad \text{where } A = 1, \ B = -\frac{1}{\sqrt{2}}, \text{ and } C = \frac{1}{\sqrt{2}}$$

Comparing this matrix with the general form in Eq. (14-5), we determine that $E_{0x} = 1$ and $E_{0y} = 1$. Making use of Eq. (14-10), we also determine that $\alpha = -45°$.

Of course, the Jones calculus can handle a case where polarized light is transmitted by a series of polarizing elements, since the product of element matrices can represent an overall *system matrix*. If light represented by Jones vector $\mathcal{V}$ passes sequentially through a series of polarizers represented by $\mathfrak{M}_1, \mathfrak{M}_2, \mathfrak{M}_3, \ldots, \mathfrak{M}_m$, that is,

$$(\mathfrak{M}_m \cdots \mathfrak{M}_3 \mathfrak{M}_2 \mathfrak{M}_1)\mathcal{V} = \mathfrak{M}_s \mathcal{V}$$

then the system matrix $\mathfrak{M}_s = \mathfrak{M}_m \cdots \mathfrak{M}_3 \mathfrak{M}_2 \mathfrak{M}_1$.

PROBLEMS

14-1. Derive the Jones matrix, Eq. (14-15), representing a linear polarizer whose transmission axis is at an arbitrary angle θ with respect to the horizontal.

14-2. Write the normalized Jones vectors for each of the following waves, and describe completely the state of polarization of each.
(a) $\mathbf{E} = \mathbf{i}E_0 \cos (kz - \omega t) - \mathbf{j}E_0 \cos (kz - \omega t)$
(b) $\mathbf{E} = \mathbf{i}E_0 \sin 2\pi \left(\frac{z}{\lambda} - ft\right) + \mathbf{j}E_0 \sin 2\pi\left(\frac{z}{\lambda} - ft\right)$
(c) $\mathbf{E} = \mathbf{i}E_0 \sin (kz - \omega t) + \mathbf{j}E_0 \sin \left(kz - \omega t - \frac{\pi}{4}\right)$
(d) $\mathbf{E} = \mathbf{i}E_0 \cos (kz - \omega t) + \mathbf{j}E_0 \cos \left(kz - \omega t + \frac{\pi}{2}\right)$

14-3. Describe as completely as possible the state of polarization of each of the following waves, including amplitude and wave direction. Be careful to distinguish between the unit vector $\mathbf{i}$ and the complex number $i = \sqrt{-1}$.
(a) $\mathbf{E} = 2E_0\mathbf{i}e^{i(kz-\omega t)}$
(b) $\mathbf{E} = E_0(3\mathbf{i} + 4\mathbf{j})e^{i(kz-\omega t)}$
(c) $\mathbf{E} = 5E_0(\mathbf{i} - i\mathbf{j})e^{i(kz+\omega t)}$

14-4. Two linearly polarized beams are given by

$$\mathbf{E}_1 = E_{01}(\mathbf{i} - \mathbf{j}) \cos (kz - \omega t) \quad \text{and} \quad \mathbf{E}_2 = E_{02}(\sqrt{3}\,\mathbf{i} + \mathbf{j}) \cos (kz - \omega t)$$

Determine the angle between their planes of polarization by (a) forming their Jones vectors and finding the vibration direction of each and (b) forming the dot product of their vector amplitudes.

14-5. Find the character of polarized light after passing in turn through (a) a half-wave plate with slow axis at 45°; (b) a linear polarizer with transmission axis at 45°; (c) a quarter-wave plate with slow axis horizontal. Assume the original light to be linearly polarized vertically. Use the matrix approach and analyze the final Jones vector to describe the product light. (*Hint*: First find the effect of the HWP alone on the incident light.)

14-6. Write the equations for the electric fields of the following waves in exponential form:
   (a) A linearly polarized wave traveling in the x-direction. The **E**-vector makes an angle of 30° relative to the y-axis.
   (b) A right elliptically polarized wave traveling in the y-direction. The major axis of the ellipse is in the z-direction and is twice the minor axis.
   (c) A linearly polarized wave traveling in the xy-plane in a direction making an angle of 45° relative to the x-axis. The direction of polarization is the z-direction.

14-7. Determine the conditions on the elements A, B, and C of the general Jones vector (Eq. 14-9), representing polarized light, that lead to the following special cases: (a) linearly polarized light; (b) elliptically polarized light with major axis aligned along a coordinate axis; (c) circularly polarized light. In each case, from the meanings of A, B, C, deduce the possible values of phase difference between component vibrations.

14-8. Write a computer program that will determine E_y-values of elliptically polarized light from the equation for the ellipse, Eq. (14-12), with input constants A, B, and C, and variable input parameter E_x. Plot the ellipse for the example given in the text,

$$E_0 = \begin{bmatrix} 3 \\ 2 + i \end{bmatrix}$$

14-9. Specify the polarization mode for each of the following Jones vectors.

   (a) $\begin{bmatrix} 3i \\ i \end{bmatrix}$ **(b)** $\begin{bmatrix} i \\ 1 \end{bmatrix}$ **(c)** $\begin{bmatrix} 4i \\ 5 \end{bmatrix}$ **(d)** $\begin{bmatrix} 5 \\ 0 \end{bmatrix}$ **(e)** $\begin{bmatrix} 2 \\ 2i \end{bmatrix}$ **(f)** $\begin{bmatrix} 2 \\ 3 \end{bmatrix}$ **(g)** $\begin{bmatrix} 2 \\ 6 + 8i \end{bmatrix}$

14-10. Linearly polarized light whose **E** is inclined at +30° relative to the x-axis is transmitted by a QWP with SA horizontal. Describe the polarization mode of the product light.

14-11. Using the Jones calculus, show that the effect of a HWP on light linearly polarized at inclination angle α is to rotate the plane of polarization through an angle of 2α. The HWP may be used in this way as a "laser-line rotator," allowing the plane of polarization of a laser beam to be rotated without having to rotate the laser.

14-12. An important application of the QWP is its use in an "isolator." For example, to prevent feedback from interferometers into lasers by front-surface, back reflections, the beam is first allowed to pass through a combination of linear polarizer and QWP, with OA of the QWP at 45° to the TA of the polarizer. Consider what happens to such light after reflection from a plane surface and transmission back through this optical device.

14-13. Light linearly polarized with a horizontal transmission axis is sent through another linear polarizer with TA at 45° and then through a QWP with SA horizontal. Use the Jones matrix technique to determine and describe the product light.

14-14. A light beam passes consecutively through (1) a linear polarizer with TA at 45° clockwise from vertical, (2) a QWP with SA vertical, (3) a linear polarizer with TA horizontal, (4) a HWP with FA horizontal, (5) a linear polarizer with TA vertical. What is the nature of the product light?

14-15. Unpolarized light passes through a linear polarizer with TA at 60° from the vertical, then through a QWP with SA horizontal, and finally through another linear polarizer with TA vertical. Determine using Jones matrices the character of the light after passing through (a) the QWP and (b) the final linear polarizer.

14-16. Determine the state of polarization of circularly polarized light after it is passed normally through (a) a QWP; (b) an eighth-wave plate. Use the matrix method to support your answer.

14-17. Show that the matrix $\begin{bmatrix} 1 & i \\ -i & 1 \end{bmatrix}$ represents a right-circular polarizer, converting any incident polarized light into right circularly-polarized light. What is the proper matrix to represent a left-circular polarizer?

14-18. Show that elliptical polarization can be regarded as a combination of circular and linear polarizations.

14-19. Derive the equation of the ellipse for polarized light given in Eq. (14-12). (*Hint*: Combine the E_x and E_y equations for the general case of elliptical polarization, eliminating the space and time dependence between them.)

14-20. **(a)** Identify the state of polarization corresponding to the Jones vector

$$\begin{bmatrix} 2 \\ 3e^{i\pi/3} \end{bmatrix}$$

and write it in the standard, normalized form of Table 14-1.

(b) Let this light be transmitted through an element that rotates linearly polarized light by $+30°$. Find the new, normalized form and describe the result.

14-21. Determine the nature of the polarization that results from Eq. (14-12) when (a) $\epsilon = \pi/2$; (b) $E_{0x} = E_{0y} = E_0$; (c) both (a) and (b); (d) $\epsilon = 0$.

14-22. A quarter-wave plate is placed between crossed polarizer and analyzer such that the angle between the polarizer TA and the QWP fast axis is θ. How does the emergent light vary as a function of θ?

REFERENCES

[1] Jones, R. Clark. "A New Calculus for the Treatment of Optical Systems." *Journal of the Optical Society* 31 (1941): 488.

[2] Walker, M. J. "Matrix Calculus and the Stokes Parameters of Polarized Radiation." *American Journal of Physics* 22 (1954): 170.

[3] Shurcliff, W. A. *Polarized Light: Production and Use*. Cambridge, Mass.: Harvard University Press, 1962.

[4] Gerrard, A., and J. M. Burch. *Introduction to Matrix Methods in Optics*. New York: John Wiley and Sons, 1975.

15

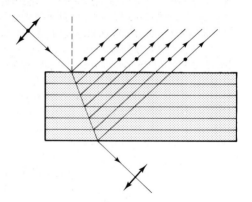

Production of Polarized Light

INTRODUCTION

Any interaction of light with matter whose optical properties are asymmetrical along directions transverse to the propagation vector provides a means of polarizing light. Indeed, if light were longitudinal rather than transverse in its nature, transverse material asymmetries along the propagation vector could not alter the sense of the oscillating **E**-vector, and the physical mechanisms to be described here would have no polarizing or spatially selective effects on light beams. The experimental observation that light can be polarized is, therefore, clear evidence of its transverse nature. The most important processes that produce polarized light are discussed in this chapter under the following general areas: (1) dichroism, (2) reflection, (3) scattering, and (4) birefringence. Optical activity is described as a mechanism that *modifies* polarized light. Finally, *photoelasticity* is briefly discussed as a useful application.

15-1 DICHROISM: POLARIZATION BY SELECTIVE ABSORPTION

A *dichroic polarizer* selectively absorbs light with **E**-vibrations along a unique direction characteristic of the dichroic material. The polarizer easily transmits light with **E**-vibrations along a transverse direction orthogonal to the direction of absorption. This preferred direction is called the transmission axis (TA) of the *polarizer*. In the ideal polarizer, the transmitted light is linearly polarized in the same direction as the transmission axis. The state of polarization of the light can most easily be tested by a second dichroic polarizer, which then functions as an *analyzer*, shown in Figure 15-1. When the TA of the analyzer is oriented at 90° relative to the TA of the polarizer, the light is effectively extinguished. As the analyzer is rotated, the light transmitted by the pair increases, reaching a maximum when their TAs are aligned. If I_0 represents the maximum transmitted intensity, then *Malus' law* states that the irradiance for any relative angle θ between the TAs is given by

$$I = I_0 \cos^2 \theta \qquad (15\text{-}1)$$

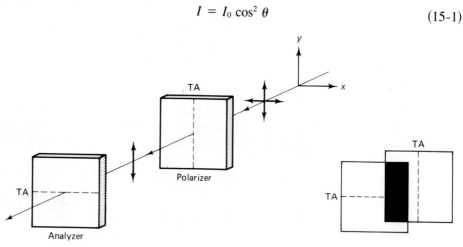

Figure 15-1   Crossed dichroic polarizers functioning as a polarizer-analyzer pair. No light is transmitted through the analyzer.

Malus' law is easily understood in conjunction with Figure 15-2. Notice that the amplitude of the light emerging from the analyzer is $E_0 \cos \theta$. The irradiance I (in W/m²) is then proportional to the square of this result.

The impressive ability of dichroic materials to absorb light strongly with **E** along one direction and to transmit light easily with **E** along a perpendicular direction can perhaps best be understood by reference to a standard experiment with microwaves, illustrated in Figure 15-3. Wavelengths of microwaves range roughly from 1 mm to 1 m. It is found that when a vertical wire grid, whose spacing is much smaller than the wavelength, intercepts microwaves with vertical linear polarization,

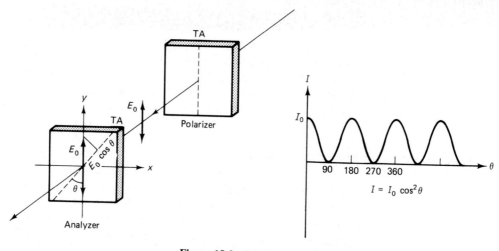

Figure 15-2   Illustration of Malus' law.

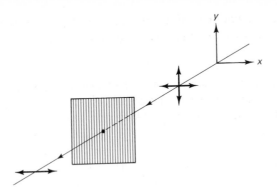

Figure 15-3 Action of a vertical wire grid on microwaves. Effective absorption of the vertical component of the radiation occurs when $\lambda \geqslant$ grid spacing.

little or no radiation is transmitted. Conversely, when the grid intercepts waves polarized in a direction perpendicular to the wires, there is efficient transmission of the waves. The explanation of this behavior involves a consideration of the interaction of electromagnetic radiation with the metal wires that operate as a dichroic polarizer. Within the metal wires, the mobile free electrons are set in oscillatory motion by the oscillations of the electric field of the radiation. We know that each electron so oscillating constitutes a dipole source that radiates electromagnetic energy in all directions, except the direction of the oscillation itself. Evidently, the superposition of an incident electromagnetic wave with vertical **E**-vibrations and the radiation of these electron oscillators leads to cancellation in the forward direction. It turns out, in fact, that the phase of the electromagnetic wave originating with the oscillating electrons is 180° out of step with that of the incident radiation,[1] so that no wave can propagate in the forward direction. In addition, the oscillation of the free electrons is not entirely free. The effective friction due to interaction with lattice imperfections, for example, constitutes some dissipation of energy, which must attenuate the incident wave. The chief reason for the disappearance of the forward wave, however, is destructive interference between the incident and generated waves. Horizontally linearly polarized light incident on the vertical wire grid would suffer the same fate, except that appreciable oscillatory motion of the electrons across the wire is inhibited. As a result, the generated electromagnetic wave is reduced in strength and effective cancellation cannot occur. If the grid is rotated by 90°, the vertical **E**-vibrations are transmitted and the horizontal **E**-vibrations are canceled. The wire grid polarizes microwaves as a dichroic absorber polarizes optical radiation.

For optical wavelengths, the conduction paths analogous to the grid wires must be much closer together. The most common dichroic absorber for light is Polaroid *H*-sheet, invented in 1938 by Edwin H. Land. When a sheet of clear, polyvinyl alcohol is heated and stretched, its long, hydrocarbon molecules tend to align in the direction of stretching. The stretched material is then impregnated with iodine atoms, which become associated with the linear molecules and provide "conduction" electrons to complete the analogy to the wire grid. Some naturally occurring materials, such as the mineral tourmaline, also possess dichroic properties to some degree. All that is required in principle is that the electrons be much freer to respond to an incident electromagnetic wave in one direction than in an orthogonal direction. In nonmetallic materials, the electrons acting as dipole oscillators are not free. In this case the wave they generate is not out of phase with respect to the incident wave, and complete cancellation of the forward wave does not occur. The energy of the driving wave, however, is gradually dissipated as the wave advances through the ab-

---

[1] The theoretical basis for this statement is presented at the end of Section 27-1, applied to the free electrons of a metal.

sorber, so that the efficiency of the dichroic absorber is a function of the thickness. The absorption follows the usual expression for attenuation,

$$I = I_0 e^{-\alpha x}$$

where I_0 is the incident irradiance and I is the irradiance at depth x of absorber. The constant α is the *absorptivity*, or *absorption coefficient*, characteristic of the material. In a good, practical dichroic absorber, α is relatively independent of wavelength, that is, the material appears transparent and yet behaves as a linear polarizer for all optical wavelengths. This ideal condition is not quite achieved in Polaroid H-sheet, which is less effective at the blue end of the spectrum. Consequently, when a Polaroid H-sheet is crossed with another such sheet acting as an analyzer, the combination contributes a blue tint to the transmitted light.

15-2 POLARIZATION BY REFLECTION FROM DIELECTRIC SURFACES

Light that is specularly reflected from dielectric surfaces is at least partially polarized. This is most easily confirmed by looking through a piece of polarizing filter while rotating it about the direction of reflected light. When the preferred **E**-direction of the reflected light is perpendicular to the TA of the filter, regions from which light is specularly reflected into the eye appear reduced in brightness. This is precisely the working principle of Polaroid sunglasses. Since the preferred **E**-vibration in light reflected from ground level into the eye turns out to be horizontal, the TA of the Polaroids in a pair of sunglasses is fixed in the vertical direction.

To appreciate the physics that underlies this phenomenon, consider Figure 15-4, which shows a narrow beam of light incident at an arbitrary angle on a smooth, flat dielectric surface. The unpolarized incident beam is conveniently represented by two perpendicular **E**-vibrations, one perpendicular (Figure 15-4a) and one parallel (Figure 15-4b) to the plane of incidence, that is, the plane of the page, including the incident ray and the normal drawn to the point of incidence. Standard notation is to refer to these components as E_s (perpendicular component) and E_p (parallel component). Alternatively, the E_s mode is called the TE (transverse electric) mode, and the E_p mode is called the TM (transverse magnetic) mode, since the **B**-component of the wave is transverse to the plane of incidence when the corresponding **E**-component is parallel.

Consider first the E_s, or TE, component (Figure 15-4a). The action of E_s on the electrons in the surface of the dielectric is to stimulate oscillations along the same direction, or perpendicular to the page. The radiation from all these electronic dipole oscillators adds up to produce a beam of light in two directions only, the di-

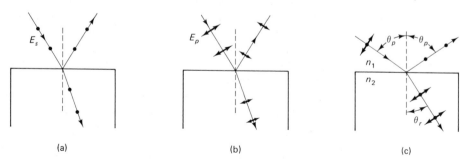

(a)   (b)   (c)

Figure 15-4 Specular reflection of light at a dielectric surface. (a) TE mode. (b) TM mode. (c) Polarization at Brewster's angle.

rection of the reflected and refracted beams, both also linearly polarized perpendicular to the plane of incidence. The reflected and refracted rays are both in a direction corresponding to maximum dipole radiation, perpendicular to the dipole axis.

Consider next the action of the E_p, or TM, component (Figure 15-4b). From the direction of the refracted beam (which may be calculated using Snell's law), we conclude that the **E**-field within the isotropic dielectric materials, and thus the axis of the dipole oscillations, is oriented perpendicular to the beam direction, as shown. Notice that the dipole oscillations include a component along the direction of the reflected beam. Recalling that a dipole oscillator radiates only weakly along directions making small angles with the dipole axis ($I \propto \sin^2 \theta$), we conclude that only a fraction of the E_p component of the original light (compared with the E_s component) appears in the reflected beam. Considering both TE and TM modes together, it follows that the reflected light is partially polarized with a predominance of the E_s mode present. Since the energy of the incident beam is equally divided between E_s and E_p components, it also follows that the refracted beam is partially polarized and richer in the E_p component.

This analysis should make it clear that when the dipole axes are in the same direction as the reflected ray, the E_p component is entirely missing from the reflected beam, and the reflected ray is linearly polarized in the E_s mode. In fact, if the dipoles radiated along the reflected ray, the electromagnetic wave could only be a longitudinal wave! This unique orientation results when the reflected and refracted rays are perpendicular to one another (Figure 15-4c). The angle of incidence that produces a linearly polarized beam by reflection is θ_p, the *polarizing angle*, or *Brewster's angle*. Combining Snell's law

$$n_1 \sin \theta_p = n_2 \sin \theta_r$$

with the trigonometric relation $\theta_r = 90 - \theta_p$, we arrive at *Brewster's law*,

$$\theta_p = \tan^{-1} \left(\frac{n_2}{n_1} \right) \qquad (15\text{-}2)$$

Polarizing angles exist for both external reflection ($n_2 > n_1$) and internal reflection ($n_2 < n_1$) and are clearly not the same. For reflection when light travels from air to glass, with $n = 1.50$, for example, $\theta_p = 56.3°$. For reflection when light travels in the opposite direction, $\theta_p = 33.7°$. These angles are seen to be precisely complementary, as required by geometry and the definition of Brewster's angle.

While reflection at the polarizing angle from a dielectric surface can be used to produce linearly polarized light, the method is relatively inefficient. For reflection from air to glass, as in the example just given, only 15% of the E_s component is found in the reflected beam. (The *Fresnel equations*, which permit calculations of this kind, are treated in Chapter 20.) This deficiency can be remedied to a degree by stepwise intensification of the reflected beam as in a *pile-of-plates* polarizer (Figure 15-5). Repeated reflections by multiple layers of the dielectric at Brewster's angle both increases the intensity of the E_s component in the integrated, reflected beam and, necessarily, purifies the transmitted beam of this component. If enough plates are assembled, the transmitted beam approaches a linearly polarized condition. Pile-of-plates polarizers are especially helpful in those regions of the infrared and ultraviolet spectrum where dichroic sheet polarizers and calcite prisms are ineffective. Multilayer, thin film coatings that show little absorption in the spectral region of interest behave in a similar manner and can be used as polarization-sensitive reflectors and transmitters.

Another interesting application of polarization by reflection is the *Brewster window*. The window (Figure 15-6) operates in the same way as a single plate of the

Figure 15-5 Pile-of-plates polarizer.

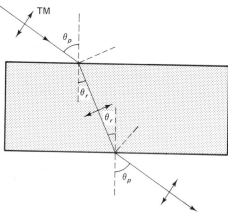

Figure 15-6 Brewster window. Brewster's law is satisfied for the TM mode at both surfaces.

pile-of-plates analyzer. TM linearly polarized light incident at Brewster's angle is fully transmitted at the first surface. The angle of incidence, θ_r, at the second surface also satisfies Brewster's law for internal reflection, so that the light is again fully transmitted. The plate acts as a perfect window for TM polarized light.

The active medium of a gas laser is often bounded by two Brewster windows, located at the ends of the gas plasma tube. The light in the cavity makes repeated passes through the windows, on its way to and from cavity mirrors positioned beyond alternate ends of the gas tube. Upon each traversal, the TM mode is completely transmitted, whereas the TE mode is partially reflected (rejected). After many such traversals in the laser cavity, the beam is essentially free of TE vibrations and the emerging laser beam is polarized in the TM mode.

15-3 POLARIZATION BY SCATTERING

Before discussing the polarization of light that occurs in scattering, we make a slight detour to discuss scattering in general, pointing out some familiar consequences of scattering that are in themselves rather interesting. By the *scattering* of light, we mean the removal of energy from an incident wave by a scattering medium and the reemission of some portion of that energy in many directions. We can think of the elemental oscillator or scattering unit as an electronic charge bound to a nucleus (a *dipole oscillator*). The electron is set into forced oscillation by the alternating electric field of incident light and at the same frequency. The response of the electron to this driving force depends on the relationship between the driving frequency ω and the natural or resonant frequency of the oscillator ω_0. In most materials, resonant frequencies lie predominantly in the ultraviolet (due to electronic oscillations) and in the infrared (due to molecular vibrations) rather than in the visible. Because atomic masses are so much larger than the electron mass, amplitudes of induced molecular vibrations are small compared with electronic vibrations and so can be neglected in this discussion. Thus for light incident on most materials, $\omega \ll \omega_0$. Calculations[2] show that in this case, the induced dipole oscillations have an amplitude that is roughly independent of the frequency of the light. The oscillating dipoles, consisting of electrons accelerating in harmonic motion, are tiny radiators, antennas that reradiate or *scatter* energy in all directions except along the dipole axis itself.

Such scattering is most effective when the scattering centers are particles whose dimensions are small compared with the wavelength of the radiation, in which case we speak of *Rayleigh scattering*. The scattering of sunlight from oxygen and nitrogen molecules in the atmosphere, for example, is Rayleigh scattering, whereas the scattering of light from dense scattering centers—like the droplets of water in clouds and fog—is not. In Rayleigh scattering, the well-separated scattering centers act independently (incoherently), so that, according to the conclusions of Section 9-3, their net irradiance is the sum of their individual irradiances. Now the radiated power can be shown to be inversely proportional to the fourth power of the wavelength of the incident radiation. Without deriving this Rayleigh scattering law, we can make the following hand-waving argument: The electric field of a dipole with a charge e accelerating back and forth along a line is proportional to the acceleration. If $r = r_0 e^{i\omega t}$, then the acceleration $d^2r/dt^2 = -\omega^2 r_0 e^{i\omega t}$ is proportional to the square of the frequency. Since the power P radiated is in turn proportional to the square of the electric field, it becomes proportional to the fourth power of the frequency. This is the Rayleigh scattering law, which is expressed by [9],

$$P = \frac{e^2 \omega^4 r_0^2}{12\pi\epsilon_0 c^3}$$

Thus the oscillating dipoles radiate more energy in the shorter wavelength region of the visible spectrum than in the longer wavelength region. The radiated power for violet light of wavelength 400 nm is nearly 10 times as great as for red light of wavelength 700 nm. Rayleigh scattering explains why a clean atmosphere appears blue: Higher frequency blue light from the sun is scattered by the atmosphere down to the earth more than the lower frequency red light. On the other hand, when we are looking at the sunlight "head-on" at sunrise or sunset, after it has passed through a good deal of atmosphere, we see reddish or yellowish light, white light from which the blues have been preferentially removed by scattering.

Scattering that occurs from larger particles[3] such as those found in clouds, fog, and powdered materials such as sugar appears as white light, in contrast to Rayleigh scattering. Here "larger particles" refers to the size of the scattering particle relative to the wavelength of light. In this case, the scattering centers (particles) are arranged—more or less—in an orderly fashion so that oscillators that are closer together than a wavelength of the incident light become coherent scatterers. The cooperative effect of many oscillators tends to cancel the radiation in all directions but the forward (refraction) direction and the backward (reflection) direction. In other words, the scattering due to these larger particles can be understood in terms of the usual laws of reflection and refraction. However, the usual departures from perfectly ordered atomic arrangement lead to some scattering in other directions as well. The

---

[2] Such calculations, giving a quantitative treatment to these ideas, are given in Section 27-1.

[3] The more general theory of scattering, including larger scattering centers, is called *Mie scattering* after its creator. Mie scattering takes into account the size, shape, refractive index, and absorptivity of the scattering particles and includes Rayleigh scattering as a special case [4].

net amplitude of the coherent scattered radiation is now the sum of the individual amplitudes, or the radiated power is proportional to N^2 when there are N coherent oscillators. Although such scattering is much less effective, *per oscillator*, than Rayleigh scattering, the density of oscillators in this case leads to considerable scattering. It can be shown that the number N of such coherent oscillators responsible for the reflected radiation is proportional to λ^2, so that the radiated power is proportional to λ^4, canceling the $1/\lambda^4$ dependence of isolated oscillators. Thus the scattered radiation is essentially wavelength independent, and fog and clouds appear white by scattered light.

Of particular interest in the context of this chapter, however, is the fact that scattered radiation may also be polarized. This phenomenon can be explained rather simply, because light is a transverse, and not a longitudinal, wave. As an example, consider a vessel of water to which is added one or more drops of milk. The milk molecules quickly diffuse throughout the water and serve as effective scattering centers for a beam of light transmitted across the medium. In pure water the light does not scatter sideways but propagates only in the forward direction. The light scattered in various directions from the milk molecules, when examined with a polarizing filter, is found to be polarized, as shown in Figure 15-7. The perpendicular **E** components of the unpolarized light incident from the left sets the electronic oscillators of a scattering center into similar forced vibrations, producing radiation reemitted in all directions. Light scattered in any direction can include only those identical **E**-vibrations executed by the oscillators, that is, along the y- and the z-directions. If scattered light is viewed from the $-y$-direction, it will be found to contain **E**-vibrations along the z-direction, but those along the y-direction are absent because they would represent longitudinal **E**-vibrations in an electromagnetic wave. Similarly, viewed along the z-direction, the z-vibrations are missing, and light is linearly polarized along the y-direction. Viewed along directions off the axes, the light is partially polarized. The forward beam shows the same polarization as the incident light.

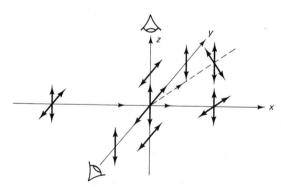

Figure 15-7 Polarization due to scattering. Unpolarized light incident from the left is scattered by a particle at the origin.

In the same way, when the sun is not directly overhead so that its light crosses the atmosphere above us, the light scattered down is found to be partially polarized. The effect is easily seen by viewing the clear sky through a polarizing filter while it is rotated. The polarization is not complete, both because we see light that is multiply scattered into the eye and because not all electronic oscillators in molecules are free to oscillate in exactly the same direction as the incident **E**-vector of the light.

Ordinary polarization by scattering is generally weak and imperfect and so is not used as a practical means of artificially producing polarized light. In the area of nonlinear optics, however, the controlled scattering of light from active media, exemplified by *stimulated Raman*, *Rayleigh*, and *Brillouin scattering*, provides much vital research in modern optics. In such cases, the scattered light is modified by the resonant frequencies of the medium. An introduction to such nonlinear applications is given in Chapter 26.

15-4 BIREFRINGENCE: POLARIZATION WITH TWO REFRACTIVE INDICES

Birefringent materials are so named because they are able to cause double refraction, that is, the appearance of two refracted beams due to two different indices of refraction for a single material.

We have already seen that anisotropy in the binding forces affecting the electrons of a material can lead to anisotropy in the amplitudes of their oscillations in response to a stimulating electromagnetic wave and hence to anisotropy of absorption. Such a material displays dichroism. For this to occur, however, the stimulating optical frequencies must fall within the absorption band of the material. Referring to Figure 15-8, we see that the dispersion $dn/d\omega$ is less than zero—or "anomalous"—over a certain frequency interval. Such intervals coincide with the existence of absorption bands in a given material. Typically, the absorption band lies in the ultraviolet, above optical frequencies, so that the material is transparent to visible light. In this case, even with anisotropy of electron-binding forces, there is little or no effect on optical absorption, and the material does not appear dichroic. Still, the presence of anisotropic binding forces along the x- and y-directions leads to different dispersion curves (like that of Figure 15-8) for refractive index n_x corresponding to E_x-vibrations and n_y corresponding to E_y-vibrations. The existence of both an n_x and an n_y for a given optical frequency is to be expected, since different binding forces along these directions produce different interactions with the electromagnetic wave and, thus, different velocities of propagation v_x and v_y through the crystal. The result is that such a crystal, while not appreciably dichroic, still manifests the property of birefringence. The critical physical properties here are the refractive index n and the extinction coefficient k (proportional to the absorption coefficient) for a given frequency of light. Each constitutes a part of the complex refractive index,

$$\tilde{n} = n + ik$$

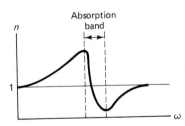

Figure 15-8 Response of refractive index as a function of frequency near an absorption band. The band in which $dn/d\omega < 0$ is said to be a region of anomalous dispersion.

Recapitulating, then, for an ideal dichroic material, $n_x = n_y$ and $k_x \neq k_y$, whereas for an ideal birefringent material, $k_x = k_y$ and $n_x \neq n_y$. Both conditions require anisotropic crystalline structures. The conditions are frequency dependent. Calcite is birefringent in the visible spectrum, for example, and strongly dichroic in certain parts of the infrared spectrum. Other common materials, birefringent in the visible region, are quartz, ice, mica, and even cellophane.

The relationship of crystalline asymmetry with refractive index and the speed of light in the medium may be understood a bit more clearly by considering the case of calcite. The basic molecular unit of calcite is $CaCO_3$, which assumes a tetrahedral or pyramidal structure in the crystal. Figure 15-9a shows one of these molecules, assumed to be surrounded by identical structures that are similarly oriented. The

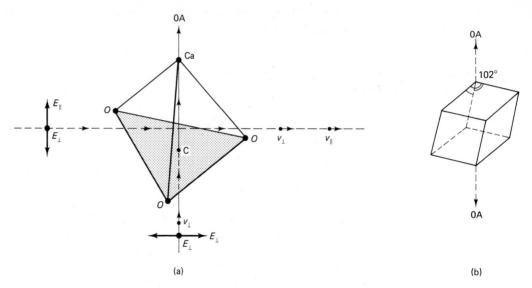

Figure 15-9 (a) Progress of light through a calcite crystal. Three oxygen atoms form the base of a tetrahedron. The optic axis is parallel to the line joining the C and Ca atoms. (b) Rhombohedron of calcite, showing the optic axis, which passes symmetrically through a blunt corner where the three face angles equal 102°.

carbon and oxygen atoms form the base of the pyramid, as shown, with carbon lying in the center of the equilateral triangle of oxygen atoms. The calcium atom is positioned at some distance above the carbon atom, at the apex of the pyramid. The figure shows unpolarized light propagating through the crystal from two different directions. First consider light entering from below, along the line joining the carbon and calcium atoms. All oscillations of the **E**-field are represented by the two transverse vectors shown. Since the molecule, and so also the crystal, is symmetric with respect to this direction, both **E**-vibrations interact with the electrons in the same way when traveling through the calcite. This direction of symmetry through the crystal is called the *optic axis* (OA) of the crystal. For the light entering from below, both **E**-components are perpendicular to the OA. Consider next the light entering the crystal from the left. From this direction the two representative **E**-vibrations have dissimilar effects on the electrons in the base plane. The component $E_\parallel$ parallel to the OA of the crystal causes electrons in the base plane to oscillate along a direction perpendicular to the plane, whereas its orthogonal counterpart $E_\perp$ causes oscillations within the plane. Oscillations within the plane—where the electrons tend to be confined due to the chemical bonding—take place more easily, that is, with smaller binding forces, than oscillations that are perpendicular to the plane. Since **E**-oscillations in the oxygen plane ($\mathbf{E} \perp \text{OA}$) interact more strongly with the electrons, the speed of these component waves is reduced most, that is, $v_\perp < v_\parallel$. No interaction at all would make $v = c$. Since $n = c/v$, we conclude that $n_\perp > n_\parallel$. The measured values for calcite are $n_\perp = 1.658$ and $n_\parallel = 1.486$ for $\lambda = 589.3$ nm. As Table 15-1 indicates, the inequality may be reversed in other materials. In materials that crystallize in the trigonal (like calcite), tetragonal, or hexagonal systems, there is one unique direction through the crystal for which the atoms are arranged symmetrically. For example, the calcite molecule of Figure 15-9a shows a threefold rotational symmetry about the optic axis. Such structures possess a single optic axis and are called *uniaxial birefringent*. Further, when $n_\parallel - n_\perp > 0$, the crystals are said to be *uniaxial positive*, and when this quantity is negative, *uniaxial negative*. Other crystalline systems, the triclinic, monoclinic, and orthorhombic, possess two

TABLE 15-1 REFRACTIVE INDICES FOR SEVERAL MATERIALS MEASURED AT SODIUM WAVELENGTH OF 589.3 nm

		$n_\parallel$	$n_\perp$	
Isotropic (cubic)	Sodium chloride		1.544	
	Diamond		2.417	
	Fluorite		1.392	
Uniaxial (trigonal, tetragonal, hexagonal)	Positive:	$n_\parallel$	$n_\perp$	
	Ice	1.313	1.309	
	Quartz (SiO$_2$)	1.5534	1.5443	
	Zircon (ZrSiO$_4$)	1.968	1.923	
	Rutile (TiO$_2$)	2.903	2.616	
	Negative:			
	Calcite (CaCO$_3$)	1.4864	1.6584	
	Tourmaline	1.638	1.669	
	Sodium Nitrate	1.3369	1.5854	
	Beryl (Be$_3$Al$_2$(SiO$_3$)$_6$)	1.590	1.598	
Biaxial (triclinic, monoclinic, orthorhombic)		n_1	n_2	n_3
	Gypsum (CaSO$_4$(2H$_2$O))	1.520	1.523	1.530
	Feldspar	1.522	1.526	1.530
	Mica	1.552	1.582	1.588
	Topaz	1.619	1.620	1.627

such directions of symmetry or optic axes and are called *biaxial crystals*.[4] Mica, which crystallizes in monoclinic forms, is a good example. Such materials then possess three distinct indices of refraction. Of course there are also cubic crystals such as salt (NaCl) or diamond (C) that are optically isotropic and possess one index of refraction. This is the case also for materials that have no large-scale crystalline structure, such as glass or fluids, and thus these materials are also optically isotropic, with a single index of refraction.

Naturally occurring calcite crystals are cleavable into rhombohedrons as a result of their crystallization into the trigonal lattice structures. The rhombohedron (Figure 15-9b) has only two corners where all three face angles (each 102°) are obtuse. These corners appear as the blunt corners of the crystal. The OA of calcite is directed through a blunt corner in such a way that it makes equal angles with the three faces there.

A birefringent crystal can be cut and polished to produce polarizing elements in which the OA may have any desired orientation relative to the incident light. Consider the cases represented in Figure 15-10. In (a), both representative direc-

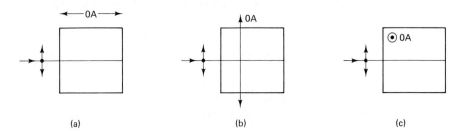

(a) (b) (c)

Figure 15-10 Light entering a birefringent plate with its optic axis in various orientations. (a) Light propagation along optic axis. (b) Light propagation perpendicular to optic axis. (c) Light propagation perpendicular to optic axis.

---

[4] For a description of such crystalline systems, see, for example, Charles Kittel, *Introduction to Solid State Physics* (New York: John Wiley & Sons, 1986).

Chap. 15 Production of Polarized Light

tions of the unpolarized light incident from the left are oriented perpendicular to the OA of the crystal. Both propagate at the same speed through the crystal with index of refraction $n_\perp$. In (b) and (c), however, the OA is parallel to one component and perpendicular to the other. In this case each component propagates through the crystal with a different index of refraction and speed. On emerging, the cumulative relative phase difference can be described in terms of the difference between optical paths for the two components. If the thickness of the crystal is d, the difference in optical paths is

$$\Delta = |n_\perp - n_\parallel| d$$

and the corresponding phase difference is

$$\Delta\varphi = 2\pi \left(\frac{\Delta}{\lambda_0}\right) = \left(\frac{2\pi}{\lambda_0}\right) |n_\perp - n_\parallel| d \qquad (15\text{-}3)$$

where λ_0 is the vacuum wavelength. If the thickness of the plate is such as to make $\Delta\varphi = \pi/2$, it is a quarter-wave plate (QWP); if $\Delta\varphi = \pi$, we have a half-wave plate (HWP), and so on. These are called *zero-order* (or sometimes *first-order*) plates. Because such plates are extremely thin, it is more practical to make thicker QWPs of higher order, giving $\Delta\varphi = (2\pi)m + \pi/2$, where $m = 1, 2, 3, \ldots$. A thicker composite of two plates may also be joined, in which one plate compensates the retardance of all but the desired $\Delta\varphi$ of the other. In this way we can fabricate optical elements that act as phase retarders. Mica and quartz are commonly used as retardation plates, usually in the form of thin, flat discs sandwiched between glass layers for added strength. Since the net phase retardation $\Delta\varphi$ is proportional to the thickness d, any device that allows a continuous change in thickness makes possible a continuously adjustable retardation plate. Such a convenient device is called a *compensator*. Figure 15-11 illustrates the working principle of a *Soleil-Babinet compensator*. Crystalline quartz is used to form a fixed lower baseplate, which is actually a wedge in optical contact with a quartz flat plate. Above is another quartz wedge, with relative motion possible along the inclined face. Notice the arrangement of the OA in this assembly. In (a) the position of the upper wedge is such that light travels through equal thicknesses of quartz with the OAs aligned perpendicular to one another. Any retardation due to one thickness is then canceled by the other, yielding zero net retardation. Sliding the upper wedge to the left increases the thickness of the first OA orientation relative to the second, yielding a continuously variable retardation up to a maximum, in position (b), of perhaps two wavelengths, or 720°. Adjustment by a micrometer screw allows small changes in $\Delta\varphi$ to be made.

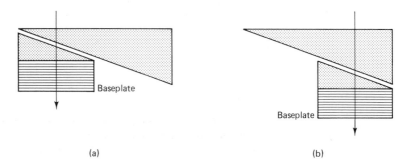

(a)

(b)

Figure 15-11 Soleil-Babinet compensator. The optic axes of the crystal are represented by dots and lines. The arrow shows the direction of light through the compensator. (a) Zero retardation. (b) Maximum retardation.

In the cases depicted in Figure 15-10b and c, the light propagating through the crystal may develop a net phase difference between **E**-components perpendicular and parallel to the crystal's OA, but the beam remains a single beam of light. If now the OA is situated so that it makes an arbitrary angle with respect to the beam direction, as in Figure 15-12, the light experiences double refraction,[5] that is, two refracted beams emerge, labeled the *ordinary* and *extraordinary* rays. The extraordinary ray is so named because it does not exhibit ordinary Snell's law behavior on refraction at the crystal surfaces. Thus if a calcite crystal is laid over a dark dot on a white piece of paper, or over an illuminated pinhole, two images are seen while looking into the top surface. If the crystal is rotated about the incident ray direction, the extraordinary image is found to rotate around the ordinary image, which remains fixed in position. Furthermore, the two beams emerge linearly polarized in orthogonal orientations, as shown. Notice that the ordinary ray is polarized perpendicular to the OA and so propagates with a refractive index of $n_0 = n_\perp = c/v_\perp$. The extraordinary ray emerges polarized in a direction perpendicular to the polarization of the ordinary ray. Inside the crystal, the extraordinary ray can be described in terms of components polarized in directions both perpendicular and parallel to the optic axis. (This situation is discussed in the following paragraph.) The perpendicular component propagates with speed $v_\perp = c/n_\perp$, as for the ordinary ray. The other component, however, propagates with a refractive index $n_e = n_\parallel = c/v_\parallel$. The net effect of the action of both components is to cause the unusual bending of the extraordinary ray shown in Figure 15-12.

Figure 15-12   Double refraction.

The situation may be clarified somewhat by reference to Figure 15-13a, which shows one Huygens' wavelet created by the extraordinary ray as it contacts the crystal surface at *P*. The incident **E**-vibration is shown resolved into components (*aa*) parallel to the OA and (*bb*) perpendicular to the OA. The parallel component propagates along the direction of $v_\parallel$, which must be perpendicular to *aa*, and the perpendicular component propagates along the direction of $v_\perp$, which must be perpendicular to *bb*. Since each component travels with a speed determined by the corresponding refractive indices, $n_\parallel$ and $n_\perp$, the speeds are unequal. For calcite, for example, $n_\perp > n_\parallel$. so that $v_\perp < v_\parallel$. The Huygens' wavelet for the extraordinary ray is not spherical as in isotropic media but ellipsoidal as shown, with major axis proportional to $v_\parallel$ and minor axis proportional to $v_\perp$. Figure 15-13b shows several such Huygens' ellipsoidal wavelets and the wavefront tangent to the wavelets. This plane wavefront, which constitutes the new surface of constant phase, is perpendicular to the propagation vector **k** for the wave. The **E** of the elliptical wavefront is intermediate between $E_\perp$ and $E_\parallel$. Notice that in this case of the extraordinary ray in an an-

---

[5] *Double refraction* is a term used to describe a manifestation of *birefringence* in materials, although it has literally the same meaning. Birefringence indicates the possession of two refractive indices, whereas double refraction refers to the splitting of a ray of light into ordinary and extraordinary parts.

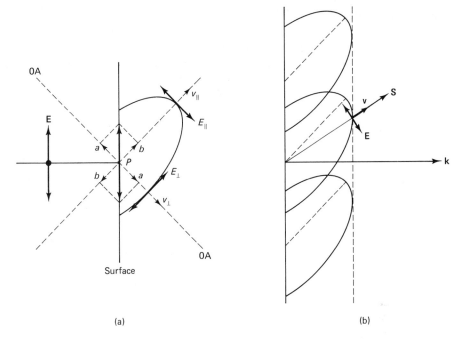

Figure 15-13 (a) Creation of an elliptical Huygens' wavelet by the extraordinary ray. The material in this case is uniaxial negative, like calcite. (b) Nonalignment of ray direction **S** and propagation vector **k** for the extraordinary ray in birefringent material.

isotropic medium, **E** is not perpendicular to **k**. Since energy propagates in the direction of the Poynting vector, $\mathbf{S} = \epsilon_0 c^2 \, \mathbf{E} \times \mathbf{B}$, and since the ray direction is the same as the direction of energy flow, the extraordinary ray with velocity **v** intermediate between $v_\perp$ and $v_\parallel$ shows the unusual refraction of Figure 15-12. The extraordinary ray is not perpendicular to the wavefront; rather, the ray direction is from the wavelet origin to the point of tangency of the elliptical wavelet with the wavefront. For the normal ray, on the other hand, due to the other **E**-component perpendicular to the OA, everything is normal; the Huygens wavelets are spheres, $\mathbf{k} \perp \mathbf{E}$, $\mathbf{k} \parallel \mathbf{S}$, and the ray is perpendicular to its wavefront.

From Figure 15-13a and the discussion above, it should be clear that the precise intermediate value of the velocity **v** of the extraordinary ray depends on the relative contributions of $v_\parallel$ and $v_\perp$, that is, on the relative orientations of the incident beam and the OA of the crystal. Thus both the velocity and index of refraction of the extraordinary ray are continuous functions of direction. On the other hand, the refractive index of the ordinary ray is a constant, independent of direction. Figure 15-14 is a plot of the refractive index versus wavelength for crystalline quartz. At any wavelength, the index for the ordinary ray is a constant, given by the lower curve, whereas the index for the extraordinary ray falls somewhere between the upper and lower curves, depending on the direction of the incident ray relative to the crystal axis.

If the two refracted rays, linearly polarized perpendicular to one another, can be physically separated, then double refraction can be used to produce a linearly polarized beam of light. There are various devices that accomplish this. One of the most commonly used is the *Glan-air prism*, shown in Figure 15-15. Two calcite prisms with apex angle θ, as shown, are combined with their long faces opposed and separated by an air space. Their optic axes are parallel, with the orientation shown. At the point of refraction out of the first prism, the angle of incidence is equal to the

Figure 15-14 Refractive indices of crystalline quartz versus wavelength at 18°C. At a given wavelength the index for the extraordinary ray may fall anywhere between the two curves, whereas the index for the ordinary ray is fixed. (Adapted from Melles Griot, *Optics Guide 3*, 1985.)

Figure 15-15 Glan-Air prism.

apex angle θ of the prisms. The critical angle for refraction into air is given as usual by $\sin \theta_c = 1/n$ and so depends on the orientation of the **E**-vibration relative to the OA. For $\mathbf{E} \parallel \text{OA}$, $n = 1.4864$ and $\theta_c = 42.3°$, while for $\mathbf{E} \perp \text{OA}$, $n = 1.6584$ and $\theta_c = 37.1°$. Thus by using prisms with apex angle intermediate between these values, the perpendicular component can be totally internally reflected while the parallel component is transmitted. The second prism serves to reorient the transmitted ray along the original beam direction. The entire device constitutes a linear polarizer. When the space between prisms is filled with some other transparent material such as glycerine, the apex angle must be modified. Several other designs for polarizing prisms constructed from positive uniaxial material (quartz) are illustrated in Figure 15-16. Notice that in these cases, the ordinary and extraordinary rays are separated without the agency of total internal reflection. In each case, the OAs of the two prisms are perpendicular to one another, so that an $E_\perp$-component in the first prism, for instance, may become an $E_\parallel$-component in the second, with corresponding change in refractive index. Different relative indices for the two components result in different angles of refraction and separation into two polarized beams. We see that birefringent materials are useful in fabricating devices that behave as linear polarizers as well as in producing phase retarders such as QWPs, considered earlier in this chapter.

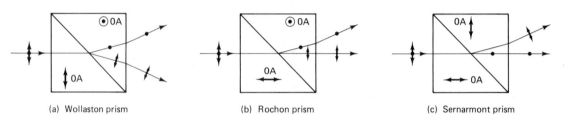

(a) Wollaston prism   (b) Rochon prism   (c) Sernarmont prism

Figure 15-16 Polarizing prisms. (a) Wollaston prism. (b) Rochon prism. (c) Sernamont prism.

Certain materials possess a property called *optical activity*. When linearly polarized light is incident on an optically active material, it emerges as linearly polarized light but with its direction of vibration rotated from the original. Viewing the beam head-on, some materials produce a clockwise rotation (*dextrorotatory*) of the **E**-field, whereas others produce a counterclockwise rotation (*levorotatory*). Optically active materials include both solids (for example, quartz and sugar) and liquids (turpentine and sugar in solution). Some materials, such as crystalline quartz, produce either rotation, traceable to the existence of two forms of the crystalline structure that turn out to be mirror images (*enantiomorphs*) of one another. Optically active materials modify the state of polarization of a beam of polarized light and can be represented mathematically by the Jones rotator matrix given in Table 14-2. Notice that the rotator mechanism involved in rotating the direction of vibration of linearly polarized light is distinct from the action of phase retarders, such as half-wave plates discussed in Sections 14-2 and 15-4, which may produce the same result.

Optical activity is easily measured using two linear polarizers originally set for extinction, that is, with their TAs crossed in perpendicular orientations (Figure 15-17). When a certain thickness of optically active material is inserted between analyzer and polarizer, the condition of extinction no longer exists because the **E**-vector of the light is rotated by the optically active medium. The exact angle of rotation can be measured by rotating the analyzer until extinction reoccurs, as shown. The rotation so measured depends on both the wavelength of the light and the thickness of the medium. The rotation (in degrees) produced by a 1-mm plate of optically active solid material is called its *specific rotation*. Table 15-2 gives the specific rotation ρ of quartz for a range of optical wavelengths. The amount of rotation caused by optically active liquids is much less by comparison. In the case of solutions, the specific rotation is defined as the rotation due to a 10-cm thickness and concentration of 1 g of active solute per cubic centimeter of solution. The net angle of rotation β due to a light path L through a solution of d grams of active solute per cubic centimeter is then

$$\beta = \rho L d \tag{15-4}$$

where L is in decimeters and d is the concentration in grams per cubic centimeter. For example, 1 dm of turpentine rotates sodium light by $-37°$. The negative sign in-

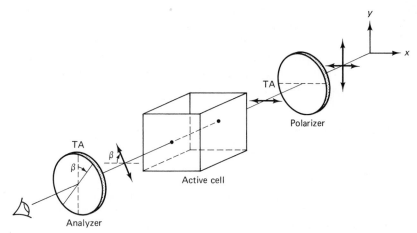

Figure 15-17 Measurement of optical activity. With the active material in place, the optical activity is measured by the angle β required to reestablish extinction.

TABLE 15-2 SPECIFIC
ROTATION OF QUARTZ

λ (nm)	ρ (degrees/mm)
226.503	201.9
404.656	48.945
435.834	41.548
546.072	25.535
589.290	21.724
670.786	16.535

dicates that turpentine is levorotatory in its optical activity. Measurement of the optical rotation of sugar solutions is often used to determine concentration, via Eq. (15-4).[6] The dependence of specific rotation on wavelength means that if one views white light through an arrangement like that of Figure 15-17, each wavelength is rotated to a slightly different degree. This separation of colors is referred to as *rotatory dispersion*.

Without giving a physical explanation of optical activity, we can, following Fresnel, offer a useful phenomenological description that enables us to relate specific rotation of an active substance to certain physical parameters. This description rests first on the fact, demonstrated in the previous chapter, that linearly polarized light can be assumed to consist of equal amounts of left- and right-circularly polarized light. Second, this description makes the assumption that the left- and right-circularly polarized components move through an optically active material with different velocities, $v_{\mathscr{L}}$ and $v_{\mathscr{R}}$, respectively. Since $v = c/n$, different refractive indices, $n_{\mathscr{L}}$ and $n_{\mathscr{R}}$, may be defined for circularly polarized light.

Consider first the case of an inactive medium for which $v_{\mathscr{L}} = v_{\mathscr{R}}$, or, equivalently, $n_{\mathscr{L}}, = n_{\mathscr{R}}$ and $k_{\mathscr{L}} = k_{\mathscr{R}}$. Here **k** is the propagation vector whose magnitude is related to wave speed by $k = \omega/v$. If the incident light is linearly polarized along the x-direction, as in Figure 15-17, it may be resolved into left- and right-circularly polarized light. Figure 15-18 makes this clear by illustrating the vector addition at three different times in an oscillation. The vector sum **E** executes oscillations along the x-axis as the $E_{\mathscr{R}}$- and $E_{\mathscr{L}}$-vectors rotate clockwise and counterclockwise, respectively, at equal rates.

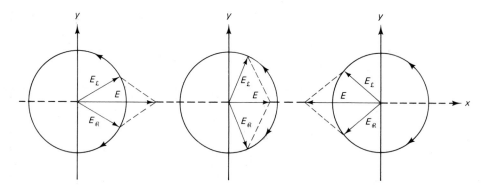

Figure 15-18 Superposition of left- and right-circularly polarized light at different instants. The light is assumed to be emerging from the page.

---

[6] Ordinary corn syrup is often used in the optics lab to demonstrate optical activity.

Next, consider the consequences of assuming $n_{\mathscr{L}} \neq n_{\mathscr{R}}$. Now the phases of the $\mathscr{L}$- and $\mathscr{R}$-components are not equal. In general, their electric fields may be expressed by

$$\tilde{\mathbf{E}}_{\mathscr{L}} = \tilde{\mathbf{E}}_{0\mathscr{L}} e^{i(k_{\mathscr{L}}z - \omega t)} \tag{15-5}$$

$$\tilde{\mathbf{E}}_{\mathscr{R}} = \tilde{\mathbf{E}}_{0\mathscr{R}} e^{i(k_{\mathscr{R}}z - \omega t)} \tag{15-6}$$

where $k_{\mathscr{L}} = (\omega/c)n_{\mathscr{L}}$ and $k_{\mathscr{R}} = (\omega/c)n_{\mathscr{R}}$. The complex amplitudes are given in vector form by

$$\tilde{\mathbf{E}}_{0\mathscr{L}} = \left(\frac{E_0}{2}\right)(1, \, i) \quad \text{and} \quad \tilde{\mathbf{E}}_{0\mathscr{R}} = \left(\frac{E_0}{2}\right)(1, \, -i) \tag{15-7}$$

corresponding to the Jones vectors for left- and right-circularly polarized modes, and the phases of the two components are

$$\theta_{\mathscr{L}} = k_{\mathscr{L}}z - \omega t$$
$$\theta_{\mathscr{R}} = k_{\mathscr{R}}z - \omega t \tag{15-8}$$

Suppose that the active medium is one for which $k_{\mathscr{L}} > k_{\mathscr{R}}$, which also means that $n_{\mathscr{L}} > n_{\mathscr{R}}$ and $v_{\mathscr{L}} < v_{\mathscr{R}}$. Then at some distance z into the medium, $\theta_{\mathscr{L}} > \theta_{\mathscr{R}}$ for all t. The situation is shown graphically at an arbitray instant in Figure 15-19a. The vector sum of $\mathbf{E}_{\mathscr{L}}$ and $\mathbf{E}_{\mathscr{R}}$ is again linearly polarized light but with an inclination angle $+\beta$ relative to the x-axis. The medium for which $n_{\mathscr{L}} > n_{\mathscr{R}}$ is therefore levorotatory. In Figure 15-19b the opposite case is also pictured, for which β is a negative angle and the optical activity is dextrorotatory. The magnitude of β can be determined by noticing that the resultant $\mathbf{E}$ that determines the angle β is always the diagonal of an equal-sided parallelogram, so that

$$\theta_{\mathscr{L}} - \beta = \theta_{\mathscr{R}} + \beta$$

or

$$\beta = \tfrac{1}{2}(\theta_{\mathscr{L}} - \theta_{\mathscr{R}}) \tag{15-9}$$

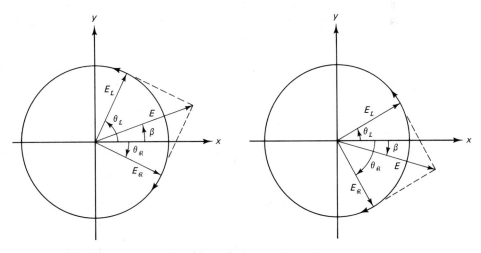

(a) Levorotatory: $n_{\mathscr{L}} > n_{\mathscr{R}}$

(b) Dextrarotatory: $n_{\mathscr{R}} > n_{\mathscr{L}}$

Figure 15-19 Optical rotation produced by left- and right-circularly polarized light having different speeds through an active medium. (a) Levorotatory: $n_{\mathscr{L}} > n_{\mathscr{R}}$. (b) Dextrorotatory: $n_{\mathscr{R}} > n_{\mathscr{L}}$.

Employing Eq. (15-8),

$$\beta = \tfrac{1}{2}(k_{\mathscr{L}} - k_{\mathscr{R}})z$$

Finally, using $k_{\mathscr{L}} = k_0 n_{\mathscr{L}}$, $k_{\mathscr{R}} = k_0 n_{\mathscr{R}}$, and $k_0 = 2\pi/\lambda_0$, where λ_0 is the wavelength in vacuum,

$$\beta = \frac{\pi z}{\lambda_0}(n_{\mathscr{L}} - n_{\mathscr{R}}) \qquad (15\text{-}10)$$

Notice that the linearly polarized light is rotated through an angle that is proportional to the thickness z of active medium, as verified experimentally. The action of the $\mathscr{L}$ and $\mathscr{R}$ modes in producing the resultant light might be visualized in the following way. At incidence the linearly polarized light is immediately resolved into $\mathscr{L}$ and $\mathscr{R}$ circular modes, which, at $z = 0$ and $t = 0$, begin together with $\theta_{\mathscr{L}} = \theta_{\mathscr{R}} = 0$. If $v_{\mathscr{R}} > v_{\mathscr{L}}$, the $\mathscr{R}$ mode reaches some point along its path before the $\mathscr{L}$ mode. Until the $\mathscr{L}$ mode arrives, $\mathbf{E}$ rotates at this point according to the circular polarization of the $\mathscr{R}$ mode acting alone. As soon as the $\mathscr{L}$ mode arrives, however, the two modes superpose to fix the direction of vibration at an angle β in a linear mode. The relative phase between the two modes at this instant determines the angle β, as expressed by Eq. (15-9). Since the frequencies of the two modes are identical, angle β remains constant thereafter.

It should be emphasized that the indices of refraction involved in optical activity characterize *circular birefringence* rather than ordinary birefringence. The indices $n_{\mathscr{R}}$ and $n_{\mathscr{L}}$ are much closer in value than $n_{\perp}$ and $n_{\parallel}$, as can be seen in the case of quartz (Table 15-3).

TABLE 15-3   REFRACTIVE INDICES FOR QUARTZ

λ (nm)	$n_{\parallel}$	$n_{\perp}$	$n_{\mathscr{R}}$	$n_{\mathscr{L}}$
396.8	1.56771	1.55815	1.55810	1.55821
762.0	1.54811	1.53917	1.53914	1.53920

Example

Determine the specific rotation produced by a 1-mm thick quartz plate at a wavelength of 396.8 nm.

Solution   From Table 15-3, at $\lambda = 396.8$ nm,

$$n_{\mathscr{L}} - n_{\mathscr{R}} = 0.00011$$

Using Eq. (15-10),

$$\beta = \frac{\pi(10^{-3})}{396.8 \times 10^{-9}}(0.00011) = 0.8709 \text{ rad} = 49.9°$$

in good agreement with Table 15-2 for the neighboring wavelength of 404.6 nm.

The above description does not explain why the velocities of the $\mathscr{L}$ and $\mathscr{R}$ circularly polarized modes should differ at all. We content ourselves for purposes of this discussion with pointing out that optically active materials possess molecules or crystalline structures that have spiral shapes, with either left- handed or right-handed screw forms. Linearly polarized light transmitted through a collection of such molecules creates forced vibrations of electrons that, in response, move not only along a spiral but necessarily around the spiral. Thus the effect of $\mathscr{L}$-circularly polarized light on a left-handed spiral would be expected to be different from its effect on a right-handed spiral and should lead to different speeds through the medium. Even if individual spiral-shaped molecules confront the light in random orientations, as in a liquid, there will be a cumulative effect that does not cancel, as long as all or most of the molecules are of the same handedness.

15-7 PHOTOELASTICITY

Consider the following experiment. Two polarizing filters acting as polarizer and analyzer are set up with a white-light source behind the pair. If the TAs of the filters are crossed, no light emerges from the pair. If some birefringent material is inserted between them, light is generally transmitted in beautiful colors. To understand this unusual effect, consider Figure 15-20, where polarizer and analyzer TAs are crossed at 45° and −45°, respectively, relative to the x-axis. Suppose that the birefringent material introduced in the light beam constitutes a half-wave plate with its fast axis (FA) vertical, as shown. Its action on the incident linearly polarized light is to convert it to linearly polarized light perpendicular to the original direction, or at −45° inclination with the x-axis. This can be understood by resolving the incident light into equal orthogonal components along the FA and SA (slow axis) and with a 180° phase difference between them. As always, the effect of the HWP on linearly polarized light is to rotate it through 2α, or, in this case, 90°. The same result follows from use of the Jones calculus:

$$\begin{bmatrix} 1 & 0 \\ 0 & -1 \end{bmatrix} \quad \begin{bmatrix} 1 \\ 1 \end{bmatrix} = \begin{bmatrix} 1 \\ -1 \end{bmatrix}$$

HWP   LP   LP
FA vertical   at 45°   at −45°

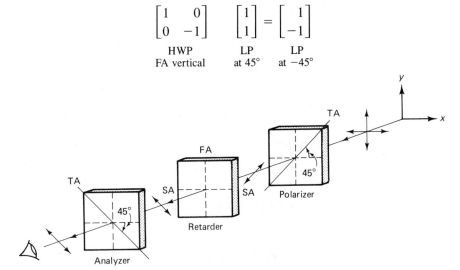

Figure 15-20 Light transmitted by cross polarizers when a birefringent material acting as a half-wave plate is placed between them.

The light from the HWP is now polarized along a direction that is fully transmitted by the analyzer. If the retardation plate introduces phase differences other than 180°, the light is rendered elliptically polarized, and some portion of the light will still be transmitted by the analyzer. Only if the phase difference is 360° or some multiple thereof, that is, if the retardation plate functions as a full-wave plate, will the character of the incident light be unmodified by the plate and the condition of extinction persist.

Now recall that the phase difference $\Delta\varphi$ introduced by a retardation plate is wavelength dependent, such that

$$\lambda_0 \, \Delta\varphi = 2\pi d (n_\perp - n_\parallel) \tag{15-11}$$

where d is the thickness of the plate. For a given plate, the right side of Eq. (15-11) is constant throughout the optical region of the spectrum, if the small variation $(n_\perp - n_\parallel)$ is neglected. It follows that the retardation is very nearly inversely proportional to the wavelength. Thus if the retardation plate acts as a HWP for red light, in the arrangement of Figure 15-20, red light will be fully transmitted, whereas shorter visible wavelengths will be only partially transmitted, giving the transmitted light a predominantly reddish hue. If the TA of the analyzer is now rotated by 90°, all components originally blocked are transmitted. Since the sum of the light transmitted under both conditions must be all the incident light, that is, white light, it follows that the colors observed under these two transmission conditions are complementary colors.

Sections of quartz or calcite and thin sheets of mica can be used to demonstrate the production of colors by polarization. Many ordinary materials also show birefringence, either under normal conditions or under stress, as in Figure 15-21. A crumpled piece of cellophane introduced between crossed polarizers shows a striking variety of colors, enhanced by the fact that light must pass through two or more thicknesses at certain points, so that $\Delta\varphi$ varies from point to point due to a change in thickness d. A similar effect is produced by wrapping glossy cellophane tape around a microscope slide, allowing for regions of overlap. Finally, $\Delta\varphi$ may also vary from point to point due to local variations in the quantity $n_\perp - n_\parallel$. Formed plastic pieces, such as a drawing triangle or safety glasses, often show such variations due to localized birefringent regions associated with strain. A pair of plastic safety goggles inserted between crossed polarizers shows a higher density of color changes in those regions under greater strain, because the difference in refractive indices changes most rapidly in such regions. The birefringence induced by mechanical stress applied to normally isotropic substances such as plastic or glass is the basis for the method of stress analysis called *photoelasticity*. It is found that in such materials,

(a)

(b)

Figure 15-21 Photoelastic stress patterns for a beam resting on two supports and (a) lightly loaded at the center, (b) heavily loaded at the center. (From M. Cagnet, M. Francon, and J. C. Thrierr, *Atlas of Optical Phenomenon*, Plate 40, Berlin: Springer-Verlag, 1962.)

an optic axis is induced in the direction of the stress, both in tension and in compression. Since the degree of birefringence induced is proportional to the strain, prototypes of mechanical parts may be fabricated from plastic and subjected to stress for analysis. Points of maximum strain are made visible by light transmitted through crossed polarizers when the stressed sample is positioned between the polarizers. Such polarized light patterns for a beam under light and heavy stress is shown in Figure 15-21.

PROBLEMS

15-1. Initially unpolarized light passes in turn through three linear polarizers with transmission axes at 0°, 30°, and 60°, respectively, relative to the horizontal. What is the irradiance of the product light, expressed as a percentage of the unpolarized light irradiance?

15-2. At what angles will light, externally and internally reflected from a diamond-air interface, be completely linearly polarized? For diamond, $n = 2.42$.

15-3. Since a sheet of Polaroid is not an ideal polarizer, not all the energy of the E-vibrations parallel to the TA are transmitted, nor are all E-vibrations perpendicular to the TA absorbed. Suppose an energy fraction α is transmitted in the first case and a fraction β is transmitted in the second.

    (a) Extend Malus' law by calculating the irradiance transmitted by a pair of such polarizers with angle θ between their TAs. Assume initially unpolarized light of irradiance I_0. Show that Malus' law follows in the ideal case.

    (b) Let $\alpha = 0.95$ and $\beta = 0.05$ for a given sheet of Polaroid. Compare the irradiance with that of an ideal polarizer when unpolarized light is passed through two such sheets having a relative angle between TAs of 0°, 30°, 45°, and 90°.

15-4. How thick should a half-wave plate of mica be in an application where laser light of 632.8 nm is being used? Appropriate refractive indices for mica are 1.599 and 1.594.

15-5. Describe what happens to unpolarized light incident on birefringent material when the OA is oriented as shown. You will want to comment on the following considerations: Single or double refracted rays? Any phase retardation? Any polarization of refracted rays?

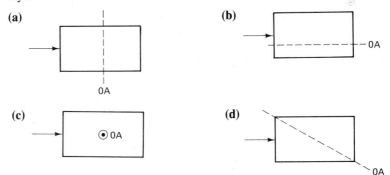

    (e) Which orientation(s) would you use to make a quarter-wave plate?

15-6. Consider a Soleil-Babinet compensator, as shown in Figure 15-11. Suppose the compensator is constructed of quartz and provides a maximum phase retardation of two full wavelengths of green mercury light (546.1 nm). Refractive indices of quartz at this wavelength are n (parallel) = 1.555 and n (perpendicular) = 1.546.

    (a) How does the total wedge thickness compare with that of the flat plate in the position of maximum retardation?

    (b) How do they compare when the emergent light is circularly polarized?

15-7. A number of dichroic polarizers are available, each of which can be assumed perfect, that is, each passes 50% of the incident unpolarized light. Let the irradiance of the incident light on the first polarizer be I_0.

(a) Using a sketch, show that if the polarizers have their transmission axes set at angle θ apart, the light transmitted by the pair is given by

$$I = \left(\frac{I_0}{2}\right) \cos^2 \theta$$

(b) What percentage of the incident light energy is transmitted by the pair when their transmission axes are set at 0° and 90°, respectively?

(c) Five additional polarizers of this type are placed between the two described above, with their transmission axes set at 15°, 30°, 45°, 60°, and 75°, in that order, with the 15° angle polarizer adjacent to the 0° polarizer, and so on. Now what percentage of the incident light energy is transmitted?

15-8. What minimum thickness should a piece of quartz have to act as a quarter-wave plate for a wavelength of 5893 Å in vaccum?

15-9. Determine the angle of deviation between the two emerging beams of a Wollaston prism constructed of calcite and with wedge angle of 45°. Assume sodium light.

15-10. A beam of linearly polarized light is changed into circularly polarized light by passing it through a slice of crystal 0.003 cm thick. Calculate the difference in the refractive indices for the two rays in the crystal, assuming this to be the minimum thickness showing the effect for a wavelength of 600 nm. Sketch the arrangement, showing the OA of the crystal, and explain why it occurs.

15-11. Light is incident on a water surface at such an angle that the reflected light is completely linearly polarized.

(a) What is the angle of incidence?

(b) The light refracted into the water is intercepted by the top flat surface of a block of glass with index of 1.50. The light reflected from the glass is completely linearly polarized. What is the angle between the glass and water surfaces? Sketch the arrangement, showing the polarization of the light at each stage.

15-12. In each of the following cases, deduce the nature of the light that is consistent with the analysis performed. Assume a 100% efficient polarizer.

(a) When a polarizer is rotated in the path of the light, there is no intensity variation. With a QWP in front of (coming first) the rotating polarizer, one finds a variation in intensity but no angular position of the polarizer that gives zero intensity.

(b) When a polarizer is rotated in the path of the light, there is some intensity variation but no position of the polarizer giving zero intensity. The polarizer is set to give maximum intensity. A QWP is allowed to intercept the beam first with its OA parallel to the TA of the polarizer. Rotation of the polarizer now can produce zero intensity.

15-13. Light from a source immersed in oil of refractive index 1.62 is incident on the plane face of a diamond ($n = 2.42$), also immersed in the oil. Determine (a) the angle of incidence at which maximum polarization occurs and (b) the angle of refraction into the diamond.

15-14. The rotation of polarized light in an optically active medium is found to be approximately proportional to the inverse square of the wavelength.

(a) The specific rotation of glucose is 20.5°. A glucose solution of unknown concentration is contained in a 12-cm-long tube and is found to rotate linearly polarized light by 1.23°. What is the concentration of the solution?

(b) Upon passing through a 1-mm-thick quartz plate, red light is rotated about 15°. What rotation would you expect for violet light?

15-15. (a) What thickenss of quartz is required to give an optical rotation of 10° for light of 396.8 nm?

(b) What is the specific rotation of quartz for this wavelength? The refractive indices for quartz at this wavelength, for left- and right-circularly polarized light, are $n_L = 1.55821$ and $n_R = 1.55810$, respectively.

15-16. (a) A thin plate of calcite is cut with its OA parallel to the plane of the plate. What minimum thickness is required to produce a quarter-wave path difference for sodium light of 589 nm?

(b) What color will be transmitted by a zircon plate, 0.0182 mm thick, when placed in a 45° orientation between crossed polarizers?

15-17. (a) Show that polarizing angles for internal and external reflection between the same two media must be complementary.

(b) Show that if Brewster's angle is satisfied for a TM light beam entering a parallel plate (a *Brewster window*), it will also be satisfied for the beam as it leaves the plate on the opposite side.

15-18. The indices of refraction for the fast and slow axes of quartz with 546 nm light are 1.5462 and 1.5553, respectively.

(a) By what fraction of a wavelength is the e-ray retarded, relative to the o-ray, for every wavelength of travel in the quartz?

(b) What is the thickness of a zeroth-order QWP?

(c) If a multiple-order quartz plate 0.735 mm thick functions as a QWP, what is its order?

(d) Two quartz plates are optically contacted so that they produce opposing retardations. Sketch the orientation of the OA of the two plates. What should their difference in thickness be such that they function together like a zeroth-order QWP?

15-19. When a plastic triangle is viewed between crossed polarizers and with monochromatic light of 500 nm, a series of alternating transmission and extinction bands is observed. How much does $(n_\perp - n_\parallel)$ vary between transmission bands to satisfy successive conditions for HWP retardation? The triangle is $\frac{1}{16}$ in. thick.

15-20. A plane plate of beryl is cut with the optic axis in the plane of the surfaces. Plane polarized light is incident on the plate such that the **E**-field vibrations are at 45° to the optic axis. Determine the smallest thickness of the plate such that the emergent light is (a) plane and (b) circularly polarized.

15-21. Find the angle at which a half-wave plate must be set to compensate for the rotation of a 1.15-mm levorotatory quartz plate using 546-nm wavelength light.

15-22. In Chapter 20, the *Fresnel equations* are derived, showing that the percent irradiance R reflected from a dielectric plane surface for the TE polarization mode is given by Eq. (20-23):

$$R = \left(\frac{\cos\theta - \sqrt{n^2 - \sin^2\theta}}{\cos\theta + \sqrt{n^2 - \sin^2\theta}} \right)^2$$

where θ is the angle of incidence and n is the ratio n_2/n_1.

(a) Calculate the percent reflectance for the TE mode when the light is incident from air onto glass of $n = 1.50$ at the polarizing angle.

(b) The reflectance calculated above is also valid for an internal reflection as light leaves the glass into air. This being the case, calculate the net percentage of the TE mode transmitted through a stack of 10 such plates relative to the incident irradiance I_0. Assume that the plates do not absorb light and that there are no multiple reflections within the plates.

(c) Calculate the *degree of polarization P* of the transmitted beam, given by

$$P = \frac{I_{TM} - I_{TE}}{I_{TM} + I_{TE}}$$

where I stands for the irradiance of either polarization mode.

15-23. A half-wave plate is placed between crossed polarizer and analyzer such that the angle between the polarizer TA and the FA of the HWP is θ. How does the emergent light vary as a function of θ?

15-24. (a) Determine the rotation produced by the optical activity of a 3-mm quartz plate on a linearly polarized beam of light at wavelength 762 nm.
(b) What is the rotation due to optical activity by a half-wave plate of quartz using the same light beam?

REFERENCES

[1] Shurcliff, W. A. *Polarized Light: Production and Use*. Cambridge, Mass.: Harvard University Press, 1962.

[2] Shurcliff, W. A., and S. S. Ballard. *Polarized Light*. Princeton, N.J.: D. Van Nostrand Company, 1964.

[3] Bennett, Jean M., and Harold E. Bennett. "Polarization." In *Handbook of Optics*, edited by Walter G. Driscoll and William Vaughan. New York: McGraw-Hill Book Company, 1978.

[4] Meyer-Arendt, Jurgen R. *Introduction to Classical and Modern Optics*, 3d ed. Englewood Cliffs, N.J.: Prentice-Hall, 1989. Ch. 4.2.

[5] Kliger, David S. *Polarized Light in Optics and Spectroscopy*. Boston: Academic Press, 1990.

[6] Azzam, R. M. A., and N. M. Bashara. *Ellipsometry and Polarized Light*. New York: North-Holland Publishing Company, 1977.

[7] Waterman, Talbot H. "Polarized Light and Animal Navigation." *Scientific American* (July 1955): 88.

[8] Wehner, Rudinger. "Polarized Light Navigation by Insects." *Scientific American* (July 1976): 106.

[9] Feynman, Richard P., Robert B. Leighton, and Matthew Sands. *The Feynman Lectures on Physics*, vol. 1. Reading, Mass.: Addison-Wesley Publishing Company, 1963. Ch. 32, 33.

[10] Weisskopf, Richard F. "How Light Interacts with Matter." In *Lasers and Light*. San Francisco: W. H. Freeman and Company Publishers, 1969.

16

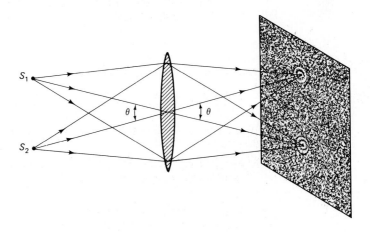

Fraunhofer Diffraction

INTRODUCTION

The wave character of light has been invoked to explain a number of phenomena, classified as "interference effects" in preceding chapters. In each case, two or more individual coherent beams of light, originating from a single source and separated by amplitude or wavefront division, were brought together again to interfere. Fundamentally, the same effect is involved in the *diffraction* of light. In its simplest description, diffraction is any deviation from geometrical optics that results from the obstruction of a wavefront of light. For example, an opaque screen with a round hole represents such an obstruction. On a viewing screen placed beyond the hole, the circle of light may show complex edge effects. This type of obstruction is typical in many optical instruments that utilize only the portion of a wavefront passing through a round lens. Any obstruction, however, shows detailed structure in its own shadow that is quite unexpected on the basis of geometrical optics.

Diffraction effects are a consequence of the wave character of light. Even if the obstacle is not opaque but causes local variations in the amplitude or phase of the wavefront of the transmitted light, such effects are observed. Tiny bubbles or imperfections in a glass lens, for example, produce undesirable diffraction patterns when transmitting laser light. Because the edges of optical images are blurred by diffraction, the phenomenon leads to a fundamental limitation in instrument resolution. More often, though, the sharpness of optical images is more seriously degraded by optical aberrations due to the imaging components themselves. *Diffraction-limited* optics is good optics indeed.

The double slit studied previously constitutes an obstruction to a wavefront in which light is blocked everywhere except at the two apertures. Recall that the irradiance of the resulting fringe pattern was calculated by treating the two openings as point sources, or long slits whose widths could be treated as points. A more complete analysis of this experiment must take into account the finite size of the slits. When this is done, the problem is treated as a diffraction problem. The results show that the interference pattern determined earlier is modified in a way that accounts for the actual details of the observed fringes.

Adequate agreement with experimental observations is possible through an application of the *Huygens-Fresnel principle*. According to Huygens, every point of a given wavefront of light can be considered a source of secondary spherical wavelets. To this, Fresnel added the assumption that the actual field at any point beyond the wavefront is a superposition of all these wavelets, taking into account both their amplitudes and phases. Thus in calculating the diffraction pattern of the double slit at some point on a screen, one considers every point of the wavefront emerging from each slit as a source of wavelets whose superposition produces the resultant field. This procedure then takes into account a continuous array of sources across both slits, rather than two isolated point sources, as in the interference calculation. Diffraction is often distinguished from interference on this basis: In diffraction phenomena, the interfering beams originate from a continuous distribution of sources; in interference phenomena the interfering beams originate from a discrete number of sources. This is not, however, a fundamental *physical* distinction.

A further classification of diffraction effects arises from the mathematical approximations possible when calculating the resultant fields. If both the source of light and observation screen are *effectively* far enough from the diffraction aperture so that wavefronts arriving at the aperture and observation screen may be considered plane, we speak of *Fraunhofer*, or *far-field*, *diffraction*, the type treated in this chapter; when this is not the case and the curvature of the wavefront must be taken into account, we speak of *Fresnel*, or *near-field*, *diffraction*, the subject of Chapter 18. In the far-field approximation, the diffraction pattern changes uniformly in size only as the viewing screen is moved relative to the aperture. In the near-field approximation, the situation is more complicated: Both shape and size of the diffraction pattern depend on the distance between aperture and screen. As the screen is moved away from the aperture, the image of the aperture passes through the forms predicted in turn by goemetrical optics, near-field diffraction, and far-field diffraction.

It should be stated at the outset that the Huygens-Fresnel principle we shall employ to calculate diffraction patterns is itself an approximation. When no light penetrates an opaque screen, it means that the interaction of the incident radiation with the electronic oscillators, set into motion within the screen, is such as to produce zero net field beyond the screen. This balance is not maintained at the edge of an aperture in the screen, where the distribution of oscillators is interrupted. The Huygens-Fresnel principle does not include the contribution to the diffraction field of the electronic oscillators in the screen material at the edge of the aperture. Such edge effects are important, however, only when the observation point is very near the aperture itself.

16-1 DIFFRACTION FROM A SINGLE SLIT

We first calculate the Fraunhofer diffraction pattern from a single slit, a rectangular aperture characterized by a length much larger than its width. For Fraunhofer diffraction, the source must be far enough away so that the wavefronts of light reaching the slit are essentially plane. Of course, this is easily accomplished in practice by placing the source in the focal plane of a positive lens. Similarly, we consider the observation screen to be effectively at infinity by using another lens on the other side of the slit, as shown in Figure 16-1. Then the light reaching any point such as P on the screen is due to parallel rays of light from different portions of the wavefront at the slit (dashed line). According to the Huygens-Fresnel principle, we consider spherical wavelets emanating from each point of the wavefront as it reaches the plane of the slit and calculate the resultant field at P by adding the waves according to the principle of superposition. As shown in Figure 16-1, the waves do not arrive at P in phase. A ray from the center of the slit, for example, has an optical path length that is an amount Δ shorter than one leaving from a point a vertical distance s above the optical axis.

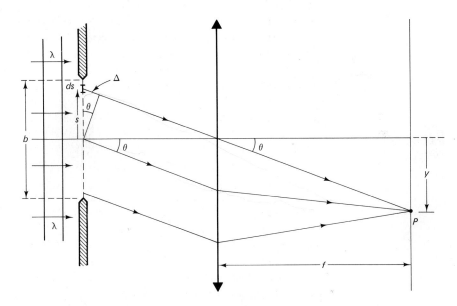

Figure 16-1 Construction for determining irradiance on a screen due to Fraunhofer diffraction by a single-slit.

The plane portion of the wavefront at the slit opening represents a continuous array of Huygens' wavelet sources. We consider each interval of dimension ds as a source and calculate the result of all such sources by integrating over the entire slit width b. Each interval ds contributes spherical wavelets at P of the form

$$dE_p = \left(\frac{dE_0}{r}\right)e^{i(kr-\omega t)} \tag{16-1}$$

where r is the optical path length from the interval ds to the point P. The amplitude dE_0 is divided by r because the spherical waves decrease in irradiance with distance, in accordance with the inverse square law, that is, $E^2 \propto 1/r^2$ and $E \propto 1/r$. The amplitude at unit distance from the source point is then dE_0. Let us set $r = r_0$ for the wave from the interval ds at $s = 0$. Then for any other wave originating at the interval ds at height s, taking the difference in phase into account, the differential field at P is

$$dE_p = \left(\frac{dE_0}{r}\right)e^{i[k(r_0+\Delta)-\omega t]} \tag{16-2}$$

In the amplitude, $dE_0/(r_0 + \Delta)$, the path difference Δ is unimportant, since $\Delta \ll r_0$, and therefore Δ can be neglected there. The phase, on the other hand, is very sensitive to small differences. For intervals ds below the axis, s is negative and the path difference is $(r_0 - \Delta)$, corresponding to shorter optical paths to P. The amplitude of the radiation from each interval clearly depends on the size of ds, so that when all such contributions are added by integration, we have the total effect at P. Accordingly, we write

$$dE_0 = E_L \, ds \tag{16-3}$$

where E_L is the amplitude per unit width of slit at unit distance away. For a point P at angle θ below the axis, relative to the lens center, the figure shows that $\Delta = s \sin \theta$. With these modifications, the differential contribution to the field at P from an arbitrary interval ds is

$$dE_P = \left(\frac{E_L \, ds}{r_0} \right) e^{i(kr_0 + ks \sin \theta - \omega t)}$$

Integrating over the width of the slit, we have

$$E_P = \left(\frac{E_L}{r_0} \int_{-b/2}^{b/2} e^{iks \sin \theta} \, ds \right) e^{i(kr_0 - \omega t)} \tag{16-4}$$

Since we are ultimately concerned with the irradiance, the square of the amplitude which we shall call E_R, we retain only the portion in parentheses and integrate:

$$E_R = \frac{E_L}{r_0} \left(\frac{e^{iks \sin \theta}}{ik \sin \theta} \right)_{-b/2}^{b/2} \tag{16-5}$$

Inserting the limits of integration into Eq. (16-5),

$$E_R = \frac{E_L}{r_0} \frac{1}{ik \sin \theta} [e^{(ikb \sin \theta)/2} - e^{-(ikb \sin \theta)/2}] \tag{16-6}$$

The phases of the exponential terms suggest we make a convenient substitution,

$$\beta \equiv \tfrac{1}{2} kb \sin \theta \tag{16-7}$$

Then

$$E_R = \frac{E_L}{r_0} \frac{b}{2i\beta} (e^{i\beta} - e^{-i\beta}) = \frac{E_L}{r_0} \frac{b}{2i\beta} (2i \sin \beta) \tag{16-8}$$

where we have applied Euler's equation to the exponential terms. Simplifying, we get

$$E_R = \frac{E_L b}{r_0} \frac{\sin \beta}{\beta} \tag{16-9}$$

The amplitude of the resultant field at P, given by Eq. (16-9), includes the sinc function $(\sin \beta)/\beta$, where β varies with θ and thus with the observation point P on the screen. We may give physical significance to β by interpreting it as a phase difference. Since a phase difference is given in general by $k\Delta$, Eq. (16-7) indicates a path difference associated with β of $\Delta = (b/2) \sin \theta$, shown in Figure 16-1. Thus β represents the phase difference between waves from the center and either endpoint of the slit, where $s = b/2$.

The irradiance at P is proportional to the square of the resultant amplitude there, or

$$I = \left(\frac{\epsilon_0 c}{2} \right) E_R^2 = \frac{\epsilon_0 c}{2} \left(\frac{E_L b}{r_0} \right)^2 \frac{\sin^2 \beta}{\beta^2}$$

or

$$I = I_0\left(\frac{\sin^2 \beta}{\beta^2}\right) \equiv I_0 \, \text{sinc}^2(\beta) \qquad (16\text{-}10)$$

where I_0 includes all constant factors. Equations (16-9) and (16-10) now permit us to plot the variation of irradiance with vertical distance from the axis at the screen. The sinc function has the property that it approaches 1 as its argument approaches 0:

$$\lim_{\beta \to 0} \text{sinc} \, (\beta) = \lim_{\beta \to 0}\left(\frac{\sin \beta}{\beta}\right) = 1 \qquad (16\text{-}11)$$

Otherwise, its zeros occur when $\sin \beta = 0$, that is, when

$$\beta = \tfrac{1}{2}(kb \sin \theta) = m\pi, \qquad \text{with } m = \pm1, \pm2, \dots$$

Equation 16-11 shows that the value $m = 0$ is excluded from this condition. The irradiance is plotted as a function of β in Figure 16-2. Setting $k = 2\pi/\lambda$, the condition for zeros of the sinc function (and so of the irradiance) is

$$m\lambda = b \sin \theta \qquad (16\text{-}12)$$

On the screen, therefore, the irradiance is a maximum at $\theta = 0$ or $y = 0$ and drops to zero at values y such that

$$y \cong \frac{m\lambda f}{b} \qquad (16\text{-}13)$$

The approximation in Eq. (16-13) comes from setting $\sin \theta \cong y/f$, usually well justified, since θ is a small angle. The irradiance pattern is symmetrical about $y = 0$.

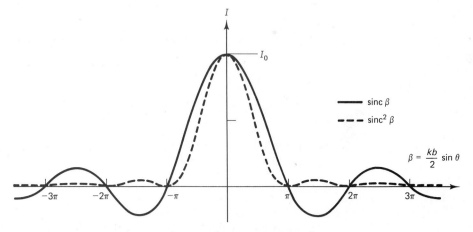

Figure 16-2 Sinc function (solid line) plotted as a function of β. The irradiance function (dashed line) for single-slit Fraunhofer diffraction is just the square of sinc β, normalized to I_0 at the center of the pattern.

The secondary maxima of the single-slit diffraction pattern do not quite fall at the midpoints between zeros, even though this condition is more nearly approached as β increases. The maxima coincide with maxima of the sinc function, points satisfying

$$\frac{d}{d\beta}\left(\frac{\sin \beta}{\beta}\right) = \frac{\beta \cos \beta - \sin \beta}{\beta^2} = 0$$

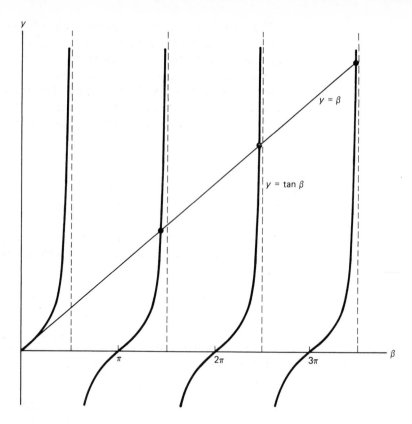

Figure 16-3 Intersections of the curves $y = \beta$ and $y = \tan \beta$ determine the angles β at which the sinc function is a maximum.

or $\beta = \tan \beta$. An angle equals its tangent at intersections of the curves $y = \beta$ and $y = \tan \beta$, both plotted in Figure 16-3. Intersections, excluding $\beta = 0$, occur at 1.43π (rather than 1.5π), 2.46π (rather than 2.5π), 3.47π (rather than 3.5π), and so on, as can be verified with a hand calculator. The plot clearly shows that intersection points aproach the vertical lines defining midpoints more closely as β increases. Thus, in the irradiance plot of Figure 16-2, secondary maxima are skewed slightly away from the midpoints toward the central peak. Most of the energy of the diffraction pattern falls under the central maximum, which is much larger than the adjoining maximum on either side.

Example

What is the ratio of irradiances at central peak maximum to the first of the secondary maxima?

Solution The ratio to be calculated is

$$\frac{I_{\beta=0}}{I_{\beta=1.43\pi}} = \frac{(\sin^2 \beta / \beta^2)_{\beta=0}}{(\sin^2 \beta / \beta^2)_{\beta=1.43\pi}} = \frac{1}{(\sin^2 \beta / \beta^2)_{\beta=1.43\pi}}$$

$$= \left(\frac{\beta^2}{\sin^2 \beta}\right)_{1.43\pi} = \frac{20.18}{0.952} = 21.2$$

Thus the maximum irradiance of the nearest secondary peak is only 4.7% that of the central peak.

The central maximum represents essentially the image of the slit on a distant screen. We observe that the edges of the image are not sharp but reveal a series of maxima and minima that tail off into the shadow surroudning the image. These effects are typical of the blurring of images due to diffraction and will be seen again in other cases of diffraction to be considered. The angular width of the central maximum is defined as the angle $\Delta\theta$ between the first minima on either side. Using Eq. (16-12) with $m = \pm 1$ and approximating $\sin\theta$ by θ, we get

$$\Delta\theta = \frac{2\lambda}{b} \qquad (16\text{-}14)$$

From Eq. (16-14) it follows that the central maximum will spread as the slit width is narrowed. Since the length of the slit is very large compared to its width, the diffraction pattern due to points of the wave front along the length of the slit has a very small angular width and is not prominent on the screen. Of course, the dimensions of the diffraction pattern also depend on the wavelength, as indicated in Eq. (16-14).

16-2 BEAM SPREADING

According to Eq. (16-14), the angular spread of the central maximum in the far field is independent of distance between aperture and screen. The linear dimensions of the diffraction pattern thus increase with distance L, as shown in Figure 16-4, such that the width W of the central maximum is given by

$$W = L\,\Delta\theta = \frac{2L\lambda}{b} \qquad (16\text{-}15)$$

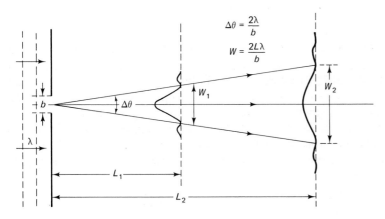

Figure 16-4 Spread of the central maximum in the far-field diffraction pattern of a single-slit.

We may describe the content of Eq. (16-15) as a linear spread of a beam of light, originally constricted to a width b. Indeed, the means by which the beam is originally narrowed is not relevant to the nature of the diffraction pattern that occurs. If one dispenses with the slit in Figure 16-4 and merely assumes an original beam of width b, all our results follow in the same way. After collimation, a "parallel" beam of light spreads just as if it emerged from a single opening.

Example

Imagine a parallel beam of 546-nm light of width $b = 0.5$ mm propagating across the laboratory, a distance of 10 m. Determine the final width of the beam due to diffraction spreading.

Solution Using Eq. (16-15),

$$W = \frac{2L\lambda}{b} = \frac{2(10)(546 \times 10^{-9})}{0.5 \times 10^{-3}} = 21.8 \text{ mm}$$

Thus even highly collimated laser beams are subject to beam spreading as they propagate, due to diffraction. It is a fundamental consequence of the wave nature of light that perfectly parallel beams of light cannot exist.

The beam spreading described by Eq. (16-14) is valid for rectangular apertures. For circular apertures, we show in the next section that a factor of 1.22 must accompany the wavelength. Furthermore, one must keep in mind that this treatment assumes a plane wavefront of uniform irradiance.[1] The spreading described by Eq. (16-15) has been deduced on the basis of Fraunhofer, or far-field, diffraction, which means here that L must remain reasonably large. If L is taken small enough, for example, the equation predicts a beam width less than b, contrary to assumption. Evidently L must be larger than some minumum value, L_{min}, which gives a beam width $W = b$, that is,

$$L_{min} = \frac{b^2}{2\lambda}$$

We may conclude that we are in the far field when

$$L \gg \frac{b^2}{\lambda}$$

A more general approach leads to the commonly stated criterion for far-field diffraction in the form

$$L \gg \frac{\text{area of aperture}}{\lambda} \tag{16-16}$$

16-3 RECTANGULAR AND CIRCULAR APERTURES

We have been describing diffraction from a slit having a width b much smaller than its length, as illustrated in Figure 16-5a. When both dimensions of the slit are comparable and small, each produces appreciable spreading, as illustrated in Figure 16-5b. For the aperture dimension a, we write, analogously, for the irradiance, as in Eq. (16-10),

$$I = I_0 \left(\frac{\sin \alpha}{\alpha} \right)^2 \qquad \text{where} \qquad \alpha \equiv \left(\frac{k}{2} \right) a \sin \theta \tag{16-17}$$

The two-dimensional pattern now gives zero irradiance for points x, y satisfied by either

$$y = \frac{m\lambda f}{b} \qquad \text{or} \qquad x = \frac{n\lambda f}{a}$$

---

[1] A laser beam usually does not have constant irradiance across its diameter. In its fundamental mode, the profile is a Gaussian function. Although its spread formula is essentially that of Eq. (16-14) with the beam diameter replacing b, the constant factor 1.22 is replaced by 1.27. Laser-beam spreading is treated in Chapter 21.

(a)

(b)

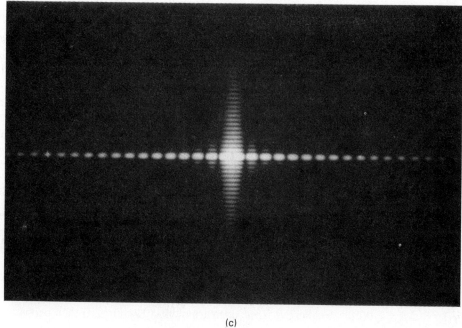

(c)

Figure 16-5 (a) Single-slit diffraction. Only the small dimension b of a narrow slit causes appreciable spreading of the light along the x-direction on the screen. (b) Single-slit diffraction. Both dimensions of the rectangular aperture are small and a two-dimensional diffraction pattern is discernible on the screen. (c) Diffraction image of a single slit, as in the representation of (a). (d) Diffraction image of a single square aperture, as in the representation of (b). (Both photos are from M. Cagnet, M. Francon, and J. C. Thrierr, *Atlas of Optical Phenomenon*, Plate 17, Berlin: Springer-Verlag, 1962.)

(d)

Figure 16-5 (continued)

where both m and n represent nonzero integral values. The irradiance over the screen turns out to be just a product of the irradiance functions in each dimension, or

$$I = I_0(\text{sinc}^2 \beta)(\text{sinc}^2 \alpha) \qquad (16\text{-}18)$$

In calculating this result, the single integration over one dimension of the slit is replaced by a double integration over both dimensions of the aperture. Photographs of single aperture diffraction patterns for rectangular and square apertures are shown in Figure 16-5c and d.

When the aperture is circular, the integration is over the entire area of the aperture since both vertical and horizontal dimensions of the aperture are comparable. Thus, in place of Eq. (16-3), we redefine $dE_0 \equiv E_A \, dA$, where E_A now represents the amplitude per unit area. The amplitude of the field at P (as in Figure 16-1) is then given by

$$E_P = \frac{E_A}{r_0} \iint_{\text{Area}} e^{isk \sin\theta} \, dA$$

We take as elemental area a rectangular strip of area $dA = x \, ds$, shown in Figure 16-6a. Using the equation of a circle, we calculate the length x at height s to be given by

$$x = 2\sqrt{R^2 - s^2}$$

where R is the aperture radius. The integral then becomes

$$E_P = \frac{2E_A}{r_0} \int_{-R}^{R} e^{isk \sin\theta} \sqrt{R^2 - s^2} \, ds$$

This integral takes the form of a standard definite integral upon making the substitutions, $v = s/R$ and $\gamma = kR \sin \theta$:

$$E_P = \frac{2E_A R^2}{r_0} \int_{-1}^{+1} e^{i\gamma v} \sqrt{1 - v^2} \, dv$$

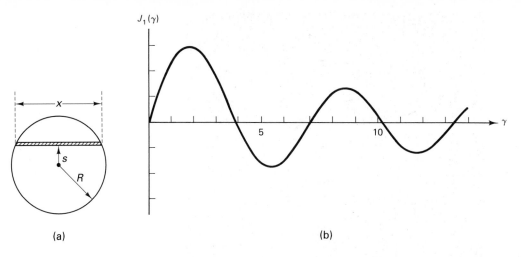

(a) (b)

Figure 16-6 (a) Geometry used in the integration over a circular aperture. (b) The Bessel function $J_1(\gamma)$. The first zero of the function occurs at $\gamma = 3.832$.

The integral has the value

$$\int_{-1}^{+1} e^{i\gamma v} \sqrt{1 - v^2}\, dv = \frac{\pi J_1(\gamma)}{\gamma}$$

where $J_1(\gamma)$ is the first-order *Bessel function of the first kind*, expressible by the infinite series,

$$J_1(\gamma) = \frac{\gamma}{2} - \frac{(\gamma/2)^3}{1^2 \cdot 2} + \frac{(\gamma/2)^5}{1^2 \cdot 2^2 \cdot 3} - \cdots$$

As can be verified from this series expansion, the ratio $J_1(\gamma)/\gamma$ has the limit $\frac{1}{2}$ as $\gamma \to 0$. Thus the circular aperture requires, instead of the sine function for the single slit, the Bessel function J_1, which oscillates somewhat like the sine function, as shown in the plot of Figure 16-6b. One difference is that the amplitude of the oscillation of the Bessel function decreases as its argument departs from zero. The irradiance for a circular aperture of diameter D can now be written as

$$I = I_0 \left(\frac{2J_1(\gamma)}{\gamma} \right)^2, \qquad \text{where } \gamma \equiv \tfrac{1}{2} kD \sin\theta \qquad (16\text{-}19)$$

and I_0 is the irradiance at $\gamma \to 0$, or at $\theta = 0$.

These equations should be compared with those of Eq. (16-17) to appreciate the analogous role played by the Bessel function. Like $(\sin x)/x$, the function $J_1(x)/x$ approaches a maximum as x approaches zero, so that the irradiance is greatest at the center of the pattern ($\theta = 0$). The pattern is symmetrical about the optical axis through the center of the circular aperture and has its first zero when $\gamma = 3.832$, as shown in Figure 16-6b. Thus the central maximum of the irradiance falls to zero when

$$\gamma = \left(\frac{k}{2} \right) D \sin\theta = 3.832 \qquad \text{or when} \qquad D \sin\theta = 1.22\lambda \qquad (16\text{-}20)$$

Equation (16-20) should be compared with the analogous equation for the narrow rectangular slit, $m\lambda = b \sin\theta$. We see that $m = 1$ for the first minumum in the slit pattern is replaced by the number 1.22 in the case of the circular aperture. Successive minima are determined in a similar way from other zeros of the Bessel function. The irradiance pattern of Eq. (16-19) is plotted in Figure 16-7a. The pattern is sim-

(a)

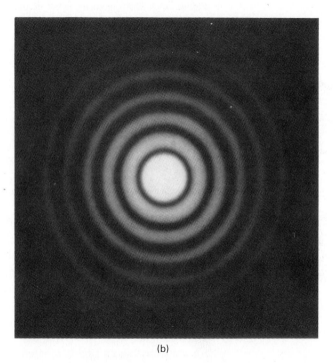

(b)

Figure 16-7 (a) Irradiance pattern of a circular aperture. By far the largest amount of light energy is diffracted into the central maximum. (b) Diffraction image of a circular aperture. The circle of light at the center corresponds to the zeroth order of diffraction and is known as the *Airy disc*. (From M. Cagnet, M. Francon, and J. C. Thrierr, *Atlas of Optical Phenomenon*, Plate 16, Berlin: Springer-Verlag, 1962.)

ilar to that of Figure 16-2 for a slit, except that the pattern for circular apertures now has rotational symmetry about the optical axis. A photograph is shown in Figure 16-7b. The central maximum is consequently a circle of light, the diffracted "image" of the circular aperture, and is called the *Airy disc*. Note that the far-field angular radius of the Airy disc, according to Eq. (16-20), is very nearly

$$\Delta\theta = \frac{1.22\lambda}{D} \qquad (16\text{-}21)$$

16-4 RESOLUTION

In forming the Fraunhofer diffraction pattern of a single slit, as in Figure 16-1, we notice that the distance between slit and lens is not crucial to the details of the pattern. The lens merely intercepts a larger solid angle of light when the distance is small. If this distance is allowed to go to zero, aperture and lens coincide, as in the objective of a telescope. Thus the image formed by a telescope with a round objective is subject to the diffraction effects described by Eq. (16-19) for a circular aperture. The sharpness of the image of a distant point object—a star, for example—is then limited by diffraction. The image occupies essentially the region of the Airy disc. An eyepiece viewing the primary image and providing further magnification merely enlarges the details of the diffraction pattern formed by the lens. The limit of resolution is already set in the primary image. The inevitable blur that diffraction produces in the image restricts the resolution of the instrument, that is, its ability to provide distinct images for distinct object points, either physically close together (as in a microscope), or separated by a small angle at the lens (as in a telescope). Figure 16-8a illustrates the diffraction of two point objects formed by a single lens. The point objects and the centers of their Airy discs are both separated by the angle θ. If the angle is large enough, two distinct images will be clearly seen, as shown in the photograph of Figure 16-8b. Imagine now that the objects S_1 and S_2 are brought closer together. When their image patterns begin to overlap substantially, it becomes more difficult to discern the patterns as distinct, that is, to resolve them as belonging to distinct object points. A photograph of the two images at the limit of resolution is shown in Figure 16-8c. *Rayleigh's criterion* for just-resolvable images—a somewhat arbitrary but useful criterion—requires that the centers of the image patterns be no nearer than the angular radius of the Airy disc, as in Figure 16-9. In this condition, the maximum of one pattern falls directly over the first minimum of the other. Thus for *limit of resolution*, we have, as in Eq. (16-21),

$$(\Delta\theta)_{min} = \frac{1.22\lambda}{D} \tag{16-22}$$

where D is now the diameter of the lens. In accordance with this result, the minimum resolvable angular separation of two object points may be reduced (the resolution improved) by increasing the lens diameter and decreasing the wavelength. We consider several applications of Eq. (16-22), beginning with the following example.

Example

Suppose that each lens on a pair of binoculars has a diameter of 35 mm. How far apart must two stars be before they are theoretically resolvable by either of the lenses in the binoculars?

Solution According to Eq. (16-22),

$$(\Delta\theta)_{min} = \frac{1.22(550 \times 10^{-9})}{35 \times 10^{-3}} = 1.92 \times 10^{-5} \text{ rad}$$

or about 4″ of arc, using an average wavelength for visible light. If the stars are near the center of our galaxy, a distance d of around 30,000 light years, then their actual separation s is approximately

$$s = d\,\Delta\theta_{min} = (30,000)(1.92 \times 10^{-5}) = 0.58 \text{ light years}$$

To get some appreciation for this distance, consider that the planet Pluto at the edge of our solar system is only about 5.5 light *hours* distant. If the stars are

(a)

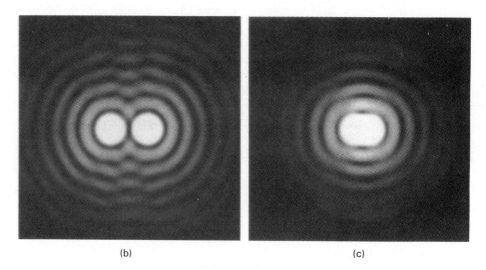

(b)   (c)

Figure 16-8 (a) Diffraction-limited images of two point objects formed by a lens. As long as the Airy discs are well separated, the images are well resolved. (b) Separated images of two incoherent point sources. In this diffraction pattern, the two images are well resolved. (c) Image of a pair of incoherent point sources at the limit of resolution. (Photos from M. Cagnet, M. Francon, and J. C. Thrierr, *Atlas of Optical Phenomenon*, Plate 16, Berlin: Springer-Verlag, 1962.)

being detected by their long-wavelength radio waves—the lenses being replaced by dish antennas—the resolution must, by Eq. (16-22), be much less.

If the lens is the objective of a microscope, as indicated in Figure 16-10, the problem of resolving nearby objects is basically the same. Making only rough estimates, we shall ignore the fact that the wavefronts striking the lens from nearby object points A and B are not plane, as required in far-field diffraction equations. The minimum separation $x_{\min}$ of two just-resolved objects near the focal plane of the lens is then given by

$$x_{\min} = f\theta_{\min} = f\left(\frac{1.22\lambda}{D}\right)$$

The ratio D/f is the *numerical aperture*, with a typical value of 1.2 for a good oil-immersion objective. Thus

$$x_{\min} \cong \lambda$$

Figure 16-9 Rayleigh's criterion for just-resolvable diffraction patterns. The dashed curve is the observed sum of independent diffraction peaks.

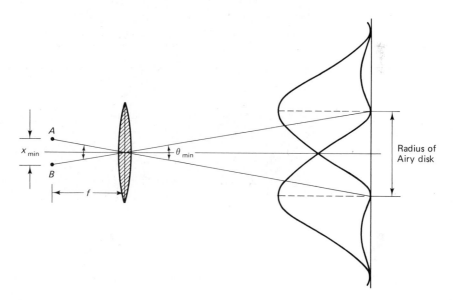

Figure 16-10 Minimum angular resolution of a microscope.

The resolution of a microscope is roughly equal to the wavelength of light used, a fact that explains the advantage of ultraviolet, X-ray, and electron microscopes in high-resolution applications.

The limits of resolution due to diffraction also affect the human eye, which may be approximated by a circular aperture (pupil), a lens, and a screen (retina), as in Figure 16-11. Night vision, which takes place with large, adapted pupils of around 8 mm, is capable of higher resolution than daylight vision. Unfortunately there is not enough light to take advantage of the situation! On a bright day the pupil diameter may be 2 mm. Under these conditions Eq. (16-22) gives $(\Delta\theta)_{min} = 33.6 \times 10^{-5}$ rad, for an average wavelength of 550 nm. Experimentally, one finds that a separation of 1 mm at a distance of about 2 m is just barely resolvable, giving $(\Delta\theta)_{min} = 50 \times 10^{-5}$, about 1.5 times the theoretical limit. One's own resolution (*visual acuity*) can easily be tested by viewing two lines drawn 1 mm apart at increasing distances until they can no longer be seen as distinct. It is interesting to

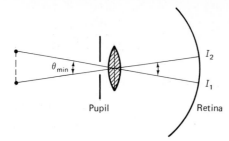

Figure 16-11 Diffraction by the eye with pupil as aperture limits the resolution of objects subtending angle $\theta_{\min}$.

note that the theoretical resolution just determined for a 2-mm-diameter pupil is consistent with the value of 1′ of arc (29×10^{-5} rad) used by Snellen to characterize normal visual acuity (see Chapter 7).

16-5 DOUBLE-SLIT DIFFRACTION

The diffraction pattern of a plane wavefront that is obstructed everywhere except at two narrow slits is calculated in the same manner as for the single slit. The mathematical argument departs from that for the single slit with Eq. (16-4), where the limits of integration are now changed to those indicated in Figure 16-12. Extracting the amplitude alone, we get

$$E_R = \frac{E_L}{r_0} \int_{-(1/2)(a+b)}^{-(1/2)(a-b)} e^{isk\sin\theta}\, ds + \frac{E_L}{r_0} \int_{(1/2)(a-b)}^{(1/2)(a+b)} e^{isk\sin\theta}\, ds \qquad (16\text{-}23)$$

Integration and substitution of the limits leads to

$$E_R = \frac{E_L}{r_0} \frac{1}{ik\sin\theta} [e^{(1/2)ik(-a+b)\sin\theta} - e^{(1/2)ik(-a-b)\sin\theta} + e^{(1/2)ik(a+b)\sin\theta} - e^{(1/2)ik(a-b)\sin\theta}]$$

Reintroducing the substitution of Eq. (16-7), involving the slit width b,

$$\beta \equiv \tfrac{1}{2}kb\sin\theta \qquad (16\text{-}24)$$

and a similar one involving the slit separation a,

$$\alpha \equiv \tfrac{1}{2}ka\sin\theta \qquad (16\text{-}25)$$

our equation is written more compactly as

$$E_R = \frac{E_L}{r_0} \frac{b}{2i\beta} [e^{i\alpha}(e^{i\beta} - e^{-i\beta}) + e^{-i\alpha}(e^{i\beta} - e^{-i\beta})]$$

Employing Euler's equation,

$$E_R = \frac{E_L}{r_0} \frac{b}{2i\beta} (2i\sin\beta)(2\cos\alpha)$$

Figure 16-12 Specification of slit width and separation for double-slit diffraction.

Finally,

$$E_R = \frac{2E_L b}{r_0} \frac{\sin \beta}{\beta} \cos \alpha \qquad (16\text{-}26)$$

The irradiance is now

$$I = \left(\frac{\epsilon_0 c}{2}\right) E_R^2 = \left(\frac{\epsilon_0 c}{2}\right)\left(\frac{2E_L b}{r_0}\right)^2 \left(\frac{\sin \beta}{\beta}\right)^2 \cos^2 \alpha$$

or

$$I = 4I_0 \left(\frac{\sin \beta}{\beta}\right)^2 \cos^2 \alpha \qquad (16\text{-}27)$$

where

$$I_0 = \left(\frac{\epsilon_0 c}{2}\right)\left(\frac{E_L b}{r_0}\right)^2$$

as defined in Eq. (16-10) for the single slit. Since the maximum value of Eq. (16-27) is $4I_0$, we see that the double slit provides four times the maximum irradiance in the pattern center as compared with the single slit. This is exactly what should be expected where the beams are in phase and amplitudes add.

On closer inspection of Eq. (16-27) we find that the irradiance is just a product of the irradiances found for double-slit interference and single-slit diffraction. The factor $[(\sin \beta)/\beta]^2$ is that of Eq. (16-10) for single-slit diffraction. The $\cos^2 \alpha$ factor, when α is written out as in Eq. (16-25), is

$$\cos^2 \alpha = \cos^2 \left[\frac{ka (\sin \theta)}{2}\right] = \cos^2 \left[\frac{\pi a (\sin \theta)}{\lambda}\right]$$

equivalent to the factor in Eq. (10-19) for Young's double slit. The sine and cosine factors of Eq. (16-27) are plotted in Figure 16-13a for the case $a = 6b$ or $\alpha = 6\beta$. Because $a > b$, the $\cos^2 \alpha$ factor varies more rapidly than the $(\sin^2 \beta)/\beta^2$ factor. The product of the sine and cosine factors may be considered a modulation of the interference fringe pattern by a single-slit diffraction envelope, as shown in Figure 16-13b. The diffraction envelope has a minimum when $\beta = m\pi$, with $m = \pm 1$, $\pm 2, \ldots$, as shown. In terms of the spatial angle θ, this condition is

$$\text{diffraction minima: } m\lambda = b \sin \theta \qquad (16\text{-}28)$$

as in Eq. (16-12). When these minima happen to coincide with interference fringe maxima, the fringe is missing from the pattern. Interference maxima occur for $\alpha = p\pi$, with $p = 0, \pm 1, \pm 2, \ldots$, or when

$$\text{interference maxima: } p\lambda = a \sin \theta \qquad (16\text{-}29)$$

When the conditions expressed by Eqs. (16-28) and (16-29) are satisfied at the same point in the pattern (same θ), dividing one equation by the other gives the condition for missing orders.

$$\text{condition for missing orders: } a = \left(\frac{p}{m}\right) b \qquad (16\text{-}30)$$

or

$$\alpha = \left(\frac{p}{m}\right) \beta$$

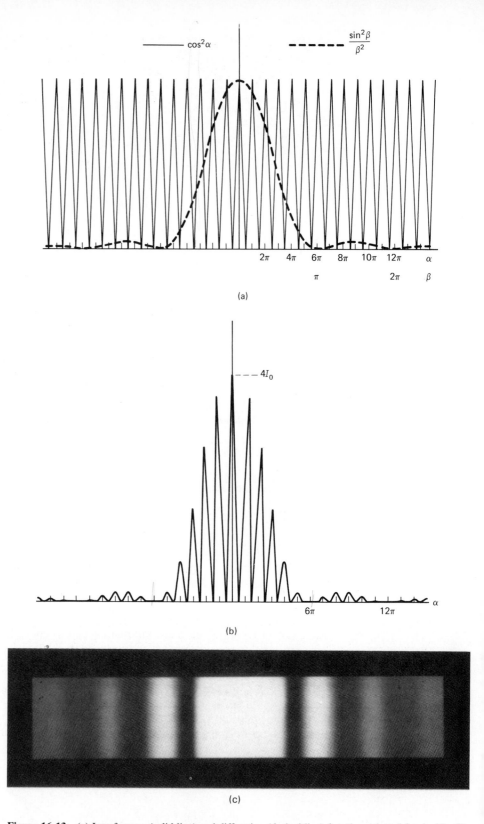

Figure 16-13 (a) Interference (solid line) and diffraction (dashed line) functions plotted for double-slit Fraunhofer diffraction when the slit separation is six times the slit width ($a = 6b$). (b) Irradiance for the double slit of (a). The curve represents the product of the interference and diffraction factors. (c) Diffraction pattern due to a single slit. (d) Diffraction pattern due to a double-slit aperture, with each slit like the one that produced (c). (Both photos are from M. Cagnet, M. Francon, and J. C. Thrierr, *Atlas of Optical Phenomenon*, Plate 18, Berlin: Springer-Verlag, 1962.)

(d)

Figure 16-13 (continued)

Thus when the slit separation is an integral multiple of the slit width, the condition is met exactly. For example, when $a = 2b$, then $p = 2m = \pm2, \pm4, \pm6, \ldots$ gives the missing orders of interference. For the case plotted in Figure 16-13a and b, $a = 6b$, and the missing orders are those for which $p = \pm6, \pm12$, and so on. Figure 16-13c and d contains photographs of a single-slit pattern and a double-slit pattern with the same slit width. (What is the ratio of a/b in this case?) Evidently, when $a = Nb$ and N is large, the first missing order at $p = \pm N$ is far from the center of the pattern. To produce a simple Young's interference pattern for two slits, one accordingly makes $a \gg b$ so that N is large. A large number of fringes then fall under the central maximum of the diffraction envelope. As a trivial but satisfying case, observe that when $a = b$, Eq. (16-30) requires that *all* orders (except $p = 0$) are missing. These dimensions cannot be satisfied, however, unless the two slits have merged into one and are unable to produce interference fringes. When $a = b$, the resulting pattern is, of course, a single-slit diffraction pattern.

16-6 DIFFRACTION FROM MANY SLITS

For an aperture of multiple slits (a *grating*), the integrals of Eq. (16-23), together with Figure 16-12, are extended by integrating over N slits. The individual slits are identified by the index j in the following expression for the resultant amplitude:

$$E_R = \frac{E_L}{r_0} \sum_{j=1}^{N/2} \left\{ \int_{[-(2j-1)a-b]/2}^{[-(2j-1)a+b]/2} e^{isk\sin\theta} \, ds + \int_{[(2j-1)a-b]/2}^{[(2j-1)a+b]/2} e^{isk\sin\theta} \, ds \right\} \quad (16\text{-}31)$$

As j increases, pairs of slits symmetrically placed below (first integral) and above (second integral) the origin are included in the integration. When $j = 1$, for example, Eq. (16-31) reduces to the double slit case, Eq. (16-23). When $j = 2$, the next two slits are included, whose edges are located at $\frac{1}{2}(-3a - b)$ and $\frac{1}{2}(-3a + b)$ below the origin and $\frac{1}{2}(3a - b)$ and $\frac{1}{2}(3a + b)$ above the origin.[2] When $j = N/2$, all slits are accounted for.

Let us first concentrate on the integrals contained within the square brackets, which we shall refer to as I, temporarily. After integration and substitution of limits, we get

$$I = \frac{1}{ik\sin\theta}\{e^{-ik\sin\theta[(2j-1)a-b]/2} - e^{-ik\sin\theta[(2j-1)a+b]/2}\} + \frac{1}{ik\sin\theta}\{e^{ik\sin\theta[(2j-1)a+b]/2} - e^{ik\sin\theta[(2j-1)a-b]/2}\}$$

---

[2] This expression is adapted to N even. For N large, one need not be concerned about the parity of N. For N small, however, N odd can be handled by taking the origin at the center of the central slit. This approach is left to the problems.

Using Eqs. (16-24) ad (16-25) again for α and β,

$$I = \frac{b}{2i\beta}[e^{-i(2j-1)\alpha}(e^{i\beta} - e^{-i\beta}) + e^{i(2j-1)\alpha}(e^{i\beta} - e^{-i\beta})]$$

With the help of Euler's equation, this can be written as

$$I = \frac{b}{2i\beta}(2i \sin \beta)\{2 \cos[(2j - 1)\alpha]\}$$

or

$$I = 2b\frac{\sin \beta}{\beta} \mathrm{Re}[e^{i(2j-1)\alpha}]$$

where we have expressed the cosine as the real part of the corresponding exponential. Returning to Eq. (16-31), we need next the sum S:

$$S = 2b\frac{\sin \beta}{\beta} \mathrm{Re} \sum_{j=1}^{N/2} e^{i(2j-1)\alpha}$$

Expanding the sum, this is

$$S = 2b\frac{\sin \beta}{\beta} \mathrm{Re}[e^{i\alpha} + e^{i3\alpha} + e^{i5\alpha} + \cdots + e^{i(N-1)\alpha}]$$

The series in brackets is a geometric series whose first term a and ratio r can be used to find its sum, given by

$$a\left(\frac{r^n - 1}{r - 1}\right) = e^{i\alpha}\left[\frac{(e^{2i\alpha})^{N/2} - 1}{e^{2i\alpha} - 1}\right] = \frac{e^{iN\alpha} - 1}{e^{i\alpha} - e^{-i\alpha}}$$

Using Euler's equation, this can be recast into the form

$$\frac{(\cos N\alpha - 1) + i \sin N\alpha}{2i \sin \alpha} = \frac{i(\cos N\alpha - 1) - \sin N\alpha}{-2 \sin \alpha}$$

whose real part is $(\sin N\alpha)/(2 \sin \alpha)$. Then

$$S = b\frac{\sin \beta}{\beta} \frac{\sin N\alpha}{\sin \alpha}$$

and

$$E_R = \frac{E_L b}{r_0} \frac{\sin \beta}{\beta} \frac{\sin N\alpha}{\sin \alpha}$$

Taking the square of this resultant amplitude gives the irradiance,

$$I = I_0\left(\frac{\sin \beta}{\beta}\right)^2\left(\frac{\sin N\alpha}{\sin \alpha}\right)^2 \tag{16-32}$$

where I_0 includes all the constants.

Although derived here for an even number N of slits, the result expressed by Eq. (16-32) is valid also for N odd (see problem 16-19). When $N = 1$ and $N = 2$, Eq. (16-32) reduces to the results obtained previously for single- and double-slit diffraction, respectively. By now we are familiar with the factor in β representing the diffraction envelope of the resultant irradiance. Let us examine the factor $(\sin N\alpha/\sin \alpha)^2$, which evidently describes interference between slits. When $\alpha = 0$ or some multiple of π, the expression reduces to an indeterminate form. We can show, in fact, that for such values, the expression is a maximum. Employing L'Hospital's rule for any $m = 0, \pm1, \pm2, \ldots ,$

$$\lim_{\alpha \to m\pi} \frac{\sin N\alpha}{\sin \alpha} = \lim_{\alpha \to m\pi} \frac{N \cos N\alpha}{\cos \alpha} = \pm N$$

Figure 16-14 (a) Interference (solid line) and diffraction (dashed line) functions plotted for multiple-slit Fraunhofer diffraction when $N = 8$ and $a = 3b$. (b) Irradiance function for the multiple slit of (a). The irradiance is limited by the diffraction envelope (dashed line).

so that the *principal maxima* of the interference pattern are proportional to N^2 in magnitude. The interference function for $N = 8$ and positive α is plotted together with the diffraction envelope in Figure 16-14a. The related irradiance is also plotted in Figure 16-14b. Notice that the principal maxima are separated by $N - 2$ small, *secondary maxima*. Furthermore, as α increases, $N\alpha$ equals an integral number of π more often than α, so that the numerator of the interference factor vanishes in instances when the denominator does not. Such instances result in $N - 1$ minima between principal maxima. The situation is precisely described by the following equations:

$$\text{with } \alpha = \frac{p\pi}{N} \text{ or } a\sin\theta = \frac{p\lambda}{N}, \qquad p = 0, \pm 1, \pm 2, \ldots$$

$$\text{principal maxima occur for } p = 0, \pm N, \pm 2N, \ldots \qquad (16\text{-}33)$$

$$\text{minima occur for } p = \text{all other values}$$

The practical device that makes use of multiple-slit diffraction is the *diffraction grating*. For large N, its principal maxima are bright, distinct, and spatially well separated. According to Eq. (16-33), with $p/N = m = 0, \pm 1, \pm 2, \ldots$ for these maxima, the *diffraction-grating equation* is

$$m\lambda = a\sin\theta \qquad (16\text{-}34)$$

where m is called the *order* of the diffraction.

Some insight into Eq. (16-34) is gained by examining Figure 16-15, which shows representative slits of a grating illuminated by plane wavefronts of monochromatic light. Wavelets emerging from each slit arrive in phase at angular deviation θ from the axis if every path difference like $AB (= a\sin\theta)$ equals an integral number m of wavelengths. When $AB = m\lambda$, the grating Eq. (16-34) follows immediately. When all waves arrive in phase, the resulting phasor diagram is formed by adding N phasors all in the same "direction," giving a maximum resultant. At such points the principal maxima of Figure 16-14a are produced. Secondary maxima result because a uniform phase difference between waves from adjoining slits causes the phase diagram to curl up with a smaller resultant. At each of the minima, the phasor diagram forms a closed figure, so that cancellation is complete. The phase difference between waves from adjoining slits and in the direction of θ can be found from Figure 16-14a by recalling that the angle α represents half the phase difference between successive

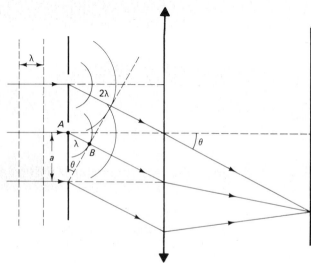

Figure 16-15 Representative grating slits illuminated by collimated monochromatic light. Formation of the first-order diffraction maximum is shown.

slits. Thus the first principal maximum from the center, at $\alpha = \pi$, occurs when the phase difference between successive waves is precisely 2π.

Photographs of diffraction fringes produced by 2, 3, 4, and 5 slits are shown in Figure 16-16. An examination of the four photographs shows that the principal maxima become narrower and secondary maxima begin to appear as the number of slits increases. For example, notice the $N - 2 = 3$ secondary maxima between the prin-

(a)

(b)

(c)

(d)

Figure 16-16 Diffraction fringes produced in turn by two, three, four, and five slits. (From M. Cagnet, M. Francon, and J. C. Thrierr, *Atlas of Optical Phenomenon*, Plate 19, Berlin: Springer-Verlag, 1962.)

cipal maxima for the case $N = 5$. The diffraction grating—for N very large—is discussed further in some detail in the next chapter.

PROBLEMS

16-1. A collimated beam of mercury green light at 546.1 nm is normally incident on a slit 0.015 cm wide. A lens of focal length 60 cm is placed behind the slit. A diffraction pattern is formed on a screen placed in the focal plane of the lens. Determine the distance between (a) the central maximum and first minimum and (b) the first and second minima.

16-2. Call the irradiance at the center of the central Fraunhofer diffraction maximum of a single slit I_0 and the irradiance at some other point in the pattern I. Obtain the ratio I/I_0 for a point on the screen that is $\frac{3}{4}$ wavelength farther from one edge of the slit than the other.

16-3. The width of a rectangular slit is measured in the laboratory by means of its diffraction pattern at a distance of 2 m from the slit. When illuminated normally with a parallel beam of laser light (632.8 nm), the distance between third minima on either side of the principal maximum is measured. An average of several tries gives 5.625 cm.
(a) Assuming Fraunhofer diffraction, what is the slit width?
(b) Is the assumption of far-field diffraction justified in this case? What is the ratio L/L_{min}?

16-4. In viewing the far-field diffraction pattern of a single slit illuminated by a discrete-spectrum source with the help of absorption filters, one finds that the fifth minimum of one wavelength component coincides exactly with the fourth minimum of the pattern due to a wavelength of 620 nm. What is the other wavelength?

16-5. Calculate the rectangular slit width that will produce a central maximum in its far-field diffraction pattern having an angular breadth of 30°, 45°, 90°, and 180°. Assume a wavelength of 550 nm.

16-6. Consider the far-field diffraction pattern of a single slit of width 2.125 μm, when illuminated normally by a collimated beam of 550 nm light. Determine (a) the angular radius of its central peak and (b) the ratio I/I_0 at points making an angle of $\theta = 5°, 10°, 15°$, and 22.5° with the axis.

16-7. The Lick Observatory has one of the largest refracting telescopes, with an aperture diameter of 36 in. and a focal length of 56 ft. Determine the radii of the first and second bright rings surrounding the Airy disc in the diffraction pattern formed by a star on the focal plane of the objective. The first two secondary maxima of the function $[J_1(\gamma)/\gamma]^2$ occur at $\gamma = 5.14$ and $\gamma = 8.42$.

16-8. A telescope objective is 12 cm in diameter and has a focal length of 150 cm. Light of mean wavelength 550 nm from a distant star enters the scope as a nearly collimated beam. Compute the radius of the central disk of light forming the image of the star on the focal plane of the lens.

8.4×10^{-4} cm

16-9. Suppose that a CO_2 gas laser emits a diffraction-limited beam at wavelength 10.6 μm, power 2 kW, and diameter 1 mm. Assume that, by multimoding, the laser beam has essentially uniform irradiance over its cross section. Approximately how large a spot would be produced on the surface of the moon, a distance of 376,000 km away from such a device, neglecting any scattering by the earth's atmosphere? What will be the irradiance at the lunar surface?

16-10. Assume a 2-mm diameter laser beam (632.8 nm) is diffraction limited and has a constant irradiance over its cross section. On the basis of spreading due to diffraction alone, how far must it travel to double its diameter?

16-11. Two headlights on an automobile are 45 in. apart. How far away will the lights appear to be just resolvable to a person whose nocturnal pupils are just 5 mm in diameter? Assume an average wavelength of 550 nm.

16-12. Assume the range of pupil variation during adaptation of a normal eye is from 2 to 7 mm. What is the corresponding range of distances over which it can detect the separation of objects 1 in. apart?

16-13. A double-slit diffraction pattern is formed using mercury green light at 546.1 nm. Each slit has a width of 0.100 mm. The pattern reveals that the fourth-order interference maxima are missing from the pattern.
  (a) What is the slit separation?
  (b) What is the irradiance of the first three orders of interference fringes, relative to the zeroth-order maximum?

16-14. (a) Show that the number of bright fringes seen under the central diffraction peak in a Fraunhofer double-slit pattern is given by $2(a/b) - 1$, where a/b is the ratio of slit separation to slit width.
  (b) If 13 bright fringes are seen in the central diffraction peak when the slit width is 0.30 mm, determine the slit separation.

16-15. (a) Show that in a double-slit Fraunhofer diffraction pattern, the ratio of widths of the central diffraction peak to the central interference fringe is $2(a/b)$, where a/b is the ratio of slit separation to slit width. Notice the result is independent of wavelength.
  (b) Determine the peak-to-fringe ratio in particular when $a = 10b$.

16-16. Calculate by integration the irradiance of the diffraction pattern produced by a three-slit aperture, where the slit separation a is three times the slit width b. Make a careful sketch of I versus $\sin \theta$ and describe properties of the pattern. Also show that your results are consistent with the general result for N slits, given by Eq. (16-32).

16-17. Make a rough sketch for the irradiance pattern from seven equally spaced slits having a separation-to-width ratio of 4. Label points on the x-axis with corresponding values of α and β.

16-18. A 10-slit aperture, with slit spacing five times the slit width of 1×10^{-4} cm, is used to produce a Fraunhofer diffraction pattern with light of 435.8 nm. Determine the irradiance of the principal interference maxima of order 1, 2, 3, 4, and 5, relative to the central fringe of zeroth order.

16-19. Show that one can arrive at Eq. (16-32) by taking the origin of coordinates at the midpoint of the central slit in an array where N is odd.

16-20. A rectangular aperture of dimensions 0.100 mm along the x-axis and 0.200 mm along the y-axis is illuminated by coherent light of wavelength 546 nm. A 1-m focal length lens intercepts the light diffracted by the aperture and projects the diffraction pattern on a screen in its focal plane.
  (a) What is the distribution of irradiance on the screen near the pattern center, as a function of x and y (in mm) and I_0, the irradiance at the pattern center?
  (b) How far from the pattern center are the first minima along the x and y directions?
  (c) What fraction of the I_0 irradiance occurs at 1 mm from pattern center along the x and y directions?
  (d) What is the irradiance at the point $(x = 2, y = 3)$ mm?

16-21. What is the angular half width (from central maximum to first minimum) of a diffracted beam for a slit width of (a) λ; (b) 5λ; (c) 10λ?

16-22. A property of the Bessel function $J_1(x)$ is that, for large x, a closed form exists, given by

$$J_1(x) = \frac{\sin x - \cos x}{\sqrt{\pi x}}$$

Find the angular separation of diffraction minima far from the axis of a circular aperture.

16-23. We have shown that the secondary maxima in a single slit diffraction pattern do not fall exactly half-way between minima, but are quite close. Assuming they are halfway:

(a) Show that the irradiance of the mth secondary peak is given approximately by

$$I_m \cong I_0 \frac{1}{[(m + \frac{1}{2})\pi]^2}$$

(b) Calculate the percent error involved in this approximation for the first three secondary maxima.

16-24. Three antennas broadcast in phase at a wavelength of 1 km. The antennas are separated by a distance of $\frac{2}{3}$ km and each antenna radiates equally in all horizontal directions. Because of interference, a broadcast "beam" is limited by interference minima. How many well-defined beams are broadcast and what are their angular half-widths?

16-25. A collimated light beam is incident normally on three very narrow, identical slits. At the center of the pattern projected on a screen, the irradiance is I_{max}.
  (a) If the irradiance I_P at some point P on the screen is zero, what is the phase difference between light arriving at P from neighboring slits?
  (b) If the phase difference between light waves arriving at P from neighboring slits is π, determine the ratio I_P/I_{max}.
  (c) What is I_P/I_{max} at the first principal maximum?
  (d) If the average irradiance on the entire screen is I_{av}, what is the ratio I_P/I_{av} at the central maximum?

16-26. Draw phasor diagrams illustrating the principal maxima and zero irradiance points for a four-slit aperture.

REFERENCES

[1] Ball, C. J. *An Introduction to the Theory of Diffraction*. New York: Pergamon Press, 1971.

[2] Born, Max, and Emil Wolf. *Principles of Optics*, 5th ed. New York: Pergamon Press, 1975. Ch. 8.

[3] Longhurst, R. S. *Geometrical and Physical Optics*, 2d ed. New York: John Wiley and Sons, 1967. Ch. 10, 11.

[4] Goodman, Joseph W. *Introduction to Fourier Optics*. New York: McGraw-Hill Book Company, 1968. Ch. 3.

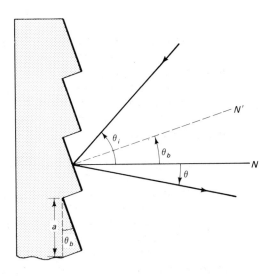

17

The Diffraction Grating

INTRODUCTION

The discussion of the preceding chapter continues here in the formal treatment of the diffraction due to a large number of slits or apertures. The diffraction grating equation is first generalized to handle light beams incident on the grating at an arbitrary angle. Performance parameters of practical interest are then developed in discussions of the *spectral range*, *dispersion*, *resolution*, and *blaze* of a grating. A brief discussion of interference gratings and several conventional types of grating spectrographs follows.

17-1 THE GRATING EQUATION

A periodic, multiple-slit device designed to take advantage of the sensitivity of its diffraction pattern to the wavelength of the incident light is called a *diffraction grating*. The grating equation developed in Chapter 16 may be generalized for the case when the incident plane wavefronts of light make an angle with the plane of the grating, as in Figure 17-1. The net path difference for waves from successive slits is then

$$\Delta = \Delta_1 + \Delta_2 = a \sin \theta_i + a \sin \theta_m \qquad (17\text{-}1)$$

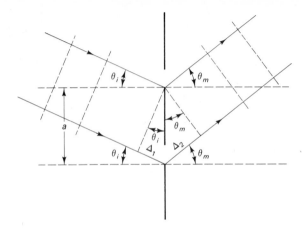

Figure 17-1 Neighboring grating slits illuminated by light incident at angle θ_i with the grating normal. For light diffracted in the direction θ_m, the net path difference from the two slits is $\Delta_1 + \Delta_2$.

The two sine terms in the path difference may add or subtract, depending on the direction θ_m of the diffracted light. To make Eq. (17-1) correct for all angles of diffraction, we need to adopt a sign convention for the angles. When the incident and diffracted rays are on the same side of the normal, as they are in Figure 17-1, θ_m is considered positive. When the diffracted rays are on the side of the normal opposite to that of the incident rays, θ_m is considered negative. In the latter case, the net path difference for waves from successive slits is the difference $\Delta_1 - \Delta_2$, as evident in a modified sketch such as Figure 17-1. In either case, when $\Delta = m\lambda$, all diffracted waves are in phase and the grating equation becomes

$$a(\sin \theta_i + \sin \theta_m) = m\lambda, \qquad m = 0, \pm 1, \pm 2, \ldots \qquad (17\text{-}2)$$

When it is not necessary to distinguish angles, the subscript on the angle of diffraction, θ_m, is often dropped. For each value of m, monochromatic radiation of wavelength λ is enhanced by the diffractive properties of the grating. By Eq. (17-2), the zeroth order of interference, $m = 0$, occurs at $\theta_m = -\theta_i$, the direction of the incident light, for all λ. Thus light of all wavelengths appears in the central or zeroth-order peak of the diffraction pattern. Higher orders—both plus and minus—produce spectral *lines* appearing on either side of the zeroth order. For a fixed direction of incidence given by θ_i, the direction θ_m of each principal maximum varies with wavelength. For orders $m \neq 0$, therefore, the grating separates different wavelengths of light present in the incident beam, a feature that explains its usefulness in wavelength measurement and spectral analysis. As a dispersing element, the grating is superior to a prism in several ways. Figure 17-2a illustrates the formation of the spectral orders of diffraction for monochromatic light. Figure 17-2b shows the angular spread of the continuous spectrum of visible light for a particular grating. Note that second and third orders in this case partially overlap. Before wavelengths of spectral lines appearing in a region of overlap can be assigned, the actual order of the line must first be ascertained so that the appropriate value of m can be used in Eq. (17-2). Unlike the prism, a grating produces greater deviation from the zeroth-order point for longer wavelengths. Thus, when the spectrum is not a simple one, the overlap ambiguity is often resolved experimentally by using a filter that removes, say, the shorter wavelengths from the incident light. In this way, the spectral range of the incident light is limited by filtering until overlap is removed and each line can be correctly identified. At other times it may be advisable to limit the wavelength range accepted by the grating by first using an instrument of lower dispersion.

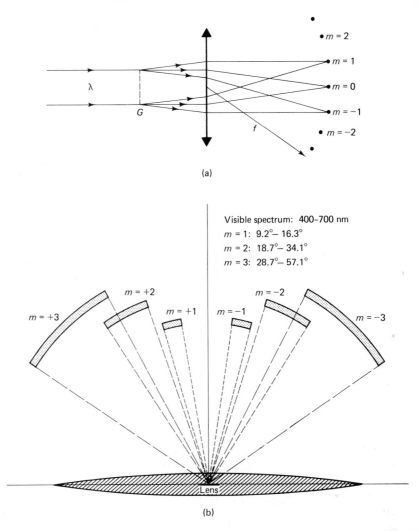

Visible spectrum: 400–700 nm
$m = 1$: 9.2°– 16.3°
$m = 2$: 18.7°– 34.1°
$m = 3$: 28.7°– 57.1°

(a)

(b)

Figure 17-2 (a) Formation of the orders of principal maxima for monochromatic light incident normally on grating G. The grating can replace the prism in a spectroscope. Focused images have the shape of the collimator slit (not shown). (b) Angular spread of the first three orders of the visible spectrum for a diffraction grating with 400 grooves/mm. Orders are shown at different distances from the lens for clarity. In each order, the red end of the spectrum is deviated most. Normal incidence is assumed.

17-2 FREE SPECTRAL RANGE OF A GRATING

The nonoverlapping wavelength range in a particular order is called the *free spectral range*, F. Overlapping occurs because in the grating equation, the product $a \sin \theta$ may be equal to several possible combinations of $m\lambda$ for the light actually incident and processed by the optical system. Thus at the position corresponding to λ in the first order, we may also find a spectral line corresponding to $\lambda/2$ in the second order, $\lambda/3$ in the third order, and so on. The free spectral range in order m may be determined by the following argument. If λ_1 is the shortest detectable wavelength in the incident light, then the longest nonoverlapping wavelength λ_2 in order m is coincident with the beginning of the spectrum again in the next higher order, or

$$m\lambda_2 = (m + 1)\lambda_1$$

The free spectral range for order m is then given by

$$F = \lambda_2 - \lambda_1 = \frac{\lambda_1}{m} \tag{17-3}$$

Notice that this nonoverlapping spectral region is smaller for higher orders.

Example

The shortest wavelength of light present in a given source is 400 nm. Determine the free spectral range in the first three orders of grating diffraction.

Solution

$$F_m = \frac{\lambda_1}{m}$$

Thus

$$F_1 = \frac{400}{1} = 400 \text{ nm (from 400 to 800 nm in first order)}$$

$$F_2 = \frac{400}{2} = 200 \text{ nm (from 400 to 600 nm in second order)}$$

$$F_3 = \frac{400}{3} = 133 \text{ nm (from 400 to 533 nm in third order)}$$

17-3 DISPERSION OF A GRATING

As Figure 16-14b shows, higher diffraction orders grow less intense as they fall more and more under the constraining diffraction envelope. On the other hand, Figure 17-2b shows clearly that wavelengths are better separated as their order increases. This property is precisely described by the *angular dispersion* $\mathscr{D}$, defined by

$$\mathscr{D} \equiv \frac{d\theta_m}{d\lambda} \tag{17-4}$$

which gives the angular separation per unit range of wavelength. The variation of θ_m with λ is described by the grating Eq. (17-2), from which we may conclude

$$\mathscr{D} = \frac{m}{a \cos \theta_m} \tag{17-5}$$

If a photographic plate is used in the focal plane of the lens to record the spectrum, it is convenient to describe the spread of wavelengths on the plate in terms of a *linear dispersion* $dy/d\lambda$, where y is measured along the plate. Since $dy = f\, d\theta$, the linear dispersion is given by

$$\text{linear dispersion} \equiv \frac{dy}{d\lambda} = f \frac{d\theta_m}{d\lambda} = f\mathscr{D} \tag{17-6}$$

The *plate factor* is the reciprocal of the linear dispersion.

Example

Light of wavelength 500 nm is incident normally on a grating with 5000 grooves/cm. Determine its angular and linear dispersion in first order when used with a lens of focal length 0.5 m.

Solution   The *grating constant* or groove separation a is

$$a = \frac{1}{5000 \text{ cm}^{-1}} = 2 \times 10^{-4} \text{ cm}$$

Clearly, for zeroth order, there is no dispersion. For first order, Eq. (17-5) requires the diffraction angle θ_1, which can be found from the grating equation (17-2):

$$\sin \theta_1 = \frac{(1)\lambda}{a} = \frac{500 \times 10^{-7}}{2 \times 10^{-4}} = 0.25$$

Thus $\theta_1 = 14.5°$ and $\cos \theta_1 = 0.968$.

The angular dispersion in the wavelength region around 500 nm can now be calculated:

$$\mathcal{D} = \frac{m}{a \cos \theta_m} = \frac{1}{(2 \times 10^{-4} \text{ cm})(0.968)} = 5164 \text{ rad/cm}$$

or

$$\mathcal{D} = 5.164 \times 10^{-4} \frac{\text{rad}}{\text{nm}} \times \frac{180°}{\pi \text{ rad}} = 0.0296°/\text{nm}$$

The linear dispersion is then found from

$$f\mathcal{D} = (500 \text{ mm})(5.164 \times 10^{-4} \text{ rad/nm}) = 0.258 \text{ mm/nm}$$

and the plate factor is $1/0.258 = 3.87$ nm/mm. One mm of film then spans a range of almost 4 nm or 40 Å.

At normal incidence, the grating equation can be incorporated with the angular dispersion relation to give

$$\mathcal{D} = \frac{m}{a \cos \theta} = \left(\frac{a \sin \theta}{\lambda}\right)\left(\frac{1}{a \cos \theta}\right)$$

or

$$\mathcal{D} = \frac{\tan \theta}{\lambda} \tag{17-7}$$

Thus the dispersion is actually independent of the grating constant at a given angle of diffraction and increases rapidly with θ. At a given angle of diffraction, the effect of increasing the grating constant is to increase the order of the diffraction there, as Eq. (17-5) clearly shows.

17-4 RESOLUTION OF A GRATING

Increased dispersion or spread of wavelengths does not by itself make neighboring wavelengths appear more distinctly, unless the peaks are themselves sharp enough. The latter property describes the *resolution* of the recorded spectrum. By the resolution of a grating, we mean its ability to produce distinct peaks for closely spaced wavelengths in a particular order. Recall that the *resolving power* $\mathcal{R}$ is defined in general by

$$\mathcal{R} \equiv \frac{\lambda}{(\Delta\lambda)_{min}} \tag{17-8}$$

where $(\Delta\lambda)_{min}$ is the minimum wavelength interval of two spectral components that are just resolvable by Rayleigh's criterion (Figure 16-9). For normally incident light of wavelength $\lambda + d\lambda$, and principal maximum of order m, we have by the grating Eq. (17-2),

$$a \sin \theta = m(\lambda + d\lambda) \tag{17-9}$$

To satisfy Rayleigh's criterion this peak must coincide (same θ) with the first minimum of the neighboring wavelength's peak in the same order, or

$$a \sin \theta = \left(m + \frac{1}{N}\right)\lambda \tag{17-10}$$

as can be deduced from both Figure 16-14a and Eq. (16-33). Equating the right members of Eqs. (17-9) and (17-10), $\lambda/d\lambda = mN$. Since $d\lambda$ here is the minimum resolvable wavelength difference, the resolving power of the grating is, by Eq. (17-8),

$$\mathscr{R} = mN \tag{17-11}$$

For a grating of N grooves, the resolving power is simply proportional to the order of the diffraction. In a given order of diffraction, the resolving power increases with the total number of grooves. It must be remembered that if N is to be increased within a given width W of grating, the grooves must be proportionately closer together. To take advantage of the maximum resolution, the light must cover the entire ruled width of the grating. If the grating in our previous example, with 5000 grooves/cm, has a width of 8 cm, then $N = 40,000$ and the resolving power in the first order is 40,000. This means, by Eq. (17-8), that in the region of $\lambda = 500$ nm, spectral components as close together as 0.0125 nm can be resolved. In the second order, this figure improves to 0.0063 nm, and so on. The best values for grating resolving power are in the range of 10^5 to 10^6. (When describing theoretical resolution, it must be remembered that the Rayleigh criterion is somewhat arbitrary and that spectral line widths also enter into the actual resolution.) A grating with 10,000 grooves/cm and 20 cm width provides a resolving power of 1 million in fifth order. For normally incident light, however, the grating equation limits the maximum wavelength (at $\theta = 90°!$) under these conditions to 200 nm. If the light is not incident along the normal, the maximum diffractable wavelength can be increased; when $\theta_i = 90°$, it is twice as much or 400 nm. Operation in high orders further severely restricts available light because of the diffraction envelope constraint, unless means are taken to redirect the central diffraction peak into the desired order. This is achieved through *blazing*, to be discussed presently. Notice that the resolving power, like the dispersion, is independent of groove spacing for a given diffraction angle. If we write $N = W/a$ for a ruled grating width W and incorporate the grating equation for normal incidence, Eq. (17-11) becomes

$$\mathscr{R} = mN = \left(\frac{a \sin \theta_m}{\lambda}\right)\frac{W}{a}$$

or

$$\mathscr{R} = \frac{W \sin \theta_m}{\lambda} \tag{17-12}$$

According to Eq. (17-12), the resolution of a grating at diffracting angle θ_m depends on the width of the grating rather than on the number of its grooves. For a fixed ratio of $(\sin \theta_m)/\lambda$, however, the grating equation also fixes the ratio m/a. Thus using

Chap. 17   The Diffraction Grating

a grating with fewer grooves and a larger grating constant requires that we work at a higher order m, where there is increased complication due to overlapping orders. Such confusion in high orders is sometimes alleviated by using a second dispersing instrument that spreads the first spectrum again but in a direction orthogonal to the first. One such instrument is described later in this chapter.

17-5 TYPES OF GRATINGS

Up to this point we have been imagining the diffraction grating to be an opaque aperture in which closely spaced slits have been introduced. Fraunhofer's original gratings were, in fact, fine wires wound between closely spaced threads of two parallel screws or parallel lines ruled on smoked glass. Later Strong used ruled metal coatings on glass blanks. Today the typical grating master is made by diamond point ruling of grooves into a low-expansion glass base or into a film of aluminum or gold that has been vacuum-evaporated onto the glass base. The base, or *blank*, itself must first be polished to closer than $\lambda/10$ for green light. The development of ruling machines capable of ruling up to 3600 sculptured grooves per millimeter over a width of 10 in. or more, with suitably uniform depth, shape, and spacing, has been an impressive and far-reaching technological achievement. Techniques involving interferometric and electronic servo-control have been used to enhance the precision of the most modern ruling engines. High-quality grating masters ruled over widths as large as 46 cm or more have become feasible.

A grating may be designed to operate either as a *transmission grating* or a *reflection grating*. In a transmission grating, light is periodically transmitted by the clear sections of a glass blank, into which grooves serving as scattering centers have been ruled, or the light is transmitted by the entire ruled area but periodically retarded in phase due to the varying optical thickness of the grooves. In the first case, the grating is a *transmission amplitude grating*, functioning like the slotted, opaque aperture; in the second case, the grating is called a *transmission phase grating*. In the reflection grating, the groove faces are made highly reflecting, and the periodic reflection of the incident light operates like the periodic transmission of waves from a transmission grating. Research-quality gratings are usually of the reflection type. A section of a plane reflection grating is shown in Figure 17-3. The path difference

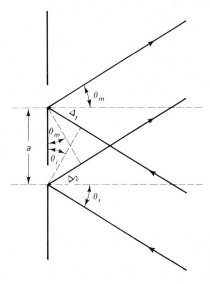

Figure 17-3 Neighboring reflection grating grooves illuminated by light incident at angle θ_i with the grating normal. For light diffracted in the direction θ_m the net path difference of the two waves is $\Delta_1 - \Delta_2$.

between equivalent reflected rays of light from successive groove reflections is just the difference

$$\Delta = \Delta_1 - \Delta_2 = a \sin \theta_i - a \sin \theta_m$$

where both rays are assumed to have the direction after diffraction specified by the angle θ_m. When $\Delta = m\lambda$ an interference principal maximum results, so that the grating equation is the same as for a transmission grating:

$$m\lambda = a(\sin \theta_i + \sin \theta_m)$$

The same sign convention also applies to the angles θ_m and θ_i: When θ_m is on the opposite side of the normal relative to θ_i, as in Figure 17-3, it is considered negative. The zeroth order of interference occurs when $m = 0$ or $\theta_m = -\theta_i$, that is, in the direction of specular reflection from the grating, acting as a mirror for all wavelengths. The metallic coating of the reflection grating should be as highly reflective as possible. In the ultraviolet range 110 to 160 nm, coatings of magnesium flouride or lithium fluoride over aluminum are generally used to enhance reflectivity, and below 100 nm, gold and platinum are often used. In the infrared regions, silver and gold coatings are both effective. The light diffracted from a plane grating must be focused by means of a lens or concave mirror. When the absorption of radiation by the focusing elements is severe, as in the vacuum ultraviolet (about 1 to 200 nm), the focusing and diffraction may both be accomplished by using a *concave grating*, that is, a concave mirror that has been ruled to form grooves onto its reflecting surface.

17-6 BLAZED GRATINGS

The *absolute efficiency* of a grating in a given wavelength region and order is the ratio of the diffracted light energy to the incident light energy in the same wavelength region. Increasing the number of rulings on a grating, for example, increases the light energy throughput. The zeroth-order diffraction principal maximum, for which there is no dispersion, represents a waste of light energy, reducing grating efficiency. The zeroth order, it will be recalled, contains the most intense interference maximum because it coincides with the maximum of the single-slit diffraction envelope. The technique of shaping individual grooves so that the diffraction envelope maximum shifts into another order is called *blazing* the grating.

To understand the effect of blazing, consider Figure 17-4 for a transmission grating and Figure 17-5 for a reflection grating. For simplicity, light is shown transmitted or reflected from a single groove, even though diffraction involves the cooperative contribution from many grooves. In each figure, (a) illustrates the situation for an unblazed grating and (b) shows the result of shaping the grooves to shift the diffraction envelope maximum ($\beta = 0$) from the zeroth-order ($m = 0$) interference or principal maximum. Recall that the diffraction envelope maximum occurs where $\beta = 0$, that is, where the far-field path difference for light rays from the center and the edge of any groove is zero. A zero path difference for these rays implies the condition of geometrical optics: For transmitted light, Figure 17-4, the diffraction peak is in the direction of the incident beam; for reflected light, Figure 17-5, it is in the direction of the specularly reflected beam. By introducing prismatic grooves in Figure 17-4 or inclined mirror faces in Figure 17-5, the corresponding zero path difference is shifted into the directions of the refracted beam and the new reflected beam, respectively, which now correspond to the case $\beta = 0$. While the diffraction envelope is thus shifted by the shaping of the individual grooves, the interference maxima remain fixed in position. Their positions are determined by the grating equation,

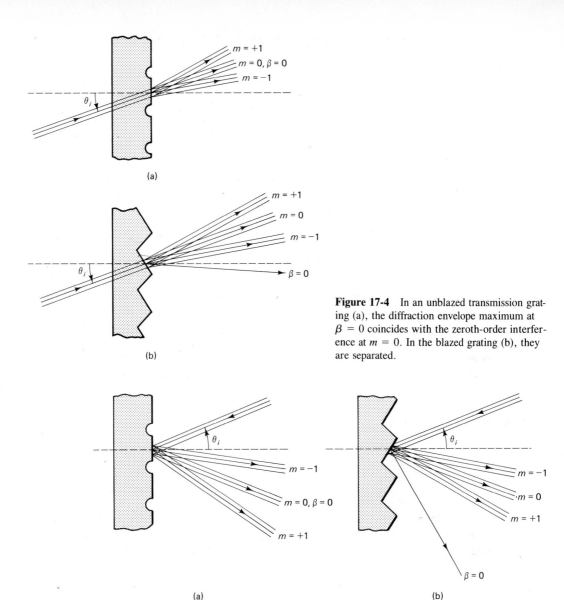

(a)

(b)

Figure 17-4 In an unblazed transmission grating (a), the diffraction envelope maximum at $\beta = 0$ coincides with the zeroth-order interference at $m = 0$. In the blazed grating (b), they are separated.

(a)

(b)

Figure 17-5 In an unblazed reflection grating (a), the diffraction envelope maximum at $\beta = 0$ coincides with the zeroth-order interference at $m = 0$. In the blazed grating (b), they are separated.

in which angles are measured relative to the plane of the grating. Neither this plane nor the groove separation have been altered in going from (a) to (b) in either Figure 17-4 or 17-5. The result is that the diffraction maximum now favors a principal maximum of a higher order ($|m| > 0$), and the grating redirects the bulk of the light energy where it is most useful.

It remains to determine the proper *blaze angle* of a grating. Consider the reflection grating of Figure 17-6, where a beam is incident on a groove face at angle θ_i and is diffracted at arbitrary angle θ, both measured relative to the normal N to the grating plane. The normal N' to the groove face makes an angle θ_b relative to N. This angle is the blaze angle of the grating. Now let us require that the diffracted beam satisfy both the condition of specular reflection from the groove face and the condition for a principal maximum in the mth order ($\theta = \theta_m$). The first condition is

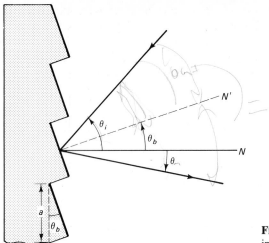

Figure 17-6 Relation of blaze angle θ_b to the incident and diffracted beams.

satisfied by making the angle of incidence equal to the angle of reflection relative to N': $\theta_i - \theta_b = \theta_m + \theta_b$, or

$$\theta_b = \frac{\theta_i - \theta_m}{2} \qquad (17\text{-}13)$$

The second condition is that the angle θ_m satisfy the grating equation,

$$m\lambda = a(\sin \theta_i + \sin \theta_m) \qquad (17\text{-}14)$$

Equation (17-13) shows that the blaze angle depends on the angle of incidence, so that various geometries requiring different blaze angles are possible. In the general case, the equation that must be satisfied by the blaze angle is found by combining Eqs. (17-13) and (17-14). Taking into account the associated sign convention, the grating equation becomes

$$m\lambda = a[\sin \theta_i + \sin (2\theta_b - \theta_i)] \qquad (17\text{-}15)$$

We consider two special cases of Eq. (17-15). In the *Littrow mount*, the light is brought in along or close to the groove face normal N', so that $\theta_b = \theta_i$ and $\theta_m = -\theta_i$, as is clear from Figure 17-6 and Eq. (17-13). For this special case, Eq. (17-15) gives

$$\text{Littrow:} \quad m\lambda = 2a \sin \theta_b \quad \text{or} \quad \theta_b = \sin^{-1}\left(\frac{m\lambda}{2a}\right) \qquad (17\text{-}16)$$

Since the quantity $a \sin \theta_b$ corresponds to the steep-face height of the groove (Figure 17-6), we see that a grating correctly blazed for wavelength λ and order m in a Littrow mount must have a groove step of an integral number m of half-wavelengths. Commerical gratings are usually specified by their blaze angles and the corresponding first-order Littrow wavelengths.

In another configuration, the light is introduced instead along the normal N to the grating itself. Then $\theta_i = 0$ and $\theta_b = -\theta_m/2$. Equation (17-15) now gives

$$\text{normal incidence:} \quad \theta_b = \tfrac{1}{2}\sin^{-1}\left(\frac{m\lambda}{a}\right) \qquad (17\text{-}17)$$

Example

(a) Consider a 1200 groove/mm grating to be blazed for a wavelength of 600 nm in first order. Determine the proper blaze angle.

(b) *An echelle grating* is a coarsely pitched grating designed to achieve high resolution by operating in high orders. Consider the operation in order $m = 30$ of a commercially available echelle grating with 79 grooves/mm, blazed at an angle of 63°26′, and ruled over an area of 406 × 610 mm. Determine its resolution when used in a Littrow mount.

Solution (a) In a Littrow mount, using Eq. (17-16), the blaze angle must be

$$\theta_b = \sin^{-1}\left[\frac{(1)(600 \times 10^{-6})(1200)}{2}\right] = 21°06′$$

On the other hand, if the grating is used in a mount with light incident normal to the grating, then from Eq. (17-17),

$$\theta_b = \tfrac{1}{2}\sin^{-1}[(1)(600 \times 10^{-6})(1200)] = 23°02′$$

(b) In a Littrow mount, the grating returns, along the incidence direction, light of wavelength

$$\lambda = \frac{2a \sin \theta_b}{m} = \frac{2(1/79) \sin (63.43)}{30} = 755 \text{ nm}$$

The total number of lines on the grating is $N = (79)(610) = 48{,}190$ so that the resolving power is $\mathcal{R} = mN = (30)(48{,}190) = 1{,}445{,}700$ at the blaze wavelength of 755 nm. The minimum resolvable wavelength interval in this region is then $\Delta\lambda_{min} = \lambda/\mathcal{R}$, or 0.0005 nm. Actual resolutions may be somewhat less than the theoretical due to grating imperfections. The high resolution is gained at the expense of a contracted spectral range of only $\lambda/m = 755/30 = 25$ nm.

17-7 GRATING REPLICAS

The expense and difficulty of manufacturing gratings prohibit the routine use of grating masters in spectroscopic instruments. Until the technique of making *replicas*—relatively inexpensive copies of the masters—was developed, few research scientists owned a good grating. To make a replica grating, the master is first coated with a layer of nonadherent material, which can be lifted off the master at a later stage. This is followed by a vacuum-evaporated overcoat of aluminum. A layer of resin is then spread over the combination, and a substrate for the future replica is placed on top. After the resin has hardened, the replica grating can be separated from the master. The first good replica grating usually serves as a submaster for the routine production of other replicas. Thin replicas made from a submaster are mounted on a glass or fused silica blank and a highly reflective overcoat of aluminum is added. This is the usual form in which the gratings are made commercially available. Replica gratings can be purchased that are as good as or better than the masters, both in performance and useful life. The efficiency of deep-groove replicas may be better than that of the master because the replication process transfers the smooth parts of the groove faces from bottom to top, improving performance.

17-8 INTERFERENCE GRATINGS

The availability of intense and highly coherent beams of light have made possible the production of gratings apart from the rulings produced by grating engines. As early as 1927, Michelson suggested the possibility of photographing straight interfer-

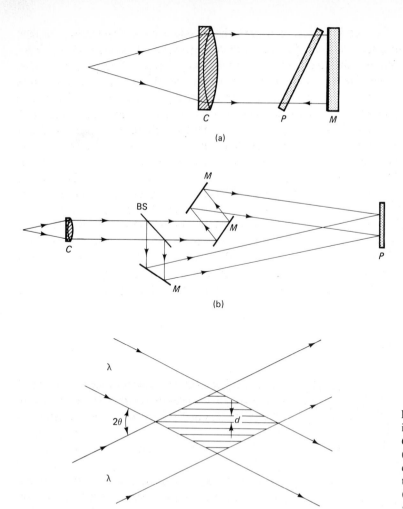

Figure 17-7 (a) Michelson system for producing interference gratings, including collimator C, mirror M, and photographic plate P. (b) Holographic system for producing interference fringes including collimator C, beamsplitter BS, mirrors M, and light-sensitive plate P. (c) Production of interference fringes in the region of superposition of two collimated and coherent beams intersecting at angle 2θ.

ence fringes using an optical system such as that shown in Figure 17-7a. Two coherent, monochromatic beams are made to interfere, producing standing waves in the region between the collimating lens and a plane mirror. The resulting straight-line interference maxima are intercepted by a light-sensitive film, inclined at an angle. When developed, straight-line fringes appear.

Interference gratings produced by such optical techniques are also called *holographic gratings*, since a grating of uniformly spaced, parallel grooves can be considered as a hologram of a point source at infinity. Other interferometric systems, such as that shown in Figure 17-7b, are essentially those used to produce holograms. Today the interfering wave fronts are photographed on a grainless film of photoresist whose solubility to the etchant is proportional to the irradiance of exposure. The photoresist is spread evenly over the surface of the glass blank to a thickness of 1 μm or less by rapidly spinning the blank. When etched, the interference pattern is preserved in the form of transmission grating grooves whose transmittance varies gradually across the groove in a sine-squared profile. A reflective metallic coating is usually added to the grating by vacuum evaporation. The fringe spacing d, as shown in Figure 17-7c, is determined by the wavelength of the light and by the angle 2θ between the two interfering beams, according to the relation $d = \lambda/(2 \sin \theta)$.

In addition to freedom from the expensive and laborious process of machine ruling, the predominant advantage of the interference grating is the absence of periodic or random errors in groove positions that produce *ghosts* and *grass*, respectively. Thus interference gratings possess impressive spectral purity and provide a high signal-to-noise advantage. On the other hand, control over groove profile, which affects the blazing and thus the efficiency of the grating, is not easily achieved. The groove profiles of normal interference gratings are sine-squared in form and so symmetrical, rather than sawtooth-shaped, as are the usual blazed gratings. Under normal incidence, a symmetrical groove profile results in an equal distribution of light in the positive and negative orders of diffraction. When used under nonnormal incidence, however, it is possible to disperse light into only one diffracted order (other than the zeroth order), and it has been shown that in this case the distribution of light does not depend to a great extent on groove shape. Efficiencies in this configuration can be comparable to those of blazed gratings. Nevertheless, various efforts are in progress to produce groove shapes more like those of ordinary blazed gratings, by exposing the photoresist to two wavelengths of radiation whose Fourier synthesis is more saw-toothed in shape, for example, or by subsequent modification of the symmetrical grooves by argon-ion etching or in a variety of other ways. Interference techniques are not practical in the production of coarse, echellelike gratings.

17-9 GRATING INSTRUMENTS

An instrument that uses a grating as a spectral dispersing element is designed around the type of grating selected for a particular application. An inexpensive transmission grating may be mounted in place of the prism in a *spectroscope*, where the spectrum is viewed with the eye, by means of a telescope focused for infinity. The light incident on the grating is rendered parallel by a primary slit and collimating lens. Research grade instruments, however, make use of reflection gratings. These may be *spectrographs*, which record a portion of the spectrum on a photographic plate, photodiode array, or other image detector, or *spectrometers*, where a narrow portion of the spectrum is allowed to pass through an exit slit onto a photomultiplier or other light flux detector. In the latter case, the spectrum may be scanned by rotating the grating. There are a number of designs possible; we describe briefly a few of the more common ones. Figure 17-8 shows the basic *Littrow mount*, where a single focusing element is used both to collimate the light incident on the plane grating and, in the reverse direction, to focus the light onto the photographic plate placed near the slit. Recall that in the Littrow configuration, light is incident along the normal to the groove faces. The Littrow condition is also used in the *echelle spectrograph* (Figure 17-9), which is designed to take advantage of the high dispersion and reso-

Figure 17-8 Littrow-mounted plane grating. Photographic plate and entrance slit are separated along a direction transverse to the plane of the drawing.

Figure 17-9 Side view of the echelle spectrograph. The echelle is positioned directly over the slit-to-mirror path, but the plate is offset in a horizontal direction.

lution attainable with large angles of incidence on a blazed plane grating. As discussed previously, the useful order of diffraction is large and the spectral free range is small, so that a second concave grating is used to disperse the overlapping orders in a direction perpendicular to the dispersion of the echelle grating. In the figure, a concave mirror collimates the light incident on the echelle, located near the slit and oriented with grooves horizontal. The light diffracted by the echelle is dispersed again by the concave grating, oriented with grooves vertical. The second grating also focuses the two-dimensional spectrum onto the photographic plate. Figure 17-10 shows a *Czerny-Turner* system in a grating spectrometer. Light from an entrance slit is directed by a plane mirror to a first concave mirror, which collimates the light incident on the grating. The diffracted light is incident on a second concave mirror, which then focuses the spectrum at the exit slit. As the grating is rotated, the dispersed spectrum moves across the slit. When the instrument is used specifically to select individual wavelengths from a discrete spectral source or to allow a narrow wavelength range of spectrum through the exit slit, it is called a *monochromator*.

Figure 17-10   Czerny-Turner spectrometer.

Other instruments dispense with secondary focusing lenses or mirrors and rely on concave gratings both to focus and to disperse the light. The grooves ruled on a concave grating are equally spaced relative to a plane projection of the surface, not relative to the concave surface itself. In this way, spherical aberration and coma are eliminated. Concave-grating instruments are used for wavelengths in the soft X-ray (1 to 25 nm) and ultraviolet regions, extending into the visible. The *Paschen-Runge* design, Figure 17-11, makes use of the *Rowland circle*. If the grating surface is tangent at its center with a circle having a diameter equal to the radius of curvature of the concave grating, it can be shown that a slit source placed anywhere on the circle gives well-focused spectral lines that also fall on the circle. If the light source and slit, grating, and plate holder are placed in a dark room at three stable positions de-

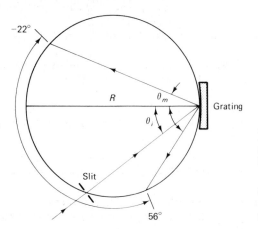

Figure 17-11   Paschen-Runge mounting for a concave grating. Diffracted slit images are formed at the Rowland circle. For a grating of 1200 grooves/mm and $\theta_i = 30°$, the first-order spectrum for wavelengths between 200 and 1200 nm fall between the angles $-22°$ and $56°$, respectively.

termined by the Rowland circle and the grating equation, the basic requirements of the Paschen-Runge spectrograph are met. Since typical radii of curvature for the grating may be around 6 m, the space occupied by this spectrograph can be quite large. The first three orders of diffraction are most commonly used. Typical angles of incidence may vary within the range 30° to 45°, and angles of diffraction may vary between 25° on the opposite side of the grating normal to 85° on the same side of the normal as the slit. Thus much of the Rowland circle is useful for recording various portions of the spectrum. In Figure 17-11, the first-order spectrum spread (200 to 1200 nm) around the Rowland circle is shown for $\theta_i = 38°$ and a grating of 1200 grooves/mm. Spectral lines formed in this way may suffer rather severely from astigmatism. The *Wadsworth spectrograph* (Figure 17-12) eliminates astigmatism by adding a primary mirror to collimate the light incident on the grating. In so doing, the spectrograph dispenses with the Rowland circle. Spectra are observed over a range making small angles to the grating normal, perhaps 10° to either side. To record different regions of the spectrum, the grating can be rotated and higher orders can be used. This version of a grating spectrograph is capable of more compact construction than is the Paschen-Runge.

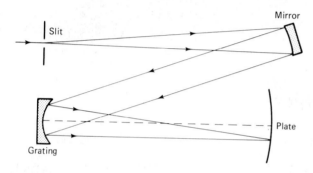

Figure 17-12 Wadsworth mount for a concave grating.

PROBLEMS

17-1. What is the angular separation in second order between light of wavelengths 400 nm and 600 nm when diffracted by a grating of 5000 grooves/cm?

17-2. (a) Describe the dispersion in the red wavelength region around 650 nm (both in °/nm and in nm/mm) for a transmission grating 6 cm wide, containing 3500 grooves/cm, when it is focused in the third-order spectrum on a screen by a lens of focal length 150 cm.

(b) Find also the resolving power of the grating under these conditions.

17-3. (a) What is the angular separation between the second-order principal maximum and the neighboring minimum on either side for the Fraunhofer pattern of a 24-groove grating having a groove separation of 10^{-3} cm and illuminated by light of 600 nm?

(b) What slightly longer (or slightly shorter) wavelength would have its second-order maximum on top of the minimum adjacent to the second-order maximum of 600 nm light?

(c) From your results in parts (a) and (b), calculate the resolving power in second order. Compare this with the resolving power obtained from the theoretical grating resolving power formula, Eq. (17-11).

17-4. How many lines must be ruled on a transmission grating so that it is just capable of resolving the sodium doublet (589.592 nm and 588.995 nm) in the first- and second-order spectra?

17-5. (a) A grating spectrograph is to be used in first order. If crown glass optics is used in bringing the light to the entrance slit, what is the first wavelength in the spectrum that may contain second-order lines? If the optics is quartz, how does this change? Assume that the absorption cutoff is 350 nm for crown glass and 180 nm for quartz.

(b) At what angle of diffraction does the beginning of overlap occur in each case for a grating of 1200 grooves/mm?

(c) What is the free spectral range for first and second orders in each case?

17-6. A transmission grating having 16,000 lines/in. is 2.5 in. wide. Operating in the green at about 550 nm, what is the resolving power in the third order? Calculate the minimum resolvable wavelength difference in the second order.

17-7. The two sodium D lines at 5893 Å are 6 Å apart. If a grating with only 400 grooves is available (a) what is the lowest order possible in which the D lines are resolved and (b) how wide does the grating have to be?

17-8. A multiple-slit aperture has (1) $N = 2$, (2) $N = 10$, and (3) $N = 15,000$ slits. The aperture is placed directly in front of a lens of focal length 2 m. The distance between slits is 0.005 mm and the slit width is 0.001 mm for each case. The incident plane wavefronts of light are of wavelength 546 nm. Find *for each case*, (a) the separation on the screen between the zeroth- and first-order maxima; (b) the number of bright fringes (principal maxima) that fall under the central diffraction envelope; (c) the width on the screen of the central interference fringe.

17-9. A reflection grating is required that can resolve wavelengths as close as 0.02 Å in second order for the spectral region around 350 nm. The grating is to be installed in an instrument where light from the entrance slit is incident normally on the grating. If the manufacturer provides rulings over a 10-cm grating width, determine (a) the minimum number of grooves/cm required; (b) the optimum blaze angle for work in this region; (c) the angle of diffraction where irradiance is maximum (show both blaze angle and diffraction angle on a sketch); (d) the dispersion in nanometers per degree.

17-10. A transmission grating is expected to provide an ultimate first-order resolution of at least 1 Å anywhere in the visible spectrum (400 to 700 nm). The ruled width of the grating is to be 2 cm.

(a) Determine the minimum number of grooves required.

(b) If the diffraction pattern is focused by a 50-cm lens, what is the linear separation of a 1-Å interval in the vicinity of 500 nm?

17-11. A concave reflection grating of 2-m radius is ruled with 1000 grooves/mm. Light is incident at an angle of 30° to the central grating normal. Determine for first-order operation the (a) angular spread about the grating normal of the visible range of wavelengths (400 to 700 nm); (b) theoretical resolving power if the grating is ruled over a width of 10 cm; (c) plate factor in the vicinity of 550 nm; (d) radius of the Rowland circle in a Paschen-Runge mounting of the grating.

17-12. How many grooves per centimeter are required for a 2-m radius, concave grating, which is to have a plate factor of around 2 nm/mm in first order?

17-13. A plane reflection grating with 300 grooves/mm is blazed at 10°.

(a) At what wavelength in first order does the grating direct the maximum energy when used with the incident light normal to the groove faces?

(b) What is the plate factor in first order when the grating is used in a Czerny-Turner mounting with mirrors of 3.4-m radius of curvature?

17-14. A reflection grating, ruled over a 15-cm width, is to be blazed at 2000 Å for use in the vacuum ultraviolet. If its theoretical resolving power in first order is to be 300,000, determine the proper blaze angle for use (a) in a Littrow mount and (b) with normal incidence.

17-15. Show that the separation d of fringes in the formation of a holographic grating, as in Figure 17-7c, is given by $\lambda/(2 \sin \theta)$, where 2θ is the angle between the coherent beams. If the beams are argon laser beams of wavelength 488 nm and the angle between beams is 120°, how many grooves per millimeter are formed in a plane emulsion ($n = 1$) oriented perpendicular to the fringes? What is the effect on the grating constant of an emulsion with a high refractive index?

17-16. A grating is needed that, working in first order, is able to resolve the red doublet produced by an electrical discharge in a mixture of hydrogen and deuterium: 1.8 Å at 6563 Å. The grating can be produced with a standard blaze at 6300 Å for use in a Littrow mount. Find (a) the total number of grooves required; (b) the number of grooves per millimeter on the grating with a blaze angle of 22°12′; (c) the minimum width of the grating.

17-17. An echelle grating is ruled over 12 cm of width with 8 grooves/mm and is blazed at 63°. Determine for a Littrow configuration (a) the range of orders in which the visible spectrum (400 to 700 nm) appears; (b) the total number of grooves; (c) the resolving power and minimum resolvable wavelength interval at 550 nm; (d) the dispersion at 550 nm; (e) the free spectral range, assuming the shortest wavelength present is 350 nm.

REFERENCES

[1] Davis, Sumner P. *Diffraction Grating Spectrographs*. New York: Holt, Rinehart and Winston, 1970.

[2] Hutley, M. C. *Diffraction Gratings*. New York: Academic Press, 1982.

[3] James, J. F., and R. S. Sternberg. *The Design of Optical Spectrometers*. London: Chapman and Hall Ltd., 1969. Ch. 5, 6.

[4] Ingalls, Albert G. "Ruling Engines." *Scientific American* (June 1952): 45.

[5] Feynman, Richard P., Robert B. Leighton, and Matthew Sands. *The Feynman Lectures in Physics*, vol. 1. Reading, Mass.: Addison-Wesley Publishing Company, 1975. Ch. 30.

18

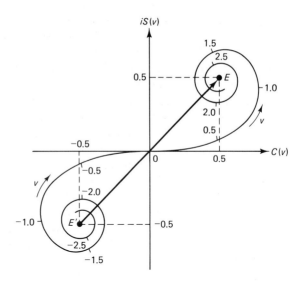

Fresnel Diffraction

INTRODUCTION

In the last two chapters we have dealt with Fraunhofer diffraction, situations in which the wavefront at the diffracting aperture may be considered planar without appreciable error. We turn now to cases where this constitutes a bad approximation, cases in which either or both source and observing screen are close enough to the aperture that wavefront curvature must be taken into account. Collimating lenses are not required, therefore, for the observation of Fresnel, or near-field, diffraction patterns, and in this experimental sense, their study is simpler. The mathematical treatment, however, is more complex and is almost always handled by approximation techniques, as we will see.

Fresnel diffraction patterns form a continuity between the patterns characterizing geometrical optics at one extreme and Fraunhofer diffraction at the other. In geometrical optics, where light waves can be treated as rays propagating along straight lines, we expect to see a sharp image of the aperture. In practice, such images are formed when the observing screen is quite close to the aperture. In cases of Fraunhofer diffraction, where the screen is actually or, through the use of a lens, effectively far from the aperture, the diffraction pattern is a fringed image that bears little resemblance to the aperture. Recall the Fraunhofer double-slit pattern, for example. In the intermediate case of Fresnel diffraction, the image is essentially an image of the aperture, but the edges are fringed.

A typical arrangement is shown in Figure 18-1. Spherical wavefronts emerge from a point source and encounter an aperture. At the aperture, the wavefront is still substantially spherical, because the aperture is not far from the source. Diffraction effects in the near field on the other side of the aperture are then of the Fresnel type. The distance from the source S to a representative point O on the wavefront at the aperture is r', and the distance from the point O to a representative point P in the field is r. Compared to Fraunhofer diffraction, this case requires special treatment in several ways. Since approaching waves are not plane, the distance r' enters into the calculations. Also, the distances r and r' are no longer so much greater than the size of the aperture that Fraunhofer diffraction applies. As a result, the variation of r and r', both with different aperture points O and with different field points P, must be taken into account. Finally, because the direction from various aperture points O to a given field point P may no longer be considered approximately constant, the dependence of amplitude on direction of the Huygens' wavelets originating at the aperture must be considered. This correction is handled by the *obliquity factor* to be discussed presently.

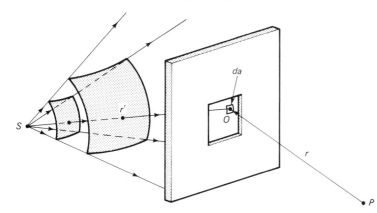

Figure 18-1 Schematic defining the parameters for a typical Fresnel diffraction.

Employing the *Huygens-Fresnel principle*, as in Fraunhofer diffraction, we seek to find the resultant amplitude of the electric field at P due to a superposition of all the Huygens' wavelets from the wavefront at the aperture, each emanating from a "point" on the wavefront, an elemental area da. Let the contribution to the resultant field at P due to such an elemental area be represented by the spherical wave,

$$dE_P = \left(\frac{dE_0}{r}\right)e^{ikr} \qquad (18\text{-}1)$$

The wave amplitude dE_0 at the aperture is proportional to the elemental area, so we can write

$$dE_0 \propto E_L\, da \qquad (18\text{-}2)$$

The amplitude E_L at point O is the amplitude of the spherical wave originating at the source, or

$$E_L = \left(\frac{E_s}{r'}\right)e^{ikr'} \qquad (18\text{-}3)$$

Except for a proportionality constant then, we have, combining these equations,

$$dE_P = \left(\frac{E_s}{rr'}\right)e^{ik(r+r')}\, da \qquad (18\text{-}4)$$

The field at P due to the secondary wavelets from the entire aperture is the surface integral

$$E_P = E_s \iint_{A_P} \left(\frac{1}{rr'}\right) e^{ik(r+r')} \, da \qquad (18\text{-}5)$$

Equation (18-5) is incomplete in two ways. First, it does not take into account the obliquity factor, which attenuates the diffracted waves according to their direction, as described earlier. For the present, we call this factor $F(\theta)$, a function of the angle θ between the directions of the radiation incident and diffracted at the aperture point O. Second, it does not take into account a curious requirement, a $90°$ phase shift of the diffracted waves relative to the primary incident wave. We will return to each of these points in the following discussion. The corrected integral is the *Fresnel-Kirchhoff diffraction formula*, given by

$$E_P = \frac{-ikE_S}{2\pi} \iint F(\theta) \frac{e^{ik(r+r')}}{rr'} \, da \qquad (18\text{-}6)$$

where the factor $-i = e^{-i\pi/2}$ represents the required phase shift, and

$$F(\theta) = \frac{1 + \cos\theta}{2}$$

limits the amplitude, E_S. This integral formula was developed by Fresnel and placed on a more rigorous theoretical basis by Kirchhoff. The ad hoc assumptions by Fresnel were shown by Kirchhoff to follow naturally by arguing from Green's integral theorem whose functions are scalar function solutions to the electromagnetic wave equation.[1] The result expressed by Eq. (18-6), however, still involves approximations, requiring that the source and screen distances remain large relative to the aperture dimensions and that the aperture dimensions themselves remain large relative to the wavelength of the optical disturbance. The integration specified by Eq. (18-6) is over a closed surface including the aperture but is assumed to make a contribution only over the aperture itself. In arriving at this result, Kirchhoff assumed as boundary conditions that the wave function and its gradient are zero directly behind the opaque parts of the aperture, and that, within the opening itself, they have the same value as they would in the absence of the aperture. These assumptions make the derivation of Eq. (18-6) possible but are not entirely justified. Furthermore, in the theory described here, the **E**-field wave function is a scalar function whose absolute square yields the irradiance. We know that, in the close vicinity of the aperture, more rigorous methods must be used that take into account the vector properties of the electromagnetic field, including polarization effects. Nevertheless, the Kirchhoff theory suffices to yield accurate results for most practical diffraction situations.

In the limiting case of Fraunhofer diffraction, Eq. (18-6) is simplified by assuming that (1) the obliquity factor is roughly constant over the aperture due to the small spread in the diffracted light and (2) the variation of distances r and r' remains small relative to that of the exponential function. When all constant (or approximately constant) terms are taken out of the integral and included in an overall constant C, Eq. (18-6) is simply

$$E_P = C \iint e^{ikr} \, da$$

---

[1] This derivation is found in many places, for example, [1] and [2], but requires mathematical ability that is beyond the stated level of this textbook.

which is a statement of the Huygens-Fresnel principle and corresponds to the integral used in Chapter 16 to calculate Fraunhofer diffraction patterns there.

For situations in which the assumptions of Fraunhofer diffraction fail, we are left with Eq. (18-6). This integration is, in general, not easy to carry out for a given aperture. Fresnel offered satisfactory methods for simplifying this task, or avoiding it altogether. We apply these methods in the simpler cases to be considered here.

18-2 CRITERION FOR FRESNEL DIFFRACTION

Before dealing with these cases, we wish to establish a practical criterion that determines when we should use Fresnel techniques rather than the simpler Fraunhofer treatment already presented. It will suffice to consider the simple case when both S and P are located on the central axis through the aperture, as in Figure 18-2. Notice that the dimension indicated by Δ is zero when the wavefront is plane. The methods of Fraunhofer diffraction suffice, however, as long as Δ is small, less than the wavelength of the light. From Figure 18-2a we may express this quantity as

$$\Delta = p - \sqrt{r'^2 - h^2} \tag{18-7}$$

or, equivalently,

$$\Delta = p - r'\left(1 - \frac{h^2}{r'^2}\right)^{1/2} \cong p - r'\left(1 - \frac{h^2}{2r'^2}\right) \tag{18-8}$$

where we have approximated the quantity in parentheses using the first two terms of the binomial expansion, $(1 - x)^{1/2} = 1 - x/2 + \cdots$. Since $p \cong r'$, the condition for significant curvature (near-field case) is

$$\Delta = \frac{h^2}{2p} > \lambda \tag{18-9}$$

and similarly, for the diffracted wave curvature in Figure 18-2b,

$$\Delta = \frac{h^2}{2q} > \lambda \tag{18-10}$$

Combining Eqs. (18-9) and (18-10), the regime of Fresnel, or near-field, diffraction may be expressed by

$$\text{near field:} \quad \frac{1}{2}\left(\frac{1}{p} + \frac{1}{q}\right)h^2 > \lambda \tag{18-11}$$

(a)

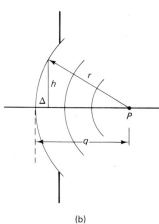

(b)

Figure 18-2 Edge view of Figure 18-1. The curvature of (a) incident and (b) diffracted wavefronts is small when Δ is small.

Of course, this condition also applies to the other dimension (transverse to h) of the aperture, not shown in Figure 18-2. When h is taken as the maximum extent of the aperture in either direction or as the radius of a circular aperture, Eq. (18-9) or Eq.(18-10) may also be expressed approximately by the condition

$$\text{near field:} \quad d < \frac{A}{\lambda} \tag{18-12}$$

where d represents either p or q and A is the area of the aperture.

18-3 THE OBLIQUITY FACTOR

The effect of the obliquity factor on the secondary wavelets originating at points of the wavefront was introduced by Fresnel. Recall that according to Huygens, a point source of secondary wavelets could radiate with equal effectiveness without regard to direction. This peculiarity would produce new wavefronts in both forward and reverse directions of a propagating wavefront, although the reverse wave does not exist. If point O in Figure 18-3 is the origin of secondary wavelets that arrive at an arbitrary point P in the field, then the correct modification of amplitude as a function of the angle θ is given by

$$a = \left(\frac{a_0}{2}\right)(1 + \cos \theta) \tag{18-13}$$

where a_0 is evidently the amplitude in the forward direction, $\theta = 0$. Notice that $a = 0$ in the reverse direction. The theoretical justification for this relation can also be found in Kirchhoff's derivation.

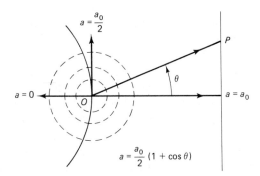

Figure 18-3   Illustration of the obliquity factor.

18-4 FRESNEL DIFFRACTION FROM CIRCULAR APERTURES

Suppose the aperture in Figure 18-1 is circular. Fresnel offered a clever technique for analyzing this special case without having to do the explicit integration of Eq. (18-6). He devised a method for dealing with the contribution from various parts of the wavefront by dividing the aperture into zones with circular symmetry about the axis SOP. The configuration is sketched in Figure 18-4a, which shows an emerging spherical wavefront centered at S. The zones are defined by circles on the wavefront, spaced in such a way that each zone is, on the average, $\lambda/2$ farther from the field point P than the preceding zone. In Figure 18-4a, then, $r_1 = r_0 + \lambda/2$, $r_2 = r_0 + \lambda, \ldots, r_N = r_0 + N\lambda/2$. This means that each successive zone's contribution is exactly out of phase with that of the preceding one. Of course, each of these half-period, or *Fresnel*, zones could be subdivided further into smaller parts,

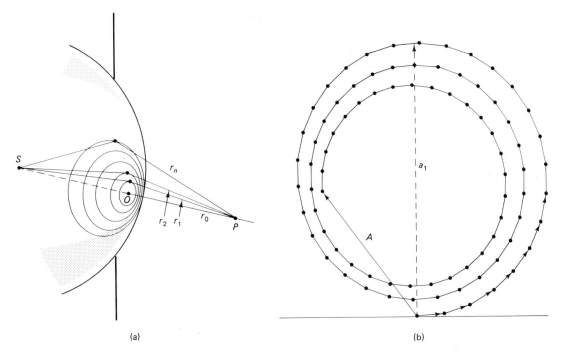

Figure 18-4 (a) Fresnel circular half-period zones on a spherical wavefront emerging from an aperture. (b) Phasor diagram for circular Fresnel zones. Each half-period zone is subdivided into 15 subzones. Individual phasors indicate the average phase angle of the subzones and are progressively shorter by 5% to simulate the effect of the obliquity factor. The amplitude a_1 represents the first half-period zone, and A represents all the zones shown, about $5\frac{1}{2}$ half-periods.

for which the phase varies from one end to the other by π. One can show that the resultant contribution from these subzones has an effective phase intermediate between the phases at the zone beginning and end, such that effective phases from successive half-period zones are π, or 180°, apart. This is also clear from Figure 18-4b, a phasor diagram in which each half-period zone is subdivided into 15 subzones. Each of the small phasors represents the contribution from one subzone. The first half-period zone is completed after a number of such phasors produce a phasor opposite in direction to the first. The amplitude a_1 (dashed line) represents the resultant of the subzones from the first half-period zone. Notice that it makes an angle of 90° relative to the reference or axial direction, so that the effective phasor from the first zone has a phase of $\pi/2$. For a large number of subzones, the phasor diagram becomes circular and the magnitude of a_1 is the diameter of the circle. The obliquity factor is taken into account in Figure 18-4b by making each succeeding phasor slightly shorter than the preceding one. Thus the circles do not close but spiral inward. The summation of the waves at P from each half-period zone can be expressed as

$$A = a_1 + a_2 e^{i\pi} + a_3 e^{i2\pi} + a_4 e^{i3\pi} + \cdots$$

or

$$A = a_1 - a_2 + a_3 - a_4 + \cdots \tag{18-14}$$

The successive zonal amplitudes are affected by three different considerations: (1) a gradual increase with N due to slightly increasing zonal areas, (2) a gradual decrease with N due to the inverse square law effect as distances from P increase, and (3) a gradual decrease with N due to the obliquity factor. With regard to the first of these, it can be shown that the surface area S_N of the Nth Fresnel zone is given by

$$S_N = \frac{\pi r_0' r_0^2}{r_0 + r_0'}\left[\frac{\lambda}{r_0} + (2N - 1)\left(\frac{\lambda}{2r_0}\right)^2\right] \qquad (18\text{-}15)$$

The quantity (λ/r_0) is very small in most cases of interest. If the second term in the brackets is accordingly neglected compared to the first, Eq. (18-15) describes zones with equal areas (independent of N), given by

$$S_N \cong \left[\frac{\pi r_0'}{(r_0 + r_0')}\right] r_0 \lambda \qquad (18\text{-}16)$$

The existence of the second term, however small, indicates increases in zonal areas with N and corresponding small increases in the successive terms of Eq. (18-14). Now one can show that these increases are canceled by the decreases that arise from the second consideration, the effect of the inverse square law. This leaves only the obliquity factor, which is responsible for systematic decreases in the amplitudes as N increases.

A phasor diagram for the amplitude terms of Eq. (18-14) is shown in Figure 18-5a, as each zone contribution is added. The corresponding resultant phasors are shown in Figure 18-5b. The individual phasors in Figure 18-5a are separated verti-cally for clarity. Each phasor is out of phase with its predecessor by 180° and is also shorter, due to the obliquity factor. The resultant phasors in Figure 18-5b begin at the start of the phasor a_1 and terminate at the end of the phasor a_N for any number N of contributing zones. Notice the large changes in the resultant phasor A_N, for small N, as the contribution from each new zone is added. For N large, the diagram shows clearly that the resultant amplitude A_N approaches a value of $A_R = a_1/2$, or half of that of the first contributing zone. The resultant amplitude is seen to oscillate be-tween magnitudes that are larger and smaller than the limiting value of $a_1/2$, de-pending on whether it represents an even or an odd number of contributing zones. A

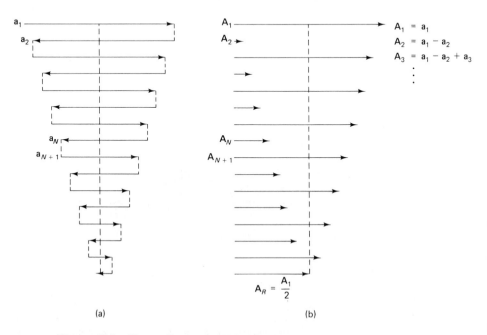

(a)

(b)

Figure 18-5 Phasor diagram for Fresnel half-period zones. Individual phasors are shown in (a) and the resultant phasors at each step in (b).

careful study of Figure 18-5 shows that for N zones, where N is even, the resultant amplitude may be expressed approximately by

$$N \text{ even:} \quad A_N \cong \frac{a_1}{2} - \frac{a_N}{2} \qquad (18\text{-}17)$$

and where N is odd by

$$N \text{ odd:} \quad A_N \cong \frac{a_1}{2} + \frac{a_N}{2} \qquad (18\text{-}18)$$

We may use either Figure 18-5 or Eqs. (18-17) and (18-18) to make the following conclusions:

1. If N is small so that $a_1 \cong a_N$, then for N odd the resultant amplitude is essentially a_1, that of the first zone alone; for N even, the resultant amplitude is near zero.
2. If N is large, as in the case of unlimited aperture, a_N approaches zero, and for either N even or odd, the resultant amplitude is half that of the first contributing zone, or $a_1/2$.

These conclusions produce some curious results, which can be verified experimentally. For example, suppose an amplitude $A_P = a_1$ is measured at P when a circular aperture coincides with the first Fresnel zone. Then by opening the aperture wider to admit the second zone as well, the additional light produces almost zero amplitude at P! Now remove the opaque shield altogether, so that all zones of an unobstructed wavefront contribute. The amplitude at P becomes $a_1/2$, or half that due to the tiny first-zone aperture alone. Since an irradiance is proportional to the square of the amplitude, the unobstructed irradiance at P is only $\frac{1}{4}$ that due to the first-zone aperture alone. Such results are surprising because they are not apparent in ordinary experience; yet they necessarily follow once Figure 18-5 is understood.

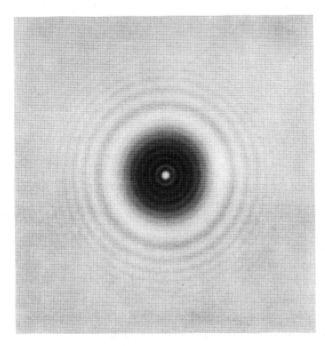

Figure 18-6 Diffraction pattern due to an opaque, circular disc, showing the *Poisson spot* at the center. (From M. Cagnet, M. Francon, and J. C. Thrierr, *Atlas of Optical Phenomenon*, Plate 33, Berlin: Springer-Verlag, 1962.)

Another conclusion that is of some historic interest follows from a consideration of the effect at P when a round obstacle or disc just covering the first zone is substituted for the aperture. The light reaching P is now due to all zones *except* the first. The first contributing zone is therefore the second zone, and by the same arguments as those used above, we conclude that light of amplitude $a_2/2$ occurs at P. Thus the irradiance at the center of the shadow of the obstacle should be almost the same as with no disc present! When Fresnel's paper on diffraction was presented to the French Academy, Poisson argued that this prediction was patently absurd and so undermined its theoretical basis. However, Fresnel and Arago showed experimentally that the spot, now known somewhat ironically as *Poisson's spot*, did occur as predicted. The diffraction pattern of an opaque circular disc, including the celebrated Poisson spot, is shown in Figure 18-6. As often happens in such cases, conclusive experimental evidence was already on hand, observed nearly a century before the argument. Here is an illustration of the need to fit experimental results into a successful conceptual framework if they are to make an impact on the scientific world.

18-5 PHASE SHIFT OF THE DIFFRACTED LIGHT

The first phasor in Figure 18-5 is drawn, rather arbitrarily, in a horizontal direction, and the other phasors are then related to it. As we have seen, however, the first phasor, due to the first Fresnel zone, has an effective phase of $\pi/2$ *behind* that of the light reaching P along the axis. The directly propagated light could therefore be represented by a phasor in the vertical direction, making an angle of $\pi/2$ with a_1. Now the resultant phasor of N zones is also in the direction of a_1. We are forced by these observations to conclude that the phase of the light at P, deduced from the Fresnel zone scheme, is at variance by 90° relative to the phase of the light reaching P directly along the axis. To remove this discrepancy and to make the results agree with the phase of the wave without diffraction, Fresnel was forced to assume that the secondary wavelets on diffraction leave with a *gain* in phase of $\pi/2$ relative to the incident wavefront. The factor of i introduced in Eq. (18-6) for this purpose appears naturally in the Kirchhoff derivation of the same equation.

18-6 THE FRESNEL ZONE PLATE

Examination of Eq. (18-14) suggests that if either the negative or the positive terms are eliminated from the sum, the resultant amplitude and irradiance could be quite large. Practically, this means that every other Fresnel zone in the wavefront should be blocked. Figure 18-7 shows a drawing of 16 Fresnel zones in which alternate zones are shaded. If such a picture is photographed and a transparency in reduced size is prepared, a *Fresnel zone plate* is produced. Let the light incident on such a zone plate consist of plane wavefronts. Then the zone radii required to make the zones half-period zones relative to a fixed field point P can be calculated. From Figure 18-8, the radius R_N of the Nth zone must satisfy

$$R_N^2 = \left(r_0 + \frac{N\lambda}{2}\right)^2 - r_0^2 \tag{18-19}$$

which can be written as

$$R_N^2 = r_0^2\left[N\left(\frac{\lambda}{r_0}\right) + \frac{N^2}{4}\left(\frac{\lambda}{r_0}\right)^2\right]$$

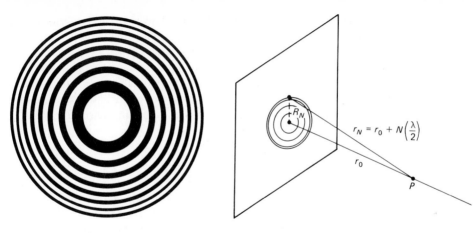

Figure 18-7   Fresnel zone plate.

Figure 18-8   Schematic for the calculation of Fresnel zone plate radii.

Since we restrict our discussion to cases where $\lambda/r_0 \ll 1$, the second term in square brackets is negligible compared with the first. For example, taking $\lambda = 600$ nm and $r_0 = 30$ cm, $\lambda/r_0 = 2 \times 10^{-6}$. The zone plate radii are thus given approximately by

$$R_N = \sqrt{Nr_0\lambda} \qquad (18\text{-}20)$$

Evidently the radii of successive zones in Figure 18-7 increase in proportion to $\sqrt{N}$. The radius of the first ($N = 1$) zone determines the magnitude of r_0, or the point P on the axis for which the configuration functions as a zone plate. If the first zone has radius R_1, then successive zones have radii of $1.41R_1$, $1.73R_1$, $2R_1$, and so on.

Example

If light of wavelength 632.8 nm illuminates a zone plate, what is the first zone radius relative to a point 30 cm from the zone plate on the central axis? How many half-period zones are contained in an aperture with a radius 100 times larger?

Solution   Using Eq. (18-20),

$$R_1 = \sqrt{(1)(30)(632.8 \times 10^{-7})} = 0.0436 \text{ cm}$$

Since $R \propto \sqrt{N}$, N increases by a factor of 10^4 when R increases by a factor of 10^2. Thus a radius of 4.36 cm encompasses 10,000 Fresnel zones.

If the first, third, fifth, ... zones are transmitting, as in Figure 18-7, then Eq. (18-14) becomes

$$A_{16} = a_1 + a_3 + a_5 + a_7 + a_9 + a_{11} + a_{13} + a_{15}$$

with 8 zones contributing. When these few zones are reproduced on a smaller scale, the obliquity factor is not very important, and we may approximate $A_{16} = 8a_1$. By comparison, this amplitude at P is 16 times the amplitude ($a_1/2$) of a wholly unobstructed wavefront. The irradiance at P is, therefore, $(16)^2$, or 256, times as great, even for an aperture encompassing only these 16 zones. If P is 30 cm away, as in the previous example, the radius of this aperture, by Eq. (18-20), is only 1.74 mm for 632.8 nm light. This concentration of light at an axial point shows that the zone plate operates as a lens with P as a focal point. Rearranging Eq. (18-20), we call the distance r_0 the first focal point ($N = 1$) with focal length f_1 given by

$$f_1 = \frac{R_1^2}{\lambda} \qquad (18\text{-}21)$$

There are other focal points as well. As the field point P approaches the zone plate along the axis, the same zonal area of radius R_1 encompasses more half-period zones. In Eq.(18-20), when R_N is fixed, N increases as r_0 decreases. Thus as P is moved toward the plate, $N = 2$ when $r_0 = f_1/2$ for the same zonal radius R_1. At this point each of the original zones covers two half-period zones and all zones cancel. When $r_0 = f_1/3$, $N = 3$, and three zones contribute from the original zone of radius R_1. Of these, two cancel, but one is left to contribute. The next three are opaque, and so on, as in Eq. (18-22):

$$A = \underbrace{a_1 - a_2 + a_3}_{a_1} - \underbrace{a_4 + a_5 - a_6}_{\text{removed}} + \underbrace{a_7 - a_8 + a_9}_{a_7} - \cdots \qquad (18\text{-}22)$$

Of the first six original zones, one zone contributes rather than three, so that the irradiance at $r_0/3$ is nearly $\frac{1}{9}$ that at r_0. The argument may, of course, be extended to $r_0/5$, when the original zone of radius R_1 includes five zones, and the irradiance is $\frac{1}{25}$ that at r_0, and so on. Thus other maximum intensity points along the axis are to be found at

$$f_N = \frac{R_1^2}{N\lambda}, \qquad N \text{ odd} \qquad (18\text{-}23)$$

Example

What are the focal lengths for the zone plate described in the preceding example?

Solution Using Eq. (18-23) together with Eq. (18-20),

$$f_N = \frac{R_1^2}{N\lambda} = \frac{r_0\lambda}{N\lambda} = \frac{r_0}{N}$$

so that $f_1 = 30/1 = 30$ cm, $f_3 = 30/3 = 10$ cm, $f_5 = 30/5 = 6$ cm, and so on.

18-7 FRESNEL DIFFRACTION FROM APERTURES WITH RECTANGULAR SYMMETRY

Diffraction by straight edges, rectangular apertures, and wires are all conveniently handled by another approximation to the Fresnel-Kirchhoff diffraction formula, Eq. (18-6). For this geometry, let the source S in Figures 18-1 and 18-2 represent a slit, so that the wavefronts emerging from S are cylindrical. Recall that cylindrical waves can be expressed mathematically in the same form as spherical waves, except that the amplitude decreases as $1/\sqrt{r}$ so that the irradiance decreases as $1/r$.

Before pursuing Fresnel's quantitative treatment of such cases, consider qualitatively what we might expect by using again the concept of Fresnel half-period zones. This time the zones are rectangular strips along the wavefront, as in Figure 18-9. We wish to show that the sum of all phasors now gives the end points of a curve called the *Cornu spiral*. As before, the average phase at P of the light from each successive zone advances by a half-period, or π. In Figure 18-9b, the rectangular strip zones are shown both above and below the axis SP. Unlike the Fresnel circular zones, the areas of the zones fall off markedly with N so that successive phasor amplitudes of the zonal contributions are distinctly shorter. A phasor diagram for the complex amplitudes from the zones above the axis might look like Figure 18-10. If the first zone is subdivided into smaller parts, which advance by equal phases, the corresponding phasors can be represented by $a_1, a_2, \ldots$, as shown.

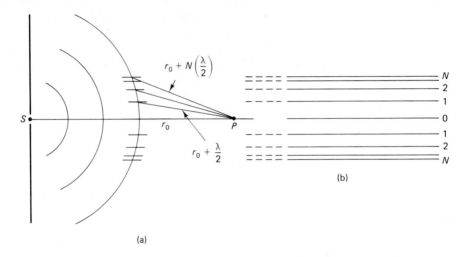

$r_0 + N\left(\dfrac{\lambda}{2}\right)$

r_0

$r_0 + \dfrac{\lambda}{2}$

S

P

N
2
1
0
1
2
N

(b)

(a)

Figure 18-9 Fresnel half-period strip zones on a cylindrical wavefront in an (a) edge view and (b) front view.

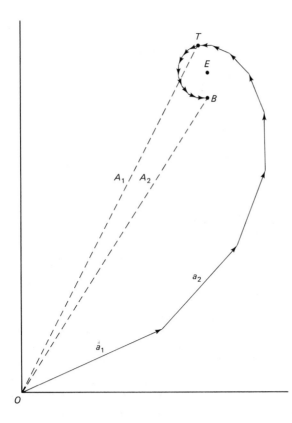

T

E

B

A_1 A_2

a_2

a_1

O

Figure 18-10 Phasor diagram for the first two half-period Fresnel zone strips, each subdivided into smaller zones of equal phase increment.

When the first half-period zone has been included, the last phasor is advanced by π relative to the first and ends at T. The sum of all these contributions is then the phasor A_1. In the case of circular zones, Figure 18-4b, the resultant phasor has a phase angle of $\pi/2$ and the point T falls on the vertical axis. Because of the rapid decrease in phasor magnitudes in this case, the phase angle is less than $\pi/2$. After advancing through the subzones of the second half-period zone, the phase changes by another 180° and the last phasor ends at B. The resultant phasor, which includes two full half-period zones, is A_2. By continuing this process, one sees that the phasors approach a smooth curve, which spirals in to a limit point E, the eye of the spiral. A phasor A_R from O to E then represents the contributions of half the unimpeded wavefront, above the axis SP in Figure 18-9a. A similar argument for the zones below the axis would lead to a twin spiral, represented in the third quadrant and connecting at the origin O. If the coordinates of all points of this Cornu spiral are known, the amplitudes due to contributions from any number of zones can be determined from such a drawing and the relative irradiances compared. The quantitative treatment that allows us to make such calculations follows.

18-8 THE CORNU SPIRAL

If we neglect the effect of the obliquity factor and the variation of the product rr' in the denominator of Eq. (18-6), the Fresnel-Kirchhoff integral may be approximated by

$$E_P = C \iint_{A_P} e^{ik(r+r')} \, da \tag{18-24}$$

where all constants are coalesced into C. We assume that the surface integral over a closed surface including the aperture is zero everywhere except over the aperture itself, so that we need perform the integration only over the aperture in the yz-plane of Figure 18-11a. A side view, which shows the curvature of the cylindrical wavefront, is shown in Figure 18-11b. The distance $r + r'$ may be determined approximately from this figure. For $h \ll p$ and $h \ll q$,

$$r' = (p^2 + h^2)^{1/2} = p\left(1 + \frac{h^2}{p^2}\right)^{1/2} \cong p\left(1 + \frac{h^2}{2p^2}\right)$$

Thus

$$r' \cong p + \frac{1}{2}\left(\frac{h^2}{p}\right)$$

and similarly,

$$r \cong q + \frac{1}{2}\left(\frac{h^2}{q}\right)$$

It follows that

$$r + r' \cong (p + q) + \left(\frac{1}{p} + \frac{1}{q}\right)\frac{h^2}{2}$$

If we abbreviate, using

$$D \equiv p + q \quad \text{and} \quad \frac{1}{L} \equiv \frac{1}{p} + \frac{1}{q} \tag{18-25}$$

we have

$$r + r' \cong D + \frac{h^2}{2L} \tag{18-26}$$

Then Eq. (18-24) becomes

$$E_P = C \iint e^{ik(D+h^2/2L)} \, da$$

(a)

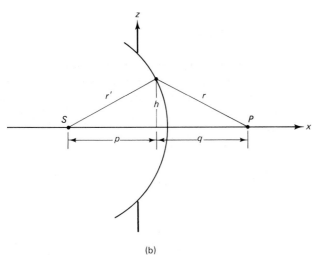

(b)

Figure 18-11 (a) Cylindrical wavefronts from source slit S are diffracted by a rectangular aperture. (b) Edge view of (a).

If the elemental area da is taken to be the shaded strip in Figure 18-11a, $da = W\,dz$, $h = z$, and

$$E_P = CWe^{ikD} \int e^{ikz^2/2L}\,dz \qquad (18\text{-}27)$$

The exponent $kz^2/2L = \pi z^2/L\lambda$. Making a change of variable, we let

$$z = \sqrt{\frac{\lambda L}{2}}\,v \quad \text{or} \quad v = \sqrt{\frac{2}{L\lambda}}\,z \qquad (18\text{-}28)$$

whereby the magnitude of E_P is

$$|E_P| = CW\left| \int e^{i\pi v^2/2} \sqrt{\frac{L\lambda}{2}}\,dv \right|$$

Using Euler's theorem on the integrand and recombining constants again, we may write

$$|E_P| = E_u\left| \int \cos\left(\frac{\pi v^2}{2}\right) dv + i \int \sin\left(\frac{\pi v^2}{2}\right) dv \right| \qquad (18\text{-}29)$$

The two integrals in this form are the *Fresnel integrals*, which we name

$$C(v) \equiv \int_0^v \cos\left(\frac{\pi v^2}{2}\right) dv \qquad (18\text{-}30)$$

$$S(v) \equiv \int_0^v \sin\left(\frac{\pi v^2}{2}\right) dv \qquad (18\text{-}31)$$

Table 18-1 provides numerical values of these definite integrals for various values of v. As we shall see in several applications, choice of the upper limit v in the Fresnel

TABLE 18-1 FRESNEL INTEGRALS

v	$C(v)$	$S(v)$	v	$C(v)$	$S(v)$
0.00	0.0000	0.0000	4.50	0.5261	0.4342
0.10	0.1000	0.0005	4.60	0.5673	0.5162
0.20	0.1999	0.0042	4.70	0.4914	0.5672
0.30	0.2994	0.0141	4.80	0.4338	0.4968
0.40	0.3975	0.0334	4.90	0.5002	0.4350
0.50	0.4923	0.0647	5.00	0.5637	0.4992
0.60	0.5811	0.1105	5.05	0.5450	0.5442
0.70	0.6597	0.1721	5.10	0.4998	0.5624
0.80	0.7230	0.2493	5.15	0.4553	0.5427
0.90	0.7648	0.3398	5.20	0.4389	0.4969
1.00	0.7799	0.4383	5.25	0.4610	0.4536
1.10	0.7638	0.5365	5.30	0.5078	0.4405
1.20	0.7154	0.6234	5.35	0.5490	0.4662
1.30	0.6386	0.6863	5.40	0.5573	0.5140
1.40	0.5431	0.7135	5.45	0.5269	0.5519
1.50	0.4453	0.6975	5.50	0.4784	0.5537
1.60	0.3655	0.6389	5.55	0.4456	0.5181
1.70	0.3238	0.5492	5.60	0.4517	0.4700
1.80	0.3336	0.4508	5.65	0.4926	0.4441
1.90	0.3944	0.3734	5.70	0.5385	0.4595
2.00	0.4882	0.3434	5.75	0.5551	0.5049
2.10	0.5815	0.3743	5.80	0.5298	0.5461
2.20	0.6363	0.4557	5.85	0.4819	0.5513
2.30	0.6266	0.5531	5.90	0.4486	0.5163
2.40	0.5550	0.6197	5.95	0.4566	0.4688
2.50	0.4574	0.6192	6.00	0.4995	0.4470
2.60	0.3890	0.5500	6.05	0.5424	0.4689
2.70	0.3925	0.4529	6.10	0.5495	0.5165
2.80	0.4675	0.3915	6.15	0.5146	0.5496
2.90	0.5624	0.4101	6.20	0.4676	0.5398
3.00	0.6058	0.4963	6.25	0.4493	0.4954
3.10	0.5616	0.5818	6.30	0.4760	0.4555
3.20	0.4664	0.5933	6.35	0.5240	0.4560
3.30	0.4058	0.5192	6.40	0.5496	0.4965
3.40	0.4385	0.4296	6.45	0.5292	0.5398
3.50	0.5326	0.4152	6.50	0.4816	0.5454
3.60	0.5880	0.4923	6.55	0.4520	0.5078
3.70	0.5420	0.5750	6.60	0.4690	0.4631
3.80	0.4481	0.5656	6.65	0.5161	0.4549
3.90	0.4223	0.4752	6.70	0.5467	0.4915
4.00	0.4984	0.4204	6.75	0.5302	0.5362
4.10	0.5738	0.4758	6.80	0.4831	0.5436
4.20	0.5418	0.5633	6.85	0.4539	0.5060
4.30	0.4494	0.5540	6.90	0.4732	0.4624
4.40	0.4383	0.4622	6.95	0.5207	0.4591

integrals is determined by the vertical dimension of the diffraction aperture, measured in terms of the aperture height z.

Now the irradiance at P, since $I_p \propto |E_P|^2$, is given by

$$I_P = I_0(C + iS)(C - iS) = I_0(C^2 + S^2) \qquad (18\text{-}32)$$

If the values of the Fresnel integrals are plotted against the variable v, as real and imaginary coordinates on the complex plane, the resulting graph is the Cornu spiral (Figure 18-12). According to Eq. (18-32), a straight line drawn between any two points of the spiral must be proportional to an amplitude of the electric field at P, since we have, by the Pythagorean relation,

$$E_P^2 \propto C^2 + S^2$$

where C and S are distances along the axes of a rectangular coordinate system. The origin $v = 0$ corresponds to $z = 0$ and therefore to the y-axis through the aperture of Figure 18-11a. The top part of the spiral ($z > 0$ and $v > 0$) represents contributions from strips of the aperture above the y-axis, and the twin spiral below ($z < 0$ and $v < 0$) represents similar contributions from below the y-axis. The limit points or eyes of the spiral at E and E' represent linear zones at $z = \pm\infty$. Furthermore, we may show that the variable v represents the length along the Cornu spiral itself. The incremental length dl along a curve in the xy-plane is given in general by

$$dl^2 = dx^2 + dy^2$$

Figure 18-12 The Cornu spiral. Intervals along the spiral measure the variable v.

Here x and y are the Fresnel integrals C and S, respectively, giving

$$dl^2 = \left[\cos^2\left(\frac{\pi v^2}{2}\right) + \sin^2\left(\frac{\pi v^2}{2}\right)\right] dv^2$$

or simply,

$$dl = dv \qquad (18\text{-}33)$$

18-9 APPLICATIONS OF THE CORNU SPIRAL

Approximate evaluations of the Kirchhoff-Fresnel integral are possible with the help of the Cornu spiral. We examine a few special cases.

Unobstructed Wavefront. The vertical dimension of the aperture ranges from $-\infty$ to $+\infty$. The resultant amplitude E_P on the Cornu spiral in this case is a phasor drawn from E' to E, as shown in Figure 18-13. These points have the coordinates $(-0.5, -0.5)$ and $(0.5, 0.5)$, as is evident by evaluating

$$\int_0^\infty \cos\left(\frac{\pi v^2}{2}\right) dv = \int_0^\infty \sin\left(\frac{\pi v^2}{2}\right) dv = 0.5$$

with the help of the definite integral formula $\int_0^\infty \cos(x^2)\, dx = \frac{1}{2}\sqrt{\pi/2}$. Then the x- and y-components of E_P are both unity and $E_P = \sqrt{2}$, or, using Eq. (18-32),

$$I_P = 2I_0 = I_u \qquad (18\text{-}34)$$

Other irradiances may conveniently be compared to this value of $I_u = 2I_0$ for the unobstructed wavefront.

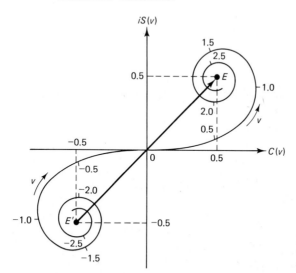

Figure 18-13 The phasor representing the unobstructed wavefront has a magnitude on the Cornu spiral of $\sqrt{2}$.

Straight Edge. Fresnel diffraction by a straight edge is pictured in Figure 18-14a. At the field point P on the axis, the edge of the geometric shadow for which $z = v = 0$, the upper half of the zones and Cornu spiral are effective. The resulting amplitude, shown as OE in Figure 18-14b, has a magnitude of $1/\sqrt{2}$ and, consequently,

$$I_P = \tfrac{1}{2}I_0 = \tfrac{1}{4}I_u \qquad (18\text{-}35)$$

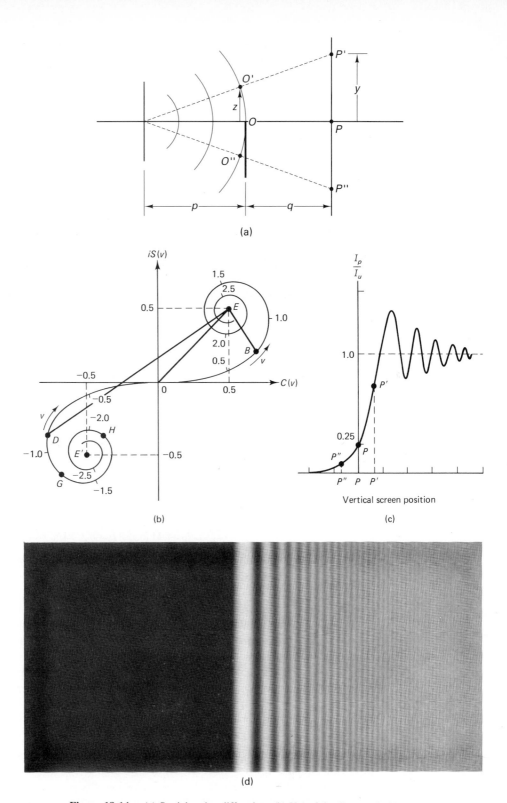

Figure 18-14 (a) Straight-edge diffraction. (b) Use of the Cornu spiral in analyzing straight-edge diffraction. (c) Irradiance pattern due to straight-edge Fresnel diffraction. (d) Diffraction fringes from a straight line. (From M. Cagnet, M. Francon, and J. C. Thrierr, *Atlas of Optical Phenomenon*, Plate 32, Berlin: Springer-Verlag, 1962.)

The irradiance is plotted as point P in Figure 18-14c. For a lower point P'' on the screen, we now consider the zones relative to the axis $O''P''$, drawn from P'' to the wavefront at the aperture. For P'', the point O'' marks the center of the wavefront, just as the point O marks the center of the wavefront relative to point P. Thus above the axis $O''P''$, the new "upper half of the wavefront," some of the first zones do not contribute to the irradiance at point P'' on the screen, due to the obstruction. Of course, all the bottom half of the wavefront is similarly blocked off. The contributing zones, beginning at a finite positive value of z and continuing to ∞, are represented by the amplitude BE on the Cornu spiral. As the observation point P'' moves from P to lower points on the screen, the representative phasor point B slides along the Cornu spiral from O, with its other end fixed at E. One sees that the amplitude, and so the irradiance, must decrease monotonically, as shown in Figure 18-14c. The edge of the shadow is clearly not sharp. On the other hand, for a point P' above P, we conclude that relative to its axis $O'P'$, all the upper zones plus some of the first lower zones contribute. The corresponding amplitude in Figure 18-14b is like DE. As P' moves up the screen, D moves down along the spiral. In this case, as D winds around the turns of the spiral, the length of DE oscillates with various maxima and minima points, as shown in Figure 18-14c.

Example

Calculate the irradiance at the first maximum above the edge of the shadow.

Solution   At the first maximum point, the tail of the phasor is at the extreme point G relative to E. From Figure 18-12, we read from the curve the value $v = 1.2$ at this point. Then from Table 18-1,

$$C(1.2) = -0.7154 \quad \text{and} \quad S(1.2) = -0.6234$$

whereas at E, $C(\infty) = S(\infty) = 0.5$. The magnitude of the phasor GE is then

$$E_P = [(-0.7154 - 0.5)^2 + (-0.6234 - 0.5)^2]^{1/2} = 1.655$$

and

$$I_P = (1.655)^2 I_0 = 2.74 I_0 = 1.37 I_u$$

The irradiance at the first maximum is 1.37 times that of the unobstructed irradiance.

Similarly, calculating the magnitude of the first minimum amplitude HE, one finds that $I_P = 0.78 I_u$, and so on. The irradiance curve approaches the value I_u due to the unobstructed wavefront. A photograph of the pattern is given in Figure 18-14d.

Notice that, from Figure 18-14a, it is possible to relate points at elevation y on the screen to corresponding points at elevation z on the wavefront, such that

$$y = \left(\frac{p + q}{p}\right) z$$

The value of z determines the length v on the Cornu spiral, permitting quantitative calculations of screen irradiance to be made.

Single Slit.   If the diffracting aperture is a single slit of width w, as in Figure 18-15a, then $z = w$, and by Eq. (18-28),

$$\Delta v = \sqrt{\frac{2}{L\lambda}} \, \Delta z = \sqrt{\frac{2}{L\lambda}} \, w \qquad (18\text{-}36)$$

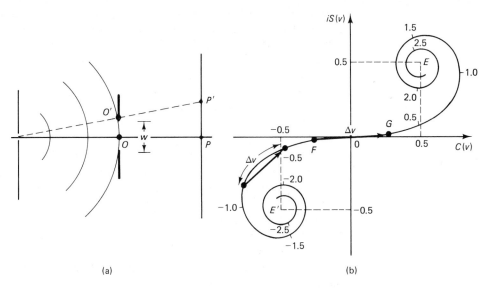

Figure 18-15 Fresnel diffraction from a single slit (a) and its amplitude representation on the Cornu spiral (b).

Once L is calculated from Eq. (18-25), the contributing interval Δv on the Cornu spiral can be determined. Note that v plays the role of a universal, dimensionless variable, allowing one Cornu spiral to serve for various combinations of p, q, and λ. For example, if $p = q = 20$ cm and $\lambda = 500$ nm, $\Delta v = 0.632$ for a slit width of 0.01 cm. To calculate the irradiance at P, a length of $v = 0.632$ along the Cornu spiral, symmetrically placed about the origin, as shown, determines the amplitude E_P. For a point like P' above P, the contributing zones form a group in the lower half of the unobstructed wavefront. Their center point z' below the axis $O'P'$ determines the center v' of the length Δv along the spiral, by Eq. (18-28). As P' moves farther above the axis, the length Δv slides toward the lower eye of the spiral, as shown. Although Δv remains fixed in length, its placement on different portions of the spiral determines a different chord length and a different E_P. When P' is below the axis, Δv is placed along the upper spiral. In this way, the irradiance of the entire pattern can be calculated. From this approach, one can reason that the diffraction pattern of the slit is symmetrical about its center and that the irradiance, while oscillatory, is never zero.

Example

Let the wavelength of the light be 500 nm and the slit of Figure 18-15a be 1 mm in width. Light emerges from a source slit, as shown, that is 20 cm from the diffracting slit. The diffraction pattern is observed 30 cm from the slit. What is the irradiance at a height of 1 mm above the axis at the screen?

Solution The parameter $L = pq/(p + q) = (20)(30)/(20 + 30) = 12$ cm. We are looking for the irradiance at a point like P' in Figure 18-15a. Contributing zones are like those included in the chord FG of Figure 18-15b, but with the spiral length FG moved somewhat toward the lower end. Fewer zones make a contribution from the upper half of the wavefront relative to P' than they do relative to P. At the screen, point P' corresponds to $y = 1$ mm. The corresponding point on the wavefront is

$$z = \left(\frac{p}{p + q}\right)y = \frac{20}{50}(1) = 0.4 \text{ mm}$$

The intersection of the dashed line (Figure 18-15a) with the wavefront at the slit thus occurs 0.1 mm below the upper slit edge. The contributing zones as "seen" from P' thus include zones from $z = +0.1$ mm to $z = 0$ in the upper half of its wavefront, and from $z = 0$ to $z = -0.9$ mm in the lower half. Contributing zones then span a continuous range from $z_1 = +0.1$ mm to $z_2 = -0.9$ mm relative to the axis from the source slit to the point P'. Corresponding endpoints on the Cornu spiral are v_1 and v_2. For example,

$$v_1 = \sqrt{\frac{2}{L\lambda}}\, z = \sqrt{\frac{2}{(0.12)(500 \times 10^{-9})}}\,(0.1 \times 10^{-3}) = 0.57735$$

Similarly, $v_2 = 5.19615$. The contributing amplitude is the chord from v_1 to v_2 on the Cornu spiral. Coordinates of these points are found by interpolation in Table 18-1, giving

For v_1: $C(0.57735) = 0.56099$ and $S(0.57735) = 0.10013$

For v_2: $C(5.19615) = -0.44016$ and $S(5.19615) = -0.50043$

The amplitude is calculated from

$$E_p^2 = [0.56099 - (-0.44016)]^2 + [0.10013 - (-0.50043)]^2 = 1.3630$$

Thus the irradiance at the point P is $I_p = 1.363\, I_0 = 0.681\, I_u$. The irradiance at the screen, 1 mm above axis, is 0.68 times the irradiance of an unobstructed wavefront there.

Wire. Suppose now the narrow slit of Figure 18-15a is replaced by a long, but thin, opaque obstacle such as a wire (Figure 18-16). If the width w of slit and wire are equal, then there is an exact reversal of the transmitting and blocking zones of the wavefront. Now all parts of the Cornu spiral are to be used in calculating the resulting amplitude E_P *except* that portion designated by Δv in Figure 18-16b. This

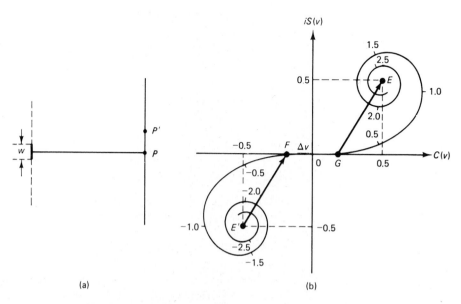

Figure 18-16   Fresnel diffraction from a wire (a) and its amplitude representation on the Cornu spiral (b).

situation clearly yields two amplitudes, as shown, which must be added together. The interval Δv *omitted* slides along the spiral as before, when different field points like P' are considered. The composite pattern is shown in the actual photograph of Figure 18-17. An example of a more complicated Fresnel pattern than those considered above is also given in Figure 18-18.

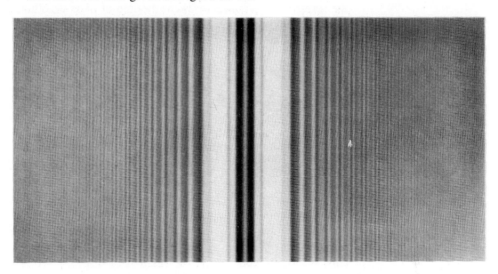

Figure 18-17   Fresnel diffraction pattern of a fine wire. (From M. Cagnet, M. Francon, and J. C. Thrierr, *Atlas of Optical Phenomenon*, Plate 32, Berlin: Springer-Verlag, 1962.)

Figure 18-18   Fresnel shadow of a screw. (From M. Cagnet, M. Francon, and J. C. Thrierr, *Atlas of Optical Phenomenon*, Plate 36, Berlin: Springer-Verlag, 1962.)

18-10 BABINET'S PRINCIPLE

Apertures like those of Figures 18-15 and 18-16, in which clear and opaque regions are simply reversed, are called *complementary apertures*. If one of the apertures, say A, and then the other, B, are put into place and the amplitude at some point of the screen is determined for each, the sum of these amplitudes must equal the unobstructed amplitude there. This is the content of *Babinet's principle*, which we express as

$$E_A + E_B = E_u \qquad (18\text{-}37)$$

with A and B representing any two complementary apertures. For example, for the slit and wire apertures we have considered, notice that in Figure 18-16b we may express the phasor addition at P on the screen by

$$E'F + FG + GE = E'E$$

where $E'F + GE$ represents the amplitude due to the wire, FG represents the amplitude due to the slit, and $E'E$ represents the unobstructed amplitude.

An interesting special case of Babinet's principle is a point where $E_u = 0$. Then, by Eq. (18-37), $E_A = -E_B$ and $I_A = I_B$ at the point. In practice, Fresnel diffraction does not produce amplitudes $E_u = 0$ without an aperture. Fraunhofer diffraction does, however, as in the case of the pattern formed by a point source and a lens. For the region outside the small Airy disc, $E_u = 0$, essentially. Complementary apertures introduced into such systems then give, outside the central image, identical diffraction patterns. Thus positive and negative transparencies of the same pattern produce the same diffraction pattern. This discussion is appropriately reminiscent of the fact that positive and negative holograms produce the same holographic image.

PROBLEMS

18-1. A 1-mm diameter hole is illuminated by plane waves of 546-nm light. According to the usual criterion, which techniques (near-field or far-field) may be applied to the diffraction problem when the detector is at 50 cm, 1 m, and 5 m from the aperture?

18-2. A 3-mm diameter circular hole in an opaque screen is illuminated normally by plane waves or wavelength 550 nm. A small photocell is moved along the central axis, recording the power density of the diffracted beam. Determine the locations of the first three maxima and minima as the photocell approaches the screen.

18-3. A distant source of sodium light (589.3 nm) illuminates a circular hole. As the hole increases in diameter, the irradiance at an axial point 1.5 m from the hole passes alternately through maxima and minima. What are the diameters of the holes that produce (a) the first two maxima and (b) the first two minima?

18-4. Plane waves of monochromatic (600 nm) light are incident on an aperture. A detector is situated on axis at a distance of 20 cm from the aperture plane.
(a) What is the value of R_1, the radius of the first Fresnel half-period zone, relative to the detector?
(b) If the aperture is a circle of radius 1 cm, centered on axis, how many half-period zones does it contain?
(c) If the aperture is a zone plate with every other zone blocked out and with the radius of the first zone equal to R_1 [found in (a)], determine the first three focal lengths of the zone plate.

18-5. The zone plate radii given by Eq. (18-20) were derived for the case of plane waves incident on the aperture. If instead the incident waves are spherical, from an axial point source at distance p from the aperture, show that the necessary modification yields

$$R_N = \sqrt{NL\lambda}$$

where q is the distance from aperture to the axial point of detection and L is defined by $1/L = 1/p + 1/q$.

18-6. Repeat parts (a) and (b) of problem 18-4 when the source is a point source 10 cm from the aperture. Take into account the results of problem 18-5.

18-7. A point source of monochromatic light (500 nm) is 50 cm from an aperture plane. The detection point is located 50 cm on the other side of the aperture plane.

  (a) The transmitting portion of the aperture plane is an annular ring of inner radius 0.500 mm and outer radius 0.935 mm. What is the irradiance at the detector, relative to the irradiance there for an unobstructed wavefront? The results of problem 18-5 will be helpful.

  (b) Answer the same question if the outer radius is 1.00 mm.

  (c) How many half-period zones are included in the annular ring in each case?

18-8. By what percentage does the area of the 25th Fresnel half-period zone differ from that of the first, for the case when source and detector are both 50 cm from the aperture and the source supplies light at 500 nm?

18-9. A zone plate is to be produced having a focal length of 2 m for He-Ne laser light of 632.8 nm. An ink drawing of 20 zones is made with alternate zones shaded in, and a reduced photographic transparency is made of the drawing.

  (a) If the radius of the first zone is 11.25 cm in the drawing, what reduction factor is required?

  (b) What is the radius of the last zone in the drawing?

18-10. A zone plate has its center half zone opaque. Find the diameters of the first three clear zones such that the plate focuses parallel light of wavelength 550 nm at 25 cm from the plate.

18-11. For an incident plane wavefront, show that the areas of the Fresnel half-period zones relative to an observation point at distance x from the wavefront are approximately constant and equal to $\pi\lambda x$. Assume that λ/x is much smaller than 1.

18-12. Light of wavelength 485 nm is incident normally on a screen. How large is a circular opening in an otherwise opaque screen if it transmits four Fresnel zones to a point 2 m away? What, approximately, is the irradiance at the point?

18-13. A single slit of width $\frac{1}{2}$ millimeter is illuminated by a collimated beam of light of wavelength 540 nm. At what observation point on the axis is $\Delta v = 2.5$?

18-14. A source slit at one end of an optical bench is illuminated by monochromatic mercury light of 435.8 nm. The beam diverging from the source slit encounters a second slit 0.5 mm wide at a distance of 30 cm. The diffracted light is observed on a screen at 15 cm farther along the optical bench. Determine the irradiance (in terms of the unobstructed irradiance) at the screen (a) on axis and (b) at one edge of the geometrical shadow of the diffracting slit.

18-15. A slit illuminated with sodium light is placed 60 cm from a straight edge and the diffraction pattern is observed using a photoelectric cell, 120 cm beyond the straight edge. Determine the irradiance at (a) 2 mm inside and (b) 1 mm outside the edge of the geometrical shadow.

18-16. Filtered green mercury light (546.1 nm) emerges from a slit placed 30 cm from a rod 1.5 mm thick. The diffraction pattern formed by the rod is examined in a plane at 60 cm beyond the rod. Calculate the irradiance of the pattern at (a) the center of the geometrical shadow of the rod and (b) the edge of the geometrical shadow.

18-17. For the near-field diffraction pattern of a straight edge, calculate the irradiance of the second maximum and minimum, using the Cornu spiral and the table of Fresnel integral values given.

18-18. Fresnel diffraction is observed behind a wire 0.37 mm thick, which is placed 2 m from the light source and 3 m from the screen. If light of wavelength 630 nm is used, compute, using the Cornu spiral, the irradiance of the diffraction pattern on the axis at the screen. Express the answer as some number times the unobstructed irradiance there.

18-19. Calculate the relative irradiance (compared to the unobstructed irradiance) on the optic axis due to a double-slit aperture that is both 10 cm from a point source of monochromatic light (546 nm) and 10 cm from the observation screen. The slits are 0.04 mm in width and separated (center to center) by 0.25 mm.

18-20. Single-slit diffraction is produced using a monochromatic light source (435.8 nm) at 25 cm from the slit. The slit is 0.75 mm wide. A detector is placed on the axis, 25 cm from the slit.

(a) Ensure that far-field diffraction is invalid in this case.

(b) Nevertheless, determine the distance above the axis at which single-slit Fraunhofer diffraction predicts the first zero in irradiance.

(c) Then calculate the irradiance at the same point, using Fresnel diffraction and the Cornu spiral. Express the result in terms of the unobstructed irradiance.

18-21. A glass plate is sprayed with uniform opaque particles. When a distant point source of light is observed looking through the plate, a diffuse halo is seen whose angular width is about 2°. Estimate the size of the particles. (*Hint*: Use Babinet's principle.)

REFERENCES

[1] Born, Max, and Emil Wolf. *Principles of Optics*, 5th ed. New York: Pergamon Press, 1975. Ch. 8.

[2] Guenther, Robert. *Modern Optics*. New York: John Wiley and Sons, 1990. Ch. 9.

[3] Reynolds, George O., John B. DeVelis, George B. Parrent, Jr., and Brian J. Thompson. *Physical Optics Notebook: Tutorials in Fourier Optics*. Bellingham, Wash.: SPIE Optical Engineering Press, 1989. Ch. 9.

[4] Longhurst, R. S. *Geometrical and Physical Optics*, 2d ed. New York: John Wiley and Sons, 1967. Ch. 13.

19

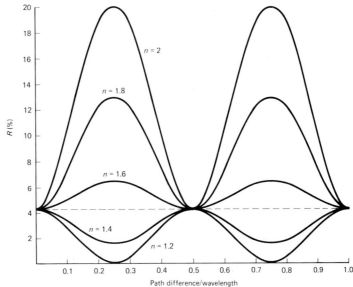

Theory
of Multilayer Films

INTRODUCTION

The physics of interference in single-layer dielectric films has been treated, in its essentials, in Chapter 10. Many useful and interesting applications of thin films, however, make use of multilayer stacks of films. It is possible to evaporate multiple layers while maintaining control over both refractive index (choice of material) and individual layer thickness. Such techniques provide a great deal of flexibility in designing interference coatings with almost any specified frequency-dependent reflectance or transmittance characteristics. Useful applications of such coatings include antireflecting multilayers for use on the lenses of optical instruments and display windows; multipurpose broad and narrow band-pass filters, available from near ultraviolet to near infrared wavelengths; thermal reflectors and cold mirrors, which reflect and transmit infrared, respectively, and are used in projectors; dichroic mirrors consisting of band-pass filters deposited on the faces of prismatic beam splitters to divide light into red, green, and blue channels in color television cameras; and highly reflecting dielectric mirrors for use in gas lasers and in Fabry-Perot interferometers.

Computer techniques have made routine the rather detailed calculations involved in the analysis of multilayer film performance. The design of a multilayer stack that will meet arbitrary prespecified characteristics, however, remains a formidable task. In this chapter we develop a *transfer matrix* to represent the film and characterize its performance. The approach differs from that used in treating multiple reflections from a thin film in Chapter 11. There we added the amplitudes of all the individual reflected or transmitted beams to find the resultant reflectance or transmittance. It will be more efficient, in the general treatment that follows, to consider all transmitted or reflected beams as already summed in corresponding electric fields that satisfy the general boundary conditions required by Maxwell's equations.

The relationships we require from electromagnetic theory, already presented in Chapter 8, are summarized here. The energy of a plane, electromagnetic wave propagates in the direction of the Poynting vector, given by

$$\mathbf{S} = \epsilon_0 c^2 \mathbf{E} \times \mathbf{B} \tag{19-1}$$

The magnitudes of electric and magnetic fields in the wave are related by

$$E = vB \tag{19-2}$$

where the wave speed can also be expressed by the refractive index,

$$n = \frac{c}{v} \tag{19-3}$$

The wave speed in vacuum is a constant, equal to

$$c = \frac{1}{\sqrt{\epsilon_0 \mu_0}} \tag{19-4}$$

where ϵ_0 and μ_0 are the permittivity and permeability, respectively, of free space. Combining Eqs. (19-2), (19-3), and (19-4), the magnitudes of the magnetic and electric fields can also be related by

$$B = \frac{E}{v} = \left(\frac{n}{c}\right)E = n\sqrt{\epsilon_0 \mu_0}\, E \tag{19-5}$$

19-1 TRANSFER MATRIX

Our analysis is carried out in terms of the quantities defined in Figure 19-1. An incident beam is shown, with $\mathbf{E}$ chosen for the moment in a direction perpendicular to the plane of incidence. (Keep in mind, however, that for normal incidence $E_\perp$ and $E_\parallel$ are equivalent since a unique plane of incidence cannot be specified.) The beam undergoes external reflection at the plane interface (a) separating the external medium of index n_0 from the nonmagnetic ($\mu = \mu_0$) film of index n_1. The transmitted portion of the beam undergoes an internal reflection and transmission at the plane interface (b) separating the film from the substrate of index n_s. Along each beam the $\mathbf{E}$-field is shown—by the usual dot notation—to be pointing out of the page ($-z$-direction), and the $\mathbf{B}$-field is shown in a direction consistent with Eq. (19-1). Notice that the y-component of $\mathbf{B}$ must reverse on reflection. The insets define a terminology for the magnitudes of the electric fields *at the boundaries* (a) and (b). For example, E_{r1} represents the sum of all the multiply reflected beams at interface (a) in the process of emerging from the film, E_{i2} represents the sum of all the multiple beams at interface (b) and directed toward the substrate, and so on. In this way, we account for multiple beams in the interference.

We assume that the film is both homogeneous and isotropic. We assume further that the film thickness is of the order of the wavelength of light, so that the path difference between multiply reflected and transmitted beams remains small compared with the coherence length of the monochromatic light. This ensures that the

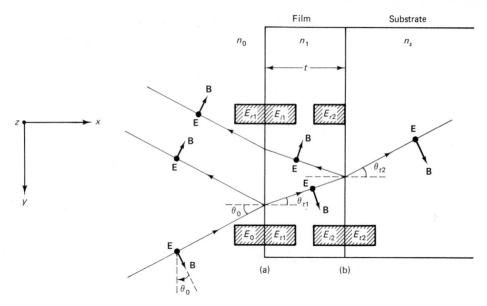

Figure 19-1 Reflection of a beam from a single layer. The diagram defines quantities used in applying boundary conditions to write Eqs. (19-6)–(19-9). Note that a bold dot is used to denote directions perpendicular to the plane of incidence.

beams are essentially coherent. The width of the incident beam, finally, is assumed to be large compared with its lateral displacement due to the many reflections that contribute significantly to the resultant reflected and transmitted beams.

Boundary conditions for the electric and magnetic fields of plane waves incident on the interfaces (a) and (b) are simply stated: The tangential components of the resultant **E**- and **B**-fields are continuous across the interface, that is, their magnitudes on either side are equal. For the case considered in Figure 19-1, **E** is everywhere tangent to the planes at (a) and (b), whereas **B** consists of both a tangential component (*y*-direction) and a perpendicular component (*x*-direction). Thus the boundary conditions for the electric field at the two interfaces become

$$E_a = E_0 + E_{r1} = E_{t1} + E_{i1} \tag{19-6}$$

$$E_b = E_{i2} + E_{r2} = E_{t2} \tag{19-7}$$

Corresponding equations for the magnetic field are

$$B_a = B_0 \cos\theta_0 - B_{r1} \cos\theta_0 = B_{t1} \cos\theta_{t1} - B_{i1} \cos\theta_{t1} \tag{19-8}$$

$$B_b = B_{i2} \cos\theta_{t1} - B_{r2} \cos\theta_{t1} = B_{t2} \cos\theta_{t2} \tag{19-9}$$

Rewriting Eqs. (19-8) and (19-9) in terms of electric fields with the help of Eq. (19-5),

$$B_a = \gamma_0(E_0 - E_{r1}) = \gamma_1(E_{t1} - E_{i1}) \tag{19-10}$$

$$B_b = \gamma_1(E_{i2} - E_{r2}) = \gamma_s E_{t2} \tag{19-11}$$

where we have written

$$\gamma_0 \equiv n_0\sqrt{\epsilon_0\mu_0}\,\cos\theta_0 \tag{19-12}$$

$$\gamma_1 \equiv n_1\sqrt{\epsilon_0\mu_0}\,\cos\theta_{t1} \tag{19-13}$$

$$\gamma_s \equiv n_s\sqrt{\epsilon_0\mu_0}\,\cos\theta_{t2} \tag{19-14}$$

Now E_{i2} differs from E_{t1} only because of a phase difference δ that develops due to one traversal of the film. Using half the phase difference calculated in Eq. (10-33) for two traversals of the film, we have

$$\delta = k_0 \Delta = \left(\frac{2\pi}{\lambda_0}\right) n_1 t \cos \theta_{t1} \qquad (19\text{-}15)$$

Thus

$$E_{i2} = E_{t1} e^{-i\delta} \qquad (19\text{-}16)$$

In the same way,

$$E_{i1} = E_{r2} e^{-i\delta} \qquad (19\text{-}17)$$

Using Eqs. (19-16) and (19-17) we may eliminate the fields E_{i2} and E_{r2} in the boundary conditions at (b), expressed by Eqs. (19-7) and (19-11), as follows:

$$E_b = E_{t1} e^{-i\delta} + E_{i1} e^{i\delta} = E_{t2} \qquad (19\text{-}18)$$

$$B_b = \gamma_1 (E_{t1} e^{-i\delta} - E_{i1} e^{i\delta}) = \gamma_s E_{t2} \qquad (19\text{-}19)$$

Disregarding for the moment the rightmost members, these equations may be solved simultaneously for E_{t1} and E_{i1} in terms of E_b and B_b, yielding

$$E_{t1} = \left(\frac{\gamma_1 E_b + B_b}{2\gamma_1}\right) e^{i\delta} \qquad (19\text{-}20)$$

$$E_{i1} = \left(\frac{\gamma_1 E_b - B_b}{2\gamma_1}\right) e^{-i\delta} \qquad (19\text{-}21)$$

Finally, substituting the expressions from Eqs. (19-20) and (19-21) into the equations (19-6) and (19-10) for boundary (a), the result is

$$E_a = E_b \cos \delta + B_b \left(\frac{i \sin \delta}{\gamma_1}\right) \qquad (19\text{-}22)$$

$$B_a = E_b (i\gamma_1 \sin \delta) + B_b \cos \delta \qquad (19\text{-}23)$$

where we have used the identities

$$2 \cos \delta \equiv e^{i\delta} + e^{-i\delta} \quad \text{and} \quad 2i \sin \delta \equiv e^{ib} - e^{-i\delta}$$

Equations (19-22) and (19-23) relate the net fields at one boundary with those at the other. They may be written in matrix form as

$$\begin{bmatrix} E_a \\ B_a \end{bmatrix} = \begin{bmatrix} \cos \delta & \dfrac{i \sin \delta}{\gamma_i} \\ i\gamma_1 \sin \delta & \cos \delta \end{bmatrix} \begin{bmatrix} E_b \\ B_b \end{bmatrix} \qquad (19\text{-}24)$$

The 2×2 matrix is called the *transfer matrix* of the film, represented in general by

$$\mathfrak{M} = \begin{bmatrix} m_{11} & m_{12} \\ m_{21} & m_{22} \end{bmatrix} \qquad (19\text{-}25)$$

If boundary (b) is the interface of another film layer, rather than the substrate, Eq. (19-24) is still valid. The fields E_b and B_b are then related to the fields E_c and B_c at the back boundary of the second film layer by a second transfer matrix. Generalizing, then for a multilayer of arbitrary number N of layers,

$$\begin{bmatrix} E_a \\ B_a \end{bmatrix} = \mathfrak{M}_1 \mathfrak{M}_2 \mathfrak{M}_3 \cdots \mathfrak{M}_N \begin{bmatrix} E_N \\ B_N \end{bmatrix}$$

An overall transfer matrix, $\mathfrak{M}_T$, representing the entire multilayer stack is the product of the individual transfer matrices, in the order in which the light encounters them,

$$\mathfrak{M}_T = \mathfrak{M}_1 \mathfrak{M}_2 \mathfrak{M}_3 \cdots \mathfrak{M}_N \qquad (19\text{-}26)$$

We return now to Eqs. (19-6), (19-7), (19-10), and (19-11) to make use of those members previously ignored in first finding the transfer matrix. Those remaining equations are

$$E_a = E_0 + E_{r1} \qquad (19\text{-}27)$$

$$E_b = E_{t2} \qquad (19\text{-}28)$$

$$B_a = \gamma_0(E_0 - E_{r1}) \qquad (19\text{-}29)$$

$$B_b = \gamma_s E_{t2} \qquad (19\text{-}30)$$

For the fields as represented by Eqs. (19-27) to (19-30), the transfer matrix, Eqs. (19-24) and (19-25), may be written as

$$\begin{bmatrix} E_0 + E_{r1} \\ \gamma_0(E_0 - E_{r1}) \end{bmatrix} = \begin{bmatrix} m_{11} & m_{12} \\ m_{21} & m_{22} \end{bmatrix} \begin{bmatrix} E_{t2} \\ \gamma_s E_{t2} \end{bmatrix} \qquad (19\text{-}31)$$

Equation (19-31) is equivalent to the two equations,

$$1 + r = m_{11}t + m_{12}\gamma_s t \qquad (19\text{-}32)$$

$$\gamma_0(1 - r) = m_{21}t + m_{22}\gamma_s t \qquad (19\text{-}33)$$

where we have used the reflection and transmission coefficients defined by

$$r \equiv \frac{E_{r1}}{E_0} \quad \text{and} \quad t \equiv \frac{E_{t2}}{E_0} \qquad (19\text{-}34)$$

Equations (19-32) and (19-33) can be solved for the transmission and reflection coefficients in terms of the transfer-matrix elements to give

$$t = \frac{2\gamma_0}{\gamma_0 m_{11} + \gamma_0 \gamma_s m_{12} + m_{21} + \gamma_s m_{22}} \qquad (19\text{-}35)$$

$$r = \frac{\gamma_0 m_{11} + \gamma_0 \gamma_s m_{12} - m_{21} - \gamma_s m_{22}}{\gamma_0 m_{11} + \gamma_0 \gamma_s m_{12} + m_{21} + \gamma_s m_{22}} \qquad (19\text{-}36)$$

Equations (19-35) and (19-36), together with the transfer-matrix elements as defined by Eq. (19-24), now enable one to evaluate the reflective and transmissive properties of the single or multilayer film represented by the transfer matrix.

Before continuing with applications of these equations, we must take into account the necessary modification of the theory that results when the incident electric field of Figure 19-1 has the other polarization, that is, in the plane of incidence. Suppose that $\mathbf{E}$ is chosen in the original direction of $\mathbf{B}$ and $\mathbf{B}$ is rotated accordingly to maintain the same wave direction. If the equations are developed along the same lines, one finds that only a minor alteration of the transfer matrix becomes necessary: In the expression for γ_1, Eq. (19-13), the cosine factor now appears in the denominator rather than in the numerator. Summarizing,

$$\mathbf{E} \perp \text{ plane of incidence:} \quad \gamma_1 = n_1 \sqrt{\epsilon_0 \mu_0} \cos \theta_{t1}$$

$$\mathbf{E} \parallel \text{ plane of incidence:} \quad \gamma_1 = n_1 \frac{\sqrt{\epsilon_0 \mu_0}}{\cos \theta_{t1}} \qquad (19\text{-}37)$$

Notice that for normal incidence, where $\mathbf{E}_\perp$ and $\mathbf{E}_\parallel$ are indistinguishable, we have $\cos \theta_{t1} = 1$, and the expressions are equivalent. For oblique incidence, however, results must be calculated for each polarization. An average can be taken for unpolarized light. For example, the reflectance becomes

$$R = \tfrac{1}{2}(R_\parallel + R_\perp) \tag{19-38}$$

19-2 REFLECTANCE AT NORMAL INCIDENCE

We apply the theory now for the case of normally incident light, the case most commonly found in practice. Results apply quite well also to cases of near-normal incidence. The beam remains normal at all interfaces, so that all angles are zero. In Eqs. (19-12) to (19-14), the cosine factors in the γ-terms are all unity. The matrix elements from Eq. (19-24), appropriately modified to become

$$
\begin{aligned}
m_{11} &= \cos \delta & m_{12} &= \frac{i \sin \delta}{n_1 \sqrt{\epsilon_0 \mu_0}} \\
m_{21} &= i n_1 \sqrt{\epsilon_0 \mu_0} \sin \delta & m_{22} &= \cos \delta
\end{aligned}
\tag{19-39}
$$

are substituted into Eq. (19-36). After cancellation of the constant $\sqrt{\epsilon_0 \mu_0}$ and some simplification, we find

$$r = \frac{n_1(n_0 - n_s) \cos \delta + i(n_0 n_s - n_1^2) \sin \delta}{n_1(n_0 + n_s) \cos \delta + i(n_0 n_s + n_1^2) \sin \delta} \tag{19-40}$$

The reflectance R, which measures the reflected exitance, is defined by

$$R = |r|^2 \tag{19-41}$$

To calculate R, first notice that the reflection coefficient r is complex and that it has the general form

$$r = \frac{A + iB}{C + iD}$$

so that

$$|r|^2 = rr^* = \frac{A + iB}{C + iD} \frac{A - iB}{C - iD} = \frac{A^2 + B^2}{C^2 + D^2}$$

By inspection then, we may write

normal incidence

$$R = \frac{n_1^2(n_0 - n_s)^2 \cos^2 \delta + (n_0 n_s - n_1^2)^2 \sin^2 \delta}{n_1^2(n_0 + n_s)^2 \cos^2 \delta + (n_0 n_s + n_1^2)^2 \sin^2 \delta} \tag{19-42}$$

Example

A 400-Å thick film of ZrO_2 ($n = 2.10$) is deposited on glass ($n = 1.50$). Determine the normal reflectance for sodium light.

Solution The phase difference is given by

$$\delta = \frac{2\pi}{\lambda}(n_1 t) = \frac{2\pi}{589.3}(2.1)(40) = 0.8956 \text{ rad}$$

so that $\cos \delta = 0.6250$ and $\sin \delta = 0.7806$. Then, substituting into Eq. (19-42),

$$R = \frac{2.1^2(1 - 1.5)^2(0.6250)^2 + [(1)(1.5) - 2.1^2]^2(0.7806)^2}{2.1^2(1 + 1.5)^2(0.6250)^2 + [(1)(1.5) + 2.1^2]^2(0.7806)^2}$$

Thus $R = 17.4\%$.

A plot of reflectance versus optical path difference Δ is shown in Figure 19-2, where the abscissa is calibrated in ratios of Δ/λ. Each curve corresponds to a different film index, but the glass substrate index has been chosen $n_s = 1.52$ in all cases. The magnitude of the film index n_1 evidently determines whether the reflectance is enhanced (for $n_1 > n_s$) or reduced (for $n_1 < n_s$) from that for uncoated glass. The curves show that quarter-wave thicknesses, or odd multiples thereof, lead either to optimum enhancement (high-reflectance coating) or to maximum reduction (antireflection coating). These minima or maxima points in R can be made to occur at various wavelengths by changing Δ through selection of the film thickness. Notice that for $\Delta = \lambda/2$ or any even multiple of a quarter-wavelength, the reflectance is just that from the uncoated glass. An antireflecting single coat, with $n_1 < n_s$, never reflects more than the uncoated glass at any wavelength. The periodic variation in R with Δ, which is proportional to the film thickness, provides a practical way of monitoring film thickness in the course of a film deposition.

The important case of quarter-wave film thickness,

$$t = \frac{\lambda}{4} = \frac{\lambda_0}{4n_1}$$

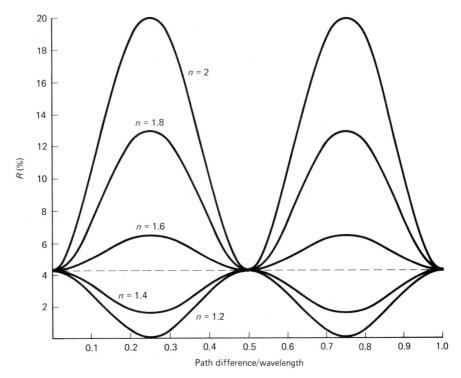

Figure 19-2 Reflectance from a single film layer versus normalized path difference. The dashed line represents the uncoated glass substrate of index $n_s = 1.52$.

makes the phase difference, Eq. (19-15), $\delta = 2\pi n_1 t/\lambda_0 = \pi/2$, so that $\cos \delta = 0$ and $\sin \delta = 1$. In this case, Eq. (19-42) reduces to

$$\text{normal incidence} \quad R = \left(\frac{n_0 n_s - n_1^2}{n_0 n_s + n_1^2}\right)^2 \qquad (19\text{-}43)$$
$$\text{quarter-wave thickness}$$

From Eq. (19-43), it follows that a perfectly antireflecting film can be fabricated with a coating of $\lambda/4$ thickness and refractive index $n_1 = \sqrt{n_0 n_s}$. If the substrate is glass, with $n_s = 1.52$, the ideal index for a nonreflecting coating is $n_1 = 1.23$, assuming an ambient with $n_0 = 1$. A compromise choice among available coating materials is a film of MgF_2, with $n_1 = 1.38$. For this film, Eq. (19-43) predicts a reflectance of 1.3% in the visible region, where the uncoated glass (set $n_1 = n_0$) would reflect about 4.3%. This difference represents a significant saving of light energy in an optical system where multiple surfaces occur. For example, after only six such interfaces, or three optical components in series, 93% of the incident light survives in the case of MgF_2 coatings, compared with 77% in the case of uncoated glass.

19-3 TWO-LAYER ANTIREFLECTING FILMS

Durable coating materials with arbitrary refractive indices are, of course, not immediately available. Practically speaking then, single films with zero reflectances cannot be fabricated. By using a double layer of quarter-wave-thickness films, however, it is possible to achieve essentially zero reflectance at one wavelength with available coating materials. At normal incidence, the transfer matrix of a single film of quarter-wave thickness is

$$\mathfrak{M}_1 = \begin{bmatrix} 0 & \dfrac{i}{\gamma_1} \\ i\gamma_1 & 0 \end{bmatrix}$$

The transfer matrix $\mathfrak{M}$ for two such layers is found, according to Eq. (19-26), by forming the product

$$\mathfrak{M} = \mathfrak{M}_1 \mathfrak{M}_2 = \begin{bmatrix} 0 & \dfrac{i}{\gamma_1} \\ i\gamma_1 & 0 \end{bmatrix} \begin{bmatrix} 0 & \dfrac{i}{\gamma_2} \\ i\gamma_2 & 0 \end{bmatrix} = \begin{bmatrix} -\dfrac{\gamma_2}{\gamma_1} & 0 \\ 0 & -\dfrac{\gamma_1}{\gamma_2} \end{bmatrix}$$

Matrix components are $m_{11} = -\gamma_2/\gamma_1$, $m_{22} = -\gamma_1/\gamma_2$, and $m_{12} = m_{21} = 0$. Using these values in Eq. (19-36), the result is

$$r = \frac{\gamma_2^2 \gamma_0 - \gamma_s \gamma_1^2}{\gamma_2^2 \gamma_0 + \gamma_s \gamma_1^2} \qquad (19\text{-}44)$$

Incorporating the refractive indices through the use of Eqs. (19-12) to (19-14) and then squaring to get the reflectance,

$$\text{normal incidence} \quad R = \left(\frac{n_0 n_2^2 - n_s n_1^2}{n_0 n_2^2 + n_s n_1^2}\right)^2 \qquad (19\text{-}45)$$
$$\text{quarter-wave thickness}$$

Zero reflectance is predicted by Eq. (19-45) when $n_0 n_2^2 = n_s n_1^2$, or

$$\frac{n_2}{n_1} = \sqrt{\frac{n_s}{n_0}} \qquad (19\text{-}46)$$

Figure 19-3   Antireflecting double layer, using $\lambda/4$-$\lambda/4$ thickness films.

For a glass substrate ($n_s = 1.52$) and incidence from air ($n_0 = 1$), the ideal ratio for the two films is $n_2/n_1 = 1.23$. The requirement is met quite well using zirconium dioxide ($n_2 = 2.1$) and cerium trifluoride ($n_1 = 1.65$), both good coating materials. The ratio of refractive indices for CeF_3 and ZrO_2 of 1.27 produces a reflectance of only 0.1% according to Eq. (19-45). The arrangement is shown in Figure 19-3 and is plotted as curve (a) in Figure 19-4. Achieving zero reflectance at some wavelength may not satisfy the very common need to reduce reflectance over a broad region of the visible spectrum. Curve (a) is rather steep on both sides of its minimum at 550 nm. Broader regions of low reflectance result for $\lambda/4$-$\lambda/4$ coatings when the substrate index is larger than that of the adjacent film layer, that is, $n_s > n_2$. In such cases the index is "stepped down" consistently from substrate to ambient. Indices high enough to satisfy this condition are possible in infrared applications where large values of n_s are available, as in the case of germanium with $n_s = 4$. A list of useful refractive indices is given in Table 19-1. Broader regions of low reflectance also be-

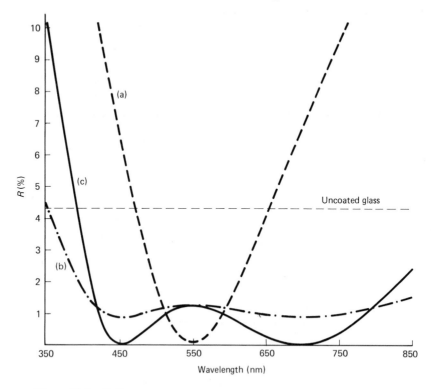

Figure 19-4   Reflectance from a double-layer film versus wavelength. In all cases $n_0 = 1$ and $n_s = 1.52$. Thicknesses are determined at $\lambda = 550$ nm. (a) $\lambda/4$-$\lambda/4$; $n_1 = 1.65$, $n_2 = 2.1$. (b) $\lambda/4$-$\lambda/2$: $n_1 = 1.38$, $n_2 = 1.6$. (c) $\lambda/4$-$\lambda/2$: $n_1 = 1.38$, $n_2 = 1.85$.

TABLE 19-1 REFRACTIVE INDICES FOR SEVERAL COATING MATERIALS

Material	Visible ($\sim$ 550 nm)	Near infrared ($\sim 2\mu$m)
Cryolite	1.30–1.33	—
MgF_2	1.38	1.35
SiO_2	1.46	1.44
SiO	1.55–2.0	1.5–1.85
Al_2O_3	1.60	1.55
CeF_3	1.65	1.59
ThO_2	1.8	1.75
Nd_2O_3	2.0	1.95
ZrO_2	2.1	2.0
CeO_2	2.35	2.2
ZnS	2.35	2.2
TiO_2	2.4	—
Si	—	3.3
Ge	—	4.0

come possible in the visible region of the spectrum, once the restriction of using equal $\lambda/4$ coatings is relaxed. For example, curves (b) and (c) of Figure 19-4 show two such solutions to the problem, where the inner coating is $\lambda/2$ thick, as illustrated in Figure 19-5. At the wavelength of 550 nm, for which the $\lambda/4$ and $\lambda/2$ thicknesses are determined, the $\lambda/2$ layer has no effect on the reflectance, and the double layer behaves like a single $\lambda/4$ layer. At other wavelengths, however, the $\lambda/2$ layer helps to keep R below values attained by a single $\lambda/4$ layer alone. For $n = 1.85$ [curve (c)], two minima near $R = 0$ appear. Although reflectance at 550 nm is 1.26%, greater than for the $\lambda/4$-$\lambda/4$ coating of curve (a), it remains at values less than this over the broad range of wavelengths from about 420 to 800 nm. For $n_2 = 1.6$ [curve (b)], the spectral response of the double layer, while more reflective, is flatter over the visible spectrum. Still other practical solutions for double-layer antireflecting films become possible if the thicknesses of the layers are allowed to have values other than multiples of $\lambda/4$.

The curves of Figure 19-4 have been calculated using the theory presented in this chapter. The overall transfer-matrix elements are first determined by forming the product of the transfer matrices of the individual layers. In these elements, the phase difference δ is expressed as a function of λ, and the film thickness is determined by the $\lambda/4$ or $\lambda/2$ requirement at a single wavelength. These matrix elements are then used in Eq. (19-36) for the reflection coefficient. When squared, the reflectance as a function of wavelength is determined. Although the calculations can be tedious, they are easily done using a programmable calculator or computer.

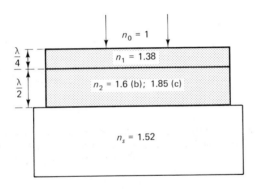

Figure 19-5 Antireflecting double layer using $\lambda/4$-$\lambda/2$ thickness films. Reflectance curves are shown in Figure 19-4.

The procedure just outlined was used to calculate the spectral reflectance of three-layer films as well. The use of three or more layer coatings makes possible a broader, low-reflectance region in which the response is flatter. If each of the three layers is of $\lambda/4$ thickness, one can show that a zero reflectance occurs when the refractive indices satisfy

$$\frac{n_1 n_3}{n_2} = \sqrt{n_0 n_s} \tag{19-47}$$

One such practical solution is shown in Figure 19-6a and plotted as curve (a) in Figure 19-7. Some improvement results when the middle layer is of $\lambda/2$ thickness, as in Figure 19-6b and curve (b) of Figure 19-7.

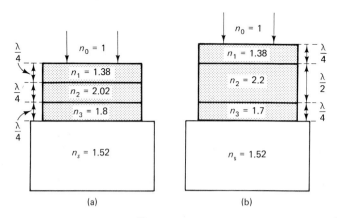

Figure 19-6 Antireflecting triple layers. (a) Quarter-quarter-quarter wavelength layers. (b) Quarter-half-quarter wavelength layers. Reflectance curves are shown in Figure 19-7.

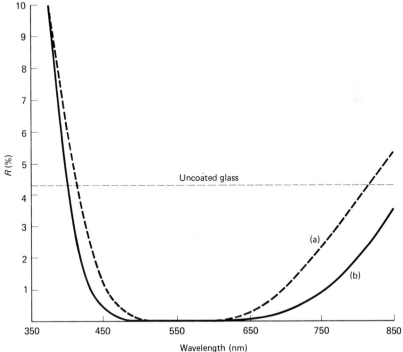

Figure 19-7 Reflectance from triple-layer films versus wavelength. In all cases $n_0 = 1$ and $n_s = 1.52$. Thicknesses are determined at $\lambda = 550$ nm. (a) $\lambda/4$-$\lambda/4$-$\lambda/4$: $n_1 = 1.38$, $n_2 = 2.02$, $n_3 = 1.8$. (b) $\lambda/4$-$\lambda/2$-$\lambda/4$: $n_1 = 1.38$, $n_2 = 2.2$, $n_3 = 1.7$.

If the order of the layers in a $\lambda/4$-$\lambda/4$ double-layer film optimized for antireflection is reversed, so that the order is air - high index - low index - substrate, all three reflected beams are in phase on emerging from the structure, and the reflectance is enhanced rather than reduced. A series of such double layers increases the reflectance further, and the structure is called a *high-reflectance stack*, or *dielectric mirror*.

We derive now an expression for the reflectance of this type of structure, shown schematically in Figure 19-8, where High and Low signify high- and low-refractive indices, respectively. The transfer matrix for one double layer of $\lambda/4$-thick coatings at normal incidence is the product of the individual film matrices, just as in the case of the double-layer antireflecting films:

$$\mathfrak{M}_{HL} = \mathfrak{M}_H \mathfrak{M}_L$$

or

$$\mathfrak{M}_{HL} = \begin{bmatrix} 0 & \dfrac{i}{\gamma_H} \\ \dfrac{i}{\gamma_H} & 0 \end{bmatrix} \begin{bmatrix} 0 & \dfrac{i}{\gamma_L} \\ i\gamma_L & 0 \end{bmatrix} = \begin{bmatrix} \dfrac{-\gamma_L}{\gamma_H} & 0 \\ 0 & \dfrac{-\gamma_H}{\gamma_L} \end{bmatrix} \qquad (19\text{-}48)$$

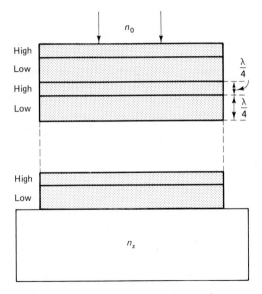

Figure 19-8 High-reflectance stack of double layers with alternating high- and low-refractive indices. Reflectance curves are shown in Figure 19-9.

For N similar double layers in series,

$$\mathfrak{M} = (\mathfrak{M}_{H1} \mathfrak{M}_{L1})(\mathfrak{M}_{H2} \mathfrak{M}_{L2}) \cdots (\mathfrak{M}_{HN} \mathfrak{M}_{LN}) = (\mathfrak{M}_H \mathfrak{M}_L)^N = (\mathfrak{M}_{HL})^N \qquad (19\text{-}49)$$

Substituting the double-layer matrix, Eq. (19-48),

$$\mathfrak{M} = \begin{bmatrix} \dfrac{-\gamma_L}{\gamma_H} & 0 \\ 0 & \dfrac{-\gamma_H}{\gamma_L} \end{bmatrix}^N = \begin{bmatrix} \left(\dfrac{-\gamma_L}{\gamma_H}\right)^N & 0 \\ 0 & \left(\dfrac{-\gamma_H}{\gamma_L}\right)^N \end{bmatrix}$$

For normal incidence,

$$\frac{\gamma_L}{\gamma_H} = \frac{n_L}{n_H} \quad \text{and} \quad \frac{\gamma_H}{\gamma_L} = \frac{n_H}{n_L}$$

so that

$$\mathfrak{M} = \begin{bmatrix} \left(\dfrac{-n_L}{n_H}\right)^N & 0 \\ 0 & \left(\dfrac{-n_H}{n_L}\right)^N \end{bmatrix} \tag{19-50}$$

The matrix elements of the transfer matrix representing N high-low double layers of $\lambda/4$ thick coatings in series are thus

$$m_{11} = \left(\frac{-n_L}{n_H}\right)^N, \qquad m_{22} = \left(\frac{-n_H}{n_L}\right)^N, \qquad m_{12} = m_{21} = 0 \tag{19-51}$$

Using these matrix elements in the expression for the reflection coefficient, Eq. (19-36), we arrive at

$$r = \frac{n_0(-n_L/n_H)^N - n_s(-n_H/n_L)^N}{n_0(-n_L/n_H)^N + n_s(-n_H/n_L)^N} \tag{19-52}$$

When numerator and denominator of Eq. (19-52) are next multiplied by the factor $(-n_L/n_H)^N/n_s$ and the result is squared to give reflectance, we have

$$R_{\max} = \left[\frac{(n_0/n_s)(n_L/n_H)^{2N} - 1}{(n_0/n_s)(n_L/n_H)^{2N} + 1}\right]^2 \tag{19-53}$$

Example

A high-reflectance stack like that of Figure 19-8 incorporates six double layers of SiO_2 ($n = 1.46$) and ZnS ($n = 2.35$) films on a glass ($n = 1.48$) substrate. What is the reflectance for light of 550 nm at normal incidence?

Solution Substituting directly into Eq. (19-53), we get

$$R = \left[\frac{(1/1.48)(1.46/2.35)^{12} - 1}{(1/1.48)(1.46/2.35)^{12} + 1}\right]^2$$

or $R = 99.1\%$.

Equation (19-53) predicts 100% reflectance when either N approaches infinity or when (n_L/n_H) approaches zero. Some data indicating these tendencies are given in Table 19-2. One sees that the reflectance quickly approaches 100% for several double layers. Since the smallest ratio of n_L/n_H yields best reflectances, high-reflectance

TABLE 19-2 REFLECTANCE OF A HIGH-LOW QUARTER-WAVE STACK

Reflectance for $N = 3$ high-low layers versus n_L/n_H		Reflectance versus N when $n_L/n_H = 0.587$ for alternating double layers of MgF_2 and ZnS	
n_L/n_H	R (%)	N	R (%)
1.0	4.26	1	39.71
0.91	21.01	2	73.08
0.83	40.82	3	89.77
0.77	57.77	4	96.35
0.71	70.44	5	98.72
0.67	79.35	6	99.56
0.625	85.48	7	99.85
0.59	89.67	8	99.95
0.56	92.55		
0.53	94.56		
0.50	95.97		

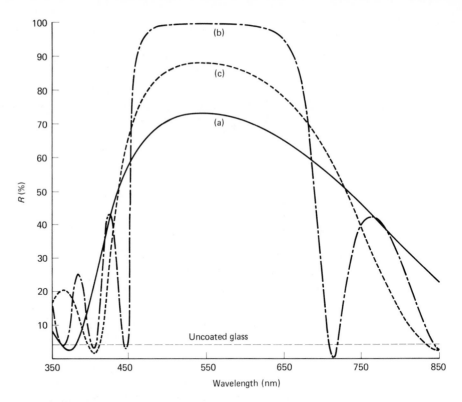

Figure 19-9 Spectral reflectance of a high-low index stack for (a) $N = 2$ and (b) $N = 6$ double layers. Curve (c) represents an $N = 2$ stack with an additional high-index layer adjacent to the substrate. Layers are $\lambda/4$ thick at $\lambda = 550$ nm. In all cases, $n_H = 2.35$, $n_L = 1.38$, $n_s = 1.52$, and $n_0 = 1.00$.

stacks may be fabricated from alternating layers of MgF$_2$ ($n_2 = 1.38$) and ZnS ($n_H = 2.35$) or TiO$_2$ ($n_H = 2.40$).

The reflectance given in Eq. (19-53) represents the maximum reflectance at the wavelength λ_0, for which the layers have optical thicknesses of $\lambda_0/4$. For other wavelengths the transfer matrix must be used in its general form, containing the wavelength-dependent phase differences. Spectral reflectance curves for $N = 2$ and $N = 6$ double-layer stacks have been calculated and plotted in Figure 19-9. Curve (c) shows the improvement in the maximum reflectance that results when an extra high-index layer is inserted between the substrate and last low-index layer. The width of the high-reflectance region in these curves is nearly independent of the number of double layers used but increases when the ratio n_L/n_H increases. This ratio is 1.70 in Figure 19-9, representing alternating MgF$_2$ and ZnS layers on glass. Outside the central *stopband*, the reflectance oscillates between a series of maxima and minima. The center of the stopband can be shifted by depositing layers whose thickness is $\lambda/4$ at another λ. Except for light energy lost by absorption and scattering during passage through the dielectric layers, the percent transmission of the structure is given by T (%) $= 100 - R$ (%). Thus such structures can be designed as *band-pass filters* whose spectral transmittance is essentially the inverse of the spectral reflectance. Narrow band-pass filters that behave like Fabry-Perot etalons can be fabricated by separating two dielectric-mirror, multilayer structures with a spacer of, say, MgF$_2$ film. Narrow wavelength regions that satisfy constructive interference can be produced far enough apart in wavelength so that all but one such region is easily filtered out by a conventional absorption color filter. The result is a filter with a pass bandwidth of perhaps 15 Å and 40% transmittance.

PROBLEMS

19-1. Show that when the incident **E**-field is parallel to the plane of incidence, γ_1 has the form given in Eq. (19-37).

19-2. A transparent film is deposited on glass of refractive index 1.50.
(a) Determine values of film thickness and (hypothetical) refractive index that will produce a nonreflecting film for normally incident light of 500 nm.
(b) What reflectance does the structure have for incident light of 550 nm?

19-3. Show from Eq. (19-42) that the normal reflectance of a single half-wave thick layer deposited on a substrate is the same as the reflectance from the uncoated substrate.

$$R = \frac{(n_0 - n_s)^2}{(n_0 + n_s)^2}$$

19-4. A single layer of SiO_2 ($n = 1.46$) is deposited to a thickness of 137 nm on a glass substrate ($n = 1.52$). Determine the normal reflectance for light of wavelength (a) 800 nm; (b) 600 nm; (c) 400 nm. Verify the reasonableness of your results by comparison with Figure 19-2.

19-5. A 596-Å-thick layer of ZnS ($n = 2.35$) is deposited on glass ($n = 1.52$). Calculate the normal reflectance of 560 nm light.

19-6. Determine the theoretical refractive index and thickness of a single film layer deposited on germanium ($n = 4.0$) such that normal reflectance is zero at a wavelength of 2 μm. What actual material could be used?

19-7. A double layer of quarter-wave layers of Al_2O_3 ($n = 1.60$) and cryolite ($n = 1.30$) are deposited in turn on a glass substrate ($n = 1.52$).
(a) Determine the thickness of the layers and the normal reflectance for light of 550 nm.
(b) What is the reflectance if the layers are reversed?

19-8. Quarter-wave thin films of ZnS ($n = 2.2$) and MgF_2 ($n = 1.35$) are deposited in turn on a substrate of silicon ($n = 3.3$) to produce minimum reflectance at 2 μm.
(a) Determine the actual thickness of the layers.
(b) By what percentage difference does the ratio of the film indices differ from the ideal?
(c) What is the normal reflectance produced?

19-9. By working with the appropriate transfer matrix, show that a quarter-wave/half-wave double layer, as in Figure 19-5, produces the same reflectance as the quarter-wave layer alone.

19-10. Write a computer program that will calculate and/or plot reflectance values for a double layer under normal incidence. Let input parameters include thickness and indices of the layers and the index of the substrate. Check results against Figure 19-4.

19-11. Prove the condition given by Eq. (19-47) for zero reflectance of three-layer, quarter-quarter-quarter-wave films when used with normal incidence. Do this by determining the composite transfer matrix for the three quarter layers and using the matrix elements in the calculation of the reflection coefficient in Eq. (19-36).

19-12. Using the materials given in Table 19-1, design a three-layer multifilm of quarter-wave thicknesses on a substrate of germanium that will give nearly zero reflectance for normal incidence of 2 μm radiation.

19-13. Determine the maximum reflectance in the center of the visible spectrum for a high-reflectance stack of high-low index double layers formed using $n_L = 1.38$ and $n_H = 2.6$ on a substrate of index 1.52. The layers are of equal optical thickness, corresponding to a quarter-wavelength for light of average wavelength 550 nm. The high-index material is encountered first by the incident light, as in Figure 19-8. Assume normal incidence and stacks of (a) 2; (b) 4; (c) 8 double layers.

19-14. A high-reflectance stack of alternating high-low index layers is produced to operate at 2 μm in the near infrared. A stack of four double layers is made of layers of germanium ($n = 4.0$) and MgF_2 ($n = 1.35$), each of 0.5-μm optical thickness. Assume a substrate index of 1.50 and normal incidence. What reflectance is produced at 2 μm?

19-15. What theoretical ratio of high-to-low refractive indices is needed to give at least 90% reflectance in a high-reflectance stack of two double layers of quarter-wave layers at normal incidence? Assume a substrate of index 1.52.

19-16. Show that R_{max} in Eq. (19-53) approaches 1 when either N approaches infinity or when the ratio n_L/n_H approaches zero.

REFERENCES

[1] Dobrowolski, J. A. "Coatings and Filters." In *Handbook of Optics*, edited by Walter G. Driscoll and William Vaughan. New York: McGraw-Hill Book Company, 1978.

[2] Knittl, Z. *Optics of Thin Films, an Optical Multilayer Theory*. New York: John Wiley and Sons, 1976.

[3] Heavens, O. S. *Thin Film Physics*. New York: Barnes and Noble, 1970.

[4] Macleod, H. A. *Thin Film Optical Filters*. New York: American Elsevier Publishing Company, 1969.

[5] Chopra, Kasturi L. *Thin Film Phenomena*. New York: McGraw-Hill Book Company, 1969.

[6] Baumeister, Philip, and Gerald Pincus. "Optical Interference Coatings." *Scientific American* (Dec. 1970): 58.

[7] Nussbaum, Allen, and Richard A. Phillips. *Contemporary Optics for Scientists and Engineers*. Englewood Cliffs, N.J.: Prentice-Hall, 1976. Ch. 8.

20

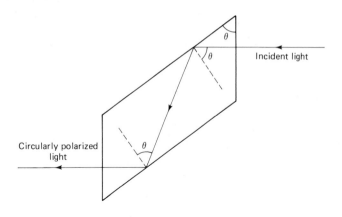

Fresnel Equations

INTRODUCTION

The basic laws of reflection and refraction in geometrical optics were derived earlier on the basis of either Huygens' or Fermat's principles. In this chapter we regard light as an electromagnetic wave and show that the laws of reflection and refraction can also be deduced from this point of view. More importantly, this approach also leads to the *Fresnel equations*, which describe the fraction of incident energy transmitted or reflected at a plane surface. These quantities will be seen to depend not only on the change in refractive index and the angle of incidence at the surface, but also on the polarization of the incident light. Finally, the important differences between internal and external reflection are clarified.

20-1 THE FRESNEL EQUATIONS

Consider Figure 20-1, which shows a ray of light incident at point P on a plane interface—the xy-plane—and the resulting reflected and refracted rays. The plane of incidence is the xz-plane. Let us assume the incident light consists of plane harmonic waves, expressed by

$$\mathbf{E} = \mathbf{E}_0 e^{i(\mathbf{k}\cdot\mathbf{r}-\omega t)} \tag{20-1}$$

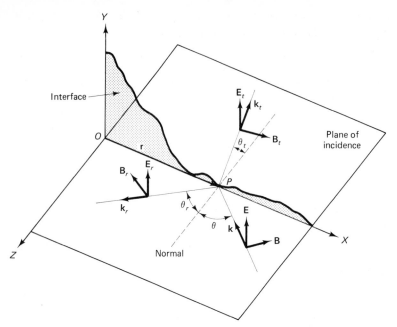

Figure diagram labels: Interface, O, Y, r, B_r, E_r, k_r, θ_r, θ, k, E, B, X, Normal, P, E_t, k_t, θ_t, B_t, Plane of incidence, Z

Figure 20-1 Defining diagram for incident, reflected, and transmitted rays at an XY-plane interface when the electric field is perpendicular to the plane of incidence, the TE mode.

where the origin of coordinates is taken to be point O. The wave vector **E** of the incident wave is chosen in the $+y$-direction, so that the wave is linearly polarized. The direction of the corresponding magnetic field vector **B** is then determined to ensure that the direction of $\mathbf{E} \times \mathbf{B}$ is the direction of wave propagation **k**. This mode of polarization, in which the **E**-field is perpendicular to the plane of incidence and the **B**-field lies in the plane of incidence, is called the *transverse electric* (TE) mode. If instead **B** is transverse to the plane of incidence, a case to be considered later, the mode is a *transverse magnetic* (TM) mode. An arbitrary polarization direction represents some linear combination of these two special cases.

The reflected and transmitted waves in Figure 20-1 can be expressed, respectively, in forms like that of the incident wave of Eq. (20-1):

$$\mathbf{E}_r = \mathbf{E}_{or}e^{i(\mathbf{k}_r \cdot \mathbf{r} - \omega_r t)} \tag{20-2}$$

$$\mathbf{E}_t = \mathbf{E}_{ot}e^{i(\mathbf{k}_t \cdot \mathbf{r} - \omega_t t)} \tag{20-3}$$

In the boundary plane xy, where all three waves exit simultaneously, there must be a fixed relationship between the three wave amplitudes (and thus their irradiances) that has yet to be determined. Since such a relationship cannot depend on the arbitrary choice of a boundary point **r** nor a time t, it follows that the phases of the three waves, which depend on **r** and t, must themselves be equal:

$$(\mathbf{k} \cdot \mathbf{r} - \omega t) = (\mathbf{k}_r \cdot \mathbf{r} - \omega_r t) = (\mathbf{k}_t \cdot \mathbf{r} - \omega_t t) \tag{20-4}$$

In particular, at the boundary point $\mathbf{r} = 0$ of Figure 20-1,

$$-\omega t = -\omega_r t = -\omega_t t$$

or

$$\omega = \omega_r = \omega_t \tag{20-5}$$

so that all frequencies are equal. On the other hand, at $t = 0$ *within the boundary plane*, Eq. (20-4) yields:

$$\mathbf{k} \cdot \mathbf{r} = \mathbf{k}_r \cdot \mathbf{r} = \mathbf{k}_t \cdot \mathbf{r} \tag{20-6}$$

Several conclusions can be drawn from the relations of Eq. (20-6). First notice that by subtracting any two members, these relations are equivalent to

$$(\mathbf{k} - \mathbf{k}_r) \cdot \mathbf{r} = (\mathbf{k} - \mathbf{k}_t) \cdot \mathbf{r} = (\mathbf{k}_r - \mathbf{k}_t) \cdot \mathbf{r} = 0 \qquad (20\text{-}7)$$

Equation (20-7) requires that the vectors $\mathbf{k}_r$ and $\mathbf{k}_t$ lie in the plane determined by the vectors $\mathbf{k}$ and $\mathbf{r}$. Thus all three propagation vectors are coplanar in the xz-plane, and we conclude that the reflected and refracted waves lie in the plane of incidence. Next, consider the first two members of Eq. (20-6), which govern the relationship between the incident and reflected waves. In terms of the angles designated in Figure 20-1, they are equivalent to

$$kr \sin \theta = k_r r \sin \theta_r$$

Since both waves travel in the same medium, their wavelengths are identical and so $k = k_r$. Therefore, we have the

$$\text{law of reflection:} \quad \theta = \theta_r \qquad (20\text{-}8)$$

Finally, the last two members of Eq. (20-6) are equivalent to

$$k_r r \sin \theta_r = k_t r \sin \theta_t \qquad (20\text{-}9)$$

Writing $k_r = \omega/v_r = n_r \omega/c$ and $k_t = n_t \omega/c$, Eq. (20-9) becomes Snell's

$$\text{law of refraction:} \quad n_r \sin \theta_r = n_t \sin \theta_t \qquad (20\text{-}10)$$

We continue now to specify further the situation at the boundary with the help of boundary conditions arising out of Maxwell's equations, and treated in texts on electricity and magnetism. We employ them here without proof. These boundary conditions require that the components of both the electric and magnetic fields parallel to the boundary plane be continuous as the boundary is crossed. In terms of the choices made for the direction of $\mathbf{E}$ in Figure 20-1, the requirement for the electric field is

$$E + E_r = E_t \qquad (20\text{-}11)$$

where we have described the total field at the incident side of the boundary as a superposition of incident and reflected waves and at the other side as the transmitted field alone. Notice that all three fields are parallel to the boundary plane and in the $+y$-direction. In the case of the corresponding magnetic fields, the parallel components satisfying the boundary condition are related by

$$B \cos \theta - B_r \cos \theta = B_t \cos \theta_t \qquad (20\text{-}12)$$

where we have made use of Eq. (20-8). The negative sign indicates that the $\mathbf{B}_r$-component is along the $-x$-direction in the reflected beam. Equations (20-11) and (20-12) are correct for the $\mathbf{E}$ and $\mathbf{B}$ vectors as chosen in Figure 20-1. If a different choice is made, for example, by reversing the $\mathbf{E}$ vector of the incident wave (and also $\mathbf{B}$ to keep the wave direction the same), Eqs. (20-11) and (20-12) appear with a change of signs. However, the physical import of these equations is the same when they are interpreted in terms of their original figures.

Before pursuing the significance of Eqs. (20-11) and (20-12) for the TE mode, we parallel their development for the TM mode pictured in Figure 20-2. Analogous to Eqs. (20-11) and (20-12), we now have

$$B + B_r = B_t \qquad (20\text{-}13)$$

$$-E \cos \theta + E_r \cos \theta = -E_t \cos \theta_t \qquad (20\text{-}14)$$

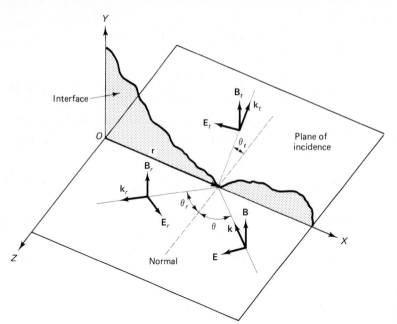

Interface

Y

Plane of incidence

B_t k_t

E_t

θ_t

B_r

k_r

θ_r

B

E_r

θ k

X

E

Normal

Z

O

r

Figure 20-2 Defining diagram for incident, reflected, and transmitted rays at an XY-plane interface when the magnetic field is perpendicular to the plane of incidence, the TM mode.

Eqs. (20-11) through (20-14) are valid for the instantaneous values of the fields at the boundary. Because of the equality of their phases, they are also valid for the amplitudes of the fields. The magnetic fields of Eqs. (20-12) and (20-13) can be expressed in terms of electric fields through the relation

$$E = vB = \left(\frac{c}{n}\right)B \qquad (20\text{-}15)$$

Writing the index of refraction for incident and refracting media as n_1 and n_2, respectively, Eqs. (20-11) through (20-14) can be recast as follows:

TE:
$$\begin{cases} E + E_r = E_t & (20\text{-}16) \\ n_1 E \cos\theta - n_1 E_r \cos\theta = n_2 E_t \cos\theta_t & (20\text{-}17) \end{cases}$$

TM:
$$\begin{cases} n_1 E + n_1 E_r = n_2 E_t & (20\text{-}18) \\ -E \cos\theta + E_r \cos\theta = -E_t \cos\theta_t & (20\text{-}19) \end{cases}$$

Next, eliminating E_t from each pair of equations and solving for the *reflection coefficient* $r = E_r/E$,

$$\text{TE:} \quad r = \frac{E_r}{E} = \frac{\cos\theta - n\cos\theta_t}{\cos\theta + n\cos\theta_t} \qquad (20\text{-}20)$$

$$\text{TM:} \quad r = \frac{E_r}{E} = \frac{n\cos\theta - \cos\theta_t}{n\cos\theta + \cos\theta_t} \qquad (20\text{-}21)$$

where we have introduced a *relative refractive index* $n \equiv n_2/n_1$. Finally, since n and θ_t are related to θ through Snell's law, $\sin\theta = n\sin\theta_t$, θ_t may be eliminated using

$$n\cos\theta_t = n\sqrt{1 - \sin^2\theta_t} = \sqrt{n^2 - \sin^2\theta} \qquad (20\text{-}22)$$

The results are then

$$\text{TE:} \quad r = \frac{E_r}{E} = \frac{\cos\theta - \sqrt{n^2 - \sin^2\theta}}{\cos\theta + \sqrt{n^2 - \sin^2\theta}} \qquad (20\text{-}23)$$

$$\text{TM:} \quad r = \frac{E_r}{E} = \frac{n^2\cos\theta - \sqrt{n^2 - \sin^2\theta}}{n^2\cos\theta + \sqrt{n^2 - \sin^2\theta}} \qquad (20\text{-}24)$$

Returning to Eqs. (20-16) through (20-19), if E_r is eliminated instead of E_t, similar steps lead to the following equations describing the *transmission coefficient* $t = E_t/E$:

$$\text{TE:} \quad t = \frac{E_t}{E} = \frac{2\cos\theta}{\cos\theta + \sqrt{n^2 - \sin^2\theta}} \qquad (20\text{-}25)$$

$$\text{TM:} \quad t = \frac{E_t}{E} = \frac{2n\cos\theta}{n^2\cos\theta + \sqrt{n^2 - \sin^2\theta}} \qquad (20\text{-}26)$$

Eqs. (20-25) and (20-26) can also be found more quickly by using Eqs. (20-16) and (20-18) written in the form

$$\text{TE:} \quad t = r + 1$$

$$\text{TM:} \quad nt = r + 1$$

into which the results expressed by Eqs. (20-23) and (20-24) can be conveniently substituted. Equations (20-23) through (20-26) are the *Fresnel equations*, giving reflection and transmission coefficients, the ratio of both reflected and transmitted **E**-field amplitudes to the incident **E**-field amplitude. In reality, measured reflection and transmission coefficients also depend on scattering losses from a nonplanar surface.

Example

Calculate the reflectance and transmittance for both TE and TM modes of light incident at 30° on glass of index 1.60.

Solution   Using Eqs. (20-23) and (20-24),

$$r_{TE} = \frac{\cos 30 - \sqrt{1.6^2 - \sin^2 30}}{\cos 30 + \sqrt{1.6^2 - \sin^2 30}} = -0.2740$$

$$r_{TM} = \frac{1.6^2\cos 30 - \sqrt{1.6^2 - \sin^2 30}}{1.6^2\cos 30 + \sqrt{1.6^2 - \sin^2 30}} = 0.1866$$

Thus

$$R_{TE} = r_{TE}^2 = 7.51\% \quad \text{and} \quad T_{TE} = 1 - R_{TE} = 92.5\%$$

$$R_{TM} = r_{TM}^2 = 3.48\% \quad \text{and} \quad T_{TM} = 1 - R_{TM} = 96.5\%$$

20-2 EXTERNAL AND INTERNAL REFLECTIONS

In dealing with the interpretation of these equations, it is necessary to distinguish between two physically different situations:

$$\text{external reflection:} \quad n_1 < n_2 \quad \text{or} \quad n = \frac{n_2}{n_1} > 1$$

$$\text{internal reflection:} \quad n_1 > n_2 \quad \text{or} \quad n = \frac{n_2}{n_1} < 1$$

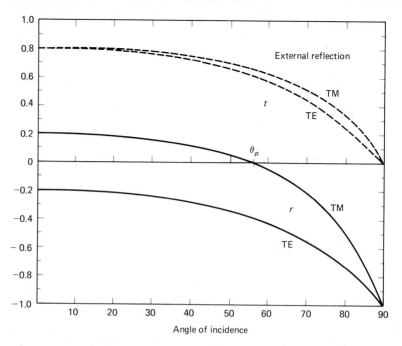

Figure 20-3 Reflection (r) and transmission (t) coefficients for the case of external reflection, with $n = n_2/n_1 = 1.50$.

Figure 20-3 is a plot of Eqs. (20-23) through (20-26) for the case of external reflection with $n = 1.50$. Notice that at both normal and grazing incidence—angles of 0° and 90°, respectively—TE and TM modes result in the same magnitudes for either the reflection or transmission coefficients. Negative values of r for both the TE and TM modes indicate a phase change of the **E**- or **B**-field vectors on reflection and will be discussed presently. The fraction of power P in the incident wave that is reflected or transmitted, called the *reflectance* and the *transmittance*, respectively, depends on the ratio of the squares of the amplitudes.

$$\text{reflectance} = R = \frac{P_r}{P_i} = r^2 = \left(\frac{E_r}{E}\right)^2 \qquad (20\text{-}27)$$

$$\text{transmittance} = T = \frac{P_t}{P_i} = n\left(\frac{\cos\theta_t}{\cos\theta_i}\right)t^2 \qquad (20\text{-}28)$$

These expressions are justified later in this chapter. Reflectance is plotted as a function of the angle of incidence in Figure 20-4. The curve for the case of external reflection, TM mode, indicates that no wave energy is reflected when the angle of incidence is near 60°. More precisely, $R_{TM} = 0$ when $\theta = \tan^{-1} n$, or *Brewster's angle,* symbolized by θ_p as the *polarizing angle*. This condition is also evident in the vanishing of r_{TM} in Figure 20-3 and of the numerator in Eq. (20-24). R_{TE} does not go to zero under this condition, so that reflected light contains only the TE mode and is linearly polarized, with $R_{TE} = 15\%$. For the case $n = 1.50$ used in Figures 20-3 and 20-4, $\theta_p = 56.31°$. At normal incidence ($\theta = 0°$), for both TE and TM modes, Eqs. (20-23) and (20-24) simplify to give

$$R = \left(\frac{1-n}{1+n}\right)^2 \qquad (20\text{-}29)$$

or a reflectance of 4% from glass of refractive index $n = 1.5$. Keep in mind, however, that n is a function of wavelength. As the angle of incidence increases to graz-

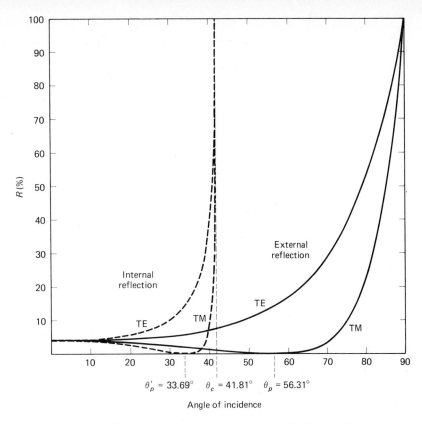

Figure 20-4 Reflectance for both external and internal reflection when $n_1 = 1$ and $n_2 = 1.50$.

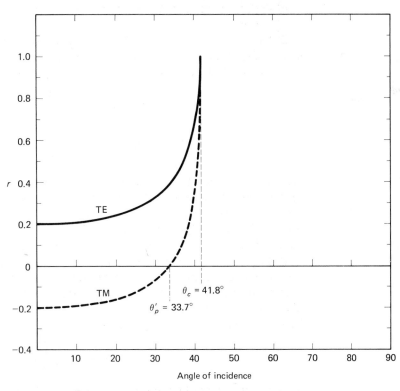

Figure 20-5 Reflection coefficient for the case of internal reflection with $n = n_1/n_2 = 1/1.5$.

ing incidence ($\theta = 90°$) both R_{TE} and R_{TM} become unity, although R_{TM} remains quite small until Brewster's angle has been exceeded.

The reflection coefficient for the case of internal reflection is shown in Figure 20-5 with $n = 1/1.5$, as when light emerges from glass to air. Evidence of phase changes and of a polarizing, or Brewster's, angle may also be seen here. However, Figure 20-5 shows that in this case both r_{TE} and r_{TM} reach values of unity before the angle of incidence reaches 90°. This is the phenomenon of *total internal reflection*, which occurs at the critical angle $\theta_c = \sin^{-1} n$. For the example of glass ($n = 1.5$) used in Figure 20-5, $\theta_p' = 33.7°$ and $\theta_c = 41.8°$. When $\sin \theta_c > n$, the radical $\sqrt{n^2 - \sin^2 \theta}$ is negative and both r_{TE} and r_{TM} are complex. Their magnitudes, however, are easily shown to be unity in this range, giving total reflection for $\theta > \theta_c$. Reflectance for internal reflection is also graphed in Figure 20-4.

20-3 PHASE CHANGES ON REFLECTION

The negative values of the reflection coefficient in Figures 20-3 and 20-5 indicate that $E_r = -|r|E$ in certain situations. Evidently the electric field vector may reverse direction on reflection. Equivalently, in such cases there is a π-phase shift of E on reflection, as the following mathematical argument demonstrates:

$$E_r = -|r|E = e^{i\pi}|r|E_0 e^{i(\mathbf{k}\cdot\mathbf{r}-\omega t)} = |r|E_0 e^{i(\mathbf{k}\cdot\mathbf{r}-\omega t+\pi)}$$

Thus in the case of external reflection, Figure 20-3, a π-phase shift of E occurs at any angle of incidence for the TE mode and for $\theta > \theta_p$ for the TM mode. When reflection is internal, Figure 20-5, we conclude that a π-phase shift occurs for the TM mode for $\theta < \theta_p'$. However, the situation in the region $\theta > \theta_c$, where r is complex, requires further investigation. When $\theta > \theta_c = \sin^{-1} n$, the radical in Eqs. (20-23) and (20-24) becomes imaginary, and the equations may be written in the form

$$\text{TE:} \quad r = \frac{\cos \theta - i\sqrt{\sin^2 \theta - n^2}}{\cos \theta + i\sqrt{\sin^2 \theta - n^2}} \tag{20-30}$$

$$\text{TM:} \quad r = \frac{n^2 \cos \theta - i\sqrt{\sin^2 \theta - n^2}}{n^2 \cos \theta + i\sqrt{\sin^2 \theta - n^2}} \tag{20-31}$$

In either equation, the reflection coefficient takes the form $r = (a - ib)/(a + ib)$. Since the real and imaginary parts of the numerator and denominator are the same, except for a sign, the magnitudes of the numerator and denominator are equal, and r has unit amplitude. The phase of r may be investigated by expressing Eq. (20-30) in complex polar form, as

$$r = \frac{e^{-i\alpha}}{e^{i\alpha}} = e^{-i(2\alpha)}$$

where, for the TE mode, $\tan \alpha = \sqrt{\sin^2 \theta - n^2}/(\cos \theta)$. If we write $r = e^{-i\phi}$, the phase of r, $\phi = 2\alpha$, and expressions enabling the calculation of ϕ are:

$$\text{TE:} \quad \tan\left(\frac{\phi}{2}\right) = \frac{\sqrt{\sin^2 \theta - n^2}}{\cos \theta} \tag{20-32}$$

$$\text{TM:} \quad \tan\left(\frac{\phi}{2}\right) = \frac{\sqrt{\sin^2 \theta - n^2}}{n^2 \cos \theta} \tag{20-33}$$

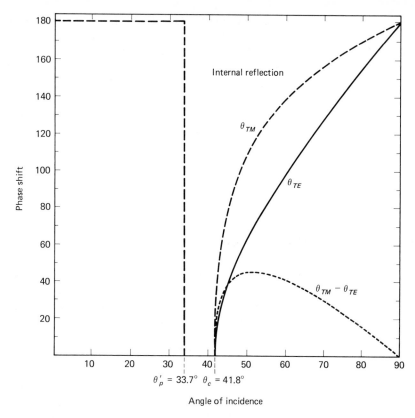

Figure 20-6 Phase shift of electric field for internally reflected rays, with $n = n_1/n_2 = 1/1.5$.

Clearly, the phase difference introduced on reflection may take on values other than 0° and 180°, depending on the angle of incidence. The phase angle ϕ, as determined from Eqs. (20-32) and (20-33), is plotted in Figure 20-6. Also plotted is the relative phase shift $\phi_{TM} - \phi_{TE}$. Notice that the relative phase is 45° at an angle of incidence near 50°. Two such consecutive internal reflections thus produce a 90° difference between the perpendicular components of the **E**-vector, resulting in circularly polarized light. This technique is utilized in the *Fresnel rhomb* (Figure 20-7). Summariz-

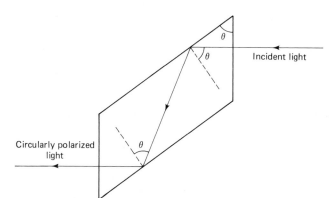

Figure 20-7 The Fresnel rhomb. With the incident light polarized at 45° to the plane of incidence, two internal reflections produce equal amplitude TE and TM amplitudes with a 90° relative phase, or circularly polarized light. For $n = 1.50$, the angle $\theta = 53°$. The device is effective over a wide range of wavelengths.

ing these results for the case of internal reflection,

$$\phi_{TM} = \begin{cases} 180°, & \theta < \theta'_p \\ 0°, & \theta'_p < \theta < \theta_c \\ 2 \arctan \left(\dfrac{\sqrt{\sin^2 \theta - n^2}}{n^2 \cos \theta} \right), & \theta > \theta_c \end{cases} \qquad (20\text{-}34)$$

$$\phi_{TE} = \begin{cases} 0° & \theta < \theta_c \\ 2 \arctan \left(\dfrac{\sqrt{\sin^2 \theta - n^2}}{\cos \theta} \right), & \theta > \theta_c \end{cases} \qquad (20\text{-}35)$$

Phase shifts for both TM and TE modes and for both internal and external reflection are summarized in Figure 20-8.

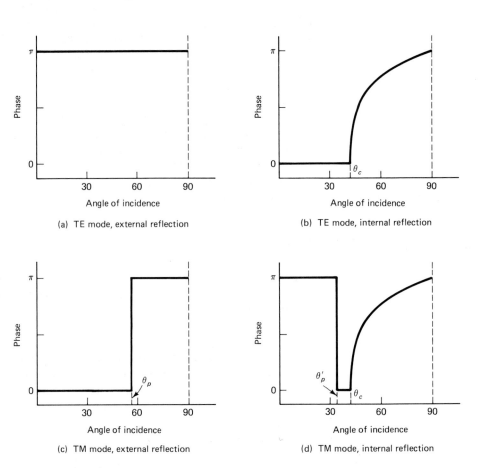

Figure 20-8 Phase changes between incident and reflected rays versus angle of incidence. Discontinuities occur at $\theta_c = 41.8°$, $\theta_p = 56.3°$, and $\theta'_p = 33.7°$ for refractive indices of $n_1 = 1$ and $n_2 = 1.50$.

Example

What is the phase shift of the TM and TE rays reflected both externally and internally in the preceding example?

Solution For this reflection,

$$\theta_c = \sin^{-1}\left(\frac{1}{1.6}\right) = 38.7°$$

$$\theta_p = \tan^{-1}(1.6) = 58.0°$$

$$\theta_p' = \tan^{-1}\left(\frac{1}{1.6}\right) = 32.0°$$

Since the angle of incidence of 30° is less than either θ_p' or θ_c, Eqs. (20-34) and (20-35) or Figure 20-8 require that for internal reflection, $\phi_{TM} = 180°$ and $\phi_{TE} = 0°$, while Figure 20-8 shows that for external reflection, $\phi_{TM} = 0°$ and $\phi_{TE} = 180°$. Thus a π-phase shift occurs between the externally and internally reflected rays for both the TE and TM modes.

A general conclusion can be drawn from the phase changes for the TE and TM modes under internal and external reflection: For angles of incidence not far from normal incidence, the phase changes that occur in internal reflection (0 for TE, π for TM) are the reverse of those at the same angles for external reflection (π for TE, 0 for TM). For both TE and TM modes, the two reflected beams experience a relative π-phase shift. For a thin film in air, we are interested in the *relative* phase shift between rays reflected from the first surface (external) and the second surface (internal). Inspection of Figure 20-8 shows that a *relative* phase shift of π occurs in the TE mode for *internal* angles of incidence less than θ_c and in the TM mode for *internal* angles of incidence less than θ_p'. The corresponding *external* angles of incidence at the first surface are 90° (TE mode) and θ_p (TM mode). Thus in the TE mode a relative phase shift of π occurs for *all* external angles of incidence, but in the TM mode this is true only for external angles less than θ_p.

20-4 CONSERVATION OF ENERGY

From the point of view of energy, it must be true that the rate of energy input or incident power must show up as the sum of reflected power and transmitted power at the boundary, or

$$P_i = P_r + P_t \tag{20-36}$$

If we represent the reflectance R as the ratio of reflected to incident power and the transmittance as the ratio of transmitted to incident power,

$$R = \frac{P_r}{P_i} \quad \text{and} \quad T = \frac{P_t}{P_i} \tag{20-37}$$

Eq. (20-36) takes the form

$$1 = R + T \tag{20-38}$$

The irradiance I is the power density (W/m²), so that we may write, in place of Eq. (20-36),

$$I_i A_i = I_r A_r + I_t A_t \tag{20-39}$$

The cross-sectional areas of the three beams (see Figure 20-9) that appear in Eq. (20-39) are all related to the area A intercepted by the beams in the boundary plane through the cosines of the angles of incidence, reflection, and refraction. We may then write

$$I_i(A \cos \theta_i) = I_r(A \cos \theta_r) + I_t(A \cos \theta_t)$$

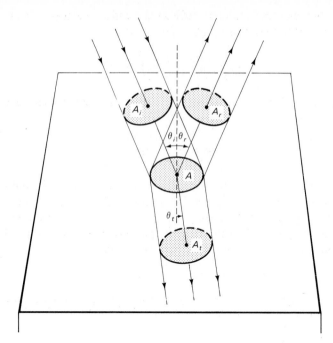

Figure 20-9 Comparison of cross sections of incident, reflected, and transmitted beams.

Of course, $\theta_i = \theta_r$ by the law of reflection. Also using the relation between irradiance and electric field amplitude,

$$I = \left(\frac{\epsilon v}{2}\right) E_0^2$$

and the facts that $v_i = v_r$, and $\epsilon_i = \epsilon_r$, since they correspond to the same medium, we arrive at the equation

$$E_{oi}^2 = E_{or}^2 + \left(\frac{v_t \epsilon_t}{v_i \epsilon_i}\right)\left(\frac{\cos \theta_t}{\cos \theta_i}\right) E_{ot}^2 \qquad (20\text{-}40)$$

The quantity $(v_t \epsilon_t / v_i \epsilon_i)$ is just a complicated way of expressing the relative refractive index n, which we can show as follows:

$$\frac{v_t \epsilon_t}{v_i \epsilon_i} = \frac{v_t}{v_i} \frac{v_i^2 \mu_i}{v_t^2 \mu_t} = \frac{v_i}{v_t} = n \qquad (20\text{-}41)$$

In arriving at this result we have used

$$\mu_i = \mu_t = \mu_0$$

for nonmagnetic materials and the relation

$$v^2 = \frac{1}{\mu \epsilon}$$

for the velocity of a plane electromagnetic wave. Incorporating Eq. (20-41) in Eq. (20-40),

$$E_{oi}^2 = E_{or}^2 + n\left(\frac{\cos \theta_t}{\cos \theta_i}\right) E_{ot}^2 \qquad (20\text{-}42)$$

Dividing the equation by the left member, it becomes

$$1 = r^2 + n\left(\frac{\cos \theta_t}{\cos \theta_i}\right) t^2 \qquad (20\text{-}43)$$

where the reflection and transmission coefficients r and t have been introduced. Now the quantity r^2 is just the reflectance R:

$$R = \frac{P_r}{P_i} = \frac{I_r}{I_i} = \left(\frac{E_{or}}{E_{oi}}\right)^2 = r^2$$

Comparing Eq. (20-43) with Eq. (20-38), it follows that the transmittance T is expressed by the relation

$$T = n\left(\frac{\cos \theta_t}{\cos \theta_i}\right)t^2 \qquad (20\text{-}44)$$

Notice that T is not simply t^2 since it must take into account a different speed and direction in a new medium. The change in speed modifies the rate of energy propagation and thus the power of the beam; the change in direction modifies the cross section and thus the power density of the beam. However, for normal incidence, Eq. (20-44) reduces to $T = nt^2$ and Eq. (20-43) becomes

$$\text{normal incidence: } 1 = r^2 + nt^2 \qquad (20\text{-}45)$$

Graphs of T versus angle of incidence are most easily drawn using Figure 20-4 together with Eq. (20-38).

20-5 EVANESCENT WAVES

In discussing the propagation of a light wave by total internal reflection (TIR) through an optic fiber, we shall mention (Chapter 24) the phenomenon of *cross-talk*, the coupling of wave energy into another medium when it is brought close enough to the reflecting wave. This loss of energy is described as *frustrated total reflection*. The theory presented in this chapter puts us in a good position to describe this phenomenon quantitatively.

The transmitted wave at a refraction is given in Eq. (20-3) as

$$E_t = E_{0t}e^{i(\mathbf{k}_t \cdot \mathbf{r} - \omega t)}$$

where, according to the coordinates chosen in Figure 20-1,

$$\mathbf{k}_t \cdot \mathbf{r} = k_t (\sin \theta_t, \cos \theta_t, 0) \cdot (x, y, 0)$$

$$\mathbf{k}_t \cdot \mathbf{r} = k_t (x \sin \theta_t + y \cos \theta_t)$$

We can express $\cos \theta_t$ as

$$\cos \theta_t \equiv \sqrt{1 - \sin^2 \theta_t} = \sqrt{1 - \frac{\sin^2 \theta}{n^2}}$$

where we have used Snell's law in the last step. At the critical angle, $\sin \theta = n$ and $\cos \theta = 0$. For angles such that $\sin \theta > n$, when TIR occurs, $\cos \theta_t$ becomes pure imaginary and we can write

$$\cos \theta_t = i\sqrt{\frac{\sin^2 \theta}{n^2} - 1}$$

Thus the exponential factor

$$\mathbf{k}_t \cdot \mathbf{r} = k_t x \frac{\sin \theta}{n} + ik_t y \sqrt{\frac{\sin^2 \theta}{n^2} - 1}$$

If we define the real, positive number,

$$\alpha \equiv k_t \sqrt{\frac{\sin^2 \theta}{n^2} - 1}$$

then the transmitted wave may be expressed as

$$E_t = E_{0t} e^{-i\omega t} e^{ixk_t \sin \theta / n} e^{-\alpha y}$$

The last factor describes an exponential decrease in the amplitude of the wave as it enters the rare medium along the y-direction. The other exponential factors include the number i, making them harmonic functions with unit amplitude. When the wave penetrates into the rare medium by an amount

$$y = \frac{1}{\alpha} = \frac{\lambda}{2\pi \sqrt{\frac{\sin^2 \theta}{n^2} - 1}} \tag{20-46}$$

the amplitude is decreased by a factor of $1/e$. The energy of this *evanescent wave* returns to its original medium unless a second medium is introduced into its region of penetration. Although detrimental in the case of cross-talk in closely bound fibers lacking sufficient thickness of protective cladding, the *frustration* of the total internal reflection is put to good use in devices such as variable output couplers, made of two right-angle prisms whose separation along their diagonal faces can be carefully adjusted to vary the amount of evanescent wave coupled from one prism into the other. Another practical application involves a prism face brought near to the surface of an optical waveguide so that the evanescent wave emerging from the prism can be coupled into the waveguide at a given angle (mode) of propagation.

Example

Calculate the penetration depth of an evanescent wave undergoing TIR at a glass ($n = 1.50$)-to-air interface, such that the amplitude is attenuated to $1/e$ its original value. Assume light of wavelength 500 nm is incident at the interface with an angle of $60°$.

Solution Since $\theta_c = \sin^{-1}(1.5) = 41.8°$, TIR occurs at $60°$. The penetration depth is given by Eq. (20-46):

$$y = \frac{0.500 \ \mu m}{2\pi \sqrt{\frac{\sin^2 60}{(1/1.5)^2} - 1}} = 0.096 \ \mu m$$

20-6 COMPLEX REFRACTIVE INDEX

We wish now to show that when the reflecting surface is metallic, the Fresnel equations we have derived continue to be valid, with one important modification: The index of refraction becomes a complex number, including an imaginary part that is a measure of the absorption of the wave. This subject is treated at greater length in Chapter 27, where we examine the frequency dependence of the real and imaginary parts of the refractive index.

When the reflecting surface is that of a homogeneous dielectric—the case we have been discussing in this chapter—the *conductivity* σ of the material is zero. The conductivity is the proportionality constant in *Ohm's law*,

$$\mathbf{j} = \sigma \mathbf{E}$$

where **j** is the *current density* (A/m²) produced by the field **E**. In such cases, both the **E**- and **B**-fields satisfy a differential wave equation of the form

$$\nabla^2 E = \left(\frac{1}{c^2}\right)\frac{\partial^2 E}{\partial t^2} \qquad (20\text{-}47)$$

as described in Chapter 8. We have written harmonic waves satisfying Eq. (20-47) in the form

$$E = E_0 e^{i(\mathbf{k}\cdot\mathbf{r}-\omega t)} \qquad (20\text{-}48)$$

Now if the material is metallic or has an appreciable conductivity, the fundamental Maxwell equations of electricity and magnetism lead to a modification of Eqs. (20-47) and (20-48). The differential wave equation to be satisfied by the **E**-field is then

$$\nabla^2 E = \left(\frac{1}{c^2}\right)\frac{\partial^2 E}{\partial t^2} + \left(\frac{\sigma}{\epsilon_0 c^2}\right)\frac{\partial E}{\partial t} \qquad (20\text{-}49)$$

Note that, compared with Eq. (20-47), the new wave Eq. (20-49) includes an additional term involving the conductivity and the first time derivative of E. As a result, when a harmonic wave in the form of Eq. (20-48) is substituted into Eq. (20-49), we find that the propagation vector **k** must have the complex magnitude

$$\tilde{k} = \frac{\omega}{c}\left[1 + i\left(\frac{\sigma}{\epsilon_0 \omega}\right)\right]^{1/2} \qquad (20\text{-}50)$$

Since the refractive index n is related to k by $n = (c/\omega)k$, the refractive index is now the complex number

$$\tilde{n} = \left[1 + i\left(\frac{\sigma}{\epsilon_0 \omega}\right)\right]^{1/2} \qquad (20\text{-}51)$$

or we write, in general,

$$\tilde{n} = n_R + in_I \qquad (20\text{-}52)$$

where Re $(\tilde{n}) = n_R$ and Im $(\tilde{n}) = n_I$. Combing Eqs. (20-51) and (20-52) and equating their real and imaginary parts, the *optical constants* n_R and n_I can be found in terms of the conductivity by the equations

$$n_R^2 - n_I^2 = 1 \qquad (20\text{-}53)$$

$$2n_R n_I = \frac{\sigma}{\epsilon_0 \omega}$$

Furthermore, if the complex character of k in the form

$$\tilde{k} = \left(\frac{\omega}{c}\right)\tilde{n} = \left(\frac{\omega}{c}\right)[n_R + in_I] \qquad (20\text{-}54)$$

is introduced into the harmonic wave, Eq. (20-48), the result is

$$E = E_0 e^{-(\omega n_I r/c)} e^{i\omega(n_R r/c - t)} \qquad (20\text{-}55)$$

We conclude from Eq. (20-55) that the wave propagates in the material at a wave speed c/n_R and is absorbed such that the amplitude decreases at a rate governed by the exponential factor $e^{-(\omega n_I r/c)}$. Thus Re $(\tilde{n}) = n_R$ must behave as the ordinary refractive index, and Im $(\tilde{n}) = n_I$, called the *extinction coefficient*, determines the rate of absorption of the wave in the conductive medium. This absorption, due to the energy contributed to the production of conduction current j in the material, is usually described by the decrease in power density I with distance given by

$$I = I_0 e^{-\alpha r} \qquad (20\text{-}56)$$

Sec. 20-6   Complex Refractive Index

By comparison with the power density as determined from Eq. (20-55), where $I \propto |E|^2$,

$$I = I_0 e^{-2\omega n_I r/c} \tag{20-57}$$

and so we see that the *absorption coefficient* α is related to the *extinction coefficient* n_I by

$$\alpha = \frac{2\omega n_I}{c} = \frac{4\pi n_I}{\lambda} \tag{20-58}$$

20-7 REFLECTION FROM METALS

Replacing n by $\tilde{n}$ in the Fresnel equations, Eqs. (20-23) and (20-24), we have for metals,

$$\text{TE:} \quad \frac{E_R}{E} = \frac{\cos\theta - \sqrt{\tilde{n}^2 - \sin^2\theta}}{\cos\theta + \sqrt{\tilde{n}^2 - \sin^2\theta}} \tag{20-59}$$

$$\text{TM:} \quad \frac{E_R}{E} = \frac{\tilde{n}^2\cos\theta - \sqrt{\tilde{n}^2 - \sin^2\theta}}{\tilde{n}^2\cos\theta + \sqrt{\tilde{n}^2 - \sin^2\theta}} \tag{20-60}$$

Introducing $\tilde{n}$ as $n_R + in_I$ into Eqs. (20-59) and (20-60), these equations take the form

$$\text{TE:} \quad \frac{E_R}{E} = \frac{\cos\theta - \sqrt{(n_R^2 - n_I^2 - \sin^2\theta) + i(2n_R n_I)}}{\cos\theta + \sqrt{(n_R^2 - n_I^2 - \sin^2\theta) + i(2n_R n_I)}} \tag{20-61}$$

$$\text{TM:} \quad \frac{E_R}{E} = \frac{[n_R^2 - n_I^2 + i(2n_R n_I)]\cos\theta - \sqrt{(n_R^2 - n_I^2 - \sin^2\theta) + i(2n_R n_I)}}{[n_R^2 - n_I^2 + i(2n_R n_I)]\cos\theta + \sqrt{(n_R^2 - n_I^2 - \sin^2\theta) + i(2n_R n_I)}} \tag{20-62}$$

In calculating the reflectance $R = |E_R/E|^2$, the complex quantity E_R/E can first be reduced to a ratio of complex numbers in the form $(a + ib)/(c + id)$, so that

$$R = (a^2 + b^2)/(c^2 + d^2)$$

In the process, we must take the square root of a complex number, which is done by first putting it into polar form. For example, if $z = A + iB$, then, in polar form,

$$z = (A^2 + B^2)^{1/2} e^{i[atn(B/A)]}$$

and the square root becomes

$$z^{1/2} = (A^2 + B^2)^{1/4} e^{i[(1/2)atn(B/A)]} \tag{20-63}$$

The complex expression in Eq. (20-63) can then be returned to the general complex form $C + iD$ using Euler's equation. These mathematical steps are easily performed with a programmable calculator or a computer. In Figure 20-10, the results of such calculations are shown for two metal surfaces, solid sodium and single crystal gallium. High reflectance in the visible spectrum is characteristic of metallic surfaces, as shown by the curves for solid sodium at a wavelength of 589.3 nm. Strong discrimination between the TE and TM modes in the incident radiation is exhibited by the curves for single crystal gallium surfaces.

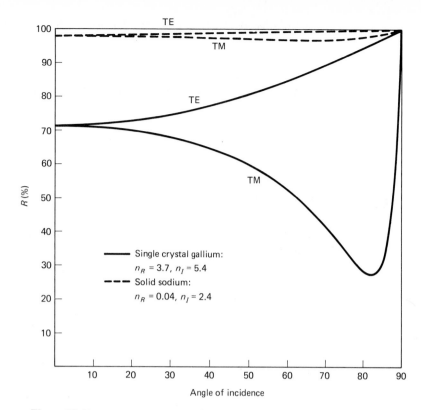

Figure 20-10 Reflectance from metal surfaces by using Fresnel's equations. The values of n_R and n_I are given for sodium light of $\lambda = 589.3$ nm.

PROBLEMS

20-1. Show that the vanishing of the reflection coefficient in the TM mode, Eq. (20-24), leads to Brewster's law.

20-2. The critical angle for a certain oil is found to be 33°33′. What are its Brewster's angles for both external and internal reflections?

20-3. Determine the critical angle and polarizing angles for (a) external and (b) internal reflections from dense flint glass of index $n = 1.84$.

20-4. For what refractive index are the critical angle and (external) Brewster angle equal when the first medium is air?

20-5. Show that the Fresnel equations, Eqs. (20-23) to (20-26), may also be expressed by

$$\text{TE:} \quad r = -\frac{\sin(\theta - \theta_t)}{\sin(\theta + \theta_t)} \qquad t = \frac{2\cos\theta\sin\theta_t}{\sin(\theta + \theta_t)}$$

$$\text{TM:} \quad r = \frac{\tan(\theta - \theta_t)}{\tan(\theta + \theta_t)} \qquad t = \frac{2\cos\theta\sin\theta_t}{\sin(\theta + \theta_t)\cos(\theta - \theta_t)}$$

20-6. Confirm Figures (20-3) and (20-5) by writing computer programs to calculate and/or plot the graphs, using Eqs. (20-23) through (20-26). Change the value of n to produce graphs for the case of external and internal reflection from diamond ($n = 2.42$).

20-7. Write a computer program to calculate and/or plot the reflectance curves of Figure 20-4. Also plot the corresponding transmittance.

20-8. Write a computer program to calculate a plot of the phase shifts as a function of angle of incidence, as in Figure 20-6, for $\theta > \theta_c$.

20-9. A film of magnesium fluoride is deposited onto a glass substrate with optical thickness equal to one-fourth the wavelength of the light to be reflected from it. Refractive indices for the film and substrate are 1.38 and 1.52, respectively. Assume that the film is nonabsorbing. For monochromatic light incident normally on the film, determine (a) reflectance from the air-film surface; (b) reflectance from the film-glass surface; (c) reflectance from an air-glass surface without the film; (d) net reflectance from the combination [see Eq. (19-43)].

20-10. Calculate the reflectance of water ($n = 1.33$) for both (a) TE and (b) TM polarizations when the angles of incidence are $0°$, $10°$, $45°$, and $90°$.

20-11. Light is incident upon an air-diamond interface. If the index of diamond is 2.42, calculate the Brewster and critical angles for both (a) external and (b) internal reflections. In each case distinguish between polarization modes.

20-12. Calculate the percent reflectance and transmittance for both (a) TE and (b) TM modes of light incident at $50°$ on a glass surface of index 1.60.

20-13. Derive Eqs. (20-25) and (20-26) for the transmission coefficient both by (a) eliminating E_r from Eqs. (20-16) to (20-19), and by (b) using the corresponding equations for the reflection coefficients, together with the relationships between reflection and transmission coefficients implied by Eqs. (20-16) and (20-18).

20-14. Unpolarized light is reflected from a plane surface of fused silica glass of index 1.458.
(a) Determine the critical and polarizing angles.
(b) Determine the reflectance and transmittance for the TE mode at normal incidence and at $45°$.
(c) Repeat (b) for the TM mode.
(d) Calculate the phase difference between TM and TE modes for internally reflected rays at angles of incidence of $0°$, $20°$, $40°$, $50°$, $70°$, and $90°$.

20-15. A Fresnel rhomb is constructed of transparent material of index 1.65.
(a) What should be the apex angle θ, as in Figure 20-7?
(b) What is the phase difference between the TE and TM modes after both reflections, when the angle is 5% below and above the correct value?

20-16. Determine the reflectance for metallic reflection of sodium light (589.3 nm) from steel, for which $n_R = 2.485$ and $n_I = 1.381$. Calculate reflectance for (a) TE and (b) TM modes at angles of incidence of $0°$, $30°$, $50°$, $70°$, and $90°$.

20-17. Determine the reflectance from tin at angles of incidence of $0°$, $30°$, and $60°$. Do this for the (a) TE and (b) TM modes of polarization. Real and imaginary parts of the complex refractive index are 1.5 and 5.3, respectively, for light of 589.3 nm.

20-18. (a) What is absorption coefficient for tin, with an imaginary part of the refractive index equal to 5.3 for 589.3 nm light?
(b) At what depth is 99% of normally incident sodium light absorbed in tin?

20-19. (a) From the power conservation requirement, as expressed by Eq. (20-43), show that for an external reflection the transmission coefficient t must be less than 1, but for an internal reflection t' may be greater than 1.
(b) Show further, using the Fresnel Eqs. (20-25) and (20-26), that as the angle of incidence approaches the critical angle, t' must approach a value of 2 in the TE mode and $2/n$ in the TM mode.
(c) Plot the transmission coefficient t' for an interface between glass ($n = 1.5$) and air.

20-20. A narrow beam of light ($\lambda = 546$ nm) is rotated through $90°$ by TIR from the hypotenuse face of a $45°–90°–45°$ prism made of glass with $n = 1.60$.
(a) What is the penetration depth at which the amplitude of the evanescent wave is reduced to $1/e$ of its value at the surface?
(b) What is the ratio of irradiance of the evanescent wave at 1 μm beyond the surface to that at the surface?

REFERENCES

[1] Dichtburn, R. W. *Light*, vol. 2. New York: Academic Press, 1976. Ch. 14.

[2] Longhurst, R. S. *Geometrical and Physcial Optics*, 2d ed. New York: John Wiley and Sons, 1967. Ch. 21.

[3] Rossi, Bruno. *Optics*. Reading, Mass.: Addison-Wesley Publishing Company, 1957. Ch. 8.

21

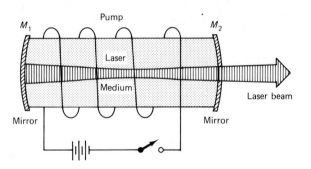

Laser Basics

INTRODUCTION

The laser is perhaps the most important optical device to be developed in the past 50 years. Since its arrival in the 1960s, rather quiet and unheralded outside the scientific community, it has provided the stimulus to make optics one of the most rapidly growing fields in science and technology today.

The laser is essentially an optical amplifier. The word *laser* is an acronym that stands for **l**ight **a**mplification by the **s**timulated **e**mission of **r**adiation. The key words here are *amplification* and *stimulated emission*. The theoretical background of laser action as the basis for an optical amplifier was made possible by Albert Einstein, as early as 1916, when he first predicted the existence of a new radiative process called stimulated emission. His theoretical work, however, remained largely unexploited until 1954, when C. H. Townes and co-workers developed a **m**icrowave **a**mplifier based on **s**timulated **e**mission of **r**adiation. It was called a *maser*. Shortly thereafter, in 1958, A. Schawlow and C. H. Townes adapted the principle of masers to light in the visible region, and in 1960, T. H. Maiman built the first laser device. Maiman's laser incorporated a ruby crystal for the laser amplifying medium and a Fabry-Perot optical cavity as the resonator. Within months of the arrival of Maiman's ruby laser, which emitted deep red light at a wavelength of 694.3 nm, A. Javan and associates developed the first gas laser, the helium-neon laser, which emitted light in both the infrared (at 1.15 μm) and visible (at 632.8 nm) spectral regions.

Following the birth of the ruby and helium-neon (He-Ne) lasers, other laser devices followed in rapid succession, each with a different laser medium and a different wavelength emission. For the greater part of the 1960s, the laser was viewed by the world of industry and technology as a scientific curiosity. It was referred to in jest as "a solution in search of a problem." In the 1960s and 1970s, however, all that changed. The laser came into its own as a unique source of intense, coherent light. Today, fueled by many sources, new laser applications are discovered almost weekly. Together with the fiber-optic cable and semiconductor optoelectronic devices, the laser has revolutionized optics and the optics industry.

In this chapter we review the essence of Einstein's prediction of the existence of stimulated emission, examine the essential elements that make up a laser, describe the operation of the laser in simple terms, and list the characteristics of laser light that make it so unique. Finally, by way of a summary, a table is provided that lists some of the popular lasers in existence today, along with their important operating parameters.

21-1 EINSTEIN'S QUANTUM THEORY OF RADIATION

In 1916, while studying the fundamental processes involved in the interaction of electromagnetic radiation with matter, Einstein showed that the existence of equilibrium between matter and radiation required a previously undiscovered radiation process called *stimulated emission*. According to Einstein, the interaction of radiation with matter could be explained in terms of three basic processes: stimulated absorption, spontaneous emission, and stimulated emission. The three processes are illustrated in Figure 21-1.

Stimulated absorption, or simply absorption, Figure 21-1a, occurs whenever radiation containing photons of energy $h\nu = E_1 - E_0$ is incident on matter with

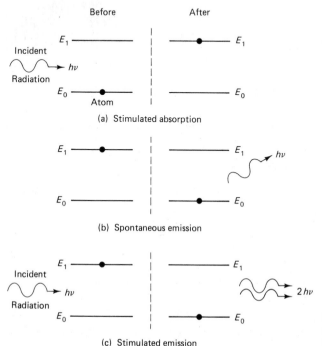

(a) Stimulated absorption

(b) Spontaneous emission

(c) Stimulated emission

Figure 21-1   Three basic processes that affect the passage of radiation through matter. Note that $h\nu = E_1 - E_0$.

ground-state energy E_0 and arbitrary excited state energy E_1. The resonant photon energy $h\nu$ raises the atom from energy state E_0 to E_1. In the process, the photon is absorbed.

Spontaneous emission, Figure 21-1b, takes place whenever atoms are in an excited state. No external radiation is required to initiate the emission. In this process, when an atom in an excited state E_1 spontaneously gives up its energy and falls to E_0, a photon of energy $h\nu = E_1 - E_0$ is released. The photon is emitted in a random direction. Even if external radiation is present, the spontaneous photon is given off in a direction that is completely uncorrelated with the direction of the external radiation.

Quite by contrast, stimulated emission, Figure 21-1c, requires the presence of external radiation. When an incident photon of resonant energy $h\nu = E_1 - E_0$ passes by an atom in excited state E_1, it "stimulates" the atom to drop to the lower state, E_0. In the process, the atom releases a photon of the *same* energy, direction, phase, and polarization as that of the photon passing by. The net effect, then, is two identical photons in the place of one, or an *increase* in the intensity of the incident "beam." It is precisely this process of stimulated emission that makes possible the amplification of light in lasers.

Einstein *A* and *B* Coefficients. Einstein's proof for the existence of stimulated emission grew out of his desire to understand the basic mechanisms involved in the interaction between electromagnetic radiation and matter. A review of his study is both interesting and informative. As a model for this study, we shall assume that matter (a collection of atoms) is in thermodynamic equilibrium with a blackbody radiation field. The atoms and the resonant radiation are contained in an enclosure at some temperature T and interact with one another. Figure 21-2 shows a simplified picture of two-level atoms and radiation (photons) bound inside of an arbitrary unit volume. If thermodynamic equilibrium exists, the number of atoms N_2 at energy level E_2, the number of atoms N_1 at energy level E_1, and the number of photons in the radiation field will all remain constant. Emission and absorption processes that add and remove photons from the radiation field occur at a constant rate, leaving the total photon number unchanged. At the same time, for every N_2 atom moving to E_1 during an emission process, there will be an N_1 atom moving to E_2 during an absorption process, so that N_1 and N_2 will not change. This condition of balance is depicted in Figure 21-3, in terms of atoms moving from level E_1 to E_2 and from E_2 to E_1.

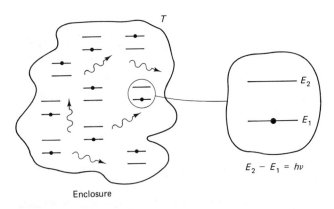

Figure 21-2 A blackbody at temperature T emits radiation that interacts with the atoms in the blackbody.

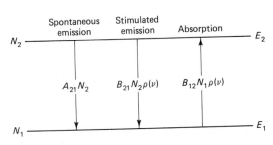

Figure 21-3 Radiative processes that affect the number of atoms at energy E_1 and E_2. The two emission processes remove atoms from level E_2 and add them to level E_1. The absorption process involves transitions from E_1 to E_2.

Of considerable importance in the quantum theory of radiation and the operation of the laser is the meaning of the *Einstein coefficients*, A_{21}, B_{21}, and B_{12}. Their significance is best appreciated in the context of Figure 21-3, where each is related to a radiative process:

Spontaneous Emission (A_{21}). Atoms at energy level E_2 decay spontaneously to level E_1, adding photons of energy $h\nu = E_2 - E_1$ to the radiation field (photon population). At the same time, the population N_2 of level E_2 decreases. The rate of decrease is proportional to the population at any time, that is,

$$\left(\frac{dN_2}{dt}\right)_{\text{spont}} = -A_{21}N_2$$

If spontaneous emission alone takes place, the solution to this equation yields

$$N_2(t) = N_{20}e^{-A_{21}t}$$

The N_2 population decreases with a time constant $\tau = 1/A_{21}$, depleting the number N_2 at level E_2 at a rate N_2/τ, or $A_{21}N_2$, and increasing the number N_1 at level E_1 at the same rate. The constant τ is referred to as the *spontaneous radiative lifetime* of level E_2; the coefficient A_{21} is referred to as the *radiative rate*, usually measured in units of s^{-1}. The coefficient A_{21} is a constant, characteristic of the atom. Note carefully that $(dN_2/dt)_{\text{spont}}$ makes no reference to the prior presence or absence of a radiation field.

Stimulated Emission (B_{21}). In this process the rate at which the N_2 atoms are stimulated by the radiation field (photons) to drop from level E_2 to level E_1 is proportional both to the number of atoms present and to the density of the radiation field, or

$$\left(\frac{dN_2}{dt}\right)_{se} = -B_{21}N_2\rho(\nu)$$

where the photon density is expressed as a function of frequency by the factor $\rho(\nu)$.

Absorption (B_{12}). Absorption is also a stimulated process, since it depends on the strength of the photon field. In effect, stimulated absorption and stimulated emission are inverse processes. The rate at which N_1 atoms are raised from energy level E_1 to E_2 is given by

$$\left(\frac{dN_1}{dt}\right)_{\text{abs}} = -B_{12}N_1\rho(\nu)$$

The coefficient B_{12} is a constant characteristic of the atom. It turns out that B_{12} and B_{21} are closely related; they are equal only under special conditions of *nondegeneracy* of the quantum states that correspond to energy levels E_1 and E_2. (Degenerate levels are those in which two or more states share the same energy.)

With the three basic processes of absorption, spontaneous emission, and stimulated emission related quantitatively to the A and B coefficients, we focus on several of Einstein's assumptions and their implications:

1. Thermodynamic equilibrium at arbitrary temperature T exists between the radiation field and the atoms.
2. The radiation field $\rho(\nu)$ has the spectral distribution characteristic of a blackbody at temperature T.

3. The atom population densities N_1 and N_2 at energy levels E_1 and E_2, respectively, are distributed according to the Boltzmann distribution at that temperature.
4. Population densities N_1 and N_2 are constant in time.

From Figure 21-3 and assumptions (1) and (4), it follows that the rate of change of atoms in level E_2 is given by

$$\frac{dN_2}{dt} = 0 = -N_2 A_{21} - N_2 B_{21}\rho(\nu) + N_1 B_{12}\rho(\nu) \tag{21-1}$$

From assumptions (2) and (3), we write for the spectral energy density $\rho(\nu)$ of blackbody radiation,[1]

$$\rho(\nu) = \frac{8\pi h\nu^3}{c^3} \frac{1}{e^{h\nu/kT} - 1} \tag{21-2}$$

and for the Boltzmann distribution of atoms between the two energy levels,

$$\frac{N_2}{N_1} = e^{-(E_2 - E_1)/kT} = e^{-h\nu/kT} \tag{21-3}$$

In Eqs. (21-2) and (21-3), ν is the frequency of radiation, such that $h\nu = E_2 - E_1$, T is the blackbody temperature, and k is the Boltzmann constant. It should be noted also that Eq. (21-3) has been written for the special case of nondegenerate energy levels. This small concession will simplify the algebra and yet not materially affect the conclusion we shall reach. Solving Eq. (21-1) for $\rho(\nu)$ and substituting for N_1/N_2 from Eq. (21-3), we obtain

$$\rho(\nu) = \frac{A_{21}}{B_{12}(N_1/N_2) - B_{21}} = \frac{A_{21}}{B_{12}e^{h\nu/kT} - B_{21}} \tag{21-4}$$

Equating this expression for $\rho(\nu)$ to that given in Eq. (21-2),

$$\frac{A_{21}}{B_{12}e^{h\nu/kT} - B_{21}} = \frac{8\pi h\nu^3}{c^3} \frac{1}{e^{h\nu/kT} - 1} \tag{21-5}$$

Rearranging to isolate multipliers of the term $e^{h\nu/kT}$,

$$\left(\frac{A_{21}}{B_{21}} - \frac{8\pi h\nu^3}{c^3}\frac{B_{12}}{B_{21}}\right)e^{h\nu/kT} - \left(\frac{A_{21}}{B_{21}} - \frac{8\pi h\nu^3}{c^3}\right) = 0 \tag{21-6}$$

We now have one equation containing the three Einstein coefficients. According to the assumptions made, Eq. (21-6) must be true for arbitrary temperature T. This can be so only if the term that multiplies $e^{h\nu/kT}$ and the remaining term in parentheses are each identically zero. Then it follows at once that

$$\frac{A_{21}}{B_{21}} = 8\pi h\nu^3/c^3 \tag{21-7}$$

and

$$B_{12} = B_{21} \tag{21-8}$$

---

[1] Eq. (21-2) for $\rho(\nu)$, spectral energy density with dimensions energy/(volume-$\Delta\nu$) follows from Eq. (2-12) for M_λ, spectral radiant exitance for a blackbody in power/(area-$\Delta\lambda$). One first changes variables in Eq. (2-12) from λ and $\Delta\lambda$ to ν and $\Delta\nu$ to get $M_\lambda d\lambda = (2\pi h\nu^3/c^2)(e^{h\nu/kT}-1)^{-1}$. Then one simply multiplies $M_\lambda d\lambda$ by $4/c$ to get the expression above for $\rho(\nu)$. See [8].

The importance of Eqs. (21-7) and (21-8) cannot be overestimated. Taken together, they tell us the following.

1. The fundamental Einstein coefficients A_{21}, B_{21}, and B_{12} are all interrelated. If one is known, by measurement or calculation, all are known.

2. The stimulated emission coefficient B_{21} and the (stimulated) absorption coefficient B_{12} are equal, at least for the case of nondegenerate energy states. This equality certainly emphasizes the observation, made earlier, that stimulated emission, the new process discovered by Einstein, and absorption are inverse processes insofar as rate of occurrence goes. Note carefully, however, that the rates $dN_2/dt = N_2 B_{21} \rho(\nu)$ and $dN_1/dt = N_1 B_{12} \rho(\nu)$ differ, depending on the population densities N_2 and N_1. If N_2 is greater than N_1 and a radiation field interacts with the atoms, stimulated emission exceeds absorption and photons will be *added* to the field. If, however, N_1 is greater than N_2, absorption exceeds stimulated emission and photons will be *removed* from the field. The first case $(N_2 > N_1)$ leads to an increase in $\rho(\nu)$, an *amplification*. The second case $(N_1 > N_2)$ leads to a decrease in $\rho(\nu)$, an *attenuation*. For the laser to operate, it is necessary that N_2 be greater than N_1. This is the condition of *population inversion*. Without a population inversion—a condition that runs contrary to the equilibrium population densities predicted by the Boltzmann distribution—laser action cannot occur.

3. Since B_{21}/A_{21} is proportional to the reciprocal of the cube of the frequency ν, the higher the frequency (the shorter the wavelength), the smaller B_{21} becomes in comparison with A_{21}. Since B_{21} is related to stimulated emission (which leads to photon amplification) and A_{21} is related to spontaneous emission (which contributes little, if any, to photon amplification), it would seem that lasers of short wavelength radiation (ultraviolet or X-ray, for example) would be more difficult to build and operate. Such has indeed been the case, even though lasers of shorter wavelengths have been developed rather extensively.

4. Although the important relations between A_{21}, B_{21}, and B_{12} were derived based on the condition of thermodynamic equilibrium, they are valid and hold under any condition. The laser, while operating, is hardly an enclosure in thermodynamic equilibrium. Yet the A and B coefficient relationships, because they are characteristic of the atom, are equally valid whether the atom is in the intense radiation field of a laser cavity or in a hot furnace that can be treated as a blackbody in thermodynamic equilibrium.

Two important ideas for the successful operation of a laser emerge from a review of Einstein's study of the interaction of electromagnetic radiation with matter. The first is that there is a process, *stimulated emission*, that leads to light amplification. The second is that a *population inversion* of atoms in energy levels must be achieved if the stimulated emission process producing coherent photons is to outrival the absorption process removing photons. We use these ideas in describing how the laser operates, but first let us examine the laser as a device and consider the individual parts essential to its operation.

21-2 ESSENTIAL ELEMENTS OF A LASER

The laser device is an optical oscillator that emits an intense, highly collimated beam of coherent radiation. The device consists basically of three elements: an external energy source or *pump*, an *amplifying medium*, and an optical cavity or *resonator*.

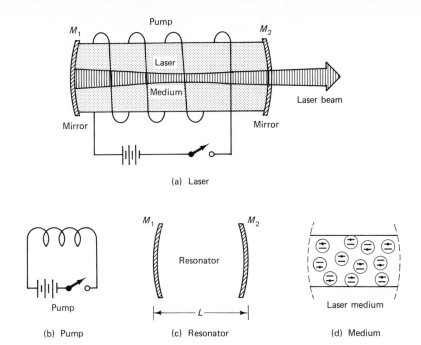

Figure 21-4 Basic elements of a laser. (a) Integral laser device with output laser beam. (b) External energy source, or *pump*. The pump creates a population inversion in the laser medium. The pump can be an optical, electrical, chemical, or thermal energy source. The battery and helix pictured are only symbolic. (c) Empty optical cavity, or *resonator*, bounded by two mirrors. (d) Active cavity, or *laser medium*. Population inversion and stimulated emission work together in the laser medium to produce amplification of light.

These three elements are shown schematically in Figure 21-4: as a unit in Figure 21-4a and separately in Figure 21-4b, c, and d.

The Pump. The pump is an external energy source that produces a population inversion in the laser medium. As explained in the previous section, amplification of a light wave or photon radiation field will occur only in a laser medium that exhibits a population inversion between two energy levels. Otherwise the light wave passing through the laser medium will be attenuated.

Pumps can be optical, electrical, chemical, or thermal in nature, so long as they provide energy that can be coupled into the laser medium to excite the atoms and create the required population inversion. For gas lasers, such as the He-Ne laser, the most commonly used pump is an electrical discharge. The important parameters governing this type of pumping are the electron excitation cross sections and the lifetimes of the various energy levels. In some gas lasers, the free electrons generated in the discharge process collide with and excite the laser atoms, ions, or molecules directly. In others excitation occurs by means of inelastic atom-atom (or molecule-molecule) collisions. In this latter approach, a mixture of two gases is used such that the two different species of atoms, say A and B, have excited states A^* and B^* that coincide. Energy may be transferred from one excited species to the other species in a process whose net effect can be symbolized by the relation $A^* + B \rightarrow A + B^*$. Atom A originally receives its excitation energy from a free electron or by some other excitation process. A notable example is the He-Ne laser, where the laser-active neon (Ne) atoms are excited by resonant transfer of energy from helium (He) atoms in a metastable state. The helium atoms receive their energy from free electrons via collisions.

Although there are numerous other pumps or excitation processes, we cite one more process which has some historical significance. The first laser, developed by T. Maiman at the Hughes Research Laboratories in 1960, was a pulsed ruby laser, which operated at the visible red wavelength of 694.3 nm. Figure 21-5 shows a drawing of the ruby laser device. To excite the Cr^{+3} impurity ions in the ruby rod, Maiman used a helical flashlamp filled with xenon gas. This particular method of exciting the laser medium is known as optical pumping. It is the only practical method that can be used to pump liquid or solid media.

Figure 21-5 Components of a ruby laser system. The shield helps to reflect light from the flashlamp back into the ruby rod.

The Laser Medium. The amplifying medium or laser medium is an important part of the laser device. Many lasers are named after the type of laser medium used, for example, helium-neon (He-Ne), carbon dioxide (CO_2), and neodymium:yttrium aluminum garnet (Nd:YAG). The laser medium, which may be a gas, liquid, or solid, determines the wavelength of the laser radiation. Because of the large selection of laser media, the range of available laser wavelengths extends from the ultraviolet well into the infrared region, sometimes to wavelengths that are a sizable fraction of a millimeter. Laser action has been observed in over half of the known elements, with more than a thousand laser transitions in gases alone. Two of the most widely used transitions in gases are the 632.8-nm visible radiation from neon and the 10.6-μm infrared radiation from the CO_2 molecule. Other commonly used laser media and their corresponding radiations are listed in Table 21-2 at the end of this chapter.

In some lasers the amplifying medium consists of two parts, the laser host medium and the laser atoms. For example, the host of the Nd:YAG laser is a crystal of yttrium aluminum garnet (commonly called YAG), whereas the laser atoms are the trivalent neodymium ions. In gas lasers consisting of mixtures of gases, the distinction between host and laser atoms is generally not made.

The most important requirement of the amplifying medium is its ability to support a population inversion between two energy levels of the laser atoms. This is accomplished by exciting (or pumping) more atoms into the higher energy level than exist in the lower level. As mentioned earlier, in the absence of pumping, there will be no population inversion between any two energy levels of a laser medium. According to the Boltzmann distribution, $N_2/N_1 = e^{-\Delta E/kT}$, where $\Delta E = E_2 - E_1$, the higher level E_2 will always be less populated than the lower level E_1. Pumping, sometimes vigorous pumping, is required to produce the "unnatural" condition of a population inversion. As it turns out though, due to the widely different lifetimes of available atomic energy levels, only certain pairs of energy levels with appropriate spontaneous lifetimes can be "inverted," even with vigorous pumping.

The Resonator. Given a suitable pump and a laser medium that can be inverted, the third basic element is a reasonator, an optical "feedback device" that directs photons back and forth through the laser (amplifying) medium. The resonator, or optical cavity, in its most basic form consists of a pair of carefully aligned plane or curved mirrors centered along the optical axis of the laser system, as shown in Figure 21-4. One of the mirrors is chosen with a reflectivity as close to 100% as possible. The other is selected with a reflectivity somewhat less than 100% to allow part of the internally reflecting beam to escape and become the useful laser output beam.

The geometry of the mirrors and their separation determine the structure of the electromagnetic field within the laser cavity. The exact distribution of the electric field pattern across the wavefront of the emerging laser beam, and thus the transverse irradiance of the beam, depends on the construction of the resonator cavity and mirror surfaces. Many transverse irradiance patterns, called *TEM modes*, are usually present in the output laser beam. By suppressing the gain of the higher-order modes—those with intense electric fields near the edges of the beam—the laser can be made to operate in a single fundamental mode, the TEM_{00} mode. The transverse variation in the irradiance of this mode is *Gaussian* in shape, with peak irradiance at the center and exponentially decreasing irradiance toward the edges.

It is helpful to compare a laser resonator and its geometry to a Fabry-Perot resonator bounded by plane, parallel mirrors. Referring back to Figure 11-8, where we show a two-parallel-plate cavity for the Fabry-Perot interferometer, the condition for resonance is given as $2t \cos \theta_t = m\lambda$. If we imagine a laser cavity made up of two parallel plane mirrors, separated by a distance $t = L$, where $\theta_t = 0$ for laser photons that reflect axially back and forth between the mirrors, the Fabry-Perot condition for resonance becomes, simply, $m\lambda/2 = L$. This is often stated as "an integral number of half-wavelengths fitting between the end mirrors," a requirement that reminds us of standing waves in a string vibrating reasonantly between two fixed ends.

Thus it is useful to think of a laser resonator as a Fabry-Perot resonator, with several modifications. In laser resonators, the cavity is generally bounded by curved rather than plane mirrors, and the empty cavity characteristic of a Fabry-Perot resonator is filled (or partially filled) with a gain medium in the laser resonator. Nevertheless, the resonance condition for the axial (or longitudinal) modes is the same for the two resonators.

21-3 SIMPLIFIED DESCRIPTION OF LASER OPERATION

We have described briefly the three basic elements that comprise the laser device. How do these elements—pump, medium, and resonator—work? We know basically that photons of a certain resonant energy must be created in the laser cavity, must interact with atoms, and must be amplified via stimulated emission, all while bouncing back and forth between the mirrors of the resonator. We can gain a reasonably accurate, though qualitative, understanding of laser operation by studying Figures 21-6 and 21-7. Figure 21-6a shows, in four steps, what happens to a typical atom in the laser medium during the creation of a laser photon. Figure 21-6b shows the actual energy level diagram for a helium-neon laser, with the four steps described in Figure 21-6a clearly identified. Figure 21-7 then shows the same four-step process while focusing on the behavior of the atoms in the laser medium and the photon population in the laser cavity. Let us now examine these figures in turn.

Figure 21-6 Four-step energy cycle associated with a lasing process, for both (a) a general four-level laser and (b) a particular laser, the helium-neon laser. (a) Four-step energy cycle for a laser atom involved in the creation of laser photons. (b) Energy level diagram for the helium-neon laser, showing the production of the 0.6328-μm laser line in terms of the four steps (circled numbers) outlined in (a).

In step 1 of Figure 21-6a, energy from an appropriate pump is coupled into the laser medium. The energy is sufficiently high to excite a large number of atoms from the ground state E_0 to several excited states, collectively labeled E_3. Once at these levels, the atoms spontaneously decay, through various chains, back to the ground state E_0. Many, however, preferentially start the trip back by a very fast (usually radiationless) decay from pump levels E_3 to a very special level, E_2. This decay process is shown in step 2. Level E_2 is labeled as the "upper laser level." It is special in the sense that it has a long lifetime. Whereas most excited levels in an atom might decay in times of the order of 10^{-8} s, level E_2 is *metastable*, with a typical lifetime of the order of 10^{-3} s, hundreds of thousands of times longer than other levels. Thus as atoms funnel rapidly from pump levels E_3 to E_2, they begin to pile up at the metastable level, which functions as a bottleneck. In the process, N_2 grows to a large value. When level E_2 does decay, say by spontaneous emission, it does so to level E_1, labeled the "lower laser level." Level E_1 is an ordinary level that decays to ground state quite rapidly, so that the population N_1 cannot build to a large value. The net effect is the population inversion ($N_2 > N_1$) required for light amplification via stimulated emission.

Once the population inversion has been established and a photon of resonant energy $h\nu = E_2 - E_1$ passes by any one of the N_2 atoms in the upper laser level (step 3), stimulated emission can occur. When it does, laser amplification begins. Note carefully that a photon of resonant energy $E_2 - E_1$ can *also* stimulate absorption from level E_1 to level E_2, thereby losing itself in the process. Since N_2 is greater than N_1, however, and $B_{21} = B_{12}$ as shown earlier, the rate for stimulated emission, $B_{21}N_2\rho(\nu)$, exceeds that for stimulated absorption, $B_{12}N_1\rho(\nu)$. Then light amplification occurs. In that event there is a steady increase in the incident resonant photon population and lasing continues. This is shown schematically in step 3, where the incident resonant photon approaching from the "left" leaves the vicinity of an N_2 atom in duplicate. In step 4, one of the inverted N_2 atoms, which dropped to level E_1 during the stimulated emission process, now decays rapidly to ground state E_0. If the pump is still operating, this atom is ready to repeat the cycle, thereby ensuring a steady population inversion and a constant laser beam output.

In Figure 21-6b, the pump energy (step 1) is supplied by an electrical discharge in the low-pressure gas mixture, thereby elevating ground state helium atoms to higher energy states, one of which is represented by the 2^1S level. Then by resonant collisional energy transfer—made possible because the 2^1S level of helium is nearly equal to the $3s_2$ level of neon—step 2 is achieved as excited helium atoms pass their energy over to ground state neon atoms, raising them to the neon $3s_2$ level. This process produces the population inversion required for effective amplification via stimulated emission of radiation.

The stimulated emission process (step 3) occurs between the neon levels $3s_2$ and $2p_4$, the transition with the highest probability[2] from $3s_2$ to any of the ten $2p$ states. This transition gives rise to photons of wavelength 0.6328 μm, photons that are amplified via stimulated emission and form the common red beam characteristic of helium-neon lasers. Finally, in step 4, the neon atom in energy state $2p_4$ decays by spontaneous emission to the $1s$ ground level. Once back in the ground state, it is again available to undergo collision with an excited helium atom and to repeat the cycle. While Figure 21-6b relates the four steps to the emission of the He-Ne 632.8-nm laser line, other transitions from the $3s$ to the $2s$ and $2p$ levels have also been made to lase. One such transition, leading to the 1.1523 μm line, is indicated in the figure.

Now let us look at Figure 21-7. It shows essentially the same action but does

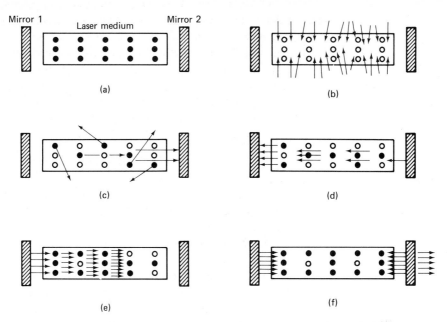

Figure 21-7 Step-by-step development of laser oscillation in a typical laser cavity. (a) Quiescent laser. (b) Laser pumping. (c) Spontaneous and stimulated emission. (d) Light amplification begins. (e) Light amplification continues. (f) Established laser operation.

so in terms of the behavior of the atoms in the laser medium and the photon population in the cavity. In (a) the laser medium is shown situated between the mirrors of the optical resonator. Mirror 1 is essentially 100% reflecting, while mirror 2 is partially reflecting and partially transmitting. Most of the atoms in the laser medium are in the ground state. This is shown by the black dots. In (b), external energy (light from a flashlamp or from an electrical discharge) is pumped into the medium, raising most of the atoms to the excited levels (E_3 in Figure 21-6). Excited states are shown by circles. During the pumping process, the population inversion is established. The light amplification process is initiated in (c), when excited atoms (some of the atoms at level E_2 in Figure 21-6) spontaneously decay to level E_1. Since this is spontaneous emission, the photons given off in the process radiate out randomly in all directions. Many, therefore, leave through the sides of the laser cavity and are lost. Nevertheless, there will generally be several photons—let us call them "seed" photons—directed *along* the optical axis of the laser. These are the horizontal arrows shown in (c) of Figure 21-7, directed perpendicularly to the mirrors. With the seed photons of correct (resonant) energy accurately directed between the mirrors and many N_2 atoms still in the inverted state E_2, the stage for stimulated emission is set. As the seed photons pass by the inverted N_2 atoms, stimulated emission adds identical photons in the same direction, providing an ever-increasing population of coherent photons that bounce back and forth between the mirrors. This buildup process, shown in (d) and (e), continues as long as there are inverted atoms and resonant energy photons in the cavity. Since output mirror 2 is partially transparent, a fraction of the photons incident on the mirror pass out through the mirror. These photons constitute the external laser beam, as shown in (f). Those that do not leave

---

[2] A readable, comprehensive discussion of the helium-neon laser, with energy level diagrams and transition probabilities, is given in [4].

through the output mirror are reflected, recycling back and forth through the cavity gain medium.

In summary then, the laser process depends on the following:

1. A *population inversion* between two appropriate energy levels in the laser medium. This is achieved by the pumping process and the existence of a metastable state.

2. *Seed photons* of proper energy and direction, coming from the ever-present spontaneous emission process between the two laser energy levels. These initiate the *stimulated emission* process.

3. An *optical cavity* that confines and directs the growing number of resonant energy photons back and forth through the laser medium, continually exploiting the population inversion to create more and more stimulated emissions, thereby creating more and more photons directed back and forth between the mirrors, and so on.

4. Coupling a certain fraction of the laser light wave (the cavity photon population) out of the cavity through the *output coupler mirror* to form the external laser beam.

Comparing Fabry-Perot and Laser Resonators.   In closing out this section, it is instructive to compare the operation of the Fabry-Perot cavity (resonator) with that of a laser oscillator (resonator) and laser amplifier. Figure 21-8 shows the comparison with three separate sketches. One can see readily that a laser oscillator (c) combines the characteristics of a Fabry-Perot cavity and laser amplifier [(a) and (b)].

Figure 21-8a shows a Fabry-Perot cavity, resonant to a wavelength λ_0, such that the resonance condition, $m\lambda_0/2 = L$, is satisfied. As the drawing indicates, with two plane parallel mirrors of equal reflectivities (say, 95%), separated by a distance L, the output power is equal to the input power, with, of course, a substantially higher circulatory power confined between the mirrors. (There is no *no gain medium* between the two mirrors.)

The next sketch (Figure 21-8b) shows the operation of a laser amplifier—a gain medium with population inversion and stimulated emission—but with no cavity (that is, *no mirrors*). In this instance, an input beam (generally a laser beam of wavelength λ_{21}) enters the gain medium, which contains an inverted population between two specific energy levels, E_2 and E_1, for which $E_2 - E_1 = hc/\lambda_{21}$. The input laser beam then gains energy via the process of stimulated emission as it passes through the gain medium, emerging finally with significantly amplified power.

The last sketch in Figure 21-8c shows how a laser oscillator combines the elements of a Fabry-Perot cavity and a bare gain medium and gives rise to a laser device and a laser beam. In this geometry the gain medium, pumped by an external source, contains inverted levels E_2 and E_1, such that $E_2 - E_1 = hc/\lambda_{21}$. In addition, the cavity is resonant to the wavelength λ_{21}, since the condition $m\lambda_{21}/2 = L$ is satisfied for this particular wavelength. Thus, with a resonant cavity and a gain medium, the essential requirements for laser oscillation are met.

Now that we have some understanding of laser operation and the origin of the laser beam, let us next look at the characteristics of laser light.

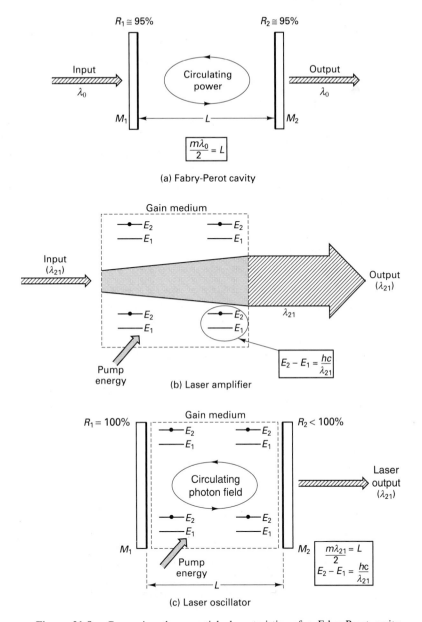

Figure 21-8 Comparing the essential characteristics of a Fabry-Perot cavity, laser amplifier, and laser oscillator. (a) Fabry-Perot cavity with identical mirrors M_1 and M_2, spaced a distance L apart. The cavity is resonant to a wavelength λ_0. Output power equals input power. (b) A bare gain medium (no end mirrors), pumped by an external energy source, has inverted population levels E_2 and E_1 such that $E_2 - E_1 = hc/\lambda_{21}$. A coherent input beam of wavelength λ_{21} is amplified as it passes through the gain medium. (c) A laser oscillator combines elements of a Fabry-Perot resonant cavity and a gain medium to produce a coherent laser beam.

Monochromaticity. The light emitted by a laser is almost pure in color, almost of a single wavelength or frequency. Although we know that no light can be truly monochromatic, with unlimited sharpness in wavelength definition, laser light comes far closer than any other available source in meeting this ideal limit.

The monochromaticity of light is determined by the fundamental emission process wherein atoms in excited states decay to lower energy states and emit light. In blackbody radiation, the emission process involves billions of atoms and many sets of energy-level pairs within each atom. The resultant radiation is hardly monochromatic, as we know. If we could select an identical set of atoms from this blackbody and isolate the emission determined by a single pair of energy levels, the resultant radiation, although weaker, would be decidedly more monochromatic. When such radiation is produced by nonthermal excitation, the radiation is often called *fluorescence*. Figure 21-9 depicts such an emission process. The fluorescence comes from the radiative decay of atoms between two well-defined energy levels E_2 and E_1. The nature of the fluorescence, analyzed by a spectrophotometer, is shown in the *lineshape* plot, a graph of spectral radiant exitance versus wavelength. Note carefully that the emitted light has a wavelength spread $\Delta\lambda$ about a center wavelength λ_0, where $\lambda_0 = c/\nu_0$ and $\nu_0 = (E_2 - E_1)/h$. While most of the light may be emitted at a wavelength λ_0, it is an experimental fact that some light is also emitted at wavelengths above and below λ_0, with different relative exitance, as shown by the lineshape plot. Thus the emission is not monochromatic; it has a wavelength spread given by $\lambda_0 \pm \Delta\lambda/2$, where $\Delta\lambda$ is often referred to as the *linewidth*. When the linewidth is measured at the half maximum level of the lineshape plot, it is called the FWHM linewidth, that is, "full width at half maximum."

(a)

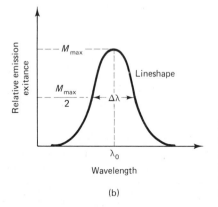

(b)

Figure 21-9 Fluorescence and its spectral content for a radiative decay process between two energy levels in an atom. (a) Spontaneous decay process between well-defined energy levels. (b) Spectral content of fluorescence in (a), showing lineshape and linewidth.

In the laser process, the linewidth $\Delta\lambda$ shown in Figure 21-9 is narrowed considerably, leading to light of a much higher degree of monochromaticity. Basically this occurs because the process of stimulated emission effectively narrows the band of wavelengths emitted during spontaneous emission. This narrowing of the linewidth is shown qualitatively in Figure 21-10. To gain a quantitative appreciation for the monochromaticity of laser light, consider the data in Table 21-1, in which the linewidth of a high-quality He-Ne laser is compared to the linewidth of the spectral output of a typical sodium discharge lamp and to the linewidth of the very narrow cadmium red line found in the spectral emission of a low-pressure lamp. The conversion from $\Delta\lambda$ to $\Delta\nu$ is made by using the approximate relationship, $\Delta\nu = c\,\Delta\lambda/\lambda_0^2$.

The data of Table 21-1 show that the He-Ne laser is 10 million times more monochromatic than the ordinary discharge lamp and about 100,000 times more so

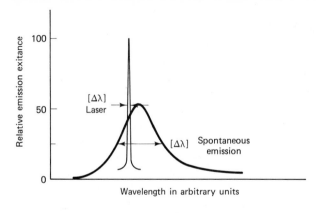

Figure 21-10 Qualitative comparison of linewidths for laser emission and spontaneous emission involving the same pair of energy levels in an atom. The broad peak is the lineshape of spontaneously emitted light between levels E_2 and E_1 *before* lasing begins. The sharp peak is the lineshape of laser light between levels E_2 and E_1 *after* lasing begins.

TABLE 21-1 COMPARISON OF LINEWIDTHS

Light source	Center wavelength λ_0 (Å)	FWHM linewidth $\Delta\lambda$ (Å)	FWHM linewidth $\Delta\nu$ (Hz)
Ordinary discharge lamp	5896	$\cong 1$	9×10^{10}
Cadmium low-pressure lamp	6438	$\cong 0.013$	9.4×10^8
Helium-neon laser	6328	$\cong 10^{-7}$	7.5×10^3

than the cadmium red line. No ordinary light source, without significant filtering, can approach the degree of monochromaticity present in the output beam of typical lasers.

Coherence. The optical property of light that most distinguishes the laser from other light sources is *coherence*. The laser is regarded, quite correctly, as the first truly coherent light source. Other light sources, such as the sun or a gas discharge lamp, are at best only partially coherent.

The subject of coherence can be treated quite rigorously and mathematically, using a statistical interpretation. Some of that was done in Chapter 12, which was devoted entirely to coherence. At this point we bypass mathematical analysis and describe coherence in a general sense, striving only for a qualitatively useful understanding of laser coherence.

Coherence, simply stated, is a measure of the degree of phase correlation that exists in the radiation field of a light source at different locations and different times. It is often described in terms of a *temporal coherence*, which is a measure of the degree of monochromaticity of the light, and a *spatial coherence*, which is a measure of the uniformity of phase across the optical wavefront. To obtain a qualitative understanding of temporal and spatial coherence, consider the simple analogy of water waves created at the center of a quiet pond by a regular, periodic disturbance. The source of disturbance might be a cork bobbing up and down in regular fashion, creating a regular progression of outwardly moving crests and troughs, as in Figure 21-11. Such a water wavefield can be said to have perfect temporal and spatial coherence. The temporal coherence is perfect because there is but a single wavelength; the crest-to-crest distance remains constant. As long as the cork keeps bobbing regularly, the wavelength will remain fixed, and one can predict with great accuracy the location of all crests and troughs on the pond's surface. The spatial coherence of the wavefield is also perfect because the cork is a small source, generating ideal waves, circular crests, and troughs of ideal regularity. Along each wave then, the spatial variation of the relative phase of the water motion is zero, that is, the surface of the

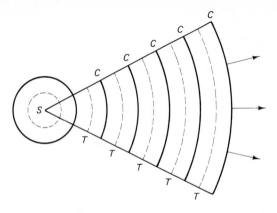

Figure 21-11 Portion of a perfectly coherent water wavefield created by a regularly bobbing cork at S. The wavefield contains perfectly ordered wavefronts, C (crests) and T (troughs), representing water waves of a single wavelength.

water all along a crest or trough is in step or in phase. Again, one can predict with great accuracy, anywhere on the pond, the vertical displacement of the water surface.

The water wavefield described above can be rendered temporally and spatially *incoherent* by the simple process of replacing the single cork with a hundred corks and causing each cork to bob up and down with a different and randomly varying periodic motion. There would then be little correlation between the behavior of the water surface at one position and another. The wavefronts would be highly irregular geometrical curves, changing shape haphazardly as the collection of corks continued their jumbled, disconnected motions. It does not require much imagination to move conceptually from a collection of corks that give rise to water waves to a collection of excited atoms that give rise to light. Disconnected, uncorrelated creation of water waves results in an incoherent water wavefield. Disconnected, uncorrelated creation of light waves results, similarly, in an incoherent radiation field.

To emit light of high coherence then, the radiating region of a source must be small in extent (in the limit, of course, a single atom) and emit light of a narrow bandwidth (in the limit, with $\Delta\lambda$ equal to zero). For real light sources, neither of these conditions is attainable. Real light sources, with the exception of the laser, emit light via the uncorrelated action of many atoms, involving many different wavelengths. The result is the generation of incoherent light. To achieve some measure of coherence with a nonlaser source, two modifications to the emitted light can be made. First, a pinhole can be used with the light source to limit the spatial extent of the source. Second, a narrow-band filter can be used to decrease significantly the linewidth $\Delta\lambda$ of the light. Each modification improves the coherence of the light given off by the source, but only at the expense of a drastic loss of light energy.

In contrast, a laser source, by the very nature of its production of amplified light via stimulated emission, ensures both a narrow-band output and a high degree of phase correlation. Recall that in the process of stimulated emission, each photon added to the stimulating radiation has a phase, polarization, energy, and direction *identical* to that of the amplified light wave in the laser cavity. The laser light thus created and emitted is both temporally and spatially coherent. In fact, one can describe or model a real laser device as a very powerful, fictitious "point source," located at a distance, giving off monochromatic light in a narrow cone angle. Figure 21-12 summarizes the basic ideas of coherence for nonlaser and laser sources.

For typical lasers, both the spatial coherence and temporal coherence of laser light are far superior to that for light from other sources. The transverse spatial coherence of a single mode laser beam extends across the full width of the beam, whatever that might be. The temporal coherence, also called "longitudinal spatial coherence," is many orders of magnitude above that of any ordinary light source. The

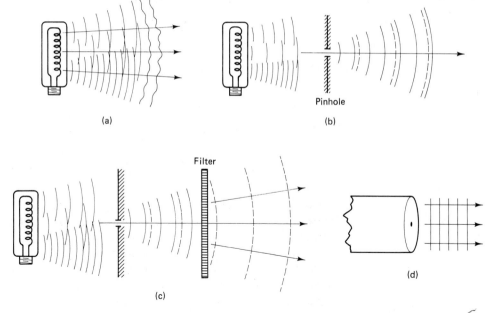

(a)

Pinhole

(b)

Filter

(c)

(d)

Figure 21-12 A tungsten lamp requires a pinhole and filter to produce partially coherent light. The light from a laser is naturally coherent. (a) Tungsten lamp. The tungsten lamp is an extended source that emits many wavelengths. The emission lacks both temporal and spatial coherence. The wavefronts are irregular and change shape in a haphazard manner. (b) Tungsten lamp with pinhole. An ideal pinhole limits the extent of the tungsten source and improves the spatial coherence of the light. However, the light still lacks temporal coherence since all wavelengths are present. Power in the beam has been decreased. (c) Tungsten lamp with pinhole and filter. Adding a good narrow-band filter further reduces the power but improves the temporal coherence. Now the light is "coherent," but the available power is far below that initially radiated by the lamp. (d) Laser. Light coming from the laser has a high degree of spatial and temporal coherence. In addition, the output power can be very high.

coherence time t_c of a laser is a measure of the average time interval over which one can continue to predict the correct phase of the laser beam at a given point in space. The *coherence length* L_c is related to the coherence time by the equation $L_c = ct_c$, where c is the speed of light. Thus the coherence length is the average length of light beam along which the phase of the wave remains unchanged. For the He-Ne laser described in Table 21-1, the coherence time is of the order of milliseconds (compared with about 10^{-11} s for light from a sodium discharge lamp), and the coherence length for the same laser is thousands of kilometers (compared with fractions of a centimeter for the sodium lamp).

Directionality. When one sees the thin, pencil-like beam of a He-Ne laser for the first time, one is struck immediately by the high degree of beam directionality. No other light source, with or without the help of lenses or mirrors, generates a beam of such precise definition and minimum angular spread.

The astonishing degree of directionality of a laser beam is due to the geometrical design of the laser cavity and to the monochromatic and coherent nature of light generated in the cavity. Figure 21-13 shows a specific cavity design and an external laser beam with an angular spread signified by the angle ϕ. The cavity mirrors shown are shaped with surfaces concave toward the cavity, thereby "focusing" the reflecting light back into the cavity and forming a *beam waist* at one position in the

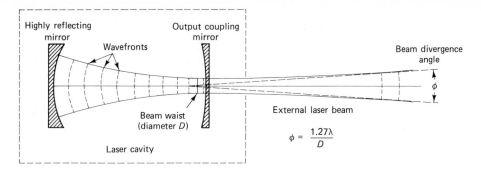

Figure 21-13 External and internal laser beam for a given cavity. Diffraction or beam spread, measured by the beam divergence angle ϕ, appears to be caused by an effective aperture of diameter D, located at the beam waist.

cavity. The nature of the beam inside the laser cavity and its characteristics outside the cavity are determined by solving the rather complicated problem of electromagnetic waves in an open cavity. Although the details of this analysis are beyond the scope of this discussion, several results are worth examining. It turns out that the beam-spread angle ϕ is given by the relationship

$$\phi = \frac{1.27\lambda}{D} \tag{21-9}$$

where λ is the wavelength of the laser beam and D is the diameter of the laser beam at its beam waist. One cannot help but observe that Eq. (21-9) is quite similar to that obtained when calculating the angular spread in light generated by the diffraction of plane waves passing through a circular aperture (Chapter 16). The pattern consists of a central, bright circular spot, the *Airy disk*, surrounded by a series of bright rings. The essence of this phenomenon is shown in Figure 21-14. The diffraction angle θ, tracking the Airy disk, is given by

$$\theta = \frac{2.44\lambda}{D} \tag{21-10}$$

where λ is the wavelength of the collimated, monochromatic light and D is the diameter of the circular aperture. Both Eqs. (21-9) and (21-10) depend on the ratio of a wavelength to a diameter; they differ only by a constant coefficient. It is tempting, then, to think of the angular spread ϕ inherent in laser beams and given in Eq. (21-9) in terms of diffraction. If we treat the beam waist as an effective circular aperture located inside the laser cavity, then by controlling the size of the beam waist we control the diffraction or beam spread of the laser. The beam waist, in

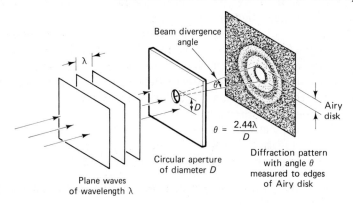

Figure 21-14 Fraunhofer diffraction of plane waves through a circular aperture. Beam divergence angle θ is set by the edges of the Airy disk.

practice, is determined by the design of the laser cavity and depends on the radii of curvature of the two mirrors and the distance between the mirrors. Therefore, one ought to be able to build lasers with a given beam waist and, consequently, a given beam divergence or beam spread in the *far field*, that is, at sufficiently great distance L from the diffracting aperture that $L \gg$ area aperture/λ. Such is indeed the case.

Several comments may be helpful in conjunction with Figures 21-13 and 21-14. Plane waves of uniform irradiance pass through the circular aperture in Figure 21-14; that is, the strength of the electric field is the same at all points along the wavefront. In Figure 21-13 the wavefronts that pass "through" the effective aperture or beam waist are also plane waves, but the irradiance of the laser light is not uniform across the plane. For the lowest-order transverse mode, the TEM$_{00}$ mode or *Gaussian beam*, the irradiance of the beam decreases exponentially toward the edges of the beam in accordance with the Gaussian form e^{-8y^2/D^2}, where y measures the transverse beam direction and D is the beam width at a given position along the beam (see Figure 21-15). The circular aperture in Figure 21-14 is a true physical aperture; the beam waist in Figure 21-13 is not. It is interpreted as an *effective aperture* when beam spread from a laser is compared with light diffraction through a real aperture.

With the help of Eq. (21-9), one can now develop a feel for the low beam spread, or high degree of directionality, of laser beams.

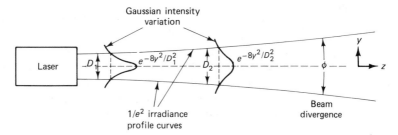

Figure 21-15　A Gaussian TEM$_{00}$ beam. The diameter (transverse width) of the beam is measured between the $1/e^2$ profile curves. These curves are the locus of points where the irradiance of the beam, on either side of the optical axis, has decreased to $1/e^2$ of its value at the center of the beam. The Gaussian irradiance variation is shown at two positions along the beam, where the beam diameters are D_1 and D_2, respectively. The diameter D_2 is greater than D_1 because the beam is spreading according to the beam divergence angle $\phi = 1.27\lambda/D$.

Example

(a) He-Ne lasers (632.8 nm) have an internal beam waist of diameter near 0.5 mm. Determine the beam divergence.

(b) Since we can control the beam waist D by laser cavity design and "select" the wavelength by choosing different laser media, what lower limit might we expect for the beam divergence? How directional can lasers be? Suppose we design a laser with a beam waist of 0.5-cm diameter and a wavelength of 200 nm. By what factor is the beam divergence improved?

Solution　(a)

$$\phi = \frac{1.27\lambda}{D} = \frac{(1.27)(632.8 \times 10^{-9} \text{ m})}{5 \times 10^{-4} \text{ m}} = 1.6 \times 10^{-3} \text{ rad}$$

This is a typical laser-beam divergence, indicating that the beam width increases about 1.6 cm every 1000 cm.

(b)

$$\phi = \frac{1.27\lambda}{D} = \frac{(1.27)(200 \times 10^{-9} \text{ m})}{5 \times 10^{-3} \text{ m}} = 5.1 \times 10^{-5} \text{ rad}$$

Thus the beam divergence angle becomes about 5×10^{-5} rad, roughly a 30-fold decrease in beam spread over the He-Ne laser described in part (a). This beam would spread about 1.6 cm every 320 m. Clearly, if beam waist size is at our command and lasers can be built with wavelengths below the ultraviolet, there is no limit to how parallel and directional the laser beam can be made.

The high degree of directionality of the laser, or any other light source, depends on the monochromaticity and coherence of the light generated. Ordinary sources are neither monochromatic nor coherent. Lasers, on the other hand, are superior on both counts, and as a consequence generate highly directional, quasi-collimated light beams.

Laser Source Intensity. It has been said that a 1-mW He-Ne laser is hundreds of times "brighter" than the sun. As difficult as this may be to imagine, calculations for luminance or visual brightness of a typical laser, compared to the sun, substantiate these claims. To develop an appreciation for the enormous difference between the radiance of lasers and thermal sources we consider a comparison of their photon output rates (photons per second).

Example

(a) Small gas lasers typically have power outputs P of 1 mW. Neodymium-glass lasers, such as those under development for the production of laser-induced fusion, boast of power outputs near 10^{14} W! Using these two extremes and an average energy of 10^{-19} J per visible photon ($E = h\nu$), determine the approximate range of photon output from lasers.

(b) For comparison, consider a broadband thermal source with a radiating surface equal to that of the beam waist of a 1-mW He-Ne laser with diameter of 0.5 mm (or area of $A = 2 \times 10^{-7}$ m²). Let the surface emit radiation at a wavelength of 633 nm with a linewidth of $\Delta\lambda = 100$ nm and temperature $T = 1000$ K. Determine the photon output rate of this thermal source.

Solution (a) The rate of photon output is given by $P/h\nu$. Thus

$$\frac{P}{h\nu} = \frac{10^{-3} \text{ J/s}}{10^{-19} \text{ J/photon}} = 10^{16} \text{ photons/s}$$

$$\frac{P}{h\nu} = \frac{10^{14} \text{ J/s}}{10^{-19} \text{ J/photon}} = 10^{33} \text{ photons/s}$$

(b) The photon output rate for the broadband source can be calculated from the equation[3]

$$\text{thermal photons/s} \cong \frac{1}{\lambda^2} \frac{1}{e^{h\nu/kT} - 1} \Delta A \, \Delta\nu \tag{21-11}$$

---

[3] One can obtain Eq. (21-11) from Eq. (21-2) by changing from $\rho(\nu)$ in units of energy per unit volume per frequency range $[E/(\text{vol-}\Delta\nu)]$ to units that give thermal photons per second $[E/(h\nu\text{-}\Delta t)]$. In this process, one must change geometry accordingly, from an isotropic radiating volume element ΔV (Eq. 21-2) to that of a radiating surface element ΔA (Eq. 21-11).

The frequency $\nu = c/\lambda = 4.74 \times 10^{14}$ Hz. The frequency bandwidth $\Delta\nu = (c/\lambda^2)\,\Delta\lambda = 7.5 \times 10^{13}$ Hz. The argument of the exponent is

$$\frac{h\nu}{kT} = \frac{(6.63 \times 10^{-34} \text{ J-s})(4.74 \times 10^{14} s^{-1})}{(1.38 \times 10^{-23} \text{ J/K})(1000 \text{ K})} = 22.77$$

Substituting all values into Eq. (21-11), the thermal photon output per second is

$$\left(\frac{1}{(633 \times 10^{-9})^2 \text{ m}^2}\right)\left(\frac{1}{7.76 \times 10^9}\right)(2 \times 10^{-7} \text{ m}^2)(7.5 \times 10^{13} s^{-1}) \cong 5 \times 10^9$$

We find a rate of only about 10^9 photons/s! This value is seven orders of magnitude smaller than the photon output rate of a low-power 1-mW He-Ne laser and 24 orders of magnitude smaller than a powerful neodymium-glass laser. The comparison is summarized in Figure 21-16.

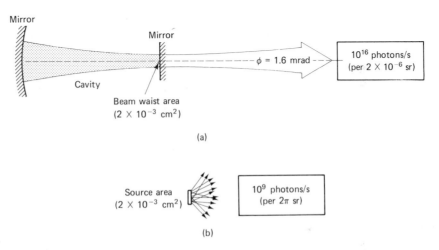

(a)

(b)

Figure 21-16 Comparison of photon output rates between a low-power He-Ne gas laser and a hot thermal source of the same radiating surface area. (a) 1-mW He-Ne laser ($\lambda = 633$ nm). (b) Broadband thermal source (Lambertian), with approximate values of $T = 1000$ K, $\Delta A = 2 \times 10^{-3}$ cm^2, $\lambda_0 = 633$ nm, and $\Delta\lambda = 100$ nm. Note that the laser emits all the photons in a small solid angle ($\sim 2 \times 10^{-6}$ sr) compared with the 2π solid angle of the thermal source.

We see also from Figure 21-16 that the He-Ne laser emits 10^{16} photons/s into a very small solid angle of about 2×10^{-6} sr, whereas the thermal emitter, acting as a Lambertian source, radiates 10^9 photons/s into a forward, hemispherical solid angle of 2π sr. If we were to ask how many thermal photons/second are emitted by the thermal source into a solid angle equal to that of the laser, we would find the answer to be a mere 320 photons/s:

$$(10^9 \text{photons/s})\left(\frac{2 \times 10^{-6} \text{sr}}{6.28 \text{ sr}}\right) = 320 \text{ photons/s}$$

The contrast between 10^{16} photons/s for the laser source and 320 photons/s for the thermal source is now even more dramatic.

Carrying our comparison of source intensity between laser and nonlaser sources one step further, we can determine how the radiance [W/(cm^2-sr)] of a high-power, nonlaser source compares with that of a low-power laser.

Example

We choose for the nonlaser source a super-high-pressure mercury lamp, capable of a source radiance L_e equal to 250 W/(cm²-sr). Such lamps were about the best high-radiance sources available before the advent of the laser. For the laser source, we choose a 4-mW He-Ne laser with beam-waist diameter of 0.5 mm and beam divergence angle of 1.6 mrad, emitting at a wavelength of 633 nm. The geometry is shown in Figure 21-17. It should be clear from the figure that the laser can be considered a radiating surface of area equal to that of the beam waist. The radiant power of the surface is equal to that of the laser, and the surface radiates only into the solid angle dictated by the laser.

Figure 21-17 Radiance of a low-power He-Ne laser. The radiance of the laser (beam waist) "seen" by the detector is about 10^6 W/cm²-sr.

Solution Using the definition for radiance L_e in terms of laser power Φ_e, source or beam-waist area A, and solid angle $\Delta\Omega$,

$$L_e = \frac{\Phi_e}{\Delta A \,\Delta\Omega} \tag{21-12}$$

the radiance of the laser is found to be 10^6 W/cm²-sr, as follows: Referring to Figure 21-17, the solid angle $\Delta\Omega = $ area B/R^2, where

$$\text{area } B = \pi\left[R \tan\left(\frac{\phi}{2}\right)\right]^2 \cong \frac{\pi R^2 \phi^2}{4}$$

Then substituting into Eq. (21-12), the radiance is

$$L_e = \frac{4\Phi_e}{\pi\phi^2\,\Delta A} = \frac{4(4 \times 10^{-3}W)}{\pi(1.6 \times 10^{-3})^2(2 \times 10^{-3}\text{ cm}^2)} = 1 \times 10^6 \text{ W/cm}^2\text{-sr}$$

Again, the comparison between laser source and high-intensity mercury lamp is dramatic. We conclude that where brightness and radiance are important in the selection of light sources, the laser stands alone.

Focusability. Focusing light to a tiny, diffraction-limited spot is a challenge. It is customary in geometrical optics to show a positive lens focusing a beam of perfectly collimated light to a "point image" (see Figure 21-18a). We know that a point image is an idealization, unattainable even in the limit of geometrical optics ($\lambda \to 0$), because perfect, aberration-free lenses do not exist. Nevertheless, the ideal of focusing light to a diffraction-limited point image has long been a goal. Now the laser, with its coherent, nearly collimated beam, has made that ideal attainable. Figure 21-18b shows the difficulty involved in focusing ordinary light to a tiny spot. First, the light emitted is incoherent. Second, the physical source of light cannot be too small, approaching a point source, for then either the light generated would be

(a) Ideal source

(b) Ordinary source

(c) Laser source

Figure 21-18 Focused beams from various sources. (a) Ideal, collimated beam is focused to a fictitious "point" in accordance with geometrical optics. (b) Incoherent radiation from a thermal source is focused to a demagnified image of size $h_i \gg \lambda$. (c) Coherent laser beam is focused to a diffraction-limited spot of diameter $d \cong \lambda$.

insufficient or the source would melt. The combination of a nonpoint source and incoherent light leads to fairly large image sizes, fixed more or less by the laws of magnification in geometrical optics.

The laser, on the other hand, radiates intense, coherent light that appears to come from a distant "point source." By its unique properties then, it overcomes the precise limitations that frustrate one's attempts to focus thermal radiation to a tiny spot. Figure 21-18c shows a laser beam focused by a positive lens to a diffraction-limited spot of incredibly small diameter, approximately equal to the wavelength of the focused light. It can be shown that a laser beam, with beam divergence ϕ, incident on a lens of focal length f, whose diameter is several times larger than the width of the incident beam, is focused to a diffraction-limited spot of diameter approximately equal to

$$d \cong f\phi \tag{21-13}$$

as shown in Figure 21-19. The beam divergence angle ϕ is equal to $1.27\lambda/D$, as given previously in Eq. (21-9). Note carefully that D in Eq. (21-9) refers to the diameter of the beam waist in the laser that "determined" the beam divergence, and d

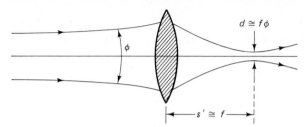

Figure 21-19 TEM$_{00}$ laser beam of beam-spread angle ϕ.

in Eq. (21-13) refers to the diameter of the laser beam focused by the positive lens. The location of the focused laser spot is essentially at the focal plane of the lens ($s' = f$), although a careful analysis shows that this is, in fact, an approximation, even if a good one.

With the help of Eq. (21-13), we can make several rough calculations to predict the spot size of focused laser beams. With a lens of focal length $f = 0.1$ m and incident laser light of beam divergence $\phi = 10^{-3}$ to 10^{-4} rad, spot sizes of the order of 10^{-4} m to 10^{-5} m (or 100 μm to 10 μm) in diameter can be obtained easily. If we compare these diameters with the wavelength of the carbon dioxide laser ($\lambda = 10.6$ μm), we see at once that CO_2 laser light—indeed all laser light—can be focused to spot sizes of the order of a wavelength.

Equation (21-13) indicates that focusing laser light down to small spots can be achieved by lenses with short focal lengths and laser beams with small beam divergences. As long as aberration-free lenses of high quality are available, the focal length can be chosen as short as is practical. The beam divergence of a laser, usually determined at the time the laser is designed, can still be reduced with the additional optics found in *beam expanders*. In Figure 21-20, a collimated laser beam of width W_i and beam divergence ϕ_i is focused by the first lens of the beam expander (focal length f_1) to a spot of diameter $d = f_1\phi_i$, in accordance with Eq. (21-13). The second lens, a distance f_2 from the focused spot, with $f_2 > f_1$, collects the light expanding from the focused spot and essentially recollimates it. The beam divergence of the expanded, recollimated beam is equal to

$$\phi_f = \left(\frac{W_i}{W_f}\right)\phi_i = \left(\frac{f_1}{f_2}\right)\phi_i \qquad (21\text{-}14)$$

where $f_2/f_1 = W_f/W_i$ is the *beam expansion ratio*. The validity of Eq. (21-14) is not difficult to show. The incident beam, focused by the first lens, has a spot diameter $d_1 = f_1\phi_i$. By the principle of reversibility of light, if the expanded beam were to be

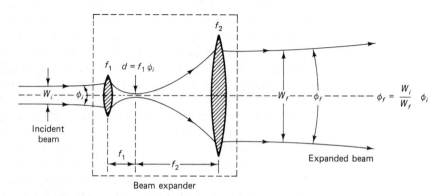

Figure 21-20 Beam expansion as a method of reducing beam divergence of a laser beam.

redirected to the left and focused by the second lens, it would form the identical spot at the same location, so that $d_2 = f_2\phi_f$. Since $d_1 = d_2$ necessarily,

$$f_1\phi_i = f_2\phi_f$$

$$\phi_f = \frac{f_1}{f_2}\phi_i$$

If the beam expansion ratio is $f_2/f_1 = 5$, the beam divergence of the expanded laser beam is $\frac{1}{5}$ that of the incident beam. Expanding the beam width by a factor of 5 achieves a reduction in beam divergence by the same factor.

What has been gained? Suppose a laser beam is expanded 10-fold by an appropriate beam expander. The outgoing beam then has had its beam divergence decreased by a factor of 10. If the expanded beam is then focused to a tiny spot with a lens of arbitrary focal length f, the diameter of the spot will be $\frac{1}{10}$ that achievable with the unexpanded beam and the same lens. This 10-fold reduction in diameter leads to a 100-fold reduction in spot size area and thus, for any given laser beam power, to a 100-fold increase in focused spot irradiance.

Laser energy focused onto small target areas makes it possible to drill tiny holes in hard, dense material, make tiny cuts or welds, make high-density recordings, and generally carry out industrial or medical procedures in target areas only a wavelength or two in size. In ophthalmology, for example, where Nd:YAG lasers are used in ocular surgery, target irradiances of 10^9 to 10^{12} W/cm^2 are required. Such irradiance levels are readily developed with the help of beam expanders and suitable focusing optics, as was discussed in Section 7-5 (see problem 7-10).

21-5 LASER TYPES AND PARAMETERS

To this point we have examined the basic assumptions that led Einstein to predict the existence of stimulated emission, identified the essential parts that make up a laser, described in a general way how a laser operates, and studied the characteristics that make lasers such a unique source of light. Now, by way of summary, we turn our attention to the identification of some of the common lasers in existence today and to parameters that distinguish them from one another.

Lasers are classified in many ways. Sometimes they are grouped according to the state of matter represented by the laser medium: gas, liquid, or solid. Sometimes they are classified according to how they are pumped: flashlamp, electrical discharge, chemical actions, and so on. Still other classifications divide them according to the nature of their output [pulsed or continuous wave (cw)] and according to their spectral region of emission (infrared, visible, or ultraviolet).

No particular classification scheme has been chosen for the lasers listed in Table 21-2. Those identified are, in a way, a cross section of the 30 or 40 common lasers on the market today. A careful examination of Table 21-2 serves as an introduction to the state of laser technology. For each laser listed, the entries include data on emission wavelength, output power (or in some cases, energy per pulse), nature of output, beam diameter, beam divergence, and operating efficiency. Table 21-2 includes examples of gas lasers (He-Ne, CO_2, nitrogen); solid state lasers (ruby, Nd:YAG, Nd-glass); liquid or dye lasers; semiconductor lasers (gallium arsenide); the excimer gas lasers (argon fluoride); chemical lasers (hydrogen fluoride); and ion lasers (argon ion). Both pulsed and continuously operating (cw) lasers are represented. Taken as a whole, Table 21-2 includes lasers whose wavelengths vary from 193 nm (deep ultraviolet) to 10.6 μm (far infrared); whose cw power outputs vary

TABLE 21-2 LASER PARAMETERS FOR SEVERAL COMMON LASERS

Type	Wavelength	Power/energy	Type of output	Beam diameter	Beam divergence	Efficiency[1]
Helium–neon (gas)	632.8 nm	0.1–50 mW	cw	0.5–2 mm	0.5–1.7 mrad	<0.1%
Ruby (solid)	694.3 nm	0.03–100 J (per pulse)	pulsed	1.5 mm–2.5 cm	0.2–10 mrad	<0.5%
Carbon dioxide (gas)	10.6 μm	3–100 W	cw	3–4 mm	1–2 mrad	5–15%
Nitrogen (gas)	337 nm	1–300 mW (average per pulse)	pulsed	2 × 3–6 × 30 mm (rectangular)	1–3 × 7 mrad	<0.1%
Nd:YAG (solid)	1.064 μm	0.04–600 W	cw	0.75–6 mm	2–18 mrad	0.1–2%
Nd–glass (solid)	1.06 μm	0.15–100 J (per pulse)	pulsed	3 mm–2.5 cm	3–10 mrad	1–5%
Argon ion (gas)	488 nm or 514.5 nm	5 mW–20 W	cw	0.7–2 mm	0.4–1.5 mrad	<0.1%
Dye (liquid)	400–900 nm (tunable)	20–800 mW	cw/pulsed (pumped by argon ion laser)	0.4–0.6 mm	1–2 mrad	10–20%
Argon fluoride (excimer)	193 nm	Up to 10 W (average per pulse)	pulsed	6 × 23–20 × 32 mm (rectangular)	2–6 mrad	<0.5%
Hydrogen fluoride (chemical)	2.6–3 μm	0.01–150 W cw or 2–600 mJ per pulse	cw/pulsed	2 mm–4 cm	1–15 mrad	0.1–1%
Gallium arsenide (semiconductor diode)	780–900 nm	1–40 mW cw or average per pulse	cw/pulsed	N/A (diverges too rapidly)	200 × 600 mrad (oval in shape)	1–20%

[1]The efficiency given here is the overall operational efficiency, often referred to as a "wall-plug" efficiency, to denote the ratio of laser-beam power out over wall-plug electrical power in. The quantum efficiency, always higher than the wall-plug efficiency, is defined as the ratio of the laser's transition energy ($E_{upper} - E_{lower}$) to the pump energy ($E_{upper} - E_{ground}$).

from 0.1 mW to 600 W; whose beam divergences vary from 0.2 mrad (circular cross section) to 200×600 mrad (oval cross section); and whose overall efficiencies (laser energy out divided by pump energy in) vary from less than 0.1% to 20%.

PROBLEMS

21-1. Beginning with the expression for rate of spontaneous decay of atom density N_2 at excited energy level E_2,

$$\left(\frac{dN_2}{dt}\right)_{spont} = -A_{21} N_2$$

show that an initial population density N_{20} decreases to a value N_{20}/e in a time τ equal to $1/A_{21}$.

21-2. Assume that an atom has two energy levels separated by an energy corresponding to a wavelength of 632.8 nm, as in the He-Ne laser. If we suppose that all the atoms are located in one or the other of these two states, what fraction of atoms is in the upper state at room temperature, $T = 300$ K, according to the Boltzmann distribution?

21-3. Show that Eq. (21-2) for the blackbody spectral energy density [energy/(vol-$\Delta\nu$)],

$$\rho(\nu) = \frac{8\pi h\nu^3}{c^3}\left(\frac{1}{e^{h\nu/kT} - 1}\right)$$

follows from the form given in Eq. (2-12) for the blackbody spectral radiant exitance [power/(area-$\Delta\lambda$)],

$$M_\lambda = \frac{2\pi hc^2}{\lambda^5}\left(\frac{1}{e^{hc/\lambda kT} - 1}\right)$$

by changing variables from λ, $d\lambda$ to ν, $d\nu$ and then multiplying the result by $4/c$. (See footnote 1.)

21-4. Treating the sun as a blackbody radiator, determine its spectral energy density $\rho(\nu)$ in the visible region near $\lambda = 550$ nm. Assume the surface temperature of the sun to be about 6000 K.

21-5. Why should one expect lasing at ultraviolet wavelengths to be more difficult to attain than lasing at infrared wavelengths? Develop your answer based on the ratio A_{21}/B_{21} and the meaning of the A_{21}, B_{21} coefficients.

21-6. Calculate the ratio of stimulated to spontaneous transitions for green light at 0.5 μm within a dense blackbody plasma at a temperature of 5000 K. That is, determine the value of the ratio

$$\frac{(dN/dt)_{se}}{(dN/dt)_{sp}} = \frac{-B_{21} N_2 \rho(\nu)}{-A_{21} N_2}, \quad \text{where} \quad \frac{B_{21}}{A_{21}} = \frac{\lambda^3}{8\pi h}$$

What does the numerical value of this ratio imply?

21-7. (a) Given the center wavelength λ_0 and linewidth $\Delta\lambda$ (FWHM) for the three entries in Table 21-1 (ordinary discharge lamp, Cd low-pressure lamp, and He-Ne laser), verify that the FWHM frequency linewidth $\Delta\nu$ (Hz) is as given in the table.

(b) How much "sharper" is a He-Ne laser line emission, for example, than a sodium line from a sodium discharge lamp?

21-8. Suppose that the coherence time of a light beam is roughly equal to the reciprocal of its frequency linewidth (FWHM). What then is the coherence time and coherence length of the He-Ne laser in Table 21-1?

21-9. A He-Ne laser has a beam waist (diameter) equal to about 1 mm. What is its beam-spread angle in the far field?

21-10. Consider a broadband thermal source with a circular radiating surface of diameter 0.5 mm (roughly the size of beam waists in He-Ne lasers). Let the surface at a temperature of 1000 K emit light at 633 nm with a linewidth (FWHM) of 100 nm. Use Eq. (21-11) to show that the thermal photon output rate is about 5×10^9 photons/s.

21-11. Consider a 1-mW He-Ne laser emitting at 632.8 nm with a FWHM frequency linewidth of $\Delta\nu = 10^4$ Hz. Assume that the full-angle beam divergence ϕ in the far field is $\phi = 1.27\lambda/D$, where D is the diameter of the diffracting aperture (beam waist). See Figure 21-17.

    (a) Show that the spectral radiance $\Delta L_e/\Delta\nu = \phi_e/(\Delta A \, \Delta\Omega \, \Delta\nu)$—where ϕ_e is the laser power, ΔA is the area of a radiating surface, $\Delta\Omega$ is the solid angle the laser radiates into, and $\Delta\nu$ is the spectral bandwidth (FWHM) of the laser emission—is independent of the value of the diameter D of the diffracting aperture.

    (b) Obtain a numerical value for the spectral radiance (in W/m²-sr-Hz) for the laser described above.

21-12. With reference to the beam expander shown in Figure 21-20, suppose $f_2/f_1 = 10$ and the beam divergence of the incident beam is 1 mrad.

    (a) What is the beam divergence of the expanded beam?

    (b) If the expanded beam is focused by a subsequent lens of power 10 diopters, what is the beam waist at the focal spot?

    (c) If the power in the incident beam is 1 mW, what is the irradiance (W/m²) at the focal spot?

21-13. Refer to Figure 21-8a. Assume that 1 W of incident laser power of resonant wavelength λ_0 is incident on the Fabry-Perot cavity.

    (a) What is the output power?

    (b) How much power is circulating within the cavity?

    (c) Describe what must be happening at mirror M_1 to make your answers to (a) and (b) valid.

21-14. For a Nd:YAG laser, there are four pump levels located at 1.53 eV, 1.653 eV, 2.119 eV, and 2.361 eV above the ground state energy level.

    (a) What is the wavelength associated with the photon energy required to populate each of the pump levels?

    (b) Knowing that a Nd:YAG laser emits photons of wavelength 1.064 µm, determine the *quantum efficiency* associated with each of the four pump levels.

21-15. To operate a Nd:YAG laser, 2500 W of "wall-plug" power are required for a power supply that drives the arc lamps. The arc lamps provide pump energy to create the population inversion. The overall laser system, from power in (to the power supply) to power out (laser output beam), is characterized by the following component efficiencies:

    80%—power supply operation
    30%—arc lamps for pump light energy
    70%—optical reflectors for concentrating pump light on laser rod
    15%—for spectral match of pump light to Nd:YAG pump levels
    50%—due to internal cavity/rod losses

    (a) Taking the efficiencies into account sequentially as they "occur," how much of the initial 2500 W is available for power in the output beam?

    (b) What is the overall operational efficiency (wall-plug efficiency) for this laser?

REFERENCES

[1] Lengyel, B. A. *Lasers*, 2d ed. New York: John Wiley and Sons, 1971.

[2] Siegman, A. E. *An Introduction to Lasers and Masers*. New York: McGraw-Hill Book Company, 1971.

[3] Siegman, A. E. *Lasers*. Mill Valley, Calif.: University Science Books, 1986.

[4] Thompson, G. H. B. *Physics of Semiconductor Laser Devices*. New York: Wiley-Interscience, 1980.

[5] Verdeyen, J. T. *Laser Electronics*. Englewood Cliffs, N.J.: Prentice-Hall, 1981.

[6] Unger, H. G. *Introduction to Quantum Electronics*. New York: Pergamon Press, 1970.

[7] O'Shea, D. C., W. R. Callen, and W. T. Rhodes. *Introduction to Lasers and Their Applications*. Reading Mass.: Addison-Wesley Publishing Co., 1978.

[8] Klein, M. V. *Optics*. New York: John Wiley and Sons, 1970.

[9] Saleh, B. E. A., and M. C. Teich. *Fundamentals of Photonics*. New York: John Wiley and Sons, 1991.

[10] Schawlow, Arthur L. "Optical Masers." *Scientific American* (June 1961): 52.

[11] Schawlow, Arthur L. "Advances in Optical Masers." *Scientific American* (July 1963): 34.

[12] Pimentel, George C. "Chemical Lasers." *Scientific American* (Apr. 1966): 32.

[13] Lempicki, Alexander, and Harold Samelson. "Liquid Lasers." *Scientific American* (June 1967): 80.

[14] Patel, C. K. N. "High-Power Carbon Dioxide Lasers." *Scientific American* (Aug. 1968): 22.

[15] Sorokin, Peter. "Organic Lasers." *Scientific American* (Feb. 1969): 30.

[16] Panish, Morton B., and Izuo Hayashi. "A New Class of Diode Lasers." *Scientific American* (July 1971): 32.

[17] Silfvast, William T. "Metal-Vapor Lasers." *Scientific American* (Feb. 1973): 88.

[18] Tsang, W. T. "The C^3 Laser." *Scientific American* (Nov. 1984): 148.

22

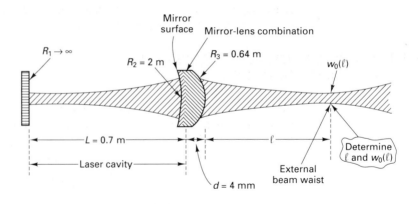

Characteristics
of Laser Beams

INTRODUCTION

We now turn our attention to the optical beam generated by the laser and the propagation characteristics of this beam. We shall see that a laser beam has characteristics of both plane and spherical waves, while retaining an identity all its own. In its simplest mode—the *fundamental mode*—the laser beam takes the form of nearly spherical wavefronts, with the electric field exhibiting a transverse *Gaussian* irradiance[1] distribution localized near the propagation axis. In other, more complex forms, referred to as *higher-order modes* or *Hermite-Gaussians*, the electric field takes on transverse irradiance distributions that depart from the simple Gaussian variation and exhibit an ordered pattern of "hot spots." In many cases, the output laser beam consists of a mixture of modes: the fundamental and several higher-order modes.

In this chapter we examine the nature of the laser beam and describe its general characteristics. We shall study its properties at positions near the laser (*near field*) and far from the laser (*far field*). We shall specify a general method for determining its characteristics—involving wavefront curvatures and transverse irradiance

---

[1] While the words *intensity* and *irradiance* are often used interchangeably in laser literature, we shall use irradiance when we mean to describe power per unit area (e.g., W/m²) at some location on a "target" plane along the beam path (see Table 2-1).

distributions—while it propagates through arbitrary optical systems. After a rather thorough study of the fundamental mode of the laser beam, the so-called TEM$_{00}$ Gaussian beam mode, we shall examine the higher-order transverse modes and their transverse irradiance distributions.

22-1 THREE-DIMENSIONAL WAVE EQUATION AND ELECTROMAGNETIC WAVES

In Chapter 8, Wave Equations, we examined the propagation of transverse waves in a homogeneous medium. We found that the one-dimensional differential wave equation describing such motion is written as

$$\frac{\partial^2 y}{\partial z^2} - \frac{1}{v^2}\frac{\partial^2 y}{\partial t^2} = 0 \tag{22-1}$$

Here y describes the transverse wave displacement, z denotes the coordinate along the wave propagation direction, t stands for time, and v represents the wave speed in the homogeneous propagation medium.

When the one-dimensional wave equation (22-1) is generalized to three dimensions for electric field displacements $E(x, y, z, t)$ in homogeneous media devoid of free charges or currents, the wave equation takes the form

$$\nabla^2 E - \frac{n^2}{c^2}\frac{\partial^2 E}{\partial t^2} = 0 \tag{22-2}$$

Here, ∇^2 is the *Laplacian* operator, n is the refractive index of the propagating medium, and c is the wave speed in vacuum.

Plane Wave Solutions to the Wave Equation.   The so-called plane wave solutions to the three-dimensional wave equation may be expressed in the complex form

$$\tilde{\mathbf{E}}(\mathbf{r}, t) = \mathbf{E}_0 e^{i(xk_x + yk_y + zk_z - \omega t)} = \mathbf{E}_0 e^{i(\mathbf{k}\cdot\mathbf{r} - \omega t)} \tag{22-3}$$

where the propagation and position vectors are given by

$$\mathbf{k} = \hat{\mathbf{i}}k_x + \hat{\mathbf{j}}k_y + \hat{\mathbf{k}}k_z \quad \text{and} \quad \mathbf{r} = \hat{\mathbf{i}}x + \hat{\mathbf{j}}y + \hat{\mathbf{k}}z$$

and $\omega = |\mathbf{k}|c/n$.[2] Plane wave solutions are useful as base functions in the Fourier analysis of arbitrary wave forms and as legitimate descriptions of electromagnetic fields reaching a detector from a distant point source. They are referred to as *plane waves* because the surfaces of constant phase ($\mathbf{k}\cdot\mathbf{r}$) are geometric planes perpendicular to the wave vector $\mathbf{k}$. The essential features of propagating plane waves have been discussed in Chapter 8 and are shown in Figure 22-1.

Spherical Wave Solutions to the Wave Equation.   The spherical solutions that satisfy the wave equation can be written as

$$\tilde{E}(r, \theta, \phi, t) = \frac{E_0(\theta, \phi)}{r} e^{i(kr - \omega t)} \tag{22-4}$$

The appropriate geometry and fundamental parameters are shown in Figure 22-2. The spherical wave solutions are useful in the formulation of diffraction theory ac-

---

[2] One needs to distinguish the unit vector $\hat{\mathbf{i}}$ from the imaginary number $i = \sqrt{-1}$ and the unit vector $\hat{\mathbf{k}}$ from the propagation constant k.

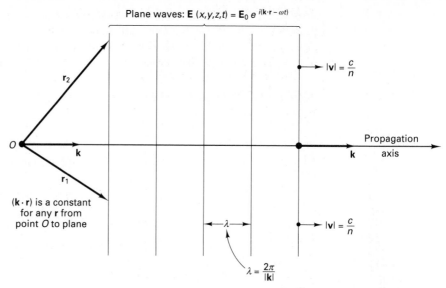

Figure 22-1 The surfaces of constant phase $\mathbf{k} \cdot \mathbf{r}$ for plane waves are geometrical planes oriented in a direction perpendicular to the wave vector $\mathbf{k}$ and moving with a speed $v = c/n$ through the homogeneous medium of refractive index n.

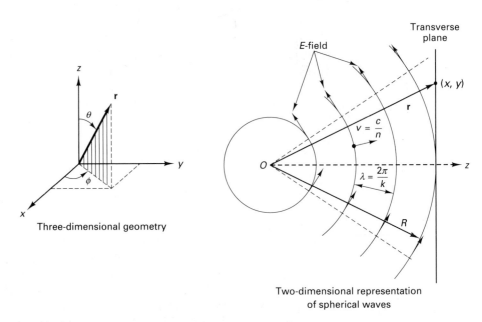

Figure 22-2 The surfaces of constant phase for spherical waves are geometrical spheres radiating outwardly from a point source O.

cording to Huygens, Fresnel, and Kirchhoff. They represent the mathematical forms used for the Huygens' secondary wavelets that radiate out from the diffraction aperture (see Section 3-1). In addition, they represent a valid description of electromagnetic energy propagating away from a point source. When the distance from the point source becomes very large, the spherical waves can be treated as plane waves, with Eq. (22-3) replacing Eq. (22-4) as a mathematical description.

22-2 PHASE VARIATION OF SPHERICAL WAVES ALONG A TRANSVERSE PLANE

To help us understand the nature of the laser beam, it will be useful at this point to determine the phase distribution of the electric field $\mathbf{E}$ associated with spherical waves along a plane perpendicular to a given direction of propagation. In Figure 22-2, this plane is shown as the x-y plane, and the direction of propagation is along the z-direction. At any point (x, y) on the transverse plane $z = R$, the spatial part of the electric field is given by

$$\tilde{E}(x, y)_{z=R} = (\text{constant})e^{ikr} \tag{22-5}$$

where $r = (R^2 + x^2 + y^2)^{1/2}$ and R is the perpendicular distance from the source to the transverse plane. It is clear from Figure 22-2 that R is also the radius of curvature of the spherical wave that is tangent to the plane at $(x, y) = (0, 0)$. Rewriting r as

$$r = R\left(1 + \frac{x^2 + y^2}{R^2}\right)^{1/2} \tag{22-6}$$

and restricting the (x, y) points in the transverse plane to positions near the axis of propagation, such that $(x^2 + y^2) \ll R^2$, we can approximate[3] Eq. (22-6) as

$$r \cong R + \frac{x^2 + y^2}{2R} \tag{22-7}$$

In turn, then, Eq. (22-5) for the spatial part of the electric field $\mathbf{E}$ at point (x, y) on the transverse plane $z = R$ takes on the explicit form

$$\tilde{\mathbf{E}}(x, y)_{z=R} \cong (\text{constant})e^{ikR}e^{i(k/2)[(x^2+y^2)/R]} \tag{22-8}$$

We shall use Eq. (22-8) later as a condition for identifying a *spherical wave* that is crossing a transverse plane, as long as we restrict the examination of the electric field $\mathbf{E}$ to points near the axis of propagation.

22-3 BASIS FOR DEFINING LASER-BEAM MODE STRUCTURES

Before we proceed further with the description of laser beams, it is necessary to say a few words about the variation of the electric field in a plane transverse to the beam, specifically, how this variation is used to characterize the mode structure of the beam. In this regard, it may be helpful to draw an analogy between the displacement of the skin on a musical drum and the value of the electric field across a transverse plane of the laser beam.

For a drum, the fundamental mode occurs when the skin fastened around the rim moves up and down as a whole, with greatest motion occurring at the center, producing a single-humped profile. Analogously, the fundamental mode for a laser beam exists when the electric field has its maximum value at the center of the beam and drops off in accordance with a *Gaussian* profile. This profile is a "hump" that decreases symmetrically in all directions from the center toward the edges of the

---

[3] Defining $(x^2 + y^2)/R^2$ as u^2, and expanding $(1 + u^2)^{1/2}$ in accordance with a binomial series for which $u^2 \ll 1$, Eq. (22-7) follows from Eq. (22-6) when smaller, higher-order terms are neglected.

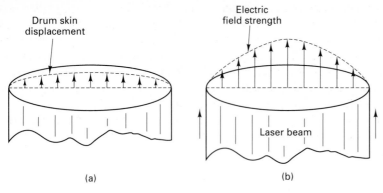

Figure 22-3 Comparison of the mechanical displacement of a vibrating drum skin and the electric field of a laser beam. (a) Instantaneous side view of a drum skin vibrating in its fundamental mode. Vertical displacement shown is the same along any *other* diameter. (b) Instantaneous view of the electric field variation along a diameter in a plane transverse to the beam. For the fundamental mode, the electric field shown is the same along any other diameter.

beam.[4] Figure 22-3 compares the displacement, at an instant, of the skin on the drum and of the electric field vector across the laser beam.

The notation used to describe the fundamental mode of the laser beam is TEM_{00}, a special case of the general notation, TEM_{mn}.[5] Here, TEM refers to "**T**ransverse **E**lectric **M**agnetic." The subscripts m and n denote the number of times the electric field goes to zero, respectively, along x- and y-coordinate axes superimposed on the beam's transverse plane. Thus TEM_{00} denotes a "single-humped profile" with no zeros along either x or y (or along any other transverse beam diameter) except, of course, at the edges where the beam *approaches* zero.

In this spirit, then, electric field variations and corresponding *burn patterns*[6] for TEM_{00}, TEM_{10}, TEM_{01}, and TEM_{11} appear as shown in Figure 22-4. For exam-

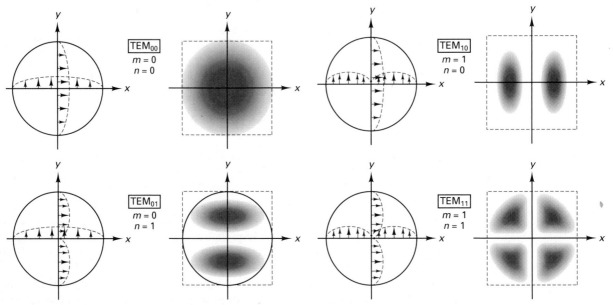

Figure 22-4 Electric-field variations across transverse beam diameters—and the corresponding burn patterns such beams would produce on a target—for different TEM_{mn} modes.

---

[4] The edges of a vibrating drum head are well-defined, while the "edges" of a laser beam are not. Still, the analogy, in principle, is useful.

[5] The *transverse* modes denoted here by the notation TEM_{mn} are generated, of course, in a mirror-bounded, open-sidewall optical resonator, the *laser cavity*.

[6] So named because of the pattern produced by a high-intensity beam when it "burns" into a target.

ple, the TEM$_{01}$ mode shows no **E**-field zeros along the *x*-direction (single-humped profile) and one **E**-field zero along the *y*-direction (double-humped profile with zero at the center point). Such a variation produces a burn pattern (on target) that contains two horizontal, dark areas centered above and below a "no-burn" region. A careful examination of the **E**-field variation and burn patterns for the other cases pictured—and, in fact, for *all* higher-order TEM$_{mn}$ modes—discloses similar correlations between **E**-field variation and "burn" areas.

For the sake of simplicity in this treatment, we shall work primarily with the TEM$_{00}$ beam, whose burn pattern is essentially a dark circle with greatest burn intensity at the center (on axis).

22-4 GAUSSIAN BEAM SOLUTION FOR LASERS

If we were to examine the electromagnetic character of a TEM$_{00}$ laser beam, we would find that its wavefronts are essentially spherical surfaces with long radii of curvature that increase as the beam advances along the propagation axis. The combined wavefront and irradiance variation of a TEM$_{00}$ laser beam passing through a converging lens appear as shown in Figure 22-5.

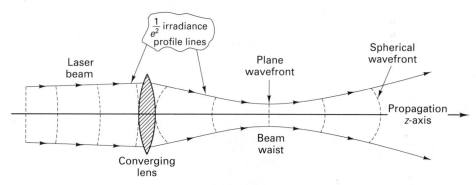

Figure 22-5 An external laser beam, confined essentially to regions within the $1/e^2$-irradiance guidelines, is focused by a converging lens. The incoming beam is highly collimated so that its wavefronts are very nearly planar. The converging lens reshapes the wavefronts and focuses the beam—forming a beam waist—to the right of the lens. The beam diverges strongly as it propagates on past the beam waist.

The solid guidelines, above and below the *z*-axis, represent the locus of points for which the beam's electric field irradiance in a transverse direction is equal to $1/e^2$ of its value on axis. Thus these lines are used to define a continuously changing *beam width*. The dashed arcs transverse to the *z*-axis indicate the wavefronts of the beam. Since the geometrical shape of common phase fronts for a highly collimated laser beam are "more-or-less" planes, we can choose a trial solution for the laser beam's electric field $E(r, t)$ of the form motivated by a one-dimensional wave, that is, a trial solution[7] of the form

$$E(r, t) = U(x, y, z)e^{i(kz-\omega t)} \tag{22-9}$$

The term $U(x, y, z)$, when determined, provides the details that accurately specify the irradiance and phase variations of the wave, primarily in the *x*, *y*-directions. The

---

[7] Equation (22-9) is cast in the spirit of the paraxial approximation. That is, a plane wave is assumed to be propagating essentially along only the *z*-direction, thereby permitting us to express the electric field $E(x, y, z, t)$ in the scalar form of Eq. (22-9).

exponential term, in Eq. (22-9), merely reflects the "more-or-less" plane wave nature of the solution. If we are successful in determining the function $U(x, y, z)$, we would then have a useful solution describing the actual laser beam.

Determination of the $U(x, y, z)$ Function. Since our trial solution in Eq. (22-9) must satisfy the wave equation, we substitute it into Eq. (22-2) and obtain a defining equation for the yet unspecified function $U(x, y, z)$:

$$e^{i(kz-\omega t)}\left[\frac{\partial^2 U}{\partial x^2} + \frac{\partial^2 U}{\partial y^2} + \frac{\partial^2 U}{\partial z^2} + 2ik\frac{\partial U}{\partial z} - \left(k^2 - \frac{\omega^2}{c^2}\right)U\right] = 0 \qquad (22\text{-}10)$$

We ignore the second-order variation of U with z since the monatonic attenuation of $E(x, y, z, t)$ with axial propagation distance is small. However, we retain the first-order variation of U with z, as well as the variation of U with x and y. The last term in Eq. (22-10) vanishes outright, since $\omega \equiv kc$. Thus Eq. (22-10) can be simplified, leaving us with the following equation that $U(x, y, z)$ must obey,

$$\frac{\partial^2 U}{\partial x^2} + \frac{\partial^2 U}{\partial y^2} + 2ik\frac{\partial U}{\partial z} = 0 \qquad (22\text{-}11)$$

Eq. (22-11) is a nontrivial partial differential equation with complex terms. To solve it, we make an "educated" guess[8] at a solution, motivated in part by the cylindrical symmetry which we expect the laser beam's electric field magnitude to have in the transverse direction, and in part by the complex nature which the solution $U(x, y, z)$ must exhibit. Thus we "guess" the form

$$\tilde{U}(x, y, z) = E_0 e^{i\{p(z)+[k(x^2+y^2)/2q(z)]\}} \qquad (22\text{-}12)$$

where $p(z)$ and $q(z)$ are general functions, as yet undetermined and subject to constraints imposed by Eq. (22-11), the defining equation for $\tilde{U}(x, y, z)$. After substituting Eq. (22-12) into Eq. (22-11), we obtain

$$\frac{2ik\tilde{U}}{q} - \frac{k^2}{q^2}(x^2 + y^2)\tilde{U} - 2k\tilde{U}\frac{\partial p}{\partial z} + \frac{k^2}{q^2}(x^2 + y^2)\tilde{U}\frac{\partial q}{\partial z} = 0$$

which, when rearranged in terms of powers of $(x^2 + y^2)$, becomes

$$\left[\left(\frac{2ik}{q} - 2k\frac{\partial p}{\partial z}\right)(x^2 + y^2)^0 + \left(\frac{k^2}{q^2}\frac{\partial q}{\partial z} - \frac{k^2}{q^2}\right)(x^2 + y^2)^1\right]\tilde{U} = 0 \qquad (22\text{-}13)$$

If Eq. (22-12) is to be a solution for all x and y, then each term multiplying a power of $(x^2 + y^2)$ must equal zero separately. Thus the "coefficients" of $(x^2 + y^2)^0$ and $(x^2 + y^2)^1$, when set equal to zero, yield

$$\frac{\partial p}{\partial z} = \frac{i}{q} \qquad (22\text{-}14)$$

and

$$\frac{\partial q}{\partial z} = 1 \qquad (22\text{-}15)$$

Let us now explore the meaning of the restrictions expressed by Eqs. (22-14) and (22-15) on the functions $q(z)$ and $p(z)$.

---

[8] A rigorous derivation of the Gaussian beam equation is given in Reference [1], Chapter 16. Those who wish to become familiar with the details of this derivation are encouraged to examine Siegman's scholarly treatment.

Definition of the Complex Radius of Curvature. We see from Eq. (22-14) that, in general, p and q may be complex. We also note, by comparing the second term in the exponent of Eq. (22-12) with the second term in Eq. (22-8), that $q(z)$ ought to resemble, in part, the radius of curvature of the wavefront, if indeed the wavefronts for the laser beam are to be spherical in nature. Having little to lose, then, we make an extremely "shrewd guess" and write q as a complex term $\tilde{q}$, calling it the *complex radius of curvature* and writing it in the form[9]

$$\frac{1}{\tilde{q}} = \frac{1}{R} + i\frac{\lambda}{\pi w^2} \qquad (22\text{-}16)$$

where R is the radius of curvature of the wavefront and w is the transverse dimension of the beam.

Quite candidly, Eq. (22-16) is an educated guess, dictated more by hindsight than foresight. Nevertheless, it bears great dividends. We see that $1/\tilde{q}$, as expressed in Eq. (22-16), is indeed complex, that its real part $1/R$ is related to the real curvature R of the wavefront, and that its imaginary part $(\lambda/\pi w^2)$ is related to the transverse dimension w of the beam. As a matter of fact, we shall see that w—called (unfortunately) the "spot size" and so confused with a diameter—is a measure of the beam's half-width (radius). The parameter w is measured from the on-axis position in a transverse plane to the off-axis position where the beam irradiance has decreased to $1/e^2$ of its on-axis value. Thus, if we introduce Eq. (22-16) into Eq. (22-12), we get

$$\tilde{U}(x, y, z) = E_0 e^{ip(z)} e^{ik(x^2+y^2)/2R(z)} e^{-(x^2+y^2)/w^2} \qquad (22\text{-}17)$$

Except for $p(z)$, whose identification we defer until later in this chapter, we now have an explicit expression for $\tilde{U}(x, y, z)$, and thus, together with Eq. (22-9), a workable form for the space and time variation of the electric field in a laser beam.[10]

Spherical Gaussian Laser Beam. At this point, we have "solved" the problem for the mathematical representation of a laser beam. Inserting Eq. (22-17) back into our original guess,

$$E(x, y, z, t) = U(x, y, z, t)e^{i(kz-\omega t)}$$

we obtain the Gaussian beam representation for the **E**-field,

$$\tilde{E}(x, y, z, t) = E_0 e^{ik(x^2+y^2)/2R(z)} e^{-(x^2+y^2)/w^2(z)} e^{i[kz+p(z)-\omega t]} \qquad (22\text{-}18)$$

The first exponential term in Eq. (22-18), identical with the second exponential term of Eq. (22-8)—developed earlier for use at this time—establishes the geometry of the wavefront for the laser beam as a *spherical surface*. The second exponential term indicates that the electric field in any transverse plane (at a fixed z) falls off as a Gaussian function,[11] with peak value on axis. The last exponential term contains phase information dependent on z. At this point, as far as $p(z)$ is concerned, we note only that it must satisfy the condition $\partial p/\partial z = i/q$.

---

[9] Some texts give Eq. (22-16) with a *negative sign* between the two terms on the right (for example, see [1], [3], and [6]). If one writes Eq. (22-9) with the term e^{ikz} as we have, Eq. (22-16) follows. If one writes Eq. (22-9) with the term e^{-ikz}, Eq. (22-16), with a negative sign, follows. In either case, in the context of a consistent set of equations, the characteristics of a Gaussian laser beam are found to be the same.

[10] We note that this is but one very useful solution to Eq. (22-11). Discussion of other solutions is deferred to Section 22-7.

[11] It is assumed that the reader is familiar with the meaning of a "Gaussian" variation of the form $\exp(-r^2/a^2)$ for a cylindrically symmetric beam where $r^2 = x^2 + y^2$.

Thus we have succeeded in finding a *beamlike solution* to the wave equation. It has explicitly the form given in Eq. (22-18) and is referred to as a *spherical Gaussian beam*. Along the beam propagation axis (*z*-direction), the electric field is harmonic in nature, but with a wavelength given approximately by $\lambda = 2\pi/k$, due to the presence of the $p(z)$ term in the last exponent. In any transverse plane (fixed *z*), the electric field drops off in a Gaussian manner with distance from the axis. These conclusions are illustrated in Figure 22-6.

(a) Axial variation

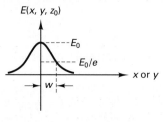

(b) Transverse variation

Figure 22-6 Approximate axial variation and transverse variation of the electric field *amplitude* $E(x, y, z)$ in the laser beam. Here *w* is the beam width at E_0/e.

22-5 SPOT SIZE AND RADIUS OF CURVATURE OF A GAUSSIAN BEAM

Suppose that at some transverse plane, $z = 0$, the Gaussian beam has a planar wavefront, that is, $R(z = 0) \to \infty$. Then, from Eq. (22-16), we have

$$\frac{1}{\tilde{q}_0} = \frac{i\lambda}{\pi w_0^2} \quad \text{or} \quad \tilde{q}_0 = -i\frac{\pi w_0^2}{\lambda} \qquad (22\text{-}19)$$

where $\tilde{q}_0$ and w_0 are defined as $\tilde{q}(z = 0)$ and $w(z = 0)$, respectively. The plane $z = 0$ locates the beam waist. The parameter w_0 is, therefore, the spot size at the beam waist, as illustrated in Figure 22-7.

For any other transverse plane $z \neq 0$, it follows from Eq. (22-15) that

$$\tilde{q}(z) = \tilde{q}(0) + z \qquad (22\text{-}20)$$

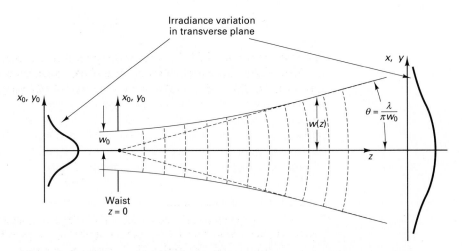

Figure 22-7 Gaussian spherical beam propagating in the *z*-direction. The spot size $w(z)$ at the beam waist (planar wavefront) is defined as w_0. The half-angle beam divergence $\theta = \lambda/(\pi w_0)$ is valid only in the far field. Note the change in transverse irradiance as the beam propagates to the right.

where z is the axial distance measured from the beam waist. This is the basic law of propagation for the laser beam. If we combine Eq. (22-20) with Eq. (22-19), we obtain

$$\tilde{q}(z) = z - i\frac{\pi w_0^2}{\lambda} \tag{22-21}$$

Writing Eq. (22-21) in reciprocal form, we obtain, after rationalizing the denominator,

$$\frac{1}{\tilde{q}(z)} = \frac{z}{\left(z^2 + \dfrac{\pi^2 w_0^4}{\lambda^2}\right)} + i\frac{\pi w_0^2/\lambda}{\left(z^2 + \dfrac{\pi^2 w_0^4}{\lambda^2}\right)} \tag{22-22}$$

Equating the real and imaginary parts in Eq. (22-22) with their counterparts in Eq. (22-16), the defining equation for $1/\tilde{q}(z)$, we obtain useful expressions for $R(z)$[12] and $w(z)$, in terms of the spot size w_0 at the beam waist and the distance z from the beam waist to the transverse plane in question. That is,

$$R(z) = z\left[1 + \left(\frac{\pi w_0^2}{\lambda z}\right)^2\right] \tag{22-23}$$

and

$$w(z)^2 = w_0^2\left[1 + \left(\frac{\lambda z}{\pi w_0^2}\right)^2\right] \tag{22-24}$$

Equation (22-23) specifies the radius of curvature of the laser-beam wavefront as a function of distance z from the reference plane, that is, from the beam waist at $z = 0$. Note that, if the spherical wavefronts were concentric about $z = 0$, then the equality $R(z) = z$ would be true. Eq. (22-23) indicates that this is not the case; the center for the spherical wavefronts is not fixed at $z = 0$ for all wavefronts. However, when the distance $z \gg \pi w_0^2/\lambda$, then $R(z) \cong z$, and we do have nearly spherical wavefronts concentric about $z = 0$.

The region, specified by

$$z \gg \frac{\pi w_0^2}{\lambda} \tag{22-25}$$

is known as the *far field*.[13] It is in this region that one can treat the diffraction from the beam waist (aperture) as purely Fraunhofer diffraction. In the limit of Eq. (22-25), Eqs. (22-23) and (22-24) take on the far-field forms given by

$$R(z) = z \tag{22-26}$$

and

$$w(z) = \frac{\lambda}{\pi w_0}z \tag{22-27}$$

---

[12] For our sign convention, a diverging or expanding wave is signified by a positive R, whereas a converging wave, heading toward some point, is signified by a negative R.

[13] In practice, the "strength" of the inequality is understood to be 20 to 50 times the value $\pi w_0^2/\lambda$. In reality, there is always a "gray area" between the *near field* (where Fresnel diffraction theory for wavefront calculations is required) and the *far field* (where Fraunhofer diffraction theory is acceptable). When in doubt, one can always perform the accurate Fresnel calculations and compare with the approximate Fraunhofer results.

Thus the target or detector is said to be in the far field relative to the laser, when the distance from beam waist to target is much greater than $\pi w_0^2/\lambda$. In the far field, it is clear from Eq. (22-27) that $w(z)$, the spot size, grows linearly with z. Since the half-angle beam spread θ, specified in Figure 22-7, is small in most applications (a milliradian or less), then $\tan \theta \cong \theta$. Thus the far-field (FF), *half-angle*[14] beam divergence for the laser is defined by

$$\theta_{FF} \cong \frac{\lambda}{\pi w_0} \tag{22-28}$$

In Figure 22-8, the implications of Eqs. (22-26) and (22-27) are evident. At large distances from the waist, the beam wavefront has a spherical shape with the center of curvature at the waist. Also, the smaller the waist w_0, the larger the beam divergence, or spread, for the beam propagating away from the beam waist, precisely as dictated by Eq. (22-28).

Figure 22-8 Two laser beams with a beam waist at $z = 0$. The smaller the beam waist w_0, the larger the beam divergence θ.

With Eqs. (22-23) and (22-24), we have expressions for the radius of curvature, $R(z)$, and the beam spot size, $w(z)$, in terms of w_0, the spot size at the beam waist, and z, the distance from the beam waist to the transverse plane, where $R(z)$ and $w(z)$ are to be found. Conversely, given $w(z)$ and $R(z)$ for a particular transverse plane, we can locate the beam waist and its spot size w_0 as follows. Consider Figure 22-9, where w_1 and R_1 in the transverse plane at z_1 are both known. Express-

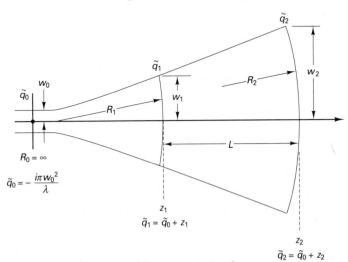

Figure 22-9 The basic law of propagation for the complex radius of curvature $\tilde{q}$. Knowing w_1, R_1 at z_1, one can determine z_1 and w_0. Or, conversely, knowing w_0 and the beam waist location, one can calculate all other $\tilde{q}$.

---

[14] Note carefully that in Chapter 21 we denoted the *full-angle* beam divergence with the letter ϕ. Here we denote the *half-angle* beam divergence with the letter θ.

ing $\bar{q}_1$ in terms of R_1 and w_1, with the aid of Eq. (22-16), and making use of Eq. (22-21), we can write

$$\left(\frac{1}{R_1} + i\frac{\lambda}{\pi w_1^2}\right)^{-1} = z_1 - i\frac{\pi w_0^2}{\lambda} \qquad (22\text{-}29)$$

Then, if we rationalize the left side appropriately and equate real parts and imaginary parts in Eq. (22-29), we obtain

$$z_1 = \frac{R_1}{1 + \left(\dfrac{\lambda R_1}{\pi w_1^2}\right)^2} \qquad (22\text{-}30)$$

and

$$w_0 = \frac{w_1}{\left[1 + \left(\dfrac{\pi w_1^2}{\lambda R_1}\right)^2\right]^{1/2}} \qquad (22\text{-}31)$$

Therefore, with w_1 and R_1 known, the location of the beam waist and its size can be determined from Eqs. (22-30) and (22-31). In addition, with Eqs. (22-23) and (22-24), one can determine $R(z)$ and $w(z)$ for any other transverse plane relative to the beam waist.

Example

Let us illustrate the use of several equations we have derived. Consider a 4-mW, TEM$_{00}$ helium-neon (He-Ne) laser ($\lambda = 632.8$ nm) with the cavity dimensions given below. The left mirror ($R_1 = 2$ m) is 100% reflecting. The right mirror ($R_2 \to \infty$) is a partially reflecting plane mirror through which the 4-mW output beam passes. The beam waist in the laser cavity ($L = 1$ m) occurs at the plane mirror, where we have chosen the reference plane $z = 0$. The dashed profiles represent the wavefronts in the cavity. For stable laser oscillation, the curvatures of the wavefronts match the curvatures of the mirrors at the mirror surfaces.

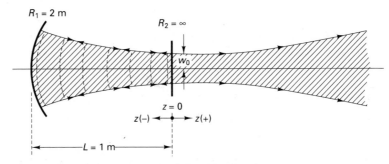

(a) Determine the spot size w_0 at the beam waist.
(b) Determine the laser-beam spot size w on the rear laser mirror.
(c) Determine the complex radius of curvature $\bar{q}(z)$ at $z = -1$ m and $z = 0$.
(d) Determine the location z_{FF} of the far field from the beam waist ($z = 0$).
(e) What is the half-angle beam divergence θ_{FF} for this laser in the far field?

Solution (a) Use Eq. (22-23), with $R(z) = -2$ m, $z = -1$ m, and $\lambda = 632.8$ nm, and solve for w_0. Substituting directly,

$$-2 = -1\left[1 + \left(\frac{3.14w_0^2}{632.8 \times 10^{-9} \times (-1)}\right)^2\right]$$

Solving for w_0 gives 4.49×10^{-4} m $\cong 0.45$ mm.

(b) Use Eq. (22-24) with $w_0 = 4.49 \times 10^{-4}$ m, $z = -1$ m, and $\lambda = 632.8$ nm:

$$w(z)^2 = (4.49 \times 10^{-4})^2\left[1 + \left(\frac{632.8 \times 10^{-9} \times (-1)}{3.14 \times (4.49 \times 10^{-4})^2}\right)^2\right]$$

giving $w(z) = 6.35 \times 10^{-4}$ m $\cong 0.64$ mm. Thus the spot size increases from a radius of $w = 0.45$ mm at the waist to $w = 0.64$ mm at the rear mirror.

(c) Using Eq. (22-21),

At $z = -1$ m: $\tilde{q}(z) = z - \frac{i\pi w_0^2}{\lambda} = -1 - \frac{i(3.14)(4.5 \times 10^{-4})^2}{632.8 \times 10^{-9}} = -1 - i$

At $z = 0$: $\tilde{q}(z) = 0 - \frac{i(3.14)(4.5 \times 10^{-4})^2}{632.8 \times 10^{-9}} = -i$

(d) Use Eq. (22-25) with w_0 and λ known: $z_{FF} \gg \pi w_0^2/\lambda$, where $w_0 = 4.49 \times 10^{-4}$ m and $\lambda = 632.8 \times 10^{-9}$ m. Then

$$z_{FF} \gg \frac{3.14 \times (4.49 \times 10^{-4})^2}{632.8 \times 10^{-9}} \quad \text{or} \quad z_{FF} \gg 1.0 \text{ m}$$

Applying the criterion that $\gg$ means 20 to 50 times larger, we conclude that the far field for this laser is 20 to 50 m from the laser. At $z = 50$ m, say, Fraunhofer diffraction calculations are valid.

(e) Use Eq. (22-28):

$$\theta_{FF} = \frac{\lambda}{\pi w_0} = \frac{632.8 \times 10^{-9}}{(3.14)(4.49 \times 10^{-4})} = 4.49 \times 10^{-4} \text{ rad} \cong 0.45 \text{ mrad}$$

22-6 LASER PROPAGATION THROUGH ARBITRARY OPTICAL SYSTEMS

With the basic law of propagation for the complex radius of curvature $\tilde{q}$ in Eq. (22-20) and the defining equations for the real radius of curvature $R(z)$ and beam width $w(z)$ in Eqs. (22-23) and (22-24), we are able to characterize the beam parameters for laser propagation in any homogeneous medium of refractive index n. We now wish to address the question of how the beam changes when it is modified by an arbitrary optical system, one that contains lenses, mirrors, prisms, and so forth.

It is helpful to note the similarity between the behavior of ordinary spherical waves encountered in geometrical optics and Gaussian spherical waves encountered here. In Figure 22-10, the basic law of propagation for each type of wave and the effect of a lens on reshaping the propagating wavefront is illustrated.

One is struck immediately with the correspondence between $R(z)$ for ordinary spherical waves and $\tilde{q}(z)$ for Gaussian spherical waves in the defining equations. For example, the law of propagation of ordinary spherical waves is given by

$$R_2 = R_1 + (z_2 - z_1) \tag{22-32}$$

Ordinary spherical waves

Gaussian spherical waves

$$R_2 = R_1 + (z_2 - z_1) \longrightarrow \tilde{q}_2 = \tilde{q}_1 + (z_2 - z_1)$$

Simple propagation

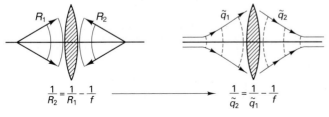

$$\frac{1}{R_2} = \frac{1}{R_1} - \frac{1}{f} \longrightarrow \frac{1}{\tilde{q}_2} = \frac{1}{\tilde{q}_1} - \frac{1}{f}$$

Simple lens effect

Figure 22-10 Correspondence between ordinary spherical waves and Gaussian spherical waves. If one knows the law for ordinary spherical waves, one can infer a similar law for Gaussian spherical waves by simply replacing $R(z)$ with $\tilde{q}(z)$ as shown above.

as we all know. Then, in accordance with a result already derived in Eq. (22-20), the basic law of propagation for the laser beam is given by

$$\tilde{q}_2 = \tilde{q}_1 + (z_2 - z_1) \tag{22-33}$$

an equation identical to Eq. (22-32) except that R has been replaced by $\tilde{q}$. Seizing the initiative, we extend this correspondence[15] to other basic laws, for example, the simple lens law. For ordinary spherical waves, a common form of the law is

$$\frac{1}{R_2} = \frac{1}{R_1} - \frac{1}{f} \tag{22-34}$$

where R_1 and R_2 are the radii of curvature at incidence and refraction, respectively, and f is the focal length (see Figure 22-10). If $\tilde{q}$ is to replace R for the laser beam, we obtain at once

$$\frac{1}{\tilde{q}_2} = \frac{1}{\tilde{q}_1} - \frac{1}{f} \tag{22-35}$$

This relationship predicts accurately the reshaping of the incident laser beam after refraction by a simple thin lens.

General Law of Laser Propagation. With the correspondence between $R(z)$ and $\tilde{q}(z)$ established, we can, with the help of the matrix methods outlined in Chapter 4, develop a simple, yet powerful, recipe for laser-beam propagation through an arbitrary optical system. Consider Figure 22-11.

An incident ray with parameters (y_1, α_1) is incident on the entrance plane of an arbitrary optical system described by the overall system matrix,

$$\begin{bmatrix} A & B \\ C & D \end{bmatrix}$$

Upon emerging from the system, the same ray has parameters (y_2, α_2). The radius of curvature R of an ordinary spherical wave can be related to its appropriate parax-

---

[15] The correspondence we refer to here can be shown to be rigorous, not just analogous.

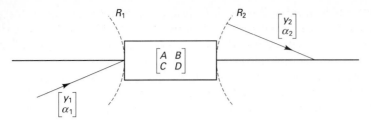

Figure 22-11 Propagation of ordinary spherical waves through an arbitrary optical system via matrix formulation.

ial ray parameters y and α simply by

$$R = \frac{y}{\alpha} \tag{22-36}$$

We know from matrix optics that when ray 1 is changed into ray 2 by an optical system, the change can be represented by the *ABCD* system matrix as follows:

$$\begin{bmatrix} y_2 \\ \alpha_2 \end{bmatrix} = \begin{bmatrix} A & B \\ C & D \end{bmatrix} \begin{bmatrix} y_1 \\ \alpha_1 \end{bmatrix} \tag{22-37}$$

Then it must be true that

$$y_2 = Ay_1 + B\alpha_1 \quad \text{and} \quad \alpha_2 = Cy_1 + D\alpha_1 \tag{22-38}$$

Dividing the first equation by the second, and using Eq. (22-36), we obtain

$$R_2 = \frac{AR_1 + B}{CR_1 + D} \tag{22-39}$$

Generalizing the basic result in Eq. (22-39) to a Gaussian spherical wave, by simply replacing $R(z)$ with $\tilde{q}(z)$, we obtain at once

$$\tilde{q}_2 = \frac{A\tilde{q}_1 + B}{C\tilde{q}_1 + D} \qquad ABCD \text{ propagation law} \tag{22-40}$$

The relationship between $\tilde{q}_2$ and $\tilde{q}_1$ in Eq. (22-40) enables one to determine the reshaping of the laser beam after it passes through an arbitrary optical system. One need only know the incident beam parameter $(\tilde{q}_1)$ as it enters the system and the overall 2×2 matrix that characterizes the optical system. Eq. (22-40) is indeed a useful and powerful result. It is often referred to as the *ABCD propagation law* for spherical Gaussian laser beams. For example, Figure 22-12 illustrates a typical laser system to which the *ABCD* propagation law can be applied.

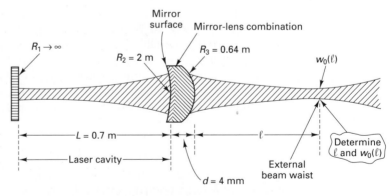

Figure 22-12 Geometry for a He-Ne laser system. With the parameters for the laser system specified, one can use the *ABCD* propagation law to determine the location ℓ and size of the beam waist $w_0(\ell)$ outside of the laser.

The problem depicted in Figure 22-12 is to determine the location ℓ of the external beam waist and its spot size w_0. Clearly, Eq. (22-40) can be used if one knows a value for $\tilde{q}_1$ somewhere in the cavity, say at the left mirror (plane[16]), and the ABCD matrix for the optical system that must, consequently, extend from the left mirror though the right mirror-lens combination and on to the transverse plane containing the external beam waist. The value of $\tilde{q}_2$ derived with the aid of Eq. (22-40) then pinpoints both ℓ and w_0 for the externally focused beam waist. The outline of the solution is given in the example that follows.

Example

For the He-Ne laser geometry given in Figure 22-12, use the ABCD propagation law to determine the spot size w_0 of the external beam waist and its distance from the outer surface R_3 of the mirror-lens combination. Determine the complex radius of curvature $\tilde{q}_1$ at the plane mirror in the cavity. Then develop an ABCD matrix for propagation to the external beam waist. Determine $\tilde{q}_2$, the complex radius of curvature at the external beam waist by using the ABCD propagation law:

$$\tilde{q}_2 = \frac{A\tilde{q}_1 + B}{C\tilde{q}_1 + D}$$

Then, from $\tilde{q}_2$, determine the spot size w_0 and location of the beam waist. The steps that follow are in outline form. Details will be assigned as problems.

(a) Determine an expression for $\tilde{q}_1$.

(b) Determine the ABCD matrix for the optical system from the plane mirror to the external beam waist.

(c) Use the ABCD propagation law to relate $\tilde{q}_2(\ell)$ to $\tilde{q}_1$.

Solution (a) Since $R_1 \to \infty$ at the plane mirror,

$$\frac{1}{\tilde{q}_1} = \frac{1}{R_1} + \frac{i\lambda}{\pi w_1^2} \longrightarrow \frac{i\lambda}{\pi w_1^2} \quad \text{or} \quad \tilde{q}_1 = -\frac{i\pi w_1^2}{\lambda}$$

Relate the wavefront curvature at R_2—where the wavefront curvature matches the mirror curvature—to the spot size w_1 at the plane mirror and the distance z from plane to curved mirror R_2. Use Eq. (22-23) to write

$$R_2 = z_2\left[1 + \left(\frac{\pi w_1^2}{\lambda z_2}\right)^2\right]$$

where $\lambda = 0.633 \times 10^{-6}$ m, $z_2 = 0.7$ m, and $R_2 = 2.0$ m. (Notice that sign conventions for laser cavity mirrors assign R positive for mirrors concave toward the cavity, R negative otherwise.) Solve for w_1 to get $w_1 = 4.38 \times 10^{-4}$ m. Then $\tilde{q}_1 = -i\pi w_1^2/\lambda = -0.952i$.

(b) From the plane mirror to the external beam waist, the system matrix is

$$\begin{bmatrix} A & B \\ C & D \end{bmatrix} = \underbrace{\begin{bmatrix} 1 & \ell \\ 0 & 1 \end{bmatrix}}_{\text{transfer}} \underbrace{\begin{bmatrix} 1 & 0 \\ \dfrac{-0.5}{0.64} & 1.5 \end{bmatrix}}_{\text{refraction}} \underbrace{\begin{bmatrix} 1 & 0 \\ \dfrac{0.5}{3} & 1.5 \end{bmatrix}}_{\text{refraction}} \underbrace{\begin{bmatrix} 1 & 0.7 \\ 0 & 1 \end{bmatrix}}_{\text{transfer}}$$

---

[16] As already alluded to and used in the previous example, it can be shown that the wavefront curvatures of the beam at the mirrors match the curvature of the mirror surfaces.

Carrying out the indicated matrix multiplication, one gets:

$$\begin{bmatrix} A & B \\ C & D \end{bmatrix} = \begin{bmatrix} 1 - 0.53\ell & 0.7 + 0.63\ell \\ -0.53 & 0.63 \end{bmatrix}$$

(c) Substitute into the propagation law given above, using the results of parts (a) and (b):

$$A = 1 - 0.53\ell \quad B = 0.7 + 0.63\ell \quad C = -0.53 \quad D = 0.63$$

$$\tilde{q}_1 = -0.952i \quad R_2(\ell) \longrightarrow \infty$$

and Eq. (22-16),

$$\frac{1}{\tilde{q}_2(\ell)} = \frac{1}{R_2(\ell)} + \frac{i\lambda}{\pi w_2(\ell)^2}$$

Equating real and imaginary parts, the solution gives $\ell = 0.06$ m and $w_2(\ell) = 0.54$ mm. Thus the external beam waist has a spot size $w_2(\ell)$ of 0.54 mm, located 6 cm from the R_3 surface of the mirror-lens combination.

Collimation of a Gaussian Beam. A collimated laser beam is a Gaussian beam with a long waist, as shown in Figure 22-13. The *collimated beam length* is arbitrarily defined as the distance between two symmetrical, transverse planes on either side of the beam waist, the two planes being those in which the spot size $w(z)$ has increased by a factor of $\sqrt{2}$ over the spot size w_0 at the waist.

Figure 22-13 A collimated laser beam and the corresponding Rayleigh range Z_R.

In Figure 22-13, the beam leaving the laser passes through a Galilean-type beam expander and forms a beam waist w_0 some distance to the right of the converging lens. The *Rayleigh range Z_R* is defined as the distance from the beam waist w_0 to the transverse plane where $w(z)$ has the value $\sqrt{2}w_0$, or, equivalently, where the beam area has doubled. The Rayleigh range can be obtained by setting $w(z) = \sqrt{2}w_0$ in Eq. (22-24) and solving for $Z = Z_R$, thus

$$w(z) = \sqrt{2}w_0 = w_0 \left[1 + \left(\frac{\lambda Z_R}{\pi w_0^2} \right)^2 \right]^{1/2} \tag{22-41}$$

from which it follows that

$$Z_R = \frac{\pi w_0^2}{\lambda} \quad \text{(Rayleigh range)} \tag{22-42}$$

Thus the collimated region extends over two Rayleigh ranges, one on either side of the beam waist. As shown earlier in Figure 22-8, the smaller the spot size w_0 of the beam at the waist, the more rapidly the beam diverges as it leaves the waist and so the smaller the Rayleigh range, or distance over which the beam remains collimated with a nearly constant diameter and planar wavefront.

Selecting Beam-Shaping Optics for Optimum Beam Propagation.

Gaussian beams are often passed through apertures such as mirrors, lenses, beam expanders, and telescopes. If such "apertures" are not to restrict severely the amount of laser energy passing through, we need to develop a criterion for the size of the aperture relative to the spot size of the beam.

To find the fraction of incident power that passes through a circular aperture (lens, diaphragm, etc.) of radius a, we return to the TEM_{00} Gaussian beam representation, Eq. (22-18), for the electric field across a plane transverse to the beam:

$$\tilde{E}(x, y, z, t) = E_0 e^{ik(x^2+y^2)/2R(z)} e^{-(x^2+y^2)/w^2(z)} e^{i[kz+p(z)-\omega t]} \qquad (22\text{-}18)$$

Since the beam irradiance is proportional to the square of the electric field amplitude, the total power Φ_{tot} in the beam is obtained by evaluating the integral

$$\Phi_{tot} = \iint_A |\tilde{E}(x, y, z, t)|^2 \, dA = \iint_A E_0^2(z) \, e^{-2(x^2+y^2)/w^2(z)} \, dA$$

where the integration is over the *entire cross-sectional area* of the beam. When this is done, one obtains for the total beam power

$$\Phi_{tot} = [E_0(z)]^2 \frac{\pi [w(z)]^2}{2}$$

Here $E_0(z)$ is the electric field amplitude at *beam center* for a transverse plane at z with spot size w. Thus,

$$E_0(z) = \sqrt{\frac{2}{\pi}} \frac{1}{w(z)} \sqrt{\Phi_{tot}}$$

Making use of the expression in Eq. (22-18) and calculating the fractional power $\Phi_{frac} = \Phi(r = a)/\Phi_{tot}$ passing through a circular aperture of radius a, one sets up the integration as follows:

$$\Phi_{frac} = \frac{\Phi(r = a)}{\Phi_{tot}} = \frac{1}{\Phi_{tot}} \iint_{Aperture} |\tilde{E}(x, y, z, t)|^2 \, dA$$

$$\Phi_{frac} = \frac{E_0^2(z)}{\Phi_{tot}} \int_0^a e^{-2r^2/w^2} (2\pi r \, dr)$$

and obtains

$$\Phi_{frac} = 1 - e^{-(2a^2)/w^2} \qquad (22\text{-}43)$$

Figure 22-14 shows the fractional power $\Phi(r = a)/\Phi_{tot}$ that passes through a circular aperture of radius a, versus the ratio a/w, where w is the spot size of the beam at the aperture location. Notice that 86% of the beam gets through (14% blocked) when the aperture radius equals the spot size, while just under 99% gets through when the aperture radius is increased to 1.5 times the spot size.

Thus, if each beam-shaping element that passes the laser beam in a given optical system has a *diameter* equal to 3 times the beam spot size ($d = 3w$), *nearly* 99% of the beam gets through. Even though nearly all the power gets through at $d = 3w$, we should note that diffraction effects caused by sharp-edged circular apertures produce ripple effects on irradiance patterns in the near field and a reduction of on-axis irradiance in the far field of about 17% [1]. To negate the diffraction effects, one enlarges the aperture so that $d \cong 4.5w$. Beam-shaping optics with diameters that satisfy the $d \cong 4.5w$ criterion transmit essentially 100% of the beam power without superimposing *additional* diffraction effects on the beam.

With Eq. (22-42) for the Rayleigh range and the aperture diameter criterion, $d = 4.5w$ (99% transmission), one can calculate the *Rayleigh range* for typical

Figure 22-14 Fractional power transmission of a Gaussian beam of spot size w through a circular aperture of radius a.

lasers as a function of aperture diameter. Results are shown in Figure 22-15 for a He-Ne, HF (hydrogen fluoride), and CO_2 laser on a log-log plot. For example, if the aperture diameter $4.5w = 1$ cm, the collimated beam length $2Z_R$ is equal to 24.5 m for He-Ne light at $\lambda = 632.8$ nm and 1.46 m for CO_2 light at 10.6 μm. For an aperture diameter $4.5w = 2$ cm, the collimated beam length becomes 98 m for the He-Ne laser and 5.8 m for the CO_2 laser, increasing as the square of the aperture diameter.

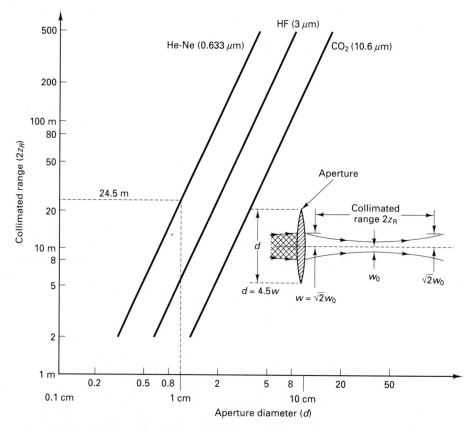

Figure 22-15 Collimated range $2z_R$ versus aperture diameter d, for a laser beam focused to a waist w_0, assuming that the aperture diameter $d = 4.5w$, where $w = \sqrt{2}\, w_0$.

Focusing a Gaussian Beam.   Let us now address the problem of focusing a Gaussian laser beam with a lens of focal length f. We take a rather general case, that of a beam with waist w_{01} located a distance Z_1 to the left of the lens, incident upon a positive thin lens and brought to a focus with beam waist w_{02} located a distance Z_2 to the right of the lens. The pertinent geometry is shown in Figure 22-16.

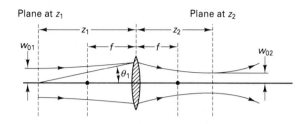

Plane at z_1 Plane at z_2

w_{01} w_{02}

Figure 22-16   A Gaussian beam of waist w_{01} and half-angle beam divergence θ_1 is focused by a thin, positive lens of focal length f. The beam is focused to a beam waist w_{02} a distance z_2 away.

The problem before us is as follows: Given the beam waist w_{01} and distance Z_1 on the incident side of the lens, determine the beam waist w_{02} and the distance Z_2 on the image side of the lens. We shall use the *ABCD* propagation law, Eq. (22-40), which involves the *ABCD* matrix from transverse plane Z_1 to transverse plane Z_2. This matrix is

$$
\begin{bmatrix} A & B \\ C & D \end{bmatrix} = \begin{bmatrix} 1 & Z_2 \\ 0 & 1 \end{bmatrix} \begin{bmatrix} 1 & 0 \\ -\dfrac{1}{f} & 1 \end{bmatrix} \begin{bmatrix} 1 & Z_1 \\ 0 & 1 \end{bmatrix} = \begin{bmatrix} 1 - \dfrac{z_2}{f} & z_1 + z_2 - \dfrac{z_1 z_2}{f} \\ -\dfrac{1}{f} & 1 - \dfrac{z_1}{f} \end{bmatrix}
\tag{22-44}
$$

Then the *ABCD* propagation law becomes

$$
\tilde{q}_2 = \frac{\left(1 - \dfrac{Z_2}{f}\right)\tilde{q}_1 + \left(Z_1 + Z_2 - \dfrac{Z_1 Z_2}{f}\right)}{-\dfrac{1}{f}\tilde{q}_1 + \left(1 - \dfrac{Z_1}{f}\right)}
\tag{22-45}
$$

Making use of Eq. (22-19), which gives us

$$
\tilde{q}_1 = -i\frac{\pi w_{01}^2}{\lambda} \text{ at } Z_1 \quad \text{and} \quad \tilde{q}_2 = -i\frac{\pi w_{02}^2}{\lambda} \text{ at } Z_2
$$

and substituting into Eq. (22-45), we obtain a rather complicated equation that contains the desired unknowns, Z_2 and w_{02}. After equating real parts and imaginary parts, we sort out the unknowns W_{02} and Z_2 in the following form:

$$
\frac{1}{w_{02}^2} = \frac{1}{w_{01}^2}\left(1 - \frac{Z_1}{f}\right)^2 + \frac{1}{f^2}\left(\frac{\pi w_{01}}{\lambda}\right)^2
\tag{22-46}
$$

$$
Z_2 = f + \frac{f^2(Z_1 - f)}{(Z_1 - f)^2 + (\pi w_{01}^2/\lambda)^2}
\tag{22-47}
$$

We note from Eq. (22-47) that for the general case, $Z_2 \neq f$. Apparently then, the Gaussian beam is not brought necessarily to a focus in the focal plane of the lens, that is, at a distance f to the right of the lens.

We can, however, make some practical assumptions that will simplify Eqs. (22-46) and (22-47) considerably. First, consider Eq. (22-46). If $w_{01} \gg w_{02}$—that is, a strong positive lens is used in the focusing process—then the first term in Eq. (22-46) is small in comparison with the second term and the focused beam waist w_{02} is given approximately by

$$w_{02} \cong \frac{f\lambda}{\pi w_{01}} = f\theta_1 \tag{22-48}$$

where $\theta_1 \equiv \lambda/\pi w_{01}$ is the defining Eq. (22-28) for the half-angle beam divergence θ_1.

Next, in Eq. (22-47), if we have a physical system for which $\pi w_{01}^2/\lambda \gg (Z_1 - f)^2$—not an uncommon situation[17]—then it follows that Eq. (22-47) reduces to

$$Z_2 \cong f \tag{22-49}$$

Eq. (22-47), in any case, reduces to Eq. (22-49) if the beam waist w_{01} is located in the focal plane of the lens ($Z_1 = f$).

If we assume that w_{01} is approximately equal to the radius of the lens, then by multiplying the numerator and denominator of Eq. (22-48) by 2 and using the definition $f\# \equiv f/d$ for the f-number of the lens, where d is the lens diameter, we obtain

$$w_{02} = \frac{2f\lambda}{\pi(2w_{01})} = \frac{2\lambda f\#}{\pi} \tag{22-50}$$

In general, then, the smaller the $f\#$ of the focusing lens, the smaller the beam waist at the focused spot.

22-7 HIGHER-ORDER GAUSSIAN BEAMS

The solution for the spherical Gaussian laser beam, derived earlier as Eq. (22-18), represents the lowest order—that is, fundamental—transverse electromagnetic mode that exists in the open-sidewall laser cavity. Other modes—higher-order modes—exist that do not have a pure Gaussian profile for their irradiance variation in the transverse plane. Let us return to our earlier development in this chapter and generalize Eq. (22-12), which we first guessed to have the form,

$$\tilde{U}(x, y, z) = E_0 e^{i\{p(z) + [k(x^2+y^2)]/2\tilde{q}(z)\}} \tag{22-12}$$

as

$$\tilde{U}(x, y, z) = E_0 g\left(\frac{x}{w}\right) h\left(\frac{y}{w}\right) e^{i[p(z) + (kr^2)/2\tilde{q}(z)]} \tag{22-51}$$

The functions $g(x/w)$ and $h(y/w)$ admit a more complicated variation of the beam irradiance with x and y. Substitution of Eq. (22-51) into Eq. (22-11), the defining equation for $\tilde{U}$, leads to

$$\underbrace{\frac{1}{w^2}\frac{g''}{g} + \frac{2ikx}{w}\left(\frac{1}{\tilde{q}} - \frac{1}{w}\frac{\partial w}{\partial z}\right)\frac{g'}{g}}_{1} + \underbrace{\frac{1}{w^2}\frac{h''}{h} + \frac{2iky}{w}\left(\frac{1}{\tilde{q}} - \frac{1}{w}\frac{\partial w}{\partial z}\right)\frac{h'}{h}}_{2}$$

$$+ \underbrace{\frac{2ik}{\tilde{q}} - 2k\frac{\partial p}{\partial z}}_{3} - \underbrace{\frac{k^2}{\tilde{q}^2}(x^2 + y^2)\left(1 - \frac{\partial \tilde{q}}{\partial z}\right)}_{4} = 0 \tag{22-52}$$

---

[17] In many cases, laser beams are first expanded in cross section before being focused to a small spot, as in the case of laser welding. In these instances, the beam waist w_{01} may be of the order of a centimeter or more, thereby ensuring that the inequality $\pi w_{01}^2/\lambda \gg (z_1 - f)^2$ is satisfied. (See, for example, problem 9 at the end of the chapter.)

where the primes denote differentiation with respect to the arguments of g and h.

Inspection of Eq. (22-52) indicates that for a given z, that is, for a given transverse plane, the first bracketed series of terms, the g-expression, is a function of x alone; the second, the h-expression, is a function of y alone; the third is independent of both x and y; and the fourth is identically zero, since $\tilde{q} = \tilde{q}_0 + z$ is the propagation law that we hold to be valid. Then Eq. (22-52) can be satisfied for all x, y at arbitrary z only if (1) the g-expression equals a constant, say $-\lambda_1$, (2) the h-expression equals a constant, say $-\lambda_2$, thereby (3) leaving the third term equal to $\lambda_1 + \lambda_2$. For the g-expression, we write

$$\frac{1}{w^2}\frac{g''}{g} + \frac{2ikx}{w}\left(\frac{1}{\tilde{q}} - \frac{1}{w}\frac{\partial w}{\partial z}\right)\frac{g'}{g} = -\lambda_1 \qquad (22\text{-}53)$$

We can show, with the help of Eqs. (22-16) and (22-24), that the term in parentheses is given by

$$\frac{1}{\tilde{q}} - \frac{1}{w}\frac{\partial w}{\partial z} = \frac{i\lambda}{\pi w^2} \qquad (22\text{-}54)$$

If we use Eq. (22-54) and a change in variable, $\xi = \sqrt{2}x/w$ in Eq. (22-53), we obtain immediately the *Hermite* differential equation,

$$\frac{\partial^2 q}{\partial \xi^2} - 2\xi\frac{\partial q}{\partial \xi} + \frac{\lambda_1 w^2}{2}q = 0 \qquad (22\text{-}55)$$

This equation is known to have a solution only if

$$\frac{\lambda_1 w^2}{2} = 2m, \quad \text{where } m = 0, 1, 2, \ldots \qquad (22\text{-}56)$$

The solutions to Eq. (22-55) are the well-known *Hermite polynomials*,

$$g(\xi) = H_m(\xi) = H_m\left(\frac{\sqrt{2}x}{w}\right) \qquad (22\text{-}57)$$

where the $H_m(\xi)$ can be obtained for each appropriate m from the generating function

$$H_m(\xi) = (-1)^m e^{\xi^2}\frac{d^m}{d\xi^m}(e^{-\xi^2}) \qquad (22\text{-}58)$$

Solution of Eq. (22-58) for $m = 0, 1, 2, \ldots$ gives

$$m = 0: \quad H_0(\xi) = 1$$

$$m = 1: \quad H_1(\xi) = 2\xi = \frac{2\sqrt{2}x}{w} \qquad (22\text{-}59)$$

$$m = 2: \quad H_2(\xi) = 4\xi^2 - 2 = \frac{8x^2}{w^2} - 2$$

Applying the second condition and solving for the h-expression in Eq. (22-52), in a manner identical to that for the g-expression, we obtain, using the change of variable $\eta = \sqrt{2}y/w$,

$$\frac{\partial^2 h}{\partial \eta^2} - 2\eta\frac{\partial h}{\partial \eta} + \frac{\lambda_2 w^2}{2}h = 0 \qquad (22\text{-}60)$$

for which the solutions are again the Hermite polynomials:

$$h(\eta) = H_n(\eta) = H_n\left(\frac{\sqrt{2}y}{w}\right), \quad \text{where } n = 0, 1, 2, \ldots \qquad (22\text{-}61)$$

obtained for all n from the generating expression, Eq. (22-58).

Before investigating the effects that the Hermite functions $g(\xi)$ and $h(\eta)$ have on the transverse nature of the beam irradiance, let us examine the consequences of the third condition imposed on Eq. (22-52),

$$\frac{2ik}{\tilde{q}} - 2k\frac{\partial p}{\partial z} = \lambda_1 + \lambda_2 \tag{22-62}$$

where $\lambda_1 = 4m/w^2$ and $\lambda_2 = 4n/w^2$. Note that if $\lambda_1 = \lambda_2 = 0$, which is necessarily true when $m = n = 0$ (see Eq. 22-56), then $g(\xi) = h(\eta) = 1$ and we are back to the pure Gaussian spherical wave solution. In this case, Eq. (22-62) reduces to $\partial p/\partial z = i/\tilde{q}$, the defining equation for p obtained earlier. However, in the general case, for which λ_1 and λ_2 are *not* equal to zero, Eq. (22-62) must remain the defining equation for the function $p(z)$. Then we must solve

$$\frac{\partial p}{\partial z} = \frac{i}{\tilde{q}(z)} - \frac{\lambda(m+n)}{\pi w(z)^2} \tag{22-63}$$

to obtain an expression for $p(z)$. Substitution of Eq. (22-16) into Eq. (22-63) for $\tilde{q}(z)$ and then Eqs. (22-23) and (22-24) for $R(z)$ and $w(z)$, we can integrate the resulting equation to obtain

$$p(z) = \frac{i}{2}\ln\left[\frac{\lambda^2 z^2 + (\pi w_0^2)^2}{(\pi w_0^2)^2}\right] - (m+n+1)\arctan\left[\frac{\lambda z}{\pi w_0^2}\right] \tag{22-64}$$

With the help of Eq. (22-64), we can form the term $e^{ip(z)}$ as

$$e^{ip(z)} = \frac{w_0}{w(z)}e^{-i(m+n+1)\arctan(\lambda z/\pi w_0^2)} \tag{22-65}$$

where we have used Eq. (22-24) for $w(z)$ to simplify the form of the expression.

The Hermite-Gaussian Beam Solutions. Now, collecting our results for p, $\tilde{q}$, $g(\xi)$, and $h(\eta)$, and putting them all together in the general expression for the electric field $E(x, y, z, t)$ as initially introduced in Eq. (22-9), we obtain finally,

$$E(x, y, z, t) = \underbrace{E_0\frac{w_0}{w(z)}H_m\left(\frac{\sqrt{2}x}{w(z)}\right)H_n\left(\frac{\sqrt{2}y}{w(z)}\right)e^{(-x^2+y^2)/w^2(z)}}_{\text{amplitude}}\underbrace{e^{ik(x^2+y^2)/2R(z)}e^{i[kz-(m+n+1)\arctan(\lambda z/\pi w_0^2)]}e^{-i\omega t}}_{\text{phase}} \tag{22-66}$$

This expression for $E(x, y, z, t)$ includes the higher-order modes as well as the Gaussian spherical wave (when $m = n = 0$) derived earlier. The first set of terms describes the amplitude variation of the electric field at any transverse plane z. The H_m and H_n Hermite polynomial terms permit a finer-tuned variation in the transverse electric field than that described previously by the pure Gaussian term alone,

$$e^{-(x^2+y^2)/w^2}$$

The variations in transverse amplitude deviate more and more from the pure Gaussian with higher integers, m and n. These integers are, of course, characteristic of the higher-order modes.

The second set of terms, a complicated phase, describes the nature of the wavefront as a function of m and n. Again, we see that for $m = n = 0$, the phase term reduces to the form characteristic of the pure Gaussian spherical wave derived earlier. It can be shown that the frequency associated with an arbitrary higher-order beam mode is given by

$$f_{mnq} = \left[q + \frac{m + n + 1}{\pi} \left(\arctan \frac{z_2}{z_R} - \arctan \frac{z_1}{z_R} \right) \right] \frac{c}{2L} \qquad (22\text{-}67)$$

where q is the *axial mode number* (not the complex radius of curvature); m and n are integers associated with the Hermite polynomials H_m and H_n; z_1 and z_2 are the distances from the beam waist inside the laser cavity to mirrors M_1 and M_2, respectively; L is the cavity length between mirrors; and Z_R is the Rayleigh range.

The general expression for $E(x, y, z, t)$, Eq. (22-66), is called the *Hermite-Gaussian* beam solution to the wave equation for a stable laser cavity. Any arbitrary beam can be expanded as a linear combination of Hermite-Gaussian beams, each of which has the same propagation law, $\tilde{q} = \tilde{q}_0 + z$.

Field and Irradiance Patterns for Hermite-Gaussian Beams.

With the help of Eq. (22-66) we can sketch the transverse electric field and irradiance variation for several of the lower-order modes—small m and n—and thus predict the nature of the "burn" pattern.

Without loss of generality, we can simplify Eq. (22-66) by setting $w(z)$ for some transverse plane $z = z_0$ to $\sqrt{2}$ units of length. This makes the arguments for the Hermite polynomials simply x and y. In addition, recognizing that the phase terms in Eq. (22-66) do not contribute to the irradiance, we can write rather simple *proportional expressions* for $E(x, y, z_0)$ and $|E(x, y, z_0)|^2$ as follows:

$$E(x, y, z_0) \propto H_m(x) H_n(y) e^{-(x^2 + y^2)/2} \qquad (22\text{-}68)$$

$$|\tilde{E}(x, y, z_0)|^2 \propto [H_m(x)]^2 [H_n(y)]^2 e^{-(x^2 + y^2)} \qquad (22\text{-}69)$$

Figure 22-17 shows the development of the irradiance patterns $|E(x, y, z_0)|^2$ for the x-variation in the transverse plane at $z = z_0$. From symmetry, it is clear that identical results would be obtained for the y-variation. Thus, for two cases, $m = 0$ and $m = 1$, Figure 22-17 shows separate sketches for $H_m(x)$, $H_m(x) \exp(-x^2/2)$, and,

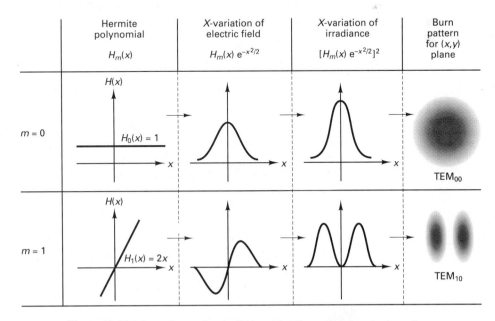

Figure 22-17 Laser-beam electric field and irradiance variations in the x-direction for two values of the Hermite integer m. Corresponding burn patterns for $m = 0$, $n = 0$ and $m = 1$, $n = 1$ are shown.

finally, $[H_m(x) \exp(-x^2/2)]^2$, the sketch that gives direct evidence of the burn pattern to be expected.[18]

After studying Figure 22-17, we see that the irradiance variation has m zeros along the x-axis, discounting the zeros at the distant tails of the Gaussian envelope. Thus, for $m = 1$, there is a single zero at $x = 0$ and peaks on either side. The same type of variation is to be expected for the irradiance $[H_n(y) \exp(-y^2/2)]^2$ in the y-direction. Putting the x- and y-variations together for any m, n pair, we see that Eq. (22-69) and its corresponding equivalent in terms of y uniquely determine the laser-beam irradiance in the transverse plane. As mentioned at the beginning of this chapter, these are commonly referred to as the TEM_{mn} modes. When $m = n = 0$, we have the Gaussian spherical beam with a pure Gaussian variation in irradiance. When $m \neq 0$ and $n \neq 0$, we have the higher-order modes. Some of the irradiance patterns for low values of the Hermite integers m and n—predictable from Eq. (22-69) and Figure 22-17—are reproduced in Figure 22-18.

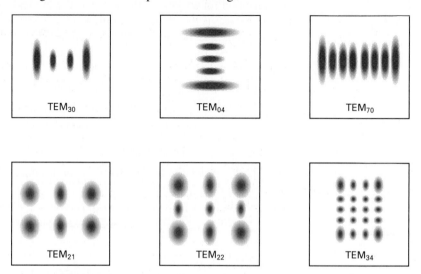

Figure 22-18   Representative sketches of irradiance or burn pattern for several different orders of Hermite-Gaussian optical resonator modes as they might be photographed in the output beam of a laser oscillator.

PROBLEMS

22-1. (a) Beginning with the expression given by Eq. (22-6) for the geometry shown in Figure 22-2, show that for points $r(x, y, z)$ near the propagation axis, where $(x^2 + y^2) \ll R^2$, one can obtain the approximate expression

$$r \cong R + \frac{x^2 + y^2}{2R}$$

(b) Then use $\tilde{E}(x, y)_{z=R} = (\text{constant})e^{ikr}$ and the approximate expression for r found above to obtain

$$\tilde{E}(x, y)_{z=R} \cong (\text{constant})e^{ikR}e^{ik(x^2+y^2)/2R}$$

---

[18] To ensure understanding of the sketch patterns shown in Figure 22-17, the reader is invited, in problem 22-17, to compute the next row, for $m = 2$.

22-2. Show that substitution of $\tilde{E}(r, t) = U(r)e^{i(kz-\omega t)}$ in the wave equation (22-2) leads to Eq. (22-10).

22-3. A TEM$_{00}$ He-Ne laser ($\lambda = 632.8$ nm) has a beam waist w_0 (at $z = 0$) of 0.5 mm and a half-angle beam divergence of $\theta = 0.4$ mrad. At a distance $z = 50$ m from the beam waist, determine

(a) the value of the complex radius of curvature $\tilde{q}_0$ at the beam waist;

(b) a numerical expression for the complex radius of curvature $\tilde{q}$, in meters, from each of the following expressions:

$$\frac{1}{\tilde{q}} = \frac{1}{R} + i\frac{\lambda}{\pi w^2(z)} \quad \text{and} \quad \tilde{q} = z - i\frac{\pi w_0^2}{\lambda}$$

(*Hint*: Is the transverse plane, 50 m from the beam waist, far enough away to be in the far field? If so, what does this say about R and z?)

22-4. (a) In problem 3, use Eqs. (22-23) and (22-24) to determine $R(z)$ and $w(z)$ at $z = 50$ m.

(b) Is it true that $R(z) \cong z$ in the far field? Is it true that $\tan \theta \cong w(z)/z$, where θ is the half-angle beam divergence, can be used as a good approximation to determine $w(z)$ at $z = 50$ m?

22-5. A TEM$_{00}$ He-Ne laser ($\lambda = 0.6328$ μm) has a cavity that is 0.34 m long, a fully reflecting mirror of radius $R = 10$ m (concave inward), and an output mirror of radius $R = 10$ m (also concave inward).

(a) From the symmetry of mirror geometries and the boundary condition that wavefront and mirror curvatures match at the mirrors, determine the location of the beam waist in the cavity. Set $z = 0$ at this location to be the reference plane.

(b) Determine the beam waist w_0.

(c) Determine the beam spot size $w(z)$ at the left and right cavity mirrors.

(d) Determine the half-angle beam divergence θ for this laser.

(e) Where is the far field for this laser if you use the criterion $z_{FF} \geq 50(\pi w_0^2/\lambda)$?

(f) If the laser emits a constant beam of power 5 mW, what is the average irradiance at the position where $z_{FF} = 50(\pi w_0^2/\lambda)$?

22-6. Relative to Figure 22-12, set up the ray-transfer matrix for the *output element* of the laser, that is, the mirror-lens combination with thickness 0.004 m, mirror surface curvature of $|R_2| = 2$ m, lens surface curvature of $|R_3| = 0.64$ m, and lens refractive index of 1.50.

(a) Using the definitions given for the refraction and translation matrices in Chapter 4, set up the $ABCD$ matrix for this element as follows:

$$\begin{bmatrix} A & B \\ C & D \end{bmatrix} = \begin{bmatrix} 1 & 0 \\ \dfrac{n-n'}{R_3 n'} & \dfrac{n}{n'} \end{bmatrix} \begin{bmatrix} 1 & L \\ 0 & 1 \end{bmatrix} \begin{bmatrix} 1 & 0 \\ \dfrac{n-n'}{R_2 n'} & \dfrac{n}{n'} \end{bmatrix}$$

Pay particular attention to the changing meaning of n and n' for the two refractions and to the sign conventions for R_2 and R_3 in the matrix formulations. Within rounding approximations, you should find

$$\begin{bmatrix} A & B \\ C & D \end{bmatrix} = \begin{bmatrix} 1.0007 & 0.0027 \\ -0.5318 & 0.9979 \end{bmatrix}$$

(b) Since $L = 0.004$ m is a very small dimension compared with $|R_2| = 2$ m or $|R_3| = 0.64$ m, repeat the $ABCD$ calculation, replacing the translation matrix with the unit matrix

$$\begin{bmatrix} 1 & L \\ 0 & 1 \end{bmatrix} \longrightarrow \begin{bmatrix} 1 & 0 \\ 0 & 1 \end{bmatrix}$$

What then is the result of the $ABCD$ matrix for this "thin lens"?

22-7. (a) Since the output element described in problem 6 is essentially a thin lens, compare the *ABCD* matrix obtained in problem 6(b) with the thin-lens matrix, namely,

$$\begin{bmatrix} 1 & 0 \\ -\dfrac{1}{f} & 1 \end{bmatrix}$$

and deduce the focal length of the output element.

(b) Use the expression for the focal length of a thin lens,

$$\frac{1}{f} = \frac{n_2 - n_1}{n_1}\left(\frac{1}{R_1} - \frac{1}{R_2}\right)$$

with careful attention to thin-lens sign conventions, and obtain the focal length of the thin-lens output element. How do the results for parts (a) and (b) compare?

22-8. Relative to the example designed around Figure 22-12, (a) determine an expression for $\tilde{q}_1$ at the plane mirror; (b) solve Eq. 22-23 for the spot size value w_1; (c) obtain a numerical value for $\tilde{q}_1$; (d) multiply $\tilde{q}_1$ by the *ABCD* matrix to obtain $\tilde{q}_2$; (e) use Eq. (22-16) and $\tilde{q}_2$ from part (d) to obtain ℓ and $w(\ell)$.

22-9. (a) Relative to the calculations in the example (refer to Figure 22-12) in which the external beam waist is focused at $\ell \cong 0.06$ m with a waist size $w(\ell) = 0.54$ mm, use Eqs. (22-46) and (22-47) together with Figure 22-16 to obtain values for w_{02} and z_2. How do these results compare with those for $w_0(\ell)$ and ℓ obtained in the example?

(b) Explain why one *cannot* use the approximations

$$w_{02} \cong \frac{f\lambda}{\pi w_{01}} = f\theta \quad \text{and} \quad z_2 \cong f$$

in this instance.

22-10. Refer to the externally focused TEM$_{00}$ laser beam shown in Figure 22-12, with beam waist $w_0(\ell) = 0.54$ mm, located at 0.06 m from the output element.

(a) Calculate the far-field distance for the externally focused beam waist.

(b) Calculate the far-field half-angle beam divergence for the laser beam that emerges from the focused beam waist.

(c) Insert a 10× beam expander in the beam at a distance $z = 30$ m past the focused beam waist. Calculate the beam spot size w at the entrance and exit faces of the beam expander.

(d) Now place a thin lens of focal length 10 cm and appropriate diameter at a distance of 20 cm from the output face of the beam expander. With reference to Figure 22-16 and Eqs. (22-46) and (22-47), calculate z_2 and w_{02} for the newly focused beam. Could you have used the approximate formulas $w_{02} \cong f\lambda/\pi w_{01}$ and $z_2 = f$ in this instance? Why? How do the calculations for the exact formulas and approximate formulas compare?

22-11. Given the expression

$$\frac{\Phi(r = a)}{\Phi_{\text{tot}}} = \frac{1}{\Phi_{\text{tot}}} \iint\limits_{\text{Aperture}} |\tilde{E}(x, y, z, t)|^2 \, dA$$

and Eq. (22-18) for $\tilde{E}(x, y, z, t)$, carry out the indicated integration over a circular aperture of radius $r = a$ to show that the fractional power of a TEM$_{00}$ laser beam passing through the aperture is given by

$$\frac{\Phi(r = a)}{\Phi_{\text{tot}}} = 1 - e^{-(2a^2)/w^2}$$

22-12. Carry out the integration for

$$\Phi_{\text{tot}} = \iint\limits_{\text{Aperture}} E_0^2(z) e^{-2(x^2+y^2)/w^2(z)} \, dA$$

over the entire cross section of a TEM_{00} beam and show that the total power Φ_{tot} in the beam is equal to

$$\Phi_{\text{tot}} = E_0^2(z) \frac{\pi w^2(z)}{2}$$

22-13. Explain how you can use an adjustable circular aperture (iris) and a power meter to determine the spot size w of a TEM_{00} laser beam at any position along the beam.

22-14. Determine collimated beam lengths $2z_R$, where z_R is the Rayleigh range, for a TEM_{00} Nd:YAG laser beam ($\lambda = 1.064 \ \mu\text{m}$) focused by lenses of aperture diameters $d = 1$ cm, 2 cm, 3 cm, and 5 cm, respectively. Assume that the lens diameter d is related to the focused beam waist w_0 by the equation $d = 4.5(\sqrt{2}w_0)$. Refer to Figure 22-15 for geometry and similar calculations made for He-Ne, HF, and CO_2 TEM_{00} lasers.

22-15. Given the generating function, Eq. (22-58), for Hermite polynomials $H_m(\xi)$, where $\xi = \sqrt{2}x/w$, verify the particular cases for $m = 0, 1, 2, \ldots$ given in Eq. (22-59).

22-16. Fill in the steps to show how Eq. (22-65) follows from Eq. (22-64).

22-17. Refer to Figure 22-17. Extend the "table" to include the case $m = 2$, $n = 0$. Thus, in a third row, sketch in curves for column 1, $H_m(x)$, column 2 for the x-variation curves of the electric field, column 3 for the x-variation curves of the irradiance, and column 4 for the expected burn pattern.

REFERENCES

[1] Siegman, Anthony E. *Lasers*. Mill Valley, Calif.: University Science Books, 1986. Ch. 16, 17.

[2] Saleh, B. E. A., and M. C. Teich. *Fundamentals of Photonics*. New York: John Wiley and Sons, 1991. Ch. 3, 9.

[3] Guenther, Robert. *Modern Optics*. New York: John Wiley and Sons, 1990. Ch. 9.

[4] Möller, K. D. *Optics*. Mill Valley, Calif.: University Science Books, 1988. Ch. 15.

[5] Gerrard, A., and J. M. Burch. *Introduction to Matrix Methods in Optics*. New York: John Wiley and Sons, 1975.

[6] Verdeyen, J. T. *Laser Electronics*. Englewood Cliffs, N.J.: Prentice-Hall, 1981.

23

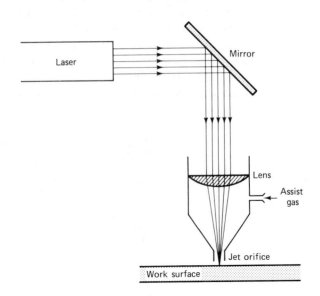

Laser Applications

When one endeavors to chronicle current applications in a rapidly evolving field such as laser technology, one accepts the inevitable fact that such information becomes outdated almost as rapidly as it is written. For information presented in textbooks, obsolescence is especially distressing. Nevertheless,[1] it seems worthwhile to describe some of the important applications of lasers as they exist today. Such applications both illustrate the usefulness of the remarkable properties of laser light and serve as a benchmark of laser technology in the mid-1980s and early 1990s.

INTRODUCTION

The variety of ways in which lasers can be applied seems to be limited only by human creativity and ingenuity. In a few short decades since its invention, the laser has found numerous applications in industry, medicine, agriculture, construction, entertainment, communications, and weaponry. In short, the laser has invaded the world of science and technology with remarkable vigor, and it continues to show great promise as a catalyst for exciting new developments in the explosive field of lightwave technology.

---

[1] For this, the second edition, the content of this chapter remains essentially as first chronicled in 1986. However, at the end of this chapter, by way of a brief update, we provide a selected glimpse of laser technology and applications in evidence at the beginning of the 1990s.

It is, of course, the unique properties of the laser—monochromaticity, directionality, coherency, and brightness—that account for its wide acceptance and usefulness. In each application, one or more of these properties is exploited to achieve a unique goal. Most laser applications can be organized into two general classes: (1) lasers and interactions and (2) lasers and information, outlined in Table 23-1. In the former, lasers interact with matter and cause desirable changes, either permanent or temporary. In the latter, lasers are used to send, detect, store, and process information.

TABLE 23-1 A CLASSIFICATION OF LASER APPLICATIONS

Lasers and interactions	Lasers and information
Materials processing	Communication
Cutting	Information processing
Welding	Optical sensing
Drilling	Optical ranging
Heat treating	Optical storage
Medical	Printers and copiers
Marking and scribing	Metrology
Microfabrication	Holography
Laser spectroscopy	Alignment
Laser-driven energy	Point-of-sale scanners
Weaponry	Entertainment and display

The broad class of applications that involve the interaction of lasers with matter includes the industrial machining and processing of materials, therapeutic and surgical uses in medicine, laser-driven energy sources, scribing, and microfabrication in semiconductor and computer technologies. The equally broad class that involves lasers and information includes information processing, optical sensing and ranging, optical communication, entertainment, printers and copiers, metrology, and alignment facilitators in construction and agriculture. In this chapter we examine several major areas of laser applications in each of the two main divisions.

23-1 LASERS AND INTERACTION

The laser beam with its high-power density, excellent directionality, monochromaticity, and purity of coherence interacts effectively with different media in many different ways. The ability of laser light to cause changes in the matter it strikes makes it ideal for many laser applications, especially those in the fields of material processing, medicine, and laser-beam fusion. In the following paragraphs we indicate some of the impressive achievements made with lasers in these areas.

Material Processing. In the industrial processing of materials, lasers are used to cut, weld, drill, etch, perforate, and heat-treat a variety of substances. The laser is an ideal source for these tasks because it delivers appropriate thermal (infrared) energy in a tightly focused beam. Flexible and easily manipulated, the laser beam can be optimally shaped to direct intense laser power, accurately and repeatedly, on almost any target or work surface. When we consider that a 500-kW laser of 1-mrad beam divergence can be focused to produce power densities of 10^{12} W/cm^2 on target, one appreciates immediately the attractiveness of lasers in material processing.

The gaseous carbon dioxide (CO_2) laser, emitting at 10.6 μm, and the solid-state neodymium (Nd:YAG) laser, emitting at 1.06 μm, are two lasers frequently used to alter materials and surfaces. The three most popular applications of lasers in material processing are usually involved with metals in cutting, welding, and heat treating. Increasingly though, lasers are used to process and fabricate plastics and other nonmetal materials. Lasers are especially compatible with automated and computer-aided manufacturing processes, because they can be programmed to move over the surface of a workpiece and perform computer-controlled cutting and welding operations. The flexible nature of lasers makes them especially suitable for prototype fabrication. An automobile manufacturer, for example, can design a one-of-a-kind plastic dashboard on a computer and later, with the help of the computer, direct a laser accurately to cut out the desired pattern.

Thousands of lasers, with output powers extending into the kilowatt levels, are used daily in materials-processing applications. Lasers are used to perform precision welding on a wide range of small components, such as hermetically sealed electronic components. They are used to weld automobile transmission parts to increase productivity. They are used to weld structural assemblies to improve quality control. They are used in job shops to cut sheet metal accurately and economically. They are also used to make precise, rapid cuts in plastics, rubber, paper, cloth, and wood.

In the automotive and aerospace industries, noncontact hole drilling by lasers is a critical manufacturing process. Accurate and reproducible holes for control of air and liquid flow in hydraulic and pneumatic components are generally made with precisely controlled laser beams. Localized heat-treating of metal surfaces exposed to wear and fatigue is accomplished by laser hardening and cladding processes. Laser scribing is used successfully to machine brittle materials such as fired ceramics and silicon wafers. Precise adjustment of microcircuits and resistor networks, a major industrial task, is handled easily with lasers. Low-power lasers are used to align, measure, and detect flaws in a variety of applications. Lasers also provide a convenient means of marking thousands of parts for identification and sorting.

The process of laser cutting is typical of laser machining in industry. Figure 23-1 shows a line drawing of a simple fixed-beam laser delivery system with gas-jet assist for a laser cutting application. The assist gas is used to "blow away" the material in the cut while at the same time enhancing the burning and melting process. The actual laser system bears little resemblance to the schematic depicted in Figure 23-1. A laser machining center that can cut, drill, and weld parts of almost any ge-

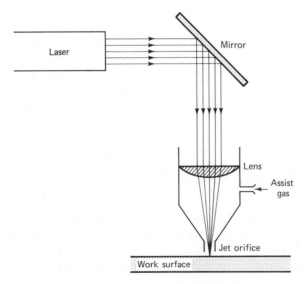

Figure 23-1 Laser fixed-beam delivery system for cutting operation. (Adapted with permission from *Lasers and Applications*.)

Figure 23-2 A heavy-duty laser machining center that permits drilling, cutting, and welding parts of almost any geometry. (Reproduced by permission from *Lasers and Applications*.)

ometry under adverse conditions is shown in Figure 23-2. The housings that accompany the laser are used to produce the high power required to operate the laser.

Figure 23-3 shows a typical curve for the cutting performance of a high-power industrial laser. The plot shows the thickness of metal cut as a function of the linear speed of the laser during the cutting operation. It is interesting to note that the laser used to generate the data in Figure 23-3 can cut through steel more easily than through aluminum, at any given cutting speed. For example, at a cutting speed of 50 mm/s, the laser in question slices through 4 mm of steel or 2 mm of aluminum.

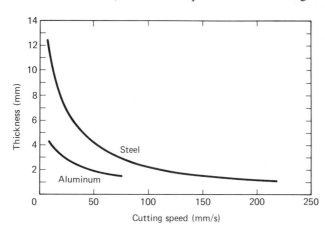

Figure 23-3 Typical performance data for a high-power industrial laser used in metal-cutting operations. (Reproduced by permission from *Laser News* 6, No. 6, November 1984, a publication of the Laser Institute of America.)

Whether lasers are used to cut, drill, weld, or perform countless other special tasks in material processing, it is apparent that lasers have become inextricably linked with industry and the manufacturing process. The integration of lasers with robots represents one of the innovative industrial applications of the laser that is only now being pioneered. Currently, the manufacturing industry is implementing heavy-duty systems that team high-power lasers—both CO_2 and Nd:YAG—with robot units. In a major automobile plant, for example, a 500-W CO_2 laser works with an assembly-line robot to trim excess material from dashboard panels, repeatedly and efficiently, with high quality control. Such applications will multiply many times over as lasers and robots continue to work together.

Medicine. It has been shown successfully that the laser beam—a scalpel of light—can cut, vaporize, rupture, weld, seal, cauterize, coagulate, open, and heal human tissue. In Chapter 7, for example, we saw in some detail how lasers are used to treat ocular disorders. One of the more recent additions to hospital operating

Figure 23-4   Lasers in the hands of doctors have entered the medical world. (Illustration by Jim Buckels.)

rooms and outpatient treatment centers, the laser beam is emerging as an effective, noninvasive surgical tool (see Figure 23-4).

Lasers that are currently applied in medicine as diagnostic and therapeutic instruments are listed in Table 23-2, with associated data on typical wavelengths, power levels, and principal medical uses. As the list indicates, the five laser types most actively used in medical fields are the excimer laser, the argon-ion laser, the Nd:YAG laser, the CO_2 laser, and the tunable dye laser. The excimer lasers [argon fluoride (ArF) and krypton fluoride (KrF)] emit invisible laser energy in the deep ultraviolet. The Nd:YAG and CO_2 lasers emit laser energy at the other end of the optical spectrum, in the near and far infrared, again invisible. The argon-ion laser emits a visible beam (blue-green) that can be seen and, therefore, more easily handled and aligned. The tunable dye laser, although generally quite low in output power, emits a beam that contains wavelengths extending from the violet end of the visible region (400 nm) to the near infrared (900 nm). For the "invisible" lasers, spotter beams— generally He-Ne lasers—are used collaboratively to facilitate handling and focusing.

The CO_2 laser, one of the first to be used extensively by medical practitioners, emits infrared (ir) energy at 10.6 μm that is almost totally absorbed by water. Since body cells and tissues are 70% to 90% water, a CO_2 laser can make a clean incision or, if focused more sharply, can vaporize malignant tissue. Because a laser cauterizes and seals capillaries as it cuts, surgery is essentially bloodless. A drawback of CO_2 lasers, to date, is their inability to be used readily with fiber optics; transmission of the beam from the laser to the target area is often accomplished with cumbersome, articulated optical arms.

Unlike the CO_2 laser, the Nd:YAG laser beam at 1.06 μm and the argon-ion laser beam at 488 nm mate well with fiber optics. The laser energy is thus transferred readily from the laser source to the operating table, without significant power loss. Using flexible systems that have emerged from the marriage of laser beams and

TABLE 23-2 LASER TYPES IN MEDICINE

Laser type	Wavelength	Power levels	Uses

Excimer
  Argon fluoride
  Krypton fluoride
	
	193 nm (ultraviolet; invisible)
	248 nm (ultraviolet; invisible)
		Pulsed, $\cong 1$ J/cm^2 with pulsewidth of about 10 ns ($\cong 10^8$ W/cm^2)	Cutting, incising, ablative photo-decomposition (ophthalmology, cardiology, and arthroscopy)

Argon-ion	488 nm (blue-green; visible)	Continuous; up to 10 W	Photocoagulation, welding, vaporization (dermatology, ophthalmology, general surgery)
Tunable dye (pumped with an argon-ion laser)	631 nm (red; visible)	Continuous; 3–4 W	Photoactivation (treatment of tumors)
Nd:YAG	1.06 μm (infrared; invisible)	Continuous; up to maximum of 60–100 W	Photocoagulation, vaporization, perforation (ophthalmology, gastroenterology, dermatology, urology, and tumors)
Carbon dioxide (CO$_2$)	10.6 μm (infrared; invisible)	Continuous; up to 80 W	Tissue vaporization, incision, excision (dermatology, gynecology, gastroenterology, and neurosurgery)

fiberoptic scopes (endoscopes), surgeons are able to perform major abdominal surgery, for example, through an incision as small as 1 inch.

The blue-green argon-ion laser beam is unique in that it is selectively absorbed by red and brown substances, such as red blood cells and melanin pigment spots. It can therefore stop retinal bleeding (photocoagulation) and weld detached retinas. It can fade port-wine birthmarks and tatoos, penetrating the nonpigmented upper layer of skin without harm.

The Nd:YAG laser, with fiberoptic scope, has also been used effectively in photocoagulative applications—controlling or stopping the bleeding of ulcers and large tumors or blood vessels deep in the body. The Nd:YAG laser has been used with some success to seal tiny bleeding vessels of the retina, thereby ameliorating a disease (retinopathy) that often leads to blindness. To accomplish this, the Nd:YAG laser is optically focused to a diameter of 1/1000 inch.

In one of its more recent dramatic applications, a Nd:YAG laser beam has been focused to a 30-μm spot within the eye. By concentrating powers of over 1 billion W/cm^2, plasma breakdown occurs in the vitreous humor. Subsequent acoustic shock waves rupture unwanted, opacified membranes along the vision axis of the eye. This unique example of noninvasive pain-free surgery was treated in more depth in Chapter 7.

Quite recently, both the Nd:YAG and the excimer lasers have been tested in an important assault on arteriosclerosis (hardening of the arteries) and critically blocked blood vessels of the heart. Again, with delicate fiber-optic scopes, threaded carefully and slowly through major blood vessels, unwanted fatty deposits and plaque are literally blasted away. This technique, while promising, remains in the experimental stage.

By way of a summary, Table 23-3 lists the major medical fields wherein lasers are currently applied and where they hold promise for the future. A quick glance reveals the many parts of the human body that come under the purview of these fields and indicates clearly the extent to which lasers have been or will be involved in the treatment of human ailments. It is apparent that almost every major part of the body, large or small, stands to benefit from some form of laser therapy.

TABLE 23-3 MEDICAL FIELDS INVOLVED WITH LASERS

Current fields of practice	
Ophthalmology	eyes
Gynecology	female reproductive organs
Dermatology	skin
Cardiology	heart and blood vessels
Gastroenterology	intestines
Oncology	tumors/cells
Future fields of practice	
Neurosurgery	nerves
Otolaryngology	ears and throat
Podiatry	feet
Urology	urinary tract
Dentistry	teeth

Laser-Induced Fusion. It has long been the goal of scientists to harness an inexhaustible source of energy. With the understanding of thermonuclear phenomena and fusion that came in the early 1950s, the goal seemed a little closer to realization. In the well-known D-T nuclear reaction that fuses two isotopes of hydrogen—deuterium and tritium—together and releases 14 MeV of kinetic energy per reaction in the process, scientists sought to emulate the sun's mechanism of producing energy. The fusion reaction, however, requires extraordinarily high pressures and temperatures. Major technologies have been developed to meet the stringent requirements of producing a confined, high-density plasma at temperatures approaching those on the surface of the sun. One of these technologies is laser-induced fusion.

The basic idea is rather simple. In laser fusion, a pellet of fusion fuel, usually a mixture of deuterium and tritium, is irradiated uniformly over its surface with high-energy laser beams, circumferentially spaced around the pellet, as shown in Figure 23-5. The rapid heating of the surface of the pellet causes a burning off (an *ablation*) of the outer material of the pellet. The rapid vaporization of the surface is accompanied, in turn, by an inward-propagating compressional wave (an *implosion*) that creates the required conditions for fusion, high density and high temperature. In the interior of the pellets, densities reach 10,000 times that of water, and temperatures rise to 100,000,000°C. These conditions, if maintained for about 1 ps, lead to the fusion reaction and the release of enormous amounts of energy per reaction.

Currently the CO_2 laser (10.6 μm) and the Nd:YAG/Nd-glass lasers (1.06 μm), in combination, are being used to irradiate the D-T pellets. The CO_2 laser fusion technology is under development at the Los Alamos National Laboratory in New Mexico; the Nd:YAG/glass technology is being pursued at the Lawrence Livermore National Laboratory in California. Laser power output with the NOVA Nd-

Figure 23-5 Symmetrically arranged lasers (*L*) irradiate a tiny deuterium-tritium pellet. Instantaneous heating of the surface causes ablation (*A*), an outward expansion of the surface. The reaction, an inward-directed compressional wave (*C*), creates the high density and temperature needed for the fusion reaction.

glass systems has reached the incredible level of 1 trillion W, making NOVA the most powerful laser in operation. Nevertheless, both experiment and theory show that wavelengths shorter than 1 μm offer significant advantages in the production of the compressional wave. Accordingly, recent laser fusion experiments have been performed at wavelengths of 0.53 μm and 0.35 μm. These experiments, in agreement with theoretical predictions, corroborate the efficacy of shorter wavelengths in the ignition and thermonuclear burn process. It is anticipated that short-wavelength lasers—the excimer lasers or the free electron lasers—will play an increasingly important role in laser fusion. If and when laser fusion becomes economically feasible, we will have realized one of our most cherished dreams, a limitless supply of energy.

23-2 LASERS AND INFORMATION

We live in an age of exploding information. Computers provide instant access to worldwide sources of information, available from a wealth of data banks. In the midst of it all, lasers are playing a leading role. They are used in systems that scan, sense, print, transmit, and store information. Laser printers and checkout scanners are already on the market. Laser sensors are used to detect pollution, wind speeds, and global weather patterns. Lasers and fiberoptic cables are changing communication technologies. Lasers and holograms are pointing the way toward an efficient system of storage and retrieval of burgeoning amounts of information. We turn now to a brief survey of laser applications in information processing.

Laser Communications. The capacity of any communication channel is proportional to the width Δf of its frequency band. Even with highly monochromatic light, where $\Delta f / f_0$ is very small, the bandwidth Δf can be enormous when f_0 is very high. Thus a communication system that uses light waves, where extremely high center frequencies f_0 near 10^{14} Hz are available, should in principle carry many times the information now carried by radio and microwave systems operating at much lower center frequencies. If, in addition, the monochromatic light can be made coherent, all the necessary ingredients for communication by light are at hand. When the laser arrived on the scene in the 1960s, a tailor-made solution to the requirements for monochromaticity, coherency, and wide bandwidth became available. Some 30 years later, with laser communication technology firmly in place, the promise is well on the way to realization.

Large volumes of information are transmitted over long distances by communication systems that involve satellites, microwave-radio relays, coaxial cables, and waveguides. Each of these systems involves electromagnetic waves of longer wavelength and lower frequency than does laser light. For example, the frequency in the center of the visible spectrum is about 100,000 times greater than the frequency of 6-cm waves used in microwave-radio relay systems. Consequently, the theoretical information capacity of a typical light wave is about 100,000 times greater than that of a typical microwave. Let us see how this comes about.

Long-distance communication systems all rely on the principle of multiplexing, the simultaneous transmission of many different messages (information) over the same pathway. The ordinary human voice (conversation) requires a frequency band from 200 to 4000 Hz, a band 3800 Hz wide. A telephone call, therefore, can be transmitted on any band that is 3800 Hz wide. It can be carried, for example, by coaxial cable in the megahertz frequency band between 1,000,200 Hz and 1,004,000 Hz, thereby occupying about 0.4% of the available coaxial carrier frequency. By contrast, it can be carried also by a He-Ne laser beam (632.8 nm, 4.738×10^{14} Hz) in the frequency band between 473,800,000,000,200 Hz and 473,800,000,004,000 Hz, where it uses less than *one billionth* of 1% of the available laser-beam frequency. For microwaves, considerably shorter in length than the waves carried by coaxial cables, the contrast is not so great, but even here, the carrying capacity of the laser exceeds that of the microwave system by a factor of 100,000.

Suppose we are able to develop a laser communication system of carrier frequency 4×10^{14} Hz and operating bandwidth of 10×10^9 Hz (10 GHz). Such a system, on a single laser beam, could carry about 2.5 million simultaneous conversations, or 2000 simultaneous TV programs. When the first telephone system was built, each wire carried only one conversation. Today each line carries hundreds of calls and each coaxial cable transmits up to 1000 calls an hour, but lines are still crowded. Fiber-optic cables, smaller in size than a finger, can be fitted with 144 fibers. Each pair of fibers can carry 672 simultaneous telephone messages on laser light pulses. For the 144-fiber cable, considerably smaller and less bulky than the copper-wire-carrying coaxial cables, this translates into 50,000 simultaneous telephone messages.

Laser-fiber-optic information-carrying systems use what is called *time-division multiplexing*. Such a system divides the voice signal into many small bits and fills the normal, short pauses in talking patterns with other conversations, increasing the carrying capacity. For a single telephone conversation, the laser-fiber-optic process works as follows. Sound waves are converted into electrical waves in the mouthpiece of the telephone. A sampler device then measures the amplitude of each electrical wave 8000 times per second and assigns to each a binary number, the same numbering system used in computer languages. In this way each sound is coded electronically. This is all accomplished at the rate of millions of bits per second. The binary-coded laser light blips are transmitted through the optical fibers and converted into sound waves by a reverse process at the opposite end of the telephone conversation. He-Ne and diode lasers (the gallium arsenide semiconductor family) are currently mated with fiber optics in today's optical communication systems (see Chapter

24, Fiber Optics). The low-loss and low-dispersion properties of silica-based single-mode optical fibers enable transmission of billions of bits of information per second over distances in excess of 100 miles. Long-wavelength light (1.3–1.5 μm), coupled with single-mode optical fibers, dominate optical communications applications, both on land and under water.

Current laser systems used in space communication transmit as high as 1 billion bits per second, a transmittal rate that could handle the entire information content of the *Encyclopaedia Brittanica* or the simultaneous program content of 14 color TV stations, all in about 1 s. The essential components that make up a typical atmospheric laser communications system are shown in Figure 23-6. Video, voice, and data information are processed by multiplexing electronics and fed to the laser. A modulated laser beam is then transmitted through the atmosphere and collected at the detection site. The photo-detector signal, with reverse multiplexing, is reconverted into the original video, voice, and data information.

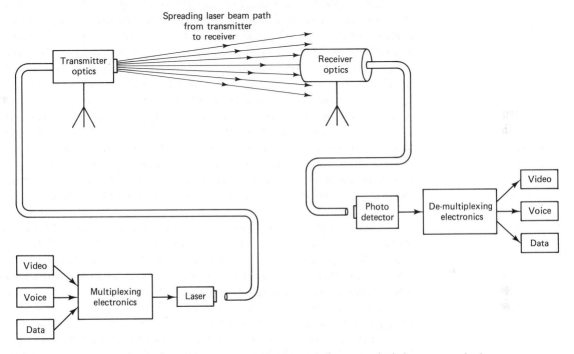

Figure 23-6 Functional arrangement of an atmospheric laser communication system. (Adapted with permission from *Lasers and Applications*.)

Information Processing.   Not only are lasers making a strong impact on communications technology, which deals with the transmission and reception of information—underwater, over land, and through space—but they are also driving new technologies in optical data storage, optical data readout or retrieval, and laser printing. Lasers are ideal in such applications because of their superior coherence and focusability. Holograms made with lasers, for example, provide highly efficient media for optical storage and were discussed in Chapter 13. The lasers most often used in information processing are the He-Ne, argon, and diode lasers.

Information storage in our society is an ongoing problem and challenge. For example, NASA alone receives 1000 trillion bits of data from satellites and space probes each year, data that must be processed, interpreted, and stored. With the seemingly exponential growth in data generated and collected, it is clear that fast, reliable systems for information processing represent one of our critical needs.

Optical data storage involves the use of lasers to store digitized information on selected surfaces by leaving a pattern of permanent or erasable imprints. The smallest unit into which digitized information can be divided, the binary digit or *bit*, is stored in binary language as a 1 or a 0. Eight bits make a *byte*, and a byte can be used to represent one alphabet character, such as the letter F, or L, or two decimal digits, such as the numbers 27 or 32. The byte 01000001 might represent A, 01000010 might represent B, and so on. Currently, high-density optical storage systems accommodate in the neighborhood of 4×10^9 bytes (4 Gbytes) on one side of a 14-in. disk. A storage capacity of 4 Gbytes translates into about 320,000 letter-sized documents! The average access time is of the order of 0.1 s to 0.5 s.

In optical data storage, a laser source is modulated appropriately by electronic input of the digital data described above. The laser writes bits of data on a photoreactive surface, usually in the shape of a 30-cm or 12-cm disk. Here it causes either a permanent or reversible change in the disk's optical behavior at precisely located, micron-size pits. The pits result from surface absorption of the laser energy with subsequent surface ablation. For playback or readout, a lower-power laser is directed onto the surface, and its light is reflected, or in some cases transmitted, according to the exact pattern of detail written on the surface. The reflected (or transmitted) laser beam is then detected by appropriately positioned photodiodes that translate the signals received back into the original electronic data pattern. Permanent recordings are currently being made for storage of archival materials, by the Library of Congress, for example. Popular compact discs (CDs) for recording or playback of audio or video information are also of this kind. Erasable recordings, while not yet available, are under development. It is only a matter of time before erasable laser recordings become part of modern data storage technology.

Laser printers, capable of producing some of the fastest, clearest printing available, are rapidly replacing conventional printers, both daisy-wheel and dot-matrix. Traditional printers create an image by forcing an ink ribbon onto paper mechanically, either with a set of small pins, as in dot-matrix printers, or with a fully formed character, as in the daisy-wheel. Dot-matrix printers produce either print or graphics but suffer from coarse resolution; daisy-wheel printers offer good text quality but cannot do graphics. Both, being mechanical printers, are noisy and slow by comparison, running at speeds of several pages a minute. By contrast, a laser printer creates images with a laser beam that scans rapidly across a photoconductive drum, selectively discharging certain areas on the drum (see Figure 23-7). Toner or ink, with charge of opposite polarity, adheres to the photoconductive surface and is then transferred to paper with both pressure and heat. In the process, a laser beam creates much smaller printed spots (coherence and focusability again) than do mechanical pins and therefore creates text of much higher quality. In addition, the laser printing process runs smoothly, quietly, and quickly. Printing rates may vary from several pages per minute to two pages per second, depending on the cost of the printer.

Remote Sensing. Recent successes of space missions, combined with advances in laser technology, have set the stage for laser remote sensing. Simply stated, remote sensing involves the detection of laser-scattered light from targets that are remotely located relative to active laser interrogating systems. The laser systems—sources and detectors—are frequently located on space platforms, whereas the remote targets, far below, may be the atmosphere, terrestrial vegetation, or the earth's crust. Ground-based systems, also prominent but not necessarily remote, are used to detect pollution and to monitor global winds.

The technique used, often referred to as *laser radar*, or LIDAR (**li**ght **d**etection **and r**anging), depends on the presence of a laser source, a suitable target for in-

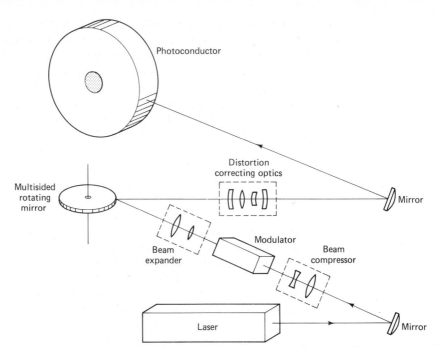

Figure 23-7 Schematic of a typical optical system for a laser printer. (Adapted with permission from *Lasers and Applications*.)

terrogation, and sensitive detectors. For any remote-sensing application, an ideal laser must have the appropriate wavelength, tunability, output power, pulse repetition rate, and amplitude and frequency stability. It should, in addition, have high operating efficiency, be mechanically rigid, and possess a long operating lifetime. Current laser sources include continuous wave carbon dioxide lasers, pulsed Nd:YAG and Nd-glass lasers, alexandrite lasers, Nd:YAG-pumped dye lasers, excimer lasers, AlGaAs-pumped Nd:YAG lasers, and an entirely new class of continuously tunable, solid-state lasers, such as the cobalt magnesium fluoride (CoMgF) lasers. Taken together, the family of special lasers provides wavelength coverage from 0.66 μm to 10.6 μm, sufficient for most remote-sensing applications.

The targets, whatever their nature, must interact with the interrogating laser light and produce a measurable change that can be detected. Such changes, returned to the detection system by the back-scattered laser radiation, include fluorescence, Raman-shifted wavelengths, and Doppler effects, each of which provides important information on specie identity, concentration, or movement. Highly specialized LIDAR systems, such as differential absorption LIDAR, Raman LIDAR, and Doppler LIDAR, are designed to detect the specific information content locked in the backscattered radiation.

The state of the art in remote sensing and the challenge ahead is immediately evident from a review of two current applications involving the measurement of wind speed and earth crust shifts. The detailed movement and speed of tropospheric winds is of considerable interest in the study of global weather patterns. To ferret out such information, a space-based CO_2 laser system would operate from an orbiting spacecraft, either an operational satellite or the space shuttle, at distances from 250 km to 800 km above the earth. From such altitudes, high initial laser power is needed, or the backscattered radiation is too weak to detect. The incident laser radiation is backscattered by dust particles carried along by the wind. Calculations of wind speed and direction are based on the Doppler frequency shifts in the reflected radiation. A CO_2 laser, to measure wind speeds with an accuracy of 1 m/s, requires 10-J pulses every second or so, with a frequency stability of several hundred kilohertz or better. Large power requirements of several kilowatts are needed to operate such laser systems. The challenge is clearly the availability of high power, stored and accessible during the space mission.

The use of a space-borne LIDAR system to detect movement of the earth's crust is of value in determining strains near earthquake zones, ground swell prior to volcanic activity, and large-area land settling. Based on time-of-flight measurements of very short (nanosecond) laser pulses from space to ground retroflectors and back, and making use of simple methods of triangulation, distances to specific parts of the earth's crust are determined to within 1 or 2 cm! By repeating the measurements at appropriate time intervals, significant earth movement is detected and monitored. Systems under consideration for earth crustal studies include a Nd:YAG laser with pulse widths less than 0.5 ns and pulse repetition rates of the order of 10 per second. Backscattered signal detection on board the spacecraft can be accomplished with existing 6-in. diameter collecting telescopes and photomultiplier electronics.

LIDAR techniques, coupled with remote-sensing laser systems, are useful in the detection of pollution, identification of toxic-particle concentrations, or global measurements of specific atmospheric gases and aerosols. Optical-sensing technology has been used successfully to monitor smokestack emissions from industrial plants or to detect gas leaks in pipelines. On a broader scale, atmospheric concentrations of ozone, water vapor, and sulfuric acid can be measured with notable precision. In particular, laser absorption spectrometers, relying on the principles of differential absorption and resonance fluorescence, have already demonstrated their usefulness. Global measurements to determine atmospheric aerosol concentrations are of much interest and are within the capability of remote-sensing systems. Such determinations illuminate the role that aerosols play in the warming or cooling of our planet as well as the formation and evolution of haze layers in the troposphere.

The recorded successes of such space missions as *Voyager*, the *Infrared Astronomical Satellite*, and several *Landsat* orbiters, coupled with advances in laser technology, indicate that spaceborne remote sensing of the earth's atmosphere and crust is a rapidly developing technology.

23-3 MORE RECENT DEVELOPMENTS[2]

During the 1990s lasers and laser applications continued to spawn strong activity in the marketplace. Laser technology moved ahead surely and steadily in well-established areas of material processing, metrology, bar-code scanning, printing, optical storage and memory, communications, diagnostic and therapeutic applications in medicine, national defense, and the entertainment/display industry.

In the early 1990s, the development of new laser materials and devices continued at a rapid pace, guaranteeing the user a wide choice of lasers operating in a continuum of wavelengths, from 0.3 μm to 1 μm, with available powers from milliwatts to tens of watts. Promising new lasers, with names such as *titanium-sapphire*, *cobalt-doped magnesium fluoride*, *alexandrite*, *Nd:YLF* (*neodymium yttrium lithium fluoride*), and *NYAB* (*neodymium yttrium aluminum borate*), joined the workhorse lasers of the 1980s: He-Ne, CO_2, Nd:YAG, argon, and gallium arsenide.

---

[2] The contents of this section are intended to present a flavor—certainly not a comprehensive coverage—of laser technology and applications in the early 1990s.

Diode lasers now compete with the old standby, the He-Ne laser. Threatening to replace the lamp-pumped lasers, diode-pumping offers an almost ideal spectral match between pump and absorber, higher operating efficiencies, better reliabilities, and less cooling requirements. In a slightly different arena, self-frequency-doubling crystals, such as *NYAB* and *MLNN* (*neodymium* and *magnesium-doped LiNbO₃*) are being used to produce directly, green light near 531 nm. In the world of optical communications, erbium-doped fiber optic amplifiers, pumped with diode lasers, are under intense development to replace costly repeaters in transoceanic, long-haul communications systems.

We shall survey in a bit more detail some of the ongoing progress in several typical fields: medicine, defense-related laser interests, and laser research in the microworld.

Medical Applications. In medicine, the laser remains extremely useful, both as a diagnostic and as a surgical tool. In operating rooms, for example, CO_2 and Nd:YAG lasers are demonstrating new successes in gall bladder operations (*laparoscopic laser cholecystectomy*). While gall bladder operations with lasers are on the increase, *laser angioplasty* (the use of a hot-tip laser to vaporize plaque in blocked arteries) is on the decrease. Instead, laser-assisted balloon angioplasty (inflating a small balloon to widen the occluded region in the artery) seems to be the currently accepted technique. Back surgery, too, is sometimes possible, using the selective heating of a laser beam to remove ruptured disk material between vertebrae of the spine.

In ophthalmology, excimer lasers (uv output) are still being considered as "scalpels" to correct nearsightedness. Two techniques are currently in use, both involving a reconfiguration of the corneal surface. One is called radial keratotomy (making carefully controlled grooves in the corneal surface); the other is corneal sculpting (judicious removal of corneal surface tissues). In medical diagnostics, research programs are under way to develop a lead-salt tunable diode laser technology for clinical testing. The goal of this research is to use stable, nonradioactive isotopes as tracers to help detect physiological dysfunctions. Instrumentation involving high-resolution infrared spectroscopy is being designed to perform isotopic analysis of exhaled human breath (a noninvasive procedure). This procedure promises to be an important future tool for clinical diagnosis of diseases such as diabetes.

Military-Related Applications. Laser-guidance techniques account for the pinpoint accuracy now possible in the delivery of weapons. In the hardware and systems of current military interest, lasers continue to be at the heart of improvements for range finders, target designators, satellite communication systems, laser radar-based navigational aids, and remote sensing. The laser systems that dominate developmental activity in these fields include the high-power CO_2 lasers, Nd:YAG lasers (diode-pumped and frequency doubled), erbium-doped YAG lasers, diode lasers, and, to a lesser extent, free-electron (FEL) and chemical (HF, DF) lasers. The wavelengths of interest include 532 nm, 693 nm, 830 nm, 1.06 μm, 1.54 μm, 3 μm, and 10.6 μm, with pulse energies ranging from millijoules to kilojoules.

Lasers and the Microworld. Extraordinary new laser devices are currently showing up at the frontiers of subatomic research. The unique properties of laser light enable biologists to examine single protein molecules attached to the inner membrane of a red blood cell. Physicists are able to levitate millions of sodium atoms in stainless-steel containers and release them simultaneously in fountains of light. Chemists can observe a single hydrogen atom pulling an oxygen atom away from a CO_2 molecule.

Femtosecond lasers (pulse lengths of 10^{-15} second) can freeze the motion of atoms and molecules, making it possible to take snapshots of chemical reactions. Beams of laser light, in so-called laser traps, are used to corral groups of atoms and to move them around at will. Techniques referred to as "optical molasses" use laser light to create electromagnetic forces that can slow down supersonic atoms. The resulting controlled atoms, observed in both free fall and oscillating motions, enable the measurement of gravitational forces with greatly increased accuracy and the development of new generations of atomic clocks.

Further, single laser beams, used as "optical tweezers," hold in place or nudge around microscopic particles, such as DNA molecules, enabling the study of their internal and external structures. Lasers are an essential part of an atomic force microscope (AFM), a remarkable "sensor" that offers a quick and accurate method for measuring the atomic details of a surface. The AFM glides over the microscopic surface, carefully sensing and tracing the outlines of individual atoms. Such microscopes offer motion control down to the angstrom level, laser spot displacement forces as low as 10^{-10} newtons, and scan sizes on the order of 100 μm^2. Understanding the atomic surface detail of insulating and conducting materials is essential, for example, in the fabrication of highly reliable fiber optics.

Other Applications.   In this chapter we have attempted to survey some of the more important applications of the laser in modern technology. The survey was organized around two general classes: lasers and information and laser interactions with materials. This afforded us the opportunity to review a number of laser applications in materials processing, medicine, laser-initiated fusion, communication, information processing, and remote sensing. As important as these are, they by no means exhaust the different ways in which the laser is being applied. Lasers also find direct applications in civil engineering and construction (the perfectly straight line that is not there); in agriculture as guidance systems for *laser planing* of the land; in the semiconductor industry, where lasers are the microtools needed for precision trimming and cutting in device-processing operations; and in entertainment, where lasers combine flashy colors and music in exotic, dynamic patterns.

PROBLEMS

23-1. Consider a pulsed infrared-type laser, with a peak power of 10 MW and beam divergence of 1 mrad, focused on a work area (target) by a lens of focal length 20 cm. What is the power density (irradiance) on target?

23-2. When a laser is focused down to a spot (beam waist), it has a certain "depth of focus" d along which the beam waist does not increase appreciably. This is given by the relation

$$d = \left(\frac{2\lambda}{\pi}\right) \sqrt{p^2 - 1} \left(\frac{2f}{D}\right)^2$$

Here f is the focal length of the focusing lens, D is the diameter of the incident laser beam at its own beam waist, λ is the wavelength, and p is a tolerance factor that sets the degree to which the focused beam is allowed to expand and still be considered "focused."

(a) Consider a CO_2, 10.6-μm laser beam, diverging from its own beam-waist diameter of 8 mm and impinging on a germanium lens of focal length 60 mm. Determine the depth of focus at the focal spot if the beam diameter is to increase no more than 10% from its minimum value at the waist ($p = 1.10$).

(b) The beam waist D' at the focal plane of the germanium lens is given by $D' = f\phi$, where $\phi = 1.27\lambda/D$ is the beam divergence of the incident laser beam. Calculate D' and the diameter at the end point of the depth-of-focus region d for the tolerance factor in (a).

23-3. Consider the cutting performance of the laser, illustrated graphically in Figure 23-3. Explain why the same laser penetrates less deeply in aluminum (2.7 g/cm^3) than it does in steel (7.8 g/cm^3) at a transverse cutting speed such as 25 mm/s. In formulating your answer, consider the importance of reflection and absorption coefficients, as well as thermal conductivity.

23-4. A pulsed Nd:YAG laser beam used in eye surgery emits a pulse of energy of about 1 mJ in 1 ns. The laser beam is focused to a tiny spot of 30-μm diameter in the interior of the eye.

(a) What is the irradiance at the focal spot?

(b) Assuming the eye to have an index of refraction of 1.33 and a permittivity of about 78 ϵ_0, estimate the electric field strength at the focal spot. (*Hint*: Use Eq. (8-42), appropriately modified.)

23-5. In the field of communication theory, it is well known that the information capacity C of a signal of average power S in the presence of additive white-noise power N in a channel of bandwidth B is given by $C = B \log (1 + S/N)$. (Note that channel capacity C is directly proportional to bandwidth B, justifying the great interest in lasers for communications purposes.) Suppose the signal-to-noise ratio, S/N, is 9 and the available bandwidth in a laser communication system is about 4000 MHz (only 0.001% of the carrier frequency, which is around 4×10^{14} Hz).

(a) What is the information capacity C?

(b) How many telephone conversations of bandwidth 4000 Hz could be carried by such a laser system?

23-6. A laser system, fixed on a geosynchronous satellite, uses a Nd:YAG laser that emits 1.06 μm radiation in a highly collimated beam of 5-μrad beam divergence.

(a) If the laser is 36,205 km above the earth, what is the minimum diameter of the laser-beam "footprint" on the surface of the earth?

(b) If the laser emits pulses of 200-MW power, what is the average electric field per pulse in the laser beam at the earth's surface? (*Hint*: See Eq. (8-42).)

23-7. Personnel who work with high-power lasers or with unshielded laser beams must take measures to protect themselves from laser damage to the skin and eyes. Towards that end, nominal hazard zones (NHZ) for given lasers can be defined. A NHZ outlines the conical space within which the level of direct, reflected, or scattered radiation—during normal operation—exceeds the assigned levels of maximum permissible exposure (MPE). Exposure levels outside the conical region are below the designated MPE level. The conical region, with tip emanating from the laser, extends to a range R_{NHZ} with sides spreading at the beam divergence angle ϕ. The range for a nonfocused laser beam is given by

$$R_{NHZ} = \frac{1}{\phi}\left[\left(\frac{4P}{\pi\,\text{MPE}}\right)^{1/2} - d\right]$$

where P is the laser power, ϕ is the beam divergence, MPE is the maximum permissible exposure in power/area, and d is the aperture diameter of the laser exit port.

(a) Draw a sketch of the NHZ (conical shape), labeling the extent of the cone, R_{NHZ}, and the full-angle beam divergence ϕ.

(b) Determine R_{NHZ} for a continuous wave Nd:YAG laser ($\lambda = 1.06$ μm) of 50 W power, 3.0 mrad beam divergence, MPE of 5.1×10^{-3} W/cm^2, a 10 s exposure, and an exit aperture diameter of 3.0 mm.

23-8. The range R_{NHZ} (refer to problem 23-7) for a laser that directs its exiting laser beam immediately onto a lens is given by

$$R_{NHZ} = \left(\frac{f}{b}\right)\left(\frac{4P}{\pi \, \text{MPE}}\right)^{1/2}$$

where R_{NHZ} defines the extent of the cone emanating from the point of beam focus, f is the focal length of the lens, b is the diameter of the beam exiting the laser, P is the power of the laser, and MPE is the maximum permissible exposure.

(a) Draw a sketch of the NHZ (conical space) for this focused laser beam.

(b) Determine R_{NHZ} for a continuous wave Nd:YAG laser of 50 W power, exit beam diameter of 5.0 mm, and capped by a lens of focal length 7.5 cm, if the MPE of this laser for an 8-hr exposure is 1.6×10^{-3} W/cm^2.

REFERENCES

[1] Berns, Michael W. "Laser Surgery." *Scientific American* (June 1991): 84.

[2] Goldman, Leon. *Applications of the Laser*. Malabar, Fla.: Robert E. Krieger Publishing Company, 1982.

[3] Ready, J. F. *Industrial Applications of Lasers*. New York: Academic Press, 1978.

[4] Sliney, D. H., and H. L. Wolbarsht. *Safety with Lasers and Other Optical Sources: A Comprehensive Handbook*. New York: Plenum Press, 1980.

[5] *Lasers and Optronics*. Morris Plains, N.J.: Elsevier Communications. Trade journal, published monthly.

[6] *Optics and Photonics*. Washington, D.C.: Optical Society of America. Published monthly.

[7] *Photonics Spectra*. Pittsfield, Mass.: Laurin Publishing Co. Trade journal, published monthly.

[8] Vali, Victor. "Measuring Earth Strains by Laser." *Scientific American* (Dec. 1969): 88.

[9] Berns, Michael H., and Donald E. Rounds. "Cell Surgery by Laser." *Scientific American* (Feb. 1970): 98.

[10] Faller, James E., and E. Joseph Wampler. "The Lunar Laser Reflector." *Scientific American* (Mar. 1970): 38.

[11] Lubin, Moshe J., and Arthur P. Fraas. "Fusion by Lasers." *Scientific American* (June 1971): 21.

[12] Feld, M. S., and V. S. Letokhov. "Laser Spectroscopy." *Scientific American* (Dec. 1973): 69.

[13] Emmett, John L., John Nuckolls, and Lowell Wood. "Fusion Power by Laser Implosion." *Scientific American* (June 1974): 24.

[14] Zare, Richard N. "Laser Separation of Isotopes." *Scientific American* (Feb. 1977): 86.

[15] Ronn, Avigdor M. "Laser Chemistry." *Scientific American* (May 1979): 114.

[16] Tsipis, Kosta. "Laser Weapons." *Scientific American* (Dec. 1981): 51.

[17] LaRocca, Aldo V. "Laser Applications in Manufacturing." *Scientific American* (Mar. 1982): 94.

[18] Anderson, Dana Z. "Optical Gyroscopes." *Scientific American* (Apr. 1986): 94.

[19] Jewell, Jack L., James P. Harbison, and Axel Scherer. "Microlasers." *Scientific American* (Nov. 1991): 86.

24

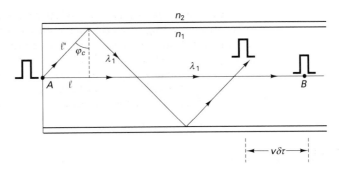

Fiber Optics

INTRODUCTION

The channeling of light through a transparent conduit has taken on great importance in recent times, as we have seen in our review of laser applications. This is especially true because of its applications in communications and laser medicine. As long as a transparent solid cylinder, such as a glass fiber, has a refractive index greater than that of its surrounding medium, much of the light launched into one end will emerge from the other end, even though somewhat attenuated, due to a large number of total internal reflections. A comprehensive treatment of fiber optics requires a wave approach in which Maxwell's equations are solved in a dielectric medium, subject to the boundary conditions at the fiber walls. In this chapter we adopt a simpler and more intuitive approach, describing the propagating wavefronts by their rays, although we appeal in some contexts to wave properties such as phase and interference.

24-1 APPLICATIONS

The simplest use of optical fibers, either singly or in bundles, is as *light pipes*. For example, a flexible bundle of fibers might be used to transport light from inside a vacuum system to the outside, where it can be more easily measured.[1] The bundle

---

[1] Interestingly, the rods and cones of the human eye have been shown to function as light pipes [1], transmitting light along their lengths, as in optical fibers.

might be divided into two or more sections at some point to act as a beam splitter. For such nonimaging applications, the fibers can be randomly distributed within the cable. When imaging is required, however, the fiber ends at the input are coordinated with the fiber ends at the output. To maintain this coordination, fibers at both ends are bonded together. The fiberscope consists of a bundle of such fibers, end-equipped with objective lens and eyepiece. It is routinely used by physicians to examine regions of the stomach, lungs, and duodenum. Some of the fibers function as light pipes, transporting light from an external source to illuminate inaccessible areas internally. Other fibers return the image.

Fibers can be bound rigidly by fusing their outer coating or *cladding*. In this way, fiber-optic faceplates are made for use as windows in cathode ray tubes. Further, when such fused-fiber bundles are tapered by heating and stretching, images can be magnified or diminished in size, depending on the relative areas of input or output faces.

The resolving power of imaging fibers depends on the accuracy of fiber alignment and, as might be expected, on the individual fiber diameter d. A conservative estimate [1] of fiber resolving power RP is given by

$$\text{RP (lines/mm)} = \frac{500}{d(\mu m)} \tag{24-1}$$

Thus a 5-μm fiber, for example, can produce a high resolution of about 100 lines/mm.

The most far-reaching applications of fiber optics lie in the area of communications, the subject of the following section.

24-2 COMMUNICATIONS SYSTEM OVERVIEW

No application has given greater impetus to the rapid development of fiber optics than has voice or video communication and data transmission. The advantages of fiber optic conduits or *waveguides* over conventional two-wire, coaxial cable or microwave waveguide systems are impressive. The replacement of microwaves and radio waves by light waves is especially attractive, since the information-carrying capacity of the carrier wave increases directly with the width of the frequency band available. Replacement of copper coaxial cable by fiber-optic cable thus offers greater communications capacity in a lighter cable that requires less space. Additionally, in contrast with metallic conduction techniques, communication by light offers the possibility of electrical isolation, immunity to electromagnetic interference, and freedom from signal leakage. The latter is especially important where security of information is vital, as in computer networks that handle confidential data.

In Figure 24-1, we give an overview of the essential components and processes involved in a fiber-optic communications system, from message source to message output. At the input end of the fiber-optic cable, the information to be transmitted is converted by some type of transducer from an electrical signal to an optical one. After transmission by the optic fiber, it is reconverted from an optical to an electrical signal. The fiber serves as an *optical waveguide* to propagate the information with as little distortion and loss of power as possible, over a distance that can range from meters to thousands of kilometers.

The message source might be audio, providing an analog electrical signal from a microphone; it might be visual, providing an analog signal from a video camera; or it might be digitally encoded information, like computer data in the form of a train of pulses. Analog and digital formats are convertible into one another, so that

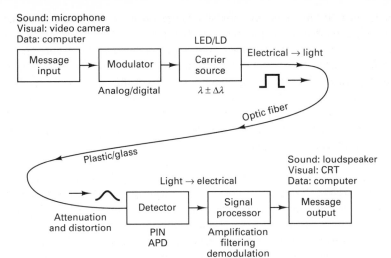

Sound: microphone
Visual: video camera
Data: computer

Message input → Modulator → Carrier source

LED/LD

Electrical → light

Analog/digital

$\lambda \pm \Delta\lambda$

Optic fiber

Plastic/glass

Attenuation and distortion

Light → electrical

Detector → Signal processor → Message output

PIN APD

Amplification filtering demodulation

Sound: loudspeaker
Visual: CRT
Data: computer

Figure 24-1 Overview of a fiber-optics communication system.

the choice of format for transmission through the fiber is always available, regardless of the original nature of the signal.

The purpose of the *modulator* is to perform this conversion when desired and to impress this signal onto the carrier wave generated by the *carrier source*. The carrier wave can be modulated to contain the signal information in various ways, usually by amplitude modulation (AM), frequency modulation (FM), or digital modulation.[2] See Figure 24-2. In fiber-optics systems, the carrier source is either a light-emitting diode (LED) or a laser diode (LD). (See Section 2-5.) The ideal carrier wave is a single-frequency wave of adequate power to propagate long distances through the fiber. In fact, no light source can produce radiation of a single frequency. The LD and LED are characterized by a frequency band with central wavelength λ and spectral width $\Delta\lambda$. The LD is preferable because it approaches the ideal

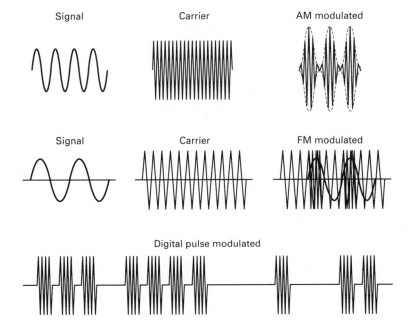

Signal   Carrier   AM modulated

Signal   Carrier   FM modulated

Digital pulse modulated

Figure 24-2 Three forms of modulation in which a carrier wave is modified to carry a signal. Top: Amplitude modulation; center: frequency modulation; bottom: digital modulation in which a pulse is either present ("on") or missing ("off").

---

[2] Some of the techniques of light modulation are discussed in Chapter 26.

single frequency more closely than the LED; however, other considerations make the LED acceptable in some systems, as we shall see.

In Figure 24-1, the carrier source output into the optic fiber is represented by a single, square pulse. As this pulse propagates through the fiber, it suffers both *attenuation* (loss of amplitude) and *distortion* (change in shape) due to several mechanisms to be discussed. The fiber may be, typically, a glass or plastic filament 50 μm in diameter. If the fiber is very long, it may be necessary to intercept the signal at one or more intermediate points with a *repeater* that amplifies and restores the original pulse shape.[3] At the remote end of the fiber, the light signal is coupled into a detector that changes the optical signal back into an electrical signal. This service is performed by a semiconductor device, most commonly a *PIN diode*, an *avalanche diode*, or a *photomultiplier* (Section 2-6). Of course, the response of the detector should be well-matched to the optical frequency of the signal received. The detector output is then handled by a signal processor, whose function is to recapture the original electrical signal from the carrier, a process that involves filtering and amplification and, possibly, a digital-to-analog conversion. The message output may then be communicated by loudspeaker (audio), by cathode-ray tube (video), or by computer input (digital).

24-3 BANDWIDTH AND DATA RATE

The more complicated the signal to be communicated, the greater is the range of frequencies required to represent it. The output of a stereo system is more faithful to the original signal than is the output of a telephone receiver because a greater frequency range is devoted to the process of reproduction. The range of frequencies required to modulate a carrier for a single telephone channel is only 4 kHz, whereas the bandwidth of an FM radio broadcasting station is 200 kHz. A commercial TV broadcasting station, which must communicate both sound and video signals, uses a bandwidth of 6 MHz. The great information-carrying potential of a light beam becomes evident when we calculate the ratio of carrier frequency to bandwidth, a measure of the number of separate channels that can be impressed on the carrier. For a TV station using a 300-MHz carrier, this ratio is 300 MHz/6 Mhz, or 50; for an optical fiber using a carrier of 1-μm wavelength (3×10^8 MHz) to carry the same information, the ratio is (3×10^8 MHz)/6 MHz or 50,000,000!

More information can be sent by optical fiber when distinct pulses can be transmitted in more rapid succession. This implies higher frequencies or, in the case of digital information, higher *bit rates*. In the latter case, suppose that 8 bits (*on* or *off* pulses) are required to represent the amplitude of an analog signal. According to the *sampling theorem*,[4] an analog signal must be sampled at a rate at least twice as high as its highest-frequency component to be faithfully represented. In the case of the TV channel with a bandwidth of 6 MHz, this means that 2×6 MHz or 12×10^6 samples must be taken each second. Since each sample is described using 8 bits, the required data rate is 96 Mbps (megabits per second). Data rates are limited by the modulator capabilities as well as by fiber distortions that prevent distinct identification of neighboring pulses, as we shall see in discussions to follow.

---

[3] Theoretical studies and applications are already well under way that use lightly erbium-doped fibers, pumped from the transmitter end so that the fiber itself serves as a distributed amplifier. As such, the fiber is able to overcome its losses and reduce or eliminate the need for repeater stations. See [11] and [18].

[4] The sampling theorem is also discussed in Section 25-2, in connection with Fourier transform spectroscopy.

We consider now the manner in which light propagates through an optical fiber. The conditions for successful propagation are developed here mainly from the point of view of geometrical optics. In addition, we consider only the meridional rays, which intersect with the fiber's central axis.[5]

Consider a short section of a straight fiber, pictured in Figure 24-3a. The fiber itself has refractive index n_1, the encasing medium (called *cladding*) has index n_2, and the end faces are exposed to a medium of index n_0. Ray A entering the left face of the fiber is refracted there and transmitted to point C on the fiber surface where it is partially refracted out of the fiber and partially reflected internally. The internal ray continues, diminished in amplitude, to D, then to E, and so on. After multiple reflections the ray will have lost a large part of its energy. Ray A does not meet the conditions for total internal reflection, that is, it strikes the fiber surface at points C, D, E, . . . such that its angle of incidence φ is less than the critical angle φ_c, or

$$\varphi < \varphi_c = \sin^{-1}(n_2/n_1) \tag{24-2}$$

Ray B, on the other hand, which enters at a smaller angle θ_m with respect to the axis, strikes the fiber surface at F in such a way that it is refracted parallel to the fiber surface. Other rays, as in Figure 24-3b, incident at angles $\theta < \theta_m$, experience total internal reflection at the fiber surface. Such rays are propagated along the fiber by a

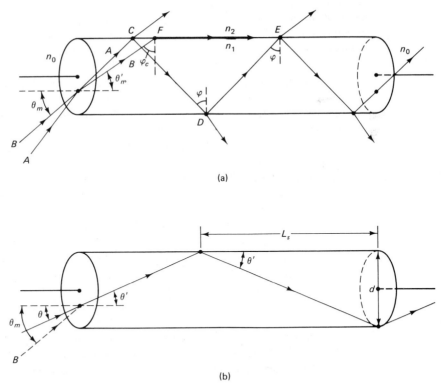

(a)

(b)

Figure 24-3 (a) Propagation of light rays through an optical fiber. Ray B defines the maximum input cone of rays satisfying total internal reflection at the walls of the fiber. (b) Propagation of a typical light ray through an optical fiber.

---

[5] Other rays, the *skew* rays, do not lie in a plane containing the central fiber axis. These rays take a piecewise spiral path through the fiber.

succession of such reflections, without loss of energy due to refraction out of the cylinder. However, depending upon the degree of transparency of the fiber material to the light, some attenuation occurs by absorption.

Ray B thus represents an extreme ray, defining the slant face of a cone of rays, all of which satisfy the condition for total internal reflection within the fiber. The maximum half-angle θ_m of this cone is evidently related to the critical angle of reflection φ_c. At the input face,

$$n_0 \sin \theta_m = n_1 \sin \theta_m' \tag{24-3}$$

and at a point like F,

$$\sin \varphi_c = \frac{n_2}{n_1}$$

Using the geometrical fact, $\theta_m' = 90° - \varphi_c$, and the trigonometric identity, $\sin^2 \varphi_c + \cos^2 \varphi_c = 1$, these relations combine to give the *numerical aperture*,

$$\text{N. A.} \equiv n_0 \sin \theta_m = n_1 \cos \varphi_c = \sqrt{n_1^2 - n_2^2} \tag{24-4}$$

If $n_0 = 1$, the numerical aperture is simply the sine of the half-angle of the largest cone of meridional rays (i.e., rays coplanar with the fiber axis) that are propagated through the fiber by a series of total internal reflections. The numerical aperture clearly cannot be greater than unity, unless $n_0 > 1$. A numerical aperture of 0.6, for example, corresponds to an acceptance cone of 74°. The light-gathering ability of an optical fiber increases with its numerical aperture.

Also from Figure 24-3b, the *skip distance* L_s between two successive reflections of a ray of light propagating in the fiber is given by

$$L_s = d \cot \theta' \tag{24-5}$$

where d is the fiber diameter. Relating θ' to the entrance angle θ by Snell's law,

$$L_s = d \sqrt{\left(\frac{n_1}{n_0 \sin \theta}\right)^2 - 1} \tag{24-6}$$

For example, in the case $n_0 = 1$, $n_1 = 1.60$, $\theta = 30°$, and $d = 50$ μm, Eq. (24-6) gives $L_s = 152$ μm. Thus in one meter of fiber, there are approximately $1/L_s$ or 6580 reflections. Table 24-1 lists various core and cladding possibilities, for which the critical angle, numerical aperture, and skip distances have been calculated. With so many reflections occurring, the condition for total internal reflection must be accurately met over the entire length of the fiber. Surface scratches or irregularities, as well as surface dust, moisture, or grease, become sources of loss that rapidly diminish light energy. If only 0.1% of the light is lost at each reflection, over a length of 1 m, this attenuation would reduce the energy by a factor of about 720. Therefore, to protect the optical quality of the fiber, it is essential that it be coated with a layer of plastic or glass called the *cladding*. Cladding material need not be highly trans-

TABLE 24-1 CHARACTERIZATION OF SEVERAL OPTICAL FIBERS

Core/cladding	n_0	n_1	n_2	φ_c	θ_{max}	N. A.	$1/L_s$
Glass/air	1	1.50	1.0	41.8°	90.0°	1	8944
Plastic/plastic	1	1.49	1.39	68.9°	32.5°	0.54	3866
Glass/plastic	1	1.46	1.40	73.5°	24.5°	0.41	2962
Glass/glass	1	1.48	1.46	80.6°	14.0°	0.24	1657

NOTE: The reciprocal of the skip distance ($1/L_s$ or skips per meter) is calculated for a fiber of diameter 100 μm and at $\theta = \theta_{max}$.

parent, but must be compatible with the fiber core in terms of expansion coefficients, for example. The index of refraction n_2 of the cladding, where $n_2 < n_1$, influences the critical angle and numerical aperture of the fiber.

The cladding around the fiber cores has another important function, which is to prevent what is called *frustrated total internal reflection* from occurring. When the process of total internal reflection is treated as the interaction of a wave disturbance with the electron oscillators comprising the medium, it becomes apparent that there is some short-range penetration of the wave beyond the boundary. Although the wave amplitude decreases rapidly beyond the boundary, a second medium introduced into this region can couple into the wave and provide a means of carrying away energy that otherwise returns into the first medium. Thus if bare optic fibers are packed closely together in a bundle, there is some leakage between fibers, a phenomenon called *cross talk* in communications applications. The presence of cladding of sufficient thickness prevents leakage, or, to put it more obliquely, negates the frustration of total internal reflection.[6]

The optic fiber cores discussed above are assumed to be homogeneous in composition, characterized by a single index of refraction n_1. Light is propagated through them by multiple total internal reflections. Such fibers are called *step-index fibers* because the refractive index changes discontinuously between core and cladding. They are *multimode* fibers if they permit a discrete number of modes (or ray directions) to propagate. When the fiber is thin enough so that only one mode (a ray in the axial direction) satisfies this condition, the fiber is said to be *single-mode*. Restrictions on possible modes will be described later. Another type of fiber, the *graded-index fiber*, is produced with an index of refraction that decreases continuously from the core axis as a function of radius. All these types are discussed in the sections that follow.

24-5 ALLOWED MODES

Not every ray that enters an optical fiber within its acceptance cone can propagate successfully through the fiber. Only certain ray directions or *modes* are allowed. To see why, we consider the simpler case of a symmetric *planar* or *slab wave guide*, shown in Figure 24-4. The waveguide core of index n_1 has a rectangular (rather than cylindrical) shape and is bounded symmetrically above and below by cladding of index n_2. A sample ray is shown undergoing two total internal reflections from the core-cladding interface at points A and B. Recalling that the ray represents plane waves moving up and down in the waveguide, it is evident that such waves overlap and interfere with one another. Only those waves are sustained that satisfy a *reso-*

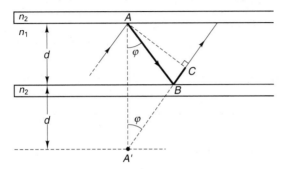

Figure 24-4 Section of a slab waveguide showing a successfully propagating ray or one of the possible modes. The geometry is used to determine the condition for constructive interference.

---

[6] This topic was treated quantitatively in Section 20-5.

nance condition. Notice that points A and C lie on a common wavefront of such waves. If the net phase change that develops between points A and C is some multiple of 2π, then the interfering wavefronts experience constructive interference and corresponding ray directions are allowed. The net phase change is made up of two parts, the optical path difference Δ and the phase change $2\phi_r$ that occurs due to the two total reflections at points A and B. Thus the self-sustaining waves must satisfy the condition

$$\frac{\Delta}{\lambda}2\pi + 2\phi_r = 2m\pi$$

where m is an integer. The geometrical extensions denoted by the dashed lines in Figure 24-4, identifying triangle ACA', make it evident that

$$\Delta = AB + BC = A'B + BC = 2n_1 d \cos \varphi \qquad (24\text{-}7)$$

so that the possible modes are given by

$$m = \frac{2n_1 d \cos \varphi}{\lambda} + \frac{\phi_r}{\pi}$$

Now, since $\phi_r < \pi$, the second term is one at most and is negligible compared with the first term. Thus each successful mode of propagation in the waveguide has an integer mode number m, related to a direction φ_m and given by

$$m \cong \frac{2n_1 d \cos \varphi_m}{\lambda} \qquad (24\text{-}8)$$

For our present purposes, the precise number of allowable modes is not as important as the qualitative dependence of the mode order m on the fiber characteristics. Notice that low-order modes—m small—correspond to $\varphi \cong 90°$, or ray directions that are nearly axial, and high-order modes—m large—correspond to rays that propagate with φ near φ_c, or at steeper ray angles. The total number of propagating modes is the value of m when $\cos \varphi_m$ has its maximum value. This occurs at the critical angle, $\varphi_m = \varphi_c$. Since from Eq. (24-4), $n_1 \cos \varphi_c = \sqrt{n_1^2 - n_2^2} = $ N. A., we can write

$$m = \frac{2d}{\lambda} \text{ N. A.} + 1 = \frac{2d}{\lambda}\sqrt{n_1^2 - n_2^2} + 1 \qquad (24\text{-}9)$$

We have added 1 to the total number of modes to account for the "straight through" mode ($m = 0$) at $\varphi = 90°$. Finally, we should point out that, because two independent polarizations are possible for the propagating plane wave,[7] the total number of modes is twice that given by Eq. (24-9) above.

This analysis for the slab waveguide has served to elucidate the physical reasons for mode restriction. The analysis giving the possible modes in a cylindrical fiber is based on the same physical principles, but is more complicated and is not developed here. It is shown [2, 3] that, in this case,

$$m_{\text{max}} = \frac{1}{2}\left(\frac{\pi d}{\lambda} \text{ N. A.}\right)^2 \qquad (24\text{-}10)$$

Notice that, as for the slab waveguide, the number of possible modes increases with the ratio d/λ. Thus larger diameter fibers are *multimode fibers*. If d/λ is small enough to make $m < 2$, the fiber allows only the axial mode to propagate. This is

---

[7] The two orthogonal polarization directions are described in Section 15-2.

the *monomode* (or *single mode*) *fiber*. The required diameter for single-mode performance is found by imposing the condition $m_{max} < 2$ on Eq. (24-10), giving

$$\frac{d}{\lambda} < \frac{2}{\pi (\text{N. A.})}$$

A more careful analysis [2] indicates that single-mode performance results even when

$$\frac{d}{\lambda} < \frac{2.4}{\pi (\text{N. A.})} \qquad (24\text{-}11)$$

Example

Suppose an optical fiber (core index of 1.465, cladding index of 1.460) is being used at $\lambda = 1.25 \ \mu$m. Determine the diameter for monomode performance and the number of propagating modes when $d = 50 \ \mu$m.

Solution   The N. A. is then $\sqrt{(1.465^2 - 1.46^2)} = 0.121$, and the required diameter for monomode performance is, using Eq. (24-11)

$$d < \frac{2.4}{\pi (0.121)} \ (1.25 \ \mu\text{m}) \quad \text{or} \quad d < 7.9 \ \mu\text{m}$$

On the other hand, if $d = 50 \ \mu$m, the fiber is multimode with

$$m_{max} = \frac{1}{2} \left[\pi \frac{50}{1.25}(0.121) \right]^2 = 115$$

giving the number of propagating modes according to Eq. (24.10).

24-6 ATTENUATION

The intensity of light propagating through a fiber invariably attenuates due to a variety of mechanisms that can be classified as *extrinsic* and *intrinsic* losses. Among the extrinsic losses are inhomogeneities and geometric effects. Inhomogeneities whose dimensions are much greater than the optical wavelength can result, for example, from inadequate mixing of the fiber material before solidification and from an imperfect interface between core and cladding. Irregularities of a geometric nature include sharp bends in the fiber as well as microbends, both of which cause radiation loss because the condition for total internal reflection is no longer satisfied (see Figure 24-5). Other extrinsic losses occur as light is coupled into and out of the fiber.

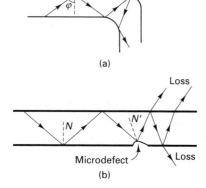

Figure 24-5   Radiation loss from an optical fiber because of (a) a sharp bend and (b) micro-defects at the fiber surface. Loss occurs where the condition for total internal reflection fails. Notice that in (b) the defect is also responsible for *mode coupling*, in this case a conversion from a lower to a higher mode.

At the fiber input there are losses due to the restrictions of numerical aperture, as well as losses due to inevitable reflections at the interface, the so-called *Fresnel losses*. The radiation pattern and size of the light source may also be ill-adapted to the fiber end, reducing input efficiency. Of course such losses also occur at the output end, where the light from the fiber is fed to a detector. Still other losses become important over longer lines wherever connectors, couplers, or splices are necessary. Losses can include mismatch of coupled fiber ends, involving core diameter and lateral and angular alignment. Separation and numerical aperture incompatibility are also possible and can lead to large losses when not properly corrected.

Intrinsic losses are due to absorption, both by the core material and by residual impurities, and by *Rayleigh scattering* from microscopic inhomogeneities, dimensionally smaller than the optical wavelength. The core material—silica, in the case of glass fibers—absorbs in the region of its electronic and molecular transition bands. See Figure 24-6. Strong absorption in the ultraviolet occurs due to electronic and molecular bands. Absorption in the infrared is due to molecular vibrational bands. Both uv- and ir-absorption decrease as wavelengths approach the visible region. Figure 24-6 shows a minimum of absorption at around 1.3 μm. Residual impurities, such as the transitional metal ions (Fe, Cu, Co, Ni, Mn, Cr, V) and, in particular, the hydroxyl (OH) ion, also contribute to absorption, the last producing significant absorption at 0.95, 1.23, and 1.73 μm. Rayleigh scattering, with its characteristic $1/\lambda^4$ dependence,[8] occurs from localized variations in the density or refractive index of the core material. For example, an optical fiber transmitting at

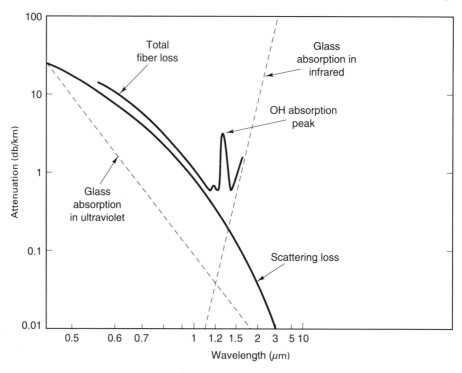

Figure 24-6 Contributions to the net attenuation of a germanium-doped silica glass fiber. [From H. Osanai, T. Shioda, T. Moriyama, S. Araki, M. Horiguchi, T. Izawa, and H. Takara, "Effects of Dopants on Transmission Loss of Low-OH-Content Optical Fibers," *Electronics Letters* 12, No. 21 (October 14, 1976): 550. Adapted with permission.]

---

[8] Rayleigh scattering was discussed in greater detail in Section 15-3.

1.3 μm rather than, say, 800 nm, represents a seven-fold reduction in Rayleigh scattering losses.

Absorption losses over a length L of fiber can be described by the usual exponential law for light irradiance I,

$$I = I_0 e^{-\alpha L} \qquad (24\text{-}12)$$

where α is an *attenuation* or *absorption coefficient* for the fiber, a function of wavelength.[9] For optical fibers, the defining equation for the absorption coefficient in *decibels* (db) is given by

$$\alpha_{\text{db}} \equiv 10 \log_{10} \left(\frac{P_1}{P_2} \right) \qquad (24\text{-}13)$$

where P_1 and P_2 refer to power levels of the light at cross sections (1) and (2), illustrated in Figure 24-7. The distance z is usually chosen to be a standard 1 km. For example, if a particular fiber experiences a loss given by $\alpha_{\text{db}} = 5$ db/km, it means that only 32% of the light energy launched into a 1-km long fiber arrives at the other end. (Negative values of α_{db} indicate amplification, rather than attenuation!) Dra-

(a)

(b)

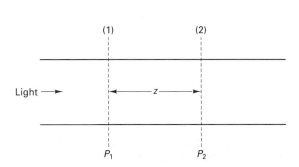

Figure 24-7 Schematic used to define the absorption coefficient for a glass fiber.

Figure 24-8   (a) Spectral attenuation for all-glass multimode fibers. (Courtesy Corning Glass Works.) (b) Spectral attenuation for all-plastic fiber cable. (Courtesy Mitsubishi Rayon America, Inc.)

---

[9] Since rays that strike the fiber wall at smaller angles of incidence travel a greater distance through the same axial length L of the absorbing medium, α is also a function of the angle of incidence.

matic advances have been made in reducing the absorption of fused silica so that today, fibers rated at 0.2 db/km (operating at 1.55 μm) are readily available. Plastic fibers are less expensive but not nearly as transparent. Their overall attenuation is at least an order of magnitude higher than for glass. Glass fibers are therefore preferable in long-distance applications. Figure 24-8 compares spectral absorption in silica and plastic fibers.

24-7 DISTORTION

Light transmitted by a fiber may not only lose power by the mechanisms just mentioned; it may also lose information through pulse broadening. When input light is modulated to convey information, the signal waveform becomes distorted due to several mechanisms to be discussed. The major causes of distortion include *modal distortion, material dispersion*, and *waveguide dispersion*, in order of decreasing severity.

Modal Distortion. Figure 24-9 indicates schematically the input of a square wave (a digital signal) into a fiber. The output pulse at the other end suffers, in general, from both attenuation and distortion. Modal distortion occurs because propagating rays (fiber modes) travel different distances in arriving at the output. Consequently, these rays arrive at different times, broadening the square wave, as shown. The shortest distance L from A to B is taken by the axial ray; the longest distance L' from A to B is taken by the steepest propagating ray that reflects at the critical angle φ_c. The distances L and L' are related, as suggested by the geometry in Figure 24-9, by

$$\sin \varphi_c = \frac{n_2}{n_1} = \frac{\ell}{\ell'} = \frac{L}{L'}$$

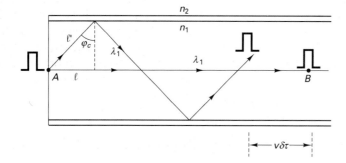

Figure 24-9 Symbolic representation of modal distortion. A square wave input arrives at the fiber end at different times, depending on the path taken. Shown are the extreme paths: an axial ray and one propagating at the critical angle. Their separation after any time interval $\Delta\tau$ is given by $v\,\Delta\tau$.

Thus the time interval $\delta\tau$ between the two rays is given by

$$\delta\tau = \tau_{\text{max}} - \tau_{\text{min}} = \frac{L'}{v} - \frac{L}{v} = \frac{L}{v}\left(\frac{n_1}{n_2} - 1\right)$$

where v is the speed of light in the fiber core. Since $v = c/n$, this result is conveniently expressed as a temporal pulse spread per unit length, in the form

modal distortion: $\delta\left(\dfrac{\tau}{L}\right) = \dfrac{n_1}{c}\left(\dfrac{n_1 - n_2}{n_2}\right)$ (24-14)
(step-index fiber)

Example

Suppose the fiber has a core index of 1.46 and a cladding index of 1.45. Determine the modal distortion for this fiber.

Solution   Using Eq. (24-14),

$$\delta\left(\frac{\tau}{L}\right) = \frac{1.46}{3 \times 10^{-4} \text{ km/ns}} \left(\frac{1.46 - 1.45}{1.45}\right) = 34 \text{ ns/km}$$

The pulse broadens by 34 ns in each km of fiber.[10]

Clearly, this broadening effect limits the possible frequency of distinct pulses. Modal distortion can be lessened by reducing the number of propagating modes. Consequently, the best solution is to use a monomode fiber, with only one propagating mode. The next best solution is to use a *graded index* (GRIN) fiber, which is described next.

The Graded Index (GRIN) Fiber.   A GRIN fiber is produced with a refractive index that decreases gradually from the core axis as a function of radius. Figure 24-10 shows the GRIN fiber profile, together with the profile of the ordinary *step-index fiber*, for comparison. In the GRIN fiber, a process of continuous refraction bends rays of light, as shown. Notice that at every point of the path, Snell's law is obeyed on a microscopic scale. Ray containment now occurs by a process of continuous refraction, rather than by total reflection. Refraction may not suffice to contain rays making steeper angles wth the axis, so GRIN fibers are also characterized by an acceptance cone. When the index profile is suitably adjusted, the rays shown in Figure 24-10c form isochronous loops, an aspect of graded-index fiber that is responsible for reducing modal distortion. Like ordinary fibers, GRIN fibers are also cladded for protection.

The variation of refractive index with fiber radius is given [4], in general, by

$$n(r) = n_1 \sqrt{1 - 2\left(\frac{r}{a}\right)^\alpha \Delta}, \qquad 0 \le r \le a \qquad (24\text{-}15)$$

where $\Delta \cong (n_1 - n_2)/n_1$ and $n_1 = [n(r)]_{max}$. The parameter α is chosen to minimize modal distortion. For $\alpha = 1$, the profile has a triangular shape; for $\alpha = 2$, it is parabolic; for higher values of α, the profile gradually approaches its limiting case, the step-index profile, as $\alpha \to \infty$. Minimizing $\delta\tau$ for all modes requires a value of $\alpha = 2$. Thus the parabolic profile shown in Figure 24-10 is optimum. It can be shown [5] that for this case, pulse broadening is given approximately by

$$\begin{array}{c} \text{modal distortion:} \\ \text{(GRIN fiber, } \alpha = 2\text{)} \end{array} \quad \delta\left(\frac{\tau}{L}\right) = \frac{n_1}{2c}\Delta^2 \qquad (24\text{-}16)$$

Comparing with modal distortion in the step-index fiber, Eq. (24-14), we can write

$$\delta\left(\frac{\tau}{L}\right)_{GRIN} = \frac{\Delta}{2}\left(\frac{n_1}{c}\Delta\right) = \frac{\Delta}{2}\delta\left(\frac{\tau}{L}\right)_{SI}$$

The factor $\Delta/2$ thus represents the improvement offered by a GRIN fiber. For the example used previously, where $n_1 = 1.46$ and $n_2 = 1.45$, we have $\Delta/2 = 1/292$. The GRIN fiber reduces the pulse broadening effect of modal distortion in this case by a factor of 292.

---

[10] Actual values are somewhat better than predicted by Eq. (24-14) due to *mode coupling* or *mixing* (rays may switch modes in transit due to scattering mechanisms that, on the average, shift power from higher and lower modes to intermediate ones) and due to preferential attenuation (higher modes taking longer paths suffer greater attenuation and so contribute less to overall pulse spreading). For longer distances, this leads to a modified dependence of the form, $\delta\tau \propto \sqrt{L}$.

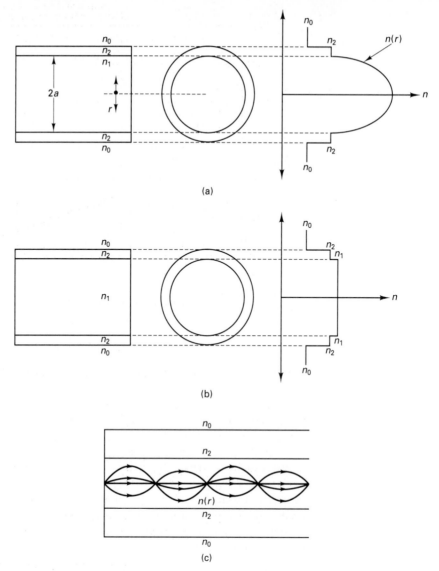

Figure 24-10 (a) Profile of a graded-index (GRIN) fiber, showing a parabolic variation of the refractive index within the core. (b) Profile of a step-index fiber, in which the core index is constant and slightly greater than that of the cladding. (c) Several ray paths within a GRIN fiber, showing their self-confinement due to continuous refraction.

Material Dispersion. Even if modal distortion is absent, some pulse broadening still occurs because the refractive index is a function of wavelength. Dispersion for a silica fiber is shown in Figure 24-11. Since no light source can be precisely monochromatic, the light propagating in the fiber is characterized by a spread of wavelengths determined by the light source. Each wavelength component has a different refractive index and therefore a different speed through the fiber. Pulse broadening occurs because each component arrives at a slightly different time. The more monochromatic the light, the less the distortion due to material dispersion. To be detected as a single pulse, the output pulse must not spread to the extent of significant overlap with neighboring pulses. Again, this requirement places a limita-

Figure 24-11 Dispersion in fused quartz.

tion on the frequency of input pulses or the rate at which bits of information may be sent.

Figure 24-12 illustrates material dispersion by showing the progress of two square pulses (initially coincident) in a fiber at wavelengths λ_1 and λ_2. If the corresponding refractive indices are n_1 and n_2, the figure implies that $n_1 > n_2$. These wavelengths are but two in a continuum described by the spectral width $\Delta\lambda$ of the source, usually chosen as the width of the source's spectral output at half-maximum, as shown.

Because the optical fiber is dispersive, we describe the speed of propagation of a pulse by its group velocity v_g (Section 9-5). The time τ required for a signal of angular frequency ω to travel a distance L along the fiber is therefore given by

$$\tau(\omega) = \frac{L}{v_g(\omega)} \quad \text{where} \quad v_g(\omega) = \frac{d\omega}{dk}$$

If the signal bandwidth is $\Delta\omega$, the spread in arrival times per unit distance is expressed by

$$\delta\left(\frac{\tau}{L}\right) = \frac{d}{d\omega}\left(\frac{1}{v_g}\right)\Delta\omega = \frac{d^2k}{d\omega^2}\Delta\omega$$

Figure 24-12 Symbolic representation of material dispersion. A square wave input arrives at the fiber end at different times, depending on wavelength. The spectral output of the light source is characterized both by a central wavelength λ and a spectral width $\Delta\lambda$.

Now the first derivative $dk/d\omega$ can be calculated from $k = 2\pi/\lambda = n\omega/c$, where n is a function of ω. This gives

$$\frac{dk}{d\omega} = \frac{1}{c}\left(n + \omega\frac{dn}{d\omega}\right) = \frac{1}{c}\left(n - \lambda\frac{dn}{d\lambda}\right) \tag{24-17}$$

where we have used the proportion $\omega/d\omega = -\lambda/d\lambda$ in the last step. Progressing to the second derivative, we write

$$\delta\left(\frac{\tau}{L}\right) = \frac{d}{d\omega}\left(\frac{dk}{d\omega}\right)\Delta\omega = \frac{d}{d\lambda}\left(\frac{dk}{d\omega}\right)\Delta\lambda$$

and substitute Eq. (24-17), giving

$$\delta\left(\frac{\tau}{L}\right) = \frac{d}{d\lambda}\left[\frac{1}{c}\left(n - \lambda\frac{dn}{d\lambda}\right)\right]\Delta\lambda = \frac{1}{c}\left(\frac{dn}{d\lambda} - \frac{dn}{d\lambda} - \lambda\frac{d^2n}{d\lambda^2}\right)\Delta\lambda$$

or simply,

$$\text{material dispersion:} \quad \delta\left(\frac{\tau}{L}\right) = -\frac{\lambda}{c}\frac{d^2n}{d\lambda^2}\Delta\lambda \equiv -M\Delta\lambda \tag{24-18}$$

where M is a property of the core material defined by the prefactor, (λ/c) $(d^2n/d\lambda^2)$, involving the second derivative of the dispersion. From Eq. (24-18), we see that M has the significance of a temporal pulse spread per unit of spectral width per unit of fiber length. Values of M (in units of ps/nm-km) for pure silica are given in Figure 24-13.

Material dispersion for pure silica

Figure 24-13 Material dispersion in pure silica. The quantity M, representing the pulse broadening (ps) per unit of spectral width (nm) per unit of fiber length (km), is plotted against the wavelength. Pulse broadening becomes zero at 1.27 μm and is negative as wavelength increases further.

Example

Using Figure 24-13, calculate the pulse spread due to material dispersion in pure silica for both a LED and a LD light source. Consider the source wavelength to be 0.82 μm, with a spectral width of 20 nm for the LED and 1 nm for the more monochromatic LD.

Solution At 0.82 μm, Figure 24-13 gives a value of near 110 ps/nm-km. Calculation then gives:

LED: $\delta(\tau/L) = (110 \text{ ps/nm-km})(20 \text{ nm}) = 2.2 \text{ ns/km}$

LD: $\delta(\tau/L) = (110 \text{ ps/nm-km})(1 \text{ nm}) = 0.11 \text{ ns/km}$

At 0.82 μm, the LD is 20 times better than the LED, as a direct result of its superior monochromaticity.

Notice also that pulse broadening due to material dispersion is much smaller than that due to modal distortion. Material dispersion therefore becomes significant only when modal distortion is greatly reduced, in both monomode and GRIN fibers. Consequently, in the presence of modal distortion, the advantage of superior monochromaticity of a LD over a LED is lost. In applications where fiber lengths are short enough, plastic fibers and LED sources may well represent the best compromise between performance and cost. Finally, notice from Figure 24-13 that M actually passes through zero at around 1.27 μm, so that material dispersion can also be reduced by finding light sources that operate in this spectral region.

We shall extend the previous numerical example to determine the bandwidth limitation due to pulse spreading. Pulse distortion limits transmission frequency and information rate in a way that we can roughly estimate. Let us use as a reasonable criterion for successful discrimination between neighboring pulses that their separation $\delta\tau$ be no less than half their period T:

$$\delta\tau > \frac{T}{2} \quad \text{or} \quad \delta\tau > \frac{1}{2\nu}$$

where ν is the frequency. It follows that the maximum frequency[11]

$$\nu_{max} = \frac{0.5}{\delta\tau} \quad \text{or} \quad \nu_{max} L = \frac{0.5}{\delta(\tau/L)} \tag{24-19}$$

For the numerical examples above, we calculate an approximate bandwidth, as follows

LED: $\nu_{max} L = \dfrac{0.5}{2.2 \text{ ns/km}} = 0.23 \text{ GHz-km}$

LD: $\nu_{max} L = \dfrac{0.5}{0.11 \text{ ns/km}} = 4.5 \text{ GHz-km}$

Waveguide Dispersion. The last pulse-broadening effect to be discussed is called *waveguide dispersion*, a geometrical effect that depends on waveguide parameters. Compared with modal distortion and material dispersion, waveguide

---

[11] This value corresponds approximately to the so-called 3-db bandwidth, the modulation frequency at which the signal power is reduced by one half due to signal distortion.

dispersion is a small effect that becomes important only when the other pulse-broadening effects have been essentially eliminated. However, its presence is important in determining the wavelength at which net fiber dispersion is zero, as we shall see.

The variation of the refractive index with wavelength leads to material dispersion, as previously discussed. An *effective refractive index* n_{eff} for the guided wave is defined by $n_{\text{eff}} = c/v_g$, where v_g is the group velocity. Waveguide dispersion leads to a variation of n_{eff} with λ for a fixed-diameter fiber, even in the absence of material dispersion. It can be shown [6] that $n_{\text{eff}} = n_1 \sin \varphi$. Since φ varies between 90° and φ_c, and $\sin \varphi_c = n_2/n_1$, it follows that n_{eff} varies between n_1 (at $\varphi = 90°$) and n_2 (at $\varphi = \varphi_c$). Thus n_{eff} for an axial ray depends only on the core index; for a ray at the critical angle, it depends only on the cladding index. The variation of n_{eff} is quite small because $n_1 - n_2$ is, in practice, quite small. Figure 24-14 suggests waveguide dispersion, in the ray representation. For a given mode, the angle between the ray and the fiber axis varies with λ. Thus the ray paths and times for two different wavelengths also vary with λ, leading to pulse broadening.

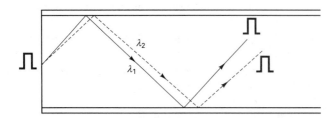

Figure 24-14 Symbolic representation of waveguide dispersion. A square wave input arrives at the fiber end at different times, depending on wavelength, even in a dispersionless medium. For any one mode, the angle of propagation is a function of the wavelength.

The variation of n_{eff} with λ simulates material dispersion and can be handled quantitatively by Eq. (24-18) by simply replacing n by n_{eff}:

$$\text{material dispersion:} \quad \delta\left(\frac{\tau}{L}\right) = -\frac{\lambda}{c}\frac{d^2n}{d\lambda^2}\Delta\lambda \equiv -M\Delta\lambda \qquad (24\text{-}18)$$

$$\text{waveguide dispersion:} \quad \delta\left(\frac{\tau}{L}\right) = -\frac{\lambda}{c}\frac{d^2n_{\text{eff}}}{d\lambda^2}\Delta\lambda \equiv -M'\Delta\lambda \qquad (24\text{-}20)$$

We can appreciate the relative contribution of material and waveguide dispersion by comparing values of M and M'. In Figure 24-13, M ranges from about 165 to -30 ps/nm-km over the spectral range of 0.7 to 1.7 μm. Values of M' for fused quartz over the same range have values of only about 1 to 4.5 ps/nm-km [6]. For example, the calculation carried out for material dispersion, using a LED source at 0.82 μm, with $M = 110$ ps/nm-km, gave a temporal pulse broadening of 2200 ps/km. For the same wavelength, $M' = 2$ ps/nm-km, giving a pulse broadening of 2/110 times as great, or only 40 ps/km.

Figure 24-13 shows that M for material dispersion becomes zero at around 1.27 μm and then becomes negative for longer wavelengths. Waveguide dispersion, on the other hand, is always positive. Combination of the two thus shifts the wavelength of zero net dispersion toward a longer wavelength of about 1.31 μm in a typical fiber. Sources operating at or near this wavelength are thus ideal in reducing pulse broadening and increasing transmission rates. In discussing attenuation earlier, we pointed out that minimum absorption in silica fibers occurs at around 1.55 μm. The closeness of the wavelengths for minimum absorption and minimum dispersion has motivated attempts to satisfy both conditions by shifting the dispersion curve towards longer wavelengths, so that it passes through zero at 1.55 μm instead of 1.31 μm. Means of modifying the dispersion curve include the use of multiple cladding layers, control of the core/cladding index difference, and variation of the profile parameter α in GRIN fibers.

By way of summary, we have discussed three principal ways of reducing pulse broadening in fibers: (1) use a monomode fiber to eliminate modal distortion, (2) use a light source of small spectral width $\Delta\lambda$ to reduce material dispersion, and (3) use a light source operating in a spectral region where both attenuation and dispersion are as small as possible. Clearly, the required length of fiber and component costs play major roles in determining the selection of the best system for a specific application.

PROBLEMS

24-1. The bandwidth of a single telephone channel is 4 kHz. In a particular telephone system, the transmission rate is 44.7 Mbps. In an actual system, some channels are devoted to housekeeping functions such as synchronization. In this system, 26 channels are so devoted. How many independent telephone channels can the system accommodate?

24-2. Determine the theoretical limit to the number of TV station channels that could transmit on a single optical beam of 1 μm wavelength.

24-3. **(a)** Show that, for a guided ray traveling at the steepest angle relative to the fiber axis, the skip distance L_s can be expressed by:

$$L_s = \frac{n_2 d}{\sqrt{n_1^2 - n_2^2}}$$

(b) How many reflections occur per meter for such a ray in a step-index fiber with $n_1 = 1.460$, $n_2 = 1.457$, and $d = 50$ μm?

24-4. Refractive indices for a step-index fiber are 1.52 for the core and 1.41 for the cladding. Determine (a) the critical angle; (b) the numerical aperture; (c) the maximum incidence angle θ_m for light that is totally internally reflected.

24-5. A step-index fiber 0.0025 in. in diameter has a core of index 1.53 and a cladding of index 1.39. Determine (a) the numerical aperture of the fiber; (b) the *acceptance angle* (or maximum entrance cone angle); (c) the number of reflections in 3 ft of fiber for a ray at the maximum entrance angle, and for one at half this angle.

24-6. **(a)** Show that the actual distance x_s a ray travels during one skip distance is given by

$$x_s = \frac{n_1 d}{\sin\theta}$$

where θ is the entrance angle and the fiber is used in air.

(b) Show that the actual total distance x_t a ray with entrance angle θ travels over a total length L of fiber is given by

$$x_t = \frac{n_1 L}{(n_1^2 - \sin^2\theta)^{1/2}}$$

(c) Determine x_s, L_s, and x_t for a 10-m-long fiber of diameter 50 μm, core index of 1.50, and a ray entrance angle of 10°.

24-7. How many modes can propagate in a step-index fiber with $n_1 = 1.461$ and $n_2 = 1.456$ at 850 nm? The core radius is 20 μm.

24-8. Determine the maximum core radius of a glass fiber so that it supports only one mode at 1.25 μm wavelength, for which $n_1 = 1.460$ and $n_2 = 1.457$.

24-9. Consider a slab waveguide of AlGaAs, for which $n_1 = 3.60$ and $n_2 = 3.55$. How many independent modes can propagate in this waveguide if $d = 5\lambda$ and $d = 50$ λ?

24-10. A flux of 5 μW exists just inside the entrance of a fiber 100 m long. The flux just inside the fiber exit is only 1 μW. What is the absorption coefficient of the fiber in db/km?

24-11. An optic fiber cable 3 km long is made up of three 1-km lengths, spliced together. Each length has a 5-db loss and each splice contributes a 1-db loss. If the input power is 4 mW, what is the output power?

24-12. The attenuation of a 1-km length of RG-19/U coaxial cable is about 12 db at 50 MHz. Suppose the input power to the cable is 10 mW and the receiver sensitivity is 1 μW. How long can the coaxial cable be under these conditions? If optical fiber is used instead, with a loss rated at 4 db/km, how long can the transmission line be?

24-13. A Ge-doped silica fiber has an attenuation loss of 1.2 db/km due to Rayleigh scattering alone, when light of wavelength 0.90 μm is used. Determine the attenuation loss at 1.55 μm.

24-14. (a) Show that the attenuation db/km is given by

$$\alpha = 10 \log_{10} (1 - f)$$

where f is the overall fractional power loss from input to output over a 1-km-long fiber.

(b) Determine the attenuation in db/km for fibers having an overall fractional power loss of 25%, 75%, 90%, and 99%.

24-15. Determine (a) the length and (b) transit time for the longest and shortest trajectories in a step-index fiber of 1 km, having a core index of 1.46 and a cladding index of 1.45.

24-16. Evaluate modal distortion in a fiber by calculating the difference in transit time through a 1-km fiber required by an axial ray and a ray entering at the maximum entrance angle of 35°. Assume a fused silica core index of 1.446. What is the maximum frequency of input pulses that produce nonoverlapping pulses on output, due to this case of modal dispersion?

24-17. Calculate the time delay between an axial ray and one that enters a 1-km-long fiber at an angle of 15°. The core index is 1.48.

24-18. Calculate the group delay between the fastest and slowest modes in a 1-km-long step-index fiber with $n_1 = 1.46$ and a *relative index difference* $\Delta = (n_1 - n_2)/n_2 = 0.003$, using a light source at wavelength 0.9 μm.

24-19. Plot the refractive index profile for a GRIN fiber of radius 50 μm and with $n_1 = 1.5$ and $\Delta = 0.01$. Do this when the profile parameter $\alpha = 2$ and repeat for $\alpha = 10$.

24-20. Calculate the delay due to modal dispersion in a 1-km GRIN fiber with $\alpha = 2$. The maximum core index is 1.46 and the cladding index is 1.44. By what factor is this fiber an improvement over a step-index fiber with $n_1 = 1.46$ and $n_2 = 1.44$?

24-21. Equation (24-19) allows calculation of bandwidth for distances less than the *equilibrium length* of fiber (see footnote 10). Assume an equilibrium length of 1 km and determine for this fiber length the 3-db bandwidth of a step-index multimode fiber whose pulse broadening is given by 20 ns/km.

24-22. Determine the material dispersion in a 1-km length of fused silica fiber when the light source is (a) a LED centered at 820 nm with a spectral width of 40 nm and (b) a LD centered at 820 nm with a spectral width of 4 nm.

24-23. The total delay time $\Delta\tau$ due to *both* modal distortion and material dispersion is given by

$$(\Delta\tau)^2 = (\Delta\tau_{\text{mod}})^2 + (\Delta\tau_{\text{mat}})^2$$

Determine the total delay time in a 1-km fiber for which $n_1 = 1.46$, $\Delta = 1\%$, $\lambda = 820$ nm, and $\Delta\lambda = 40$ nm.

24-24. Waveguide dispersion is measured in a silica fiber at various wavelengths using laser diode sources with a spectral width of 2 nm. The results are:

$\lambda\,(\mu\mathrm{m})$	$\Delta(\tau/L)$ (ps/km)
0.70	1.88
0.90	5.02
1.10	7.08
1.40	8.40
1.70	8.80

(a) Plot the waveguide parameter M' versus λ in the range 0.70 to 1.70 μm.

(b) Determine the waveguide dispersion in ps/km at $\lambda = 1.27$ and 1.55 μm for a source with a spectral width of 1 nm.

24-25. Compare pulse broadening for a silica fiber due to the three principal causes—modal distortion, material dispersion, and waveguide dispersion—in a step-index fiber. The core index is 1.470 and the cladding index is 1.455 at $\lambda = 1$ μm. Assume a LED source with a spectral width of 25 nm. The values of the parameters M and M' are 43 ps/nm-km and 3 ps/nm-km, respectively.

(a) Determine each separately by calculating $\Delta\tau$ for a 1-km length of fiber.

(b) Determine an overall broadening $\Delta\tau$ for a 1-km length of fiber, using

$$(\Delta\tau)^2 = (\Delta\tau_{\mathrm{mod}})^2 + (\Delta\tau_{\mathrm{mat}})^2 + (\Delta\tau_{\mathrm{wg}})^2$$

REFERENCES

[1] Siegmund, Walter P. "Fiber Optics." In *Handbook of Optics*, edited by Walter G. Driscoll and William Vaughan. New York: McGraw-Hill Book Company, 1978.

[2] Yariv, Amnon. *Optical Electronics*, 3d ed. New York: Holt, Rinehart and Winston, 1985. Ch. 3.

[3] Cheo, Peter K. *Fiber Optics Devices and Systems*. Englewood Cliffs, N.J.: Prentice-Hall, 1985. Ch. 4.

[4] Gloge, D., and E. A. J. Marcatili. "Multimode Theory of Graded-Core Fibers." *Bell Syst. Tech. J.* 52 (Nov. 1973): 1563.

[5] Miller, Stewart E., Enrique A. J. Marcatili, and Li Tingye. "Research toward Optical-Fiber Transmission Systems." *Proc. IEEE* 61, no. 12 (Dec. 1973): 1703.

[6] Palais, Joseph C. *Fiber Optic Communications*. Englewood Cliffs, N.J.: Prentice-Hall, 1988.

[7] Guenther, Robert. *Modern Optics*. New York: John Wiley and Sons, 1990. Ch. 5.

[8] Karim, Mohammad A. *Electro-Optical Devices and Systems*. Boston: PWS-Kent Publishing Company, 1990. Ch. 9.

[9] Nerou, Jean Pierre. *Introduction to Fiber Optics*. Sainte-Foy, Quebec: Les Editions Le Griffon D'Argile, 1988.

[10] Boyd, Waldo T. *Fiber Optics Communications, Experiments and Projects*. Indianapolis: Howard W. Sams and Co., 1982.

[11] Urquhart, Paul, and Timothy J. Whitley. "Long Span Fiber Amplifiers." *Applied Optics* 29, no. 24 (Aug. 20, 1990): 3503.

[12] Kapany, Narinder S. "Fiber Optics." *Scientific American* (Nov. 1960): 72.

[13] Busignies, Henri. "Communication Channels." *Scientific American* (Sept. 1972): 98.

[14] Cook, J. S. "Communication by Optical Fiber." *Scientific American* (Nov. 1973): 28.

[15] Yariv, Amnon. "Guided Wave Optics." *Scientific American* (Jan. 1979): 64.

[16] Mandoli, Dina F., and Winslow R. Briggs. "Fiber Optics in Plants." *Scientific American* (Aug. 1984): 90.

[17] Katzir, Abraham. "Optical Fibers in Medicine." *Scientific American* (May 1989): 120.

[18] Desurvire, Emmanuel. "Lightwave Communications: The Fifth Generation." *Scientific American* (Jan. 1992): 114.

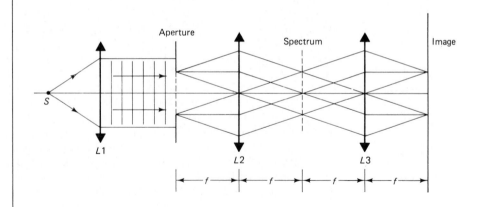

Fourier Optics

INTRODUCTION

Two rather extensive areas in which the Fourier transform is central to applications in optics are treated in this chapter, although, necessarily, somewhat cursorily. The first is included under the general heading of *optical data imaging and processing* and the second, *Fourier-transform spectroscopy*. Both are included within a branch of physics referred to generally as *Fourier optics*, in which the *Fourier transform, convolution,* and *correlation* are central concepts of mathematical analysis.

Optical data processing takes advantage of the fact that the simple lens constitutes a Fourier-transform computer, capable of transforming a complex two-dimensional pattern into a two-dimensional transform at very high resolution and at the speed of light. The diffraction pattern of a spatial object formed by the lens is shown to be a two-dimensional Fourier transform, or *spectrum*, of the input. This pattern may be manipulated in turn, using masks or filters to modify the final image produced by a second lens in a process called *spatial filtering*. Since various details of the image can be modified by appropriate filtering, this technique is exploited in such areas as *contrast enhancement* and *image restoration*. If the image is compared directly with a second object, the two may be *optically correlated*. Such correlation is applied, for example, in the problem of *pattern recognition*. By such optical means, two-dimensional pictures or text are processed at once, without the necessity of sequential scanning of the object. Optical data processing represents a fruitful convergence of the fields of optics, information science, and holography. As in many other fields, the availability of the laser as a coherent light source has insured rapid growth.

Fourier-transform spectroscopy capitalizes on the fact that the spatial or temporal variations of an irradiance pattern due to polychromatic radiation can be Fourier-transformed into a spectral decomposition of the radiation. This technique makes possible another application of interferometry with distinct advantages for spectroscopy. Fourier-transform spectroscopy is the subject of the second part of the present chapter.

25-1 OPTICAL DATA IMAGING AND PROCESSING

Fraunhofer Diffraction and the Fourier Transform. We wish to show that the Fraunhofer diffraction pattern is, within certain approximations, the Fourier transform (Section 12-1) of the **E**-field amplitude distribution in the object plane. Recall the one-dimensional Fourier transforms presented in Chapter 12:

$$f(x) = \frac{1}{2\pi} \int_{-\infty}^{+\infty} g(k)e^{-ikx} \, dk \tag{25-1}$$

$$g(k) = \int_{-\infty}^{+\infty} f(x)e^{ikx} \, dx \tag{25-2}$$

Equation (25-1) states that an arbitrary, nonperiodic function $f(x)$ can be synthesized by summing a continuous distribution of plane waves with amplitude distribution $g(k)$ given by Eq. (25-2). The functions $f(x)$ and $g(k)$ are said to be a Fourier-transform pair, so that they are Fourier transforms of one another. Symbolically,

$$g(k) = \mathscr{F}\{f(x)\} \tag{25-3}$$

$$f(x) = \mathscr{F}^{-1}\{g(k)\} \tag{25-4}$$

The transform of the inverse transform of a function returns the function, that is,

$$f(x) = \mathscr{F}^{-1}\{g(k)\} = \mathscr{F}^{-1}\mathscr{F}\{f(x)\} = f(x) \tag{25-5}$$

In two dimensions, the transforms take the form

$$f(x, y) = \frac{1}{(2\pi)^2} \int\!\!\int_{-\infty}^{+\infty} g(k_x, k_y)e^{-i(xk_x + yk_y)} \, dk_x \, dk_y \tag{25-6}$$

$$g(k_x, k_y) = \int\!\!\int_{-\infty}^{+\infty} f(x, y)e^{i(xk_x + yk_y)} \, dx \, dy \tag{25-7}$$

The nonperiodic function of two variables $f(x, y)$ can thus be synthesized from a distribution of plane waves, each with amplitude $g(k_x, k_y)$ and constant phase, such that

$$xk_x + yk_y = \text{constant} \tag{25-8}$$

The quantities k_x and k_y are the *spatial frequency* $(2\pi/\lambda)$ components needed in the expansion to represent the desired function $f(x, y)$. The individual plane waves in the continuous distribution intersect the xy-plane along the straight lines defined by Eq. (25-8). As k_x and k_y vary, the slopes of these lines vary. Thus the synthesis involves plane waves that vary in direction.

Consider the Fraunhofer diffraction pattern due to an arbitrary aperture situated in an xy-plane, as shown in Figure 25-1. Plane monochromatic waves diffract

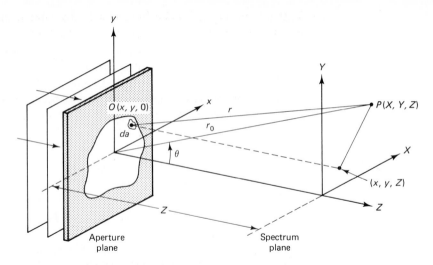

y

Y

O(x, y, 0)

da

θ

r

r₀

P(X, Y, Z)

X

(x, y, Z)

Z

Z

Aperture
plane

Spectrum
plane

Figure 25-1 Fraunhofer diffraction in the spectrum XY-plane due to an aperture in the xy-plane.

from the *aperture* (xy) *plane*. The diffraction pattern is observed in the XY-plane, which we shall call the *spectrum plane*, a distance Z along the axis. The contribution dE_p at an arbitrary point P due to the light amplitude from an elemental area da surrounding point O in the aperture is given by

$$dE_P = \left(\frac{E_s da}{r}\right) e^{i(\omega t - kr)} \tag{25-9}$$

where r is the distance from point O to point P. Neglecting the obliquity factor for small angles θ, Eq. (25-9) represents a spherical wave whose amplitude decreases with distance r. The quantity E_s is the *source strength*, or amplitude per unit area of aperture in the neighborhood of point O. The combination $E_s da$ is then the amplitude at unit distance $(r = 1)$ from O due to the elemental aperture area da. If the aperture is not uniformly illuminated or is not uniformly transparent, then $E_s = E_s(x, y)$ and is called the *aperture function*. In Eq. (25-9), ω and k refer to the properties of the incident and diffracted radiation. The point P in the spectrum plane is a distance r_0 from the origin of the xy-coordinate system in the aperture plane. The distance r may be referred to the distance r_0 as follows. From the geometry apparent in Figure 25-1,

$$r^2 = (X - x)^2 + (Y - y)^2 + (Z - 0)^2$$

and

$$r_0^2 = X^2 + Y^2 + Z^2$$

so that

$$r^2 = r_0^2 - 2xX - 2yY + (x^2 + y^2) \tag{25-10}$$

Although the dimensions X and Y in the spectrum plane may be appreciable, the dimensions x and y are typically negligible in comparison with r_0 for far-field diffraction. Accordingly, the terms x^2 and y^2 are ignored and Eq. (25-10) is rewritten as

$$r = r_0 \left[1 - 2\frac{(xX + yY)}{r_0^2}\right]^{1/2} \tag{25-11}$$

In this form, Eq. (25-11) is immediately adaptable to approximation by the binomial expansion $(1 + u)^2 = 1 + (\frac{1}{2})u + \cdots$. Retaining only the first two terms.

$$r = r_0 \left[1 - \frac{(xX + yY)}{r_0^2} \right] \tag{25-12}$$

In Eq. (25-9), the distance r appears in both the amplitude and the phase. In the amplitude it can be safely approximated by the distance Z between planes, but in the phase we use the approximate expression just derived. Then

$$dE_P = \left(\frac{E_s \, dx \, dy}{Z} \right) e^{i\omega t} e^{-ik[r_0 - (xX + yY)/r_0]} \tag{25-13}$$

so that, upon integration over the area of the aperture, we have

$$E_P = \left[\frac{e^{i(\omega t - kr_0)}}{Z} \right] \iint E_s(x, y) e^{ik(xX + yY)/r_0} \, dx \, dy \tag{25-14}$$

If we are interested in the relative *amplitude* distribution of the electric field in the spectrum plane, we may set $Z = 1$ and write

$$E_P = \iint E_s(x, y) e^{ik(xX + yY)/r_0} \, dx \, dy \tag{25-15}$$

Next, introducing the *angular spatial frequencies,*

$$k_x \equiv \frac{kX}{r_0} \quad \text{and} \quad k_Y \equiv \frac{kY}{r_0} \tag{25-16}$$

corresponding to each point (X, Y) in the spectrum plane, Eq. (25-15) may be expressed as

$$E_P(k_X, k_Y) = \iint E_s(x, y) e^{i(xk_X + yk_Y)} \, dx \, dy \tag{25-17}$$

In this form, Eq. (25-17) may be compared directly with Eq. (25-7), and our goal is established. We see that E_P and E_s are related through a Fourier transformation. The inverse transform, as in Eq. (25-6), is

$$E_s(x, y) = \frac{1}{(2\pi)^2} \iint E_P(k_X, k_Y) e^{-i(xk_X + yk_Y)} \, dk_X \, dk_Y \tag{25-18}$$

Within the approximations made, we have shown that the Fraunhofer diffraction pattern described by $E_P(k_X, k_Y)$ is just the two-dimensional Fourier transform of the aperture function described by $E_s(x, y)$. The continuous distribution of constituent multidirectional plane waves is responsible for the "bending" of the light into the various regions of the two-dimensional diffraction pattern.

Optical Spectrum Analysis. The Fraunhofer diffraction pattern of a given aperture is most conveniently displayed using a positive lens, as in Figure 25-2. Light from a monochromatic (temporally coherent) point source (spatially coherent) is collimated by lens $L1$ and illuminates, in the *input* or *aperture plane*, a two-dimensional pattern whose transmittance varies across the aperture. Lens $L2$ forms the Fraunhofer pattern in the spectrum plane. We shall neglect lens aberrations and also assume that the aperture is large enough so that its own boundaries do not appreciably modify the diffraction pattern. The *aperture function* may thus be formed

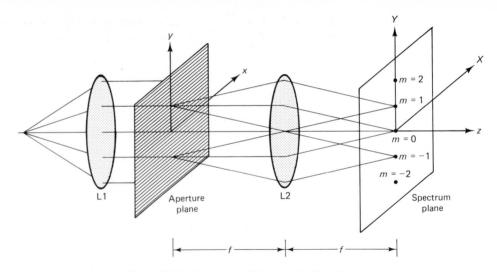

Figure 25-2 Fraunhofer diffraction of a Ronchi ruling.

by any photographic negative. For simplicity we shall imagine the aperture function to vary like a square wave, such as would be produced by a *Ronchi ruling*, a grating of parallel straight lines with large grating space, whose opaque and transparent regions are of equal width. Since the Fourier transform is an amplitude (not an irradiance) transform, we describe the square wave in Figure 25-3 by the amplitude of the transmitted light. We refer to the ratio of transmitted to incident amplitudes E_t/E_0 as the *transmission*, in contrast with the ratio of irradiances I_t/I_0, which we have called the *transmittance*. Transmittance is then just the square of the transmission. The aperture function, involving amplitudes, may also be called the *transmission function*. Lens $L2$ acts as a *Fourier-transform lens*. With a transmission function in its first focal plane, the Fraunhofer diffraction pattern, which is its Fourier transform, is produced in the second focal plane, the *spectrum*, or *output*, *plane*. The Ronchi ruling acts as a coarse grating, producing a series of bright spots that correspond to the various orders of diffraction. Since the Ronchi rulings are aligned parallel to the x-axis in the aperture plane, the spectrum of bright spots in the output plane occurs along the Y-direction, as shown. Now, according to the grating equation,

$$m\lambda = d \sin \theta = d \frac{Y_m}{f} \qquad (25\text{-}19)$$

where d is the spatial period of the ruling. Spots appear at distances Y_m from the optical axis given by

$$Y_m = m\left(\frac{\lambda f}{d}\right) \qquad (25\text{-}20)$$

We wish to show now that this series of bright spots is, in fact, the spectrum of frequencies required in a Fourier representation of the aperture or transmission function

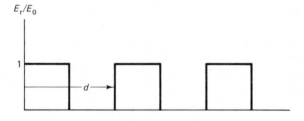

Figure 25-3 Transmission function of period d due to a Ronchi ruling, in which opaque and transmitting widths are equal.

of Figure 25-3. The angular spatial frequencies required in the Fourier integral were introduced in Eq. (25-16). In the Y-direction these are given by

$$k_Y = \frac{kY}{f} \qquad (25\text{-}21)$$

Since the transmission function for the Ronchi ruling is a periodic square function, it is represented by a discrete set of frequencies in a Fourier series rather than by a continuous distribution of frequencies in a Fourier integral. Let us introduce a wave number or "normalized" form of the spatial frequencies in the Fourier series by

$$\nu_Y \equiv \frac{1}{\lambda_Y} = \frac{k_Y}{2\pi} \qquad (25\text{-}22)$$

Then, substituting for k_Y from Eq. (25-21) and for Y from Eq. (25-20), we have, for the spectrum of spatial frequencies displayed in the diffraction pattern,

$$\nu_Y = \frac{m}{d} \qquad (25\text{-}23)$$

The central spot with $m = 0$ thus corresponds to a frequency $\nu_Y = 0$, the *DC component*, in analogy with electrical frequencies. The first-order ($m = 1$) spots above and below the central spot represent the fundamental frequency $\nu_{Y1} = 1/d$. Higher-order ($m > 1$) spots represent higher harmonics given by $m\nu_{Y1}$. We see that when the frequency of the square wave is larger (more closely spaced rulings with smaller d), the fundamental frequency in the Fourier spectrum is also larger, and the separation $Y_1 = \lambda f/d$ is increased—a fact that should already be familiar from our study of the diffraction grating.

A Fourier analysis of the square function (also calculated in Chapter 12 for a square wave as an even function) gives the Fourier series

$$f(Y) = \frac{1}{2} + \frac{2}{\pi} \left(\cos kY + \frac{1}{3} \cos 3kY + \frac{1}{5} \cos 5kY + \cdots \right) \qquad (25\text{-}24)$$

Here we find a constant ($k = 0$) term of $\frac{1}{2}$ corresponding to the DC component or central spot of the diffraction pattern; a term with fundamental (spatial) frequency $k_1 = 2\pi/d$; and terms with higher *odd* harmonics, $3k_1$, $5k_1$, ... The absence of the even harmonics might at first be puzzling, on the basis of Eq. (25-23), since it woud lead us to expect all the higher harmonics in the representation. The even harmonics, however, are just those corresponding to the missing orders in the grating diffraction. These missing orders are expected when the ratio of slit separation is twice the width of the slit opening, precisely the case in the Ronchi ruling. The squares of the coefficients in the Fourier series are proportional to the irradiances of the corresponding diffraction spots.

Suppose now that the transmission function is not a square wave but a sine wave. If the lines of the Ronchi ruling have gradually changing opacity, such that the amplitude transmitted varies sinusoidally, we have the *sinusoidal grating*. Arguing from the Fourier series required to represent this kind of aperture function, it is clear that orders in the diffraction spectrum higher than $m = 1$ do not appear. Clearly, only one frequency is required to represent a sine wave. Why then does the spectrum also show a central spot, the DC component with $m = 0$? A little thought will make clear that an amplitude aperture function cannot be produced with both positive and negative portions, like the pure sine wave of Figure 25-4a. A photographic negative, at points of ideal opacity, may produce an amplitude $E = 0$ but

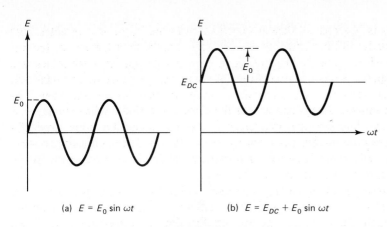

(a) $E = E_0 \sin \omega t$ (b) $E = E_{DC} + E_0 \sin \omega t$

Figure 25-4 Sinusoidal amplitude or transmission functions including negative displacements as in (a), and with all displacements positive, as in (b). Actual aperture functions do not have negative amplitudes.

cannot provide negative values. Thus the sinusoidal grating produces a transmission function like that of Figure 25-4b, in which the sine wave is offset by a DC bias. It is precisely the component E_{DC} in the figure that accounts for the zeroth-order diffraction signal.

The inverse of the Fourier transformation of sinusoidal grating is the diffraction pattern produced by two slits. As shown earlier, when considering the interference of two point sources, the resultant diffraction pattern is a series of fringes whose irradiance pattern across the fringes is cosine-squared or sinusoidal in form. A photographic film or other type of detector records the irradiance pattern in the form shown in Figure 25-4b.

Optical Filtering. We have seen that the back focal plane of the transform lens is the spectrum plane in which a Fourier transform of the aperture or transmission function is located. If this spectrum plane now serves in turn as a new aperture function for a second lens $L3$, a focal length away (Figure 25-5), the back focal plane of the second lens recieves the Fourier transform of the new aperture function. This second Fourier transform is thus the transform of the transform of the original aperture function and so returns the original aperture function, that is, an image of the original aperture is formed there. This conclusion also follows from a simple application of the laws of geometrical optics, evident from the ray diagram included in Figure 25-5.

Each diffraction spot in the spectrum plane, with coordinates (X, Y), represents spatial frequencies present in the aperture function, as we have pointed out. Each of these diffraction spots now helps to illuminate the image of the aperture in the image plane. How is this image affected if the light from one or more of these

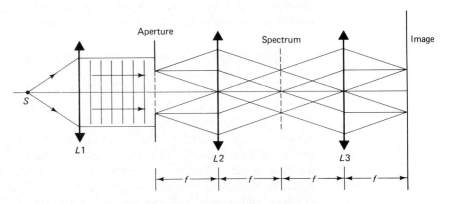

Figure 25-5 Optical filter.

diffraction spots is blocked, so that its contribution to the image is subtracted out? From our knowledge of Fourier series, we conclude that the finer features of the image disappear when spots corresponding to the higher spatial frequencies are blocked. If all spots are blocked except the DC component, or undeviated diffraction beam—say by an iris diaphragm centered on the central spot—the image plane is illuminated but no image details appear. As the circular opening of the diaphragm is gradually widened, higher spatial frequencies are admitted and the image gradually sharpens. The physical operation of opening the diaphragm is thus analogous mathematically to the systematic inclusion of higher and higher frequency terms in the Fourier series representing the aperture function.

Optical filtering is the process of intentionally blocking certain portions—that is, certain spatial frequencies—present in the diffraction pattern, to manipulate the image. Suppose, for example, that the aperture function is the superposition of two sine waves that are produced by back-to-back sinusoidal gratings with parallel rulings but different line spacings or spatial frequencies. The diffraction pattern consists, in addition to the direct beam, of two pairs of light spots, each pair due to one of the spatial frequencies present. If one of these pairs is blocked, that frequency is eliminated, or *filtered* from the illumination. The image is a sinusoidal pattern of the other frequency.

This example shows how optical filtering is applied to the extraction of desired periodic signals from background noise or, on the other hand, to the elimination of periodic noise from a desirable signal. As another example, suppose the aperture function is a television picture in which horizontal raster lines are visible. The diffraction pattern due to this function may be quite complicated, but the raster lines, like a Ronchi ruling, produce a series of diffraction spots along the vertical direction in the spectrum plane. If a rectangular-shaped, opaque shield is used to block the contribution of these spots, the raster line frequencies are filtered out and the final image is a reproduction of the TV picture but without the raster lines present. This technique was used to remove video scan lines from the video micrograph of a diatom frustule, as shown in Figure 25-6.

From the point of view of optical filtering, then, it should be clear that a diaphragm, which blocks all but those frequencies near the direct beam, functions as a low-pass optical filter; a diaphragm, which blocks only those frequencies near the direct beam, functions as a high-pass optical filter; and an annular ring, which blocks the lowest and the highest frequencies, functions as a band-pass filter. A case in point is the suppression of low spatial frequencies, or high-pass optical filtering, to enhance the contrast in a photograph. (Recall the importance of the high-frequency components in a Fourier series when synthesizing the fine features of a function, like the corners of a square wave; Section 12-1.) More complex filtering has also been used in image restoration, for example, in the deblurring of lunar photographs.

Optical Correlation.   As we have seen, an image of the two-dimensional object situated in the aperture plane is formed in the image plane of the optical filter (Figure 25-5). Suppose now that in the position of the image plane we insert a second object, that is, another photographic transparency containing an image, so that the image of the original object is superimposed over that of the second. Then the amount of light passed by the second object at any point depends both on the amount of light available in the image and the transparency of the second object. Let the light so transmitted be intercepted by an additional lens and, in its second focal plane, be monitored by a film or light detector, as shown in Figure 25-7. We have, in effect, added an *optical spectrum analyzer* to the optical filter of Figure 25-5. In the output plane where the detector is placed, we expect to measure the spectrum or

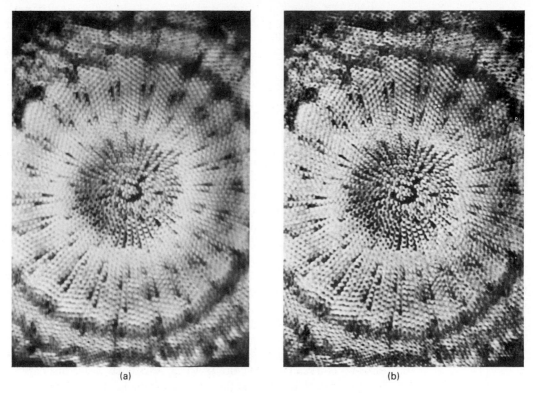

(a)   (b)

Figure 25-6 (a) Video micrograph of a diatom frustule including video scan lines (shown vertically). (b) Video micrograph of (a), spatially filtered to remove the video scan lines. (Photos by Gordon W. Ellis, University of Pennsylvania.)

Fourier transform of the transmission function represented by the light transmitted through the second object. This system provides an experimental means of comparing, or *correlating*, the two objects or the two transmission functions they determine. If the two functions are identical, for example, and so situated that the image of the first coincides with the second, then maximum light throughput occurs, a case of maximum *correlation*. If one object is translated relative to the other, however, the bright points of the image no longer all coincide with the transparent regions of

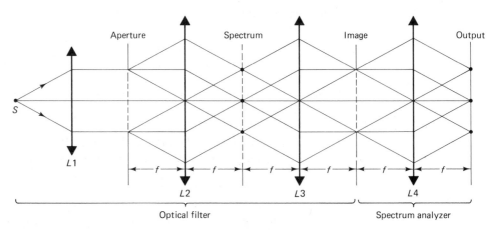

Figure 25-7 Optical correlator formed by the combination of an optical filter and a spectrum analyzer.

the second object, and light throughput and correlation are reduced. If the first object in the aperture plane is a photographic image of the block letter *A* and the second object in the image plane is of a similar shape, a high degree of correlation should be obtained when the objects are properly positioned; on the other hand, if the second object is a photographic image of the letter *B*, the maximum light throughput and correlation should be significantly reduced. This technique of *pattern recognition* is applied, for example, to the recognition and counting of small particles with different shapes, as in the case of blood cells, or to the search for characteristic patterns in aerial photographs, medical X-rays, and fingerprint files.

Let us express the situation more precisely in mathematical terms. Let the first object in the aperture plane be illuminated uniformly by light of unit amplitude, and let its transmission function be described by $E_1(x, y)$. The transmitted light, amplitude modulated and imaged at the position of the second object in the image plane, is then represented by $E_1(-x, -y)$. The change to negative coordinates is required by the inversion of the real image relative to the object. If now the second transmission function is $E_2(x, y)$, the light transmitted is the product function $E_1(-x, -y) E_2(x, y)$. The Fourier transform or spectrum of this composite transmission function is formed in the output plane, that is, the diffraction pattern there is described by

$$\mathscr{F}[E_1(-x, -y)E_2(x, y)] = \int\!\!\int_{-\infty}^{\infty} E_1(-x, -y)E_2(x, y)e^{i(xk_x + yk_y)}\, dx\, dy \qquad (25\text{-}25)$$

To concentrate only on the direct beam, or DC component, in the pattern, we set the spatial frequencies k_x and k_y equal to zero, so that

$$\mathscr{F}[E_1(-x, -y)E_2(x, y)]_{DC} = \int\!\!\int_{-\infty}^{\infty} E_1(-x, -y)E_2(x, y)\, dx\, dy \qquad (25\text{-}26)$$

Both transmission functions $E_1(x, y)$ and $E_2(x, y)$ have been referred to xy-coordinate system origins that differ only by translation along the z-, or optical, axis. If the first object is shifted in the aperture plane by an arbitrary translation given by components (q_x, q_y), for instance, its transmission function must reflect a translation of origin within the xy-plane, and Eq. (25-26) is expressed more generally by

$$\mathscr{F}[E_1(-x, -y)E_2(x, y)]_{DC} = \int\!\!\int_{-\infty}^{\infty} E_1(q_x - x, q_y - y)E_2(x, y)\, dx\, dy \qquad (25\text{-}27)$$

The integral in Eq. (25-27) is an example of the two-dimensional *convolution function*,

$$\rho_{12}(q_x, q_y) = \int\!\!\int_{-\infty}^{\infty} f_1(q_x - x, q_y - y)f_2(x, y)\, dx\, dy \qquad (25\text{-}28)$$

If the transmission function possesses inversion symmetry, that is, if

$$f_1(-x, -y) = f_1(x, y)$$

then the negative signs in the integrand of Eq. (25-28) may be written as positive signs, and the integral is instead the *correlation function*,

$$\Phi_{12}(q_x, q_y) = \mathscr{F}[\, f_1(x, y)\, f_2(x, y)]_{DC} \qquad (25\text{-}29)$$

$$\Phi_{12}(q_x, q_y) = \int\!\!\int_{-\infty}^{\infty} f_1(x + q_x, y + q_y)\, f_2(x, y)\, dx\, dy$$

Further, when f_1 and f_2 are merely shifted versions of the *same* function, we speak instead of the *autocorrelation function*,

$$\Phi_{11}(q_x, q_y) = \int\limits_{-\infty}^{\infty}\!\!\int f(x + q_x, y + q_y)\, f(x, y)\, dx\, dy \qquad (25\text{-}30)$$

Transmission functions with inversion symmetry are imaged in such a way that the actual image inversion due to the lens is not apparent. Let us briefly examine the autocorrelation integral of Eq. (25-30). The integrand is a product of two functions and is nonzero only at those (x, y) points where both functions have nonzero values. With (q_x, q_y) fixed, the integral is the area under a curve representing the product of the two functions. This area, which we call the correlation, clearly depends on the choice of (q_x, q_y). If (q_x, q_y) are large enough so that there is no overlap of the functions, the area and correlation are zero. When q_x and q_y are both zero, the functions coincide, yielding a product curve with the maximum area and correlation. As an example, Figure 25-8, we have chosen as a function the top half of a circle. As one such curve is translated along the x-axis relative to the other, their autocorrelation varies as a function of the parameter B_x, the displacement of their y-axes. The example illustrates a one-dimensional correlation.

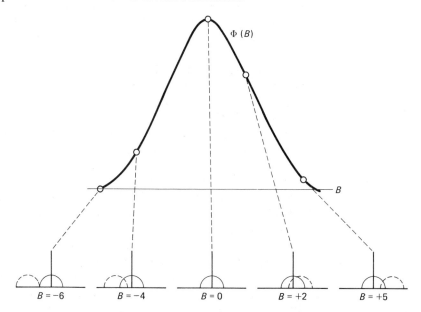

Figure 25-8 One-dimensional autocorrelation $\Phi(B)$ of a semicircle with radius $= 3$ as a function of the displacement parameter B. Several points on the correlation curve are referred to the specific translations that produce them.

We see from Eq (25-29) that the correlation is given by the DC component, or zeroth-order spectral point of the Fourier transform, or spectrum. Thus a detector, placed on axis at the output plane in the optical correlation system of Figure 25-7, measures the correlation. More precisely, since it is sensitive only to irradiances, the detector measures a quantity proportional to the square of the correlation. As the first object in the aperture plane is translated along its x-axis, the light energy in the direct beam varies, producing the correlation function $\Phi_{12}(q_x)$. A given function (first object) can be simultaneously correlated with many other reference functions (second object) by separating the reference functions as horizontal strips, or *channels*, at the position of the second object. The DC spectral components corresponding to each reference channel are kept separated in the output plane by using a cylindrical lens as the final transform lens.

The method of pattern recognition just described is perhaps the simplest to understand, but many other techniques with various advantages have been developed. In this introduction, we can only mention briefly one other approach that makes use of a hologram as a spatial filter. The technique was introduced by Vander Lugt in 1963. First, a hologram is made of a particular pattern to be "recognized." Let us refer to its amplitude distribution by the function f. The holographic plate is situated in the Fourier-transform plane. The resulting hologram is called a *matched filter* or a *Vander Lugt filter*, after its originator. This holographic filter is subsequently used in the Fourier plane, together with various test patterns having amplitude distributions $g_1, g_2, \ldots$, in the object plane. It can be shown [1, 2] that, in general, three angularly distinct beams result, so that three distinct spots appear in the image plane. One of these is centered on the optical axis, and the other two, off-axis, represent, respectively, the convolution and the correlation of the f and g pattern functions. When the test pattern g matches the desired pattern f, the correlation image appears with a bright, central spot, and pattern recognition is achieved.

Optical correlation techniques have been developed that allow recognition of a pattern independently of its size or orientation. Matched filtering techniques using incoherent light have also been devised to reduce the signal-to-noise background that is typical in coherent light systems.

Another Model of Imaging: Convolution.   In the preceding sections, we have presented imaging as (1) a result of diffraction or Fourier analysis, producing a spectrum of spatial frequencies, and (2) their subsequent recombination, or Fourier synthesis, to form the image. We wish now to introduce some of the mathematical formalism and terminology commonly used to discuss Fourier transformations in another approach to imaging.

Consider a two-dimensional aperture (xy plane) and its image (XY plane) formed by some intervening optical system. We assume that the two sets of axes are similarly oriented, as in Figure 25-1. In the case of a perfect optical system, there is established a one-to-one correspondence between conjugate object and image points. For simplicity, we shall assume a lateral magnification of 1. Let the irradiance of such a (hypothetical) perfect image be given by $I_0(X, Y) \equiv I_0(x, y)$. In reality, light from each object point is spread out over its conjugate image point, due to diffraction and aberration. In this model, the resultant image is considered to be the overlapping of such "blurred" image points. In a *linear system*, these elementary irradiance patterns are simply additive. Let the actual irradiance over the image plane be given by $I_i(X, Y)$. The transformation from $I_0(X, Y)$ to $I_i(X, Y)$ clearly characterizes the optical system and is accomplished by a third function, called the *point spread function*, $G(x, y, X, Y)$. For example, in the case of an aberration-free system, G is just the function describing the Airy pattern (Section 16-3).

Now, if we assume the point spread function to be space-invariant (independent of object point coordinates), it can only depend on the relative displacement of conjugate points:

$$G(x, y, X, Y) = G(X - x, Y - y)$$

Further, if the light from the object plane is incoherent, irradiances add,[1] and we

---

[1] If the light is coherent, the sum is a vector sum of complex electric field amplitudes.

can write for the irradiance at the image point (X, Y) due to all object points (x, y):

$$\underbrace{I_i(X, Y)}_{\substack{\text{image} \\ \text{irradiance}}} = \iint \underbrace{I_0(x, y)}_{\substack{\text{object} \\ \text{irradiance}}} \underbrace{G(X - x, Y - y)}_{\substack{\text{point spread} \\ \text{function}}} dx\, dy \qquad (25\text{-}31)$$

The integral in Eq. (25-31) is called the *convolution*[2] of the functions I_0 and G, usually abbreviated by

$$I_i = I_0 \otimes G \qquad (25\text{-}32)$$

Suppose that we calculate the Fourier transform of each of these functions, represented by $\mathcal{F}(I_0)$, $\mathcal{F}(I_i)$, and $\mathcal{F}(G)$. The *convolution theorem* (see problems) states that the Fourier transform of the convolution of two functions is equal to the product of their individual transforms. Symbolically,

$$\mathcal{F}(I_i) = \mathcal{F}(I_0 \otimes G) = \mathcal{F}(I_0) \times \mathcal{F}(G) \qquad (25\text{-}33)$$

The content of Eqs. (25-32) and (25-33) can be succinctly summarized by stating that convolution in *real space* corresponds to multiplication in *Fourier space*. Combining this result with our understanding of the equivalence of Fourier transform and spatial frequency spectrum (or Fraunhofer diffraction function) of an aperture function, we can read Eq. (25-33) as follows: The spatial frequency spectrum of image irradiance is equal to the product of the spatial frequency spectrum of object irradiance *times* the spatial frequency spectrum of the point spread function. The last of these, $\mathcal{F}(G)$, is called the *optical transfer function* (OTF), because it *transfers* or changes the object spectrum into the image spectrum. Thus the OTF is used to characterize the performance of an optical system.

As an example of the convolution theorem, recall the results given in Eq. (16-32) for the Fraunhofer diffraction of a grating. There we found that the two product functions could be interpreted separately as diffraction from a single slit and interference from multiple (negligible width) slits. Since these functions are the Fourier transforms of their respective aperture functions, we can say either that (1) in Fourier space, the grating diffraction pattern is given by the product of the Fourier transform of the single-aperture function and the transform of the array of line sources defining the grating; or (2) in real space, the grating aperture function is a convolution of the slit-array aperture function with the single-slit aperture function. The second formulation is suggested by Figure 25-9. Practically speaking, if one knows the Fourier transform of simple aperture functions, one can more easily cal-

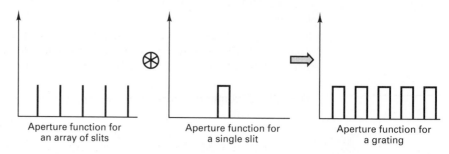

| Aperture function for an array of slits | Aperture function for a single slit | Aperture function for a grating |

Figure 25-9 Symbolic representation of the convolution theorem for a grating.

---

[2] This integral has other important applications in physics. It requires the multiplication of one function at each point by the whole of another function and then the summation of the results. Hence it is also called a *folding* or *superposition* integral.

culate the Fraunhofer pattern that results from more complicated aperture functions by using the convolution theorem.

System Evaluation Using the Transfer Function. Characterization of the imaging capacity of an optical system by simply citing its resolving power does not give an adequate assessment of the system's performance. The preferred criterion of performance is the optical transfer function (OTF). To test an optical system properly, objects having both high and low spatial frequencies are required. As usual, low spatial frequencies are sufficient to image the gross details of an object, while high spatial frequencies are required to reproduce the finer details.

One technique of testing an optical system is to use a series of test patterns with sinusoidally varying darkness, each at a different spatial frequency K. When illuminated, incoherent imaging of the test pattern takes place. Let us assume a system magnification of 1. The image produced by a linear optical system is also sinusoidal at the same spatial frequency, but with a modification of amplitude and phase, as shown in Figure 25-10a. The OTF encompasses both modifications when it is written in complex form:

$$\text{OTF} = (\text{MTF})\, e^{i(PTF)} \tag{25-34}$$

where MTF is the modulus and PTF is the phase. When either is known as a function of frequency, MTF is the *modulation transfer function* and PTF is the *phase transfer function*. The sinusoids corresponding to object and image are described by their *contrast modulation* γ, given by

$$\gamma_0 = \frac{O_{\max} - O_{\min}}{O_{\max} + O_{\min}} \quad \text{and} \quad \gamma_I = \frac{I_{\max} - I_{\min}}{I_{\max} + I_{\min}} \tag{25-35}$$

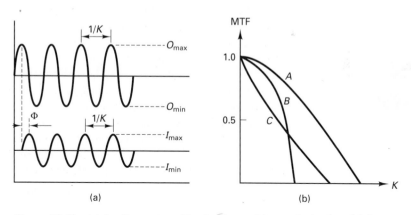

Figure 25-10 (a) Irradiance sinusoids of object and image, both of spatial frequency K. The optical system has unit magnification. (b) Modulation transfer function (MTF) for three optical systems plotted against spatial frequency.

Then the MTF and OTF are given, simply, by

$$\text{MTF} = \gamma_I/\gamma_0 \quad \text{and} \quad \text{PTF} = \Phi \tag{25-36}$$

Determination of MTF at various spatial frequencies, like the curves shown in Figure 25-10b, allows a more complete evaluation of system performance than resolution alone. Three systems are shown characterized. All approach a MTF of 1 as the spatial frequency approaches zero, but indicate different resolution limits as MTF becomes zero. System A clearly shows the best performance. System B has a lower frequency limit than system C, but better performance at lower frequencies.

Fourier-transform spectroscopy represents an elegant alternative to traditional methods of spectrum analysis. The special advantages of this technique have led to widespread applications in research and industry. Employing as a spectrometer an instrument such as the Michelson interferometer, these advantages derive both from the use of a large aperture at signal input and from the presence of the entire spectrum at signal output. The large energy throughput that results from the use of a large aperture is called the *Jacquinot advantage*, whereas the simultaneous processing of the entire spectral range during a single scan of the instrument is referred to as the *Fellgett*, or *multiplex*, *advantage*. Thus the Fourier-transform spectrometer is not limited, as are prism and grating spectrometers, by the presence of narrow slits that restrict both the wavelength interval and irradiance available at any one time. In addition, the technique is capable of high resolution, limited in principle only by the sample width of the input data and the wavelength region under analysis.

The large aperture and integrated throughput of the Michelson interferometer make it useful as a Fourier-transform spectrometer. It will be shown in the following treatment that the spectral distribution, or *spectrogram* (irradiance versus wave number), of the light incident on a Michelson interferometer is just the Fourier transform of the irradiance distribution, or *interferogram* (irradiance versus path difference), of its two-beam interference as a function of mirror movement. Figure 25-11 schematically shows the Michelson interferometer, which uses a beam splitter SP to separate equal-amplitude portions of a spectral input beam from source S and reunite them again after reflection from mirrors $M1$ and $M2$. The interfering beams are collected at detector D. Let the electric fields of the interfering beams for a particular wave number $k \, (= 2\pi/\lambda)$ component in the light source, on arrival at the detector, be represented by

$$E_1 = E_0 \cos (kx_1 - \omega t) \tag{25-37}$$

and

$$E_2 = E_0 \cos (kx_2 - \omega t) \tag{25-38}$$

where the two beams have experienced a physical path difference of $x = x_2 - x_1$ between separation and recombination. The time-averaged irradiance for the k component at the detector is then

$$I_k = \langle (E_1 + E_2)^2 \rangle$$

which gives, as also calculated in Chapter 10,

$$I_k = 2I_0(1 + \cos kx) \tag{25-39}$$

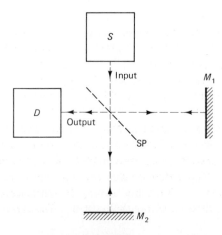

Figure 25-11 Elements of a Michelson interferometer used as a Fourier-transform spectrometer.

where I_0 represents the time-averaged irradiance of one beam. Since there will be a spread of k values in the source, I_k can be interpreted as irradiance $I(k)$ per unit k interval at k, giving an integrated irradiance over all wavelengths of

$$I = \int_0^\infty I(k)\ dk = \int_0^\infty 2I_0(k)\ dk + \int_0^\infty 2I_0(k)\ \cos(kx)\ dk \qquad (25\text{-}40)$$

The first term in the result behaves as a bias term, representing the constant integrated irradiance due to all wavelength components in the two noninterfering beams added together. The second term represents interference between the two beams and can be considered as a positive or negative deviation from the constant term, dependent upon the path difference x. Irradiance fluctuations about the constant bias comprise the spectral distribution (interferogram) given by

$$I(x) = \int_0^\infty I(k)\ \cos(kx)\ dk \qquad (25\text{-}41)$$

which is the Fourier transform of the spectrogram,

$$I(k) = \left(\frac{2}{\pi}\right) \int_0^\infty I(x)\ \cos(kx)\ dx \qquad (25\text{-}42)$$

Thus detection of the interferogram output $I(x)$, as a function of path difference x, at a point on the optical axis of the system enables one to calculate the spectral irradiance distribution $I(k)$ as a function of wavenumber by the Fourier-transform integration indicated in Eq. (25-42). In Figure 25-12 three experimental sample interferograms are shown, produced by a Michelson interferometer using various spectral inputs. Such interferograms are approximated for the purposes of Fourier-transform calculations by periodic sampling. When the function $I(x)$ is such a discrete set of sample points, the continuous Fourier transform is allowed to go over into sums and is referred to as a *discrete Fourier transform*. The use of finite sampling intervals across a finite total sample width or window leads to limitations both in the resolving power of the instrument and in the minimum wavelength that is unambiguously handled by the transform calculation. It can be shown that the restriction of data to a finite window x_w limits the resolution of the spectral distribution so that the minimum resolvable wavelength interval is given by

$$\Delta\lambda = \frac{\lambda^2}{x_w} \qquad (25\text{-}43)$$

yielding a resolving power of

$$\mathcal{R} \equiv \frac{\lambda}{\Delta\lambda} = \frac{x_w}{\lambda} \qquad (25\text{-}44)$$

One sees that the resolution is improved by using large sample widths. For example, a mirror movement of 0.5 cm, producing a total path difference or window of 1 cm, results in a resolving power at 500 nm of 20,000 and a resolution of 0.025 nm. Spectrometers have been built with relative mirror displacements of a meter or more, yielding resolving powers of 10^5 or greater. However, another important limitation must be taken into account. Because the true interferogram is only approximated at a specific sampling interval (nm/reading), a well-known phenomenon in sampling theory called *aliasing* places a limit on the smallest wavelength that can be unambiguously processed by this method. Wavelengths present in the input radiation, which are smaller than a particular λ_{min}, show up as longer wavelengths in the transformed spectrum. Such overlapping of wavelengths can be avoided by observing the *Nyquist criterion* of sampling theory: The signal must be sampled at a rate at

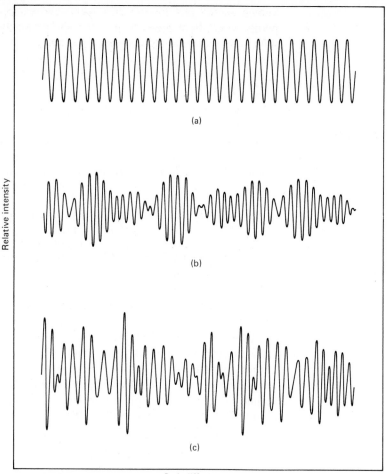

Figure 25-12 Interferograms produced by a Michelson interferometer using different light sources. (a) He-Ne laser. (b) Hg source, violet filter. (c) Hg source, unfiltered.

least twice as high as its highest-frequency component. It is interesting to note that this criterion is also used in the production of modern digital audio recordings, where an audio signal sampling rate of 50 kHz ensures accurate reproduction of the maximum audio frequency of 20 kHz. Expressed in terms of our experimental parameters, the criterion states equivalently that, to avoid aliasing, the sampling interval must be less than half the smallest wavelength present in the source. Thus the minimum wavelength is given by

$$\lambda_{\min} = \frac{2x_w}{N - 1} \qquad (25\text{-}45)$$

where N is the total number of samples, giving $N - 1$ sampling intervals. One sees now that a large x_w, which is beneficial in producing good resolution, may also be detrimental in limiting the spectral range of the spectrometer, unless N is also suitably large. The maximum number of data points, however, is limited by computer data-storage requirements and by computer time in handling the calculations. The number of operations performed by a computer in calculating the spectral distribution $I(k)$ is roughly equal to N^2. Use of the *Cooley-Tukey algorithm* for doing this series of calculations reduces the number of calculations to about $N \log_2 N$ and is known as the *fast Fourier transform*. For example, a transform using 1000 data points would be reduced from 1,000,000 operations to around 10,000, a considerable saving of computer time and expense. In the example just discussed, if the input radiation includes wavelengths in the visible and near ultraviolet, then N could not be less than about 67,000 without jeopardizing the correct analysis of wavelengths as small as 300 nm.

PROBLEMS

25-1. (a) Calculate the distances from the axis of the first three bright spots produced by a Ronchi ruling with transmitting slits of width 0.25 mm, as in Figure 25-2. Assume laser irradiation of 632.8 nm and a 50-cm focal length lens.
(b) What is the wavelength corresponding to the fundamental frequency?
(c) Determine the three lowest angular spatial frequencies, apart from the DC component, required in a Fourier representation of the Ronchi aperture function.
(d) What are the ratios of irradiance of the first three spots, relative to the irradiance of the "fundamental"?

25-2. (a) When two transmission functions are put together, by physically placing two transparencies back to back in the aperture plane, how must the combined transmission function relate to the individual transmission functions?
(b) Consider an aperture function formed by two perpendicularly crossed Ronchi rulings. What would you expect to see in the spectrum plane?

25-3. The optical density of film is defined as the common logarithm of its opacity. The opacity, in turn, is just the reciprocal of the transmittance T.
(a) Thus, show that optical density is equal to $-\log_{10} T$.
(b) Show that the total optical density of several film layers is just the sum of their individual optical densities.
(c) What is the transmittance of five layers of film, each with an opacity of 1.25? What is the net optical density of the combined layers?

25-4. The sinusoidal transmission of a grating varies as 5 sin (ay), in arbitrary units.
(a) To produce faithfully the sinusoidal variation in the transmittance of the grating, what bias is required in the transmission function, assuming 100% maximum transmission?
(b) Sketch the aperture function with and without the bias term.
(c) What is the irradiance function at the detector for unit irradiance incident at the grating?

25-5. Prove the convolution theorem, that is, prove that if

$$h(x) = f(x) \otimes g(x)$$

then

$$\mathcal{F}[h(x)] = \mathcal{F}[f(x)]\mathcal{F}[g(x)]$$

25-6. Plot the convolution in one dimension of two identical square pulses, of unit height and of 6 units length.

25-7. Determine the one-dimensional autocorrelation function $\Phi_{11}(\tau)$ for the sinusoidal function $y = A \sin(\omega t + \alpha)$.

25-8. (a) The output of a Michelson spectrometer is fed to a photodetector. The input is mercury green light of 546.1 nm. If one mirror translates at a speed of 5 mm/s, what is the frequency of modulation of the photocurrent?
(b) What is the beat frequency of the photocurrent when the input is the yellow light of sodium, at 5889.95 Å and 5895.92 Å? [*Hint:* Recall Eq. (11-14).]

25-9. The mirror translation in a Michelson spectrometer is 5 cm. What is the minimum resolvable wavelength at (a) 632.8 nm and at (b) 1 μm?

25-10. Light from a mercury lamp falls on the beam splitter of a student Michelson spectrometer. Wavelengths shorter than 360 nm are filtered from the light. The mirror translation rate is 71.5 nm/s. The rate at which spectrogram data is sampled is 1.28 readings/s. A total of 256 data points is fed to the computer for Fourier-transform analysis. Find the (a) window width x_w; (b) minimum resolvable wavelength interval at 400 nm; (c) minimum wavelength that is not subject to aliasing; (d) minimum sampling rate according to the Nyquist criterion.

25-11. The total path difference executed by a Fourier-transform spectrometer operating in the infrared is 2.78 mm. Its range is from 4400 to 400 cm^{-1}.

(a) What is its resolution in wave number?

(b) How many data points must be taken over the scan to avoid aliasing within this range?

(c) What is the scan rate if one run is completed in 30 s?

REFERENCES

[1] Steward, E. G. *Fourier Optics: An Introduction*, 2d ed. New York: Halsted Press, 1987. Ch. 4, 5.

[2] Goodman, Joseph W. *Introduction to Fourier Optics*. New York: McGraw-Hill Book Company, 1968.

[3] Almeida, Silverio P., and Guy Indebetouw. "Pattern Recognition via Complex Spatial Filtering." *Applications of Optical Fourier Transforms*, edited by Henry Stark. New York: Academic Press, 1982.

[4] Duffieux, P. M. *The Fourier Transform and Its Applications to Optics*, 2d ed. New York: John Wiley and Sons, 1983.

[5] Françon, M. *Optical Image Formation and Processing*. New York: Academic Press, 1979.

[6] Lee, S. H. ed. *Optical Information Processing, Fundamentals*. New York: Springer-Verlag, 1981.

[7] Bell, Robert John. *Introductory Fourier Transform Spectroscopy*. New York: Academic Press, 1972.

[8] Griffiths, Peter, and James A. de Haseth. *Fourier Transform Infrared Spectrometry*. New York: John Wiley and Sons, 1986. Ch. 1–4.

[9] Matthys, D. R., and F. L. Pedrotti. "Fourier Transforms and the Use of a Microcomputer in the Advanced Undergraduate Laboratory." *American Journal of Physics* 50 (1982): 990.

[10] Strong, John. *Concepts of Classical Optics*. San Francisco: W. H. Freeman and Company Publishers, 1958. Appendix F.

[11] Mertz, Lawrence. *Transformations in Optics*. New York: John Wiley and Sons, 1965. Ch. 1, 2.

[12] Williams, Charles S., and Orville A. Becklund. *Introduction to the Optical Transfer Function*. New York: John Wiley and Sons, 1989.

[13] Smith, F. Dow. "How Images Are Formed." *Lasers and Light, Readings from Scientific American*, p. 59. San Francisco: W. H. Freeman and Company Publishers, 1969.

[14] Campbell, Fergus W., and Lamberto Maffei. "Contrast and Spatial Frequency." *Scientific American* (Nov. 1974): 106.

[15] Bracewell, Ronald N. "The Fourier Transform." *Scientific American* (June 1989): 86.

[16] James, J. F., and R. S. Sternberg. *The Design of Optical Spectrometers*. London: Chapman and Hall Ltd., 1969. Ch. 3.

[17] Reynolds, George O., John B. DeVelis, George B. Parrent, Jr., and Brian J. Thompson. *Physical Optics Notebook: Tutorials in Fourier Optics*. Bellingham, Wash.: SPIE Optical Engineering Press, 1989.

26

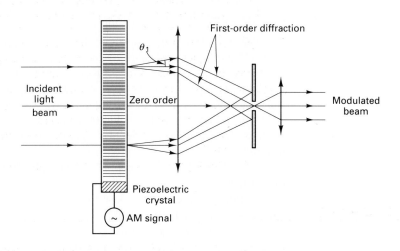

Incident light beam

Zero order

θ_1

First-order diffraction

Piezoelectric crystal

~ AM signal

Modulated beam

Nonlinear Optics and the Modulation of Light

INTRODUCTION

The optics treated in most of this text, including the processes of transmission, reflection, refraction, superposition and birefringence, fall in the category of what is called *linear optics*. When we speak of linear optics, we assume that an optical disturbance propagating through an optical medium can be described by a linear wave equation. As a consequence of this assumption, two harmonic waves in the medium obey the principle of superposition, traveling without distortion due to the medium itself or as a result of the mutual interference of the waves, regardless of the intensity of the light. Only the wavelength and velocity of a light beam in a transparent material are required to describe its behavior.

We now know that when the light intensity becomes great enough, linear optics is not adequate to describe the situation. With the advent of the more intense and coherent light made available by the laser, we find that the optical properties of the medium, such as its refractive index, become a function of the intensity of the light. When two or more light waves interfere within the medium, the principle of superposition no longer holds. The light waves interact with one another and with the medium. These *nonlinear* phenomena require an extension of the linear theory that allows for a nonlinear response of optical materials to the electromagnetic radiation.

In this chapter we define more precisely the area of nonlinear optics, describe some of the new properties that are found, and discuss some of their practical applications in the modulation of light beams.

Nonlinear phenomena are due ultimately to the inability of the dipoles in the optical medium to respond in a linear fashion to the alternating **E**-field associated with a light beam. Atomic nuclei are too massive and inner-core electrons too tightly bound to respond to the alternating **E**-field at the frequency of light ($\sim 10^{14}$–10^{15} Hz). Thus the outer electrons of the atoms in a material are primarily responsible for the polarization of the optical medium by the beam's **E**-field.[1] When the oscillations of these electrons in response to the field are small, the polarization is proportional to the **E**-field, as described later in Section 27-1. However, as the **E**-field increases with the intensity of the beam, strict proportionality begins to fail, just as the harmonic oscillations of a simple spring become increasingly anharmonic as the amplitude of the oscillations increases. Another means of exciting nonlinear behavior without using high beam intensities is to choose the exciting optical frequency near a resonant frequency of the oscillating dipoles, a technique widely utilized in *nonlinear spectroscopy* and known as *resonance enhancement* [1].

The polarization of a linear medium by an electric field **E** is usually written in the form

$$\mathbf{P} = \epsilon_0 \chi \mathbf{E} \tag{26-1}$$

where χ is the *susceptibility* and ϵ_0 is the vacuum permittivity. When departures from linearity are small, it is possible to represent the modification of the susceptibility in a nonlinear medium by a power series in the form

$$\chi = \chi_1 + \chi_2 E + \chi_3 E^2 + \cdots \tag{26-2}$$

When substituted into Eq. (26-1), the polarization takes the form

$$P = \epsilon_0(\chi_1 E + \chi_2 E^2 + \chi_3 E^3 + \cdots) \tag{26-3}$$

or

$$P = \underbrace{P_1}_{\text{linear}} + \underbrace{(P_2 + P_3 + \cdots)}_{\text{small nonlinear terms}}$$

where the subscripts on χ match the powers of E and reflect the decreasing magnitude of the higher-order terms. The linear and nonlinear susceptibility coefficients characterize the optical properties of the medium, and this relation between P and E completely characterizes the response of the optical medium to the field.

The first term P_1 in Eq. (26-3) represents linear optics in which the polarization of the medium is simply proportional to the **E**-field. Unless the **E**-field is very large, the coefficients of the higher-power terms are too small to allow higher-power terms to influence the polarization appreciably. Only with the availability of intense, coherent light have these higher-order terms become important. The high coherence of laser light allows the beam to be focused onto small spot sizes of the order of a wavelength,[2] producing **E**-fields of around 10^{10} V/m, on the order of the fields binding electrons to nuclei in the optical medium. Peter Franken and his associates are credited with the first nonlinear coherent optics experiment [2], conducted at the University of Michigan in 1961. The team focused the coherent 694.3-nm output from a pulsed ruby laser onto a quartz crystal and detected *second harmonic generation*, the presence in the output of a weak ultraviolet coherent radiation component at 347.15 nm, twice the frequency or half the wavelength of the exciting light. This nonlinear phenomenon is discussed in the next section.

---

[1] We speak here of the *electric* polarization, rather than the *wave* polarization that was the subject of Chapters 14 and 15.

[2] See the discussion on focusing laser beams in Section 21-4.

Second harmonic generation results from the contribution of the second-order term in Eq. (26-3):

$$P_2 = \epsilon_0 \chi_2 E^2 \qquad (26\text{-}4)$$

in which the second-order polarization term P_2 of the optical medium is proportional to the square of the **E**-field. Figure 26-1 shows the polarization as a function of the **E**-field for the linear case and the deviation from linearity due to this second-order term.

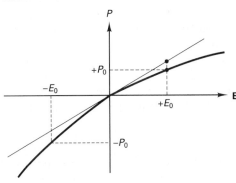

Figure 26-1 Linear and typical nonlinear response of polarization P to an applied electric field **E**. For equal positive and negative fields, the response of the optical medium is not symmetrical in the case of the nonlinear (curved line) response. In this case, the negative field E_0 produces a greater polarization than a positive field of the same magnitude.

It can easily be shown that the second-order term makes no contribution to polarization in an isotropic optical material, or one having a center of symmetry. A crystal having a center of symmetry is characterized by an inversion center, such that if the radial coordinate r is changed to $-r$, the crystal's atomic arrangement remains unchanged and so the crystal responds in the same way to a physical influence. In such a crystal, reversing the applied field should not—except for a change in sign—change any physical property, such as its polarization. Thus we should have both

$$P_2 = \epsilon_0 \chi_2 (+E)^2 \quad \text{and} \quad -P_2 = \epsilon_0 \chi_2 (-E)^2$$

Because the **E**-field is squared, $P_2 = -P_2$, which can only be true if $P_2 = 0$. The quartz crystal used by Franken, and many other crystals as well, do not possess inversion symmetry. They can, therefore, manifest second harmonic generation in addition to other second-order phenomena to be described presently.

The appearance of a *second harmonic* in the polarization is expected from the following mathematical argument. If the applied **E**-field, or one of its Fourier components, is of the form

$$E = E_0 \cos \omega t$$

substitution into Eq. (26-4) gives

$$P_2 = \epsilon_0 \chi_2 E_0^2 \cos^2 \omega t = \epsilon_0 \chi_2 E_0^2 \left[\tfrac{1}{2}(1 + \cos 2\omega t) \right]$$

where we have substituted the double-angle identity for $\cos^2 \omega t$. Then

$$P_2 = \tfrac{1}{2}\epsilon_0 \chi_2 E_0^2 + \tfrac{1}{2}\epsilon_0 \chi_2 E_0^2 \cos 2\omega t \qquad (26\text{-}5)$$

The second-order polarization P_2 consists of a term of twice the frequency of the applied optical field as well as a constant or DC component that represents *optical rectification*. Thus the dipole oscillations constituting the time dependence of the polarization generate electromagnetic radiation of angular frequency 2ω, which is present in the resultant field, together with the much stronger first-order effects at

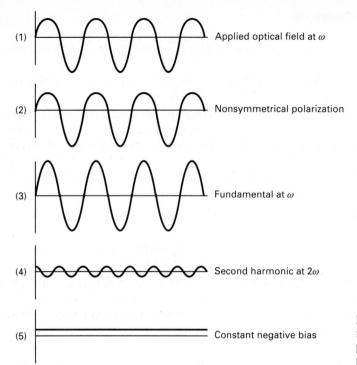

(1) Applied optical field at ω

(2) Nonsymmetrical polarization

(3) Fundamental at ω

(4) Second harmonic at 2ω

(5) Constant negative bias

Figure 26-2   Exciting E-field (1) and its effects in a nonlinear medium: a nonsymmetrical polarization (2) and its radiative Fourier components (3, 4, 5).

the fundamental frequency ω. Figure 26-2 illustrates the Fourier components of the nonsymmetrical polarization that produces them.

Because of dispersion, the light continuously generated at frequency 2ω travels at a different speed in the optical material than the light at frequency ω. The two interfering waves are periodically in and out of step as they traverse the crystal. One can show [3] that the output intensity is proportional to

$$\text{sinc}^2 \left(\frac{L \Delta k}{2} \right)$$

where L is the distance into the crystal and k is the wave propagation constant, equal to $n\omega/c$. Here, $\Delta k = k_{2\omega} - 2k_\omega$. When $\Delta k = 0$, therefore, the crystal shows no dispersion and the intensity factor above is a maximum. Because dispersion is present in materials, $\Delta k \neq 0$ and the intensity factor describes the resultant interference between the fundamental and second harmonic. Reference to the discussion of the sinc function (see Figure 16-2) shows that the change in argument of the function from a minimum to a maximum is $\pi/2$. The corresponding change in L is called the *coherence length*.[3] Thus

$$L_c \Delta k/2 = \pi/2 \quad \text{and} \quad L_c = \pi/\Delta k$$

The coherence length can also be expressed by

$$L_c = \frac{\pi}{\Delta k} = \frac{\pi}{\dfrac{2\omega}{c} \Delta n} = \frac{\pi}{\dfrac{4\pi}{\lambda_0} \Delta n}$$

$$L_c = \frac{\lambda_0}{4 \Delta n} \tag{26-6}$$

---

[3] This *coherence length* is not to be confused with the quantity of the same name presented in Chapter 12.

where we have used

$$k = \frac{n\omega}{c} \quad \text{and} \quad \Delta k = \frac{n_{2\omega}(2\omega)}{c} - \frac{2n_\omega(\omega)}{c} = \frac{2\omega}{c}\,\Delta n$$

Here λ_0 is the vacuum wavelength of the fundamental and Δn the difference in refractive index for fundamental and second harmonic.

Example

Suppose the fundamental $\lambda = 0.8\ \mu m$ and the refractive indices in KDP (potassium dihydrogen phosphate) are 1.4802 for the second harmonic and 1.5019 for the fundamental. What is the maximum crystal thickness useful in generating second harmonic light.

Solution   Substitution into Eq. (26-6) gives

$$L_c = \frac{0.8\ \mu m}{(4)(1.5019 - 1.4802)} = 9.2\ \mu m$$

This calculation shows that the maximum crystal thickness useful in generating second harmonic light is typically quite small, in this case around 10 times the wavelength of the fundamental. Crystals with thicknesses equal to their coherence length are impractically small.

A technique that can circumvent the small coherence length of nonlinear crystals involves making use of their birefringence. As described in Section 15-5, the refractive index (and so the velocity) of the extraordinary (E) ray varies with direction through the crystal. If a direction through the crystal is chosen such that $n_{2\omega}$ for the E-ray equals n_ω for the O-ray, the fundamental and second harmonic waves remain in step and the crystal can be a centimeter or so thick. This technique is called *index matching* or *phase matching* and is clarified by Figure 26-3, which shows how the ellipsoids representing the velocity versus crystal direction for the E- and O-rays intersect along the direction of matching.

Second harmonic generation is not the only nonlinear phenomenon that results from the quadratic dependence of the polarization on the electric field. Table 26-1 lists others, as well as several that depend on the next higher order of approximation, where $P_3 = \epsilon_0 \chi_3 E^3$. For example, notice that for third-order nonlinear proc

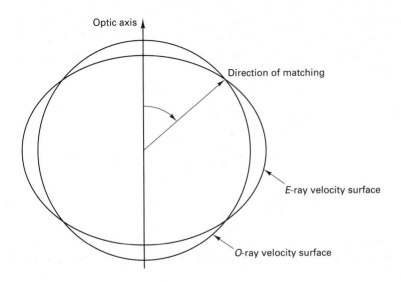

Optic axis

Direction of matching

E-ray velocity surface

O-ray velocity surface

Figure 26-3   Velocity ellipsoids for orthogonally polarized light beams in a birefringent medium. The O-ray ellipsoid is spherical and intersects the E-ray ellipsoid along a direction (shown relative to the optic axis) for which both rays have the same velocity.

TABLE 26-1   LINEAR AND NONLINEAR PROCESSES

Linear first order: $P_1 = \epsilon_0 \chi_1 E$	Nonlinear second order: $P_2 = \epsilon_0 \chi_2 E^2$	Nonlinear third order: $P_3 = \epsilon_0 \chi_3 E^3$
Classical optics:	Materials lacking inversion	Third harmonic generation
Superposition	symmetry:	Four-wave mixing
Reflection	Second harmonic generation	Kerr effect
Refraction	Three-wave mixing	Raman scattering
Birefringence	Optical rectification	Brillouin scattering
Absorption	Parametric amplification	Optical phase conjugation
	Pockels effect	

esses, *third harmonic generation* can occur. Some of the higher-order nonlinear phenomena are discussed in the paragraphs that follow.

26-3 FREQUENCY MIXING

When two or more incident beams with different frequencies are allowed to interfere within a nonlinear dielectric material, *frequency mixing* can occur. Let two interfering incident waves of frequencies ω_1 and ω_2 be represented in the form

$$E = E_{01} \cos \omega_1 t = E_{02} \cos \omega_2 t$$

or, in the equivalent exponential form,

$$E = \tfrac{1}{2}E_{01}(e^{i\omega_1 t} + e^{-i\omega_1 t}) + \tfrac{1}{2}E_{02}(e^{i\omega_2 t} + e^{-i\omega_2 t})$$

Second-order polarization $P_2 = \epsilon_0 \chi_2 E^2$ requires the square of the incident field. Thus, by observation, we can see that the square of E produces harmonics in $2\omega_1$, $2\omega_2$, $\omega_1 - \omega_2$, and $\omega_1 + \omega_2$. Similarly, third-order polarization produces frequency mixing in all possible combinations of two or more incident beams.

A special case of frequency mixing is the process known as *parametric amplification*. Instead of $2\omega_1 \rightarrow \omega_3$, as in second harmonic generation, it is possible to achieve a situation in which $\omega_3 \rightarrow \omega_1 + \omega_2$, with power flowing from a *pump wave* at ω_3 into *signal* and *idler* waves at ω_1 and ω_2. (Notice that in second harmonic generation the power flow is reversed, from the wave at ω_1 to a wave of double the frequency at ω_3.) To visualize this effect more clearly, suppose a small signal wave at ω_s and a powerful pump wave at ω_p interact within a nonlinear medium. A difference or idler frequency $\omega_i = \omega_p - \omega_s$ is produced as a beat frequency. This idler frequency can in turn beat with the pump frequency to enhance the signal frequency, $\omega_s = \omega_p - \omega_i$. In this process, therefore, both idler and signal waves are amplified, drawing power from the pump wave. When signal and idler frequencies correspond to resonant frequencies in the nonlinear crystal acting as a tuned Fabry-Perot cavity, the parametric oscillator is a tunable source of coherent radiation. Tuning the cavity is accomplished by varying the refractive index of the cavity through control of temperature or an applied DC field.

Some of the nonlinear processes itemized in Table 26-1 are discussed in the following paragraphs, together with several of their device applications in the production of modulated light beams. By *light modulation*, we refer to any means of modifying the amplitude (AM), frequency (FM), phase, polarization, or direction of a light wave. The purpose of modulation is to render the wave capable of carrying information, as described in Chapter 24, for example, where light beams carry information through optical fibers. Light choppers and shutters can accomplish some modulation mechanically. We are interested here in describing the modulation that is accomplished by varying the refractive index of a material through the use of electric, magnetic, or acoustic means.

The basic equation describing nonlinear behavior was given as Eq. (26-3), a relation between the polarization of the medium and the applied **E**-field. In dealing with crystalline media, which represent most of the useful acousto-optic materials, it is customary to express nonlinearity of the refractive index n by an equation [4] analogous to Eq. (26-2) for the susceptibility:

$$\frac{1}{n^2} = \frac{1}{n_0^2} + rE + RE^2 \qquad (26\text{-}7)$$

where r and R are the linear and quadratic *electro-optic coefficients*,[4] respectively, and we assume that there is no other effect present (like crystal strain) that can modify n. The refractive index in the absence of an applied field is n_0. In general, the refractive index depends on the propagation direction and wave polarization relative to the crystal axes. Since E is a vector field, the coefficients r and R are tensors that reflect the crystal symmetry. Depending on the degree of symmetry, many tensor components may vanish or become equal to others, reducing the total number of independent elements[5] required to represent a particular crystalline material [5].

26-4 THE POCKELS EFFECT

The *Pockels effect* results from the linear term in Eq. (26-7), where E is an applied DC field. This effect can be considered a special case of two-wave mixing, where one of the waves is the incident optical wave and the other a field of zero frequency. The optical **E**-field can be small, since the DC field is itself large enough to produce nonlinear behavior. In general, the DC field redistributes electrons in such a way that birefringence is induced in an otherwise isotropic material, or new optic axes appear in naturally birefringent crystals. Since the Pockels effect is a second-order effect relative to the polarization (Eq. 26-3), it is not found in isotropic materials having inversion symmetry. All crystalline materials exhibiting a Pockels effect[6] are also *piezoelectric*, that is, they show induced birefringence due to mechanical strain. Since the effect was discovered in 1893, long before the discovery of the laser, it was well known even before intense optical fields became available.

In one configuration of the Pockels cell, the natural optical axis of the crystal is aligned parallel to the applied field. Fast and slow axes are induced in a plane normal to the applied field, as shown in Figure 26-4. If the Pockels cell crystal is rotated until the FA and SA are at 45° to the x- and y-axes, a vertically polarized light wave E_0 incident on the crystal along the field direction has equal amplitude components on FA and SA. These components experience different refractive indices and different speeds through the crystal. The crystal therefore behaves as a phase retarder (Section 14-2), and the component waves emerge with a phase difference.

---

[4] It is important to distinguish the "order" of an electro-optic process as determined from Eq. (26-3) for the polarization and as determined from Eq. (26-7) for $1/n^2$ or Δn. For example, the Pockels effect is a second-order effect (involving χ_2) by the first criterion and a first-order effect (involving r) by the second. Both criteria are used.

[5] The linear electro-optic tensor p_{ij} is defined by the relation $\Delta(1/n^2)_i = \Sigma_j p_{ij} E_j$, with $i = 1, 2, 3, \ldots, 6$ and $j = x, y, z$. For example, in crystals of triclinic symmetry, all the 18 possible tensor elements are required; in zinc blende (GaAs), only one is required; in centrosymmetric crystals, all elements are zero.

[6] Single crystals can be divided into 32 symmetry classes; of these, 20 show the Pockels effect.

Figure 26-4 Pockels cell schematic. The retardation action due to an applied voltage V is suggested by a separation on axis of the polarized components transmitted by the fast and slow axes of the crystal.

One component advances in phase by $\Delta\varphi$ and the other lags in phase by $\Delta\varphi$ while traversing the crystal of length L, so that their relative phase on emerging is given by $\Phi = 2\Delta\varphi$. Now $\Delta\varphi = (2\pi/\lambda_0)L\,\Delta n$, where λ_0 is the vacuum wavelength and $L\,\Delta n$ represents the optical-path difference. We find Δn from Eq. (26-7), which, for small changes, can be approximated by $d(1/n^2) = rE$. The E^2 term is considered negligible in the Pockels effects. Thus

$$-2\frac{dn}{n^3} = rE$$

$$|\Delta n| \cong \frac{r}{2}n_0^3 E$$

Substituting into the phase equations above, we find

$$\Phi = \frac{2\pi}{\lambda_0}rn_0^3 EL = \frac{2\pi}{\lambda_0}rn_0^3 V$$

where $V = EL$ is the voltage applied across the length of the cell. Notice that the phase difference Φ is independent of the crystal length.

If the Pockels cell is to behave as a half-wave plate, we need to make $\Phi = \pi$. The half-wave voltage required is then

$$V_{HW} = \frac{\lambda_0}{2rn_0^3} \tag{26-8}$$

Example

Suppose the cell is made from a KD*P crystal of 1-cm thickness and the optical wave has a wavelength of 1.06 μm. What half-wave voltage is required?

TABLE 26-2 LINEAR ELECTRO-OPTIC COEFFICIENTS
FOR REPRESENTATIVE MATERIALS

Material (wavelength if not 633 nm)	Linear electro-optic coefficient[a] r (pm/V)	Refractive index n_0
KH₂PO₄ (KDP)	11	1.51
KD₂PO₄ (KD*P)	24.1	1.51
(NH₄)H₂PO₄ (ADP) $\lambda = 0.546$ nm	8.56	1.48
LiNbO₃ (lithium niobate)	30.9	2.29
LiTaO₃ (lithium tantalate)	30.5	2.18
GaAs (gallium arsenide) $\lambda = 10.6$ μm	1.51	3.3
ZnS (zinc sulfide) $\lambda = 0.6$ μm	2.1	2.36
Quartz	1.4	1.54

[a] Depending on crystalline symmetry, materials have more than one electro-optic coefficient. Only one has been listed here for use in a Pockels cell. These and others may be found in [3].

Solution From Table 26-2, we find $r = 24.1 \times 10^{-12}$ m/V and a refractive index of 1.51. Then

$$V = \frac{1.06 \times 10^{-6}}{2(24.1 \times 10^{-12})(1.51)^3} = 6390 \text{ V}$$

Thus an applied voltage of 6.4 kV transforms the crystal into a half-wave plate.

Recall (Section 15-7) that in this case the plane of polarization of the emergent light is rotated by 90° relative to the plane of polarization of the incident light. If a linear polarizer with TA along the x-direction intercepts the emergent light, as in Figure 26-5, the transmittance of the system is zero when $V = 0$ and maximum when $V = V_{HW}$. Variations in V therefore modify the polarization state of the emergent light, rendering it elliptical, in general, with an x-component that can be trans-

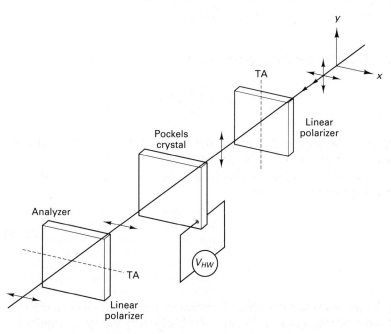

Figure 26-5 The effect of applying a half-wave voltage to a Pockels cell when the system includes a crossed polarizer and analyzer. The vertically polarized incident beam is transmitted as a horizontally polarized beam. For other values of applied voltage, the beam incident on the analyzer is elliptically polarized and is only partially transmitted.

mitted by the analyzer. In effect, the polarizer-analyzer pair transforms phase modulation into amplitude modulation. Thus we see that the transmittance of the system can be modulated by variations in the applied voltage. Variations of a signal voltage superimposed on V are transformed into variations in light intensity in such a device, known as a *Pockels electro-optic modulator*.

The Pockels cell can be used also with the field oriented orthogonally to the beam direction, an arrangement that simplifies placement of the electrodes. In the geometry we have been describing, the electrodes are usually end-rings that allow the light beam to pass through and still provide a reasonably uniform field in the crystal.

The transmittance of the beam can be expressed [6] by the relation:

$$I = I_{\text{max}} \sin^2 \left(\frac{\pi}{2} \frac{V}{V_{HW}} \right) \tag{26-9}$$

and is plotted in Figure 26-6. To take advantage of the more linear region of the transmittance, a quarter-wave plate is often inserted between the initial polarizer and the Pockels crystal. This has the effect of producing 50% transmittance when $V = 0$ so that the operating point is located at P in the figure, rather than at the origin. Variations in modulating voltage, if not too large, then occur with a system response that is linear.

Two other closely related applications of the Pockels cell are illustrated in Figures 26-7 and 26-8. In Figure 26-7, the cell is used as a *Q-switch* that allows sudden dumping of the energy stored in a laser. Without an applied voltage, the Pockels crystal transmits the linearly polarized beam from a laser cavity without changing its state of polarization. A Glan-laser prism, tuned to pass vertically polarized light, re-

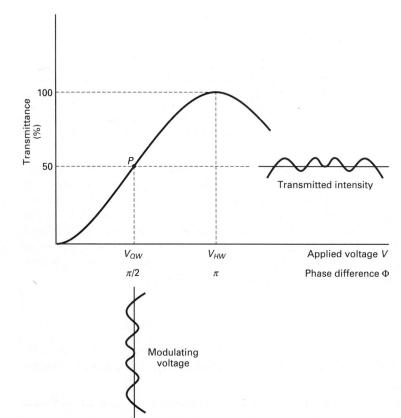

Figure 26-6 Transmittance curve for the Pockels cell modulator. Without a quarter-wave plate, the transmittance is zero when the applied voltage is zero. Using a quarter-wave plate between polarizer and modulator, the transmittance is 50% at operating point P when the applied voltage is zero. Under these conditions, the modulator responds more linearly to an input signal.

(a)

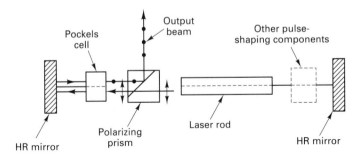

(b)

Figure 26-7 Light-controlling action of a Pockels cell, used as a Q-switch. The configuration in (a) produces low transmission at zero-cell voltage and in (b) high transmission at half-wave voltage. In (b), the incident and reflected beams are separated for clarity. Repeated re-entries of the beam into the laser cavity initiates stimulated emission that produces the laser pulse.

Figure 26-8 Light-controlling action of a Pockels cell, used as a cavity dumper. When half-wave voltage is applied to the Pockels cell, it rotates the linear polarization of the laser beam so that it can be dumped by the polarizing prism.

jects the polarized beam. With half-wave voltage applied, the linearly polarized beam from the laser cavity is rotated by 90° and is accordingly passed by the Glan-laser prism, allowing the beam to be back-reflected from a high-reflectance mirror. The configuration now permits rapid traversal of the beam back and forth through the laser cavity and stimulated emission occurs, producing an energetic pulse of laser radiation. In Figure 26-8, the Pockels cell is used to initiate cavity dumping. When half-wave voltage is applied, the polarization state of the laser radiation is rotated by 90°, so that it can be extracted with the help of the polarizing prism.

26-5 THE KERR EFFECT

When the optical medium is isotropic, as in the case of liquids and glasses, the Pockels effect is absent and the polarization is modified by the third-order[7] electro-

---

[7] In Eq. (26-7), the Kerr effect arises from the E^2 term involving the quadratic electro-optic coefficient.

optic effect, better known as the *Kerr effect*. This effect, like all third-order effects, occurs whether or not a material possesses inversion symmetry. Actually, the Kerr effect was the first electro-optic effect to be discovered (1875), despite its third-order dependence. Kerr cells usually contain nitrobenzene or carbon disulfide in the space between two electrodes across which a voltage is applied, as indicated in Figure 26-9. The applied electric field induces birefringence with an optic axis parallel to the applied field. Light traversing the cell thus encounters two refractive indices for polarizations parallel and perpendicular to the optic axis, and phase retardation results. In this case, Eq. (26-7) becomes

$$\frac{1}{n^2} = \frac{1}{n_0^2} + RE^2 \quad \text{or} \quad |\Delta n| = \frac{R}{2}n_0^3 E^2 \tag{26-10}$$

Figure 26-9 Kerr cell. The applied voltage creates a field that is perpendicular to the beam direction. As with the Pockels cell, a modulating voltage produces phase modulation that is converted to amplitude modulation by the polarizer-analyzer pair.

Experimentally, the difference between n_E and n_0 is found to obey a relation of the form,

$$\Delta n = KE^2\lambda \tag{26-11}$$

where K is the *Kerr constant*.

Equating Eqs. (26-10) and (26-11), we find the relationship between K and R to be

$$K = \frac{Rn_0^3}{2\lambda}$$

As explained for the Pockels cell, the relative phase retardation for the ordinary and extraordinary components is

$$\Phi = \frac{2\pi}{\lambda}L\,\Delta n$$

Introducing the Kerr constant through Eq. (26-11),

$$\Phi = \frac{2\pi K V^2 L}{d^2}$$

where we have set $V = Ed$ and d is the interelectrode distance. To function as a half-wave plate, $\Phi = \pi$, and we find that the required voltage V_{HW} is given by

$$V_{HW} = \frac{d}{\sqrt{2KL}} \tag{26-12}$$

TABLE 26-3 KERR CONSTANT
FOR SELECTED MATERIALS

Material (λ = 589 nm, RT)	K (pm/V^2)
Nitrogen (STP)	4×10^{-6}
Glass (typical)	0.001
Carbon disulfide (CS_2)	0.036
Water (H_2O)	0.052
Nitrotoluene ($C_5H_7NO_2$)	1.4
Nitrobenzene ($C_6H_5NO_2$)	2.4

Example

Consider a nitrobenzene Kerr cell for which $K = 2.4 \times 10^{-12}$ m/V^2 (see Table 26-3) at room temperature and $\lambda = 589.3$ nm. If $d = 1$ cm and $L = 3$ cm, what half-wave voltage is needed?

Solution   Substituting into the equation above,

$$V_{HW} = \frac{0.01}{\sqrt{(2)(2.4 \times 10^{-12})(0.03)}} = 26.4 \text{ kV}$$

Thus the Kerr cell behaves as a half-wave plate at a voltage of around 26 kV, considerably higher than for a typical Pockels cell.

Kerr cells can be used as modulators in the manner described for Pockels cells. Because of the higher voltages required, and because of the toxic and explosive nature of nitrobenzene, Pockels cells are usually preferred. Nevertheless, Kerr cells find application as high-speed shutters and as a substitute for mechanical light choppers, being capable of response to frequencies in the range of 10^{10} Hz. They can often be found operating as Q-switches in pulsed lasers.

26-6 THE FARADAY EFFECT

In contrast to the electro-optic effects discussed to this point, the *Faraday Effect* is a first-order (i.e., $\Delta n \propto B$) *magneto*-optic interaction.[8] When a transparent material is placed in a magnetic field and linearly polarized light is passed through it along the direction of the magnetic field, the emerging light is found to remain linearly polarized, but with a net rotation β of the plane of polarization that is proportional both to the thickness d of the sample and the strength of the magnetic field B, according to the empirical relation,

$$\beta = VBd \tag{26-13}$$

Here V is the *Verdet constant* for the material, usually expressed in minutes of angle per Gauss-cm (G-cm). The Verdet constant is both temperature and wavelength dependent. Its dependence on the wavelength is explained in what follows.

An interesting aspect of the Faraday rotation is that the sense of rotation relative to the magnetic field direction is, for a given material, independent of the prop-

---

[8] The magnetic field analogues of the Kerr effect (where $\Delta n \propto E^2$) are the *Voigt effect* (in gases) and the *Cotton-Mouton effect* (in liquids), in which a constant magnetic field is applied normal to the light-beam direction. Both are very small effects and will not be discussed further here.

agation direction of the light. Thus, repeated forward and backward traversals of the material by a light beam has a cumulative effect on the angle of rotation β. This behavior contrasts with that exhibited in the closely related phenomenon of *optical activity,* discussed in Section 15-6.

The optical rotation of the polarized light can be understood as *circular birefringence,* the existence of different indices of refraction for left-circularly and right-circularly polarized light components (Section 15-6). Recall that linearly polarized light is equivalent to a combination of right- and left-circularly polarized components. Each component is affected differently by the applied magnetic field and traverses the sample with a different speed, since the refractive index is different for the two components. The end result consists of left- and right-circular components that are out of phase and whose superposition, upon emerging from the Faraday rotator, is linearly polarized light with its plane of polarization rotated relative to its original orientation.

A classical derivation [7] of the angle of rotation β predicts a relation of the form

$$\beta = \left(\frac{e}{2m} \frac{\lambda}{c} \frac{dn}{d\lambda} \right) Bd \tag{26-14}$$

with e and m the electronic charge and mass, c the speed of light, λ the wavelength, and $dn/d\lambda$ the *rotatory dispersion.* By comparison with Eq. (26-13), the theory predicts a dependence of the empirical Verdet constant V given by

$$V = \frac{e}{2mc} \lambda \frac{dn}{d\lambda} \tag{26-15}$$

When the constants are evaluated and V is expressed in the standard units of min/(G-cm), Eq. (26-15) becomes

$$V = 1.0083\lambda \frac{dn}{d\lambda}$$

We note that the Verdet constant is proportional to both the wavelength of the light and the induced rotatory dispersion in the medium. Measured values for V at 589.3 nm are given in Table 26-4.

The Faraday effect can be used for light modulation, although it is difficult, practically speaking, to modulate a magnetic field at very high frequencies. Figure 26-10 shows schematically the Faraday rotator, a crystal or liquid cell whose axis of symmetry is aligned with a magnetic field. The figure shows a field B established by

TABLE 26-4 VERDET CONSTANT FOR SELECTED MATERIALS

Material	V (min/G-cm) λ = 589 nm
H_2O	0.0131
Crown glass	0.0161
Flint glass	0.0317
CS_2	0.0423
CCl_4	0.0160
NaCl	0.0359
KCl	0.02858
Quartz	0.0166
ZnS	0.225

Figure 26-10 Faraday effect producing rotation of the plane of polarization.

current windings and indicates a rotation of the polarization in the same sense as the current producing the field. This particular geometry defines a *positive* Verdet constant. Figure 26-11 illustrates the principal application of the Faraday rotator as an *optical isolator*. The isolator consists of a Faraday rotator situated between a polarizer-analyzer pair. As shown, the incident vertically polarized light is rotated 45° counterclockwise by the Faraday rotator and in this orientation is fully transmitted by the analyzer. Optical elements (not shown) farther down the line are responsible for undesirable back reflections (retropulses) of this radiation along the optical axis. In traversing the Faraday rotator a second time, the polarization vector of the reflected light is rotated an additional 45° *in the same rotational sense*, so that it emerges horizontally polarized and encounters the polarizer at an angle of 90° with the polarization direction of the original beam. In this state, it is rejected by the polarizer, preventing it from continuing back into the optical system, where, in high-power laser systems, it can damage optical components. Thus the optical isolator effectively *isolates* the optical system from stray retropulses.

Figure 26-11 Faraday rotator used between a polarizer-analyzer pair to produce optical isolation of the optical system providing the incident light.

Example

Let us calculate the required length of SF58 flint glass, having a Verdet constant of 0.112 min/G-cm for 543.5-nm light, if it is to produce the 45° rotation of the polarization vector required in an optical isolator when the magnetic field has a value of 9 kG.

Solution Using Eq. (26-13), we have

$$d = \frac{\beta}{VB} = \frac{45° \times 60 \text{ min/}°}{(0.112 \text{ min/G-cm}) \times 9000 \text{ G}} = 2.68 \text{ cm}$$

Photoelasticity (Section 15-7) is the change in refractive index of a crystal due to mechanical stress. This phenomenon makes possible the *AO* or *acousto-optic effect*—the interaction of optical and acoustic waves—in which a longitudinal acoustic wave launched by means of a piezoelectric transducer produces a periodic mechanical stress in the crystal. The acoustic wave (see Figure 26-12) consists of a series of compressions and rarefactions (longitudinal vibrations) in atomic density and so a periodic—although small—variation of the refractive index about its normal value. Light incident on this structure is scattered to a greater extent from regions of higher refractive index. The scattering is called *Brillouin scattering* and is another third-order effect that does not require a medium possessing inversion symmetry.

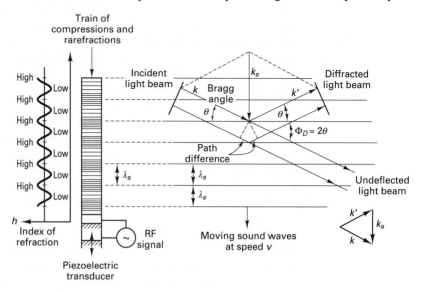

Figure 26-12 Variations in the refractive index of a medium due to the passage of a harmonic acoustic wave (left) and the scattering of an incident optical beam by the induced "planes" (right). The inset shows the relationship of wave vectors required by momentum conservation when the acoustic wave has the direction indicated.

If the crystal is thin enough, variations in refractive index along its length lead to corresponding variations in the speed of light so that the crystal behaves as a transmission phase grating (Section 17-5). Some of the incident light beam is diffracted into various orders, according to the diffraction grating equation, $m\lambda = d \sin \theta_m$, where the grating constant $d = \lambda_s$, the acoustic wavelength, and θ_m is the angle of diffraction in mth order. This is the so-called *Raman-Nath* regime [8]. If the crystal is thicker, regions of higher refractive index represent planes normal to the direction of the acoustic wave, as suggested in Figure 26-12. In this case, the light wave is diffracted in a way similar to that of X-rays from crystalline planes in Bragg scattering (the *Bragg* regime). The light wave traverses the crystal with a speed that is around five orders of magnitude greater than the speed of the acoustic wave. This means that the induced grating appears essentially stationary to the light wave.

A diffracted light beam appears in any direction in which portions of the wavefront reflected (1) from different parts of a given plane and (2) from successive planes obey the usual condition for constructive interference, that is, the path difference must be an integral number of wavelengths. The first requirement is satisfied by an angle of scatter that is equal to the angle of incidence. This condition is shown satisfied in Figure 26-12. The second requirement is the *Bragg condition*. The path difference between the incident and diffracted waves is made up of the two segments indicated in the figure. By geometry, each has a magnitude of $\lambda_s \sin \theta$. This leads to the equation

$$m\lambda = 2\lambda_s \sin \theta = 2\lambda_s \sin \frac{\Phi_D}{2} \qquad (26\text{-}16)$$

a relation worked out in this context by Brillouin in 1921 and identical to the Bragg equation for X-ray diffraction.[9] In Eq. (26-16), the angles and optical wavelength are those measured within the medium. (However, see problem 26-15.)

The Bragg condition for diffraction maxima can be found by an alternative argument that makes use of the particle nature of waves (Chapter 1). The incident light beam with wave vector **k** can be considered a flux of photons, each of energy $\hbar\omega$ and momentum $\hbar\mathbf{k}$, where ω is the angular frequency and $\hbar$ is the Planck constant divided by 2π. In the same way, the acoustic wave with wave vector $\mathbf{k}_s$ can be considered a flux of quantized particles called *phonons*, each of energy $\hbar\omega_s$ and momentum $\hbar\mathbf{k}_s$. The acousto-optic interaction then consists of the interaction or collision of these particles in which both energy and momentum are conserved. The two wave vectors **k** and $\mathbf{k}_s$, as well as the wave vector $\mathbf{k}'$ of the diffracted light beam, are also shown in Figure 26-12. The acoustic waves are shown propagating down in the figure. Conservation of momentum in a collision between a photon and a phonon requires that $\mathbf{k}' = \mathbf{k} - \mathbf{k}_s$, as indicated in the inset vector triangle. On the other hand, if the acoustic wave is reversed in direction, the corresponding vector triangle is satisfied by $\mathbf{k}' = \mathbf{k} + \mathbf{k}_s$. In general,

$$\mathbf{k}' = \mathbf{k} \pm \mathbf{k}_s \qquad (26\text{-}17)$$

With the plus sign, we interpret this to mean that an incident photon combines with a phonon to produce the diffracted photon; with the negative sign, the incident photon is considered to yield an additional phonon to the acoustic wave, as well as a photon to the diffracted beam. Thus Eq. (26-17) describes both the emission and the absorption of a phonon by the crystal lattice.

We now show that Eq. (26-17) is equivalent to the Bragg diffraction condition. Since light frequencies are of the order of 10^{14} Hz, while acoustic frequencies are generally less than 10^{10} Hz, both ω and ω' are much greater than ω_s, that is,

$$\omega' \cong \omega \quad \text{and accordingly} \quad \mathbf{k}' \cong \mathbf{k}$$

The vector triangle (Figure 26-13) for the wave vectors then shows that, in magnitude, $\mathbf{k}_s = 2\mathbf{k} \sin \theta$, or, in terms of wavelength,

$$\lambda = 2\lambda_s \sin \theta \qquad (26\text{-}18)$$

which is Bragg's equation, with $m = 1$.

Conservation of energy of the interacting particles requires that $\hbar\omega' = \hbar\omega \pm \hbar\omega_s$, or

$$\omega' = \omega \pm \omega_s \qquad (26\text{-}19)$$

---

[9] An important difference between X-ray diffraction and light diffraction in the Bragg regime is that light scattering occurs in a continuous manner from a thick, sinusoidal grating, rather than from discrete planes. As a consequence, only the first-order diffraction, $m = 1$, occurs in the acousto-optic effect. In this connection, see Section 13-2 (Hologram of a Point Source) and Section 25-1 (Optical Data Imaging and Processing).

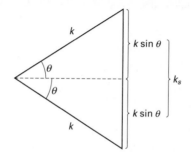

Figure 26-13. Wave vector triangle in the approximation $|k'| = |k| = k$. The angle θ is the Bragg angle. The geometrical relationship of sides is equivalent to the Bragg diffraction condition.

This result shows that the diffracted photon differs in frequency—however little—from the incident photon by the amount ω_s, greater or less, depending on the direction of the acoustic wave. This turns out to be another way of arriving at the *Doppler effect* for light. When the incident light encounters the acoustic wave approaching, the scattered frequency is greater, and when departing (as in Figure 26-12), the scattered frequency is less.

Example

To get an idea of the magnitude of the diffraction angle in first order, let us consider a typical case in which the incident light has a wavelength of 550 nm and the acoustic wave has a frequency of 200 MHz and a speed of 3000 m/s.

Solution   Then

$$\lambda_s = \frac{V_s}{\nu_s} = \frac{3 \times 10^3}{2 \times 10^8} = 1.5 \times 10^{-5} \text{ m}$$

Using Eq. (26-16),

$$\sin \theta = \frac{\lambda}{2\lambda_s} = \frac{550 \times 10^{-9}}{(2)(1.5 \times 10^{-5})} = 0.0183$$

so that

$$\theta = 1.05°$$

The acousto-optic (AO) effect can be applied to the modulation of a light beam by controlling its amplitude (AM), its frequency (FM), or its direction. Figure 26-14 illustrates one means of achieving AM modulation using a thin AO material in the Raman-Nath regime. The fraction of light removed from the zero-order diffracted beam depends on the magnitude of the induced stress and so on the amplitude of the modulating RF signal. The slit allows only the modulated zero-order beam to be transmitted. Other applications make use of the beam deflection capabilities of the AO effect. Referring back to Figure 26-12, it should be evident that both a change in frequency and a change in direction of the acoustic wave cause a change in the direction of the diffracted beam. If the frequency-sensitive position of the output beam is detected by a photodetector array, the AO device can be used as a spectrum analyzer. Again, because the frequency of the diffracted beam is shifted by an amount equal to the acoustic frequency, the beam can be frequency modulated. In this case, the design aims to minimize the angular spread of the diffracted light; when used as a spectrum analyzer, the design aims instead to maximize it.

Another application of the device as a beam deflector to initiate laser-cavity dumping is illustrated in Figure 26-15. When no acoustic wave is applied, the beam (1) bounces back and forth in the laser cavity, building up energy to a maximum

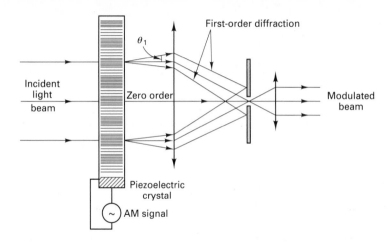

Figure 26-14 Modulation of a light beam by an acousto-optic grating in the Raman-Nath regime. The modulated signal driving the piezoelectric crystal is transferred to the output beam in zero order. Only the zero- and first-order diffracted beams are shown.

Figure 26-15 Cavity dumping of a laser using an acousto-optic beam deflector. Turning on the acoustic wave deflects the beam outside the laser cavity and initiates cavity dumping.

value. Turning on the acoustic wave causes a deflection of the beam (2) out of the cavity, thereby dumping the energy stored in the cavity.

26-8 NONLINEAR OPTICAL PHASE CONJUGATION

Optical phase conjugation (OPC) represents another third-order, nonlinear phenomenon with some fascinating applications. It was observed relatively recently, the first publications occurring in 1971–1972 by researchers in the Soviet Union. It is so named because the nonlinear interaction produces a light beam that, mathematically, is the *spatial complex conjugate* of the incident wave. This means that a new wave is produced that exactly reverses the direction and overall phase factor of the primary beam. Thus the phase conjugate wave precisely retraces the path of the original beam and, at each position, reproduces the exact shape of the original wavefront.

We shall, for the moment, view the process as a unique type of reflection and refer to the nonlinear medium that creates the phase conjugate wave as a *phase conjugate mirror* (PCM). To appreciate the uniqueness of the process, consider Figure 26-16, which shows reflections from ordinary mirrors. In (a), a plane wave is reflected from an ideal (infinite) plane mirror. The reflected wave is also a plane wave. To express mathematically the reversal in direction of the incident wave, the sign of the kz term is changed from minus to plus. Notice then that, except for the sign of the ωt term, the reflected wave is the complex conjugate of the original wave and has the properties of a phase conjugate wave as described above: It retraces the path of the incident beam and is its phase-reversed replica. Figure 26-16b shows the same process for an incident spherical wave. All that is needed to produce the phase-reversed replica on reflection is a concave spherical mirror whose curvature exactly matches that of the wavefront at incidence. For an arbitrary incident wavefront, as

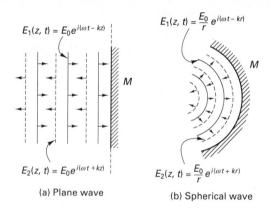

$E_1(z, t) = E_0 e^{i(\omega t - kz)}$

$E_2(z, t) = E_0 e^{i(\omega t + kz)}$

(a) Plane wave

$E_1(z, t) = \dfrac{E_0}{r} e^{i(\omega t - kr)}$

$E_2(z, t) = \dfrac{E_0}{r} e^{i(\omega t + kr)}$

(b) Spherical wave

$E_1(r, t) = Re[\Psi(r)e^{i(\omega t - kz)}]$

$E_2(r, t) = Re[\Psi^*(r)e^{i(\omega t + kz)}]$

(c) "Nearly plane" wave

Figure 26-16 Three examples of a phase-reversed replica of an incident wavefront produced by an ordinary mirror. In each case, the rays corresponding to the wavefront are everywhere normal to the mirror surface on reflection. A phase-conjugate mirror handles all cases and also responds to instantaneous changes in the incident wavefront.

shown in (c), the amplitude $\Psi(r)$ is complex and includes the amplitude and phase factors that describe its deviation from a plane wave. We imagine such a wave as a plane wave that has been shaped by passing through a distorting medium, by diffraction, or by modulation. Its phase reversed replica is expressed by taking the complex conjugate of both $\Psi(r)$ and ikz, in other words, the spatial part of the wave equation. To produce this wave with ordinary mirrors, we would need to construct a mirror surface that matched exactly the wavefront of the incident wave at the instant of reflection. The uniqueness of the PCM is that the phase conjugate replica is produced *regardless* of the shape of the incident wavefront, as long as the PCM has an aperture large enough to receive the entire wavefront. Unlike an ordinary mirror, the PCM is able to respond immediately to varying spatial and temporal features of an incident wave, such that a phase conjugate wave is continually produced.

Notice that the equations for the phase conjugate waves of Figure 26-16 imply waves that are equivalent physically to those that result by leaving the spatial part unchanged and reversing the sign of t. Because of this, the phase conjugated wave is often referred to as the "time reversed replica" of the incident wave. If one could make a video of the incident wave and then play it backwards, one would see an exact reproduction of the behavior of the conjugate wave itself.

Figure 26-17a further contrasts the ordinary mirror with a PCM. With an ordinary mirror, a diverging wavefront continues its divergence after reflection, whereas with a PCM, the reflected wavefront converges so that it exactly matches the curvature of the incident wave at every point and returns to its point of origin. To elaborate this difference, consider what one would see looking into an ordinary mirror and a PCM, as suggested in Figure 26-17b. In the first case, the image of the tip of the nose can be seen by means of light diverging from that point, reflecting from the mirror, and entering the eye. With a PCM, such a diverging wavefront would be returned as a converging wavefront that collapsed at the original point on

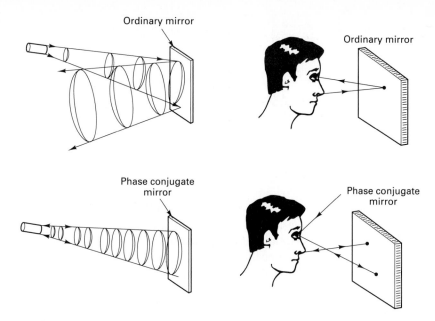

Figure 26-17 Differences between the action of an ordinary and a phase-conjugate mirror in (a) reflection of a diverging wavefront and (b) formation of an image.

the nose. Thus the eye looking into a PCM does not see an image of the face but views only the light that scatters from the pupil. The eye sees a uniformly lit mirror whose brightness depends on the intensity of the light scattered toward the PCM from the cornea covering the pupil.

One way of understanding OPC is to see it as *real time holography*. First, recall with the help of Figure 26-18 the procedure for making a conventional hologram (Section 13-3). A photographic emulsion is exposed (1) simultaneously to a reference beam and an object beam, which interfere to produce a complex fringe pattern within the emulsion. In the next step (2), the film is developed to bring out the interference pattern and produce the hologram. In the third step (3), the object beam is reproduced as a phase-reversed replica of the original by again illuminating the

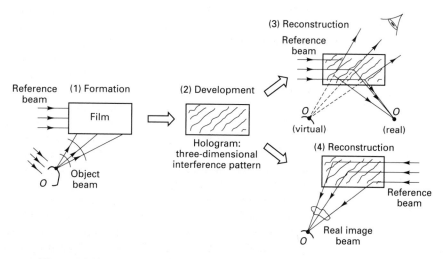

Figure 26-18 Conventional holography. Steps (3) and (4) are alternative techniques. The three steps are sequential.

(1) + (3) : Real time hologram
(2) + [(1) + (3)] : Bragg reflection; PC beam (4)

Figure 26-19 Real-time holography or phase conjugation. The four-wave mixing that produces the output beam occurs simultaneously. The output beam (4) corresponds to the real image beam in conventional holography.

hologram with the reference beam. When the reference beam has the original direction, both real and virtual images of the original object appear, with the virtual image located at the position of the original object. If the reference beam is reversed (4), the real image appears in the position of the original object, as shown. In OPC, the processes (1), (2), and (4) occur simultaneously, or in "real time." Figure 26-19 shows the PCM exposed to counter-propagating beams (1) and (2), which take the place of the reference beams in Figure 26-18. Similarly, the input and output beams (3) and (4) take the place of the object and image beams. We can consider beams (1) and (3) as interfering to produce a real-time hologram, complex regions of alternating refractive index induced by alternating regions of high and low light intensity. As pump beam (2) encounters this structure, Bragg scattering occurs, producing the phase conjugate beam (4), a phase-reversed replica of the input beam (3). Notice that the roles of beams (1) and (2) could just as well be interchanged in this description, so that Bragg scattering occurs when beam (1) encounters the interference patterns produced by beams (2) and (3).

This technique of producing an OPC wave is called *four wave mixing*, involving as it does four interfering waves in a nonlinear medium.[10] Refer to Figure 26-20. Two of the beams are the pump or reference beams that are incident on the medium from opposite directions. Usually these beams are plane waves of the same wavelength as that of the third input beam, the signal beam whose OPC is desired. The signal beam can have any direction, but its direction determines that of its phase conjugate. The amplitudes of the pump beams, A_1 and A_2, are much larger than those of the signal and phase conjugate beams, A_3 or A_4, so that they can be considered constant within the volume of the nonlinear medium. Analysis of this interaction [9] shows that the PC output beam has an amplitude proportional to the amplitudes of the other beams:

$$A_4 \propto A_1 A_2 A_3^*$$
(26-20)

Figure 26-20 Conventional geometry for phase conjugation by four-wave mixing. Pump beams A_1 and A_2 are directed antiparallel and are much stronger than the signal beam A_3. The phase conjugate output of the signal beam is the beam shown as A_4.

---

[10] Media used for nonlinear optical phase conjugation depend upon the wavelength region of interest but are wide ranging, including liquids, vapors, and solids. In a small sampling of these materials, we might list alcohol, carbon disulfide, sodium and iodine vapors, glasses, liquid crystals, ruby, lithium niobate, and semiconductors such as Ge, Si, CdTe, and HgCdTe.

This proportionality implies that the phase conjugate beam can be modulated by appropriately encoding—spatially or temporally—one or more of the input beams.

Several properties of the OPC beam, implied in the preceding, have interesting applications. These include aberration correction, pointing, and tracking. These properties follow from the basic nature of the PC wave, as described above. First, consider a transparent, distorting medium like frosted glass, which is placed in the path of a light beam on its way to a PCM. The wave is modified in a nonuniform manner, but its phase conjugate exactly retraces its path and reverses its modifications, so that the return passage through the distorting medium undoes or "heals" the original distortion. This means that a beam distorted by passing through an optical system with severe aberrations can be recovered by "reflection" from a PCM that sends it back through the same system. It also means that, if the light beam originates from a point source, divergence and diffraction effects are reversed when the beam is returned through the system, so that the PC beam converges to the original point. Furthermore, if the source point moves, the returning beam adjusts so that it continues to point to the source, the pointing and tracking property.

A few examples[11] of the usefulness of these properties will now be given. Others can be found in the references provided at the end of the chapter. Figure 26-21 illustrates the basic action of self-targeting, using a PCM. It is desired to concentrate

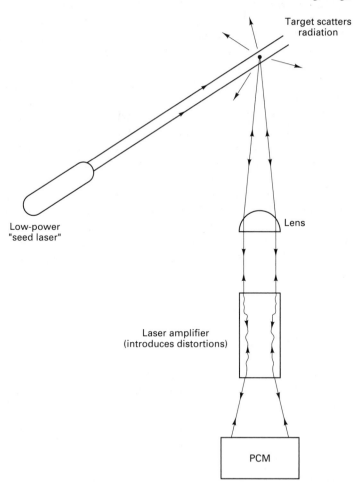

Figure 26-21 Pointing and targeting properties of a PC beam are illustrated in this scheme. A highly directed and powerful beam is returned to the scattering center that initiates it.

[11] Although the applications described here are theoretically possible, they have not all been implemented commercially.

a high-energy laser beam onto a small target, precisely the goal of laser fusion, for example, where the target is a hydrogen pellet. A laser amplifier that can deliver high energy invariably introduces distortions in the beam that impair the directivity of the beam and the ability to focus it accurately onto a small area. This problem can be circumvented by using a "seed" laser that has high directivity at the expense of power. Some of the light from the low-power laser that scatters from the particle enters the laser amplifier, triggering stimulated emission and amplification. Distortions produced in this process are removed by using a PCM as a reflector. The result is a high-power directed laser beam that retraces its route to converge precisely on the original scatterer. In this process, it is fairly assumed that the light traverses the amplifier so quickly that time-varying inhomogeneities in the laser medium are not detrimental to the healing process. To increase the power on target beyond the capability of a single laser amplifier, it is possible to arrange a number of laser amplifiers to pick up some of the scattered, pulsed radiation and to synchronize their return pulses to the target.

A slight change in perspective serves to indicate the wide variety of applications of this basic scheme. Suppose the seed laser is aboard a satellite and the PCM laser amplifier is situated at a ground station. The satellite can initiate a return PC beam from the ground station that heals the distortions introduced both by atmospheric turbulence and by the amplifier. Not only does this make possible satellite tracking, but it also allows for communication between ground and satellite. If one of the pump beams is pulsed, the PC beam will turn on and off accordingly; if one of the pump beams is information-modulated, this modulation is transferred to the return PC beam, as Eq. (26-20) implies.

Another interesting application of OPC lies in the area of fiber-optic transfer of information. Recall (Section 24-3) that a light signal sent through an optic fiber is gradually degraded by both modal and spectral dispersion. Figure 26-22 shows a length of fiber (fiber 1) that is used to transmit an image, in this case, the arrow object. If light reflected downward by the beam splitter is focused, the distortions introduced by the fiber result in a degraded image. To correct this situation, light transmitted by the beam splitter is reflected from a PCM and directed as shown to a

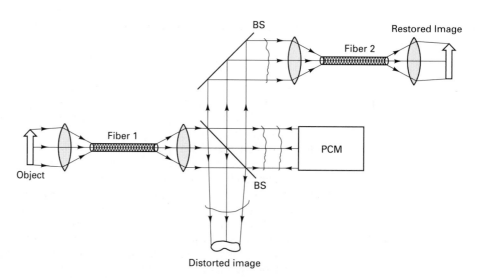

Figure 26-22 Restoration of image quality to light transmitted by an optical fiber. The healing of the PC beam is accomplished by returning the beam through a second, identical fiber.

second length of fiber (fiber 2), which we assume to be identical to fiber 1. This arrangement allows the second fiber to compensate for the distortions introduced by the first, so that it is possible to produce a high-quality image at a remote location.

PROBLEMS

26-1. Write out the third-order terms of the polarization for a single beam described by a plane wave with amplitude E_0 and frequency ω. What frequencies appear in the polarization wave?

26-2. Write out the third-order terms of the polarization for two-beam interaction, where the beams are plane waves having amplitudes E_{01} and E_{02} and frequencies ω_1 and ω_2, respectively. What frequencies are radiated by the polarization wave?

26-3. Write out the second-order terms of the polarization for three-beam interaction, where the beams are plane waves having amplitudes E_{01}, E_{02}, and E_{03}, and frequencies ω_1, ω_2, and ω_3, respectively. What frequencies are radiated by the polarization wave?

26-4. Arguing from Eq. (26-7), show that the linear electro-optic effect is found only in crystals lacking inversion symmetry.

26-5. (a) Determine the coherence length for second harmonic generation in KDP when submitted to pulsed ruby laser light at $\lambda_0 = 694$ nm. Appropriate refractive indices are n (694 nm) $= 1.505$ and n (347 nm) $= 1.534$.
(b) The measured coherence length of barium titanate at $\lambda_0 = 1.06$ μm is 5.8 μm. Calculate the expected change in refractive index at $\lambda = 0.53$ μm.

26-6. Determine the half-wave voltage for a longitudinal Pockels cell made of ADP (ammonium dihydrogen phosphate) at $\lambda = 546$ nm. What is its length?

26-7. A longitudinal Pockels cell is made from lithium niobate. Determine the change in refractive index and the phase difference produced by an applied voltage of 426 V when the light beam is from a He-Ne laser at 632.8 nm. The length of the crystal is 1 cm.

26-8. Using Eq. (26-9), show that the transmittance of a Pockels cell can also be written as $I = I_{max} \sin^2 (\Phi/2)$.
(a) At what values of V and Φ (greater than zero) is the transmittance zero?
(b) If the Pockels cell is preceded by an ordinary half-wave plate, what is the irradiance when $V = 0$ and when $V = V_{HW}$?

26-9. In what kinds of media are both longitudinal Pockels and Kerr effects present? To get some idea of their relative strengths, compare them by calculating the ratio of retardations produced by an appropriately applied 10 kV. Derive an expression for this ratio. Then do a numerical calculation by assuming a hypothetical medium with "typical" values of $r = 10$ pm/V, $K = 1$ pm/V^2, $L = 2$ cm, $d = 1$ cm, and $n_0 = 2$. Take $\lambda = 550$ nm.

26-10. Calculate the length of a Kerr cell using carbon disulfide required to produce half-wave retardation for an applied voltage of 30 kV. The electrodes of the cell have a separation of 1.5 cm. Is this cell practical?

26-11. Show that Eq. (26-19) is equivalent to the Doppler effect for light. Use the fact that the Doppler frequency shift $\Delta \nu$ for light reflected from a moving object is twice that of light emanating from a moving object, or $\Delta \nu = 2\nu u_p/v$, where ν is the light frequency, v its velocity in the medium, and u_p is the component of the object velocity parallel to the light wave's propagation direction. Use the geometry of Figure 26-12 and the Bragg condition.

26-12. The speed of sound in glass is 3 km/s. For a sound wave having a width of 1 cm, calculate the advance of the sound wave while it is traversed by a light wave. Take $n = 1.50$ for the glass. What is the significance of this result?

26-13. (a) Show that a small change in angle $\Delta\theta$ around the direction of the diffracted beam can be expressed approximately by $\Delta\theta = \Delta k_s/k$.

(b) Show that this result can be expressed as

$$\Delta\theta = (\lambda/v_s)\Delta\nu_S$$

where λ is the wavelength in the medium.

(c) The factor by which $\Delta\theta$ exceeds the beam divergence is a practically useful number N called "number of resolvable spots." This serves as a figure of merit, giving the number of resolvable positions that can be addressed by the beam deflector. If the beam divergence is expressed by the diffraction angle $\theta_D = \lambda/D$, with D the beam diameter, show that

$$N = \frac{\Delta\theta}{\theta_D} = \tau\Delta\nu_s$$

where τ is the time for the sound to cross the optical beam diameter.

(d) As a numerical example, consider modulation of the sound frequency in the range 80–120 MHz in fused quartz, where $v_s = 5.95 \times 10^5$ cm/s. If the beam diameter is 1 cm, determine the number of resolvable spots.

26-14. What acoustic frequency is required of a plane acoustic wave, launched in an acousto-optic crystal, so that a He-Ne laser beam is deflected by 1°? The speed of sound in the crystal is 2500 m/s and its refractive index at 632.8 nm is 1.6.

26-15. In Bragg's equation (26-18), the wavelength of the light and the angle are those measured within the medium. Show that, if the medium is isotropic and its sides are parallel to the direction of a plane acoustic wave, the equation also holds for the wavelength and angle of diffraction measured *outside* the medium.

26-16. Determine the difference in deflection angle for a He-Ne laser beam that is Bragg-scattered by an acoustic plane wave when the frequencies are 50 MHz and 80 MHz. The acoustic crystal is sapphire, with $n = 1.76$ and a sound speed of 11.00 km/s.

26-17. Design an optical isolator, as in Figure 26-11, that uses ZnS as the active medium. Let the magnetic field be produced by winding a solenoid directly onto the ZnS crystal at a turn density of 60 turns/cm. Assume $\lambda = 589$ nm.

26-18. A sample of SF57 glass with polished, parallel sides and 2.73 cm in length is placed between the tapered poles of an electromagnet. A small, central hole is drilled through the pole pieces to allow passage of a linearly polarized He-Ne laser beam through the sample and parallel to the magnetic field direction. The magnetic field is set at 5.098 kG.

(a) When red He-Ne laser light (632.8 nm) is used, the measured rotation is 900 min. Determine the Verdet constant for the glass.

(b) When green He-Ne laser light (543.5 nm) is used, the measured rotation is 1330 min. Determine the Verdet constant for the glass.

26-19. A 5-cm long liquid cell is situated in a magnetic field of 4 kG. The cell is filled with carbon disulfide and linearly polarized sodium light is transmitted through the cell, along the B-field direction. Determine both the net rotation of the light and the circular dispersion of CS_2 at this wavelength.

26-20. Sketch the shape of a nonsymmetrical pulse before and after reflection from an ordinary mirror and before and after reflection from a PCM. In the latter case, assume that the PCM is "turned on" by initiating the pump beams at the instant the entire pulse has moved inside the PC medium. Show how this effect might be used to correct dispersion broadening in an optical fiber. (If necessary, consult [12].)

26-21. Sketch an arrangement using a PCM to project a sharp, high-intensity image of a mask onto the photo-resist layer on a semiconducting chip without using lenses. This provides a means of doing photolithography without placing a mask in direct contact with the chip. (If necessary, consult [12]).

REFERENCES

[1] Butcher, P. N., and D. Cotter. *The Elements of Nonlinear Optics*. New York: Cambridge University Press, 1990. Ch. 6.

[2] Franken, P. A., A. E. Hill, C. W. Peters, and G. Weireich. "Generation of Optical Harmonics." *Phys. Rev. Letters* 7 (1961): 118.

[3] Yariv, Ammon. *Optical Electronics*, 3d ed. New York: Holt, Rinehart and Winston, 1985. Ch. 8.

[4] Kaminov, Ivan P. *An Introduction to Electrooptic Devices*. New York: Academic Press, 1974. Ch. 3.

[5] Yariv, Ammon. *Optical Electronics*, 3d ed. New York: Holt, Rinehart and Winston, 1985. Ch. 9.

[6] Wilson, J., and J. F. B. Hawkes. *Optoelectronics: An Introduction*. London: Prentice-Hall International, 1983. Ch. 3.

[7] Pedrotti, Frank L., and Peter Bandettini. "Faraday Rotation in the Undergraduate Advanced Laboratory." *Am. J. Phys.* 58 (June 1990): 542.

[8] Guenther, Robert. *Modern Optics*. New York: John Wiley and Sons, 1990. Ch. 14.

[9] Yariv, Ammon. *Optical Electronics*, 3d ed. New York: Holt, Rinehart and Winston, 1985. Ch. 16.

[10] Yariv, Ammon, and Robert A. Fisher. In *Optical Phase Conjugation*, edited by Robert A. Fisher, Ch. 1. New York: Academic Press, 1983.

[11] Giuliano, Concetto R. "Applications of Optical Phase Conjugation." *Physics Today* 34, no. 4 (April 1981): 27.

[12] Shkunov, Vladimir V., and Boris Ya. Zel'dovich. "Optical Phase Conjugation." *Scientific American* (Dec. 1985): 54.

[13] Pepper, David M. "Applications of Optical Phase Conjugation." *Scientific American* (Jan. 1986): 74.

[14] Karim, Mohammad A. *Electro-optical Devices and Systems*. Boston: PWS-Kent Publishing Company, 1990. Ch 7.

[15] Boyle, W. S. "Light Wave Communications." *Scientific American* (Aug. 1977): 40.

[16] Giordmaine, J. A. "The Interaction of Light with Light." *Scientific American* (Apr. 1964): 38.

[17] Miller, Stewart E. "Communication by Laser." *Scientific American* (Jan. 1966): 19.

[18] Nelson, Donald F. "The Modulation of Laser Light." *Scientific American* (June 1968): 17.

[19] Pepper, David M., Jack Feinberg, and Nicolai V. Kukhtarev. "The Photorefractive Effect." *Scientific American* (Oct. 1990): 62.

27

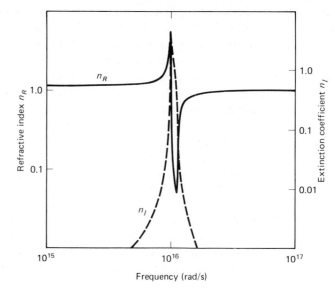

Optical Properties of Materials

INTRODUCTION

Electromagnetic waves that encounter materials create a complex of interactions with the charged particles of the medium. Forces are exerted on the charges by the electric field of the waves and, because of the motions of the charges, also by the magnetic field of the waves. In responding to these oscillating fields, the charges themselves oscillate and act as radiators of secondary electromagnetic waves. Thus in determining the net field at some point, the fields of both the source waves and the charged oscillators must be taken into account. In the case of ordinary fields, smaller than those now attainable with high-energy lasers, the net fields are assumed to be a linear superposition of the constituent fields. The complicated effects of all the microscopic contributions to the resultant field by the charges in the material can, for certain purposes, be simply described by macroscopic material parameters, the *optical constants* of the material. In this chapter we show in particular how the *refractive index* and the *absorption coefficient* for isotropic conducting (metals) and nonconducting (insulators or dielectrics) materials can be understood. In order to do this we make use of Maxwell's equations and the mathematical techniques of vector calculus.

27-1 POLARIZATION OF A DIELECTRIC MEDIUM

We take as our model a *simple dielectric*, that is, a nonconducting material whose properties are isotropic. By *nonconducting* we mean that the medium, unlike a metal, contains no free charges. Positive charges are associated with the constituent nuclei and negative charges with the electrons bound to such nuclei. By *isotropic* we mean that the relevant physical properties we consider are independent of direction in the medium, so that we may treat the physical constants as scalar quantities. Application of an electric field to such a medium causes charge displacement, in which the negative charge distribution bound to the nuclei shifts in a direction opposite to the electric field. The shift may occur in a *polar molecule*, like H_2O, because the molecule has a *permanent electric dipole*, that is, the effective centers of its positive and negative charge distributions do not coincide. In this case application of the field produces some reorientation of the molecules so that, on the average, the positive end of the dipole is in the direction of the field. The tendency toward alignment is counteracted by the thermal motions of the molecules. The shift in charge distribution may also occur in *nonpolar molecules*, such as O_2, in which positive and negative charge distributions normally have the same effective center. Application of the field results in a slight shift of the electron cloud relative to its nucleus, producing an *induced dipole*. In either case, the *dipole moment* **p** due to each atom or molecule is given by the product of the displaced charge q and the effective separation of the positive and negative charge in the atomic dipole, or

$$\mathbf{p} = -q\mathbf{r} \tag{27-1}$$

as indicated in Figure 27-1a. The direction of the dipole moment is from the negative toward the positive charge. The magnitude of the dipole moment for a given material depends on how easily charge is displaced under the influence of a given electric field. The *polarization* **P** of the medium is then said to be the collective dipole moment per unit volume, the sum of dipole moments given by

$$\mathbf{P} = -Ne\mathbf{r} \tag{27-2}$$

where N is the number elementary dipoles per unit volume and e is the magnitude of the electronic charge.

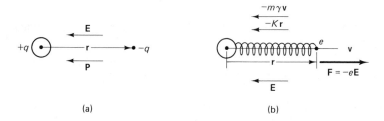

Figure 27-1 The elementary electric dipole. (a) Alignment with the field. (b) Forces acting on a dipole when the electric field has the direction indicated.

Electrons behave as though the forces binding them to the nuclei are elastic forces given by Hooke's law, where the restoring force is proportional to the displacement and oppositely directed. The more massive nuclei can be considered stationary since they are unable to respond to the rapid changes in the field representing an electromagnetic wave in the optical region of the spectrum. A simple model in which electrons are held by springlike forces to a fixed nucleus is therefore applicable. In an alternating electric field, however, forced oscillations of electrons remove a certain amount of energy from the incident radiation, the energy that the electrons radiate in turn and the energy of interaction with neighboring atoms that shows up as heat. The model of the oscillating electron is therefore that of a damped, harmonic oscillator, with a frictional force proportional to the velocity. Newton's second law, applied to the electron in the model of Figure 27-1b, then leads to the equation of motion,

$$-K\mathbf{r} - m\gamma \frac{d\mathbf{r}}{dt} - e\mathbf{E} = m\frac{d^2\mathbf{r}}{dt^2} \tag{27-3}$$

In Eq. (27-3), K is the force constant of the effective spring, m is the electronic mass, and γ is a frictional constant with dimensions of reciprocal time. Notice that the force $(e\mathbf{v} \times \mathbf{B})$ on the electron due to the magnetic field of the radiation is omitted; it is, in fact, negligible compared with the force $(e\mathbf{E})$ due to the electric field.

When the applied $\mathbf{E}$-field is static, there is no oscillation of the dipoles, so that both velocity and acceleration of the electron vanishes. In this special case, Eq. (27-3) reduces to

$$-K\mathbf{r} = e\mathbf{E}$$

or, eliminating $\mathbf{r}$ with the help of Eq. (27-2), the *static polarization* is given by

$$\mathbf{P} = \frac{Ne^2\mathbf{E}}{K} \tag{27-4}$$

Suppose now that $\mathbf{E}$ is the field of a harmonic wave with a time dependence given by $\mathbf{E} = \mathbf{E}_0 e^{-i\omega t}$ and that the oscillations respond with a similar dependence, $\mathbf{r} = \mathbf{r}_0 e^{-i\omega t}$. Inserting the corresponding derivatives $d\mathbf{r}/dt = -i\omega\mathbf{r}$ and $d^2\mathbf{r}/dt^2 = -\omega^2\mathbf{r}$ into Eq. (27-3) gives

$$\mathbf{r} = \frac{-e\mathbf{E}}{-m\omega^2 - im\omega\gamma + K} \tag{27-5}$$

which, when substituted in Eq. (27-2), now yields a time-dependent polarization given by

$$\mathbf{P} = \left(\frac{Ne^2}{-m\omega^2 - im\omega\gamma + K}\right)\mathbf{E} \tag{27-6}$$

Note that Eq. (27-6) agrees with Eq. (27-4) in the case of a static $\mathbf{E}$-field specified by the conditions $\omega = 0$ and $\gamma = 0$. In all other cases, polarization is a function of the radiation frequency ω and, because the coefficient of $\mathbf{E}$ is complex, the polarization may possess a frequency-dependent phase relative to $\mathbf{E}$, as we will see. The field $\mathbf{E}$ in Eq. (27-6) should represent the actual field at the dipole, in the interior of the medium. This local field $\mathbf{E}_{\text{loc}}$ is a superposition of the applied field $\mathbf{E}_{\text{app}}$ and the field that results from all the other dipoles aligned in a polarized medium. It is shown in standard texts on electricity and magnetism that the latter contribution is given by $\mathbf{P}/3\epsilon_0$, where ϵ_0 is the permittivity of free space. Thus

$$\mathbf{E}_{\text{loc}} = \frac{\mathbf{P}}{3\epsilon_0} + \mathbf{E}_{\text{app}} \tag{27-7}$$

Retaining the symbol $\mathbf{E}$ for the applied field and substituting Eq. (27-7) into Eq. (27-6),

$$\mathbf{P} = \left(\frac{Ne^2}{-m\omega^2 - im\omega\gamma + K}\right)\left(\mathbf{E} + \frac{\mathbf{P}}{3\epsilon_0}\right) \tag{27-8}$$

$\mathbf{P}$ now appears twice in the equation, and we can solve for it explicitly. Setting the prefactor in Eq. (27-8) equal to F for the moment, we conclude

$$\mathbf{P} = \left(\frac{F}{1 - F/3\epsilon_0}\right)\mathbf{E} \tag{27-9}$$

The multiplier of $\mathbf{E}$ can now be determined to be

$$\frac{F}{1 - F/3\epsilon_0} = \frac{Ne^2/m}{(K/m - Ne^2/3m\epsilon_0) - \omega^2 - i\omega\gamma}$$

Defining ω_0^2 as the quantity in parentheses, that is,

$$\omega_0^2 \equiv \frac{K}{m} - \frac{Ne^2}{3m\epsilon_0} \tag{27-10}$$

Eq. (27-9) becomes

$$\mathbf{P} = \frac{Ne^2/m}{\omega_0^2 - \omega^2 - i\omega\gamma}\mathbf{E} \tag{27-11}$$

Taking the magnitude of this complex expression for $\mathbf{P}$,

$$|\mathbf{P}| = \frac{Ne^2/m}{\sqrt{(\omega_0^2 - \omega^2)^2 + \omega^2\gamma^2}}|\mathbf{E}|$$

Clearly, $\mathbf{P}$ can increase dramatically as $\omega \to \omega_0$, so that ω_0 represents a *resonance frequency* for the dipoles of the medium. Equation (27-11) has the same form as the equation of motion of a driven harmonic oscillator with damping. As the driving frequency approaches the resonance frequency ω_0 of the oscillator, the amplitude of the vibrations becomes very large and subsides again as the frequency increases beyond ω_0. In the case of a dielectric medium, the increase of dipole moments at resonance results in a large maximum polarization. Equation (27-11) also illustrates a frequency-dependent phase shift between the applied field and the polarization. Far from resonance, we may set $\gamma = 0$, corresponding to negligible damping. Then for $\omega \ll \omega_0$, $\mathbf{P}$ and $\mathbf{E}$ have the same sign and the dipoles are oscillating in phase with the field. Beyond resonance, however, when $\omega \gg \omega_0$, $\mathbf{P}$ and $\mathbf{E}$ have opposite signs, indicating a phase difference of 180°. Free electrons respond in this manner. When $\omega \cong \omega_0$, near resonance, the vibrations are large. The damping term in the denominator in this case is not negligible, and the division by $-i$, equivalent to multiplication by i, indicates a 90° phase shift between $\mathbf{E}$ and $\mathbf{P}$.

The dependence of $\mathbf{P}$ on $\mathbf{E}$, as given in Eq. (27-11), can now be used to discover the conditions under which plane waves are able to propagate in a dielectric. The fundamental wave equation for electromagnetic waves in the dielectric is a consequence of the Maxwell equations.

27-2 PROPAGATION OF LIGHT WAVES IN A DIELECTRIC

The four Maxwell equations may be written in the general form

$$\mathbf{\nabla} \cdot \mathbf{E} = \frac{\rho}{\epsilon_0} \tag{27-12}$$

$$\mathbf{\nabla} \times \mathbf{E} = -\frac{\partial \mathbf{B}}{\partial t} \tag{27-13}$$

$$\mathbf{\nabla} \cdot \mathbf{B} = 0 \tag{27-14}$$

$$c^2\mathbf{\nabla} \times \mathbf{B} = \frac{\partial \mathbf{E}}{\partial t} + \frac{\mathbf{J}}{\epsilon_0} \tag{27-15}$$

In these equations ρ is the charge density, which in general includes both free and bound charge, as indicated by $\rho = \rho_b + \rho_f$. In a dielectric, however, $\rho_f = 0$. It is standard in a course in electricity and magnetism to show that the bound-charge density is related to the polarization by

$$\rho_b = -\mathbf{\nabla} \cdot \mathbf{P} \tag{27-16}$$

The quantity $\mathbf{J}$ similarly represents the current density and can arise from both free and bound charge, as indicated by $\mathbf{J} = \mathbf{J}_b + \mathbf{J}_f$. In a dielectric where $\rho_f = 0$, $\mathbf{J}_f = 0$ also. Furthermore, it can be shown that

$$\mathbf{J}_b = \frac{\partial \mathbf{P}}{\partial t} \tag{27-17}$$

With these constraints, the four Maxwell equations for a dielectric can be written

$$\nabla \cdot \mathbf{E} = \frac{-\nabla \cdot \mathbf{P}}{\epsilon_0} \tag{27-18}$$

$$\nabla \times \mathbf{E} = -\frac{\partial \mathbf{B}}{\partial t} \tag{27-19}$$

$$\nabla \cdot \mathbf{B} = 0 \tag{27-20}$$

$$c^2 \nabla \times \mathbf{B} = \frac{\partial \mathbf{E}}{\partial t} + \frac{1}{\epsilon_0} \frac{\partial \mathbf{P}}{\partial t} \tag{27-21}$$

Now we take the curl of both sides of Eq. (27-19), giving

$$\nabla \times (\nabla \times \mathbf{E}) = \nabla \times \left(-\frac{\partial \mathbf{B}}{\partial t} \right) = -\frac{\partial}{\partial t} (\nabla \times \mathbf{B}) \tag{27-22}$$

where we have interchanged the order of differentiation with respect to space and time in the last step. The left member of Eq. (27-22) can be reexpressed by the identity

$$\nabla \times (\nabla \times \mathbf{E}) \equiv \nabla(\nabla \cdot \mathbf{E}) - \nabla^2 \mathbf{E} \tag{27-23}$$

In a *homogeneous* dielectric, the effect of polarization is to produce a net surface charge density, while leaving the internal charge density $\rho_b = 0$ unchanged. The internal charge density is zero because, in any internal closed surface, every bit of charge that moves into the enclosed volume in response to a polarizing field is balanced by an equal bit of charge that moves out. The surface-charge density appears because such balancing is not possible there. Thus by Eqs. (27-16) and (27-18) we conclude that $\nabla \cdot \mathbf{E} = 0$ and substitute the remainder of Eq. (27-23) into Eq. (27-22), giving

$$\nabla^2 \mathbf{E} = \frac{\partial}{\partial t} (\nabla \times \mathbf{B}) \tag{27-24}$$

For the right member we may make use of Maxwell's equation (27-21) and write

$$c^2 \nabla^2 \mathbf{E} = \frac{\partial^2 \mathbf{E}}{\partial t^2} + \frac{1}{\epsilon_0} \frac{\partial^2 \mathbf{P}}{\partial t^2} \tag{27-25}$$

The last term is expressible in terms of $\mathbf{E}$ using Eq. (27-11), so we have

$$c^2 \nabla^2 \mathbf{E} = \left[1 + \frac{Ne^2}{m\epsilon_0 (\omega_0^2 - \omega^2 - i\omega\gamma)} \right] \frac{\partial^2 \mathbf{E}}{\partial t^2} \tag{27-26}$$

For a harmonic wave expressed as $\mathbf{E} = \mathbf{E}_0 e^{i(kz - \omega t)}$, in which case $\nabla^2 \mathbf{E} = -k^2 \mathbf{E}$ and $\partial^2 \mathbf{E}/\partial t^2 = -\omega^2 \mathbf{E}$, Eq. (27-26) solved for k^2 becomes

$$k^2 = \frac{\omega^2}{c^2} \left[1 + \frac{Ne^2}{m\epsilon_0} \frac{1}{(\omega_0^2 - \omega^2 - i\omega\gamma)} \right] \tag{27-27}$$

We conclude that the analysis of plane waves propagating in a homogeneous dielectric requires in general that the propagation constant k be a complex number.

Defining the real and imaginary parts of $\tilde{k}$ by

$$\tilde{k} = k_R + ik_I \qquad (27\text{-}28)$$

and inserting this form into the expression for a harmonic wave, we have

$$\mathbf{E} = \mathbf{E}_0 e^{i(k_R z + ik_I z - \omega t)} = \mathbf{E}_0 e^{-k_I z} e^{i(k_R z - \omega t)} \qquad (27\text{-}29)$$

The exponential factor in k_I represents a depth-dependent absorption of an otherwise harmonic wave, and k_I measures the *amplitude attenuation* of the wave. By taking the square of the magnitude of both sides of Eq. (27-29), the result describes instead the energy flux density, giving

$$I = I_0 e^{-\alpha z}$$

where $\alpha = 2k_I$ is the *absorption coefficient* of the mediun. If the propagation constant is complex, so must be the refractive index, for we can write

$$k = \frac{2\pi}{\lambda} = \frac{2\pi f}{v} = \left(\frac{\omega}{c}\right) n \qquad (27\text{-}30)$$

If we identify the real and imaginary parts of the complex refractive index by

$$\tilde{n} = n_R + in_I \qquad (27\text{-}31)$$

where n_R is the usual refractive index and n_I is called the *extinction coefficient*, it follows from Eqs. (27-28) and (27-30) that

$$k_R + ik_I = \left(\frac{\omega}{c}\right)(n_R + in_I)$$

yielding the relations

$$k_R = \left(\frac{\omega}{c}\right) n_R \qquad (27\text{-}32)$$

and

$$k_I = \left(\frac{\omega}{c}\right) n_I \qquad (27\text{-}33)$$

Writing n^2 as

$$n^2 = (n_R + in_I)^2 = \left(\frac{ck}{\omega}\right)^2$$

and relating this equation to Eq. (27-27) gives

$$(n_R + in_I)^2 = 1 + \left(\frac{Ne^2}{m\epsilon_0}\right) \frac{1}{\omega_0^2 - \omega^2 - i\omega\gamma} \qquad (27\text{-}34)$$

Expressions for the real and imaginary parts of the refractive index can be found by equating real and imaginary parts in Eq. (27-34). The left member

$$(n_R + in_I)^2 = (n_R^2 - n_I^2) + i(2n_R n_I) \qquad (27\text{-}35)$$

The right member can also be written as the sum of a real and imaginary part. The complex term is first modified by multiplying numerator and denominator by the conjugate of the denominator. The result, after simplification, is

$$(n_R + in_I)^2 = 1 + \left(\frac{Ne^2}{m\epsilon_0}\right)\left(\frac{\omega_0^2 - \omega^2}{(\omega_0^2 - \omega^2)^2 + \omega^2\gamma^2}\right) + i\left(\frac{\omega\gamma}{(\omega_0^2 - \omega^2)^2 + \omega^2\gamma^2}\right) \qquad (27\text{-}36)$$

Now by comparing the right members of Eqs. (27-35) and (27-36),

$$n_R^2 - n_I^2 = 1 + \frac{Ne^2}{m\epsilon_0}\left[\frac{\omega_0^2 - \omega^2}{(\omega_0^2 - \omega^2)^2 + \gamma^2\omega^2}\right] \tag{27-37}$$

and

$$2n_I n_R = \frac{Ne^2}{m\epsilon_0}\left[\frac{\gamma\omega}{(\omega_0^2 - \omega^2)^2 + \gamma^2\omega^2}\right] \tag{27-38}$$

The equations can be solved simultaneously for n_I and n_R. The appearance of the mass m in the denominator of these equations shows that electronic oscillations are more important than ionic oscillations in determining the index of refraction. Ionic polarization may be significant in the region of resonance, however, where the large bracketed term balances the small prefactor containing the mass.

Figure 27-2 shows both n_R and n_I calculated from Eqs. (27-37) and (27-38) as a function of driving frequency ω. The absorption described by the extinction coefficient is seen to peak at the resonant frequency ω_0. The real refractive index experiences a sharp rise and fall as ω increases toward and passes through resonance, after which it increases again, approaching the value $n_R = 1$ at high frequencies. The narrow region where n_R decreases with frequency is contrary to the usual dispersion of transparent media and is called the region of *anomalous dispersion*. A resonance frequency such as ω_0 for the dielectric means that, for incident photons of frequency ω_0, there is a high probability of absorption. Absorption of such a photon corresponds to a transition of $E_0 = \hbar\omega_0 = hf_0$ in the energy-band structure of the material. As ω is varied, there will be a series of resonance frequencies characteristic of the material. If such a resonance occurs in the visible range of frequencies, for example, the material absorbs a portion of the spectrum and appears colored, while transmitting the remainder. Transparent materials like glass have resonance frequencies in the infrared and ultraviolet regions but not in the visible. In terms of our simplified model of a dielectric, we interpret the existence of a number of resonance frequencies to mean that electrons experience different degrees of freedom in response to the applied field. To take this into account formally, Eq. (27-34) is usually generalized to include a number of terms summed over the resonant frequencies ω_j, given by

$$n^2 = 1 + \frac{Ne^2}{m\epsilon_0}\sum_j \frac{f_j}{\omega_j^2 - \omega^2 - i\gamma_j\omega} \tag{27-39}$$

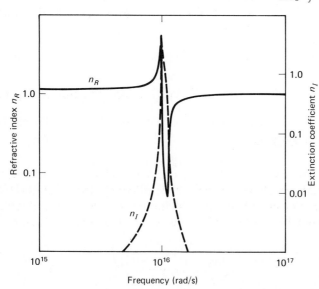

Figure 27-2 Angular frequency dependence of the refractive index n_R and the extinction coefficient n_I for a dielectric. Assumed values are $\omega_0 = 1 \times 10^{16}$ s^{-1}, $\gamma = 10^{14}$ s^{-1}, and $N = 1 \times 10^{28}$ m^{-3}.

where f_j, called the *oscillator strength* for the resonance ω_j, represents the fraction of dipoles having this resonant frequency. The rigorous treatment of this problem requires the application of quantum theory.

The Dispersion Equation. The variation in refractive index with frequency, described by Eq. (27-39), is what we mean by *dispersion*. We wish to show that the Cauchy dispersion equation, introduced in Eq. (6-17) as an empirical formula, can be deduced from Eq. (27-39) under certain simplifying assumptions. We shall assume a single resonant frequency ω_0 in the ultraviolet, such that frequencies in the visible obey the inequality, $\omega \ll \omega_0$. We shall also assume that damping is negligible by setting $\gamma = 0$. Since all dipoles have the same resonant frequency, $f = 1$ and Eq. (27-39) takes the form:

$$n^2 = 1 + \frac{Ne^2}{m\epsilon_0}\left(\frac{1}{\omega_0^2 - \omega^2}\right)$$

Notice that, for $\omega \ll \omega_0$, as for a gas, the refractive index is nearly constant. As ω approaches ω_0 from below, the refractive index increases slightly, as shown in Figure 27-2. The slowly increasing index with frequency (decreasing with wavelength) is characteristic of normal dispersion.

To derive the Cauchy dispersion equation, let us first expand the frequency factor in a binomial series:

$$\frac{1}{\omega_0^2 - \omega^2} = \frac{1}{\omega_0^2}\left(1 - \frac{\omega^2}{\omega_0^2}\right)^{-1} = \frac{1}{\omega_0^2}\left(1 + \frac{\omega^2}{\omega_0^2} + \frac{\omega^4}{\omega_0^4} + \cdots\right)$$

so that

$$n^2 = 1 + \frac{Ne^2}{m\epsilon_0\omega_0^2}\left(1 + \frac{\omega^2}{\omega_0^2} + \frac{\omega^4}{\omega_0^4} + \cdots\right)$$

Writing $\omega = 2\pi c/\lambda$ and gathering constants A', B', and C' appropriately, we can express

$$n^2 = A' + \frac{B'}{\lambda^2} + \frac{C'}{\lambda^4} + \cdots$$

We may take the square root of both sides and, since each higher-order term in the expression is less than A', use the binomial expansion again on the right member. After re-collecting constants, we get

$$n = A + \frac{B}{\lambda^2} + \frac{C}{\lambda^4} + \cdots$$

which is the Cauchy relation introduced earlier to describe normal dispersion.

27-3 CONDUCTION CURRENT IN A METAL

In metals, the existence of "free" electrons, not bound to particular nuclei, modifies the treatment outlined above for dielectrics. Although there are also bound electrons, the response of the free electrons dominates the electrical and optical properties of the medium. So, in Eq. (27-3), we set $K = 0$, and the equation of motion becomes

$$m\frac{d\mathbf{v}}{dt} + m\gamma\mathbf{v} = -e\mathbf{E} \tag{27-40}$$

The equation may be conveniently expressed in terms of the *conduction current density* **J**, defined by

$$\mathbf{J} = -Ne\mathbf{v} \tag{27-41}$$

where **J** has (SI) units of amperes per square meter. Writing Eq. (27-40) in terms of **J** rather than **v**,

$$\frac{d\mathbf{J}}{dt} + \gamma\mathbf{J} = \left(\frac{Ne^2}{m}\right)\mathbf{E} \tag{27-42}$$

In the case where the applied field is the harmonic wave $\mathbf{E} = \mathbf{E}_0 e^{-i\omega t}$, we expect the current density to vary at the same rate and write $\mathbf{J} = \mathbf{J}_0 e^{-i\omega t}$. Equation (27-42) then takes the form

$$(-i\omega + \gamma)\mathbf{J} = \left(\frac{Ne^2}{m}\right)\mathbf{E} \tag{27-43}$$

In the static, or DC, case specified by $\omega = 0$,

$$\mathbf{J} = \left(\frac{Ne^2}{m\gamma}\right)\mathbf{E} \tag{27-44}$$

The static *conductivity* σ, defined by Ohm's law,

$$J = \sigma E \tag{27-45}$$

then takes the theoretical form

$$\sigma = \frac{Ne^2}{m\gamma} \tag{27-46}$$

Since conductivities are usually measured, we rewrite Eq. (27-43) in terms of σ, giving

$$\mathbf{J} = \left(\frac{\sigma}{1 - i\omega/\gamma}\right)\mathbf{E} \tag{27-47}$$

27-4 PROPAGATION OF LIGHT WAVES IN A METAL

An electromagnetic wave propagating in the conducting medium satisfies Maxwell's equations (27-12) through (27-15). Although free charge exists in the metal, the internal free-charge volume density ρ_f is zero. The free charge is so mobile that it quickly redistributes in response to an applied field, preventing the buildup of local charge densities. The appropriate Maxwell equations are then

$$\nabla \cdot \mathbf{E} = 0 \tag{27-48}$$

$$\nabla \times \mathbf{E} = -\frac{\partial \mathbf{B}}{\partial t} \tag{27-49}$$

$$\nabla \cdot \mathbf{B} = 0 \tag{27-50}$$

$$c^2 \nabla \times \mathbf{B} = \frac{\partial \mathbf{E}}{\partial t} + \frac{\mathbf{J}}{\epsilon_0} \tag{27-51}$$

As before, $\nabla \times (\nabla \times \mathbf{E}) = -\nabla^2 \mathbf{E}$ because $\nabla \cdot \mathbf{E} = 0$ in the identity of Eq. (27-23). Calculating the same quantity by taking the curl of Eq. (27-49), we have

$$-\nabla^2 \mathbf{E} = \nabla \times \left(-\frac{\partial \mathbf{B}}{\partial t}\right) = -\frac{\partial}{\partial t}(\nabla \times \mathbf{B}) = -\frac{1}{c^2}\frac{\partial^2 \mathbf{E}}{\partial t^2} - \frac{1}{\epsilon_0 c^2}\left(\frac{\partial \mathbf{J}}{\partial t}\right)$$

where we have used Eq. (27-51) in the last step. Representing **J** with the help of Eq. (27-47), we conclude

$$\nabla^2 \mathbf{E} = \frac{1}{c^2}\left(\frac{\partial^2 \mathbf{E}}{\partial t^2}\right) + \frac{1}{\epsilon_0 c^2}\left(\frac{\sigma}{1 - i\omega/\gamma}\right)\frac{\partial \mathbf{E}}{\partial t} \tag{27-52}$$

For plane, harmonic waves given by $\mathbf{E} = \mathbf{E}_0 e^{i(kz - \omega t)}$, the appropriate space and time derivatives required by Eq. (27-52) can be calculated to give

$$k^2 = \frac{\omega^2}{c^2} + i\left(\frac{\sigma\omega\mu_0}{1 - i\omega/\gamma}\right) \tag{27-53}$$

where we have also made use of the fact that $c^2 = 1/\epsilon_0\mu_0$, with μ_0 the permeability of vacuum. Again, we find that the propagation constant must be a complex number to properly describe the propagation of the wave in a metal.

27-5 SKIN DEPTH

Before proceeding with the general case described by Eq. (27-53), we pause to consider the special case in which the frequency ω of the incident radiation is small enough to allow as a good approximation to Eq. (27-53),

$$k^2 = i\omega\sigma\mu_0$$

Expressing i as $e^{i\pi/2}$ and taking the square root of both sides,

$$k = (1 + i)\left(\frac{\sigma\mu_0\omega}{2}\right)^{1/2} \tag{27-54}$$

Writing k as the complex number $\bar{k} = k_R + ik_I$, as before, we can identify the real and imaginary coefficients by

$$k_R = k_I = \left(\frac{\sigma\mu_0\omega}{2}\right)^{1/2} \tag{27-55}$$

and the real and imaginary refractive indices by

$$n_R = \frac{c}{\omega}k_R = \left(\frac{c^2\sigma\mu_0}{2\omega}\right)^{1/2} = \left(\frac{\sigma}{2\omega\epsilon_0}\right)^{1/2} \tag{27-56}$$

and

$$n_I = \frac{c}{\omega}k_I = \left(\frac{\sigma}{2\omega e_0}\right)^{1/2} \tag{27-57}$$

The complex character of k, when introduced into the plane, harmonic wave equation, leads as in Eq. (27-29) to

$$\mathbf{E} = \mathbf{E}_0 e^{-k_I z} e^{i(k_R z - \omega t)}$$

The real exponential factor $e^{-k_I z}$ describes absorption. When the radiation has penetrated a depth of $z = 1/k_I$, therefore, the amplitude has decreased to $1/e$ of its surface value. This particular distance is called the *skin depth*, δ, where

$$\delta \equiv \frac{1}{k_I} = \sqrt{\frac{2}{\sigma\mu_0\omega}} \tag{27-58}$$

and is evidently smaller for better conductors with larger σ. For 3-cm microwaves, for example, the skin depth in copper, with conductivity of $5.8 \times 10^7/\Omega$-m, is only about 6.6×10^{-5} cm.

Returning to the general case of Eq. (27-53) and introducing there the complex refractive index, we write

$$n^2 = \left(\frac{c}{\omega}k\right)^2 = 1 + \frac{i\sigma c^2 \mu_0}{\omega(1 - i\omega/\gamma)}$$

After multiplying the complex term by $i\gamma/i\gamma$, this becomes

$$n^2 = 1 - \frac{\mu_0 \sigma c^2 \gamma}{\omega^2 + i\omega\gamma} \qquad (27\text{-}59)$$

The numerator in the second term must have the same dimensions as ω^2 and is identified as the square of a *plasma frequency* given by

$$\omega_p^2 = \mu_0 c^2 \gamma \sigma = \mu_0 c^2 \gamma \left(\frac{Ne^2}{m\gamma}\right) = \frac{Ne^2}{m\epsilon_0} \qquad (27\text{-}60)$$

where we have made use of both Eq. (27-46) and the relation $c^2 = 1/\epsilon_0\mu_0$. The plasma frequency is a resonant frequency for the free oscillations of the electrons about their equilibrium positions. Inserting it into Eq. (27-59),

$$n^2 = 1 - \frac{\omega_p^2}{\omega^2 + i\omega\gamma} \qquad (27\text{-}61)$$

the plasma frequency turns out to be a critical frequency whose value determines whether the refractive index is real or imaginary. This can be seen by neglecting the γ-term, valid for high enough frequency ($\omega \gg \gamma$), in which case Eq. (27-61) is simply

$$n^2 = 1 - \frac{\omega_p^2}{\omega^2} \qquad (27\text{-}62)$$

Equation (27-62) now shows that for $\omega < \omega_p$, the refractive index of the metal is complex and radiation is attenuated, whereas for $\omega > \omega_p$, the index is real and the metal is transparent to the radiation.

Returning to Eq. (27-61), we find, as before, two equations from which the real and imaginary parts of the refractive index can be calculated. Equating real and imaginary parts in

$$n^2 = (n_R + in_I)^2 = (n_R^2 - n_I^2) + i(2n_R n_I) = 1 - \left(\frac{\omega_p^2}{\omega^2 + i\omega\gamma}\right) \qquad (27\text{-}63)$$

we find

$$n_R^2 - n_I^2 = 1 - \left(\frac{\omega_p^2}{\omega^2 + \gamma^2}\right) \qquad (27\text{-}64)$$

$$2n_R n_I = \frac{\gamma}{\omega}\left(\frac{\omega_p^2}{\omega^2 + \gamma^2}\right) \qquad (27\text{-}65)$$

These equations, solved simultaneously, permit calculation of curves such as those in Figure 27-3. The curves cross at $\omega = (\omega_p^2 - \gamma^2)^{1/2}$, as is evident from Eq. (27-64). Since typically $\omega_p \gg \gamma$, the crossover occurs at $\omega \cong \omega_p$, dividing the transparent and the opaque (and highly reflecting) regions. The plasma frequency for metals falls in the visible to near-ultraviolet regions, so that they are opaque to visible and transparent to ultraviolet radiation at sufficiently high frequency.

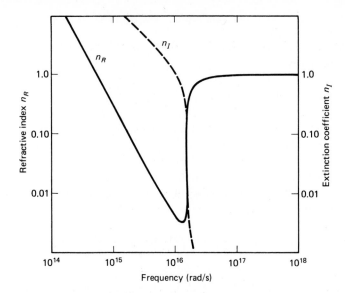

Figure 27-3 Angular frequency dependence of the refractive index n_R and the extinction coefficient n_I for copper. Values assumed are $\omega_p = 1.63 \times 10^{16}$ s^{-1} and $\gamma = 4.1 \times 10^{13}$ s^{-1}. The crossover point of the curves coincides with the plasma frequency.

Intermediate to the good insulator and good conductor we have treated separately are materials, like semiconductors, for which neither of these extreme cases suffices to explain the properties. Such materials manifest appreciable contributions to their optical properties from both free and bound charges and accordingly must be treated by allowing for both types of behavior.

PROBLEMS

27-1. In general, the "electrical constant" K, the *dielectric constant*, is related to the refractive index by

$$K = n^2$$

(a) Show that if K_R and K_I are the real and imaginary parts of the dielectric constant, then

$$n_R = \left[\frac{K_R + (K_R^2 + K_I^2)^{1/2}}{2} \right]^{1/2}$$

and

$$n_I = \left[\frac{-K_R + (K_R^2 + K_I^2)^{1/2}}{2} \right]^{1/2}$$

(b) Calculate n_R and n_I for a dielectric, in terms of K_I, at frequencies high enough such that $K_I = K_R$.

27-2. Show that in a nearly transparent medium, the absorption coefficient is related to the conductivity and refractive index by

$$\alpha = \frac{377\sigma}{n_R}$$

27-3. Write a computer program to calculate and/or plot real and imaginary parts of the refractive index for a dielectric given the frictional parameter γ, the resonant frequency ω_0, and the dipole density N. Check your program against Figure 27-2.

27-4. Assume that aluminum has one free electron per atom and a static conductivity given by $3.54 \times 10^7/\Omega$-m. Determine (a) the frictional constant γ; (b) the plasma frequency; (c) the real and imaginary parts of the refractive index at 550 nm.

27-5. Show that Eq. (27-58) for the skin depth at low frequency is an adequate approximation when $\omega \ll \gamma$ and $\omega \ll \sigma / \epsilon_0$.

27-6. Calculate the skin depth in copper for radiation of (a) 60 Hz and (b) 3 m. First ensure that the approximations of problem 27-5 are satisfied. (Handbook data for copper: $\sigma = 5.76 \times 10^7 / \Omega\text{-m}$.)

27-7. Compare the skin depth of (a) aluminum, with conductivity of $3.54 \times 10^7 / \Omega\text{-m}$ and (b) seawater, with conductivity of $4.3 / \Omega\text{-m}$, for radio waves of 60 kHz.

27-8. Calculate the skin depth of a solid silver waveguide component for 10-cm microwaves. Silver has a conductivity of $3 \times 10^7 / \Omega\text{-m}$. Explain why a more economical silver-plated brass component will work as well.

27-9. The energy density of red light of wavelength 660 nm is reduced to one-quarter of its original value by passage through 342 cm of seawater.
(a) What is the absorption coefficient of seawater for red light of this wavelength?
(b) At what depth is red light reduced to 1% of its original energy density?

27-10. Write a computer program to calculate and/or plot the real and imaginary parts of the refractive index for a metal, given the frictional parameter γ and the plasma frequency. Check your results against Figure 27-3.

27-11. Determine the theoretical content of the constants A, B, and C used to express the Cauchy dispersion equation.

REFERENCES

[1] Feynman, Richard P., Robert B. Leighton, and Matthew Sands. *The Feynman Lectures on Physics*, vol. 1. Reading, Mass.: Addison-Wesley Publishing Company, 1975. Ch. 31.

[2] Sommerfeld, Arnold. *Optics*. New York: Academic Press, 1964. Ch. 3.

[3] Javan, Ali. "The Optical Properties of Materials." *Scientific American* (Sept. 1967): 238.

[4] Longhurst, R. S. *Geometrical and Physical Optics*, 2d ed. New York: John Wiley and Sons, 1967. Ch. 20.

[5] Strong, John. *Concepts of Classical Optics*. San Francisco: W. H. Freeman and Company Publishers, 1958. Ch. 4.

[6] Rossi, Bruno. *Optics*. Reading, Mass.: Addison-Wesley Publishing Company, 1957. Ch. 8.

Suggestions
for Further Reading

ARTICLES ON OPTICS FROM SCIENTIFIC AMERICAN
(chronological order)

Kirkpatrick, Paul. "X-ray Microscope." (Mar. 1949): 44.

Evans, Ralph M. "Seeing Light and Color." (Aug. 1949): 52.

Wald, George. "Eye and Camera." (Aug. 1950): 32.

Wilson, Albert G. "The Big Schmidt." (Dec. 1950): 34.

Kelner, Albert. "Revival by Light." (May 1951): 22.

Muller, Erwin W. "A New Microscope." (May 1952): 58.

Ingalls, Albert G. "Ruling Engines." (June 1952): 45.

James, T. H. "Photographic Development." (Nov. 1952): 30.

LaMer, Victor K., and Milton Kerker. "Light Scattered by Particles." (Feb. 1953): 69.

Waterman, Talbot H. "Polarized Light and Animal Navigation." (July 1955): 88.

Rush, J. H. "The Speed of Light." (Aug. 1955): 62.

Sperry, R. W. "The Eye and the Brain." (May 1956): 48.

Milne, L. J., and M. J. Milne. "Electrical Events in Vision." (Dec. 1956): 113.

Gordon, James P. "The Maser." (Dec. 1958): 42.

Land, Edwin H. "Experiments in Color Vision." (May 1959): 84.

Wald, George. "Life and Light." (Oct. 1959): 92.

McClain, Edward, Jr. "The 600-Foot Radio Telescope." (Jan. 1960): 45.

Bloom, Arnold L. "Optical Pumping." (Oct. 1960): 72.

Kapany, Narinder S. "Fiber Optics." (Nov. 1960): 72.

Arnon, Daniel. "The Role of Light in Photosynthesis." (Nov. 1960): 104.

Butler, W. L., and R. J. Downs. "Light and Plant Development." (Dec. 1960): 56.

Hess, Eckhard H. "Shadows and Depth Perception." (Mar. 1961): 138.

Schawlow, Arthur L. "Optical Masers." (June 1961): 52.

Pritchard, Roy M. "Stabilized Images on the Retina." (June 1961): 72.

Evans, Ralph M. "Maxwell's Color Photograph." (Nov. 1961): 118.

Wallach, Hans. "The Perception of Neutral Colors." (Jan. 1963): 107.

Schawlow, Arthur L. "Advances in Optical Masers." (July 1963): 34.

Giordmaine, J. A. "The Interaction of Light with Light." (Apr. 1964): 38.

Leith, Emmett N., and Juris Upatnieks. "Photography by Laser." (June 1965): 24.

Miller, Stewart E. "Communication by Laser." (Jan. 1966): 19.

Pimentel, George C. "Chemical Lasers." (Apr. 1966): 32.

Rock, Irvin, and Charles Harris. "Vision and Touch." (May 1967): 96.

Morehead, Fred F., Jr. "Light-Emitting Semiconductors." (May 1967): 108.

Hubbard, Ruth, and Allen Kropf. "Molecular Isomers in Vision." (June 1967): 64.

Lempicki, Alexander, and Harold Samelson. "Liquid Lasers." (June 1967): 80.

Javan, Ali. "The Optical Properties of Materials." (Sept. 1967): 238.

Pennington, Keith S. "Advances in Holography." (Feb. 1968): 40.

Hartmann, Sven R. "Photon Echoes." (Apr. 1968): 32.

Nelson, Donald F. "The Modulation of Laser Light." (June 1968): 17.

Patel, C. K. N. "High-Power Carbon Dioxide Lasers." (Aug. 1968): 22.

Feiberg, Gerald. "Light." (Sept. 1968): 50.

Weisskopf, Victor F. "How Light Interacts with Matter." (Sept. 1968): 60.

Connes, Pierre. "How Light is Analyzed." (Sept. 1968): 72.

Smith, F. Dow. "How Images Are Formed." (Sept. 1968): 96.

Jones, R. Clark. "How Images Are Detected." (Sept. 1968): 110.

Schawlow, Arthur L. "Laser Light." (Sept. 1968): 120.

Herriott, Donald R. "Applications of Laser Light." (Sept. 1968): 140.

Oster, Gerald. "The Chemical Effects of Light." (Sept. 1968): 158.

Hendricks, Sterling B. "How Light Interacts with Living Matter." (Sept. 1968): 174.

Fitch, James Marston. "The Control of the Luminous Environment." (Sept. 1968): 190.

Neisser, Ulric. "The Processes of Vision." (Sept. 1968): 204.

Sorokin, Peter. "Organic Lasers." (Feb. 1969): 30.

Michael, Charles R. "Retinal Processing of Visual Images." (May 1969): 104.

Metherell, Alexander F. "Acoustical Holography." (Oct. 1969): 36.

Vali, Victor. "Measuring Earth Strains by Laser." (Dec. 1969): 88.

Berns, Michael H., and Donald E. Rounds. "Cell Surgery by Laser." (Feb. 1970): 98.

Faller, James E., and E. Joseph Wampler. "The Lunar Laser Reflector." (Mar. 1970): 38.

Drexhage, Karl H. "Monomolecular Layers and Light." (Mar. 1970): 108.

Heilmeier, George H. "Liquid Crystal Display Devices." (Apr. 1970): 100.

Young, Richard W. "Visual Cells." (Oct. 1970): 80.

Burbidge, E. Margaret, and C. R. Lynds. "The Absorption Lines of Quasi-Stellar Objects." (Dec. 1970): 22.

Baumeister, Philip, and Gerald Pincus. "Optical Interference Coatings." (Dec. 1970): 58.

Crewe, Albert V. "A High-Resolution Scanning Electron Microscope." (Apr. 1971): 26.

Lubin, Moshe J., and Arthur P. Fraas. "Fusion by Lasers." (June 1971): 21.

Panish, Morton B., and Izuo Hayashi. "A New Class of Diode Lasers." (July 1971): 32.

Ingham, M. F. "The Spectrum of the Airglow." (Jan. 1972): 78.

Ashkin, Arthur. "The Pressure of Laser Light." (Feb. 1972): 62.

Kellermann, K. I. "Intercontinental Radio Astronomy." (Feb. 1972): 72.

Busignies, Henri. "Communication Channels." (Sept. 1972): 98.

Silfvast, William T. "Metal-Vapor Lasers." (Feb. 1973): 88.

Cook, J. S. "Communication by Optical Fiber." (Nov. 1973): 28.

Feld, M. S., and V. S. Letokhov. "Laser Spectroscopy." (Dec. 1973): 69.

Tien, P. K. "Integrated Optics." (Apr. 1974): 28.

Metelli, Fabio. "The Perception of Transparency." (Apr. 1974): 90.

Emmett, John L., John Nuckolls, and Lowell Wood. "Fusion Power by Laser Implosion." (June 1974): 24.

Campbell, Fergus W., and Lamberto Maffei. "Contrast and Spatial Frequency." (Nov. 1974): 106.

Govindjee, and Rajni Govindjee. "The Absorption of Light in Photosynthesis." (Dec. 1974): 68.

Wurtmann, Richard J. "The Effects of Light on the Human Body." (July 1975): 68.

Beck, Jacob. "The Perception of Surface Color." (Aug. 1975): 62.

Ross, John. "The Resources of Binocular Perception." (Mar. 1976): 80.

Wehner, Rudiger. "Polarized Light Navigation by Insects." (July 1976): 106.

Price, William H. "The Photographic Lens." (Aug. 1976): 72.

Leith, Emmett N. "White-Light Holograms." (Oct. 1976): 80.

Zare, Richard N. "Laser Separation of Isotopes." (Feb. 1977): 86.

Nussenzveig, H. Moyses. "The Theory of the Rainbow." (Apr. 1977): 116.

Horridge, G. Adrian. "The Compound Eye of Insects." (July 1977): 108.

Boyle, W. S. "Light-Wave Communications." (Aug. 1977): 40.

Land, Edwin H. "The Retinex Theory of Color Vision." (Dec. 1977): 108.

Dickinson, Dale F. "Cosmic Masers." (June 1978): 90.

Spiller, Eberhard, and Ralph Feder. "The Optics of Long-Wavelength X Rays." (Nov. 1978): 70.

Land, Michael F. "Animal Eyes with Mirror Optics." (Dec. 1978): 88.

Yariv, Amnon. "Guided Wave Optics." (Jan. 1979): 64.

Ronn, Avigdor M. "Laser Chemistry." (May 1979): 114.

Quate, Calvin. "Acoustic Microscope." (Oct. 1979): 62.

Nassau, Kurt. "The Causes of Color." (Oct. 1980): 124.

Thomas, D. E. "Mirror Images." (Dec. 1980): 206.

Yellott, John I., Jr. "Binocular Depth Inversion." (July 1981): 148.

Tsipis, Kosta. "Laser Weapons." (Dec. 1981): 51.

LaRocca, Aldo V. "Laser Applications in Manufacturing." (Mar. 1982): 94.

Bahcall, J. N., and L. Spitzer, Jr. "The Space Telescope." (July 1982): 40.

Abraham, Eitan, et al. "The Optical Computer." (Feb. 1983): 85.

Poggio, Tomaso. "Vision by Man and Machine." (Apr. 1984): 106.

Mandoli, Dina F., and Winslow R. Briggs. "Fiber Optics in Plants." (Aug. 1984): 90.

Habing, Harm J., and Gerry Neugebauer. "The Infrared Sky." (Nov. 1984): 48.

Tsang, W. T. "The C^3 Laser." (Nov. 1984): 148.

Brookner, Eli. "Phased-Array Radars." (Feb. 1985): 94.

Shkunov, Vladimir V., and Boris Ya. Zel'dovich. "Optical Phase Conjugation." (Dec. 1985): 54.

Pepper, David M. "Applications of Optical Phase Conjugation." (Jan. 1986): 74.

Hoskin, Michael. "William Herschel and the Making of Modern Astronomy." (Feb. 1986): 106.

Anderson, Dana Z. "Optical Gyroscopes." (Apr. 1986): 94.

Ramachandran, Vilayanur S., and Stuart M. Anstis. "The Perception of Apparent Motion." (June 1986): 102.

Craxton, R. Stephen, Robert L. McCrory, and John M. Soures. "Progress in Laser Fusion." (Aug. 1986): 68.

Brou, Philippe, Thomas R. Sciascia, Lynette Linden, and Jerome Y. Lettvin. "The Colors of Things." (Sept. 1986): 84.

Rowell, J. M. "Photonic Materials." (Oct. 1986): 146.

Abu-Mostafa, Yaser S., and Demetri Psaltis. "Optical Neural Computers." (Mar. 1987): 88.

Schnapf, Julie L., and Denis A. Baylor. "How Photoreceptor Cells Respond to Light." (Apr. 1987): 40.

Winick, Herman. "Synchrotron Radiation." (Nov. 1987): 88.

Livingstone, Margaret S. "Art, Illusion and the Visual System." (Jan. 1988): 78.

Slusher, Richard E., and Bernard Yurke. "Squeezed Light." (May 1988): 50.

Loretz, Jane F., and George H. Handelman. "How the Human Eye Focuses." (July 1988): 92.

Letokhov, Vladilen S. "Detecting Individual Atoms and Molecules with Lasers." (Sept. 1988): 54.

Nathans, Jeremy. "The Genes for Color Vision." (Feb. 1989): 42.

Katzir, Abraham. "Optical Fibers in Medicine." (May 1989): 120.

Bracewell, Ronald N. "The Fourier Transform." (June 1989): 86.

Romer, Grant B., and Jeannette Delamoir. "The First Color Photographs." (Dec. 1989): 88.

Pepper, David M., Jack Feinberg, and Nicolai V. Kukhtarev. "The Photorefractive Effect." (Oct. 1990): 62.

Howells, Malcolm R., Janos Kirz, and William Sayre. "X-ray Microscopes." (Feb. 1991): 88.

Winston, Roland. "Nonimaging Optics." (Mar. 1991): 76.

Berns, Michael W. "Laser Surgery." (June 1991): 84.

Robinson, Glen M., David M. Perry, and Richard W. Peterson. "Optical Interferometry of Surfaces." (July 1991): 66.

Jewell, Jack L., James P. Harbison, and Axel Scherer. "Microlasers." (Nov. 1991): 86.

Desurvire, Emmanuel. "Lightware Communications: The Fifth Generation." (Jan. 1992): 114.

Answers
to Selected Problems

Chapter 1

1-1. (a) 6.6×10^{-34} m (b) 3.9 Å
1-2. 3.6×10^{-17} W
1-3. 3.10 and 1.77 eV
1-4. 0.024 Å; 2.7×10^{-22} kg-m/s
1-9. (a) 1.49×10^{-18} kg-m/s (b) 4.45×10^{-16} m (c) 4.22×10^{-16} m
1-10. 3.75×10^{17}

Chapter 2

2-1. $3.9 - 7.9 \times 10^{14}$ Hz
2-2. (a) 2050 lm (b) 39.8 W/sr; 163 cd (c) 10^5 W/m^2; 4.1×10^5 lm/m^2
(d) 9.95 W/m^2; 40.8 lx (e) 0.0195 W; 0.0803 lm
2-3. (a) He-Cd appears about 1.3× brighter (b) about 2.4 mW
2-4. (a) 900 cd (b) 85.4 lm/m^2 or lx
2-5. 1.055:1
2-6. 320 lx
2-7. (a) 1.7×10^9 cd/m^2 (b) $2\pi L$
2-8. 0.97 lm
2-11. 5800 K
2-12. (a) 0.4830 μm (b) 0.0756 W
2-13. 6266 K; 462.5 nm
2-14. 6105 K

Chapter 3

3-1. $t = (\Sigma_i \, n_i x_i)/c$
3-2. $1.25(x^2 + y^2) + 70(x^2 + y^2)^{1/2} - 135x + 800 = 0$

3-3. 4.00 mm

3-4. 3 ft, with top edge of mirror at a height halfway between the person's eye level and the top of the person's head

3-5. The ray emerges from the bottom at 45°.

3-6. Reflection from the bottom surface; 1.60

3-7. 1.55

3-8. 1.153 cm

3-9. 8 cm

3-10. Light from the bubble is refracted through the plane surface, both directly and after reflection from the spherical mirror; 3.33 cm and 10 cm.

3-11. 12.5 cm; 75 cm

3-12. 10 cm behind the near surface; 3×

3-13. (a) $f = n_1 R/(n_2 - n_1)$ (b) $R > 0$ (convex) and $R < 0$ (concave), respectively

3-14. (a) center, $\frac{4}{3}$ actual size (b) 6.4 cm behind the glass, $\frac{8}{7}$ actual size

3-15. Virtual, inverted, 15 cm from the window, twice the object size

3-16. 13.0 cm

3-17. +20 cm or −20 cm

3-18. 22.5 cm behind the lens; 1.50 times the actual size

3-19. (a) −6.7 cm (b) −10 cm or −60 cm

3-20. −50 cm

3-21. 3.33 mm in front of the objective; erect and magnified

3-22. Final image between lens and mirror at 21/34 f from lens, virtual, inverted, and $\frac{1}{17}$ the original size

3-23. (a) 33.3 cm, 2× (b) 86.67 cm, 2× (c) 7.37 cm, −0.316×

3-24. 1.63

3-25. 150 cm and 600 cm; inverted

3-26. (a) 10, 5, −2.5 diopters; 12.5 diopters (b) 8.33 m^{-1}, 4.17 m^{-1}; 24 cm

3-30. $f_1/f_2 = n/(n - 1)$

3-31. At $s = s' = 2f$

3-32. (b) $\Sigma_i(t_i \tan \theta_i)$

3-33. Incident on plane side: 8 cm beyond lens; on curved side: 5.33 cm beyond lens

3-34. +40 cm, +30 cm; −30 cm, −40 cm

3-35. 25,000 ft

Chapter 4

4-1. $f_1 = -62.05$ cm; $f_2 = 46.66$ cm; $r = 2.91$ cm; $s = -0.98$ cm

4-2. (a) $f_1 = 14.06$ cm $= -f_2$; $r = 1.17$ cm; $s = -0.73$ cm
(b) 8.92 cm left of lens center
(c) 9.78 cm left of lens center, with 9.6% error

4-3. Erect, virtual image at 6.67 cm left of second vertex, 0.556 in. high

4-4. $f_1 = -11.51$ cm; $f_2 = 15.31$ cm; $r = 0.400$ cm; $s = -1.16$ cm; $v = 4.20$ cm; $w = 2.64$ cm; $s_i = 18.9$ cm from H_2; $h_i = -1.18$ cm

4-5. (a) $f_1 = -20.15$ cm $= -f_2$; $r = 10$ cm $= -s$; $v = 10$ cm $= -w$
(b) Image is inverted, real, 61.38 cm from sphere center and $m = -2.05\times$

4-6. (a) $y = 1$ cm; $\alpha = -5.73°$
(b) $A = 1 - x/10$; $B = 10/3 + 2x/3$; $C = -1/10$; $D = 2/3$
(c) $x = 10$ cm

4-8. $p = -4.17$ cm, $q = +2.17$ cm, $r = -0.83$ cm, $s = -2.17$ cm, $f_1 = -3.33$ cm, $f_2 = 4.33$ cm

4-9. $f_1 = -20$ cm, $f_2 = +20$ cm, $p = -30$ cm, $q = +10$ cm, $r = -10$ cm, $s = -10$ cm

4-10. $f_1 = -16.7$ cm, $f_2 = +23.3$ cm, $q = +18.7$ cm, $p = -18.3$ cm, $r = -1.67$ cm, $s = -4.67$ cm

4-11. (a) $A = -\frac{1}{2}$, $B = 0$, $C = -\frac{1}{10}$, $D = -2$ (b) Input and output planes fall at conjugate object and image positions; A is identical with the linear magnification.

4-12. (a) $p = -2$, $q = +2$, $f_1 = -6$, $f_2 = +6$, $r = 4$, $s = -4$ in. (b) 2 in. beyond ball

4-13. (a) Elements of system matrix: $A = \frac{16}{15}$, $B = \frac{2}{3}$, $C = -\frac{1}{150}$, $D = \frac{14}{15}$

    (b) $p = -140$, $q = 160$, $r = s = 10$, $f_1 = -150$, $f_2 = 150$, all in cm

4-14. (a) $A = 0.9764$, $B = 0.9676$, $C = 0.009182$, $D = 1.033$ (b) $f_1 = 108.9$ cm,

    $f_2 = -108.9$ cm, $p = 112.5$ cm, $q = -106.3$ cm, $r = 3.62$ cm, $s = 2.57$ cm

    (c) -100 cm

4-15. (a) $A = \frac{2}{3} - s'/6$; $B = \frac{25}{3} + 2 - ss'/6 + s'$; $C = -\frac{1}{6}$; $D = -s/6 + 1$

    (b) $s' = (4s + 12)/(s - 6)$; $m = \frac{2}{3} - s'/6$ (c) $s' = 6\frac{4}{7}$ cm; $m = -0.429$

    (d) $s' = 4$ cm corresponds to second focal plane; $s = 6$ cm corresponds to first

    focal plane

4-16. $A = 0.93935$; $B = 22.2212$; $C = -0.009284$; $D = 0.8448$; $r = v = 16.72$ mm;

    $s = w = -6.53$ mm; $p = -90.99$ mm; $q = 101.18$ mm; $f_1 = -f_2 =$

    -107.71 mm; the film plane is a distance $q = 101.2$ mm behind the last lens

    surface.

4-17. $A = \dfrac{n - n_L}{n_L R_1}t + 1$ $B = \dfrac{nt}{n_L}$

$$C = \frac{n_L - n'}{n'R_2} - \frac{n_L - n}{n'R_1} - \frac{(n_L - n)(n_L - n')}{n'n_L}\frac{t}{R_1 R_2} \qquad D = \frac{n}{n'} + \left(\frac{n_L - n'}{n'R_2}\right)\frac{nt}{n_L}$$

4-21. For $\alpha = 0°$, $s' = 3.180$ cm and $\alpha' = -23.51°$; for $\alpha = -20°$, $s' = 16.104$ cm

    and $\alpha' = 6.081°$

4-22. $s' = -49.525$ cm; $\alpha' = 3.371°$; $Q = 2.912$

4-23. For $h = 1$ mm, $s' = 98.20$ mm, $\alpha' = -0.567°$; for $h = 5$ mm, $s' = 102.45$ mm,

    $\alpha' = -2.723°$

Chapter 5

5-2. (d) $s' = 0$, $+8$, $+14.4$ cm

5-3. $a = +0.0954$ mm; $b_y = +1.22$ mm; $b_z = +3.91$ mm

5-5. $a = 0.015$ mm; $b_y = 0.49$ mm; $b_z = 3.9$ mm

5-6. (a) 0.0296 mm (b) 0.021 mm

5-7. (a) 0.60 mm (b) 1.2 mm

5-8. $b_z = 1.48$ mm; $b_y = 0.15$ mm

5-9. $b_z = 0.974$ mm at $h = 1$ cm, 3.84 mm at $h = 2$ cm, 8.44 mm at $h = 3$ cm,

    14.53 mm at $h = 4$ cm, 21.81 mm at $h = 5$ cm

5-10. $b_z = 1.82$ cm; $b_y = 0.970$ mm

5-12. For $\sigma = 0.7$, $r_1 = 17.65$ and $r_2 = -100$ cm; for $\sigma = 3$, $r_1 = 7.50$ and

    $r_2 = 15.0$ cm

5-13. $r_1 = 18.62$ cm, $r_2 = -33.75$ cm

5-14. optimum $\sigma = +0.714$, closer to $+1$ than to -1

5-15. (a) $+0.714$ (b) $r_1 = 17.5$ cm; $r_2 = -105$ cm (c) -0.714, reverse the lens

5-16. (a) 0.8 (b) $r_1 = 16.7$ cm; $r_2 = -150$ cm (c) -0.8, reverse the lens

5-17. $+20$ and -20 cm

5-18. answers the same

5-19. -17.7 cm

5-20. (a) $R = 15.7$ cm (b) $f_2 = -3.476$ cm

5-21. $r_{11} = 8.5168$ cm, $r_{22} = -434.89$ cm; $f_D = 20.0000$ cm, $f_C = 20.0096$ cm,

    $f_F = 20.0096$ cm

5-22. (a) $r_{11} = 3.4535$ cm, $r_{22} = -12.6576$ cm

    (b) $f_D = 5.0000$ cm, $f_C = 5.0026$ cm, $f_F = 5.0026$ cm

    (c) $P_{1D} = 0.3695$ cm^{-1}, $P_{2D} = -0.1695$ cm^{-1}, $\Delta_{1D} = 0.01802$, $\Delta_{2D} = 0.03928$

    (d) yes

5-23. (a) $r_{11} = -5.2415$ cm, $r_{22} = 53.1840$ cm (b) $f_{1D} = -4.5770$ cm, $f_{2D} =$

    8.4399 cm (c) $f_D = -10.0000$ cm, $f_C = -10.0050$ cm, $f_F = -10.0050$ cm

Chapter 6

6-1. Entrance pupil is the stop; exit pupil is 3.33 cm in front of the lens, with an

    aperture of 3.33 cm; image is 10 cm behind the lens, inverted and 2 cm long.

6-2. Exit pupil is the stop; entrance pupil is 4.29 cm behind the lens, with an aperture of

    3.43 cm; image is 10.5 cm behind the lens, inverted, and 3 cm long.

6-3. Entrance pupil is the stop; exit pupil is 12 cm in front of the lens, with an aperture of 6 cm; image is 10.5 cm behind the lens, inverted, and 1.5 cm long.

6-4. (b) 20 cm right of L_2 (c) both at L_1 (d) 8.57 cm beyond L_2 and $\frac{4}{7}$ cm in diameter (e) field stop at A, entrance window in object plane with 1 cm diameter, exit window in image plane with 1 cm diameter   (f) $\alpha = 1.43°$, $\alpha' = 5.01°$

6-6. 53′

6-7. (a) crown: $A = 1.511$, $B = 4240$ nm^2, $n_D = 1.523$; flint: $A = 1.677$, $B = 13{,}190$ nm^2, $n_D = 1.715$ (b) crown: -4.146×10^{-5} nm^{-1}; flint: -1.290×10^{-4} nm^{-1} (c) crown: 3110, 1.9 Å; flint: 9675, 0.61 Å

6-8. (a) $50.0°$ (b) $1/55.5$ (c) $A = 1.6205$, $B = 6073.7$ nm^2; 4.297×10^{-5} nm^{-1} (d) 1.12 m

6-9. 0.01909

6-10. $5.99°$, 2.16′

6-11. $4.82°$, $4.37°$

6-12. (a) $M_e = 1 \times 10^4$ W/m^2, $I_e = 3.98$ W/sr, $L_e = 1.59 \times 10^3$ W/m^2-sr (b) 7.81×10^{-5} W   (c) 18 W/m^2

6-14. 5.7 cm

6-15. (a) $f = 53.3$ cm, 13.33 cm, 1.86 cm   (b) $r = -100$ cm; $s = -40$ cm

6-16. 5.3 to 7.0 ft

6-17. 1.3×10^7 lx

6-18. (a) 0.90 cm   (b) 5.45 cm, $3\times$

6-19. (a) 27.8 mm   (b) $f/3.1, f/5.4, f/9.4$ (c) 16.0, 9.26, 5.35 mm   (d) 0.03, 0.09, 0.27 s

6-21. (a) 2.8 cm   (b) $10\times$

6-22. (a) $320\times$ (b) 0.516 cm

6-23. (a) $46.7\times$ (b) 8.68 cm

6-24. 5 cm

6-25. 14.9 cm

6-26. (a) $7\times$ (b) 2 cm   (c) 5 mm   (d) 2.3 cm   (e) 337 ft

6-27. (b) $7.50\times$; $8.70\times$

6-28. 1.05 cm

6-29. (a) 8 cm, $3\times$ (b) 7.38 cm, $2.6\times$

6-30. 1.25 cm farther from the objective

6-31. (a) $12.5\times$ (b) $15\times$ (c) 0.13 cm, 3 mm   (d) $3.8°$

6-33. -2.5 ft; $-180\times$

Chapter 7

7-1. $+41.6$ D

7-2. (a) 8.33 mm; $+120$ D   (b) 42.86 mm; $+23.33$ D   (c) 43.65 mm, measured from its second principal plane, or 42.38 mm from its second surface; $+22.9$ D

7-3. (a) 22.34 mm from cornea   (b) 21.60 mm from cornea

7-4. (a) $A = 0.75846$, $B = 5.1050$, $C = -0.05011$, $D = 0.65180$ (b) Focal points are 13.01 mm in front and 22.34 mm behind the cornea; principal points are 1.96 mm behind and 2.38 mm behind the cornea.

7-5. Block-letter sizes are 1.309 in. for 20/300; 0.436 in. for 20/100; 0.262 in. for 20/60; 0.087 in. for 20/20; 0.065 in. for 20/15. Letter details are $\frac{1}{5}$ block letter size in each case.

7-6. (a) $+3.2$ D   (b) yes

7-7. -2.0 D; 21.4 cm

7-8. (a) argon ion; doubled YAG   (b) CO_2 (c) Nd:YAG; krypton red

7-9. (a) myopia; astigmatism   (b) myopia   (c) hyperopia   (d) hyperopia; astigmatism

7-10. (a) Glass absorbs 10.6 μm light.   (b) 4.4×10^{-4} rad   (c) 14.5 μm (d) 3×10^6 W/cm^2

7-11. (a) 10^6 W   (b) 5 μm   (c) 5.1×10^{12} W/cm^2

7-12. (a) 1.04×10^{-2} W/cm^2 (b) 783 W/cm^2 (c) 75,000

7-13. (a) 0.21 J/cm^2 (b) 4.6

Chapter 8

8-1. $y = Ae^{-b(x + 10t)^2}$

8-2. (a) $y = \dfrac{4}{(x + 2.5t)^2 + 2}$

8-3. (a) (1) and (2) qualify because they satisfy the wave equation; more simply, if $w = z + vt$, they are functions of w: $y = A \sin^2 (4\pi w)$ and $y = Aw^2$.
(b) (i) $v = 1$ m/s in $-z$-direction; (ii) $v = 1$ m/s in $+x$-direction.

8-4. 10 m/s in $+x$-direction

8-5. (a) $\psi = 2 \sin 2\pi (z/5 + t/3)$ (b) $\psi = 2 \sin (2\pi/5)(z + \tfrac{5}{3}t)$
(c) $\psi = 2 \exp [(2\pi i(z/5 + t/3)]$

8-6. (a) $y = 5 \sin (\pi x/25)$ (b) $y = 5 \sin (\pi/25)(x + 8)$

8-7. (a) 0.01 cm   (b) 1000 Hz   (c) 628.3 cm^{-1} (d) 6283 s^{-1} (e) 1 ms
(f) 10 cm/s   (g) 10 cm

8-8. (a) $+1$ in y-direction   (b) $-C/B$ in x-direction   (c) C in z-direction

8-9. $y = 15 \sin (kx + \pi/3)$

8-10. (b) 90°, 60°, 0°, −90°, 108°   (c) Subtract 90° from each.

8-11. (a) $A \sin (2\pi/\lambda)(z - vt)$ (b) $A \sin (2\pi/\lambda)(\sqrt{2}x \pm vt)$
(c) $A \sin (2\pi/\lambda)[(\sqrt{3}/3)(x + y + z) \pm vt]$

8-15. $E = 1028$ V/m; $B = 3.43 \times 10^{-6}$ T

8-16. (a) 5×10^{-7} T   (b) 19.88 W/m^2

8-17. (a) 1.01×10^3 V/m, 3.37×10^{-6} T   (b) 4.76×10^{21}/m^2-s
(c) $E = 1010 \sin 2\pi (1.43 \times 10^6 r + 4.28 \times 10^{14}t)$, r in m, t in s

8-18. (a) 8.75×10^{-3} W/m^2, 2.57 V/m
(b) 2×10^{13} W/m^2, 1.23×10^8 V/m, 0.410 T

8-21. $v = 0.168c$

8-22. $v = -0.917c$

8-23. $2\,\Delta\lambda = 0.12$ Å

Chapter 9

9-1. (a) The waves move in opposite directions along the x-axis, E_1 to the right, E_2 to the left, with equal speeds of $\tfrac{4}{3}$ m/s.   (b) $t = \tfrac{3}{4}$ s   (c) $x = 1$ m

9-2. (b) $E_R = 8.53 \sin (\omega t + 0.2\pi)$

9-3. $E_R = 6.08 \sin (2\pi t + 1.13)$

9-4. $y = 11.6 \sin (\omega t + 0.402\pi)$

9-5. $E = 1.71 \sin (\pi t + 4.57)$

9-6. (a) 2 V/m   (b) 0.2 V/m

9-7. $\psi(t) = 2.48 \sin (20t + 0.8\pi)$

9-8. (a) $v_g = v_p [1 - (\omega/n)(dn/d\omega)]$ (b) $v_g < v_p$

9-9. $c/1.56$

9-10. $v_p = c/1.5$; $v_g = c/1.73$

9-12. $v_g = A = $ constant

9-14. $2(v/c)\nu_0$

9-15. 14 cm; 1.57 cm; 0.785 cm; 0 cm/s; T seconds

9-16. (a) 1.5 cm; 25 Hz; 20 cm; 5 m/s; opposite directions   (b) 10 cm   (c) −3 cm;
0 cm/s; 7.40×10^4 cm/s^2

Chapter 10

10-1. (a) 11,945 and 21,235 W/m^2 (b) 18,723 W/m^2 (c) 51,903 W/m^2 (d) 0.96

10-2. 0.86, 0

10-3. 0.8; 3.73/1

10-4. 1.78, 2.55, 4.00, 13.9

10-5. Lloyd's mirror interference fringes are produced, aligned parallel to the slit, and separated by 0.273 mm. The irradiance of the pattern is given by
$I = 4I_0 \sin^2 (115y)$, with y measured in cm from the mirror surface.

10-6. 509 nm

10-7. 514.5 nm

10-8. To acquire coherent beams; 560 nm

10-9. (a) 83.3 cm (b) 83.3 fringes (c) 150 nm
10-10. 556 nm, 455 nm
10-11. 20.3′
10-12. 6′5″
10-13. 35′40″
10-14. 9.09×10^{-5} cm; orders 4 and 3, respectively
10-15. 498 nm
10-16. 1.33; 103 nm
10-17. (a) 2.78% (b) 89.3 nm (c) 1%
10-18. Soap film becomes wedge-shaped under gravity; the angle of the wedge is 1′14″.
10-19. 15
10-20. 1.16×10^{-3} cm
10-21. 1.09 mm; 184
10-23. 3 m
10-25. 603.5 nm; 2.39 mm; 2.87×10^{-4} cm
10-26. 928 nm

Chapter 11

11-1. 436 nm
11-2. One mirror makes a wedge angle of 0.0172° with the image of the other, reflected through the beam splitter. Fizeau fringes result.
11-3. 23.75 μm
11-4. (a) 80,000; (b) 79,994
11-5. (a) $n = 1 + N\lambda/2L$ (b) 153
11-6. (a) 11.2° (b) 45.9°
11-7. 79.1 nm or $\lambda/8$
11-8. (a) 980 V/m (b) 30° (c) $r' = 0.28$; $tt' = 0.9216$ (d) 274, 253, 19.8 V/m; 7.8%, 6.7%, 0.041% (e) 903, 70.8 V/m; 85%, 0.52% (f) 258 nm
11-10. (a) 48,260 (b) 0.01013 cm
11-11. (a) 3.996×10^6 (b) 3.16×10^6 (c) 0.318 mm (d) 6.29 Å (e) 0.002 Å
11-12. (a) 329,670 (b) 361 (c) 9.8×10^6
11-13. 2.18 cm
11-14. 16
11-15. 0.161 mm
11-16. (a) 360° (b) 180° (c) 2
11-17. 1; 0.47

Chapter 12

12-1. $f(x) = (4/\pi)(\sin kx + \frac{1}{3} \sin 3kx + \frac{1}{5} \sin 5kx + \cdots)$
12-2. $f(t) = \dfrac{E_0}{\pi} + \dfrac{E_0}{2} \cos \omega t + \dfrac{2E_0}{3\pi} \cos 2\omega t - \dfrac{2E_0}{15\pi} \cos 4\pi t + \cdots$
12-3. $g(\omega) = \dfrac{\sigma h}{\sqrt{2\pi}} e^{(-\sigma^2 \omega^2)/2}$

If the width of the first is σ, the width of the second is $1/\sigma$. Thus the spectrum broadens as the original Gaussian narrows, and vice versa.
12-4. $|g(\omega)|^2 = (A^2 \tau_0^2/4\pi^2)(\sin u/u)^2$, where $u = \omega\tau_0/2$
12-5. The narrow-band filter has a coherence length better by one order of magnitude: 3.48×10^{-5} m
12-6. 0.013 nm; 10^{10} Hz; 3 cm
12-7. 0.0243 mm
12-8. (a) 0.00138 nm (b) 1 ns
12-9. 2.5 mm
12-10. 0.0625 cm; 2.08×10^{-12} s
12-11. 0.144 cm
12-12. 4×10^{-7} Å; 3×10^4 Hz

12-13. (a) 2.08×10^{-12} s, 0.0625 cm   (b) 0.36, 0.36   (c) 53

12-14. 1.01×10^{-4} cm, 2.90×10^{-6} cm^2; 1.8, 35

12-15. (b) 2.55

12-17. 0.998, 0.63

12-18. 0.937, 0.686, 15.95 cm

12-19. (a) 0, 0, 0.596 cm   (b) 0.895 mm

Chapter 13

13-2. (b) 0.866

13-4. 1.82 μm

13-5. 63.3 m/s

13-7. 1.8×10^{10} bits

13-8. (a) 250 nm   (b) 500 nm   (c) 433 nm

13-9. (b) 365 nm   (c) 38°

13-10. 365 nm; blue components shift into ultraviolet and are missing.

13-11. (a) $1.88\times$ (b) $6330\times$

Chapter 14

14-2. (a) $\dfrac{1}{\sqrt{2}}\begin{bmatrix} 1 \\ -1 \end{bmatrix}$: linearly polarized at $-45°$

(b) $\dfrac{1}{\sqrt{2}}\begin{bmatrix} 1 \\ 1 \end{bmatrix}$: linearly polarized at $+45°$

(c) $\dfrac{1}{\sqrt{2}}\begin{bmatrix} 1 \\ \dfrac{1}{\sqrt{2}}(1-i) \end{bmatrix}$: right-elliptically polarized at $+45°$

(d) $\dfrac{1}{\sqrt{2}}\begin{bmatrix} 1 \\ i \end{bmatrix}$: left-circularly polarized

14-3. (a) linearly polarized along x-direction, traveling in $+z$-direction with amplitude of $2E_0$ (b) linearly polarized at 53.1° relative to the x-axis, traveling in the $+z$-direction with amplitude of $5E_0$ (c) right-circularly polarized, traveling in $-z$-direction, with amplitude of $5E_0$

14-4. 75°

14-5. right-circularly polarized light

14-6. (a) $\mathbf{E} = E_0(\sqrt{3}\,\mathbf{j} + \mathbf{k})e^{i(kx-\omega t)}$ (b) $\mathbf{E} = E_0(2\mathbf{k} - i\mathbf{i})e^{i(ky-\omega t)}$
(c) $\mathbf{E} = \mathbf{k}E_0 \exp\{i[(x+y)k/\sqrt{2} - \omega t]\}$

14-7. (a) $C = 0$, $m\pi$ (b) $B = 0$, $(m + \tfrac{1}{2})\pi$ (c) $B = 0$, $A = \pm C$, $(m + \tfrac{1}{2})\pi$

14-9. (a) linearly polarized, $\alpha = 18.4°$, $A = \sqrt{10}$ (b) right-circularly polarized, $A = 1$ (c) right-elliptically polarized; semimajor axis = 5 along y-axis, semiminor axis = 4 along x-axis   (d) linearly polarized, horizontal, $A = 5$ (e) left-circularly polarized, $A = 2$ (f) linearly polarized, $\alpha = 56.3°$, $A = \sqrt{13}$ (g) left-elliptically polarized, $\epsilon = 53.1°$, $\alpha = -7°$, $E_{0x} = 2$, $E_{0y} = 10$

14-10. right-elliptically polarized, symmetrical with x- and y-axes, $E_{0x}/E_{0y} = \sqrt{3}$

14-13. right-circularly polarized light

14-14. no light emerges

14-15. (a) right-elliptically polarized, major axis along x-axis
(b) vertically linearly polarized

14-16. (a) linearly polarized at $\pm 45°$ (b) elliptically polarized

14-17. $\begin{bmatrix} 1 & -i \\ i & 1 \end{bmatrix}$

14-20. (a) Elliptical polarization with inclination angle $\alpha = -25.097°$:

$$\frac{1}{\sqrt{13}}\begin{bmatrix} 2 \\ \dfrac{3}{2} + \dfrac{i3\sqrt{3}}{2} \end{bmatrix}$$

(b) Elliptical polarization with inclination angle $\alpha = 4.903°$, or rotated by 30°:

$$\frac{1}{\sqrt{34.475}}\begin{bmatrix} 2.6519 \\ -0.6651 + i(5.1962) \end{bmatrix}$$

14-21. (a) elliptically polarized with semi-axes E_{0x} and E_{0y} aligned with coordinate axes
(b) elliptically polarized with principal axes at 45° to coordinate axes
(c) circularly polarized, centered at origin, radius of E_0
(d) linearly polarized with slope E_{0y}/E_{0x}

14-22. $I = I_0(2 \sin^2 \theta \cos^2 \theta)$

Chapter 15

15-1. 28.1%
15-2. 67.5°; 22.5°
15-3. (a) $I_0\{0.5(\alpha^2 + \beta^2) \cos^2 \theta + \alpha\beta \sin^2 \theta\}$
(b) $0.4525I_0$ versus $0.5I_0$; $0.351I_0$ versus $0.375I_0$; both $0.25I_0$; $0.0475I_0$ versus 0
15-4. 0.0633 mm
15-5. (a) single refraction, phase retardation, any polarization possible (b) single refraction, no phase retardation, unpolarized (c) same as (a) (d) double refraction, no phase retardation in each separated beam, each beam linearly polarized (e) cases (a) and (c)
15-6. (a) difference of 0.121 mm (b) 0.015 mm
15-7. (b) 0% (c) 33%
15-8. 0.0162 mm
15-9. 20°
15-10. 0.005
15-11. (a) 53.12° (b) 11.5°
15-12. (a) mixture of unpolarized and circularly polarized light (b) elliptically polarized light
15-13. (a) 56.2° (b) 33.8°
15-14. (a) 0.05 g/ml (b) about 46°
15-15. (a) 0.200 mm (b) 50°
15-16. (a) 8.57×10^{-5} cm (b) green
15-18. (a) 0.0091 (b) 15 μm (c) 12 (d) 15 μm
15-19. 3.15×10^{-4}
15-20. (a) 36.8 μm (b) 18.4 μm
15-21. 14.7°
15-22. (a) 14.8% (b) 2.03% of I_0 (c) 0.92
15-23. $I_0 \sin^2 (2\theta)$
15-24. (a) 42.5° (b) 0.60°

Chapter 16

16-1. (a) 0.218 cm (b) 0.218 cm
16-2. 0.090
16-3. (a) 0.135 mm (b) 139
16-4. 496 nm
16-5. 2.125, 1.44, 0.778, and 0.55 μm
16-6. (a) 15° (b) 0.678, 0.166, 0, 0.0461
16-7. 1.68×10^{-3} cm; 2.75×10^{-3} cm
16-8. 8.4×10^{-4} cm
16-9. 9725 km in diameter; 2.69×10^{-11} W/m^2
16-10. 5.2 m
16-11. 5.3 miles
16-12. 75.7 to 265 m
16-13. (a) 0.400 mm (b) 0.8106, 0.4053, 0.09006
16-14. (b) 2.10 mm
16-15. (b) 20
16-18. 0.875, 0.573, 0.255, 0.0547, 0

16-20. (a) $I = I_0 \dfrac{\sin^2(1.151y)\,\sin^2(0.575x)}{0.438x^2y^2}$ (b) 5.46 mm along x; 2.73 mm along y
(c) 0.895 along x; 0.629 along y (d) 0.005

16-21. (a) 90° (b) 11.5° (c) 5.7°

16-22. $\Delta\theta = (\lambda/D)(1/\cos\theta)$

16-23. (b) 4.7%, 1.8%, 0.84% for $m = 1, 2, 3$, respectively

16-24. $m = 0$; $\theta_{1/2} = 30°$

16-25. (a) 120° (b) $I_p = (\tfrac{1}{9})I_{max}$ (c) $I_p = I_{max}$ (d) $I_p = 3I_{av}$

Chapter 17

17-1. 13°18′

17-2. (a) 0.0823°/nm; 0.464 nm/mm   (b) 63,000

17-3. (a) 8.66′ (b) 612.5 nm (or 587.5 nm)   (c) 48; 48

17-4. 987; 494

17-5. (a) 700 nm, 360 nm   (b) 57.1°, 25.6°   (c) 350 nm and 175 nm for crown glass; 180 nm and 90 nm for quartz

17-6. 120,000; 0.069 Å

17-7. (a) third order   (b) any width smaller than light beam

17-8. (a) 21.8 cm, in each case   (b) 9, in each case   (c) 21.8 cm, 4.37 cm, and 0.0029 cm, respectively

17-9. (a) 8750 grooves/cm   (b) 18.89°   (c) 37.77°   (d) 7.88 nm/deg

17-10. (a) 7000   (b) 0.018 mm

17-11. (a) $-5.7°$ to $+11.5°$ (b) 100,000   (c) 10 Å/mm   (d) 1 m

17-12. about 5000 grooves/cm

17-13. (a) 1.16 μm   (b) 18.4 Å/mm

17-14. (a) 11.5°   (b) 11.8°

17-15. 3550 grooves/mm; reduces it

17-16. (a) 3647   (b) 1200 grooves/mm   (c) 3.04 mm

17-17. (a) 557 to 318   (b) 960   (c) 388,800; 0.014 Å   (d) 0.41°/nm   (e) 5.5 Å

Chapter 18

18-1. near, near, far

18-2. maxima: 409, 136, 81.8 cm; minima: 204.5, 102, 68 cm

18-3. (a) 1.88 and 3.26 mm   (b) 2.66 and 3.76 mm

18-4. (a) 0.0346 cm   (b) 833   (c) 20 cm, 6.67 cm, 4 cm

18-6. (a) 0.02 cm   (b) 2500

18-7. (a) 4×   (b) very nearly zero   (c) 5; 6

18-8. 0.0012%

18-9. (a) 1/100   (b) 50.31 cm

18-10. 1.05, 1.48, 1.82 mm

18-12. 1.97 mm radius; zero

18-13. 14.8 cm

18-14. (a) $0.066I_u$ (b) $1.53I_u$

18-15. (a) $0.0119I_u$ (b) $1.23I_u$

18-16. (a) $0.018I_u$ (b) $0.0538I_u$

18-17. $1.19I_u$; $0.86I_u$

18-18. $0.55I_u$

18-19. 21%

18-20. (b) 0.145 mm   (c) $0.645I_u$

18-21. 19 μm

Chapter 19

19-2. (a) 102 nm, 1.22   (b) 0.084%

19-4. (a) 2.81%   (b) 3.17%   (c) 4.26%

19-5. 32.3%

19-6. 2; 0.25 μm; ZrO_2

19-7. (a) 859 Å of aluminum oxide, 1058 Å of cryolite; 0.0003%   (b) 15.6%

19-8. (a) 227 nm and 370 nm   (b) 10%   (c) 1.2%

19-12. For example, from surface to substrate: MgF_2 ($n = 1.35$), SiO ($n = 1.5$), ZnS ($n = 2.2$)

19-13. (a) 81.1%   (b) 98.4%   (c) 99.99%

19-14. 99.96%

19-15. 2.24

Chapter 20

20-2. $61°4'$; $28°56'$

20-3. $\theta_c = 32.9°$, $\theta_p = 61.5°$, $\theta_p' = 28.5°$

20-4. 1.272

20-9. (a) 2.55%   (b) 0.233%   (c) 4.26%   (d) 1.26%

20-10. (a) 2.01%, 2.10%, 5.23%, 100%   (b) 2.01%, 1.91%, 0.274%, 100%

20-11. (a) TM: $\theta_p = 67°33'$, no θ_c; TE: no θ_p, no θ_c
(b) TM: $\theta_p = 22°27'$, $\theta_c = 24°24'$; TE: no θ_p, $\theta_c = 24°24'$

20-12. (a) $R = 13.85\%$, $T = 86.15\%$ (b) $R = 0.62\%$, $T = 99.38\%$

20-14. (a) $\theta_c = 43.3°$, $\theta_p = 55.6°$, $\theta_p' = 34.4°$
(b) $R = 3.47\%$, $T = 96.53\%$; $R = 8.21\%$, $T = 91.79\%$
(c) $R = 3.47\%$, $T = 96.53\%$; $R = 0.67\%$, $T = 99.33\%$
(d) $180°$, $180°$, $0°$, $41.0°$, $27.9°$, $0°$

20-15. (a) $59°51'$ (b) $97.0°$ and $82.3°$

20-16. (a) 29.3%, 34.5%, 45.4%, 65.7%, 100%
(b) 29.3%, 24.2%, 14.9%, 5.4%, 100%

20-17. (a) 82.5%, 84.7%, 90.9%   (b) 82.5%, 80.1%, 69.5%

20-18. (a) 0.113 nm^{-1} (b) 41 nm

20-20. (a) 0.164 μm   (b) 5.1×10^{-6}

Chapter 21

21-2. $N_2/N_1 = 1.2 \times 10^{-33}$

21-4. 1.30×10^{-15} J/m^3-Hz

21-6. 0.00318; spontaneous emission is about 314 times larger than stimulated emission. This is to be expected since the ratio calculated is greater than 1 whenever $h\upsilon \gg kT$.

21-7. (a) See Table 21-1.   (b) Linewidth is about one ten-millionth as wide.

21-8. 0.13 ms; 40 km

21-9. 0.80 mrad

21-11. (b) 2.5×10^5 W/m^2-sr-Hz

21-12. (a) 10^{-4} rad   (b) 10 μm   (c) 12.7 MW/m^2

21-13. (a) 1 W   (b) 20 W

21-14. (a) 0.81 μm; 0.75 μm; 0.585 μm; 0.525 μm   (b) 76%; 70%; 55%; 49%

21-15. (a) 31.5 W   (b) 1.26%

Chapter 22

22-3. (a) $-i(1.24$ m)   (b) for each, approximately, 50 m $-i(1.24$ m)

22-4. (a) $R_{50} = 50.03$ m; $w_{50} = 20.15 \times 10^{-3}$ m

22-5. (a) center of cavity   (b) 0.51 mm   (c) 0.51 mm   (d) 0.4 mrad   (e) ≥ 64.6 m
(f) 245 μW/cm^2

22-6. (b) $\begin{bmatrix} 1 & 0 \\ -0.53125 & 1 \end{bmatrix}$

22-7. (a) 1.88 m   (b) 1.88 m

22-8. (a) $-i\pi w_1^2/\lambda$ (b) 0.438 mm   (c) $-0.952i$
(d) $\dfrac{AC\tilde{q}_1\tilde{q}_1^* + BD + AD\tilde{q}_1^* + BC\tilde{q}_1}{A^2\tilde{q}_1\tilde{q}_1^* + B^2} = \dfrac{1}{R(\ell)} + i\dfrac{\lambda}{\pi w_2^2(\ell)}$
(e) $\ell \cong 6$ cm; $w_2(\ell) \cong 0.54$ mm

22-9. (a) $w_{02} = 0.543$ mm; $z_2 = 0.0663$ m

22-10. (a) $z_{FF} \gg 1.46$ m   (b) 0.371 mrad   (c) $w_{in} = 1.113$ cm; $w_{out} = 11.13$ cm
(d) $z_2 \cong 0.1$ m; $w_{02} \cong 1.81 \times 10^{-7}$ m

22-14. 14.6 m; 58.3 m; 131.2 m; 364.5 m

Chapter 23

23-1. 3.18×10^{10} W/cm^2

23-2. (a) 0.7 mm (b) $D' = 101$ μm; diameter at edge = 111 μm

23-4. (a) 1.4×10^{15} W/m^2 (b) 1.35×10^8 V/m

23-5. (a) 4×10^9 Hz, or the channel bandwidth (b) 10^6

23-6. (a) 181 m (b) 2.42×10^3 V/m

23-7. (b) $R_{NHZ} = 371$ m

23-8. (b) $R_{NHZ} = 29.9$ m

Chapter 24

24-1. 672

24-2. 50 million

24-3. (b) 1284

24-4. (a) 68.1° (b) 0.567 (c) 34.5°

24-5. (a) 0.64 (b) 79.5° (c) 3281

24-6. (c) 432 μm; 429 μm; 10.07 m

24-7. 159

24-8. 10.2 μm

24-9. 12 and 120, counting both polarizations

24-10. -70 db/km

24-11. 0.080 mW

24-12. 3.33 km; 10 km

24-13. 0.136 db/km

24-14. (b) -1.25 db, -6.02 db, -10 db, -20 db

24-15. (a) 1.0069 km; 1 km (b) 4.900 μs; 4.867 μs

24-16. 431 ns; 2.32 MHz

24-17. 77.2 ns

24-18. 14.6 ns/km

24-20. 457 ps; 1/146

24-21. 25 MHz

24-22. (a) 4 ns (b) 0.4 ns

24-23. 48.9 ns

24-24. (b) 3.9 ps/km; 4.3 ps/km

24-25. (a) 50.5 ns; 1.075 ns; 0.075 ns (b) 50.5 ns

Chapter 25

25-1. (a) 0.633, 1.898, 3.164 mm (b) 0.50 mm (c) 12.57, 37.70, and 62.83 cycles/mm (d) $1 : \frac{1}{9} : \frac{1}{25}$

25-2. (a) product

25-3. (c) 32.8%; 0.48

25-4. (a) 5 units of amplitude (c) $25[1 + \sin{(ay)}]^2$

25-7. $(\pi A^2 / \omega) \cos{(\omega \tau)}$

25-8. (a) 18.3 kHz (b) 17.2 Hz

25-9. (a) 0.04 Å (b) 0.1 Å

25-10. (a) 2.86×10^{-3} cm (b) 5.59 nm (c) 224 nm (d) 0.80 reading/s

25-11. (a) 3.6 cm^{-1} (b) 2450 (c) 0.093 mm/s

Chapter 26

26-1. ω, 3ω

26-2. $3\omega_1$, $3\omega_2$, $2\omega_1 + \omega_2$, $2\omega_1 - \omega_2$, $2\omega_2 + \omega_1$, $2\omega_2 - \omega_1$, ω_1, ω_2

26-3. 0, $2\omega_1$, $2\omega_2$, $\omega_1 \pm \omega_2$, $\omega_1 \pm \omega_3$, $\omega_2 \pm \omega_3$

26-5. (a) 5983 nm (b) 0.046

26-6. 9.84 kV; V_{HW} is independent of the length

26-7. 7.9×10^{-6}; $\pi/2$

26-8. (a) $\Phi = 2\pi, 4\pi, \ldots$; $V = 2V_{HW}, 4V_{HW}, \ldots$

(b) At $V = 0$, $I = I_{\max}$; at $V = V_{HW}$, $I = 0$

26-9. $\Phi_{\text{pock}}/\Phi_{\text{kerr}} = (r/K)(n_0^3/\lambda_0 \, VL)$; 73

26-10. 3.47 m; not practical

26-12. The sound wave advances 150 nm, which is $\lambda/33$ for $\lambda = 500$ nm. The ultrasonic wave thus appears practically stationary to the light beam.

26-13. (d) 67

26-14. 221 MHz

26-16. 2.97′

26-17. For a 5-cm length, the current is 31.8 A.

26-18. (a) 0.0647 min/G-cm   (b) 0.0956 min/G-cm

26-19. 14.1°; 0.0712 μm^{-1}

Chapter 27

27-1. (b) $n_I = 0.455 \sqrt{K_I}$; $n_R = 1.099 \sqrt{K_I}$

27-4. (a) 4.80×10^{13} s^{-1} (b) 1.38×10^{16} s^{-1} (c) $n_R = 0.0292$; $n_I = 3.92$

27-6. (a) 0.856 cm   (b) 6.63 μm

27-7. (a) 0.35 mm   (b) 1 m

27-8. 1.7 μm

27-9. (a) 0.405 m^{-1} (b) 11.4 m

27-11. $A = (1 + \alpha)^{1/2}$; $B = 2(\pi c/\omega_0)^2(\alpha + 1)^{-1/2}$
$C = 2(\pi c/\omega_0)^4 \alpha (3\alpha + 4)(\alpha + 1)^{-3/2}$ where $\alpha = Ne^2/m\epsilon_0\omega_0^2$

Index

Abbe number, 120
Abbe sine condition, 96
ABCD propagation law, laser, 470–71
Absorption coefficient, 301, 422,
    511, 573
Accommodation, eye, 155
Achromatic doublet, 102–6
Achromatic prism, 123
Acousto-optic (AO) effect, 556–59
   Bragg regime, 556
   Brillouin scattering, 556
   cavity dumping, 558–59
   light modulation, 558–59
   Raman-Nath regime, 556, 558
Adaptation, eye, 155
Airy disc, 334–36, 444
Airy function, 238, 533
Aliasing, 537
Amplitude modulation, 503
Anastigmat objective, 100
Angular dispersion, 352
Angular frequency, 175
Angular magnification, 131
   eyepiece, 131, 134
   microscope, 136
   telescope, 140–41
Anomalous dispersion, 119, 306, 574
Antireflecting coating, 212, 397–401
Aperture stop, 110
   table, 116

Aplanatic optical system, 97
Apochromatic, 138
Aqueous humor, 152
Argon-ion laser, 488–89
Astigmatism, 93, 98–100
   eye, 162–63
Autocorrelation, 532, 539
Avalanche diode, 26

Babinet's principle, 388, 390
Band-pass filter, 404
Bandwidth, 504, 515
Barrel distortion, 100–102
Beam:
   expander, 450–51
   Gaussian, 445, 456–60
   higher order, 476–80
   splitter, 224
   spreading, 329–30, 443–46
   waist, 443–46, 464–67
   width, 461
Beat frequency, 196
Bessel function, 333–34, 347
Binoculars, 141–42
Binocular vision, 156
Birefringence, 306–10
   circular, 316, 554
Bit rate, 504
Blackbody radiation, 15–17, 28, 428–30

Blaze angle, grating, 357
Blind spot, 153
Bolometer, 24
Boltzmann distribution, 430, 433
Boundary conditions, **E** and **B**, 393, 409
Bragg condition, 274, 557
Brewster angle, 301–2 (*See also*
   Polarizing angle)
Brewster window, 302–3
Brillouin scattering, 556
Burn patterns, laser, 460–61, 479–80

Calcite, 306–12, 318
Camera, 125–30
   depth of field, 128–29
   exposure, 127
   irradiance, 127–28
   lens design, 129–30
   pinhole, 125–26
   relative aperture (*f*-number), 127–28
   35-mm, 130
Candela, 10, 15
Carbon arc lamp, 20
Cardinal points, 63, 74–79
   eye, 154
   locations, *table,* 77
Cartesian surface, 41–43
Cassegrain telescope, 143–44, 149
Catadioptric telescope, 139, 144

Cataract, 163, 166
Cauchy dispersion formula, 119, 122, 575
Cavity dumper, 551
Charge-coupled device, 26–27
Chief ray, 112
  table, 116
Chromatic aberration, 88, 102–6
  lateral, longitudinal, 105
Chromatic dispersion, fiber, 514–18
Chromatic resolving power, prism, 121–23
CIE response curve, 14
Circle of confusion, 128
Circular birefringence, 316, 554
Circular polarization, 280, 285, 292–94, 314–16, 415, 554
Cladding, fiber optic, 502, 505–7
CO_2 laser, 433, 451–52, 486, 488–90, 496–97
Coddington shape factor, 95, 97
Coefficience of finesse, 238
Coherence:
  correlation function, 255
  degree of, 256–58
  laser, 441–43
  length and time, 253, 443
  partial, 247, 254–59
  relations, summary, 258
  spatial, 247, 259–63, 441–43
  temporal, 247, 253–54, 441–43
Coherence length, nonlinear crystal, 544–45
Coherent sources, 192, 449
Color temperature, 17
Coma, 93, 96–97
Comatic circle, 96–97
Compensator, polarization, 309
Complementarity principle, 4
Complementary aperture, 388
Complex numbers, 177–78
Complex radius of curvature, 463
Concave grating, 356
Conductivity, 420–21, 576–78
Cones, eye, 153, 156, 501
Conjugate points, 41
Constructive interference, 200, 203, 205–7 214
Convolution, 522, 533–35
Cooke triplet lens, 85, 130
Cornea, 152, 154, 162–66
Cornu spiral, 376–86
Correlation, 522, 529–33
Correlation function, coherence, 255
Critical angle, 40, 413–14
  optical fiber, 505–6, 508
Current density, **J**, 572, 575–76
Curvature of field, 93, 98–100
Czerny-Turner spectrometer, 362

Davisson-Germer experiment, 3
de Broglie wavelength, 3
Degree of coherence, 256–58, 262
Depth of field, 126, 128–29
Destructive interference, 200, 203, 205–7, 214
Detectivity, 27
Detector noise, 27
Detectors, optical radiation, 24–27

Deuterium arc lamp, 21–22
Dextrorotatory, 313
Dichroism, 298–301
Dielectric, simple, 568–69
Dielectric constant, 198, 579
Dielectric film, 211–21
  multilayer, 391–404
Dielectric mirror, 213, 402
Diffraction, 31, 43, 323 (See also Fraun-hofer diffraction; Fresnel diffraction)
  double slit, 338–41
  single slit, 324–29
  grating, 341–46, 349–63, 534
Diffraction grating, 341–46, 349–63
  blazing, 354, 356–59
  concave, 356
  dispersion, 352–53
  efficiency, 356
  equation, 344, 349–50
  free spectral range, 351–52
  holographic, 360
  instruments, 361–63
  interference, 359–61
  irradiance, 342–43
  resolving power, 353–55
  types, 355–56
Diffuse reflector, 11, 37
Digital modulation, 503
Diode:
  avalanche, 504
  laser, 23, 503, 517
  LED, 23, 503, 517–18
  PIN, 504
Diode laser, 497, 451–52
Diopter, 55, 58
Dipole oscillator, 302–5, 543, 569–71
  induced, 569
  moment, 569
  resonance frequency, 571
Direct vision prism, 123
Dirichlet theorem, 248
Discharge lamp, 20
Dispersion:
  anomalous, 119, 306, 574
  diffraction grating, 352–53
  harmonic generation, 544
  optical fiber, 514–18
  prism, 116, 118–120
  rotatory, 314, 554
Dispersive power, 104, 120, 198
Distortion:
  barrel, 100–102
  optical fiber, 504, 512–19
  pincushion, 100–102
Doppler effect, 183–84, 186
  light, 495, 558, 565
Double refraction, 2, 306, 310–12 (See also Birefringence)
Double-slit diffraction, 338–41
  missing orders, 339–41
  irradiance, 339–40
Double-slit (Young's) experiment, 2, 259–60
Dove prism, 124

Echelle grating, 359, 361, 365
Einstein coefficients, 428–31
Electromagnetic spectrum, 8–10

Electromagnetic wave, 181–83
Electro-optic coefficients, 547
  table, 549
Elliptical polarization, 280, 285–87, 295
Emissivity, 17
Entrance pupil, 110–12
  table, 116
Entrance window, 115
  table, 116
Equal inclination fringe, 215, 226, 237
  (See also Haidinger fringe)
Equal thickness fringe, 215–17, 229–30
  (See also Fizeau fringes)
Etalon, 236, 243
Euler formula, 177, 249
Evanescent waves, 419–20
Excimer laser, 488–89, 451–52
Exitance, 10
Exit pupil, 112
Exit window, 115
  table, 116
Exposure, film, 127
External reflection, 211–12, 411–17
Extinction coefficient, 421–22, 573, 577–79
Extraordinary ray, 310–12, 545, 552
Eye:
  astigmatism, 162–63
  hyperopia, 160–61
  laser treatment, 163–68, 489–90
  lens, 152–54, 166
  myopia, 158–60
  resolution, 157–58, 337–38
  response to radiant energy, 13–14
  schematic, 154
  structure, 152–53
  visual acuity, 157–58
Eye chart, Snellen, 157–58
Eyepiece:
  angular magnification, 131, 134
  chromatic aberration, 132
  Huygens, 133–35
  Kellner, 134
  Ramsden, 133–35, 149
Eye relief, 135, 141

Fabry-Perot interferometer, 236–44
  resonator, 434, 438–39
Faraday effect, 553–55
  optical isolator, 555
  Verdet constant, table, 553–54
Far-field diffraction, 324, 330, 445, 465
  (See also Fraunhofer diffraction)
Far point, eye, 158
Far-sighted eye (See Hyperopia)
Fast Fourier transform, 538
  Cooley-Tukey algorithm, 538
Fellgett (multiplex) advantage, 536
Fermat's principle, 34–36, 41, 89, 91
Fiber optics, 501–19
  attenuation, 504, 509–12
  bandwidth, 504, 515
  characteristics, table, 506
  communications, 502–4
  distortion, 504, 512–19
  modes, 507–9
  propagation optics, 505–7

Fibers, optic:
  graded-index (GRIN), 507, 513–14
  multimode, 507–8
  planar (slab), 507
  single-mode (monomode), 507, 509
  step-index, 507, 513–14
Field of view, 113–16
Field stop, 115
Film:
  antireflecting, 212, 397–401
  high reflectance, 213, 397, 402–4
  interference, 211–12
  multilayer, 213, 391–404
  refractive index, *table*, 400
  thickness measurement, 219–21
Filter, band-pass, 404
Finesse, 242–43
  coefficient of, 238–42
Fizeau fringes, 215–17, 219
Fluorescence, 440
f-number, 127–28, 476
Focal length:
  refracting surface, 59–60
  spherical mirror, 45
  system, 63, 77
  thin lens, 51
  thin lenses:
    Abbe's method, 60
    Bessel's method, 60
Fourier integral transform, 250–52, 523
Fourier series, 174, 248–50, 527
Fourier transform, 250–52
  discrete, 537
  fast, 538
  spectroscopy, 536–39
Fraunhofer diffraction, 324, 366
  circular aperture, 332–34, 444
  criterion, 324, 330, 465
  double-slit, 338–41
  and Fourier transform, 523–25
  grating, 341–46, 349–63
  multiple slits, 341–46
  rectangular aperture, 330–32
  single slit, 324–29
Fraunhofer wavelength, 103–5, 120
Free spectral range:
  diffraction grating, 351–52
  Fabry-Perot, 242–44
Frequency mixing, 546–47
Frequency modulation, 503
Fresnel biprism, 210
Fresnel diffraction, 324, 366
  Babinet's principle, 388
  circular aperture, 370–76
  Cornu spiral, 376–86
  criterion, 369–70, 465
  rectangular aperture, 376–87
  single slit, 384–86
  straight edge, 382–84
  wire, 386–87
  zone plate, 374–76
Fresnel equations, 2, 407
  dielectric, 411
  metal, 422
Fresnel integrals, *table*, 380
Fresnel-Kirchhoff diffraction integral, 367–69
Fresnel loss, 510

Fresnel mirror, 209–10
Fresnel rhomb, 415
Fresnel zone, 370–71 (*See also* Half-period zone)
Fresnel zone plate, 269, 374–76
Fringe contrast, 203, 238, 258, 262, 265, 269
Fringes, interference:
  contrast, 203
  types, 215–16
Fusion, laser induced, 490–91

Gabor zone plate, 267–69, 277
Gaussian, laser mode, 434, 456–57
Gaussian beam, 445, 456–60
  collimation, 472
  focusing, 475–76
  higher order, 476–80
  radius of curvature, 464–67
  solution, 461–67
  spherical, 463–64, 469
  spot size, 464–67
Gaussian optics, 44
Glan-air prism, 311–12, 551
Glasses, optical, *table*, 104
Glaucoma, 163
Globar, 17, 19
Golay cell, 2
Graded index optical fiber, 507, 513–14
Grating:
  constant, 353
  interference (holographic), 359–61, 268–69
  replica, 359
Graybody, 17
Gregorian telescope, 143–44
Group velocity, 194–97

Haidinger fringe, 215
Hale telescope, 144
Half-period zone, 370–77
Half-wave plate, 290, 292–93, 309, 317–18, 548–49
Half-wave voltage:
  Kerr cell, 552
  Pockels cell, 548
Harmonic wave, 174–76
Heisenberg principle, 5
Helium-neon laser, 434–35, 441, 445–48, 452, 470–71, 474, 488
Hermite-Gaussian beam, 456, 479–80
Hermite polynomials, 477–80
High-reflectance film, 243, 397, 402–4
  (*See also* Dielectric mirror)
  stack, 266–67, 402–4
Hologram:
  data storage, 276–77
  grating, 268
  storage, 276–77
  point source, 267–69
  real time, 561–62
  reconstruction, 271
  ultrasonic, 275–76
  Vander Lugt (matched), 533
  volume, 273–74, 279
  white-light (rainbow), 273–74
Holographic grating, 268–69, 360

Huygens eyepiece, 133–34
Huygens-Fresnel principle, 324
Huygens' principle, 31–34
Hyperopia, 160–61

Illuminance, 10, 15
Imaging, 40–42
Incandescent source, 17–18
Index of refraction, 34 (*See also* Refractive index)
Index (phase) matching, 545
Intensity, 10–11, 446, 456
Interference:
  coherence requirement, 202–3, 247, 254
  dielectric film, 211–15
  double slit, 205–8
  Fabry-Perot interferometer, 236–44
  grating, 359–61
  Michelson interferometer, 225–30, 259–60
  multilayer film, 212–13
  multiple beam, 233–36
  Newton's rings, 217–19
  objective, microscope, 219–21
  two-beam, 200–204
  virtual sources, 209–10
Interferogram, 237, 536–38
Interferometer:
  classification, 224–25
  Fabry-Perot, 236–44
  Mach-Zehnder, 232
  Michelson, 225–30
  stellar, 262–63
  Twyman-Green, 230–32
Internal reflection, 212, 411–17
Inverse square law, 11
Irradiance, 10–11, 15
  camera, 127–28
  circular aperture, 333
  coherent and random source, 192
  Cornu spiral, 381
  diffraction grating, 342–43
  double slit, 339–40
  electromagnetic wave, 183
  hologram, 271
  interferogram, 536–37
  multiple beam, 234–35
  polarized light, 299
  rectangular aperture, 330–32
  single slit, 327, 384–86
  two-beam, 201–4
Isochronous rays, 41

Jacquinot advantage, 536
Jones matrices, 289–95
  summary, 293
Jones vectors, 281–89
  normalized form, 282
  summary, 288

Lambertian surface, 12, 147, 447
Lambert's cosine law, 11
Laplacian operator, 180, 457
Laser:
  applications, *table*, 485
  beam characteristics, 456–80

Laser (*cont.*)
  characteristics:
    coherence, 441–43
    directionality, 443–46
    focusability, 448–51
    intensity, 446–48
    monochromaticity, 440–41
  communications, 491–93
  Einstein radiation theory, 427–31
  essential elements, 431–34
  eye therapy, 163–68, 489–90
  fiber optics, 503, 517, 488–89, 492–93
  information processing, 493–94
  medical, *tables*, 487–90
  modes, 434
  operation, 434–38
  types and parameters, *table*, 451–53
  propagation, 468–76
    general law, 469–70
  spot size, 464–67
  stimulated emission, 426–31
Laser diode, 23, 503, 517
Lateral coherence, 247 (*See also* Spatial coherence)
Lens:
  camera, 126–30
  Cooke triplet, 85, 130
  design, 129–30
  eye, 152–54, 166
  Fourier transform, 526
  Protor, 86
  Rapid landscape, 83–84
  relative aperture, 127–28
  resolution, 335–37
  simple magnifier, 131–32
  telephoto, 126
  wide-angle, 126
Lensmaker's formula, 51, 95
Lens matrix, 68–71
Levorotatory, 313
LIDAR, 494–96
Light-emitting diode (LED), 23, 503
Light pipe, 501–2
Linear dispersion, 352
Linear polarizer, 282–83, 289–92
Linewidth, spectral, 440–41
  comparison, *table*, 441
Lissajous figure, 283–84
Littrow mount, grating, 358, 361
Lloyd's mirror, 209
Localized fringes, 215–16, 229, 231
Longitudinal coherence, 247 (*See also* Temporal coherence)
Lumen, 10, 14
Luminance, 10
Luminous efficacy, 14
Luminous efficiency, 14–15
Lux, 10, 15

Mach-Zehnder interferometer, 232
Magnification:
  angular, 131
  eyepiece, 131, 134
  holographic, 275
  lateral, 38, 47, 49–50
  microscope, 136
  spherical mirror, 45–47
  spherical refracting surface, 48–49
  summary, 54
  telescope, 140–41
  thin-lens, 53–54
Malus' law, 299
Material dispersion, 514–18
Maxwell equations, 2, 392, 571–72, 576
Mercury arc lamp, 20, 448
Mercury-xenon arc lamp, 21–22
Meridional rays, 80
Michelson interferometer, 225–30, 259–60
  Fourier transform spectroscopy, 536–38
  stellar, 262–63
Michelson-Morley experiment, 3
Microscope, 135–39
  interference objective, 219–21
  magnification, 136
  numerical aperture, 137
  objective types, 137–38
  resolution, 336–37
Mie scattering, 304–5
Minimum deviation, prism, 117–18
Mirror:
  dielectric, 213
  Fresnel, 209–10
  Lloyd's, 209
  plane, 37–38
  spherical, 43–47
Missing orders, 339–41
Modal distortion, 512–13
Modes:
  fiber optics:
    allowed, 507–9
    coupling, 509, 513
  laser beam:
    electric field profile, 459–61
    Gaussian (fundamental), 434
    higher order, 456, 476–80
Modulation, light, 546–59
Modulator, fiber optics, 503
Monochromator, 362
Multilayer films, 213, 391–404
Multiplexing, 492
Myopia, 158–60, 164

Natural line width, 253–54
Nd:YAG laser, 433, 451–52, 486, 488–90, 495–97
Near-field diffraction, 324, 465 (*See also* Fresnel diffraction)
Near point, eye, 131, 158
Near-sighted eye (*See* Myopia)
Neodymium-glass laser, 446, 452, 490
Nerst glower, 17–18
Newtonian telescope, 143, 149
Newtonian thin-lens equation, 57
Newton's rings, 2, 217–19
Nodal points, 63, 74–77
  locations, *table*, 77
Noise-equivalent power, 27
Nonlinear optics, 188, 541–65
  acousto-optic effect, 556–59
  Faraday effect, 553–55
  frequency mixing, 546–47
  Kerr effect, 551–53
  phase conjugation, 559–65
  Pockels effect, 547–51
  second harmonic generation, 542–45
Nonlocalized fringes, 208, 215, 230
NOVA laser, 490–91

Numerical aperture, 137, 336
  fiber optics, 506, 508–9
Nyquist criterion, 537–38

Obliquity factor, 367, 370
Ohm's law, 420
Oil immersion objective, 137
Opacity, 539
Opera glass, 140–41
Optical activity, 313–17
  measurement, 313–14
  quartz, 313–14
  rotation, 313–14
  rotatory dispersion, 314
Optical constants, 421, 568
Optical data imaging, 522–35
  convolution, 533–35
  correlation, 529–33
  filtering, 528–29
  pattern recognition, 522, 531
  point spread function, 533
  spectrum analysis, 525–28
  transfer function, 534–35
Optical density, 267, 539
Optical glasses, *table*, 104
Optical isolator, 555
Optical path length, 42
Optical phase conjugation (OPC), 559–65
  four-wave mixing, 562
  mirror (PCM), 559–61
  nuclear fusion, 563–64
  optical fiber, 564–65
  pointing and tracking, 563–64
  real time holography, 561–62
Optical transfer function (OTF):
  modulation (MTF), 534–35
  phase (PTF), 534–35
Optic axis (OA), 307–12
Ordinary ray, 310–12, 545, 552

Parametric amplification, 546
Paraxial ray, 39, 65, 87
Partial coherence, 247, 254–59
Paschen-Runge mounting, 362, 364
Pattern recognition, 522, 531
Pellin-Broca prism, 123–24
Petzval lens, 130
Petzval surface, 98
Phase, wave, 176, 188–89
Phase angle, 176
Phase retardation, 290, 292–93, 547–48, 552
Phase velocity, 194–97
Phasor, 189–90, 344, 372, 376–78
Phonon, 557
Photodiode, 25–26
Photoelasticity, 317–19
Photoelectric effect, 3
Photometric quantities, *table*, 10
Photometry, 13–15
Photomultiplier, 24, 504
Photon, 3–5, 428–29, 557–58
Photopic vision, 13, 156
Photovoltaic effect, 25–26
Piezoelectric, 547, 556
Pile-of-plates polarizer, 302–3
Pincushion distortion, 100–102

Pinhole camera, 125–26
Pixel, 26
Planckian source, 17
Planck's constant, 3, 16
Plane of incidence, 31
Plane wave, 178–80, 457–58
Plasma frequency, 578
Plate factor, grating, 352
p-n junction, 25–26
Pockels effect, 547–51
    cavity dumper, 551
    half-wave voltage, 548
    Q-switch, 550–51
Point spread function, 533
Poisson spot, 373–74
Polarization, electric, **P**, 542–43,
    569–72
Polarization, optical:
    birefringence, 306–9
    circular, 280, 285, 292–93, 554
    dichroism, 298–301
    double refraction, 310–12
    elliptical, 280, 285–87, 295
    Jones matrices, *table*, 293
    Jones vectors, *table*, 288
    linear, 280, 282–83, 289–92
    optical activity, 313–17
    partial, 280
    photoelasticity, 317–19
    by reflection, 301–3
    by scattering, 303–5
    system matrix, 295
Polarizer-analyzer pair, 299
Polarizing angle, 302, 412
Polar molecule, 569
Polaroid *H*-sheet, 300–301
Population inversion, 431, 436–38
Porro prism, 124
Posterior capsulotomy, 166–68, 170
Power spectrum, 252
Poynting vector, **S**, 183, 392
Presbyopia, 155
Principal maximum, 344–46, 351
Principle of complementarity, 4
Principle of reversibility, 36
Principal points, 63, 74–77
    locations, *table*, 77
Prism:
    angular deviation, 116–18
    designs, 123–25
    dispersion, 116, 118–20
    minimum deviation, 117–18
    polarizing, 311–12
    resolving power, 121–23
    spectrometer, 120–21
Propagation constant, 175
Protor photographic lens, 86
Pump, laser, 431–32

Q-switch, 550–51
Quantum, 3
Quantum electrodynamics, 1, 6
Quantum mechanics, 3, 4, 197, 254
Quarter-wave plate, 290, 292–94, 309
Quartz, 308–9, 312–14

Radial keratotomy, 164–66
Radiance, 10–12

laser, 448
Radiometric quantities, *table*, 10
Radiometry, 10–13
Rainbow hologram, 273
Ramsden eyepiece, 133–34
Rapid landscape lens, 83–84
Ray aberrations, 88–89
Ray diagrams:
    optical system, 63, 75–76
    spherical mirror, 45–47
    thin lens, 52
Rayleigh criterion, 122, 239, 335–37, 354
Rayleigh range, 472–73
Rayleigh scattering, 304, 510
Ray tracing, 79–84
    equations, *table*, 82
    software, 80
Ray transfer matrix, 66
    summary, 70
Real image, 38, 53, 268, 272
Reconstruction, hologram, 267–68, 271–72
Reflectance, 232, 396–404, 412–13, 417,
    419
Reflection:
    diffuse, specular, 37
    external, internal, 211–12, 411–17
    formulas, *summary*, 54
    law, 31, 35, 409
    from metals, 422–23
    phase change, 209, 210–11, 214, 233,
        414–17
    plane mirror, 37–38
    spherical surface, 43–47
    total internal, 40, 414
Reflection coefficient, 212, 232–33,
    410–14, 419
Reflection grating, 355–56
Reflection matrix, 67–68
Refracting power, 55
Refraction:
    formulas, *summary*, 54
    plane surface, 38–40
    Snell's law, 31, 34, 36, 38–40
    spherical surface, 47–50
    thick lens, 62–65
    thin lens, 50–57
Refraction matrix, 66–67
Relative aperture, 127–28, 476
Refractive index, 34
    complex, 306, 420–23, 573–74, 577–79
    eye, 152–54
    function of frequency, 119, 306, 574–75,
        578–79
    gas, measurement, 228–29
    optical fiber, 505–7
    polarizing materials, *table*, 308
    prism, 118–19
    quartz versus wavelength, 312
    relative, 212, 410
Resolution:
    eye, 157–58, 337–38
    lens, 335–37
    microscope, 336–37
    telescope, 335–36
Resolving power:
    eye, 157–58
    Fabry-Perot, 239–41
    Fourier spectrometer, 537
    grating, 353–55

optical fiber, 502
    prism, 121–23
Resonator, laser, 431–32, 434, 438–39
Responsivity, detector, 27
Retina, 152
Rochon prism, 312
Rods, eye, 153, 156, 501
Ronchi ruling, 526–27
Rotator, polarization, 290–91, 293
Rotary dispersion, 314, 554
Rowland circle, 362
Ruby laser, 433

Sampling theorem, 504, 537–38
Scalar theory, 188
Scattering, 303–5
Schematic eye, 154
Schmidt telescope, 144–45
Scotopic vision, 13, 156
Secondary maxima, 344
Second harmonic generation, 542–45
Seidel aberrations, 87, 93
    astigmatism, 98–100
    coma, 96–97
    curvature of field, 98–100
    distortion, 93, 100–102
    spherical aberration, 43, 91–96
Sernamount prism, 312
Sign convention, imaging, 45, 48
Simple magnifier, 131–32
Sinc function, 252–53, 326–27, 544
Single-slit diffraction, 324–38
    beam spreading, 329–30
    central maximum, 329
    Fresnel, near-field, 384–86
    irradiance, far-field, 327
    rectangular and circular apertures,
        330–35
    resolution, 335–38
Sinusoidal grating, 527–28
Skew rays, 80
Skin depth, 577
Skip distance, 506
Snellen eye chart, 157–58
Snell's law, 31, 34, 36, 38–40, 302, 409
Sodium arc lamp, 20
Solar constant, 18, 186
Solar radiation, 18
Soleil-Babinet compensator, 309
Sources, optical radiation, 17–23
Spatial coherence, 259–63, 441–43
    width, 261
Spatial filtering, 522, 528–29
Spatial frequency, 250, 523, 527
    angular, 525
Specific rotation, polarization, 313
Spectrogram, 536
Spectrometer:
    Czerny-Turner, 362
    grating, 361–63
    prism, 120–21
    Wadsworth, 363
Spectrum analysis, 525–30, 536
Specular reflection, 37
Spherical aberration, 43, 91–96, 129,
    143–44
Spherical wave, 180–81, 457–59
Spontaneous radiation lifetime, 429

Standing waves, 192–94
Stefan-Boltzmann law, 17, 28
Stellar interferometer, 262–63
Stereoscopic vision, 156
Stimulated emission, 426–31, 436–38
Stokes relations, 232
Sunlight, 17–18
Superposition principle, 187–88
System matrix, polarization, 295
   laser propagation, 469–70
System ray-transfer matrix, *summary*,
   71–74

Talbot, 10, 15
Telescope, 139–45
   angular magnification, 140–41
   astronomical, 139–40, 149
   binoculars, 141–42
   eye relief, 141–42
   Galilean, 140–41
   reflecting, 143–45
   resolution, 335–36
TEM beam, 450, 461
TEM modes, 434, 460–61
Temporal coherence, 247, 253–54, 259,
   441–43
Tessar lens, 129–30
Thermistor, 24
Thermocouple, 24
Thick lens, 62–65
   cardinal points, 63
Thin lens, 50–57
   combinations, 56–57, 77–78, 103–6

focal length:
   Abbe's method, 60
   Bessel's method, 60
formulas, *summary*, 54
lateral magnification, 38, 47, 49–50
lensmaker's equation, 51, 95
matrix, 68–71
Newtonian equation, 57
refracting power, 55–56
Third-order aberration, 87
Total internal reflection, 40, 414
   fiber optics, 505–7
   frustrated, 419–20, 507
Transfer matrix, 391–95, 398, 403
Translation matrix, 66
Transmission axis, TA, 298
Transmission coefficient, 411–12, 232, 419
Transmission function, 526, 530
Transmission grating, 355
Transmittance, 232, 238–39, 269, 412,
   419, 526, 550
Transverse electric (TE), magnetic (TM),
   301, 408–17
Tungsten lamp, 19–20
Twyman-Green interferometer, 230–32

Van Cittert-Zernike theorem, 262
Vander Lugt filter, 533
Vergence, 55
Vignetting, 115
Virtual image, 37–38, 53, 268, 272
Virtual object, 50, 53
Visibility, fringe, 203

Vision (*See also* Eye)
   acuity, 157–58, 337–38
   defects, 158–63
   laser treatment, 163–68, 489–90
   photopic, scotopic, 13, 156
   stereoscopic, 156–57

Wadsworth spectrograph, 363
Wave aberration, 88–89
Wave equation, 172–74
   complex, 178
   electromagnetic, 181–83, 457
   harmonic, 174–76
   plane, 178–80, 457
   spherical, 180–81, 457–59
   standing, 192–94
Wavefront, 179
   reconstruction, 267, 271
Waveguide dispersion, 517–18
Wave number, 175
Wave-particle duality, 1, 3
Wien displacement law, 16, 28
Wollaston prism, 312

Xenon arc lamp, 21

Young's double slit, 205–8 [*See also*
   Double-slit (Young's) experiment]

Zone plate:
   Fresnel, 374–76
   Gabor, 267–69, 277